D1448759

THE CHIRONOMIDAE

THE CHIRONOMIDAE

Biology and ecology of non-biting midges

Edited by

P.D. Armitage

Institute of Freshwater Ecology, Wareham, UK

P.S. Cranston

CSIRO Division of Entomology, Canberra, Australia

and

L.C.V. Pinder

Institute of Freshwater Ecology, Monks Wood, Cambridgeshire, UK

CHAPMAN & HALL

London · Glasgow · Weinheim · New York · Tokyo · Melbourne · Madras

Published by Chapman & Hall, 2–6 Boundary Row, London SE1 8HN, UK

Chapman & Hall, 2–6 Boundary Row, London SE1 8HN, UK

Blackie Academic & Professional, Wester Cleddens Road, Bishopbriggs, Glasgow G64 2NZ, UK

Chapman & Hall GmbH, Pappelalle 3, 69469 Weinheim, Germany

Chapman & Hall USA, One Penn Plaza, 41st Floor, New York NY 10119, USA

Chapman & Hall Japan, ITP-Japan, Kyowa Building, 3F, 2-2-1 Hirakawacho, Chiyoda-ku, Tokyo 102, Japan

Chapman & Hall Australia, Thomas Nelson Australia, 102 Dodds Street, South Melbourne, Victoria 3205, Australia

Chapman & Hall India, R. Seshadri, 32 Second Main Road, CIT East, Madras 600 035, India

First edition 1995

Typeset in 10/11 Bembo by Type Study, Scarborough
Printed in Great Britain by Hartnolls Ltd, Bodmin, Cornwall

ISBN 0 412 45260 X

A catalogue record for this book is available from the British Library

Library of Congress Catalog Card Number: 94–72675

(∞) Printed on permanent acid-free text paper, manufactured in accordance with ANSI/NISO Z39.48–1992 and ANSI/NISO Z39.48–1984 (Permanence of Paper).

Contents

Contributors

Arshad Ali, Central Florida Research and Education Center, University of Florida, 2700 East Celery Avenue, Sanford, Florida 32771–9608, USA.

Patrick D. Armitage, Institute of Freshwater Ecology, The River Laboratory, East Stoke, Wareham, Dorset BH20 6BB, UK.

Martin B. Berg, Department of Biological Sciences, University of Notre Dame, Notre Dame, Indiana, 46556, USA. Present address: Department of Biology, Loyola University, Chicago, IL 60626, USA.

Bill (W.P.) Coffman, Department of Biological Sciences, A-2343 Langley Hall, University of Pittsburgh, Pittsburgh, PA 15260, USA.

Peter S. Cranston, Taxonomy and General Biology Section, CSIRO Division of Entomology, GPO Box 1700, ACT 2601, Australia.

Peter H. Langton, 3 St Felix Road, Ramsey Forty Foot, Huntingdon PE17 1YH, UK.

Claus Lindegaard, Freshwater Biological Laboratory, 51 Helsingørsgade, DK-3400 Hillerød, Denmark.

Clive (L.C.V.) Pinder, Institute of Freshwater Ecology, Eastern Rivers Laboratory, c/o ITE, Monks Wood, Abbots Ripton, Huntingdon PE17 2LS, UK.

Mutsunori Tokeshi, School of Biological Sciences, Queen Mary and Westfield College, University of London, Mile End Road, London E1 4NS, UK.

Ian Walker, Biology Department, Okanagan College, 1000 K.L.O. Road, Kelowna, BC V1Y 4X8 Canada.

Preface

The dipteran family Chironomidae is the most widely distributed and frequently the most abundant group of insects in freshwater, with representatives in both terrestrial and marine environments. A very wide range of gradients of temperature, pH, oxygen concentration, salinity, current velocity, depth, productivity, altitude and latitude have been exploited, by at least some chironomid species, and in grossly polluted environments chironomids may be the only insects present. The ability to exist in such a wide range of conditions has been achieved largely by behavioural and physiological adaptations with relatively slight morphological changes.

It has been estimated that the number of species world-wide may be as high as 15 000. This high species diversity has been attributed to the antiquity of the family, relatively low vagility leading to isolation, and evolutionary plasticity. In many aquatic ecosystems the number of chironomid species present may account for at least 50% of the total macroinvertebrate species recorded. This species richness, wide distribution and tolerance to adverse conditions has meant that the group is frequently recorded in ecological studies but taxonomic difficulties have in the past prevented non-specialist identification beyond family or subfamily level. Recent works, including genetic studies, have meant that the family is receiving much more attention globally.

The short life cycles and high densities of many species have provided much basic information on productivity and population dynamics. The family plays an important role in detritus processing and trophic cycles and interactions in standing and running waters. Their wide distribution and the tolerance of some species to extreme conditions has led to the use of Chironomidae to assess and monitor water quality, and the preservation of larval material in lake sediments has provided clues to historical environmental change. In the terrestrial phase, their behavioural activities interact in a complex way with the local environment to maximize survival and reproduction. Occasionally very high densities of emerging adults cause a severe nuisance and may result in allergic response in humans. All these aspects make Chironomidae a central component of theoretical studies on population dynamics and productivity in freshwater environments. In the applied field they have considerable importance in biological monitoring and in providing a food resource for a wide range of animals.

The only detailed attempt to summarize studies on Chironomidae was made by Thienemann in his classic work published in 1954, *Chironomus. Leben, Verbreitung und wirtschaftliche Bedeutung der Chironomiden*. Since then there have been short reviews of aspects of chironomid biology and since 1964 regular publications from the triennial symposia on chironomid studies, but no comprehensive book covering both pure and applied aspects of chironomid research has appeared for 40 years. In this period there have been many advances in all areas of chironomid research and it is the aim of this present book to provide a 'state of the art' account of the biology and ecology of Chironomidae which will be of interest to a wide range of readers including entomologists, ecologists, researchers and workers active in biological monitoring of freshwaters, fishery biologists, public health workers and students of the environment.

Patrick Armitage
Wareham

Peter Cranston
Canberra

Clive Pinder
Monks Wood

Acknowledgements

The editors of this work wish to acknowledge the assistance of their respective Institutions in providing facilities, including libraries that provided excellent service and friendly assistance with even the most obscure queries: our thanks therefore go to the library staff at The Ferry House, Windermere, England, and at CSIRO Black Mountain library, Canberra, Australia. We are very grateful to Donald Webb at Illinois Natural History Survey for providing many references from the North American Benthological Society bibliography. Sue Briggs at CSIRO Wildlife, Canberra, generously provided many references to research on birds that feed on Chironomidae. Odwin Hoffrichter kindly perused a late draft of the references and found many errors – any remnant missing accents are not his fault!

In Canberra the willing assistance of Eva Bugledich in converting all electronic contributions to a single word processor form and chasing up references and author names was greatly valued. One or more chapters were refereed in full or in part by the following Australian colleagues (from Canberra unless noted): Tom Bellas, Andrew Calder, Geoff Clarke, Nick Drayson, Don Edward (Perth), Penny Gullan, Terry Hillman (Albury-Wodonga), Sam Lake (Melbourne), Jon Martin (Melbourne), Chris Reid and John Trueman. Declan Murray and Paddy Ashe (Ireland), Ulrike Nolte (Brazil), Matt Colloff (Glasgow, Scotland), Don Oliver (Canada), Pat Charlebois and Rich Merritt (USA) kindly gave critical comment upon certain chapters and R.W. Lawrenz, P.H. Kansanen and E. Haworth are thanked for providing constructive critiques of Chapter 16.

J.A.T. Boubée generously gave us permission to reproduce the unpublished Lake Maratoto chironomid stratigraphy (Chapter 16) which constituted an important part of his dissertation research. Dr Carlo Giacomin generously allowed us to use his photograph of skin tests of a Venetian midge-allergy patient. CSIRO and Melbourne University Press gave permission for the use of several illustrations in Chapter 4, taken from *Insects of Australia*.

1

Introduction

P.S. Cranston

1.1 INTRODUCTION TO THE CHIRONOMIDAE

Members of the family Chironomidae are true flies (order Diptera). They are the most widely distributed and frequently the most abundant insects in freshwater. Under certain conditions, such as at low levels of dissolved oxygen, larval chironomids may be the only insects present in benthic sediments. Extremes of temperature, pH, salinity, depth, current velocity, and productivity have been exploited by the immature stages of at least some chironomid species. Chironomids live in the glaciated areas of the highest mountains, including at elevations of up to 5600 m in the Himalaya (Kohshima, 1984; Sæther and Willassen, 1987) and are active at temperatures of −16 °C. Larvae of *Sergentia* live at over 1000 m depth in the abyssal of the world's deepest body of freshwater, Lake Baikal (Linevich, 1963).

Geographically, chironomids are the most widely distributed free-living holometabolous insects, with a range paralleled only by some fleas and lice, and exceeded only by a few collembolans and mites. At Ellesmere Island, at Lake Hazen (81° 49′ N), a diverse community exists (Oliver and Corbett, 1966). At the other extreme, the Antarctic mainland is home to three species, with *Parochlus steinenii* (Gercke) and *Belgica antarctica* Jacobs reaching 62° S and 68° S respectively (Usher and Edwards, 1984; Edwards and Usher, 1985). Inclusion of the chironomid fauna of subantarctic islands gives at least ten species in the region. For geo-political reasons, these southern, cold-hardy midges are amongst the best-studied of chironomids. American, British, French, German and South African researchers have published studies on these few subantarctic species. Thus more is known about the physiological mechanisms of cold-tolerance of *Belgica antarctica*, for example, than is known of the mechanisms that permit tolerance of low oxygen, heavy metals, low pH or high conductivity in common, pollution-indicative species in temperate regions.

The Chironomidae: Biology and ecology of non-biting midges
Edited by P.D. Armitage, P.S. Cranston and L.C.V. Pinder
Published in 1995 by Chapman & Hall. ISBN 0 412 45260 X

1.2 CHIRONOMID SPECIES RICHNESS

There are estimated to be as many as 15 000 species of chironomid world-wide; however, the basis for such estimates ought to be examined critically. As with most groups of insects, the northern hemisphere temperate fauna is known best – after all, most chironomid taxonomists have been European, and more recently, temperate North America and Japanese scientists have described chironomids. However, even in these relatively well-documented areas, previously undescribed species are still discovered sporadically. These midges may be distinctive in morphology, or, more often, so-called cryptic species that are recognized on internal (chromosomal) features, but with an external morphology very similar or apparently identical to their close relatives. Outside the well-known western Palaearctic fauna, distinctive new species (that is, those undescribed scientifically) are commonplace, and there is cytological evidence that cryptic species are just as ubiquitous (Chapters 3 and 18).

Tempering the estimated incidence of novelty is the widespread nature of some chironomid species that are virtual cosmopolitan 'weeds'. Failure to recognize this has led to redescription of some of these species as 'new' many times over in different countries. Similar contributions to overestimation of species totals has come from some taxonomists emphasizing certain minute differences that subsequently have been realized to be insignificant – some currently recognized single species have had more than ten names applied to them. There are undoubtedly other similar cases, as yet unrevealed by taxonomic revisionary studies. These problems apply more particularly to temperate faunas, as a result of the attentions of numerous taxonomists who are sometimes without the means to make adequate reference to the studies of their predecessors.

The less-studied, poorly-known faunas, particularly of the southern continents, are little affected by this form of taxonomic overestimation; estimation of species richness based on published data is distorted more by lack of knowledge. In these regions, species richness estimates can come from extrapolations and comparisons with better-known regions, or from semi-formal or informal discovery curves, in which the frequency or proportion of novel taxa in successive samples is used to predict an upper limit. One perhaps unexpected finding from recent studies (e.g. Coffman *et al.*, 1993 and other unpublished findings) is that the described fauna and proportion of novelty discovered in one particular part of a poorly-known continent cannot be extrapolated to the remainder of the region without risk of substantial error. High species richness may be very localized, perhaps amidst a broadly homogeneous area of low diversity. For these reasons, we are hardly yet in a position to make realistic estimates of true global species richness in Chironomidae, though a total of greater than 10 000 species is well supported.

Some of the suggested reasons for high chironomid species richness, and for variations in richness with geographic distribution are discussed in more detail in Chapter 4 and particularly in Chapter 18.

1.3 GENERIC RICHNESS

Lack of knowledge of species taxonomy and biology often means that information is presented at the generic level. Many identification keys attempt to include all genera known from a region (e.g. Oliver, 1981; Coffman and Ferrington, 1984; Wiederholm, 1983, 1986, 1989) when, for various reasons, species-level identification cannot be made. We might ask what kind of biological information can be represented at the generic level. It is tempting to answer that some aspects of diversity can be associated with this rank. Ashe (1983) documented 355 valid generic (and subgeneric) names that had been proposed up until 1983, and these were the ranks used in the synthesis of the global distribution (zoogeography) of chironomids by Ashe *et al.* (1987). However, the generic rank is not applied coherently across all chironomid species – nor, for that matter, across any sizeable taxonomic grouping of other organisms. The reasons for this are expanded on in Chapter 3, but basically generic limits (and therefore their total number) are necessarily arbitrary in their delimitation (Lindeberg, 1971b), despite much discussion to the contrary. Thus genera are not strictly comparable in number of included species, phylogenetic history or evolutionary and biogeographical radiation.

A critique of the use of the generic rank in selected ecological studies, such as the role of competition in structuring communities, is given in section 12.6.1. Tokeshi (1991b) has drawn attention elsewhere to the necessity for species-level ecological knowledge, because, as Cranston (1990) and others have stated, the linkage between chironomid genera and the ecology of their constituent species is tenuous at best.

1.4 ECOLOGICAL DIVERSITY

Chironomids are most familiar as adults that swarm beside productive standing waters, and especially to fishermen as larvae ('bloodworms') in these water bodies. Certainly chironomids are ubiquitous inhabitants of organically enriched places, but the often-expressed popular view that these represent the major or even the sole habitat for chironomids is far from the truth. The following chapters in this book demonstrate the range of biotopes in which chironomids occur.

Certainly there is a strong preference for aquatic habitats for the development of the immature stages, as seen in Chapter 6. These range from the 'conventional' flowing waters (trickles, torrents, streams, rivers) and standing waters (puddles, pools, lakes) to more unexpected habitats, such as temporary rain-pools (section 6.6.2), plant-held waters (phytotelmata, section 6.7) and even the thin film of water on high-altitude glaciers (Kohshima, 1984). Some chironomids tolerate high osmotic levels in brackish and shoreline saline pools and, unusually amongst the insects, several genera include intertidal and marine species (section 6.8.2).

Chironomid larvae are not restricted to sediments and other surfaces exposed to free-water: many can develop in marginal and interstitial aquatic habitats and a substantial number of genera include species that are considered to be fully terrestrial (Table 6.7). The terrestrial habitats used by chironomid larvae predominantly are humic soils (section 6.9), but decaying vegetation and even, exceptionally, living greenhouse vegetation (Cranston, 1987) may be used. Perhaps the most unusual biotope adopted by any of these terrestrial chironomids is fresh cow dung, the sole habitat of immature *Camptocladius stercorarius* DeGeer. However, outside the temperate northern hemisphere, the larval ecologies of terrestrial chironomids are poorly known and other curious life histories are likely to await discovery.

1.5 PHYSIOLOGICAL DIVERSITY

Concomitant with ecological diversity must come a range of physiological abilities to tolerate environmental extremes. Perhaps the best understood is the ability to tolerate reduced levels of dissolved oxygen, even to the point of occasional anoxia (section 6.4.4). Amongst the aquatic insects, only certain larval chironomids and a few notonectid bugs possess haemoglobin. Vertebrate haemoglobins have 'low affinity' for oxygen; that is, oxygen is obtained from high oxygen environments and unloaded in an acid (carbonic acid from dissolved CO_2) environment – the **Bohr effect**. Where environmental oxygen concentrations are consistently low, as in the virtually anoxic and often acidic sediments of lakes, the Bohr effect would be counterproductive. Thus, in contrast to vertebrate haemoglobins, chironomid haemoglobins have a 'high affinity' for oxygen. Chironomid midge larvae can saturate their haemoglobins through undulating their bodies within their silken tubes or substrate burrows to permit oxygenated water to flow over the cuticle. Oxygen is unloaded when undulations stop, or when recovery from anaerobic respiration is needed. The respiratory pigments allow a much more rapid oxygen release than is available by diffusion alone. Haemoglobins are present in certain larval Tanypodinae (none of which are Pentaneurini), all Chironominae and *Propsilocerus* and *Tokunagayusurika* amongst the Orthocladiinae. There has been much study of the multiple haemoglobins of *Chironomus*, and more recently of *Tokunagayusurika*. Selections from the extensive literature are reviewed in section 14.6.1 in relation to the well-established allergenicity of chironomid haemoglobins.

The best-known chironomid from a perspective of a curious physiology is *Polypedilum vanderplanki* Hinton (section 6.6.2). This sub-Saharan African midge lives in temporary pools that are natural depressions in granite rock outcrops, or those formed by human activities such as grinding of millet. Pools are filled by sporadic rainfall and, because of high prevailing temperatures, they often fail to remain wet for long enough for *P. vanderplanki* to complete development. The larvae desiccate and in this condition they await the re-filling of the pool, at which time they rehydrate and continue their development. Experimentally, up to ten cycles can be tolerated but evidence is lacking for the natural number of cycles experienced. This extraordinary cryptobiotic midge can withstand temperatures from −270° C to 102° C when desiccated – a unique physiological adaptation in the insects that deserves further study.

Elsewhere in this book, chironomid physiology is discussed predominantly in relation to ecological issues, such as oviposition site selection, larval habitat and tolerance of physiological extremes such as salinity, temperature and reduced oxygen. Studies using chironomids as model systems for insect physiology and biochemistry are not expressly surveyed and readers are referred to Schmidt (1989) for a review of molecular biological studies involving Chironomidae.

1.6 CHIRONOMIDS AND HUMANS

Given their wide distribution and frequent abundance as larvae and as adults, chironomids might be expected to impinge in many ways on human activities. Historically, adult midges have provided human food beside central/east Africa lakes (section 17.3.4), and for many centuries carp ponds in central Europe have been manipulated to enhance chironomid larval production (section 17.4.3) as a food source for the fish. More recently, with the growth of fish-keeping as an indoor hobby, bloodworms have become known as a favourite fish food amongst the aquarist community. Commercial rearing of chironomids has been attempted in several countries, with Hong Kong and Thailand currently providing large quantities of these delicacies for tropical fish. The important role of chironomids in molecular biological studies also has led to demands for mass-reared chironomid larvae.

A large body of people in developed nations that are affected in some way by chironomids are aquatic biologists. There are few water bodies anywhere in the world in which midges do not occur, and whatever the interests of the limnologist, chironomids are unavoidable. Their ubiquitous presence has led to the inclusion of chironomids in nearly all programmes of biological monitoring of water quality (Chapter 15). Furthermore, since aquatic ecosystems possess many qualities that encourage their use in ecological theoretical studies, chironomids have been the subject of frequent autecological and community ecological studies (see Chapters 10, 11, and 12). The International Biological Programme of the mid 1960s provided a major stimulus to these studies, many of which have continued since that decade. The programme, and the dramatically increased number of aquatic-system researchers, provided the stimulus for the systematic community to transfer interest to the immature stages of aquatic insects (in place of a previous adult-bias) and led further to the development of identification manuals reviewed in section 3.8. Hand in hand has been a proliferation of morphological (Chapter 2), taxonomic and phylogenetic studies (Chapter 3) and biogeographic studies (Chapter 4). It is of particular interest that much of the research described in Chapters 2–4 has been undertaken by scientists with training primarily in aquatic ecology rather than in traditional systematics. The marriage of ecology and systematics advocated by Oliver (1979) was displayed *par excellence* throughout their careers by August Thienemann and Lars Brundin, for example. Their students and many successors continue to provide exemplars of interdisciplinary study to other research scientists.

An increasing human use of Chironomidae is in climatic reconstruction, using palaeontological evidence. This arises from recovery of chironomid residues from dated sediments and using the faunal assemblage to reconstruct past changes in aquatic ecosystems. Interpretation is somewhat controversial, as is clear from Chapter 16.

Recent research has revealed some negative aspects of the biology of chironomids. As documented in Chapter 13, adult midges can occur in such large numbers that they cause severe nuisance. The circumstances often involve human activity, generally the provision of nutrient-rich effluents in water bodies that lead to eutrophication. These polluted water-bodies allow the development of huge populations of one or a few species of tolerant midges.

Nuisance populations of midges not only result from direct human impacts on water quality but may also be a consequence of human population pressure. This leads to the desire or necessity to live in areas that are exposed naturally to high midge densities. In Australia, for example, certain *Tanytarsus* species have probably always been abundant in the many otherwise species-poor saline lakes. Naturally, when houses are built around such lakes, people become exposed to the midges. The problems of human-induced salinization of standing waters and nutrient enrichment by faulty sewage treatment and fertilizer run-off exacerbate an existing problem. Likewise, *Kiefferulus longilobus* Kieffer, a saline-tolerant Pacific/Indian Ocean species, becomes a nuisance when it invades salt-water canal estates made by damming and dredging coastal mangrove swamps, even though water quality remains high. Elsewhere, water quality deterioration, natural eutrophication and elevated temperatures conspire to encourage nuisance numbers in places such as desirable lake frontages in Perth, Western Australia.

An ironic twist to our deleterious impacts upon aquatic systems comes from Japan, where there is increasing documentation of midge nuisance. The rapid onset of Japanese industrialization in the absence of environmental safeguards led to eutrophication, anoxia and frequent abiosis of urban and peri-urban waters. Two factors have conspired to promote midge nuisance in Japan. The first is the recovery in water quality as environmental consciousness has provoked remedial actions. This has resulted in previous abiotic waters improving to become suitable for a limited range of species such as *Chironomus yoshimatsui* Martin and Sublette, *C. plumosus* (Linnaeus) and *Tokunagayusurika akamusi* (Tokunaga) to develop in nuisance numbers. The second factor is the proximity of dense habitation to chironomid-rich waters, such as rice fields and urban rivers where *Polypedilum kyooense* (Tokunaga) and *Tanytarsus oyamai* Sasa occur in nuisance numbers.

Scenarios similar to those in Australia and Japan are described in Chapter 13 for many other parts of the world, and notably in Florida and California in the warmer parts of the United States.

One difficulty with assessment of midge nuisance is the variation of threshold of complaint – what constitutes a nuisance in a developed country may be of little or no public note in a less developed country. Such was the case in the Sudan, where Nilotic midges appear to have swarmed in significant numbers for many years. Interest from outside the Sudan came when it was shown that the midges were responsible for the induction of human allergy (section 14.3.1.). The development of our knowledge of the medical significance of nuisance midges is surveyed in Chapter 14. This allergenicity appears to be both environmental and occupational, where those that handle bloodworms are particularly prone to debilitating disease such as urticaria, dermatitis, allergic rhinitis and even fatal attacks of asthma. The principal allergens have been shown to be haemoglobins, present in the larva and partially retained in the pupa and adult.

Lest this introduction should finish with an overemphasis of the negative side of the chironomid impact on humans, mention should be made of a community with a great interest in chironomids and no small observational skill – the anglers. It is commonly recognized that trout, in particular, not only rise to caddis and mayflies but also may feed exclusively for a period on the larger chironomids, particularly species of *Chironomus*. The rising pupa and newly-emerged adult thus form frequent models for fly-fishers' lures. The accuracy of these lures varies, but it was a salutary lesson to see a fly fisher patrolling a windswept London reservoir using *Chironomus* lures representing the pupa and newly-emerged adult, both of which portrayed the red pigmentation of retained larval haemoglobin. The scientific debate over whether this was so had taken much longer than an astute angler's observation.

Part One

Taxonomy, Morphology and Biogeography

2

Morphology

P.S. Cranston

2.1 INTRODUCTION

The newcomer to the study of midges will not be impressed by the variety of Chironomidae – all stages appear at first sight to be highly homogeneous. Even for the experienced observer, most species cannot be recognized by eye without the aid of magnification. Furthermore, for the aquatic larval stage that is most often encountered by the biologist, few accurate identifications can be made without examination of slide-mounted specimens with high power microscope optics. Increasingly, this is true also of adults, although once they were examined dry or in alcohol. Only the cast pupal cuticle (exuviae) may be identified with consistent accuracy at lower magnifications of about ×100.

The need for great magnification does not derive solely from the small size of the flies (though the smallest may be less than 2 mm long), but more from the often minute dimensions of the subtle features needed to discriminate genus and species. In this chapter, the detailed description of morphology of each life history stage is not intended to illustrate the skills of microanatomists, undoubted as they are. It is intended instead as a practical guide to the characters required to identify chironomids, whether as larva, pupa or adult.

Since some of the important features of chironomids are in the range of 10–100 microns, and may often be hyaline, use of the magnification and resolution of the scanning electron microscope (SEM) in identification seems evident (Sublette, 1979). SEM has been used to illustrate highly complex features, such as the microarchitecture of larval silk-spreading apparatus (e.g. Webb and Scholl, 1987) and in permitting interpretation of highly three-dimensional features which are often difficult to interpret with the narrow focal range of a conventional microscope. The SEM has not been used as a tool in identification, partially because the equipment is expensive and of restricted availability, and fundamentally because the opacity of the cuticle to electrons prevents visibility of structures beneath the cuticle. Visibility of many morphological features is improved with optical microscopes if phase contrast or interference optics are used. However, scanning electron

The Chironomidae: Biology and ecology of non-biting midges
Edited by P.D. Armitage, P.S. Cranston and L.C.V. Pinder
Published in 1995 by Chapman & Hall. ISBN 0 412 45260 X

microscopy does produce some excellent images and this section is illustrated with two of them.

As with all fields of comparative biology, new observations give rise to new hypotheses of character evolution and homology. Ideally there should be uniform morphological nomenclature over all Diptera, and indeed for insects as a whole. As this goal is gradually attained, morphological terminology changes as new homologies are identified and others are refuted. Changes to long-standing morphological terminology may be as troublesome as changes in taxonomic nomenclature of favourite taxa. However, an invaluable foundation for stable terminology is provided by Sæther's excellent (1980a) review and by the glossary in Wiederholm (1983, 1986, 1989).

2.2 THE EGG

There is little, if any, tradition of the use of egg morphology in chironomid studies and elucidation of egg features in chironomids lags far behind studies in, for example, Culicidae (mosquitoes) and Odonata (dragonflies and damselfies). However, following Nolte (1993), some generalizations can be made. Of all subfamilies in which egg laying has been observed, only in the Telmatogetoninae are eggs laid singly, without benefit of a gelatinous matrix (Nolte, 1993; P.S. Cranston *pers. obs.*). *Telmatogeton (Paraclunio)* eggs are ovoid, 400 × 200 μm and have a chorion (Ring, 1989). In other subfamilies, the eggs, numbering up to several hundred, are laid in a gelatinous matrix which expands on contact with water. The egg mass may be an elongate ribbon or a more compact cylindrical to tear-shaped globule. Within the gelatinous matrix the eggs may be randomly placed, or more typically, arranged linearly or helically.

There is phylogenetic significance in shape of the egg mass and arrangement of eggs, as noted by Thienemann in his 1954 synthesis and elaborated upon by Nolte (1993). Chironomine eggs are predominantly arranged helically and tanypod eggs arranged more irregularly in a more or less globular matrix. Orthoclad and diamesine eggs tend to be arranged more linearly in a ribbon-like mass. Nolte and Hoffman (1992) found the eggs of *Diamesa incallida* (Walker) to be uni- or biserial in a secondarily folded, linear cord consisting of two different types of mucilage. The same arrangement is found in *Paratanytarsus grimmii* (Schneider) but the wider phylogenetic distribution of this double mucilage layer remains to be investigated. Another common arrangement is for the egg string to be looped alternately to right and left.

Further details concerning chironomid eggs are provided in Chapter 5, including egg number (section 5.3.1), morphology of the egg (section 5.3.2), egg-mass (section 5.3.3) and gelatinous sheath (section 5.3.4).

2.3 THE LARVAL INSTARS

2.3.1 RECOGNITION

Larval Chironomidae are typical nematocerous Dipterans, having a well-developed, exposed, complete, non-retractile head capsule, with mandibles

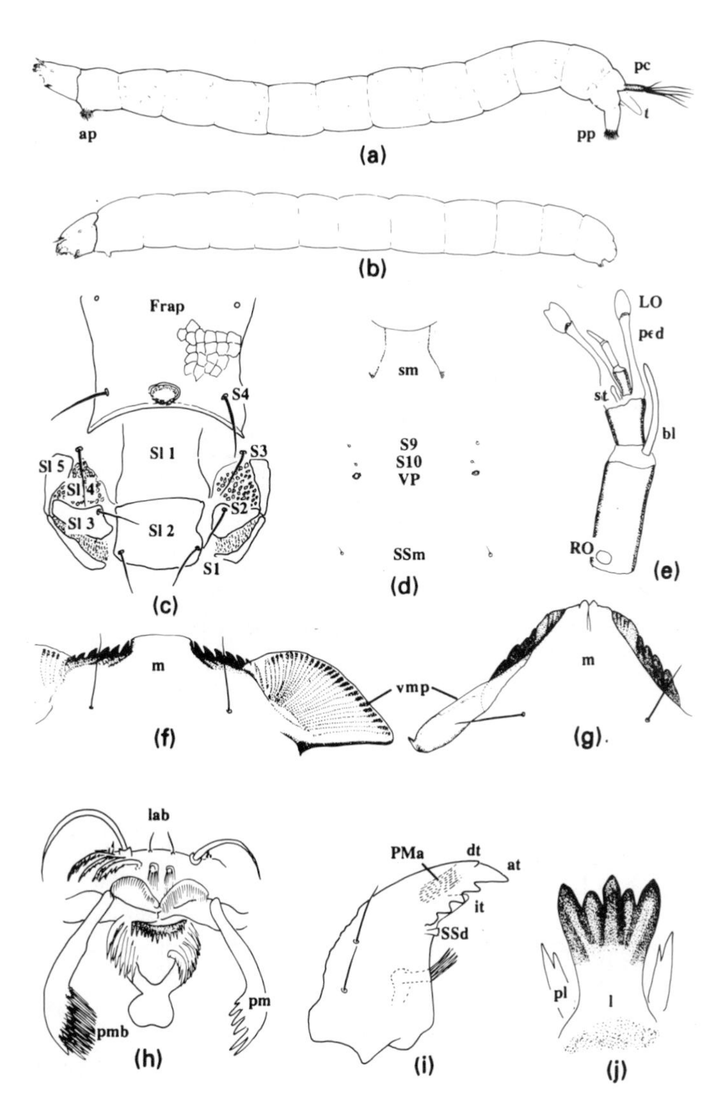

Figure 2.1 Larval morphology. (a) *Parochlus bassianus* Brundin; (b) *Gymnometriocnemus* sp.; (c) frontal apotome and labral sclerites of *Kiefferulus disparilis* (Goetghebuer); (d) ventral head of *Nilotanypus* sp.; (e) antenna of *Neozavrelia* sp.; (f) mentum of *Paracladopelma* sp.; (g) mentum of *Nanocladius* sp.; (h) labrum and epipharynx of *K. disparilis*; (i) mandible of *K. disparilis*; (j) ligula and paraligulae of *Nilotanypus* sp. Key: ap = anterior parapod, at = apical tooth, bl = blade, dt = dorsal tooth, Frap = frontal apotome, it = inner teeth, l = ligula, lab = labrum, LO = Lauterborn organ, m = mentum, pc = procercus, ped = pedestal, pl = paraligula, pm = premandible, PMa = pecten mandibularis, pmb = premandibular brush, pp = posterior parapod, RO = ring organ, SI–IV = labral sensillae I–IV, Sl 1–5 = head sclerites 1–5, sm = submentum, SSd = seta sub-dentalis, SSm = seta submenta, st = style, t = anal tubule, vmp = ventromental plate, VP = ventral pit.

operating in an oblique to horizontal plane, and a narrow, elongate, segmented body that lacks jointed thoracic legs (Figure 2.1a). Chironomid larvae may be distinguished from other nematocerous larvae only by a combination of features:

- The spiracles predominantly are absent (**apneustic**), although the **metapneustic** condition (with posterior spiracles only) occurs in some Podonominae.
- The prolegs, which occur on the first thoracic and terminal abdominal segments, are paired, although variable fusion and/or reduction occurs, particularly in the terrestrial species.
- The terminal abdominal segment bears paired procerci, each bearing apically a tuft of setae.

In larvae in which these abdominal features are reduced (Figure 2.1b), differentiation from larval Ceratopogonidae or Thaumaleidae may be less easy. Thaumaleid larvae, which have fused anterior prolegs, are metapneustic and hypognathous (with the mouthparts directed ventrally), whereas all chironomid larvae are prognathous, with the mouthparts directed anteriorly. Terrestrial orthocladiine chironomid larvae may resemble closely some ceratopogonid larvae. However, the uniquely modified pharynx of ceratopogonid larvae, consisting of two strongly sclerotized, divergent arms and a complex of food-sorting combs, contrasts with the poorly sclerotized and unmodified pharynx of chironomids.

Chironomids have four larval instars, with unconfirmed reports of a fifth instar in a tanypod. Although most morphological and taxonomic observations are made on the final instar, most structures appear to be present in earlier instars (Olafsson, 1992). However, many features of final instar larvae, particularly ratios and shapes, do not apply to earlier instars and do not permit differentiation (Mozley, 1979; P.S.Cranston, *pers. obs.*). Larval morphology has been considered by numerous authors since Miall and Hammond (1900), with Zavřel (1941a) creating the framework for modern interpretation. This has been followed by several thorough studies, notably by Gouin (1959), Sæther (1971), Hirvenoja (1973) and Müh (1985). Many different terminologies have been used, some congruent with prevailing views of insect morphology, while others have been more idiosyncratic: the terminology used in this review follows Sæther's (1980a) detailed assessment of morphological homology.

2.3.2 THE HEAD CAPSULE

The chironomid head capsule predominantly consists of a fully sclerotized cranium that comprises a dorsal apotome and a pair of lateral genae. These three sclerites are separated by ecdysial lines of weakness, along which the integument may split during moulting. There is variation between subfamilies in the splitting at ecdysis: for example, orthoclads and chironomines open a dorsal flap, with the orthoclads also completing a ventral split (along the median suture); tanypods often lack the dorsal splitting. Morphological subdivisions of the dorsal head by sutures are seen in most Chironominae: in the most divided state the clypeus is discrete and up to five labral sclerites lie anterior to a delimited frontal apotome (Figure 2.1c). Variations include fusion of the clypeus and frontal apotome to form a frontoclypeal apotome. In some instances, some or all labral sclerites may become incorporated into a single frontoclypeolabral apotome. Homologies of the various patterns of dorsal sclerites are assessed through recognition of the location of the five anterior-most pairs of cephalic setae: S1 and S2 lie on the labrum (labral setae), S3 on the clypeus (clypeal setae) and S4 and S5 on the frontal apotome (frontal setae).

The remainder of the sclerotized head is made up of the genae. The genae form the ventral and lateral walls, which may extend to meet dorsally in larvae in which the frontal apotome fails to contact the postoccipital margin (the posterior margin of the head capsule) because it is tapered posteriorly. Lying on the genae are the remaining cephalic setae, S6 (suborbital), S7 (supraorbital), S8 (parietal), S9, S10 (genal) and S11, S12 (coronal), all located as their names imply. Variation in the placement of certain of these setae (S9, S10) relative to a ventral pore (VP) and the seta submenta (SSm) (e.g. Figure 2.1d) has proven to be of great value in generic delimitation within the Tanypodinae (Kowalyk, 1985).

In contrast to many other Diptera, chironomid larvae have little internal skeletal structure to the head, with no more than slight internal ridges and the tentorium recognizable only from a relictual posterior pit.

Posterior to the mouthparts, the dominant ventral feature of the larval head is a plate, which is usually toothed (Figure 2.1f,g). The terminology for this plate depends upon hypotheses concerning its origin, with earlier disparate views giving rise to the older terms labium or hypostomium (hypostoma). However, Sæther (1971) argued convincingly for the use of Craig's (1969) term **mentum**, and he established that the plate is double-walled and of dual origin. The two components are a median ventral wall (**ventromentum**) and a dorsal wall (**dorsomentum**) that extends more laterally and curves dorsally behind the ventromentum. The tanypod ventromentum is hyaline with the dorsomentum delimited as a distinct toothed row on each side of the ventromentum only in the more phylogenetically basal genera. Although at first inspection the mentum, especially of Orthocladiinae and Chironomini, appears to be a single structure, the line of fusion between the dorsal and ventral components can often be recognized.

The lateral or posterolateral expansions of the mentum, called paralabial plates by earlier workers, are contiguous extensions of the ventromentum and therefore are correctly termed **ventromental plates** (Sæther, 1971). Orthocladiine ventromental plates vary from scarcely indicated to very well developed, as, for example, in *Nanocladius* (Figure 2.1g). In the Chironominae, the dorsal (inner) surface of the ventromental plates is variously striated, often with attendant hooks and similar structures that give rise to fan-like plates with highly complex microarchitecture (Figures 2.1f, 3.3d). The counter surface to the striations of the ventromental plates are on the cardo of the maxilla, and together the surfaces extrude 'spun' silk that is produced by the salivary glands. In the Orthocladiinae, the role of the

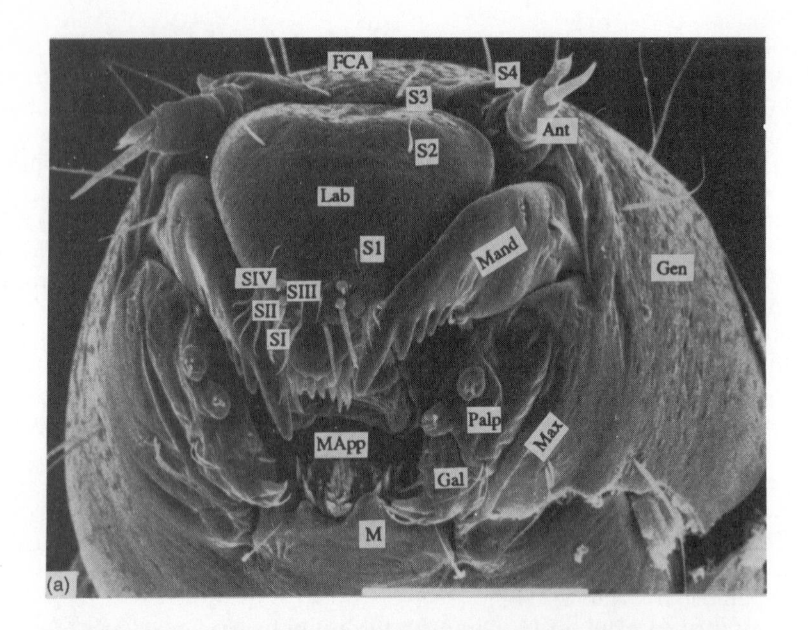

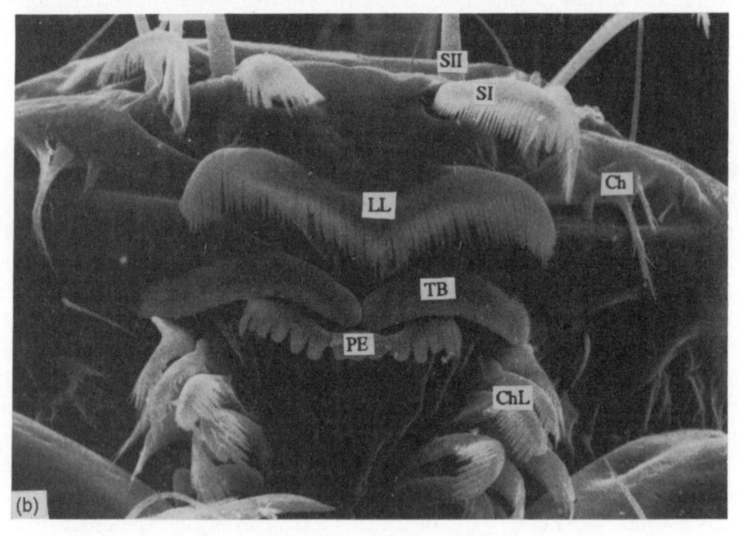

Figure 2.2 Larval morphology. Scanning electron micrographs of larval heads of (a) *Paraheptagyia tonnoiri* Freeman; (b) *Chironomus* sp. Key: Ant = antenna, Ch = chaeta, ChL = chaetulae laterales, FCA = frontoclypeal apotome, Gal = galea (of maxilla), Gen = gena, Lab = labrum, LL = labral lamella, M = mentum, MApp = M appendage of hypopharynx, Max = maxilla, Palp = palp and palpiger of maxilla, PE = pecten epipharyngis, SI–IV = setae of anterior labrum, S1–4 = anterior dorsal cephalic setae (S1,2 labral, S3 clypeal, S4 frontal), TB = tormal bar.

ventromental plates (and perhaps even their homology to those of the Chironominae) is not understood. Extrusion of silk is unlikely, since the opening of the salivary gland is associated with the hypopharynx, within the oral cavity, rather than associated with the plates. The mentum and associated ventromental plate morphology provide useful features for recognition of larval chironomid taxa at taxonomic levels from subfamily to species.

Terminology for the area of the head capsule posterior to the mentum is unclear, with the terms **gula** and **submentum** being used for essentially the same structure. The position of the posterior tentorial pits has been used to mark a boundary between an anterior submentum and posterior gula, but no such anterior markers for the submentum have been recognized. Essentially the posterior mentum grades into a submentum that extends to the occipital margin. Although the term post-mentum (mentum plus submentum) may be more appropriate, the frequently differentiated dark sclerotisation justifies the retention of the term mentum.

The **labrum** is the anterior extension of the frontal apotome (Figure 2.2a,b). Delimitation may be unclear due to fusion of sclerites including the clypeus; however, the labrum may be recognized by the presence of labral setae S1 and S2. At its most distinctive, the labrum comprises two unpaired medial sclerites (Sl 1 and 2) and 3 pairs of lateral sclerites (Sl 3–5) (Figure 2.1c). The ventral surface of the labrum is the epipharynx or palatum, which bears setae, chaetae and scales and is involved in sensing the anterior environment and in feeding. Lying astride the suture delimiting the anterior of the labrum are the posterior-most labral setae, termed SIVA and SIVB. These are actually paired sensilla basiconica, with SIVA a larger bisensillum and SIVB the smaller peg, which may be missing. Anterior to these lie three distinctive pairs of sensilla chaetica. The usually simple and fine SIII lie more medial to, and slightly posterior to, the paired SII setae. The SII also are usually simple, but may be broadened or even pectinate. The anterior-most labral setae, the SI, display a range of structure, from simple to bifid, plumose or pectinate. The shape is often of phylogenetic significance, particularly in the Orthocladiinae. The Tanypodinae appear different superficially and an alternative terminology has been used for many structures, including the labral setae; however, Sæther (1980a) and Müh (1985) recognize homologies that permit uniform terminology.

The anterolateral labral margin bears chaetae and anteromedially there may be a single or divided plate, the labral lamella(e), overhanging the labral margin. The margin itself comprises sclerotised structures, including the torma, and delimits the palatum beneath.

The major features of the ventral labral surface are the **premandibles** and **palatum** (Figures 2.1h, 2.2a,b). Premandibles are paired, movable and toothed, and attach ventrally to the tormal bar. They are lacking in Tanypodinae, Podonominae and Aphroteniinae. The palatum comprises a median epipharynx, bounded by a U-shaped ungula and terminating in a basal sclerite. Three groups of scales or spines occur on the epipharynx: an anteromedian pecten epipharyngis and lateral and basal chaetulae. The pecten epipharyngis is of taxonomic significance: it may be a fused pectinate plate as in Chironomini, three scales, sometimes basally fused or even lacking completely in all Tanypodinae.

Amongst the most prominent features of the mouthparts are paired, articulated **mandibles** (Figure 2.2a). These operate in an oblique plane between the labrum and the maxilla, though some wood-mining larvae have a narrowed labrum and the mandibles thus may operate in an almost vertical plane. The mandible (Figure 2.1i) is toothed, typically with an outer dorsal tooth, a dominant apical tooth and a variable number (usually two or three) of inner teeth, although the dorsal tooth is lost in many taxa. Three setae or groups of setae can be identified: a comb-like pecten mandibularis on the mesal subapical surface; a seta subdentalis on the mola internal to the inner teeth; a branched seta interna basally, usually on the inner mandibular surface. These setae, in conjunction with those on the epipharynx, are used to direct food into the mouth.

The **maxilla** lies dorsolateral to the mentum (Figure 2.2a) and comprises a base of variable breadth bearing a ventrolateral maxillary palp, dorsomedial galea, and posteromedial lacinia. Detailed morphological studies on the maxilla by, for example, Strenzke (1960) and Mozley (1971) have shown that variation in relative proportions and distinctive arrangements of setae, sensilla, ring organs and sometimes combs (pectens) have taxonomic value. In some Chironomini the lacinia apicomedially bears long setae directed behind the mentum. Maxillary movement is important in feeding and particularly in the extension of silk.

Most chironomid larvae have well developed, multi-segmented **antennae** (Figure 2.1e) placed anterodorsally on the upper genae (Figure 2.2a). In Aphroteniinae and some Podonominae the antennae are sited more mid-dorsally on the head. The antenna is retractile into the head in all Tanypodinae (Figure 3.3a). If non-retractile, the antenna may be mounted on an elevated pedestal, which can be elaborated with spines or combs, or the basal antennal segment may arise directly from the genae. The primitive number of antennal segments appears to be five, with variations including reduction to four or three, or an increase through division to six or an apparent seven segments. In some taxa, portions of elongate segments may be hyaline and poorly sclerotized, allowing the antenna to trail posteriorly rather than being the usual anteriorly directed sensory organ. Terrestrial larvae tend to show reduction in antennal segment number and length. Segment homology can be recognized by the consistent placement of a blade (often accompanied by an auxiliary blade) at the apex of the first segment, and also by the **Lauterborn organs** on the second segment. These paired organs are usually sited apically and opposite, but one may be subapical and they may be alternate. These organs are compound, comprising a peg sensillum and paired, thin-walled, fan-like sensilla, which in Tanytarsini may be located apically on pedestals. Reduction of the Lauterborn organs to pegs or styles occurs, notably in terrestrial taxa.

The head also has internal structures associated with feeding. The most obvious is the **premento–hypopharynx complex**, which lies dorsal to the mentum, but is often partially or completely obscured in ventral view by the mentum. The complex is double lobed, consisting of a ventral prementum and dorsal hypopharynx, with the salivary gland outlet between the lobes. The hypopharynx is never strongly developed, usually being a scaly lobe, but with clear rows of teeth in some Tanypodinae (e.g. Figure 3.3b) or brushes in Diamesinae (Figure 3.3k). The prementum is best developed in the Tanypodinae, where it forms a principal part of the feeding apparatus. In this subfamily, the median, articulated, toothed plate is the **ligula**, which is composed of fused glossae, and bears paired, toothed paraligula laterally (Figure 2.1j). The **M appendage**, which lies dorsal to the ligula, is a hyaline, triangular plate adorned with a median strip, the **pseudoradula**, in the Tanypodinae. In other subfamilies all elements of the prementum are quite differently constructed (e.g. Figures 2.2a, 3.3k) or often reduced.

Chironomid larval **eyes** are typically simple (Figure 2.1a,b), being areas of pigment lying beneath the cuticle. There are, however taxonomic differences in eye shape: Chironominae larvae have double eye spots in contrast to the single spot in the Orthocladiinae and most other subfamilies.

2.3.3 THE LARVAL BODY

The chironomid larval body is nearly always demarcated into three somewhat broader thoracic segments, followed by nine narrower abdominal segments. Pseudosegmentation, giving rise to an apparent duplication of segments, is seen in some rare Chironomini. Late fourth instar larvae can be recognized by the swollen thoracic segments with the pharate pupa within: this thoracic swelling may occur in earlier (non-terminal) instars in wood-mining taxa and in *Corynoneura* and *Thienemanniella*. Thoracic and abdominal segments have a setal pattern which may be homologous with that of the pupa; however, the use of such patterns in larval taxonomy is limited to recognition of the diagnostic value of plumosity of some lateral setae, for example amongst *Cricotopus* species (Hirvenoja, 1973). Other abdominal cuticular features include feathered extensions and plates in the Aphroteniinae (Figure 3.3h) (Brundin, 1966; Cranston and Edward, 1992).

The only appendages of the thorax are the anterior **parapods** (Figure 2.1a) which are paired, fleshy, unsegmented 'false legs' that bear claws and are placed ventrolateral on the first thoracic segment. Very similar posterior parapods are found ventrolaterally on the terminal abdominal segment (Figure 2.1a). Parapods are found in nearly all chironomid larvae, though they may be partially or completely fused; even more rarely, parapods and claws may be reduced or absent, notably in some terrestrial taxa (Figure 2.1b). Pre-anally, the **procerci**, which are paired tubercles bearing an apical tuft of setae, are usually prominent (Figure 2.1a).

Lying behind and between the bases of the posterior parapods, surrounding the terminal anus, are one to three (usually two) pairs of anal tubules. These function in ionic regulation (Strenzke and Neumann, 1960) and vary in length intraspecifically with inorganic ion concentration (McLachlan, 1976a). *Chironomus* and some relatives often have haemolymph-filled abdominal tubules associated with respiration (Harnisch, 1954; Nagell and Orrhage, 1980) which therefore are appropriately termed **tracheal gills**.

Usually there are two ventral pairs of these tubules on abdominal segment VIII, and often a lateral pair arise near the posterior margin of abdominal segment VII. A few Podonominae are spiraculate, but all apneustic chironomid larvae have a tracheal system with an anastomosing structure (Zavřel, 1918). Variation of the tracheal system, for example in the number of thoracic commissures and the segregation of thoracic from abdominal tracheal patterns, appears to follow phylogenetic and ecological lines (Tait-Bowman, 1978; Goetz, 1980).

2.4 THE PUPA

2.4.1 RECOGNITION

The pupal stage of chironomids is short-lived, particularly when compared with the larval stage. The duration may be for just a few hours up to several days for long-lived Podonominae (P.S. Cranston *pers. obs.*). Though short-lived, the pupal stage involves major changes in morphology as the larva metamorphoses to the adult.

The pupa is somewhat comma-shaped, with a swollen cephalothorax and dorsoventrally flattened abdomen (Figure 2.3a, b). Because this general shape is very similar to that of pupae of many aquatic nematoceran families, these are difficult to distinguish by any individual character or even by concise combinations of characters. Like the mosquitoes and their relatives, the pupae of Chironomidae may be free-living, either hanging beneath the water surface, as in Tanypodinae and Podonominae, or mobile in the benthos, as in the Aphroteniinae. The pupae of the remaining subfamilies live generally in some kind of tube or covering film produced by the final instar larva. The paired thoracic respiratory organs of free-living pupae characteristically are tubular with a distinctive apical respiratory surface, the plastron plate (Figure 2.3d). In contrast, tubicolous chironomid pupae have variously shaped respiratory organs that lack a plastron plate and act as gills (Figure 2.3a, b). Features of the terminal abdominal segments assist in recognition of chironomid pupae, with tergite and sternite IX being modified as a swimming lobe (Figures 2.3a, b; 3.5b, c). The lack of division of the anal lobe and lack of any supporting buttressing distinguishes the anal lobe of chironomids from those of most other aquatic nematocerans.

Although all morphological features are visible on a pupa with the developing (or fully developed) adult within, they are more readily seen on a cast pupal skin. The cast skin is known as the **exuviae**, a Greek word meaning 'cast clothes' and for which there is no singular, though contrivances such as exuvia and exuvium have been coined.

2.4.2 THE PUPAL CEPHALOTHORAX

Three areas are distinguishable on the pupa: head (cephalic area), thorax and abdomen. On the cast exuviae, the cephalic area and thorax are treated as one unit, the **cephalothorax**.

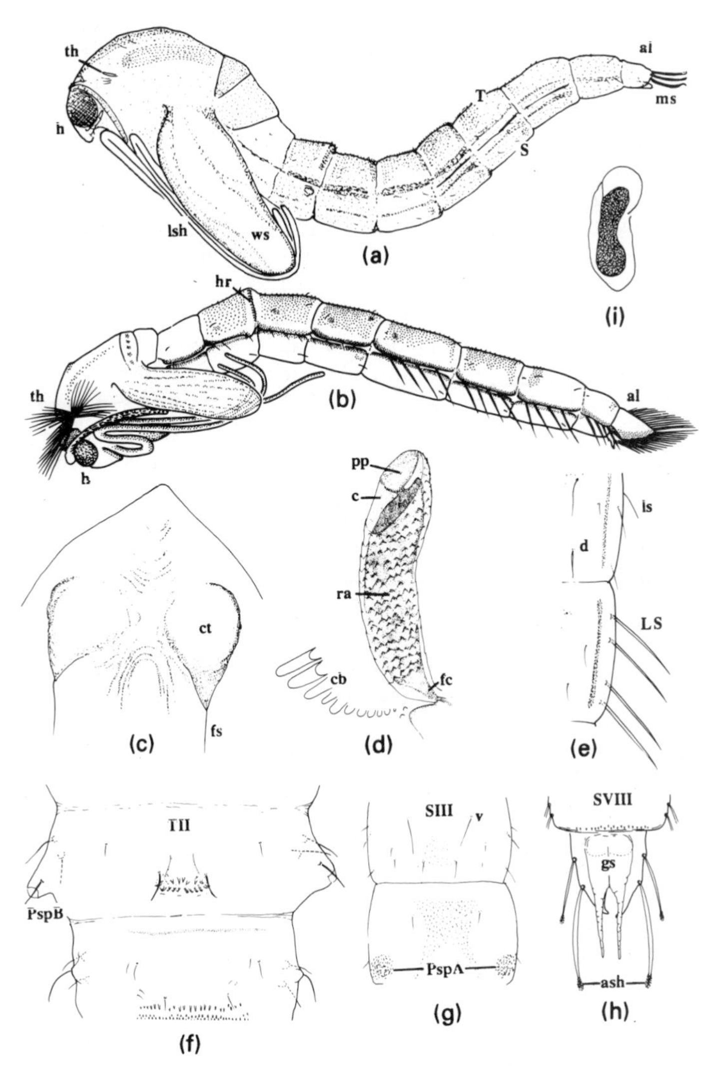

Figure 2.3 Pupal morphology. (a) *Cricotopus* sp.; (b) *Chironomus* sp.; (c) cephalic tubules and frontal setae, *Chironomus nepeanensis* Skuse; (d) thoracic horn, Pentaneurini; (e) lateral setae, Chironomini; (f) tergite II, *Nanocladius* sp.; (g) sternite III and IV, *Paracladopelma* sp.; (h) terminal segments, *Nanocladius*; (i) base of thoracic horn, *Chironomus*. Key: al = anal lobe, ash = anal macrosetae sheath, c = corona, cb = thoracic comb, ct = cephalic tubule, d = dorsal seta, fc = felt chamber, fs = frontal seta, gs = genital sac, h = head, hr = hook row, LS (ls) = filamentous (fine) lateral seta, lsh = leg sheaths, ms = anal macrosetae, pp = plastron plate, PspA = pedes spurii A, PspB = pedes spurii B, ra = respiratory atrium, T = tergite, tc = tracheal connection, th = thoracic horn, S = sternite, v = ventral seta, ws = wing sheath.

The emergent adult splits the dorsal pupal thorax together with a Y-shaped anterior extension around the posterior cephalic area, which opens as a forward projecting flap. Features of taxonomic significance on the cephalic area include location, number and strength of setae on the frons, vertex, postorbit and ocular area and occurrence of tubercles and warts, particularly on the frons and prefrons. The number, length and location of paired frontal setae and the position of any tubercles on the head (Figure 2.3c) are variable features of taxonomic significance, particularly in the Chironomini.

The thorax bears groups of diagnostic setae on the antepronotum (median and lateral), precorneal, prealar, dorsocentral, supralar and metanotal areas, whose number and position are of phylogenetic significance (Coffman, 1983). The wing sheaths, prealar area and area along the median suture may have species-specific microsculpturing, including rugosity and tubercles. The most prominent variable thoracic feature is the respiratory organ, termed a **thoracic horn**, which lies mediolaterally on the anterior thorax. At its most complex (and probably the plesiomorphic condition) (Figure 2.3d), this is a tube directly connected to the developing adult tracheal system, containing a felt chamber (the tracheal continuation into the horn) forming the basal part of a respiratory atrium. The atrium connects distally to the porous plastron plate which makes direct contact to the aerial environment at the water surface. The higher phylogeny of the Chironomidae involves numerous changes in the thoracic horn, notably loss of the direct contact with the adult spiracle and re-establishment of a secondary indirect contact (Coffman, 1979). When the plastron plate is lost, as in some Podonominae and Aphroteniinae, the remnant inner chamber is termed the atrium or horn sac. Further loss of the atrium and even of the complete organ occurs in some Orthocladiinae. Elaboration of the simple tracheal gill is found in Chironominae in which the structure is multifilamentous and plumose. In a few Orthocladiinae there may be no evidence of a thoracic horn. and some terrestrial orthoclads clearly rely exclusively on cuticular respiration. The functional significance of various arrangements of the respiratory apparatus is discussed in more detail in Chapter 9.

The arrangement of **leg sheaths** can be diagnostic amongst the Nematocera. In the Chironomidae, each leg sheath is separate (Figure 2.3a, b) and frequent patterns are for some or all leg sheaths to be recurved under the wing sheath, or with only the hindleg sheath so recurved (Figure 3.5o). However, in the Chironomidae the leg sheath organization includes virtually every pattern seen amongst Nematocera. Brundin (1966) discusses variations in these patterns.

2.4.3 THE PUPAL ABDOMEN

The abdomen is highly and specifically variable in its patterns of distribution of spines, spinules and tubercles. These patterns are particularly strongly elaborated in tubicolous pupae, with free-living pupae less obviously adorned. This tergal pattern is often called **shagreen**, but as Soponis (1977) observed, the term should probably be restricted to granulation (rounded tubercles) rather than spinules. The term **armament** is also used to describe the spine/spinule/hooklet pattern. There are few shared elements to the patterns but some generalizations are discussed below and in Chapter 8.

On each tergite there are three primary fields of posteriorly-directed cuticular projections: an anterior transverse band, a median quadrate patch and a posterior transverse band. In addition, on some segments there is an apical transverse band just posterior to the posterior band, the points, hooks or spinules of which are directed forwards. Between these last two bands is a narrow band of thin cuticle that allows the apical band to be tucked under the posterior margin of the tergite.

Each field may be medially separated into two blocks. Though such a pattern will be repeated on successive tergites, there is no correlation between fields on a tergite: only one of them may be divided, or any combination of them. Amalgamation and reduction of one or more fields also occurs; the reduction may be in the overall extent of the field, or in the density of the armament. The fields are generally best developed on the third and fourth tergites, diminishing in extent and strength posteriorly. Frequently tergite II has a pattern that is different from that of the following segments, but tergite I is rarely embellished. In addition, the paratergites are frequently armed but usually more weakly than the tergites. A variety of points and spines may occur in each of the fields.

Similar fields of cuticular projections occur also on the sternites. In some species there are anterior and/or posterior transverse rows or bands of long spines, although these are usually less strongly developed than on the tergites. The parasternites also may be armed with points and spines.

The abdominal segments have characteristic patterns of setae, with a basic five pairs of dorsal setae, five of ventral setae and four lateral setae (ls), of which some may be filamentous (LS) (Figure 2.3e). Close to the intersegmental membranes, or conjunctives, between segments there are often minute 'O'-setae (Coffman, 1979). In many Podonominae setae on segment VIII and the anal lobe may be bent sharply twice, the so-called **wavy setae** (Brundin, 1966) (Figure 3.5i). Chironominae pupae usually bear a characteristic and often specifically diagnostic spur or comb on the posterolateral margin of sternite VIII (Figure 3.5d,e), and some Orthocladiinae (*Zalutschia* and relatives) may bear a somewhat similar, but doubtfully homologous, 'embedded spine' at the same site (Figure 8.9e). The mechanical roles of pupal abdominal armament are discussed in greater detail in Chapter 8.

The pupal legs are not used in locomotion; what limited powers of locomotion that pupae possess derive from abdominal flexion. In tubicolous species locomotory (or undulatory) structures include the two types of 'false legs', the pedes spurii A and B (Figure 2.3f,g). The pedes spurii A are paired swellings posterolaterally situated on some sternites, particularly IV, with each swelling crowned by whorls of spines (Figure 2.3g). Pedes spurii B are swellings on the posterolateral corner of segments II, and sometimes III, and are also non-muscular (Figure 2.3f). The role of these structures in pupal mobility is discussed in Chapter 8.

The posterior segments are modified as an **anal lobe**, formed from the paratergites of several posterior segments (Figure 2.3a,b,h). In tubicolous species, the anal lobe is undulated to provide respiratory current and to aid in swimming to the water surface at emergence. In free-living forms the lobe has a predominant role in swimming. Again, there are many variations in structure, ranging from the near fusion of tergites VIII and IX into a unique, subcircular, terminal disc in the Telmatogetoninae (Figure 3.5a) to flattened paratergites fringed with elongate filamentous setae in the anal lobe of many Orthocladiinae and virtually all Chironominae. In certain Tanypodinae, the setae of the anal lobe may have adhesive sheaths (Figure 2.3h). On the ventral surface of the anal lobe, the adult genitalia are enveloped in genital sacs (Figure 2.3h). Many terrestrial Orthocladiinae are exceptional in having little or no such specialized development of the anal segments.

2.5 THE IMAGO

The earliest descriptions of Chironomidae concern the adult insect and traditionally Dipteran taxonomists have concerned themselves more with this life history stage than any other. Morphological nomenclature of the adult stage in the Chironomidae is quite similar to that used for the Diptera generally (McAlpine, 1981).

2.5.1 THE IMAGINAL HEAD

The adult head is rounded, somewhat flattened in an anteroposterior plane, and has shortened mouthparts. The **antenna** typically is strongly sexually dimorphic: males have a prominent, generally plumose antenna comprising a narrow scape, globular pedicel and a number (usually 11–14) of cylindrical flagellomeres (Figure 2.4a), but the antenna may resemble that of the female (Figure 2.4b). Although primitively the male and female antennae were probably relatively monomorphic, the antenna of most females has a smaller pedicel, lacks any plume and often has fewer, more flask-shaped flagellomeres (Figure 2.4c). The **eyes** are dichoptic, round, kidney-shaped or sometimes with dorsomedial extensions that may almost meet medially. The ommatidia may have microtrichia between them, which, when extended beyond the ommatidia, give rise to the term 'hairy' eyes.

The **clypeus** is relatively well developed (Figure 2.4d) and is larger than the bilobed labrum, excepting in the few species where there is a marked development of a proboscis. The palp is elongate and primitively five-segmented, with the basal segment usually very small, and with a sensory pit on the third segment. The mouthparts have been termed 'reduced', and it is true that functional mandibles occur only in females of the southern hemisphere podonomine *Archaeochlus*. Apart from mandibles, however, the remaining elements of the mouthparts are present, namely the hypopharynx, labium and soft labella, which together form a food canal that leads to the cibarial pump (Burtt *et al.*, 1986). Adults are observed feeding on nectar or honeydew, but the diet and mode of feeding of the mandibulate *Archaeochlus* remains unknown.

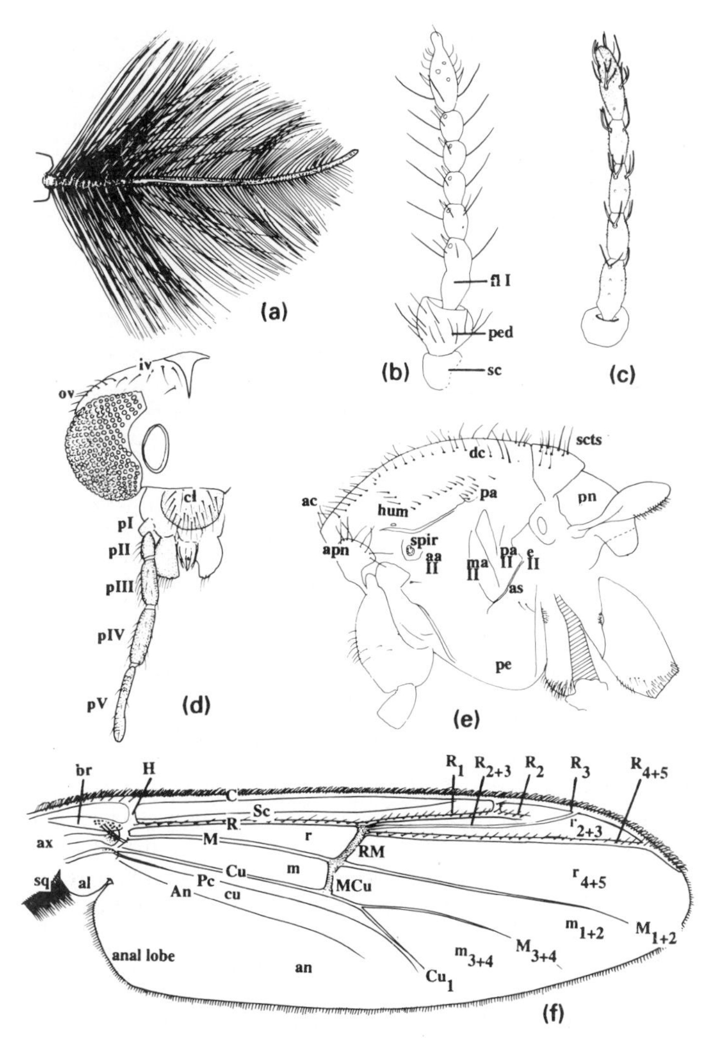

Figure 2.4 Adult morphology. (a) Male antenna (*Conochironomus* sp); (b) male antenna (*Thalassomyia frauenfeldi* Schiner); (c) female antenna (*Bryophaenocladius furcatus* (Kieffer); (d) head (*Procladius paludicola* Skuse); (e) lateral thorax (*T. frauenfeldi*); (f) wing (*Coelopynia pruinosa* Freeman). Key: aaII = anterior anepisternum II, ac = acrostichal setae, al = alula, An (an) = anal vein (cell), apn = antepronotum, as = anapleural suture, ax = axillary area, br = brachiolum, C = costa, cl = clypeus, Cu (cu) = cubital vein (cell), dc = dorsocentral setae, eII = epimeron II, fl = flagellomere, H = humeral crossvein, hum = humeral setae, iv = inner vertical setae, M (m) = medial vein (cell), maII = median anepisternum II, ov = outer vertical setae, pI–V = palpomeres, pa = prealar setae, paII = posterior anepisternum II, Pc = postcubitus, ped = pedestal, pe = pre-episternum, pn = postnotum, R (r) = radial vein (cell), sc = scape, Sc = subcosta, scts = scutellar setae, spir = spiracle, sq = squama.

Primitively the head bears many setae and, particularly in more derived taxa, setal groups can be delimited, namely the temporals (comprising postorbitals, inner and outer verticals, orbitals, frontals), coronals and clypeals. The sclerotized internal structure of the head associated with muscle attachment (the tentorium) and with feeding (the torma linking labrum to cibarial pump and the maxillary stipes) all vary and may provide taxonomic data (Sæther, 1971).

2.5.2 THE IMAGINAL THORAX

The chironomid **thorax** (Figure 2.4e) is dorsally convex in winged species, but more flattened in brachypterous and apterous taxa which do not require the space for bulky flight muscles. The anterior thoracic segment, the pronotum, is divided anteriorly into an often setose antepronotum, a collar formed from paired, medially fused sclerites, and a small, bare postpronotum fused dorsally to the scutum and laterally to the mesopleuron. As in all Diptera, the thorax is dominated by the **mesonotum**, the second thoracic segment that contains the flight muscles. Dorsally the scutum is the major component, with the scutellum a rounded lobe lying posteriorly and a quite well developed postnotum posterior to this. The scutum may have a median suture or a tubercle, and often has a median longitudinal row of setae (the acrostichals) and a mediolateral longitudinal row of dorsocentral setae. The anterolateral scutum may have paler humeral pits and humeral setae sited close to the anterior end of the parapsidal suture. This suture delimits the lateral scutum from the pleuron.

Homology and hence terminology of the pleural parts have been somewhat inconsistent between different families of Diptera. In the Chironomidae, the dorsal episternum is a tripartite anepisternum II divided into anterior, median and posterior parts, with the anterior spiracle in the centre of anterior anepisternum II. More ventrally an anapleural suture divides the anepisternum from the large pre-episternum, which is linked beneath the mid-venter with the same sternite on the other side. Posterior to the episternum, and dorsal to the base of the midleg, is the epimeron which abuts the basal wing articulation. Setae and groups of setae derive their names from the sclerites that bear them: variation in presence, number and strength are of great taxonomic value. The final, posterior, thoracic segment, the metathorax, is a scarcely developed strip connecting to the abdomen.

Taxonomic and phylogenetic information can be derived from many features of the chironomid **wing**, particularly the setosity of veins and membrane, the presence and relative proportions of veins and various patterns of pigmentation of membrane and setae. Interpretation of wing venation requires recognition of homology across many families, often possessing quite disparate venations. Chironomid venation (Figure 2.4f) is reduced relative to the basic Dipteran pattern, and hence there are still some minor uncertainties concerning terminology. The six primary veins are present: the costa is single and usually extends no further than the wing apex; the subcosta is also single, but short; the radius is usually three branched, but may be two-branched through loss of R_{2+3}, e.g. in Podonominae, Aphroteniinae and Telmatogetoninae, or four branched through division of R_2 from R_3, e.g. in some Tanypodinae; the median sector is two-branched; the cubitus and closely associated postcubitus are single; the anal vein short or extending to the margin. Humeral, radial-medial, and mediocubital cross-veins may be present. The wing membrane is divided into cells which are named after the anterior vein that delimits them. Structures involved in the articulation at the wing base, including the brachiolum and arculus, may vary taxonomically, as may the shape of the anal lobe and the setosity of alula, squama and veins. There is sexual dimorphism in wing shape, with the female wing being relatively broader than the male, but generally wing shape has not been well quantified.

Many variable features are also found in the **legs**: from the earliest days of chironomid taxonomy the ratio of foreleg tibia to tarsomere 1 (the Leg Ratio) was recognized as allowing subfamily segregation. Ratios such as this may be functionally associated with the chironomid habit of holding its forelegs aloft as sensory organs. Several further ratios of lengths of some or all of femur, tibia and five tarsomeres may prove distinctive. The apex of the tibia may bear several taxonomic features, including the presence of one or more (usually two) apical spurs, a scale on the fore tibia and combs of varying complexity on the hindleg. The tibia and tarsomeres, usually of the foreleg, may bear lengthy setae of many times the diameter of the leg segment, forming a beard. The claws may be simple or more rarely toothed, with some taxa having swollen pulvilli, in addition to the empodium. Commonly there are sensilla chaetica on some leg segments, particularly on the mid tarsi, whose distribution and number is of taxonomic significance, though less so for phylogenetic study since their development is susceptible to homoplasy (Säwedal, 1982).

The anterior seven abdominal segments are dorsoventrally flattened without prominent pleurae: the female abdomen is shorter and more swollen than that of the male. Hirvenoja (1973) found the distribution of tergal setae was valuable in species segregation in *Cricotopus*. However, beyond the observation that abdominal setosity is reduced in more derived taxa, the character has a limited history of use in chironomid systematics. It is the more terminal abdominal segments that form the internal and external genitalia that provide a wealth of systematic characters. This has been known for a century for male chironomids, but the realization that there is almost comparable detail in the females derives from more recent studies of Wensler and Rempel (1962), Rodova (e.g. 1978) and particularly Sæther (1974, 1977).

2.5.3 THE MALE GENITALIA

In the male **genitalia** (Figure 2.5a–c), segment VIII is modified only in *Polypedilum* and many Tanytarsini, in which it is anteriorly tapered to a narrow waist where connected to segment VII. Sporadically, the hypopygium may be rotated, variably up to 180°, with rotation occurring at the junction of segments VII and VIII. This rotation occurs notably in species

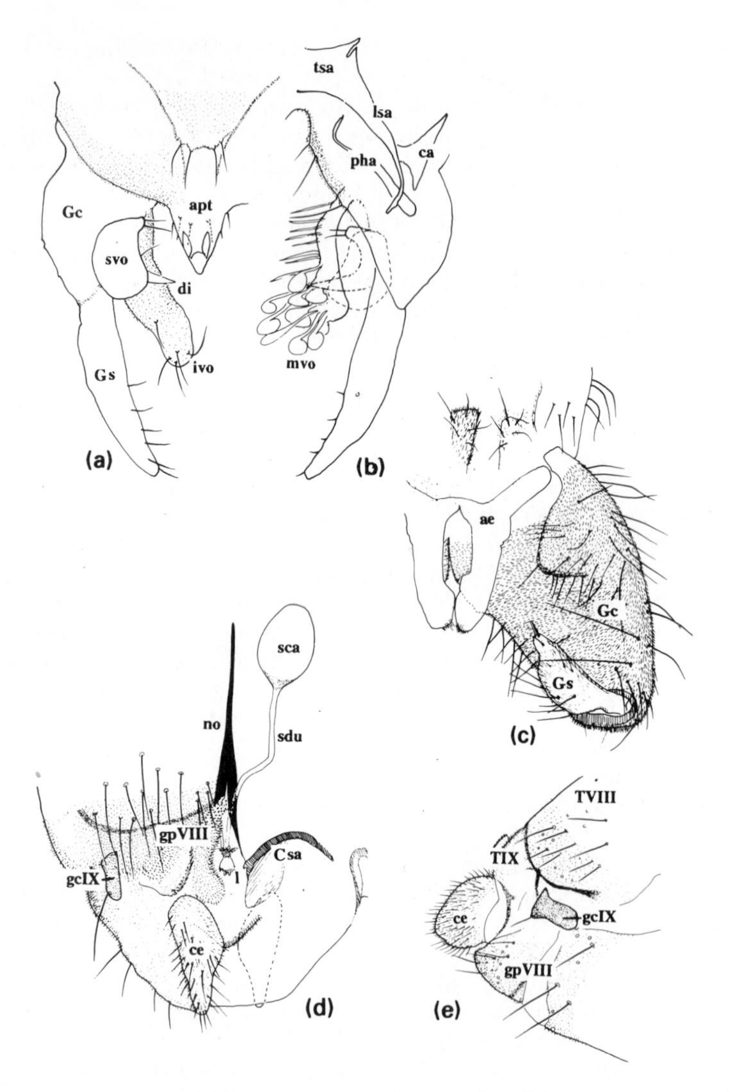

Figure 2.5 Adult morphology. (a) (b) Male genitalia of *Paratanytarsus* sp.– (a) left half, (b) right half, internal; (c) right half of male genitalia (*Thalassomyia frauenfeldi* Schiner); (d) (e) female genitalia of *Conochironomus* sp.– (d) ventral, right internal, (e) lateral. Key: ae = aedeagus, apt = anal point, ca = coxapodeme, ce = cercus, Csa = coxosternapodeme, di = digitus, ivo = inferior volsella, Gc = (gono)coxite, gcIX = gonocoxite IX, gpVIII = gonapophysis VIII, Gs = (gono)style, l = labium, lsa = lateral sternapodeme, mvo = median volsellus, n = notum, pha = phallapodeme, sca = seminal capsule, sdu = seminal duct, svo = superior volsellus, T = tergite, tsa = transverse sternapodeme.

that have lost the aerial mating habit and adopted substrate mating (Fittkau, 1971b).

Posterior to segment VIII lies the major male external genitalia. Features of the genitalia have been subject to several conflicting theories of their origins, with resultant differing terminologies, outlined and mostly reconciled by Sæther (1971).

Tergite IX has a frequent posteromedian extension, forming a specifically distinctive **anal point** (Figure 2.5a). Segment IX laterally is a simple laterosclerite, often fused to the tergite. One of the most distinctive features of the male genitalia are the paired claspers, comprising a basal **gonocoxite** and apical or subapical **gonostylus**. Quite exceptionally, in Chilenomyiinae, there is a second, well developed, pair of gonocoxites (Figure 3.1j) . The gonocoxite appears to be homologous with the female gonocoxite IX, and is thus identified as a homologue of the laterosternite IX of males of other chironomid subfamilies. These homologies necessitate reinterpretation of the theory that claspers originate from sternite X (Brundin, 1983a).

The chironomid gonocoxite bears a variable number of mesal appendages, termed **volsellae** and named for their relative positions (median, inferior, superior, etc) and shape (cuspis and digitus) (Figure 2.5a,b). Furthermore, there may be additional lobes associated with the aedeagus and penis (Figure 2.5c). However, the establishment of homologies for volsellae across rather widely differing subfamilies has not proved easy and a uniform terminology remains elusive. Excepting perhaps many Pentaneurini (Tanypodinae) and some Podonominae, the external structures of the male genitalia provide the most valuable information for species delimitation in most Chironomidae.

The gonostylus is inserted at the apex or immediate subapex of the gonocoxite, where it articulates against the gonocoxite. In Chironominae, the gonostylus appears fused to the gonocoxite, but a variable amount of movement is possible, and the adductor and abductor muscles of the gonostylus are present in the gonocoxite (Wensler and Rempel, 1962). In the Chilenomyiinae, the gonostylus is subapical and rotated through 90° to articulate against the apex of the gonocoxite (Brundin, 1983a).

The internal supporting structures or apodemes (Figure 2.5b) vary, though less markedly than the external structures. Terminology follows Schlee (1968) and Sæther (1971). Important and ever-present structures are a **sternapodeme** (the sternal apodeme), which may be divided into one transverse sternapodeme and paired lateral sternapodemes, each of which is associated with a **coxapodeme**, the apodeme within the cavity of the gonocoxite. The **phallapodeme**, the anterior margin of a hyaline aedeagal lobe, is a pivoted lever which everts the endophallus (penis). The **aedeagus** itself, a complex intromittent organ which passes a spermatophore (sperm package) to the female spermathecae, is formed from gonapophysis IX. In the Orthocladiinae, the penis cavity may contain a complex sclerotized virga, which may be associated with the distal endophallus and is everted by the movement of the phallapodemes.

The Chilenomyiinae are aberrant, as in many genitalic features, in lacking any coxapodeme, possessing an arched sternapodeme that ends free in the gonocoxite cavity and does not articulate with the phallapodeme (Figure 3.1j).

2.5.4 THE FEMALE GENITALIA

The terminology and homology of the female genitalia (Figure 2.5d,e) derives from the study of Smith (1969) as understood and applied by Sæther (1974 *et seq.*). Fundamental to the interpretation is the recognition of homologies with the primitive **telopodites** (or **gonostyli**) and **endites** (**gonapophyses**). These are variously modified lobes on abdominal gonocoxites VIII and IX, some of which are reduced to sclerotized apodemes.

In the female, tergite VIII is modified only in the Telmatogetoninae, in which it is strongly reduced. Sternite VIII may be unmodified or may incorporate gonocoxapodeme VIII as thickened internal ridges. However, it is gonapophysis VIII that provides particularly diagnostic morphological features, notably when divided into two lobes. These lobes, termed dorsomesal and ventrolateral, vary diagnostically in their relative size and orientation, shape and surface microtrichia. Gonapophysis VIII (or, if divided, its ventrolateral lobes) forms the walls of the medially placed vagina.

Tergite IX is well developed, with a frequent tendency to median division. Sternite IX is membranous except for the sclerotized coxosternapodeme. The gonapophysis of segment IX forms an anterior projecting notum, a caudal rami, posterior paired or fused labia and, in Chironomoinae, a variably distinctive apodeme lobe that links with gonapophysis VIII (Sæther, 1989). Gonocoxite IX is variably developed but is often a laterally protruding bulge. Phylogenetically significant fusions of some or all the sclerites of segment IX may occur. Segment X is not differentiated into tergites and sternites, but bears a posterior postgenital plate and a pair of one-segmented cerci.

Internally, there are three sclerotized seminal capsules primitively, with frequent reduction to two, and they are lacking altogether in the Telmatogetoninae (Figure 3.2b). The spermathecal ducts are usually separate for much or all of their length, terminating in separate or common openings on a spermathecal eminence, which lies anterior to the labia, within the vagina. In the Telmatogetoninae, the dilated spermathecal ducts (Figure 3.2b) act as sperm storage organs.

With this introduction to the morphology of the Chironomidae, we have the background to discuss the systematics in the next chapter.

3

Systematics

P. S. Cranston

3.1 INTRODUCTION

There is a voluminous published literature on Chironomidae (see bibliographies by Fittkau *et al.* (1976) and Hoffrichter and Reiss (1981). Many papers provide taxonomic descriptions. There are good reasons for this predominance: current estimates of insect diversity imply that many, perhaps even a substantial majority, are undescribed. The Chironomidae resemble other insects in that documentation of the world fauna is far from fulfilled. An enormous and perhaps insurmountable task remains. The target of complete world inventory is made more elusive by the requirement of users of chironomid taxonomy for descriptions of all lifehistory stages.

The responsibility for documentation lies with systematists, who describe, name, classify, identify and elucidate phylogeny and biogeography. Taxonomy, the theory and practice of describing, naming and classifying, is distinguished from systematics, which includes taxonomy but also involves the study of the diversity of and relationships between different organisms. Chironomid studies demonstrate a continuum from purely descriptive to investigations of phylogeny and related matters such as historical biogeography.

In this chapter, the theoretical basis for classification and phylogenetic determination is outlined. This is followed by the results of the application of phylogenetic methods to the Chironomidae, with contentious areas highlighted.

3.2 CLASSIFICATION AND NOMENCLATURE

The division of chironomids into groups – that is, classification – reflects ideas on the concepts of resemblance and relatedness. These two concepts may be the same but this is not necessarily so. This basic conflict in systematics gives rise to opposing methods and varying conclusions. Before

The Chironomidae: Biology and ecology of non-biting midges
Edited by P.D. Armitage, P.S. Cranston and L.C.V. Pinder
Published in 1995 by Chapman & Hall. ISBN 0 412 45260 X

outlining these methods, the distinction between classification and identification should be clarified: identifications can only be made within a pre-existing classificatory framework.

Biological **classifications** are hierarchical arrangements of groups within groups, in which the levels (categories or ranks) represented in this hierarchy bear names. The application of these names are governed by rules of nomenclature (ICZN, 1985). These rules endeavour to maintain stability of names by reconciling 'priority', the historically correct name, with consistent 'usage' when there is conflict. The fundamental nomenclatural category is the **species**, which is placed into increasingly inclusive categories, both obligatory and optional. Categories used in chironomid nomenclature are given below, with their standard suffixes in bold:

- Superfamily: Chironom**oidea**
- Family: Chironom**idae**
- Semifamily: Chironom**oinae**
- Subfamily: Chironom**inae**
- Tribe: Chironom**ini**
- Genus: *Chironomus*
- Subgenus: *Chironomus*
- Species: *plumosus*.

The subtribe category, which has been recognized rarely, is given the suffix **-ina** as in Zavrelina. The subspecies category was used widely in early descriptive studies for forms that differed subtly in morphology or colour. Many of these infraspecific taxa have proved subsequently to be either morphologically distinctive species, or to fall within the previously unrecognized range of variation of a polytypic species. Thus, although the use of subspecific categories continues in some groups (e.g. birds, butterflies) there is little use of infraspecific concepts in contemporary chironomid studies.

Some nomenclatural activity in the Chironomidae appears to conflict with the desire for stability of names. Changes often result from the discovery of **synonymy**, when a species is found to have been described as new (and named as such) on more than one occasion – indeed on as many as a dozen times for some common European species. Synonymy and commensurate name-changes also occur when the concept of a genus is believed to be unsubstantiated and is identical to another, with an earlier name. Changes occur when a generic name is found to have been applied to another animal taxon, as when Beck's chironomid studies were commemorated in the name *Beckiella* (Sæther, 1977b) although the name was already in use for an oribatid mite genus (Grandjean, 1964).

Another reason to change a name is when two different species in the same genus come to have the same species name (**homonymy**), either through original description (primary homonymy) or through subsequent union in the same genus (secondary homonymy). Perhaps the most confusing nomenclatural matter concerning students of Chironomidae was the prolonged use of the family name Tendipedidae by some workers. Following over a century of the use of *Chironomus* (and *Tanypus*), names that originated with Meigen (1803), an earlier, previously unknown paper by Meigen (1800) was discovered by Hendel (1908) in which the names *Tendipes* and *Pelopia* had been proposed for identical concepts to *Chironomus* and *Tanypus*. It took 55 years before the matter was resolved (see Ashe, 1983 for an excellent discussion of this and other historical nomenclatural matters).

Rules govern which name is to be applied to a taxon. Is the 'correct' name only a matter of nomenclatural pedantry that affects only a small clique of taxonomists? Since a major function of the biological classification system is to provide an information storage and retrieval system of great capacity, a species name provides a label for all biological knowledge pertaining to that animal. Taxon names of various ranks, but particularly that of the species, allow access to all literature concerning the species, from behaviour to ecology and physiology. If this information is stored (or hidden) under numerous different names that in reality apply to the same animal, synthesis is made very difficult.

The second major function of classifications is to allow prediction of biologically significant features. Entomologists are faced with many little-known or even undescribed taxa. Species-level information is often lacking. Classifications ought to be predictive, allowing integration and extrapolation of knowledge from well-known species to their lesser-known relatives. However, classifications do not always make sensible predictions: one based upon geographic distribution, pest status or a single life history stage is limited in prediction of other features. Classifications that reflect evolutionary history, or the pattern of nature, have greater predictive value.

Several methods are used to deduce pattern and relationships and implement the results into classification. Since different assumptions may lead to different estimates of relationships, altered classification (and names) can result. Likewise the incorporation of new or different data may have a significant effect upon classification. Continued study of the Chironomidae and implementation of current and new methods will undoubtedly lead to novel and perhaps dissenting views on classification.

3.3 CHARACTERS FOR SPECIES RECOGNITION IN CHIRONOMIDS

Discrimination of individual taxa from their relatives involves recognition of **characters**, which are features of organisms that may be expressed in different ways, called **states**. Often the most useful (and non-controversial) characters are those whose states are presence or absence, such as a particular wing vein, or seta. Other useful features may exhibit discontinuous variation in states (such as the number of antennal segments, or of setae in a particular area) or continuous variation between limits, such as absolute length or the shape of a structure. Ratios of two independent features may constitute a discriminatory feature, but rarely have these been examined for allometric influences, which would threaten their validity.

Traditionally, gross features of external morphology were examined, with features of adults such as wing venation, hairiness and colour patterns important in providing discrimination. Morphometrics were used early in chironomid studies, with the discovery that the relative lengths of the segments of legs and antennae often varied in a species-specific manner. As microscope techniques grew, more subtle features such as structures of the genitalia and setal arrangements were found to aid in species discrimination and these provided evidence of increased numbers of species amongst adult chironomids.

With time, and arising from demands of users of taxonomy, the immature stages were examined in more detail and a wealth of novel discriminatory characters were found in the pupal exuviae. Features of similar significance in larvae have been acquired with greater difficulty, and the external morphology of this stage is the least likely to allow discrimination from close relatives. Eggs can provide useful characters at supraspecific rank and even at species level (Nolte, 1993).

It is beyond the scope of this chapter to investigate and document the range of characters used in modern systematics. The morphology from which taxonomically discriminatory characters are drawn is outlined in Chapter 2. Examination of any recent taxonomic paper shows the wealth of characters in all developmental stages from which diagnostic and descriptive information can be obtained. These data may be expressed either in cursive form, or more concisely and comparably in a tabular manner. Whether diagnostic features should be accompanied by uncritically assessed extensive morphometrics is a matter of personal opinion (e.g. Cranston and Oliver, 1988a), constrained by the increasing demands for economy and speed of delivery of the products of taxonomists' labours.

3.3.1 CYTOTAXONOMY

Although features derived from external anatomy provide the characters used most often in chironomid taxonomy, there is a wealth of information provided by the chromosomes. As in all Diptera, there are relatively few chromosomes ($2n = 4$–8), paired as double structures, and often forming polytene chromosomes through **endomitosis**. This is mitosis unaccompanied by nuclear (or cell) division, thus allowing association of 3000–4000 chromatids. In certain cells, polytenization can lead to 2^{13}-fold increase in DNA (Daneholt and Edström, 1967). Polytenization and despiralled state of the DNA leads to very elongate interphase chromosomes, up to 800 μm in chironomids, over 100 times the metaphase length. These 'giant' polytene chromosomes, within equivalently large nuclei, are notable in the salivary glands. However, they are not restricted to these tissues, being found also in the epithelium of midgut, hindgut and the Malpighian tubules. In fresh chromosomal preparations, but even more evident in appropriately fixed and stained material, morphological structures of taxonomic and potential phylogenetic significance are evident.

Michailova (1989) in her review of chironomid polytene chromosomes, lists these features as follows:

- Banding pattern: each polytene chromosome has a species-specific pattern of bands.
- Centromere shape and position.

- Puffs and giant puffs (Balbiani rings): number and location of functionally active bands (puffs).
- Nucleolus organizer: the number and location of nucleoli.
- Chromosomal rearrangements: i.e. homozygous reciprocal translocations (in which whole arms are exchanged between non-homologous chromosomes), chromosome fusions and homozygous inversions (in which sections of chromosome, identified by specific banding patterns, are turned around between two break points).

Cytology has proved to be of greatest value in the taxonomic discrimination of morphologically similar (sibling) species. The technique is valuable in direct assessment of species status, through observation of the chromosomes of laboratory-induced hybrids, and monitoring of the actual occurrence (or lack thereof) in nature. The summary of Michailova (1989) of the use of cytotaxonomy in chironomid systematics should be consulted for cytotaxonomic keys for many taxa and more detail of the methods and their application. Cytology is used in phylogenetics (Martin, 1979) but, with some exceptions, methods of interpretation appear to lag a little behind those used by morphologists.

3.4 PHYLOGENETIC RECONSTRUCTION

Before outlining current consensus views on the phylogenetic classification of the Chironomidae, it is appropriate to outline how phylogenies (and hence classifications) are estimated.

The human need to classify came long before scientists had ideas on how classifications originated. For example, the English name 'midge', 'mygg' of Swedish and 'myg' of Dutch derives from 'muggia' in old Norse. As early as Aristotle, classification was a search for 'natural' groupings. Linnaeus's uniform nomenclatural system in place of vernacular names continued and revived interest in classification. With the recognition of the importance of homology, the outline of a natural hierarchy of taxa was well under way when Darwin provided a theory to explain the origins of the pattern. Darwin understood that the natural system and 'true' classification should be based on genealogy, that is, relationship by descent. In reality, the process of taxonomy changed little, but what did change was the justification for the classifications. These were now erected through some intuitive recognition of the course of evolution. The evolutionary changes of some 'important' characters were traced with justification for use of characters ranging from evolutionary conservativeness to postulated adaptive significance. Fossils could be incorporated, although often recognized inappropriately as actual ancestors of extant taxa. The resultant 'evolutionary' classifications were often similar to current ones, but the methods used in their construction were essentially untestable and unrepeatable.

Dissatisfaction with the subjectivity of this 'evolutionary' approach led to two contrasting explicit methods of assessing phylogeny. These approaches developed into the schools of **phenetics**, or numerical taxonomy, and phylogenetic systematics, which has come to be known as **cladistics**.

3.4.1 PHENETICS

Some systematists, including some studying Chironomidae, regard the calculation of overall similarity as an objective means of grouping and classification. This has been termed numerical taxonomy, but since numerical methods are used by most systematists, phenetics is a more appropriate term. Pheneticists argue that patterns based on resemblance are identifiable, whereas the pathways of the evolutionary process are unknowable.

Pheneticists adhere to the following principles:

- The more characters used, the more informative and, up to a point, the better the resulting classification.
- *A priori* every character is of equal weight.
- Overall similarity between two comparable taxa is an additive function of the similarities in individual characters.
- Degrees of similarity are non-uniform, thereby allowing recognition of distinct clusters.
- Taxonomy should be empirical.
- Classification should be based on phenetic resemblance.

Furthermore, it is implicit that overall similarity does estimate phylogenetic relationship, as far as that is possible. The advent of computers allows the handling of the many characters required to estimate similarities, which are expressed in the form of a **data matrix** of characters (allocated to states) and taxa studied. After compilation of the data matrix, distance (or similarity) coefficients are estimated between each taxon and every other, based upon all characters. There are many measures of dissimilarity and similarity available for calculation of coefficients. The calculated values are then tabulated into a **distance matrix** of each taxon against every other and two principal types of analysis, **clustering** and **ordination**, can be applied to the data or distance matrix (Sneath and Sokal, 1973). The results can be shown either as a **dendrogram**, a tree-like portrayal of phenetic proximity, or as **cluster (scatter) diagrams** of distances between organisms.

Although several groups of Diptera have been subjected to phenetic analysis, chironomid studies using this method are very limited and only Roback and Moss (1978), Roback (1979) and Rossaro (1979, 1981, 1989) have employed phenetic methods of analysis. Roback and Moss (1978) delimited species and higher category relationships within the non-pentaneurine Tanypodinae and Rossaro has examined relationships within the *Cricotopus* and *Diamesa* groupings.

3.4.2 PHYLOGENETIC SYSTEMATICS OR CLADISTICS

The German dipterist Willi Hennig also expressed concern about the subjectivity inherent in biological classification. Hennig differed from those that adopted phenetics in arguing that relationship could be clearly defined only in terms of recency of common ancestry (genealogy) and that genealogy could be identified by a set of procedures (Hennig, 1950, 1966). The first is the detection of derived homologous character states, termed **apomorphies**. These character states, when shared among taxa, are termed **synapomorphies** and indicate shared (common) ancestry. However, advanced character states restricted to one taxon only, termed **autapomorphies**, convey no information on the relationships of that taxon. Primitive character states, termed **plesiomorphies** (**symplesiomorphies** when shared) cannot indicate relationship. Actually these are relative terms: an autapomorphy of a genus is a synapomorphy for all the species of the genus. Likewise, a generic-level plesiomorphy is an apomorphy at some higher level, perhaps at the tribal or familial level.

This fundamental distinction between derived and ancestral character states allows the recognition of sets of nested synapomorphies of greater or lesser generality as reflections of nature's hierarchy. From these sets it can be seen that every taxon at any rank uniquely possesses a **sister group** – a taxon which is its closest relative. Thus, reconstruction of the phylogeny can be seen as a search for successive sister groups, through recognition of increasingly more general synapomorphies.

The hierarchical pattern derived by cladists from these analyses are branching diagrams called **cladograms** (e.g. Figure 4.3). A cladogram reflects the distributions of character states, and systematists usually take this to be a representation of the likely evolutionary relationships of the taxa. The **node** (point of bifurcation) in a cladogram represents the character state homologies shared by taxa united by the node. It does not represent an actual ancestor but may be termed a **hypothetical ancestor**. The term **groundplan** refers to a set of predominantly plesiomorphic character states that define the basal node of that group.

A hypothetical ancestor and all of its descendants, the products of cladogenesis, form a **monophyletic** group or **clade**. If either a single taxon or a larger monophyletic group descendent from a hypothetical ancestor is excluded from a larger group, then the residual taxa form a **paraphyletic** group, lacking some descendants of the ancestor. A group that is broader than that derived from a single common ancestor is **polyphyletic**. These terms for groupings can be defined in terms of the character states that define them: a monophyletic assemblage should be united by synapomorphy, a paraphyletic assemblage by only symplesiomorphy and a polyphyletic assemblage by convergent character state(s). Cladists seek monophyletic groupings by recognition and analysis of apomorphic character states, distinguishing these from **homoplasious** (non-homologous or convergently acquired) and plesiomorphic states.

Early cladists handled relatively few characters and constructed cladograms by hand; many contemporary exponents, particularly those studying Chironomidae, follow this practice. With the advent of computers, cladistic analyses can utilize extensive data matrices derived from numerous characters across many taxa.

Cladistics differs from phenetics in considering **polarity** of the character states. The assignment of character polarity, that is distinguishing derived (**apomorphic**) from ancestral (**plesiomorphic**) character states, is crucial to the cladistic method. Hennig (1950, 1966) recommended several ways to assess the direction of change of characters: the three most widely accepted are outgroup comparison, ontogenetic transformation and the use of fossils. Although the palaeontological record reveals several crucial fossils (Chapter 4), the general use of palaeontological evidence in phylogenetic reconstruction is limited and alternative methods have wider applicability.

The **outgroup** method may be applied *a priori*, with polarity assigned before analysis, or *a posteriori*, with polarity inferred after analysis. In contrast, polarities derived from ontogeny, are applied only prior to analysis. The outgroup method depends upon comparison of the character states observed in the study group (the **ingroup**) with those seen in sister groups (outgroups). Assessment of the polarity of each ingroup character state is made by wider reference: if two states are present for a character, then it is assumed that the state occurring in the outgroup(s) is more ancestral (Watrous and Wheeler, 1981). Polarity assignment using this outgroup method assumes some prior knowledge of the phylogeny of the broader group under study in order to establish appropriate outgroups for analysis.

Ontogeny emphasizes the transformation of characters during the growth and development of an individual. The general utility of ontogenetic data is restricted and metamorphosis in holometabolous insects clearly complicates ontogeny. Features from all developmental stages of chironomids have long been used, but they do not appear to have been applied ontogenetically for deducing character polarities for phylogenetics. The discrete developmental stages (the instars) provide excellent opportunity to test, for example, congruence between phylogenies derived from ontogenetic evidence and those using other criteria, and to examine character homologies throughout life cycles. Descriptions of all four larval instars have been made (e.g. Roback, 1989) and the changes in larval morphology through development have been examined by Mozley (1979) and reviewed and extended by Olafsson (1992). Such data clearly allows determination of the plesiomorph condition of, for example, the tanypod mentum and prementum, and the 'predatory-type' mandible with apical tooth dominant.

Hennig believed that **fossils** could be used to determine character polarity with the geologically earliest appearance of a character state being plesiomorphic relative to a homologue appearing later in the fossil record. However, although an ancestral character state must precede the derived, the fossil record may not reflect the temporal sequence because of imperfect preservation. With the exception of amber materials, the paucity of chironomid fossil representation and the imperfect nature of preservation also means that vital characters are obscured or lost. None-the-less, fossil material has proved vital in chironomid biogeographic studies (section 4.5.2).

Although it is conventional to ascribe the cladistic methods of phylogenetic deduction to Hennig, there is a substantial earlier history of systematists who recognized problems with classification by 'archaic' features, interpreted the importance of parallel loss of character states and argued the need to use recently derived characters in establishing relationships. Edwards

(1926b: 113), for example, made quite explicit statements when addressing the 3rd International Entomological Congress, stating that 'resemblances between species, even in striking features, are not necessarily indicative of a close relationship'. In continuing, he discussed how misleading resemblances arise, including 'parallel retention in this stage of numerous archaic features'.

3.5 CONFLICT IN PHYLOGENETIC RECONSTRUCTION

3.5.1 CONFLICTS IN DATA

One might ask, given Hennig's explicit method, why there are such polemical views on phylogenetics amongst chironomid workers. One may cite Schlee's (1975) questioning of the seminal work of Brundin (1966), and criticism of aspects of Sæther's phylogenetic methods by Cranston and Humphries (1988) and Murray and Ashe (e.g. 1985). The problems basically derive from conflicts in data. If all character states were observed to change (transform) in concert, with novel characters from any or all life-history stages providing support for a single phylogeny with no incongruence, then there would be no dispute. However, this is not the case. Some characters will support one particular phylogeny, while others support one or more alternative phylogenies. In fact, from observations on many data sets from many different sources, but including substantial ones concerning Chironomidae, it seems that characters whose transformations are fully consistent are very rare. Parallel and convergent development, phenomena subsumed under the term **homoplasy**, are frequent. The gain of a novel character state, but subsequent loss and possible regain seems to be prevalent, in contravention of Dollo's law of irreversibility. Despite the implicit views of Hennig that character incongruence could only be due to faulty assessment of homology and therefore was amenable to reconciliation by more intensive re-examination, homoplasy is common and may not be susceptible to more detailed analysis.

(a) Weighting

Two solutions have been proposed to deal with homoplasy, each with differing philosophical foundations. The first involves the specialist judging the relative probabilities that transformations of particular characters accurately reflect phylogeny. Thus some characters are judged to give more reliable estimates of phylogeny whilst others, deemed to be unreliable, are permitted to have high variability or are excluded from consideration. Arguments put forward in favour of recognition of highly reliable characters include:

- Complexity (Dollo's law: once evolved, never lost, or lost only with difficulty).

- Non-adaptive characters in which selection is deemed to be reduced (e.g. Sæther, 1989).

Selection is believed to increase the likelihood of **parallelism**, i.e. multiple evolutionary origins of apparently identical character states. Conflict (homoplasy) is restricted where possible to the 'unreliable' characters. Postulated reliability is used to weight characters, with minimum weight being zero – i.e. the characters are eliminated completely from consideration. This method is amenable to analysis by hand, especially if a restricted set of the many possible characters is assessed.

(b) Parsimony

The second approach is to treat change (transformation) from one state to another in any character as equally likely to reflect phylogeny. Analysis involves minimizing the number of character state changes across all characters measured in relation to a cladogram. This method is termed **parsimony analysis**, derived from application of parsimony to a character matrix and the minimization of the number of evolutionary origins (or alternatively, maximizing the number of characters whose transformation is fully congruent with the proposed phylogeny). Parsimony analysis does not infer that evolution acts in a parsimonious manner, but only that the principle of parsimony is the only means of choosing between alternative hypotheses of phylogeny (alternative cladograms).

Unless few characters and taxa are assessed, computation is generally required for analysis, with different programmes involving variable, but often undiscussed, assumptions.

3.5.2 INCONGRUENCE OF CLASSIFICATIONS FROM DIFFERENT STAGES

Amongst the persistent difficulties with classifications, particularly phenetic ones, is their susceptibility to different life-history stages giving incongruent phylogenetic estimates. For Chironomidae, these problems and how they might arise are addressed in some detail by Roback and Moss (1978). However, the basic problem is not discussed: a species has only one phylogeny and each life stage (and part of that stage) shares an identical phylogeny – incongruence between phylogeny estimates for different life-history stages arises only through faulty hypotheses of phylogeny. The use of phenetics, which identifies patterns of resemblance rather than phylogenetic proximity, risks misleading results – and incongruence (or faulty resolution) may arise as an artifact of the phenetic method itself.

One means to establish the relative contribution of information from different life-history stages was provided by Cranston *et al.* (1990) in cladistic analyses of *Kiefferulus* and its relatives. Tests were designed to reveal the relative contributions of cladistic information from each stage and to identify whether some classificatory conflicts could be attributed to reliance upon few characters in a single life history stage. Furthermore, a subset of the total data (ventromental plate data alone) allowed assessment of the reliability of information derived from a highly functional suite of characters – those associated with silk production. Phylogenetic analyses performed on data subsets, namely matrices derived separately from adult, pupal, larval and larval ventromental plate characters, showed that little reliance could be placed on minimum-length trees for any subset, using statistical tests derived from randomization of the data (Cranston *et al.*, 1990). In contrast, the cladogram derived from the combined data appeared to be robust under the same tests and this was taken as justification for viewing the total character set as representing the best estimate of phylogeny.

It was instructive to compare trees derived from cladistic analyses of each subset (larva, pupa, adult and larval ventromental plate character matrices) with that derived from the total characters set. Analysis of the larval data matrix alone supported the 'total evidence' tree, with a sister-group relationship between *'Carteronica' longilobus* and *Kiefferulus*, these two being the sister-group of *Nilodorum*. The adult character matrix gave a similar pattern to the larvae, but, in contrast, pupal evidence produced uninformative and partially incongruent phylogenetic trees in comparison with the 'total evidence' tree.

Cranston *et al.* (1990) believed these results cast doubt on pupal evidence in elucidating relationships, thus contradicting the views of Roback and Moss (1978) and Coffman (1973) who argued, based largely upon theoretical considerations, that lack of selection on pupae actually ought to lead to better reflection of phylogeny. Langton (1989 and Chapter 8) makes a good case for parallel selection leading to repetitive expression of pupal morphologies from an otherwise suppressed part of the genome (i.e. homoplasy is rampant). This may explain the high levels of cladistically-revealed homoplasy and frequent apparent inability to detect any strong phylogenetic signals from pupal evidence (Cranston, unpubl.). None-the-less, the view that identifiable selection disguises any phylogenetic signal may be countered by the finding that ventromental plate structure, which is highly selected for silk production, tracks the putative phylogeny within *Chironomus* (Webb *et al.*, 1990), *Kiefferulus* (Cranston *et al.*, 1990) and of the Chironominae in general (Webb and Cranston, unpubl.).

The difficulty with discussions of which characters should be included or excluded according to whether they appear to be under selection or not is that we have little or no idea of how selection operates, despite views to the contrary (e.g. Sæther, 1986, 1989). How are we to distinguish which (if any) characters are unselected, which are under selection and which are pleiotropic effects of others? This has a direct bearing on the question of whether weight should be placed on any character to represent its likelihood of being congruent with phylogeny. The logical answer is to use total evidence (i.e. all characters obtainable, unweighted) (Kluge, 1989) to derive phylogenetic hypotheses, and use cladistic permutation tests (Faith and Cranston, 1991; Faith, 1991) to establish the reliability of estimates of phylogeny, in the manner used by Cranston *et al.*, (1990). Furthermore, such analyses to detect the pattern of relationships must be made prior to the genesis of assumptions about processes such as selection and its effect on the distribution of homoplasy.

As usual, Edwards (1926b:112) had addressed these issues, succinctly stating:

> It is inconceivable that, rightly understood, the evidence as to ancestry provided by the larvae should be in conflict with that given by the adult forms. Wherever such appears to be the case, it can only be due to a mis-reading of the facts.

In the ensuing paragraphs Edwards pointed out that an incongruence between larval and adult mosquito classifications arose because certain very obvious characters (in this case, of the adult) were not of phylogenetic significance. However, small and largely overlooked characters of the adult 'were in perfect accord with the arrangement based on . . . the larvae'. This, and continuous confirmatory evidence from the Chironomidae, is the strongest warning against the *a priori* inclusion or exclusion (*i.e.* weighting) of characters – we simply do not know in advance which characters provide the phylogenetic message.

3.5.3 DATA MATRICES

A fundamental problem remaining in phylogenetics, possibly at the root of many dissenting opinions, lies in debates concerning the distribution of characters across a suite of taxa. It may seem evident that analysis requires comparable information across all taxa under study: i.e. theories of homology must be made explicit. A major contribution of phenetics has been the demand for formal presentation of the evidence for conclusions in the form of a data matrix, usually accompanied by discussion of the reasons for allocation of particular states to each character. Presented with this rigorously acquired evidence, others can verify, test and, if necessary, refute character assignment and homology assessments.

The difficulties in deriving a character/taxa matrix for further cladistic analysis based only on discursive published information was exemplified and acknowledged by Cranston and Humphries (1988) in seeking to derive a matrix from Sæther (1976). Sæther (1990) justifiably complained about the erroneous nature of the matrix that Cranston and Humphries had derived solely from publications. However, the previous failure to provide any data matrix had required those who wished to explore the significance of phylogenetic conclusions to compile their own, from less than adequate literature. The corrected matrices (Sæther, 1990a: Appendices 3 and 4) represent the first and, thus far, only data matrices published in 25 five years of 'qualitative Hennigian' analyses (Sæther, 1990a). Describing cladistics as 'the vagaries of quantitative phenetics' or 'neocladistics' is a poor substitute for explicit documentation of character states across all taxa investigated. Data matrices are explicit statements about characters observed and their putative homologies and they allow comparisons between data sets under different methods of analysis. Chironomid systematics can only benefit from such information.

3.6 THE FAMILY STATUS AND PHYLOGENETIC POSITION OF THE CHIRONOMIDAE

The Chironomidae belong to the suborder Nematocera, the phylogeny of which has been most recently examined by Wood and Borkent (1989), to which readers are directed for detailed discussion. Briefly, the Chironomidae and Ceratopogonidae combined form the sister group to the Simuliidae (black flies). These three families form a unit that Hennig (1973) referred to as the family group Chironomidea. Inclusion of the small family Thaumaleidae as sister to the Chironomidea forms the superfamily Chironomoidea. Together with the superfamily Culicoidea (mosquitoes and relatives), the Chironomoidea form the infraorder Culicomorpha.

Chironomidae have been called colloquially 'non-biting midges' since Malloch (1917) and Edwards (1926a, 1929) recognized separate family status for the biting midges (Ceratopogonidae). Edwards (1926a: 390), in his listing of the differences between the two families, contrasted the ceratopogonid mouthparts, complete with toothed mandibles and maxillary blades (=laciniae), with the reduced chironomid mouthparts lacking these structures. In fact, these 'reduced' chironomid mouthparts lack only the toothed mandibles; the labrum, hypopharynx and labium are present and, in conjunction with the soft labella, form a food canal leading to a cibarial pump. Furthermore, laciniae appear to be present in many species (Burtt *et al.*, 1986; Cranston *et al.*, 1987). In addition, the apical armature of the labrum, hypopharynx and laciniae are much reduced in non-biting species of Ceratopogonidae and in virtually all Chironomidae.

In questioning whether the genus *Archaeochlus* belonged to the Chironomidae, Cranston *et al.* (1987) reviewed the list of 20 characters of adult midges that Edwards (1926a) believed to separate Ceratopogonidae from Chironomidae. *Archaeochlus* resembles a ceratopogonid in some of these features; for example, the possession of functional mandibles, the presence of alular setae, and the foreleg being subequal in length to the mid and hindlegs. However, these, and some other similarities are plesiomorphic character states in the superfamily Chironomoidea: they do not indicate internal relationships. Apomorphic character states, including loss of wing vein M_2 on a relatively large wing, male hypopygia lacking strongly sclerotized parameres, well-developed five-segmented larval antenna and larval procerci indicated that *Archaeochlus* could be assigned to the Chironomidae (Cranston *et al.*, 1987). These features thus remain as defining characteristics of a monophyletic Chironomidae.

Although many of the characters used to define the sister group, the Ceratopogonidae, are in the plesiomorphic state, separation of the taxa at family level has been universally accepted. With recognition of an apomorphy for larval Ceratopogonidae, namely the pharyngeal apparatus being strongly developed with paired divergent arms and rows of combs (Borkent *et al.*, 1987, Wood and Borkent, 1989), the monophyly of this family is assured.

As Wood and Borkent (1989) observe, the family-level relationships amongst the aquatic nematocerans are the least contentious in dipteran phylogeny.

3.7 SYSTEMATIC SURVEY

3.7.1 TELMATOGETONINAE

The Telmatogetoninae, with world-wide distribution, is a small grouping of predominantly marine intertidal midges, formerly placed with several other genera of inter-tidal midges centred on *Clunio* in a subfamily (or tribe) Clunioninae (or Clunionini). Strenzke (1960a), in an early explicit application of Hennig's phylogenetic methods, demonstrated that the resemblance between these intertidal midges was convergent through similar life styles. The *Clunio* grouping was postulated to be the sister group of some terrestrial Orthocladiinae, centred on *Smittia*, a placement confirmed by Sæther (1977a). Strenzke observed that *Telmatogeton* and allies was a plesiomorphic grouping but, beyond dismembering the Clunioninae, he made no further estimation of its phylogenetic position.

Brundin (1966: 370) examined this problem in his assessment of Diamesinae relationships (section 3.7.7). The general similarity in habitus of *Telmatogeton* to *Diamesa* was noted, but Brundin (1966) confirmed that many characters, including all those of the very generalized larva, the pupal spiraculate thoracic horn, and those of the setose adult (including lack of a postnotal median furrow) were in the plesiomorph state. In an extensive discussion of character states, Brundin recognized modifications, including elements of the distinctive pupa (features discussed in Chapter 8) and the elongate female ovipositor, as apomorphic states. Together with the common presence of 'bunches of hair-like processes on the ventral hypopharynx' (the median ligula brush and lateral paraligulae brushes of the premento-hypopharyngeal complex), these indicated to Brundin a linking of the *Telmatogeton* group to the Diamesinae. Brundin erected the subfamily Telmatogetoninae for the grouping, asserting monophyly and a sister-group relationship to the Diamesinae.

In a study involving a more thorough examination of female genitalic structures, Sæther (1977a) noted: the lack of seminal capsules; the swollen spermathecal ducts; reduced tergite VIII; well developed, lobed tergite IX; ovipositor formed from elongate cerci; and large gonapophyses VIII. Although a placement as sister group to the remaining Chironomidae was proposed, Sæther argued that, were the female alone known, the uniqueness could justify familial rank for the Telmatogetoninae. Ashe *et al.* (1987) took exception to the arguments used in support of this position, observing that plesiomorphies (gonostylus IX present, gonapophysis VIII large and elongate) were used to support the placement. Furthermore, most features of the female were unique apomorphies (autapomorphies) associated with alteration in oviposition behaviour in the intertidal environment. They further raised ecological reasoning to counter the Telmatogetoninae being sister to the remaining Chironomidae, discussing the unlikelihood of a marine–intertidal sister taxon to all extant Chironomidae. Whilst certainly there are some freshwater torrenticolous *Telmatogeton* in Hawaii, as stated by Sæther, this is clearly a radiation from a marine ancestor (Newman, 1977), not the reverse. Sæther (1989) addressed some of the other issues including observing that the hypopharyngeal 'brush' structures of Diamesinae and Telmatogetoninae were not homologous. At present the placement ought to be accepted until further evidence is assessed.

The chromosomal evolution of Hawaiian *Telmatogeton* is discussed by Newman (1977), but the characters shed no light on higher relationships.

3.7.2 CHILENOMYIINAE

The Chilenomyiinae is the most recently erected subfamily of Chironomidae, stemming from the recognition of a high rank for an enigmatic Chilean species, *Chilenomyia paradoxa* (Brundin, 1983a). Although the species is known only from the male and female adult, the unusual morphology warranted systematic placement without reference to the immature stages. The principal unusual features are from the male genitalia, where there are two pairs of well developed gonocoxites (Figure 3.1j). The anterior pair are large, simple lobes which Brundin assesses as homologous with laterosternite IX of other male chironomids and with gonocoxite IX of females. This feature is clearly plesiomorphous within the Chironomidae (and the Diptera in general). The gonocoxite plus gonostylus typical of all other chironomid males is actually the posterior pair in Chilenomyiinae, which implies a derivation from sternite X, rather than IX as previously believed. In the Chilenomyiinae the gonostylus is autapomorphically modified by rotation through 90° and insertion subapically, thus clasping by movement against the apex of the gonocoxite. A further autapomorphy is the sternapodeme which forms a semicircular arch, with sharp points free in the gonocoxite cavity, rather than connected to the coxapodeme at the anterodorsal of the gonocoxite.

Utilizing evidence from these male genitalic features, supplemented by cibarial structure, Brundin proposed that Chilenomyiinae form the sister group of the remaining Chironomidae less the Telmatogetoninae.

The remaining Chironomidae, which form the bulk of the genera and species, fall predominantly into two discrete groupings ranked as semifamily by Sæther (1983). One semifamily, the Tanypodoinae, comprises the subfamilies Podonominae, Aphroteniinae and Tanypodinae. The other semifamily, the Chironomoinae, comprises Diamesinae, Prodiamesinae, Orthocladiinae and Chironominae. The placement of the subfamily Buchonomyiinae in the semifamily system is controversial (section 3.7.5).

3.7.3 APHROTENIINAE

The Aphroteniinae was recognized by Brundin (1966) for three species-poor genera distributed exclusively in the southern hemisphere. All except one of the presently known species were discovered during Brundin's researches on the austral chironomid biota (Brundin, 1966). Five synapomorphies for the included species in the subfamily can be derived from the very distinctive pupae (Brundin, 1966, 1983b): displacement of dorsal setae to the posterior margin; reduction of one dorsal seta to a finger-shaped lobe; prolonged, slender, thin-walled, blunt-tipped tubular thoracic horn; anterior dorsocentral seta (dc_1) stout, arising from tubercle (Figure 3.5k); lateral seta (L_3) on segment VIII spine-shaped and dominant. Additional evidence from the adult includes wing with R_{2+3} and MCu absent (Figure 3.1f), absence of tibial spurs and comb on the hindleg, and absence of a megaseta on the gonostylus. Knowledge of the larva was very limited in 1966, but Brundin's suggestions of the diagnostic uniqueness of the sickle-shape and location on a pedestal of the SII seta, reduction in other labral setae, loss of epipharyngeal features of pecten and chaetulae and approximation of premandibular setae are all upheld with further larval material (Cranston and Edward, 1992). The presence of sclerotized plates on the body cited as diagnostic by Murray and Ashe (1981) is inconsistently distributed through the subfamily, although there are some cuticular modifications in all taxa (Cranston and Edward, 1992).

Schlee (1975) identified some of Brundin's (1966) characters, principally of the adult Aphroteniinae, as occurring outside the clade and argued that Brundin had not rigorously established monophyly as he had claimed. Brundin's (1976) restatement and elaboration incontestably asserts the monophyly of the group.

Sæther (1977a) discussed the female genitalia of the Aphroteniinae, Brundin (1966) outlined the internal phylogenetic relationships and Cranston and Edward (1992) provided additional records and estimated the age of the grouping on vicariance grounds.

3.7.4 PODONOMINAE

This grouping was established for a few northern hemisphere species by Thienemann (1937). The unexpected high diversity in the southern continents was discovered by Brundin and documented in his 1966 monograph. The position regarding the monophyly of the Podonominae resembles that of the Aphroteniinae – Brundin's 1966 observations were criticized by Schlee (1975), to which Brundin (1976) responded in detail. The structure of larval labral sensilla SI and SII being enlarged, sickle-shaped and on elongate pedestals (Figure 3.3g), known since Zavřel (1941b), are synapomorphic features. The pupa provides two putative synapomorphic features in the 'wavy setae' – the characteristic sinuous lateral setae of segments VIII and IX (Figure 3.5i) – and the reduced anal lobe. These pupal character states are widespread but by no means universal within the Podonominae (Cranston *et al.*, 1987). Sæther (1977a) identifies the fusion of the labia of the female genitalia as a synapomorphy of the subfamily, although this feature occurs 'in different type' in the Telmatogetoninae.

The internal relationships within the Podonominae are discussed by Brundin (1966), Schlee (1975) and Cranston *et al*, (1987).

3.7.5 BUCHONOMYIINAE

This subfamily is based upon a single genus *Buchonomyia* Fittkau, with three included species: *thienemanni* Fittkau, from the western Palaearctic and Iran, *burmanica* Brundin and Sæther, from Burma and an unnamed species from Costa Rica (Anderson and Sæther, in press). The genus has had a chequered history, with Fittkau (1955) making a provisional placement in the Podonominae and Brundin (1966) suggesting it to be a possible plesiomorph Orthocladiinae. Following recognition of subfamily status for Buchonomyiinae by Brundin and Sæther (1978), dissent concerning the phylogenetic position was reopened when they argued its position as sister group to the Diamesinae + Prodiamesinae + Orthocladiinae + Chironominae clade. Debate, notably between Murray and Ashe (1981, 1985) and Sæther (1979a,b, 1983), centred on interpretation of genitalic features, since the larva was unknown at that time and the strongly plesiomorphic pupa (Figure 3.5l) appeared to provide little assistance in assessment of relationships. Sæther's (1977a) studies of chironomid female genitalia had provided vital evidence for higher phylogenetic relationships, into which the female of *Buchonomyia* might be expected to be placed without difficulty. However, subfamily relationships relied heavily upon subjectively assessed non-universal characters with limited reliability ('underlying synapomorphies'). Furthermore, some homologies in the male and female genitalic morphology appeared to have been misinterpreted by Brundin and Sæther (1978) and Sæther (1979a,b). Three crucial misinterpreted features – the fusion of tergal and sternal elements of male segment IX, the undivided female gonapophysis VIII and absence of an apodeme lobe – are argued by Murray and Ashe (1985) as supporting their previous (1981) placement of Buchonomyiinae within the Tanypodoinae (Podonominae + Aphroteniinae + Tanypodinae clade). No further information from the pupa is used to support this position. However, knowledge of non-terminal instar larvae of *Buchonomyia* (Ashe, 1985, in press) shows that the absence of a premandible, presence of a labral rod and insertion of stout labral SII setae on well-developed pedestals strongly supports this placement. Within this clade, Buchonomyiinae is sister to Podonominae + Aphroteniinae, based predominantly upon characters of the wing, with the Tanypodinae a sister to this grouping. Sæther's (1989) response to this proposal was to restate and reinterpret some ambivalent features, but without using consistent parsimonious reasoning for character state derivations (viz. discussion on premandibles).

3.7.6 TANYPODINAE

The subfamily Tanypodinae was established by Thienemann and Zavřel (1916) primarily on the basis of the immature stages. The larval feeding apparatus differs strongly from other Chironomidae, with the strong development of premental elements as a ligula and paired paraligulae (Figure 2.1j). The derived (apomorphic) nature of these features, and of the curiously developed mentum, is evident from Roback's (1989) study of the ontogeny of *Djalmabatista*. The monophyly of the Tanypodinae is strongly supported.

The female genitalia of the Tanypodinae resemble those of the Aphroteniinae and Podonominae in which a gonotergite IX is formed from the fusion of gonocoxite IX with tergite IX. However, a synapomorphy for Tanypodinae is the reduction of tergite IX to a narrow, usually bare band (Figure 3.2d) (Sæther, 1977a).

Fittkau (1962) examined the internal relationships, establishing the tribes Anatopynyiini, Macropelopiini and Pentaneurini. Roback and Moss (1978), using phenetic reasoning, erected the tribes Procladiini and Natarsiini. The female genitalia provided Sæther (1977a) with some evidence on internal relationships within the tribe Pentaneurini, but these were not presented formally and deserve further investigation.

3.7.7 DIAMESINAE

The subfamily Diamesinae was established by Kieffer (1923). However, in the following period the included genera tended to be treated within the Tanypodinae (as in older works) or the Orthocladiinae. Even following the listing of their special features by Edwards (1929) there has been no consensus on the subfamily ranking. Pagast (1947) treated the *Diamesa* group as primitive Orthocladiinae and he and Brundin (1956) considered it to be of tribal rank (together with Protanypini, Orthocladiini and Metriocnemini) within the subfamily Orthocladiinae.

Revisiting the problem, Brundin (1966) asserted the monophyly of each of the southern hemisphere tribes (Heptagyini, Lobodiamesini) and the northern Boreoheptagyini, combined as a monophyletic Heptagyiae. Brundin considered the Heptagyiae to be the sister to the Diamesae (Pagast's Diamesini), which Brundin recognized for three tribes, the Diamesini, Protanypini and the enigmatic southern African Harrisonini. Previous larval and pupal features that led Thienemann (1934, 1952) and Pagast (1947) to view the '*Diamesa* group' as Orthocladiinae were symplesiomorphies according to Brundin (1966), as was the possession of MCu in the wing of the imago. Despite finding no synapomorphy for the Diamesinae (Heptagyiae + Diamesae), subfamily status was advocated, and furthermore the Telmatogetoninae were proposed as sister group of subfamily rank (Brundin, 1966)(see also section 3.7.1). Reservations were expressed, however, that the status and monophyly of the Diamesinae could only be resolved by a better understanding of the Orthocladiinae (Pagast, 1947; Brundin, 1966).

Analysis of characters of the female genitalia did not allow Sæther (1977a) to identify any synapomorphies for the Diamesinae and it remains possible that the Diamesinae are non-monophyletic. Internal relationships within the

Heptagyiae are considered by Brundin (1966) and the tribe Diamesini are discussed by Serra-Tosio (1971).

3.7.8 PRODIAMESINAE

The Prodiamesinae was established by Sæther (1976) for three anomalous genera previously treated either as Diamesinae or Orthocladiinae: *Prodiamesa*, *Odontomesa* and *Monodiamesa*. The subfamily is treated as a sister group to the Orthocladiinae + Chironominae, based on the synapomorphic parallel-sided rami of the female gonapophysis IX and the plesiomorphic retention of wing vein MCu. As with the Diamesinae, it is possible that the grouping is paraphyletic to the Orthocladiinae.

3.7.9 ORTHOCLADIINAE

The Orthocladiinae is probably the least understood subfamily of Chironomidae. It is species- and genera-rich, and is essentially the residue after the removal of other, possibly paraphyletic groupings (Prodiamesinae, Diamesinae). The authorship of the subfamily is due to Edwards (1929), with the concept derived largely from Kieffer's Orthocladiariae. With the exception of *Clunio*, it bears a great resemblance to the current view. However, in the intervening period species now assigned to Prodiamesinae and Diamesinae have been included (e.g. Brundin, 1956), the marine genera around *Clunio* have been excluded (e.g. Wirth, 1949) and subsequently restored on well-argued phylogenetic grounds (Strenzke, 1960a) and the subfamily Corynoneurinae has been raised and sunk.

The monophyly of the Orthocladiinae cannot be established on any single character state or even any acceptable combination of characters – exceptions abound and homoplasy is rampant. The only attempts to handle the bewildering array of character combinations by parsimony methods (Sæther, 1989; Rossaro, 1989) were preliminary attempts, restricted by computers and programmes incapable of handling efficiently the substantial data matrices that the authors had generated.

3.7.10 CHIRONOMINAE

The subfamily Chironominae was established by Macquart (1838) and the concept has remained stable since that time. The monophyly of the subfamily is evident, based on the larval synapomorphy of striated ventromental plate and cardo. Although these features are strongly reduced in *Harrisius* and *Stenochironomus*, some elements remain detectable.

Three tribes are recognized, the long-recognized Chironomini, Tanytarsini and the Pseudochironomini, of more recent designation (Sæther, 1977b). The exact composition of the latter tribe is uncertain, however. The monophyly of each tribe is evident, with the Pseudochironomi (comprising at least *Pseudochironomus*, *Manoa* and *Riethia*) at the base, forming a sister group to the Tanytarsini. The phylogeny of the Chironominae is estimated by Sæther (1977a), a cladistic structure largely confirmed by parsimony analyses using features of all life history stages in addition to characteristics derived from the ventromental plates (Webb and Cranston, unpubl.). The relationships of *Kiefferulus* and relatives are assessed by Cranston *et al.* (1990).

3.8 CATALOGUES AND IDENTIFICATION GUIDES

Section 3.7 outlines the subfamily arrangement recognized currently by those publishing on chironomid systematics. For those whose main demands of taxonomy are the provision of distributional information and means of identification generated from these studies, section 3.8 provides an entry into available information. Included is the first key to all subfamilies, world-wide, in all life history stages.

3.8.1 CATALOGUES

A prerequisite for any biological study is to know the research organisms – what is known already, how to recognize them, what is known of their relatives and where to find published information. For some chironomids, this information is compendious; however, the overwhelming number of species means that any such detailed knowledge is patchy and almost arbitrarily distributed. For the vast majority of described chironomids we barely know their natural distributions, let alone what effects we are having on their existence and they on ours.

One of the more humble tasks of taxonomists is 'housekeeping' – the provision and maintenance of records of what we do know about described species. These records are variously termed checklists or catalogues according to some unwritten feelings about how much information should be included (or how diligent the compiler wishes to be). In dipterology, the term 'catalogue' is used most often. The basic information contained is the species name, its current generic placement, the genus to which it belonged when first described, the author, date and publication in which the description was made first and the recorded distribution. Citation of subsequent taxonomic and ecological publications adds value to any such work but is usually beyond the scope of catalogues of chironomid names. The task of compiling these guides tends to be underestimated by the wider biological community. However, the products are long-lasting, authoritative works that are fundamental tools for many scientists.

The following works are the current catalogued references to the Chironomidae fauna based on world-wide or major biogeographical regions:

- Genera and subgenera described until 1983 – Ashe (1983).
- Species and higher taxa from the Palaearctic region described until the start of 1983 – Ashe and Cranston (1990) (with corrections by Ashe, 1993).

- Species and higher taxa for the Nearctic region described until 1989 – Oliver *et al.* (1990).
- Species and higher taxa for the Afrotropical region described until 31 July 1978 – Freeman and Cranston (1980).
- Species and higher taxa for the Australian and Oceanian regions and the Antarctic described until 1988 – Cranston and Martin (1989).
- Species and higher taxa for the Oriental region described until 1970 – Sublette and Sublette (1973).

Notably the Neotropical chironomid fauna lacks a catalogue but the works of Edwards (1931), Fittkau and Reiss (1970), Roback and Coffman (1983) and Watson and Heyn (1993) provide bibliographical and taxonomic entry to the sparse literature.

Major publications updating the chironomid faunal lists for subcontinental or substantial country areas smaller than the major biogeographical areas are listed in section 4.2.4. Papers published since 1983 concerning Chironomidae of the Palaearctic region render Ashe and Cranston (1990) somewhat out of date concerning some nomenclature and many distributions. With the ability to maintain computer data bases, formally published catalogues may become superfluous. None-the-less, the amount of effort required to maintain and update entries should not be underestimated, whether or not computers are used as aids.

3.8.2 IDENTIFICATION GUIDES

Many systematists have produced guides to the identification of chironomids on geographic scales that range from a single drainage system, to continental or, rarely, world-wide overviews. Taxonomic coverage given by keys range from single genera to subfamily or all described chironomid taxa in a geographic area; keys may include all life history stages or, more usually, only selected stages. Many keys are produced as working manuals with little or no intention to publish formally in refereed journals. In fact the difficulty of publishing substantial, authoritative works, and the recognition of the ephemeral nature of some restricted-use keys, means that there are probably more 'grey literature' guides than there are within the sphere of traditional publications.

It is beyond the scope of this section to survey all such keys – there are too many – but an entry is provided into recent major literature. An obvious starting point is the collaborative work that produced the keys to larvae, pupae and adult males of the genera of Holarctic Chironomidae (Wiederholm, 1983, 1986, 1989). These keys to subfamilies and genera allow access to lower-level taxonomic publications by reference to species-level keys published prior to 1989. Adult females, not covered by the Wiederholm (*op. cit.*) works, are keyed to subfamily, tribe and genera by Sæther (1977a).

For Holarctic (Nearctic plus Palaearctic) regional Chironomidae, recent keys are provided by Sasa (1989) for adult males of all Japanese species. The chironomid pupal exuviae from Britain and the West Palaearctic have been keyed by Langton (1984, 1991 respectively). The genus *Limnophyes* in the Holarctic and Afrotropical regions are keyed in all known stages (including female adults) by Sæther (1990b) and the genus *Boreoheptagyia* by Serra-Tosio (1989a). Adult male Chironomidae of the Saudi Arabian peninsula were keyed by Cranston and Judd (1989).

In comparison with the Holarctic region and its African periphery, other parts of the world are much less well endowed with identification guides. The neotropical region is particularly poorly served with keys, with Edwards (1931) providing keys to subfamilies and genera of Patagonian chironomids, and only isolated species descriptions, redescriptions and generic reallocations since. Where keys are available, they usually cover the adults alone, such as Freeman (1959) for New Zealand; Freeman (1961) for Australia, amplified for the Australian Tanytarsini by Glover (1973); and Freeman (1955, 1956, 1957, 1958) for the sub-Saharan African (Afrotropical) Chironomidae. The lack of keys, for example, to the immature stages of the Afrotropical fauna does not indicate lack of recent study, however: for example, Lehmann, Harrison, Dejoux, McLachlan, Sæther, Willassen, Edward and Cranston have contributed to our knowledge of all stages of many elements of the fauna.

There are some exceptions to adult-only keys for extra-Holarctic areas. There are keys to all species known as larvae in New Zealand (Stark, in Winterbourne and Gregson, 1989) and to the immature stages of Chironomidae from the Alligator Rivers Region of monsoonal northern Australia (Cranston, 1991). Brundin (1966) provided seminal keys to all known stages of all species of the subfamilies Aphroteniinae, Podonominae and austral Heptagyiae (Diamesinae), notably across the southern continents. Extension to the keys of the Australian Aphroteniinae are made by Cranston and Edward (1992).

Whilst the Wiederholm (*op. cit.*) keys allow correct identification for many elements of faunas outside the Holarctic region, there is a clear need for comparably detailed, illustrated keys in other geographic areas. The provision of these guides should be a priority for chironomid systematists.

3.9 IDENTIFICATION KEYS TO SUBFAMILIES OF CHIRONOMIDAE, WORLD-WIDE

Identification keys are based on materials preferably examined and drawn by the author or compiled from other published sources. Whilst the key attempts to cover all known material, there will be exceptions unrecorded in the literature or unseen by the author. Furthermore, there are some exceptions to virtually all diagnostic characters used and, as with all keys, these should be used with these limitations in mind.

In the key that follows here, adult wing venation is used widely; however, wing reduction (aptery and brachyptery) occurs in some Telmatogetoninae, Diamesinae, Orthocladiinae and Chironominae. These will have to be identified to subfamily on genitalic features. Furthermore, there are some missing life-history stages, notably the immature stages of Chilenomyiinae, and the larvae of Buchonomyiinae are known in first instar only, although the larval key otherwise concerns features best seen in the final (fourth) instar.

KEY TO SUBFAMILIES OF CHIRONOMIDAE – ADULT WINGED* MALES

1 Wing with crossvein MCu present (Figure 3.1a–d) 2
– Crossvein MCu lacking (Figure 3.1e–g) 7
2 R_1 and R_{4+5} widely separated, R_{2+3} absent (Figure 3.1a,b) 3
– R_{2+3} present; if absent, then R_1 and R_{4+5} narrowly separated by no more than width of a vein 5
3 Male sternite IX with 'non-clasper gonocoxites IX' – large lateral lobes extending further posterior than gonostylus and gonocoxites (Figure 3.1j) .... Chilenomyiinae
– Male sternite IX with at most small lateral lobes, never extending over the genitalia (Figure 3.1k,l) 4
4 Crossvein MCu close to wing base (Figure 3.1a) .... Buchonomyiinae
– Crossvein MCu more distal (Figure 3.1b) .... Podonominae
5 R_{2+3} usually forked into R_2 and R_3 (Figure 3.1c); if this fork is absent, then wing macrotrichiose .... Tanypodinae
– R_{2+3} simple, unforked. Wing with few apical setae or bare 6
6 Wing with FCu usually proximal to crossvein MCu (Figure 3.1d), except on shortened wing, where it may be opposite or distal to MCu. Tarsomere 4 usually cordiform .... Diamesinae
– Wing with FCu opposite or distal to crossvein MCu, wing never shortened. Tarsomere 4 never cordiform .... Prodiamesinae
7 R_{2+3} absent. R_1 and R_{4+5} widely separated (Figure 3.1e,f) 8
– R_{2+3} present (Figure 3.1g); if absent, then R_1 and R_{4+5} narrowly separated by no more than width of a vein 9
8 Fourth tarsal segment cordiform (Figure 3.1h); male genitalia enlarged and variably rotated .... Telmatogetoninae
– Fourth tarsal segment cylindrical (Figure 3.1j); male genitalia small, not rotated .... Aphroteniinae
9 Foreleg with basitarsus longer than or subequal to tibia; male gonostylus less flexibly attached or fused to gonocoxite, directed backwards (Figure 3.1k). Spurs of mid and hind tibiae often comb-like, comprising fused spines (Figure3.1n) .... Chironominae
– Foreleg with basitarsus shorter than tibia; male gonostylus flexibly attached to gonocoxite (Figure 3.1l). Tibial spurs, if comb-like, composed of free spines (Figure 3.1m) 10
10 Midleg longest, hindleg shortest with hind tarsomere 1 thickened, bearing paired thick spines on oblique apex, no longer than tarsomere 2 . ... (tribe Harrisonini) Diamesinae
– Midleg shortest, hindleg never as above .... Orthocladiinae

* Aptery and brachyptery occur in some Telmatogetoninae, Diamesinae, Orthocladiinae and Chironominae, which will have to be identified on genitalic features.

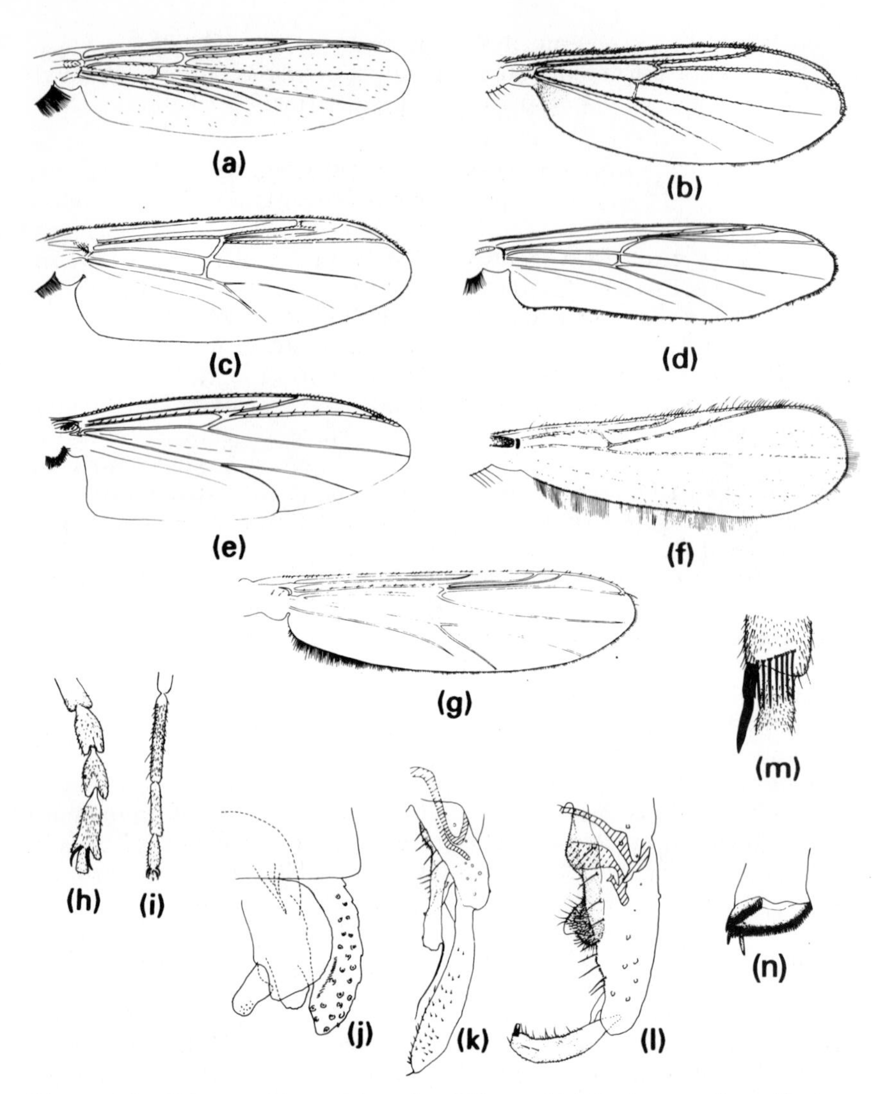

Figure 3.1 Chironomidae: adult males. Wings of (a) Buchonomyiinae (*Buchonomyia thienemanni* Fittkau); (b) Podonominae (*Podonomus evansi* Brundin); (c) Tanypodinae (*Coelopynia pruinosa* Freeman); (d) Diamesinae (*Paraheptagyia tonnoiri* Freeman); (e) Telmatogetoninae (*Telmatogeton japonicus* Tokunaga); (f) Aphroteniinae (*Aphroteniella filicornis* Brundin); (g) Chironominae (*Skusella* sp.). Distal tarsomeres of (h) Telmatogetoninae, *Telmatogeton australicus* Womersley; (i) Chironominae, *Polypedilum* sp., right half of hypopygia. Ventral view of (j) Chilenomyiinae (*Chilenomyia paradoxa* Brundin); (k) Chironominae (*Phaenopsectra flavipes* (Meigen)); (l) Orthocladiinae (*Paratrichocladius micans* (Kieffer)). Hind tibial apex of (m) Orthocladiinae (*Cricotopus* sp.); (n) Chironominae (*Einfeldia pagana* (Meigen)).

KEY TO SUBFAMILIES OF CHIRONOMIDAE – ADULT WINGED* FEMALES
(Modified after Sæther 1977)

1 Ovipositor formed from gonostyli, posterior lobes of tergite IX, elongate cerci and large gonapophysis VIII (Figure 3.2a). Seminal capsules absent; swollen spermathecal duct acts as seminal receptacle (Figure 3.2b). Tergite VIII reduced .... Telmatogetoninae
– Gonostyli and ovipositor absent. 2–3 well-developed seminal capsules present; spermathecal duct not swollen (Figure 3.2e,f). Tergite VIII well developed 2
2 R_1 and R_{4+5} widely separated, R_{2+3} absent (Figure 3.1a,b) 3
– R_{2+3} present; if absent, then R_1 and R_{4+5} narrowly separated by no more than width of a vein 5
3 Gonocoxite IX separate, strongly developed, reaching as far posterior as the cerci (Figure 3.2c) .... Chilenomyiinae
– Gonocoxite IX fused with tergite IX to form an annular gonotergite (Figure 3.2e) 4
4 Crossvein MCu close to wing base (Figure 3.1a). First axillary sclerite setose .... Buchonomyiinae
– Crossvein MCu more distal (Figure 3.1b). First axillary sclerite bare .... Podonominae
5 Gonocoxite IX fused with tergite IX to form an annular gonotergite (Figure 3.2d,e) 6
– Gonocoxite IX separated from tergite IX (Figure 2.5e) 7
6 Gonotergite IX weak, narrow and strap-like, with few or no setae (Figure 3.2e) .... Tanypodinae
– Gonotergite IX well developed, hood-shaped, setose (Figure 3.2d) .... Aphroteniinae
7 Wing with crossvein MCu present (Figure 3.1d) 8
– Crossvein MCu lacking (Figure 3.1e–g) 9
8 Wing with FCu usually proximal to crossvein MCu (Figure 3.1d), except on shortened wing, where it may be opposite or distal to MCu. Tarsomere 4 usually cordiform .... Diamesinae
– Wing with FCu opposite or distal to crossvein MCu, wing never shortened. Tarsomere 4 never cordiform .... Prodiamesinae
9 Foreleg with basitarsus longer than or subequal to tibia. Spurs of mid and hind tibiae often comb-like, comprising fused spines (Figure 3.1n) .... Chironominae
– Foreleg with basitarsus shorter than tibia. Tibial spurs, if comb-like, composed of free spines (Figure 3.1m) 10
10 Midleg longest, hindleg shortest. Three spermathecae .... (tribe Harrisonini) Diamesinae
– Midleg shorter than hindleg. Usually with 2 spermathecae, though 3 in several genera .... Orthocladiinae

* Aptery and brachyptery occur in some Telmatogetoninae, Diamesinae, Orthocladiinae and Chironominae, which will have to be identified on genitalic features.

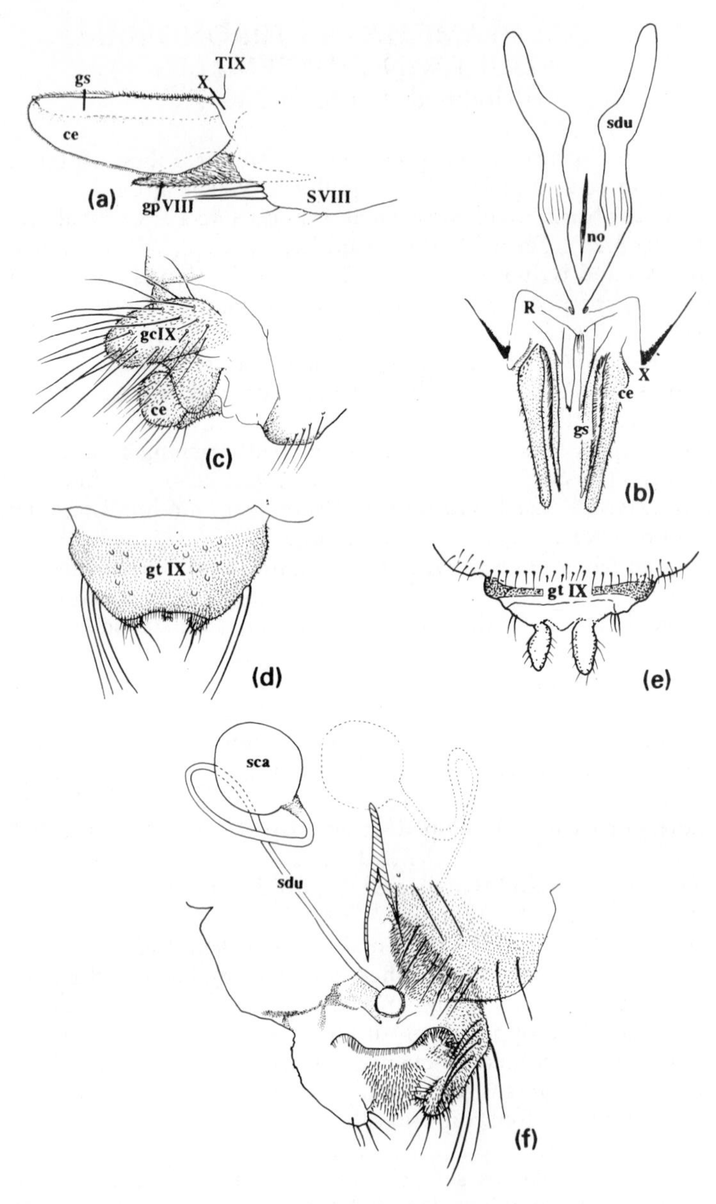

Figure 3.2 Chironomidae: adult females. Genitalia of (a) (b) Telmatogetoninae (*Telmatogeton australicus* Womersley) – (a) lateral, (b) dorsal; (c) Chilenomyiinae, lateral; (d) Aphroteniinae (*Paraphroteniella fascipennis* Brundin), dorsal; (e) Tanypodinae (*Procladius paludicola* Skuse), dorsal; (f) Orthocladiinae, ventral. Key: ce = cercus, gc = gonocoxite, gs = gonostylus, gt = gonotergite, S = sternite, sca = seminal capsule, sdu = seminal ducts, T = tergite.

KEY TO SUBFAMILIES OF CHIRONOMIDAE – LARVAE*

1 Antenna retractile into head (Figure 3.3a). Hypopharynx with distinctive toothed ligula (Figure 3.3b). Mentum usually weakly sclerotized .... Tanypodinae

– Antenna non-retractile (Figure 3.3c) Ligula never developed as in Tanypodinae. Mentum nearly always a dark sclerotized toothed plate (e.g. Figures 3.3d,e; 3.4a–c) 2

2 Mentum associated with variably developed but always broad, and nearly always striated, ventromental plates (Figure 3.3d) [Exceptionally *Stenochironomus* (Figure 3.3e) and *Harrisius* lack striations, but some hooks are present] .... Chironominae

– Mentum with, at most, relatively small, non-striate ventromental plates (e.g. Figure 3.4b,c) 3

3 Mentum untoothed (Figure 3.3f). Sensory setae of anterior labrum elongate (Figure 3.3g). Body covered with tubercles and hairs (Figure 3.3h) .... Aphroteniinae

– Mentum nearly always toothed. Labral setae small. Body cuticle smooth but may be strongly setose. 4

4 Premandibles absent (Figure 3.3g) 5

– Premandibles present (Figure 3.4e) 6

5 Second or second/third antennal segments often annulate (Figure 3.3i). Procercus predominantly elongate, many times as long as wide (Figure 3.3j) .... Podonominae

– Antenna non-annulate. Procercus absent .... Buchonomyiinae

6 Central ligula and lateral paraligulae of prementum resembling three brushes (Figure 3.3k). Third antennal segment often annulate .... Diamesinae

– Prementum never developed as three brushes, although a single median brush (M-appendage) may be present (Figure 3.4d). No annulate antennal segments 7

7 Ventromental plates elongate, with beard beneath (Figure 3.4a) .... Prodiamesinae

– Ventromental plates variable; if large (Figure 3.4b,c), then never with beard beneath 8

8 Prementum with brush-like M appendage divided into many fine branches, lying ventral to the mentum (Figure 3.4d). Premandible short and broad with strong inner brush (Figure 3.4e). Antenna short but distinctly 4-segmented (Figure 3.4f) .... Telmatogetoninae

– No brush-like development of prementum. Premandibular brush absent or weak. If the antenna is 4-segmented and short (less than half mandible length), then the apical segmentation is indistinct .... Orthocladiinae

* Larvae of Chilenomyiinae unknown; that of Buchonomyiinae based upon non-terminal instar only.

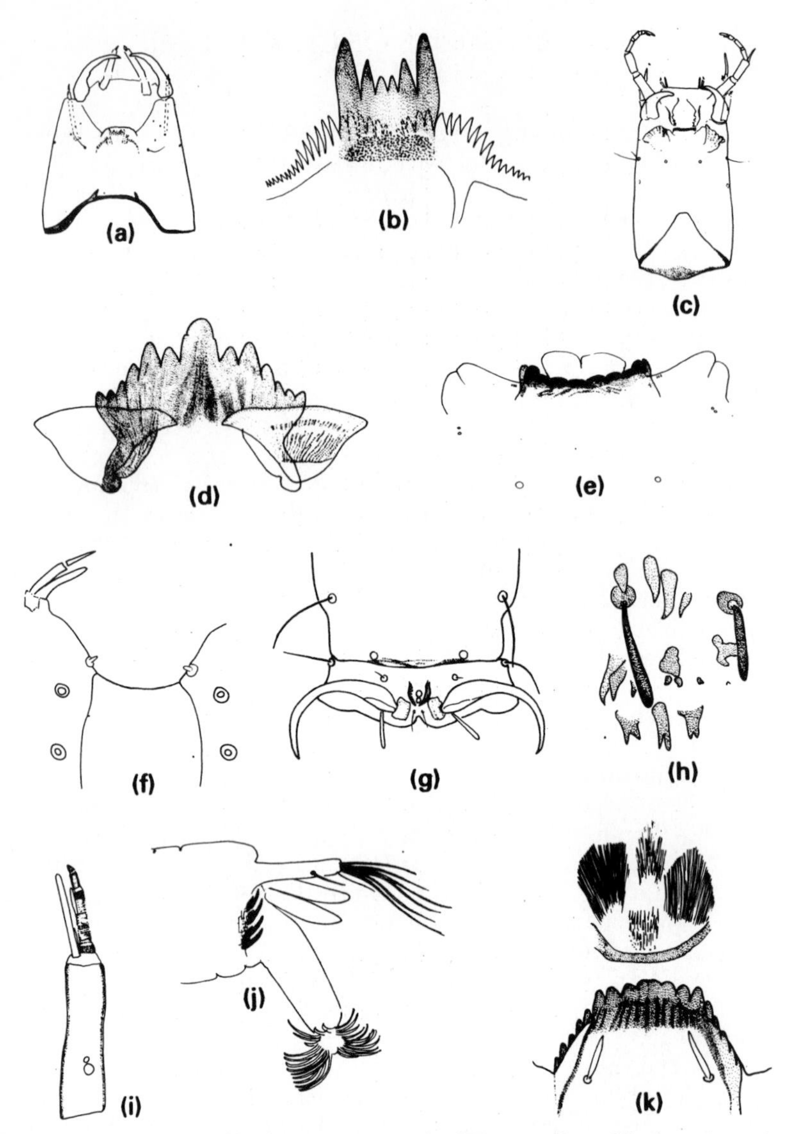

Figure 3.3 Chironomidae: larvae. (a) Head of Tanypodinae (*Coelopynia pruinosa* Freeman); (b) ligula and paraligulae of Tanypodinae (*C. pruinosa*); (c) head of Chironominae (*Robackia* sp.). Menta of (d) Chironominae (*Dicrotendipes* sp.); (e) *Stenochironomus* sp.; (f) Aphroteniinae (*Aphroteniella filicornis* Brundin). (g) Labium of Aphroteniinae (*A. filicornis*); (h) cuticular sculpturing of Aphroteniinae; (i) antenna of Podonominae (*Podochlus australiensis* Brundin); (j) posterior segments of Podonominae (*P. australiensis*); (k) prementum and mentum of Diamesinae (*Paraheptagyia tonnoiri* Brundin).

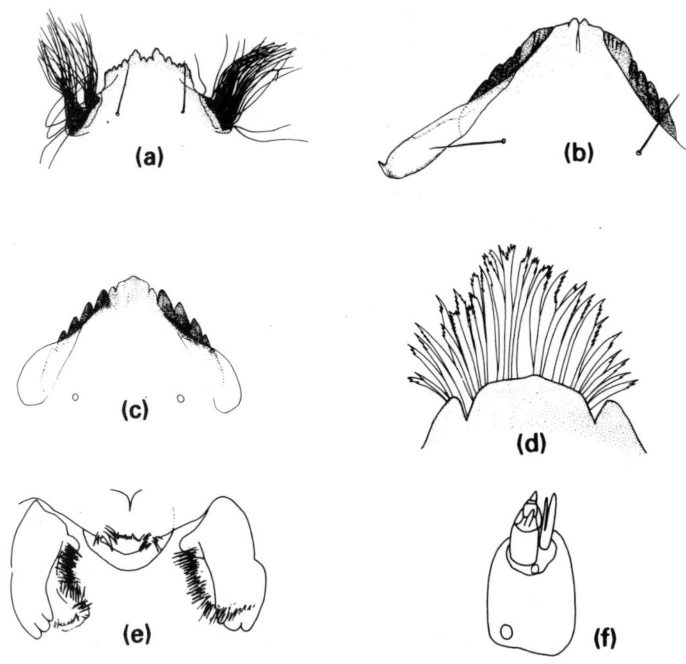

Figure 3.4 Chironomidae: larvae. Menta of (a) Prodiamesinae (*Prodiamesa olivacea* (Meigen)); (b) (c) Orthocladiinae – (b) *Nanocladius* sp., (c) *Parakiefferiella* sp. (d) Prementum of Telmatogetoninae; (e) epipharynx and premandibles of Telmatogetoninae; (f) antenna of Telmatogetoninae (all *Telmatogeton japonicus* Tokunaga).

KEY TO SUBFAMILIES OF CHIRONOMIDAE – PUPAE*

1 Anal lobe comprising tergites VIII and IX fused into a circular disc, fringed with short hair-like spines around posterior disk (Figure 3.5a) .... Telmatogetoninae

– Anal lobe comprising segment IX alone (Figure 3.5b,c) 2

2 Anal lobe fringed with setae but lacking more distinctive macrosetae (Figure 3.5c). Postero-lateral corner of segment VIII usually with spur or comb (Figure 3.5d,e). Thoracic horn often complex, bi- to multibranched, often plumose, simple only in tribes Tanytarsini and some Pseudochironomini .... Chironominae

– Anal lobe with or without setal fringe; if fringed, then nearly always 3 distinctive macrosetae apical/subapical among the fringe (Figure 3.5f). Postero-lateral corner of segment VIII never with comb or spur. Thoracic horn simple or absent, never two-branched from base or plumose. 3

* Pupae of Chilenomyiinae unknown.

3 Thoracic horn present (except when damaged, when remnant of stem often visible), with horn sac (Figure 3.5g) and usually plastron (Figure 3.5h) 4
– Thoracic horn may be absent; when present, without horn sac or plastron (Figure 3.5p) 5

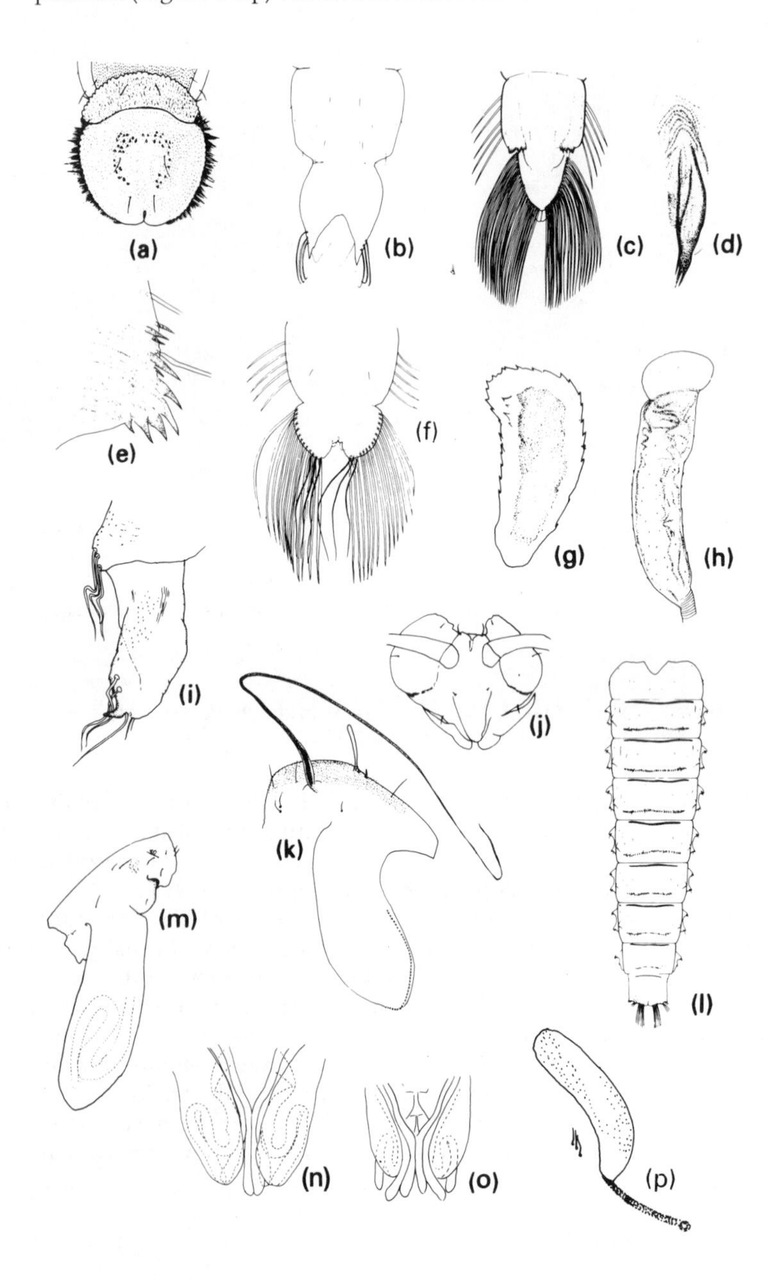

4 Two pairs of frontal setae (Figure 3.5j). Some lateral and anal lobe setae may be 'wavy' (Figure 3.5i). Sheaths of fore- and midlegs straight, terminating beside recurved hindleg sheath at apex of wing sheath .... Podonominae
– One pair of frontal setae. No 'wavy' setae. All leg sheaths recurved beneath wing sheath .... Tanypodinae
5 Thoracic horn very elongate, at least half length of thorax (Figure 3.5k) . ... Aphroteniinae
– Thoracic horn variable but never as long as half thorax 6
6 Anal lobe with postero-lateral spurs, 4–6 postero-lateral setae and 7–8 pairs of anal macrosetae (Figure 3.5l). Fore- and midleg sheaths extend directly backward, hindleg sheath recurved beneath wing sheath. Thoracic horn absent .... Buchonomyiinae
– Anal lobe never with postero-lateral spurs or such an arrangement of setae and macrosetae. Leg sheaths variably arranged; fore- and midleg sheaths straight and directed posteriorly only in Diamesinae (Figure 3.5o). Thoracic horn typically present, very variable in shape. 7
7 Dorsomedian area of thorax with 3 setae, dc3 typically in supra-alar position, dc4 absent (Figure 3.5m), or all dorsocentral setae absent. Fore- and midleg sheaths extend directly backward, hindleg sheath recurved beneath wing sheath (Figure 3.5n) or (tribe Harrisonini) also directed backward (Figure 3.5o) .... Diamesinae
– Dorsomedial area of thorax with 4 setae, with neither dc3 nor dc4 in supra-alar position. Typically all leg sheath recurved beneath wing sheath; exceptions never like Diamesinae 8
8 Thoracic horn with indirect connection to adult spiracle (Figure 3.5p). Anal lobe with either (i) full fringe of long setae and usually 4–5 marginal macrosetae; or (ii) 3 marginal macrosetae; or (iii) short fringe and 2 small median setae. Tergite II without hookrow; no tergites with transverse bands of stronger spines. .... Prodiamesinae
– Thoracic horn without connection to adult spiracle. Anal lobe quite variable: with or without fringe, usually with 0 or 3 macrosetae; when with more macrosetae, then tergite II with hookrow and/or tergites spinose .... Orthocladiinae

---

Figure 3.5 Chironomidae: pupae. Posterior segments of (a) Telmatogetoninae (*Telmatogeton japonicus* Tokunaga); (b) Orthocladiinae (*Parakiefferiella* sp.); (c) Chironominae (*Stenochironomus watsoni* Freeman). (d) (e) Posterolateral combs of Chironominae – (d) *Chironomus nepeanensis* Skuse; (e) *Kiefferulus intertinctus* Skuse). (f) Posterior segments of Orthocladiinae (*Nanocladius* sp.). Thoracic horn of (g) Podonominae (*Podochlus australiensis* Brundin); (h) Tanypodinae (*Coelopynia pruinosa* Freeman). (i) Lateral posterior segments of Podonominae (*P. australiensis*); (j) head of Podonominae (*Archaeochlus* sp.); (k) thorax of Aphroteniinae (*Aphroteniella filicornis* Brundin); (l) tergites of Buchonomyiinae (*Buchonomyia thienemanni* Fittkau); (m) lateral thorax of Diamesinae (*Diamesa freemani* Willassen and Cranston); (n) (o) leg sheaths of Diamesinae – (n) *Diamesa* sp., (o) *Boreoheptagyia* sp.; (p) thoracic horn of Prodiamesinae (*Odontomesa fulva* Meigen).

4

Biogeography

P.S. Cranston

4.1 INTRODUCTION

Biogeography is the study of the distribution patterns of plants and animals, and the explanations of such patterns. Some fields of biogeographical study have been particularly influenced by chironomid studies, notably those of Brundin (1966, *et seq.*). Biogeography, like other subdisciplines of biology, has unique concepts and a terminology that must be explained before making further progress.

Every taxon, whether species, genus or whatever rank, occupies a particular geographical area, called its **distribution**, **range** or **area of endemism**. This area of distribution need not be continuous, but might be discontinuous (**disjunct**), as when interrupted by unsuitable habitat. Within a range, a taxon may be scarce or abundant, permanent, seasonal or ephemeral. Furthermore, distributions may change over shorter or longer periods of geological time (Chapter 16). Ranges may expand, contract or shift. Long-distance dispersal may be followed by establishment in a new area. Increasingly, human activity causes extinction or redistribution (e.g. Hoffrichter, 1973), leading to global impoverishment and homogeneity.

A variable number of regions of distinctive endemism are recognized. These biogeographical regions number from 20 (Candolle 1820) to 40 (Candolle 1838) to the smaller number of six regions (Sclater and Sclater, 1899) widely recognized today, namely the Afrotropical, Australasian, Nearctic, Neotropical, Oriental and Palaearctic regions (Figure 4.1). The Afrotropical region was formerly called the Ethiopian region, but confusion with the nation of Ethiopia encouraged the use of a new name. This name also more accurately indicates the actual region, which comprises Africa south of the Sahara. The Nearctic and Palaearctic regions are often united as the Holarctic region, a view supported by extensive shared generic and species-level chironomid distributions and relatively low levels of endemism in each area (Cranston and Oliver, 1987; Ashe *et al.*, 1987).

The discipline is commonly referred to as biogeography (as in this

The Chironomidae: Biology and ecology of non-biting midges
Edited by P.D. Armitage, P.S. Cranston and L.C.V. Pinder
Published in 1995 by Chapman & Hall. ISBN 0 412 45260 X

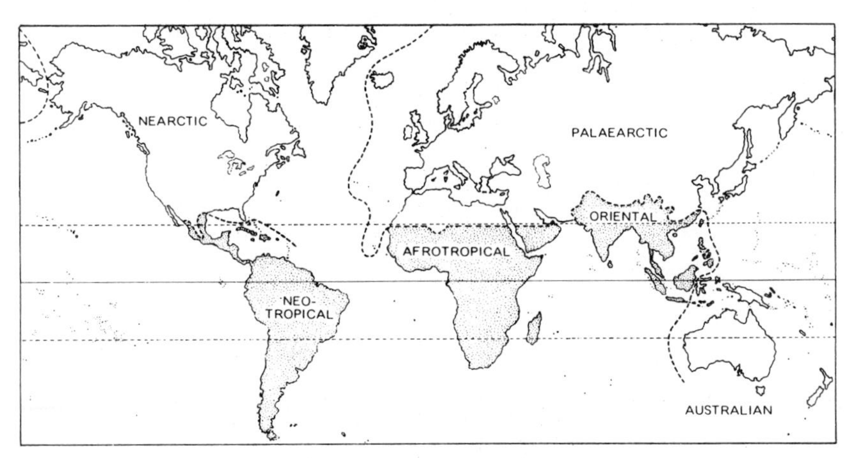

Figure 4.1 The biogeographical regions of the world (modified after Wallace, 1876).

chapter), rather than zoogeography, since distributions of both plants and animals may have similar causes. Faunistic surveys, combined with recognition of areas of endemism, explanation of their genesis and interpretation of the interrelationship of faunas and the areas they occupy, is all part of biogeography. Biogeography unites biology with geography and geology and is intimately linked to ecology.

In this chapter the data sources for chironomid distributions are considered, followed by an outline of the different approaches to biogeographical analysis together with a review of some case studies. A section considering the value of palaeontological evidence in biogeography precedes a final summary and suggestions for future biogeographical studies.

4.2 DISTRIBUTIONAL DATA

A fundamental requirement for any biogeographical analysis is to understand the distribution of the study organisms. The detail depends upon scale, related to the questions being considered. In this section it is convenient to separate local scale, which for chironomids is likely to be affected by current ecological factors, from broader scale patterns in which a historic component may be more influential (or more readily detectable).

4.2.1 DISTRIBUTIONAL DATA FOR LOCAL SCALE ECOLOGICAL BIOGEOGRAPHY

A biogeographical comparison of several types of lake would require that faunas be known in detail, on a type-of-lake by type-of-lake basis. Likewise, inter- and intra-drainage comparisons require detailed knowledge of the fauna, probably at many sites throughout the drainage(s). All types of such local comparisons ought to involve detailed survey to elucidate seasonal and inter-annual variability. Sampling and analysis must be statistically valid, to allow testing of correlation of faunal composition with the environmental factors that are postulated to control their distribution in space. The faunal classifications for lakes and rivers, discussed in section 15.3, are ecological biogeographical schemes. Biogeography might also be taken to include observations on local responses of species of Chironomidae to anthropogenic pollutants; however, conventionally the scale under consideration is greater.

4.2.2 DISTRIBUTIONAL DATA FOR BROAD-SCALE BIOGEOGRAPHY

Moving up the geographical scale, inventories of larger areas are necessary foundations of biogeographic knowledge. Unlike local-scale ecological biogeography, in which the boundaries are chosen to include pertinent geographical features such as a single drainage, or a suite of lakes, at the broader scale the region predominantly is chosen on a non-biological basis. Regions and countries are political constructs that often fail to coincide with significant biogeographical entities. Probably, we ought to disdain biogeographical documentation of the fauna of, for example, Britain that excluded Wales or Scotland, or of Eire that failed to take cognizance of the Northern Ireland fauna because it belongs to a different political unit. Our scepticism might extend to other surveyed areas: for example, Austria is unlikely to have a fauna not found generally in central Europe, and Maine probably has no species not found in similar contiguous habitat in the United States or in adjacent eastern Canada. Exceptions to the biological arbitrariness of such political division are islands that are single political units, such as Hawaii, Iceland and some islands of the Caribbean or the Pacific. Even Australia, which represents uniquely a single political unit and a continent, ought to include New Guinea in its biogeographical region. Some countries cover more than one biogeographical region; for example, China and Japan combine some of the Palaearctic and Oriental regions, Mexico contains Nearctic and Neotropical areas, and Saudi Arabia links the southern Palaearctic and Afrotropical regions (section 4.4.1).

Despite these reservations concerning the circumscribed units of spatial distribution that are imposed on biologists, country-by-country inventory data are fundamental to the documentation of diversity at wider geographical scales. Data acquisition for these broader scale studies may derive from local ecological biodiversity studies, but differs usually in the statistical rigour of sampling. Furthermore, such studies often seek congruence of distributions only with current ecological environmental data. Compilers of regional inventories seek to include all taxa ever recorded from an area – an exercise that must include interpretation of historical data. An excellent model of an annotated inventory is that of Serra-Tosio and Laville (1991) for the chironomids of France and Corsica.

Serra-Tosio and Laville (1991) limit their interpretation of the inventory to observing that France (including Corsica) has one of the richest chironomid faunas in Europe, with virtually 600 species. Although this may be unsurprising in view of the habitat heterogeneity and size of the country, comparisons of variations in species richness through space (regionally) and across taxonomic groups (usually within and between subfamilies) are often made from such inventories. Indeed, this seems to be the most frequent biogeographical 'pattern' that is detected and interpreted in chironomid studies. Major ecological input into these discussions centres on ratios of taxon richness between subfamilies, notably Tanypodinae, Orthocladiinae, Diamesinae and Chironominae. These ratios are taken to reflect variation in proportions of cold stenothermic taxa (Diamesinae, Orthocladiinae) to warm eurytherms (Chironominae). As outlined in Chapter 15, these summaries have some ecological validity. Furthermore, they may be valid worldwide in reflecting trends of the responses of chironomid assemblages to physico-chemical parameters such as nutrient availability and temperatures.

Elaborations on the ecological biogeographical theme include the perceptive commentaries by Coffman (1989) and Coffman *et al.* (1993) concerning patterns of species richness. In surveying the mechanisms that support overall patterns of lotic chironomid diversity, Coffman (1989) proposed maximal environmental heterogeneity as a predictor of maximal species richness. In contrast to other similar interpretations that reflect only current ecology and take little or no cognizance of past processes, Coffman used the term 'biogeographical potential' – that is, the historical component governing local faunal taxonomic richness – as an important factor in species richness. In speculative mode, Coffman (1989) proposed a 3rd order lowland stream in the temperate Holarctic region as likely to provide the highest diversity.

As is often the fate of such speculations, more recent information, in this case from non-Holarctic locations, soon overturned this hypothesis. Pupal exuvial collections from the New and Old World Tropics revealed unexpectedly high alpha-diversity (community level) in both West African and Costa Rican streams. Furthermore, West African collections revealed unexpectedly high beta-diversity (regional level), refuting earlier authors such as Fittkau (1980) and provoking Coffman *et al.* (1993) to address the inadequacy of previous data-sources and to re-emphasize the importance of historic processes in providing the pool of species from which alpha-diversity is derived. These themes are revisited in the sections that follow.

4.2.3 SAMPLING BIASES AND ERRORS

Between-area comparisons (beta-diversity) may be seriously compromised by variation in sampling intensity between regions. As Ashe *et al.* (1987) observed, the Palaearctic region is thoroughly studied, followed by the Nearctic, Afrotropical and Australasian, with the Neotropical and Oriental regional faunas poorly known. Although the period since the mid-1980s has seen greater activity concerning the less well-known regional faunas, the observation remains generally true. Extrapolations from what is known are fraught with danger, as observed by Coffman *et al.* (1993).

Another instance concerns the Nearctic fauna, which might have been considered well-known in view of the number and quality of North American chironomid systematists. Deficiencies in the knowledge were illustrated by Sæther (1981), in describing Orthocladiinae from the Antilles (British West Indies). Sæther recognized several genera as new to science (*Antillocladius*, *Lipurometriocnemus*, *Compterosmittia* and *Diplosmittia*) and suggested that these endemic elements bore phylogenetic and biogeographical relationships to Afrotropical elements. Subsequent survey of continental North America reveals that all the novel genera are now known to occur in the eastern Nearctic region (Cranston and Oliver, 1988a; Hudson *et al.*, 1990). Whilst the biogeographical ideas (Sæther, 1981) are not necessarily refuted, the implications of the new findings are addressed in more detail in section 4.6.

An additional complication is that compilation and publication of faunal lists is often considered to be outside mainstream science. The biogeographer thus has difficulties even when regional faunas are better known but unpublished. Biogeographical databases such as regional catalogues are generally in ink-on-paper form and updating is infrequent, despite the continuing acquisition of novel information.

Any attempt in this chapter to synthesize chironomid distributional data would be negated by the frequency with which such distributional information becomes out-of-date through new discoveries. Furthermore, there are limitations on the length of this contribution, and there is general availability of such data in other publications (section 4.2.4). In this chapter, literature sources are reviewed, the methods of interpretation are detailed and selected case histories are considered.

4.2.4 EXISTING DISTRIBUTIONAL INFORMATION

(a) Regional catalogues

Regional faunal catalogues are primary sources of distributional data for use in biogeographical studies. Current catalogues containing distributional information (predominantly at nation level, within a specified biogeographical region) with their cut-off dates for inclusion are discussed in section 3.8.1.

(b) National inventories

Major papers that add significant information to the above catalogues, based on a multi-taxon regional approach, include:

- **Afrotropical Region** – Ethiopia (Harrison, 1987, 1991, 1992); Western Africa (Dejoux, 1984a); Guinea (Dejoux, 1984b); Zaire (Lehmann, 1979, 1981); Nigeria (Hare and Carter, 1987).
- **Nearctic region** – South-eastern United States (Hudson *et al.*, 1990).
- **Oriental Region** – Thailand (Hashimoto *et al.,* 1981); Sumatra (Indonesia) (Kikuchi and Sasa, 1990); India and Bhutan (Chaudhuri and

Guha, 1987); southern India (Coffman *et al.*, 1988); Indonesia (Sulawesi) (Ashe, 1990); Thailand (Moubayed, 1988); Oriental China (Wang and Zheng, 1993); Oriental Japan: Ryuku Islands (Sasa and Hasegawa, 1983; Hasegawa and Sasa, 1987); Nansei Islands (Sasa, 1990).

• **Palaearctic region** – Fennoscandia (Tuiskunen and Lindeberg, 1986); Norway (Schnell, 1988, 1991); Netherlands (Buskens and Moller-Pillot, 1993); France (Laville and Vinçon, 1986; Laville and Serra-Tosio, 1987; Serra-Tosio, 1989b; Serra-Tosio and Laville, 1991); Italy (Rossaro, 1988); Spain (Vilchez and Casas, 1987; Casas and Vilchez, 1989); Portugal (Reiss, 1989); Canary Islands (Cranston and Armitage, 1988); Turkey (Reiss, 1985a); Syria (Reiss, 1986); Morocco (El Medzi and Guidicelli, 1986; Azzouzi *et al.*, 1992); Lebanon (Moubayed and Laville, 1983; Moubayed, 1987); Mediterranean (Laville and Reiss, 1993); Arabian Peninsula with new records from much of the Middle East (Cranston and Judd, 1989); USSR (Diamesinae only) (Makarchenko, 1989); Japan (Sasa, 1989); and Palaearctic China (Wang and Zheng, 1993).

In contrast to the above papers which report national-based inventories, there have been a number of taxon-based studies which contain some discussions of biogeography. These include reviews of:

• world-wide Aphroteniinae (Brundin, 1966);
• world-wide Podonominae (Brundin, 1966);
• austral Diamesinae (Brundin, 1966);
• Afrotropical Diamesinae (Willassen and Cranston, 1986);
• Afro-Australian Podonominae (Cranston *et al.*, 1987); Diamesinae: *Boreoheptagyia* (Serra-Tosio, 1989c); Aphroteniinae (Cranston and Edward, 1992);
• Gondwanan Chironomidae in south-west Australia (Edward, 1989).

Recent attempts to document the distributions of chironomids on a broad scale include *Limnofauna Europaea* (Fittkau and Reiss, 1978). In this work Europe is divided for biogeographical purposes into 25 regions, with all pre–1978 published chironomid records tabulated. As a compendium of knowledge the publication is very valuable, with several provisos:

• Many nomenclatural and other taxonomic changes have been made since 1978.

• The distributions recorded represent somewhat biased data collection, with the Mediterranean countries poorly sampled at that time.

• There is difficulty in interpreting records from the then Soviet Union, because of somewhat differing taxonomic systems, and an emphasis on naming of unreared (and sometimes unassociated) larvae.

From this author's perspective, the documentation of pattern alone, with no explanation of the basis for the regions chosen, nor interpretation of the patterns, is a necessary but only first step towards biogeographical analysis.

(c) World-wide overviews

Finally, in this review of publications on chironomid biogeography, there are a few excellent reviews on world-wide aspects. Brundin's (1966) publication is notable in addressing the causes of the distributions of the predominantly austral Podonominae, Aphroteniinae and austral Diamesinae (Heptagyiae). Aspects of this seminal study are discussed elsewhere in this chapter.

Another comprehensive work is that of Ashe *et al.* (1987), who addressed distributions on a world-wide basis, using the generic rank as a vehicle for very extensive review and commentary. Many previously unreported extensions of distributions are provided and all records are tabulated on a biogeographical regional basis. The detailed documentation is accompanied by a review of selected ecologies on a subfamily-by-subfamily basis. If there is an explanation of the patterns revealed, ecological reasoning predominates, although the importance of past history is recognized.

The next sections of the chapter examine explanations of distribution patterns: section 4.3 covers the principal methods, followed in section 4.4 by some examples of their application in chironomid studies.

4.3 METHODS IN BIOGEOGRAPHICAL ANALYSIS

Examination of just a few of the papers cited in section 4.2.4 shows that there are many approaches to biogeography. Four partially overlapping categories can be recognized: ecological (physical) biogeography; phenetic biogeography; dispersal biogeography; and vicariance biogeography. Dispersal and vicariance biogeography often are grouped together as 'historical biogeography'. It is increasingly obvious that current ecological explanations must take cognizance of historical biotic and environmental change.

Ideally, biogeography requires a phylogeny of a group under study, although it is possible (and even prevalent) for speculations to lack hypotheses of taxonomic relationship. It is important to note that distributional data must not be incorporated into genesis of a phylogenetic hypothesis, to avoid compromise if there is to be any subsequent biogeographical analysis.

4.3.1 ECOLOGICAL BIOGEOGRAPHY

Different species of chironomid thrive in different environments: some eurytopic species have widespread ecological tolerances whereas stenotopic ones are more restricted. It may be possible to determine the optimal conditions for some populations and, by inference, for a species. For example, environmental parameters can be manipulated experimentally to demonstrate how individuals and populations respond physiologically. However, in the unlikely event that we could to do this for more than a very small number of species of chironomid, it is difficult to extrapolate from laboratory to field conditions. Amongst other untested effects, species interactions, such as competition, may restrict particular taxa to a narrow band of their potential physico-chemical, edaphic and other abiotic environmental tolerances. None-the-less, even without quantitative laboratory-based studies, it is clear that chironomids live in preferred ranges governed by biotic and abiotic environmental variables. For example, many *Chironomus* Meigen live as larvae in nutrient-rich, oxygen-deprived sediments; all *Clunio* Haliday live in the marine littoral zone, and all *Rheotanytarsus* Kieffer live in flowing waters. If a chironomid lives in a closely restricted mutualistic, phoretic or ectoparasitic host association (Chapter 12), then the distribution will reflect that of the host, though the chironomid need not fully occupy the host range.

Ecological biogeographers examine the broad requirements of study organisms in terms of physical and climatic factors and discuss how different regions of the globe fulfil these criteria. However, these models explain only a limited number of biogeographical observations and do not address the frequent phenomenon of areas with comparable climate, soil and topography supporting quite disparate plant and animal communities. Furthermore, the relationship between ecological factors and the distribution patterns of higher taxa (for example, species groups, genera, families) is far from clear. If species diversification in monophyletic lineages (clades, section 3.4.2) is predominantly allopatric, with speciation in isolation (Bush, 1975) and little or no ecological divergence, then models of ecology, particularly incorporating climatic influences, might predict the distributions of these multi-taxon clades. However, if cladogenesis produces sister taxa that diverge in their ecologies, or speciation is not allopatric and not concurrent with ecological differentiation, environmental parameters would fail to correlate with distribution.

Several of the chapters in this book detail the postulated relationships between selected environmental factors and the local distribution of chironomid species and communities. Readers are directed to these sections for further discussion of this aspect of biogeography. However, although environmental determinants clearly play an important rôle in constraining the distribution of organisms, historical factors are vital in determining more fundamental patterns.

4.3.2 PHENETIC BIOGEOGRAPHY

Process explanations in chironomid faunistic studies, where they are given, typically utilize taxa shared between areas. In such interpretations, widespread taxa are understood to convey biogeographical information, because they allow assessment of estimates of overall similarity between regions. Endemics, particularly narrow-endemics, are not considered relevant. By analogy with phenetic estimates of phylogenetic similarity (section 3.4.1), Nelson and Platnick (1978) and Grisvold (1991) refer to this prevalent method as 'phenetic biogeography'. This term is used because of the primacy of similarity as a measure of geographical relatedness. Typically, widespread species are considered to indicate common history of areas, with the number of taxa in common being related proportionately to recency of faunal connection between areas. Phylogenetic relationships between taxa in the areas under study either are ignored, or, more usually, are unavailable.

Although such studies allow description and comparison of biotas, they are argued to reveal little or nothing about the history of the biotas (Farris, 1981; Grisvold, 1991). (However, for a possible counter-example, see section 4.4.1.) The major difficulty concerns the emphasis on widespread species, which may be taxa that have failed to respond to earth-history events, and the ignoring of the restricted-distribution, often allopatric, endemics that actually have tracked that history.

Extending the analogy between phylogenetics and biogeography further, it can be argued (e.g. Nelson and Platnick, 1978) that reliance on widespread species in historical reconstruction is the equivalent of the exclusive use of plesiomorphic character states in the reconstruction of hypotheses of phylogeny (section 3.4.2). Since this is inadvisable, other methods of historical biogeography might be pursued profitably.

4.3.3 DISPERSAL BIOGEOGRAPHY

The frequently observed disjunct distributions of related organisms are interpreted in dispersal biogeography as having arisen in the following manner:

- Groups originate in one place.
- They diffuse (range expand) until some kind of barrier is reached.
- They disperse ('jump') across these pre-existing barriers.
- They then differentiate subsequently in isolation (Figure 4.2).

Darwin (1859), in *The Origin of Species*, and subsequently Wallace (1876), promoted this view and it was important in the formulation of Darwin's and Wallace's ideas on species formation.

The view that organisms disperse from centres of origin to other parts of the globe is central to the dispersalist approach. Criteria suggested for recognizing a centre of origin of a group of organisms include the areas that:

1. are most ecologically suitable; or
2. harbour the greatest number of extant species of the group; or
3. circumscribe the greatest morphological diversity for the group; or
4. contain the greatest number of advanced forms with primitive forms 'pushed' to the periphery; or
5. contain the greatest number of primitive forms as advanced forms 'progressively' disperse from the origin (the reverse of (4) and known as **Hennig's progression rule**).

Palaeontological evidence may be sought, in the belief that the site of the oldest fossil indicates the origin of the group. However, reliance upon the vagaries of the fossil record is fraught with danger.

Following identification of the centre(s) of origin, the next task of the dispersalist biogeographer is to trace a route for the colonization of the remainder of the range. This involves assessment of the relative **vagility** of

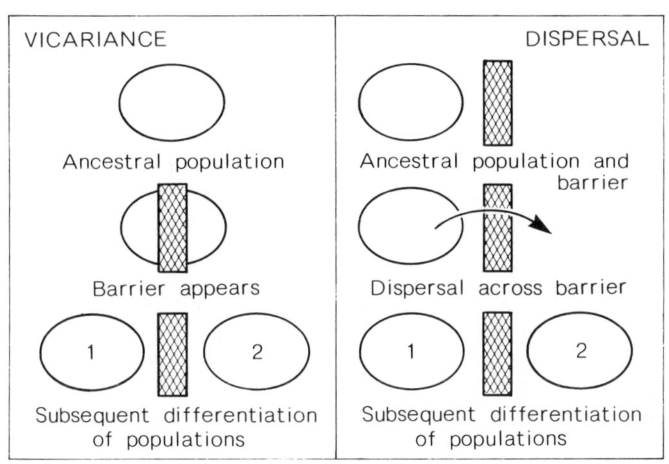

Figure 4.2 The significance of barriers in vicariance and dispersalist explanations of disjunction (modified after Nelson and Platnick, 1984).

the group, i.e. the tendency or propensity and ability to disperse or diffuse from a centre of origin. Subsequently, barriers that prevent the organisms from becoming universal must be identified or postulated.

Assessment of the propensity to disperse tends to be somewhat subjective and arbitrary, but entomologists make some common observations. Firstly, brachypterous and apterous adult insects are presumed to have limited powers of dispersal relative to winged forms. However, evidence of the ability of other life history stages (such as eggs) to survive dispersal can counter evidence of a sedentary adult. The occurrence of particular insects in the aerial planktonic drift is often cited as evidence of dispersal powers. The discovery of particular insects on newly formed islands or faunistically denuded areas, such as those following volcanic activity or marine retreat after inundation of terrestrial systems, is taken to indicate their ready dispersivity. For non-volant insects where a distribution pattern transcends physical barriers such as oceans, dispersivity is often inferred to be through contiguous land connections either extant, ancient or postulated, in combination with various *ad hoc* scenarios such as propensity to trans-oceanic rafting, elevation in cyclonic winds and transport in more recent times by humans, either intentionally or involuntarily.

Given powers of dispersal, it is necessary to postulate restrictions that prevent or restrict radiation of all taxa from their point of origin. Commonly three kinds of filters or barriers are recognized (cf. Cranston and Naumann, 1991):

- **Corridors** or 'bottlenecks' that are variably narrow constrictions of suitable habitats that may be impediments to dispersal between larger areas at each end. Ecologically similar areas connected by corridors may have very similar biotas.

- **Filter bridges** that allow limited transgressions compared with corridors but more than the sweepstake routes described below. Chains of oceanic or ecological islands are often cited as examples of filter bridges.
- **Sweepstake routes** are major barriers, such as great expanses of ocean for terrestrial and freshwater biota, or tropical regions for cold stenothermic organisms. Generally it is held to be a matter of chance which organisms will survive a sweepstakes route and successfully colonize the new area. Dispersal via sweepstakes routes is commonly invoked to explain the depauperate, unbalanced mixture of taxa found on remote islands.

Dispersal has always been central to explanations of island biotas. The **theory of island biogeography** of Macarthur and Wilson (1967) is a mathematical elaboration of dispersal biogeography. Among other things, this theory postulates a relationship between species and area, such that larger islands support a more diverse biota than do smaller ones. Macarthur and Wilson (1967) and others visualize an island biota in or approaching a dynamic equilibrium. At equilibrium, immigration and extinction rates are equal, so that the total number of species on an island is constant. However, the actual composition of the biota changes with time, through species interactions often in relation to geological and/or climatic changes.

Theories concerning explanations of island biotas, whether dispersalist or vicariant (see below) have wide application in biogeography. There is little difference between a remote coral atoll, an isolated spring, a lake or the glacial streams of an isolated high mountain peak – all can be envisaged as 'islands' isolated by surrounding unsuitable habitats.

(a) Problems with dispersalist biogeography

Dispersalist explanations of biotic relationships between disjunct areas originated when the present land masses were believed to be fixed. If land masses were immobile, then disjunctions must have arisen by biotic movements – the mechanisms outlined above were postulated. All observed distributions became explicable by one or a combination of the above scenarios of faunal relocation, with differential extinctions explaining absences.

Although dispersal scenarios have been made for groups with understood phylogenies, few attempts were made to relate phylogeny with distribution – each taxon could be treated in isolation and, generally, congruence between distribution and phylogenetic relationship was not sought. Repeated patterns might imply similar dispersal histories for the organisms investigated, but repeated *ad hoc* explanations could account for any distribution pattern.

The dispersalist view concerning the pre-existence of barriers and taxonomic radiation by crossing of barriers postulates speciation occurring in subsequent isolation. Just how does a barrier that allows a founder population to cross, subsequently act as a total barrier to further exchange that would prevent allopatric speciation?

A most cogent criticism concerning 'jump' dispersal across a barrier followed by establishment is that this implies dispersal into some kind of ecological 'vacuum'. Ecological 'vacuums' do occur and we have clear recent evidence that newly formed or denuded areas such as post-volcanic Mount St Helens (USA: Washington State) (Anderson, 1992) or Krakatoa (Pacific Ocean) (Thornton and New, 1988) are being colonized rapidly. Furthermore, on a rather longer geological time-scale, dispersal must be involved in the acquisition of characteristic post-glacial floras and faunas by the British Isles, New Zealand and vast areas of the cool temperate zones of the Northern Hemisphere by post-Pleistocene colonization from non-glaciated areas. Range changes have occurred in the past (see, for example, Coope *et al.*, 1981; also Chapter 16 of this volume) and such mechanisms are clearly important in governing current distributions. However, biogeography must take more cognizance of ecological theory (Myers and Giller, 1988) and the concept of potential colonists arriving at a void is unlikely – usually there will be a pool of existing species exploiting and locally adapted to conditions. Incomers clearly face difficulty in establishment. Increasingly biogeographers view dispersal alone as inadequate to account for regularly structured distribution patterns.

4.3.4 VICARIANCE BIOGEOGRAPHY

Disjunct distributions of taxa that are interpreted by mobile biota between static continents can be viewed equally in terms of mobile land masses and static biotas. Although the ideas do not originate with Leon Croizat (1958), he is credited with promulgating the view that present biotic distributions represent ancient patterns that have been disrupted (vicariated) in the past by factors such as altered geology and climate. Croizat compendiously documented distributions of many organisms, mapping them and connecting distributions of related taxa by lines (**tracks**). Repeated lines connecting occurrences (generalized tracks) across the globe bore little or no direct relationship to modern geography. In the southern hemisphere, generalized tracks cross the oceans, linking distant land masses. In the northern hemisphere tracks traverse mountain ranges. Croizat (1964) argued that the congruence (or regularity) of so many distributions to form generalized tracks could not be explained by dispersal. Instead, they were evidence for the appearance of impassable barriers that divided many taxa, thus producing congruent patterns of speciation, whose geographical relationships were identical. In the terminology of historical biogeography, widespread ancestral species are divided into vicariant populations (incipient species) by a vicariance event. This may arise through geological or climatological occurrences such as sea level alteration, ocean formation, orogeny, aridity or glaciation. In dispersal biogeography, the barrier predates the disjunction, but in vicariance biogeography the barrier is the cause of the disjunction (Figure 4.2).

Croizat gave no explicit means of assessing taxonomic relationship but his studies were incorporated within the cladistic method (Croizat *et al.*, 1974), thereby becoming exposed to a wider audience. The melding of cladistic phylogenetics and vicariance biogeography became the framework espoused by many subsequent historical biogeographers.

Hennig's formulation of an explicit method for phylogenetic reconstruction (Chapter 3) provided a framework within which biogeographic patterns could be examined. For example, a phylogenetic background to biogeography was included in the works of Hennig himself (1960), Illies (1961a, 1965) and Brundin (1966). When Croizat's views were melded to cladistics, biologists including several European entomologists, were receptive. However, ideas of centres of origin were retained and dispersalist views persisted rather than the exclusively vicariant paradigm. For example, Hennig's progression rule stemmed from a belief that taxic plesiomorphy indicated a centre of origin, with increasing apomorphy expected towards the periphery of the range.

The two vicariance methods, cladistic and Croizatian (**panbiogeographic**), have moved in separate directions, with cladists stressing the primacy of phylogeny and the panbiogeographers emphasizing distribution pattern and proximity of taxa. The vicariant biogeographical method can be summarized as follows. Taxon names in a cladogram (Figure 4.3a) are replaced by their respective areas of endemism to give an area cladogram. This shows the historical relationships between the areas, as evidenced by the biota studied. Identical geographical areas with repetitious presence of taxa ('redundant areas') are then removed to give a reduced area cladogram (Figure 4.3b). Comparison of the reduced area cladogram with theories of earth history may reveal explanations of the processes that determined the patterns observed. Such explanations have greater generality when the same patterns recur many times based on different taxa.

Interest in vicariance as providing an explanation of biogeographical pattern, rather than reliance upon dispersal scenarios depended upon acceptance of Wegener's (1915) theory that continents were not fixed but were mobile. **Continental drift** implies that land masses move and therefore disjunct distributions can be explained by ancient contiguity or at least proximity, with subsequent disruption. Croizat's generalized tracks can be interpreted as links between faunas that were originally united. Instead of mobile faunas dispersing on a fixed geography, immobile faunas were dispersed on moving plates of land. Brundin, notably in 1966 (but see also, for example, 1988), was able to use chironomid distributions (section 4.4.2(a)) to provide an explanation for early observations concerning the similarity of the plant faunas of all southern continents (e.g. Hooker, 1853). The dispersalist faunal interpretations of Darwin and Wallace, and many successors who believed in a stable geography, could be refuted.

Just as the distribution of a single organism or pair of species of organism can be interpreted by past dispersal, so a single cladistic biogeographical analysis can always be interpreted. The likelihood of a common explanation increases as more geographically congruent patterns are discovered. As with Croizat's generalized tracks, the more similarity in pattern that is discovered, the stronger is the evidence that disjunct distributions are deterministic rather than stochastic (through jump dispersal). Vicariance biogeographers are concerned with detecting patterns for as many groups as possible and seeking congruence.

(a) Problems with vicariance biogeography

Relationships postulated between areas are only as good as the phylogenetic hypotheses from which they are derived. Although the systematist may have great confidence in a phylogeny derived for their own particular study organisms, vicariance biogeography demands comparisons of area relationships derived from many unrelated groups. These may be incongruent with each other, and some incongruence may derive from the different approaches to deriving the phylogenies compared. Furthermore, few cladograms have been tested for robustness and it could well be the case that an apparently incongruent area relationship is no more robust than a fully congruent one.

Further difficulties in interpretation arise when well-substantiated area relationships remain contradictory. Incongruence can arise when studies lack comparable areas of endemism due to the effects of redundancy of data, extinction, existence of unique taxa (or areas) and/or the occurrence of widespread taxa that overlap endemic areas defined by others. A common finding is that compared area relationships are often trivial, providing no more than documentation of the geophysicists' view of the breakup of the main plates of Gondwana. Fine level relationships, for example concerning the biota of terranes of Gondwanan south-east Asia, are scarce; when they are available, their complexities make comparisons very difficult.

Vicariance paradigms assume that the earth's history is reflected in the patterns of evolutionary relationships and geographical distribution. Dispersalists argue that congruences of dispersal could explain similar patterns, although the mechanism that might produce such concerted movements through history is not evident. Vicariance biogeographers do not ignore dispersal, but reject its primacy. Dispersal that produces incongruence in area cladograms is the equivalent of homoplastic character states in a phylogenetic analysis of characters. However, dispersal is recognized, particularly when there is evidence of the historical removal of barriers. Thus range expansion provides the ancestral distributions which are disrupted by vicariance events.

4.4 EXAMPLES OF CHIRONOMID BIOGEOGRAPHICAL STUDIES

4.4.1 EXCLUSIVELY DISPERSAL EXPLANATIONS

A predominantly dispersalist explanation using a phenetic biographical paradigm was invoked by Cranston and Judd (1989) for the colonization of lowland Saudi Arabia. This country was predominantly beneath the Tethys Sea, excepting high ground that followed the Rift Valley, along the north-west shore of the Red Sea. Sea-level change, associated with the retreat of the Tethyian Sea in the late Palaeogene, exposed much of the present arid lowland landscape. The chironomid fauna of this once-inundated land contains many species that occur more widely in the region and are therefore understood to have colonized the new habitats. Colonizers were identifiable predominantly as having diffused (range-expanded) from proximate, non-inundated Palaearctic areas. Two predominant groups were evident: those with Mediterranean distributions and those with wider Palaeotropical distributions.

Afrotropical species appeared to remain restricted to the historically non-indundated montane areas along the Red Sea. Endemism was low and in only one case might there have been vicariant speciation across the Red Sea opening – all other Afrotropical species were undifferentiated from populations on the opposite (Ethiopian) side of the Red Sea.

4.4.2 EXCLUSIVELY VICARIANCE EXPLANATIONS

(a) *Parochlus araucanus* group

The relationships among 16 species of the monophyletic *Parochlus araucanus* group of podonomine midges are depicted in the cladogram shown in Figure 4.3a, derived from data of Brundin (1966). Substitution of the areas of endemism for the terminal taxon names gives an area cladogram. The species are distributed over the southern land masses, except that one species (*P. kiefferi*) occurs in North America. One clade (*P. novaezelandiae*, *longicornis* and *glacialis*) contains redundant information since all are New Zealand species. Removal of this and further redundancy provides the reduced area cladogram shown in Figure 4.3b. This depicts several New Zealand–South America vicariance events and one Australia–South America vicariance event occurring after the initial event involving Australia–South America–New Zealand. These sequences are not incompatible with the known earth history concerning late Cretaceous Gondwanan fragmentation. Accumulating evidence on the probable multiple origins of New Zealand (e.g. Craw 1988) suggest these patterns might profitably be re-examined.

There are three possible explanations for the position of *P. kiefferi*:

- Faulty phylogenetic relationships. (However, even if true, the problem recurs wherever *kiefferi* lies within the otherwise austral clade.)
- Dispersal from a cold stenothermic southern hemisphere location to its present North American localities.
- The clade predates the early Cretaceous Gondwana, having originated in Pangaea, in which North and South America were contiguous.

Recent evidence that may contribute to the problem is the discovery of males of a species of the *P. araucanus* clade from high elevations in Costa Rica, central America (Watson and Heyn, 1993). Unfortunately, males provide little or no evidence for phylogenetic interpretation, with the pupa providing informative characters. If the unknown pupa of the Costa Rican species were to be identical, or have a sister-group relationship to *P. kiefferi*, this might indicate northward diffusion from South America rather than a more archaic pattern.

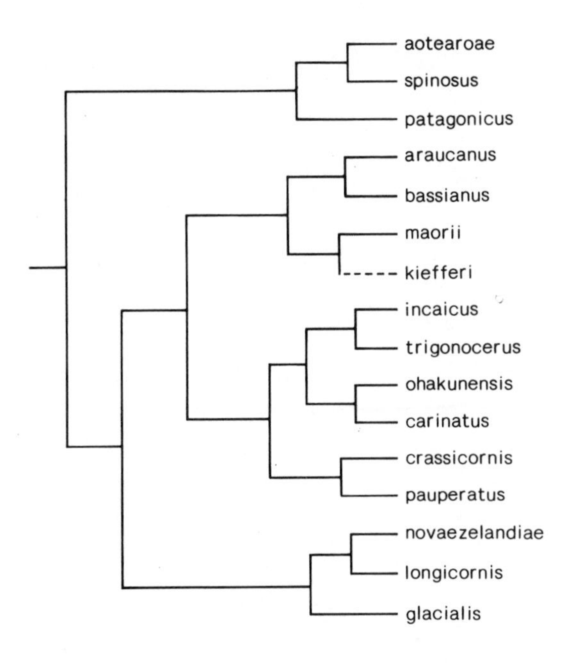

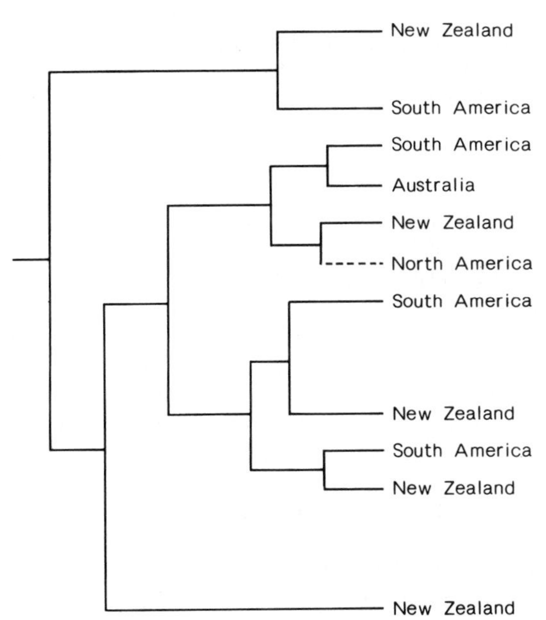

Figure 4.3 (a) Estimate of the phylogeny of the *Parochlus araucanus* group (Podonominae) (after Brundin, 1966); (b) the reduced area cladogram derived from (a).

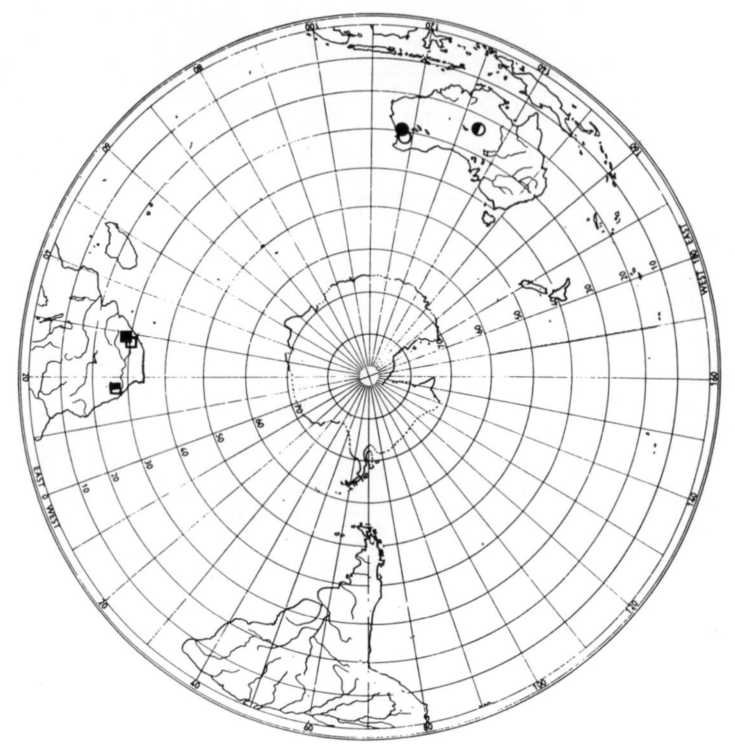

Figure 4.4 Distribution of *Archaeochlus* (Podonominae). Key: (◨) = *A. biko* Cranston, Edward & Colless, (■) = *A. bicirratus* Brundin, (□) = *A. drakensbergensis* Brundin, (●) = *A. brundini* Cranston, Edward & Colless, (○) = *A.* sp. nov., (◑) = *A.* sp. nov.

(b) *Archaeochlus*

In 1966, Brundin described the podonomine genus *Archaeochlus* from two southern African species, and a monotypic genus *Afrochlus* from a little further north in Zimbabwe. Gradually it became clear that a curious mandibulate podonomine from western Australia was related to the South African clade. The group was revised by Cranston *et al.* (1987) and the phylogeny and biogeography were considered in some detail.

The present bi-continental distribution of *Archaeochlus* (with or without *Afrochlus*) (Figure 4.4) was ascribed to a Gondwanan distribution of the ancestral taxon. The clade was presumed to have been distributed at least in Australia, east and west Antarctica and southern Africa prior to the breakup of Gondwana. Whether an expanding earth (Owen, 1981) or constant dimension globe is modelled, the scenario of the breakup of Gondwana is similar – African–Australian separation starting by rift faulting between south-east Africa and Antarctica in the Oxfordian Upper Jurassic (c. 155 Ma).

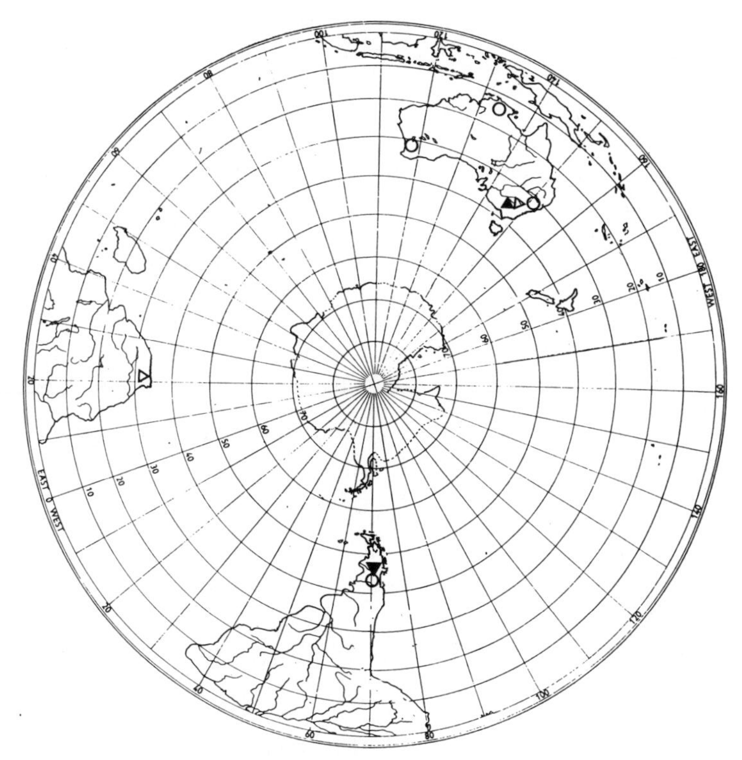

Figure 4.5 Distribution of Aphroteniinae. Key: (○) = *Aphroteniella*, (▼) = *Paraphrotenia*, (▽) = *Aphrotenia*.

Movement apart occurred throughout the Kimmeridgian Upper Jurassic and by the Hauterivian stage of the Lower Cretaceous (c. 120 Ma) Africa was separated from the Antarctic by oceanic crust. Cranston *et al.* (1987) argued that this vicariance event, over 120 million years ago, divided *Archaeochlus* into two, with three species (and the sister group *Afrochlus* Brundin) remaining on the African plate. At that time, it was believed that only one Australian species remained, which they suggested was a descendant of what might have been a more diverse fauna spread through a temperate Mesozoic to mid-Tertiary Antarctic–Australian continent. Recent unpublished observations have confirmed the suspicions of a wider Australian *Archaeochlus* radiation, with a second species partially sympatric with *A. brundini* on western Australian granite outcrops (D.H.D Edward, *pers. obs.*), and a third species in relictual seepages in arid central Australia (P.S.Cranston *pers. obs.*).

In this example, a vicariant explanation of the distribution of a clade allowed a minimum age to be applied to the group. Applying similar logic to cladograms such as those of Brundin (1966), one can arrive at minimum ages for other clades. However, Jurassic datings can be derived only for southern African–Australian clades, since other southern continental connections persisted very much longer, notably that between Magellania (the southern Neotropics) and Australia via Antarctica.

(c) Aphroteniinae

The subfamily Aphroteniinae, erected by Brundin (1966), is exclusively Gondwanan in its current distribution, with three genera distributed amongst Australia, southern South America and southern Africa (Figure 4.5). The discovery of more extensive Australian distributions prompted Cranston and Edward (1992) to re-examine the systematics and biogeography. Australian range extensions did not refute the view that the Aphroteniinae are a group of Gondwanan origin, but provided evidence that the taxa need not be cool stenotherms restricted to the elevated south-eastern areas. Particularly notable, however, was the documentation of the genus *Aphrotenia* in Australia – previously the genus had been known only from the Drakensberg Mountains of South Africa.

Once again, a vicariance paradigm of continental faunal disjunction allowed the estimation of divergence times of lineages. Brundin (1966) had few examples of southern African–Australian generic or species group disjunctions. The Aphroteniinae study therefore presented another analogous case to *Archaeochlus* (section 4.4.2(b)) of intra-generic southern African–Australian disjunction in a second subfamily.

4.5 THE SIGNIFICANCE OF PALAEONTOLOGY TO BIOGEOGRAPHY

4.5.1 THE ROLE OF PALAEONTOLOGY

Just as palaeontology was once considered to be central to phylogenetics, it has been argued that it is central to biogeography in the determination of historical distribution patterns and processes. Many biologists state that the biogeography (and evolutionary history) of their study group cannot be reconstructed because of the lack of fossils. As seen above, biogeographical reconstruction does not rely on, or even require, a knowledge of fossils. Although palaeontology provides the only direct evidence of past biotic distributions, the imperfections and incompleteness of the fossil record is well known. With the development of phylogenetic reconstruction through cladistics and the development of vicariance methods of biogeographical analysis, some palaeontologists such as Patterson (1981) admitted that fossils were too often uninformative, providing too few characters for analysis, or too vague in their relationships to other fossil and extant taxa. Furthermore, assumptions that the site of the oldest fossil of a group or the most 'primitive' member marked its centre of origin were challenged as untenable.

However, it is evident that fossils can:

- provide new morphological and ontogenetic data for systematics;

- provide additional taxa to add to the known range of a higher taxon; and (of significance for our discussion below):
- establish the minimum age for a taxon; and
- allow analysis of biogeographic patterns from a different time than the Recent.

As far as insects (including chironomids) are concerned, some of Patterson's reservations are removed with the excellent preservation of fossils in amber (e.g. Schlee and Dietrich, 1970; Brundin, 1976; Poinar, 1993). Fossiliferous amber dates from the Lower Cretaceous (Neocomian) of Lebanon through the Upper Cretaceous of Canada, Eocene/Oligocene of the Baltic, Miocene of Saxony, Sicily and Mexico to more Recent copal. Amber is rarely present in the southern hemisphere, but Lebanon may have been part of the northernmost margin of Cretaceous Gondwana (Brundin 1976; Cranston *et al.* (1987).

4.5.2 FOSSILS AND VICARIANCE-BASED DATING

The two vicariance-based biogeographical dating proposals, suggesting that *Aphrotenia*, an aphroteniine, and *Archaeochlus*, a podonomine, both date from at least the Jurassic (section 4.4.2(b), (c)), can be substantiated with fossil evidence.

Substantiation for the dating of *Archaeochlus* by South African–Australian vicariance comes from the presence of a more derived podonomine, †*Libanoclites* Brundin, 1976, in lower Cretaceous Lebanese amber. Despite the reservations of Schlee (1975) and Cranston *et al.* (1987) concerning the tribes of Podonominae, there is no doubt that †*Libanoclites* is a close relative of *Boreochlus* and more particularly *Paraboreochlus*. However, it is unclear what geographical relationship the present Lebanon had to Gondwana in the Upper Jurassic–Lower Cretaceous. The maps of Owen (1981) and Howarth (1981) show this area to lie on the northern margin of the African plate, but differ in the presence or extent of the Tethys that separated it from Laurasia. None-the-less, as Brundin (1976) stated, †*Libanoclites* confirms that the Podonominae already showed radiation in morphology and distribution by the Lower Cretaceous.

Kalugina (1980) described an amber fossil, †*Electrotenia brundini* from the Siberian Upper Cretaceous, that clearly possesses all the adult synapomorphic defining features of the Aphroteniinae (section 3.7.3). The fossil evidence confirms that Aphroteniinae had already shown morphological radiation in Pangaea. The evidence strongly supports the argument (section 4.4.2(c)) concerning continental fragmentation, with the added elucidation of an earlier Pangaean aphroteniine clade prior to Pangean breakup, subsequent to which the northern Laurasian part appears to have become extinct and the southern section of the genus *Aphrotenia* was vicariated by the Jurassic rifting of Gondwana.

The third chironomid subfamily for which amber fossil material allows a

Cretaceous dating is the Diamesinae. Kalugina (1976) described *Cretodiamesa taimyrica* from Upper Cretaceous amber from Siberia, allocating it to the new tribe Cretodiamesini. The phylogenetic placement of this tribe relative to the modern Diamesinae is unknown.

The earliest dating for the subfamily Orthocladiinae derives from Canadian Upper Cretaceous amber (Boesel, 1937), with Kalugina (1976) mentioning 623 specimens of Orthocladiinae in her Siberian (presumed Upper Cretaceous) amber. The subfamilies Tanypodinae, Chironominae and Buchonomyiinae are not seen until the Upper Eocene Baltic amber (Brundin, 1966; Ashe *et al.*, 1987) and there are no fossil records of Chilenomyiinae, Telmatogetoninae or Prodiamesinae.

Fossils have another role in biogeography: they may refute vicariance-based biogeographic scenarios. For example, fossil evidence of a presently exclusively Gondwanan-distributed taxa from post-Cretaceous (i.e. post breakup) non-Gondwanan facies would require re-examination of an exclusively austral vicariance hypothesis.

4.6 SUMMARY AND SUGGESTIONS FOR FUTURE DIRECTIONS

The hazards of an inadequate biogeographic database were mentioned in section 4.2.3. However, we would make no progress if we required absolute knowledge of the distributions of all Chironomidae before speculating on the reasons for their distributions. Problems arise from human activity:

- Species are being lost from their native ranges because of anthropogenic impacts, meaning that we may well be interpreting recent 'palaeontological' patterns rathern than current ones.
- Increasingly we cannot recognize natural distributions, because of the substantial number of synanthropic species that appear to be becoming world-wide as part of the universal faunal homogenization that goes hand in hand with enhanced extinction rates.

A further difficulty lies in the variation in taxonomic knowledge between zoogeographic regions: a single species may reside under different names in different places. Few taxonomists can take a broad perspective and review faunas on a world-wide basis, although this approach is fundamental to biogeographical studies.

Accepting the limitations of our present knowledge, at least we ought to frame biogeographical hypotheses in a scientific manner, making them amenable to testing against further data. This implies being explicit about the relevance of new data, allowing new data to be sought in such a way as to provide an explicit test of a well-formulated hypothesis. An example is Willassen and Cranston (1986), who used phylogeny in postulating past dispersal from the north for East African montane *Diamesa* and suggested that the scenario was testable in two ways: by congruence of the observed pattern with other organisms; and particularly by the predicted phylogenetic relationship between other Gondwanan *Diamesa* species. In the latter test, a

Gondwanan but non-African *Diamesa* with a closer relationship to Afrotropical taxa than to a northern (Laurasian) group was seen as a potential falsifier of their biogeographical hypothesis.

Such tests may be derived from other publications, as for example for the Antillean orthoclad taxa, whose relationships were postulated as Afrotropical (Sæther, 1981) (section 4.2.3). Here the now-known Nearctic distributions of several relevant taxa could be seen as a failure of the hypothesis of biogeographical relationship, but might also be taken as evidence for an expanded relationship. Thus the Antillean fauna extends northwards through eastern North America (basically the Appalachian subregion): the Afrotropical relationship is maintained with this broader distributed grouping. Again, this scenario is testable by specific phylogenetic relationships and also by new data, such as distributional extensions.

In an analagous manner, distributions postulated as reflecting the temporal sequence of Gondwanan fragmentation may be amenable to testing. Firstly, fossil evidence may demonstrate the previous presence of the taxon in an area not derived from a Gondwanan region. Secondly, the discovery of crucial taxa lying outside the regions of Gondwanan origin might serve to refute a particular hypothesis. However, three problems arise in defining testable hypotheses concerning Gondwanan distributions. Some of these are exemplified by *Parochlus kiefferi*, as discussed in section 4.4.2(a):

- Dispersal may have occurred from a Gondwanan area at any time in the past, thereby disrupting a purely vicariance-derived pattern.
- The 'non-Gondwanan' area may actually be of Gondwanan origin: the fragmentation history of small blocks (terranes) on the margins of Gondwana is complex, and still somewhat unclear (e.g. Burrett *et al.*, 1991).
- The clade may predate an early Cretaceous Gondwana, perhaps having originated in geologically earlier time in megacontinental Pangaea.

Besides erecting testable hypotheses concerning biogeographical scenarios, certain other matters could be profitably pursued. From the austral perspective of the author, the generally accepted Gondwanan fragmentation of cool temperate species (leaving vicariant daughter taxa in Magellanic South America, New Zealand, southern Australia and, sporadically, upland southern Africa) is a repeated and well-established pattern amongst predominantly cool stenothermic taxa. However, another pattern that may be Gondwanan and bears further examination concerns the warm eurythermic species that are commonly distributed through central Africa, the Indian Ocean (including Sri Lanka and parts of India), south-east Asia and northern Australia. This occurs at the generic level (e.g. *Djalmabatista*, *Conochironomus*, *Skusella*) and even at species level, with several previously unpublished examples including *Polypedilum (Pentapedilum) convexum* Johannsen, as well as the well known *Polypedilum nubifer* (Skuse), *Microchironomus tener* (Kieffer) and *Microtendipes umbrosus* Freeman. It will be difficult to separate natural distributions from those caused by anthropogenic agencies, but the absence of some of these taxa from otherwise appropriate habitats in Saudi Arabia may allow elimination of dispersal as a mechanism.

Finally, there are known to be many excellently preserved chironomids in

Lower Cretaceous Lebanese amber (Brundin, 1976, P.S. Cranston, *pers. obs.*) and Siberian Upper Cretaceous amber (Kalugina, 1976) which have already provided influential data in chironomid biogeographical studies. Further effort by palaeontologists familiar with the present faunas world-wide would undoubtedly reveal more fascinating scenarios, not to say earlier dates for more clades.

Part Two

Biology, Behaviour and Ecology

5

Biology of the eggs and first-instar larvae

L.C.V. Pinder

5.1 INTRODUCTION

Egg-laying behaviour and selection of oviposition sites, coupled with any subsequent redistribution of egg-masses through drift and active dispersal of first instar larvae, are primarily responsible for determining patterns of larval distribution. It is therefore both logical and convenient that they should be considered in combination. The relative importance of each factor, of course, varies between species. All may be significant in species that deposit their egg-masses on to the surface of lakes and rivers, whereas oviposition behaviour alone determines the distribution of species that colonize certain specialized, patchy habitats such as phytotelmata or cattle dung.

The diffuse literature dealing with the eggs of Chironomidae has been reviewed excellently by Nolte (1993) who, personally, has contributed substantially to knowledge of this, the least well studied, part of the chironomid life cycle. Early studies included those of Miall (1895) whose splendid illustrations of the eggs of *Chironomus* (Figure 5.1) have often been reproduced, for example, in the early treatise on the natural history of Chironomidae, *The Harlequin Fly*, by Miall and Hammond (1900), and more recently in Hinton's (1981) three-volume work on insect eggs. Prior to the publication of Nolte's work the most comprehensive study of chironomid eggs was that of Munsterhjelm (1920), who described 62 different egg-masses. Unfortunately the specific identity of only about 35 of these can be determined with confidence (Nolte, 1993).

First-instar larvae of Chironomidae are usually quite distinct, both morphologically and in their behaviour, from the later stages. The differences are so marked that the distinguishing term **larvula** has sometimes been applied to the first instar (Dorier, 1933). As would be expected, first-instar larvae are often present in enormous numbers in freshwater habitats but, because of their small size and partly pelagic behaviour, they are not easily

The Chironomidae: Biology and ecology of non-biting midges
Edited by P.D. Armitage, P.S. Cranston and L.C.V. Pinder
Published in 1995 by Chapman & Hall. ISBN 0 412 45260 X

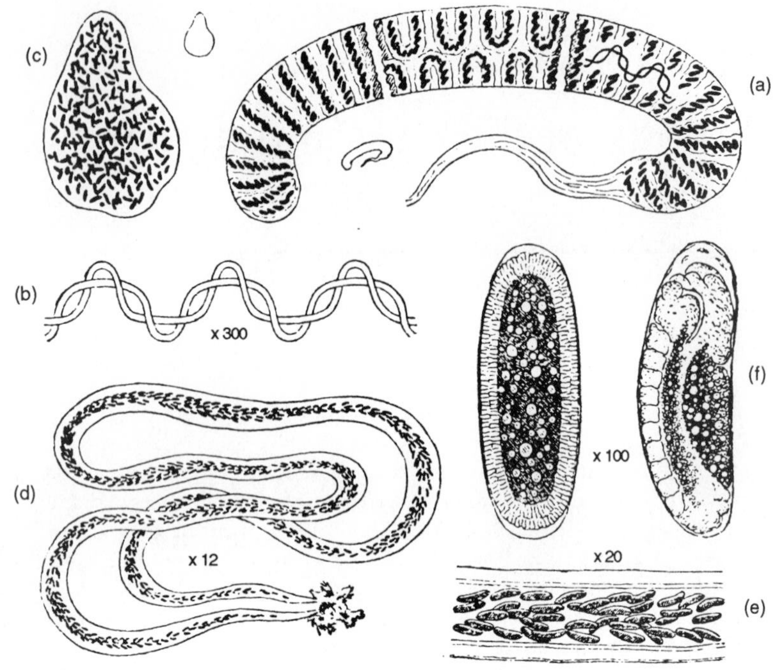

Figure 5.1 Egg-masses of *Chironomus*. (a) Egg-rope of *C. dorsalis*, divided into sections to show both sides; (b) twisted fibres which traverse the egg-rope; (c) egg-mass of another species of *Chironomus*; (d) egg-mass of a third species; (e) part of the last, more highly magnified; (f) developing eggs in two stages. (From Miall, 1895.)

sampled by conventional benthic sampling methods. Storey and Pinder (1985) demonstrated that the vast majority of live first instars, as well as a substantial proportion of second instars, of *Eukiefferiella ilkleyensis* (Edwards), *E. claripennis* (Lundbeck) and *Tvetenia calvescens* (Edwards) will pass through a 125 μm sieve with ease. In most autecological studies, therefore, first-instar larvae have been overlooked and in general little is known about the habits of individual species. Studies that have been concerned with first instars (e.g. Davies, 1973) indicate that a major function of this stage is one of dispersal.

5.2 OVIPOSITION

In most species of Chironomidae the females emerge with ovarian follicles one third to one half developed (Anderson and Hitchcock, 1968; Fischer, 1969; Oliver, 1968) but some Tanytarsini and *Clunio* eclose with fully developed eggs (Hashimoto, 1957; Oliver, 1968). Males of the marine tanytarsine, *Pontomyia pacifica* Tokunaga (Chapter 9) will attempt to mate even before the female, which lies on the surface of the sea, has fully eclosed.

OVIPOSITION

The eggs are laid immediately following copulation after which the female soon dies. The whole process may be completed in 10 minutes or an hour at the most (Tokunaga, 1932). Some parthenogenetic species may even lay their eggs without ever fully eclosing (Chapter 9).

In many species oviposition appears to be triggered by changing light intensity, especially at dusk. Williams (1982) noted a peak in numbers of drifting egg-masses in a stream in southern England immediately after dark, with a period of substantial drift for about two hours either side of this. A second but smaller peak occurred just after dawn. Several species were involved, although he made particular reference to *Polypedilum convictum* (Walker). Several other studies, covering a range of species, tend to confirm that most eggs are laid at dusk or during the night (Chapter 9). Oviposition at such times probably reduces predation by visual predators (Williams, 1982) but by no means all chironomids conform to this diurnal pattern of egg-laying activity. Nolte and Hofmann (1992) noted *Diamesa incallida* (Walker) ovipositing in shaded situations at the source of the Breitenbach (central Germany) between 08.00 and 10.00 in the morning. The author has observed another Diamesinae, *Potthastia gaedii* (Meigen), laying eggs on the water surface during early morning (but several hours after sunrise) while communally ovipositing *Thienemanniella vittata* (Edwards) continued laying throughout the day (L.C.V. Pinder, *pers. obs.*).

Although a few species of Chironomidae lay their eggs singly, the vast majority produce them in batches, often of several hundreds encased in a mucilaginous coating, and extrusion of eggs is a rapid process. *Chironomus plumosus* (Linnaeus) extrudes six or seven eggs per second, permitting an entire egg-rope of around 1600 eggs to be laid in about five minutes (Munsterhjelm, 1920, as reported by Nolte, 1993).

According to Nolte (1993), most chironomids deposit their eggs on firm substrata such as macrophytes, stones or leaf litter, close to the water's edge. Miall and Hammond (1900) described the sequence of events as follows:

> The fertile female skims over the surface of the water touching it lightly from time to time with her legs. This is a preliminary to the laying of the eggs, which commonly takes place in the late evening or early morning. She settles at last on the margin of a pool or stream, and brings the tip of the abdomen close to the surface of the water. A dark gelatinous mass, consisting of eggs thinly covered with mucilage, is then protruded until it touches the water, when it at once begins to swell up. After all the eggs are passed out, the whole mass, which forms a gelatinous cylinder, is secured by the female to some fixed object close to the water's edge. The attachment varies according to the species of the fly but often takes the form of a double cord, which traverses the egg-mass and projects beyond it at one end.

In contrast to certain other aquatic insects, such as some species of Ceratopogonidae and Simuliidae, there are no recorded instances of chironomids venturing beneath the water surface to oviposit. However, circumstantial evidence suggests that this may be the case with *Tanytarsus neoflavellus* (Malloch), since Davis (1966) found eggs of this species (cited as *Calopsectra neoflavellus*) on the underside of *Nymphaea* leaves. Alternatively, these egg-masses may have become attached to the lower surface of the leaves after drifting for a short distance at or just beneath the water surface.

Some species lay their eggs directly on to the surface of the water while in flight. The author has observed this process in *Potthastia gaedii* (Diamesinae). During the early calm of a summer morning, over a chalk stream in southern England, many female *P. gaedii* (their identity was confirmed by hatching some of the eggs in the laboratory) were noticed flying low over the smooth, gently flowing water of a deep pool. The insects moved slowly upstream, frequently dipping to touch the water surface with the tips of their abdomens as long strings of eggs, encased in gelatin, were gradually extruded. If the mass had not been released by the time the insects reached the upper limit of the pool, marked by the more disturbed water of a riffle, they quickly dropped back to the tail of the pool, bounded by another riffle, and repeated the process. Eventually, after about three to five minutes the green, string-like mass was released on to the water surface and large numbers were visible, slowly sinking, as they drifted downstream. When first laid the masses were extremely sticky, which caused them to become firmly attached to the first solid surface with which they came into contact. This was usually a submerged plant. After a few minutes the gelatinous matrix, in which the eggs were encased, had swollen to several times its initial thickness and was no longer sticky.

5.2.1 COMMUNAL OVIPOSITION

The usual egg-laying behaviour of *Thienemanniella vittata* in rivers in southern England is similar to that described for *P. gaedii* (L.C.V. Pinder, *pers. obs.*) but on occasions the females indulge in a different type of activity that produces enormous composite masses, consisting of what on occasions must amount to the egg production of millions of females. Female *T. vittata*, apparently all of them gravid, swarmed just above the water surface and close to the river bank. The first to lay eggs crawled out along trailing leaves and attached their egg-ropes to vegetation at the water surface. Later-laying females crawled out along the growing, composite egg-mass and added their egg-ropes to it. Over a matter of one or two days this resulted in tapering clumps of eggs up to 20–30 cm long and 3–4 cm wide at the base (Pinder, 1992).

Although this type of behaviour is apparently not common, similar examples have been recorded for a range of species and habitats (Nolte, 1993) (Table 5.1). A most spectacular example of communal oviposition was described by Dinulesco (1932). For a distance of hundreds of metres along the margins of the Danube, and attached to protruding boulders, were large numbers of composite masses, 10–15 cm in diameter and yellowish in colour. So conspicuous were these masses that, from a boat, they were visible at a distance of 100–150 m. The masses were the result of egg-laying by vast numbers of *Cardiocladius capucinus* Zetterstedt [cited as *leoni* Goetghebuer] (Chapter 9).

The significance of communal egg-laying behaviour is not known. In the

Table 5.1 Chironomid taxa for which large aggregations of egg masses have been observed (after Nolte 1993)

Taxon	Habitat
Telmatogetoninae	
Telmatogeton torrenticola Terry	Mountain stream
Tanypodinae	
? species	Lake
Diamesinae	
Diamesa alpina (Goetghebuer)	Mountain stream
Diamesa incallida (Walker)	Upland stream
Orthocladiinae	
Cardiocladius capucinus Zetterstedt (as *leoni*)	River
Cricotopus sylvestris (Fabricius)	Lake and pond
Halocladius variabilis (Staeger)	Baltic Sea
Thienemanniella vittata (Edwards)	Chalk stream
Chironominae	
Chironomus riparius Meigen	Sewage canal

T. vittata example described above, the masses disintegrated within two to three days and fragments of varying dimensions drifted away from the oviposition site (Pinder, 1992). Their ultimate fate is unknown but eggs in fragments that were reared in the laboratory hatched over a prolonged period. At 15 °C, peripheral eggs hatched after about 6 days while those in the middle of the mass took up to 58 days (Williams, 1981). Larvae that hatched early had completed their development and pupated while others from the same batch were still in the first instar or not yet hatched. Pinder (1992) surmised that the differences resulted from a gradient of oxygen concentration within the egg-masses and quoted Hynes's (1976) suggestion that the delayed hatching of a proportion of eggs may act as a form of insurance, in an unpredictable habitat, thereby increasing the probability that a proportion of eggs will hatch under optimal environmental conditions for larval development.

5.3 EGGS AND EGG-MASSES

5.3.1 EGG NUMBER

Typical egg-masses of freshwater chironomids contain from 20 or 30 eggs, in the case of some smaller species, to 2000 or more for the larger species, although there may be substantial intraspecific variation. The highest recorded number of eggs in a single mass is 3300, for *Chironomus* (*Camptochironomus*) *tentans* (Fabricius) (Nolte, 1993).

In general, chironomids have been found to produce a single egg mass, although Wensler and Rempel (1962) deduced from a study of ovarian structure that *Chironomus plumosus* is capable of producing up to three. In some species a second batch is produced (Fischer, 1969; Oliver, 1968) and Fischer also noted that a third batch began to develop in *Chironomus nuditarsis* Keyl, although no third oviposition was recorded. *Pseudochironomus prasinatus* (Staeger), however, lays up to six separate egg-ropes, each containing around 70–80 eggs (Nolte, 1993), and repeated oviposition also occurs in the parthenogenetic species, *Paratanytarsus grimmii* (Schneider) (Langton *et al.*, 1988).

P. richardsoni Malloch also oviposits repeatedly, but with only between one and five eggs in each batch (Nolte, 1993). This pattern of oviposition is uncommon among Chironomini but the Telmatogetoninae, some Tanytarsini and some terrestrial Orthocladiinae also lay eggs singly or in numerous small batches.

Telmatogeton contrasts with all other taxa studied by Nolte in that the eggs are always laid individually. *T. japonicus* Tokunaga lays its eggs attached to stones in the intertidal zone of the southern Baltic Sea. It may be surmised that, unlike freshwater species, females of intertidal species are able to select specific oviposition sites, during low tide, that are suitable for larval development. This eliminates the necessity for young larvae to disperse in order to find suitable habitats, in an environment where large numbers might otherwise be washed out to sea and lost. However, the only known freshwater species of the same genus, *T. torrenticola* (Terry), also lays its eggs individually on rocks, although the activities of a number of females sometimes result in accumulations of several thousand eggs in a single layer (Terry, 1913).

The terrestrial species, *Bryophaenocladius virgo* Thienemann and Strenzke, *Pseudosmittia gracilis* (Goetghebuer) and *Pseudosmittia holsata* Thienemann and

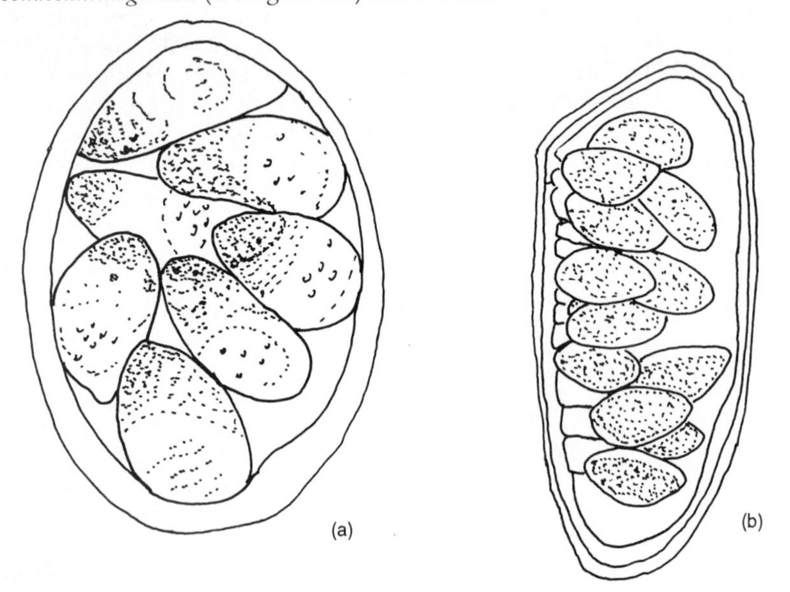

Figure 5.2 Egg balls of two terrestrial species. (a) *Bryophaenocladius virgo* Strenzke; (b) *Pseudosmittia gracilis* (Goetghebuer). (From Thienemann and Strenzke, 1940.)

Strenzke (Thienemann and Strenzke, 1940) produce several minute (<1 mm) egg-masses, each containing a low number of eggs (Figure 5.2). This seems to be typical of terrestrial orthoclads since small egg-masses with few eggs were also described by Williams (1944) in respect of a species of *Smittia* from Hawaii. In terrestrial habitats, dispersal by larvae, other than very locally, is impossible. However, like *Telmatogeton japonicus*, females of terrestrial species are able to place their eggs much more precisely, in relation to the habitat requirements of their larvae, than is possible for riverine or lacustrine species. Locally high densities and intraspecific competition are avoided by laying eggs singly or in batches of very few eggs.

5.3.2 MORPHOLOGY OF CHIRONOMID EGGS

Chironomid eggs are most commonly elliptical or reniform in shape, although some (for example, those of *Telmatogeton japonicus*, *Eukiefferiella claripennis* and some *Orthocladius* sensu stricto) are described by Nolte (1992) as deltoid (roughly triangular) (Figure 5.3).

Egg size varies considerably depending on the species. The smallest eggs, those of *Corynoneura* and *Thienemanniella*, are around 170 μm long and 70 μm wide while *Tanypus punctipennis* Meigen, a large tanypodine, produces eggs that may be as much as 612 μm long and 135 μm wide (Nolte, 1992). These are unusually elongate eggs with a length:width ratio of about 4.5. At the other extreme are the eggs of *Telmatogeton japonicus*, which have a length:width ratio of only 1.6. The majority of eggs, however, are about 2.5 to 3 times as long as wide (Nolte, 1992).

Little is known about the structure of the chorion. In some species, including *Cardiocladius capucinus* Zetterstedt (Dinulesco, 1932), *Telmatogeton torrenticola* (Terry, 1913), *Telmatogeton* sp. (Williams, 1944) and *Buchonomyia thienemanni* Fittkau (Ashe and Murray, 1983), a large, distinct micropyle is present. However, scanning electron micrographs of the eggs of *B. thienemanni* (Ashe and Murray, 1983) and *Tanytarsus barbitarsis* Freeman (Kokkinn and Williams, 1988) indicate that the chorion is otherwise smooth

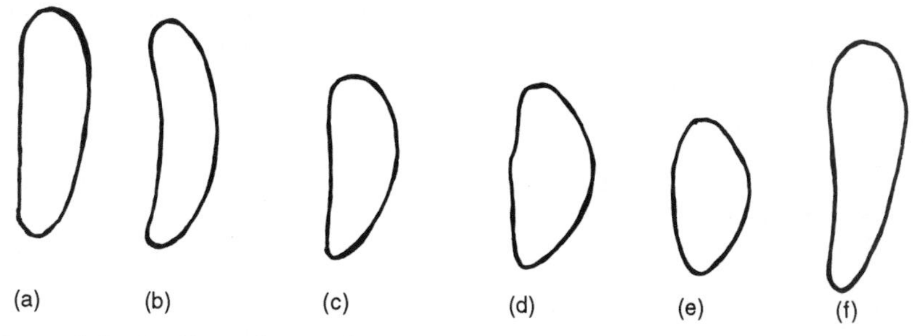

Figure 5.3 Outlines of some chironomid eggs. (a) *Chironomus tentans* Fabricius; (b) *Dicrotendipes nervosus* (Staeger); (c) *Cricotopus sylvestris* (Fabricius) (d) *Metriocnemus fuscipes* (Meigen); (e) *Paratanytarsus tenuis* (Meigen); (f) *Procladius culiciformis* (Linnaeus). (After Munsterhjelm, 1920.)

and featureless. In *Telmatogeton* and *Diamesa*, as well as some Orthocladiinae, the colour of the chorion changes from an initial pale colour to brownish olive or dark brown but in the majority of species there is no such change or only a slight darkening (Nolte 1993). The eggs of Telmatogetoninae are apparently unique among the Chironomidae in possessing a thickened chorion, which presumably affords the eggs, laid singly or in pairs on intertidal rocks, some protection from desiccation. Even this relatively tough chorion retains some degree of elasticity, however, since the eggs increase in size by around 50% during development (Nolte, 1993) and shrink again following eclosion of the larva.

Newly laid eggs are most often white or yellowish, frequently but not always darkening to brown as they develop. Some Tanypodinae, including *Apsectrotanypus trifascipennis* Zetterstedt and *Psectrotanypus varius* (Fabricius), produce eggs that are brownish-orange. Among the Chironominae, the eggs of *Microtendipes pedellus* (de Geer), *Paratendipes* sp. and *Pseudochironomus richardsoni* were described by Nolte (1993) as golden-brown, while those of *Pseudochironomus prasinatus* are bright green.

5.3.3 THE FORM OF THE EGG-MASS

Although the majority of chironomid species lay eggs in batches that are encased in a clear gelatinous mass, there is substantial interspecific variation in the morphology of the mass. The topic was extensively studied by Nolte (1993) and the following account owes much to her work.

Tanypodine egg-masses are usually globular or club-shaped and, in the majority of species, are attached to a firm substratum by a slender stalk at one end. However, in the tribe Macropelopiini the masses are attached, not by a stalk, but by a flattened region along one side. *Tanypus punctipennis* is unusual in that its egg-masses have been found lying freely in muddy littoral sediments (Zilah, 1932).

Eggs of Diamesinae are laid in long strings. The usual practice, among rheophilous species of this subfamily, is to oviposit while crawling over a firm substratum at the water's edge, where movements of the abdomen during oviposition give the egg-mass a characteristic coiled or looped appearance. *Potthastia gaedii* is an exception (L.C.V. Pinder, *pers. obs.*), as is the limnophilous species, *Pseudodiamesa arctica* (Malloch) (Oliver, 1959), as both species deposit their eggs on the water surface.

Egg-masses of Orthocladiinae are either linear or, to use Nolte's (1993) terminology, 'bale shaped'. Nolte used this description to encompass those masses whose shape could not be described as linear, globular, club-shaped or cylindrical, and about half of the orthoclad egg-masses that she considered fell into this general category.

The most conspicuous and therefore the most familiar egg-masses are the cylindrical masses associated with species of the tribe Chironomini (subfamily Chironominae), especially those of *Chironomus* spp. The masses laid by *Chironomus plumosus*, for example, may be as much as 36 mm in length (Munsterhjelm, 1920). Smaller masses, otherwise similar in form, are laid by a number of genera, including *Glyptotendipes*, *Microtendipes*, *Parachironomus*, *Polypedilum* and *Zavreliella*. Globular, club-shaped, string-like or bale-shaped masses are also produced by some Chironominae but all appear to be derived from the basic cylindrical form. The egg-masses of *Polypedilum convictum* (Walker) were described by Williams (1982) as being formed by a variable number (between 1.5 and 6) of helical turns and, unusually, may contain a gas vacuole when first laid.

The mechanics of producing these variations from a basically cylindrical egg mass is unknown. Williams (1982) suggested that the spiral form of *P. convictum* masses results from properties of the gelatin, which cause it to spiral as it is progressively extruded and exposed to water. However, the female of *Pontomyia pacifica* (Tanytarsini) holds its abdomen above the surface of the sea while the egg-mass is extruded, but the mass nevertheless coils up regularly 'just like a watch spring', according to Tokunaga (1932).

In contrast to Chironomini, the egg-masses of Tanytarsini exhibit a much greater range of morphology. Nolte (1993) considered that a continuous range of types exists within this tribe, extending from string-shaped egg-masses, through loosely associated groups of eggs to single eggs. Intraspecific variation is not uncommon. For example, *Tanytarsus barbitarsis*, may lay its eggs singly or in numerous batches of only a very small number (Kokkinn and Williams, 1988); *Paratanytarsus dissimilis* (Johannsen) has been described variously as laying string-shaped egg-masses (Cavanaugh and Tilden, 1930), eggs joined loosely to form a string, or single eggs (Munsterhjelm, 1920). The disparity between these reports may relate to a misidentification or confusion of the parthenogenetic *P. grimmii* Schneider with *P. dissimilis* (Langton *et al.*, 1988). In contrast, the majority of species of Tanytarsini produce linear, string-like eggmasses, although these may be coiled to form an irregular 'bale' as in *Paracladius nana* (Meigen) (Munsterhjelm, 1920), a globular mass as in *Tanytarsus neoflavellus* (Davis, 1966) or wound spirally, giving rise to club-shaped masses such as those produced by *Paratanytarsus laccophilus* Edwards (Lindeberg, 1958).

5.3.4 NATURE OF THE GELATINOUS SHEATH

Among the simplest egg-masses are those of the very distinctive small subfamily, Buchonomyiinae. The eggs of this family are known from a single observation, by Ashe and Murray (1983), of a mass produced in the laboratory and therefore possibly atypical. This mass was roughly tubular in shape and 'open at both ends' with a 'seam' along its entire length. The eggs were apparently freely mobile within the fluid gelatin of the central lumen of the tube. Usually, however, egg-masses are more structured, with more than one type of gelatin being distinguishable by differences between their refractive indices.

Tanypodinae egg-masses all have a basic matrix, constituting the bulk of the mass, in which the eggs are suspended. Stalked masses have a second type of gelatin, which extends throughout the stalk (Nolte, 1993). When freshly laid, the egg masses of Macropelopiini have a third type of streaky gelatin

(Munsterhjelm, 1920; Nolte, 1993) which disappears as the eggs develop and, according to Koreneva (1959), the eggs of *Procladius rufovittatus* (van der Wulp) are arranged around a distinct skein of gelatin that runs through the basic matrix. The arrangement of tanypod eggs within the matrix is, according to Nolte (1993), basically linear but the rows may be spirally or irregularly coiled or folded into loops, producing substantial variation in their ultimate appearance.

Nolte also distinguished two types of gelatin in the egg-masses of *Diamesa* spp.: an inner tube containing the eggs and a surrounding sheath. In *Pseudodiamesa*, however, only one type of gelatin is distinguishable and the eggs are arranged in a single row (Oliver, 1959; Nolte, 1993).

The linear egg-masses of Orthocladiinae usually consist of three components: an inner layer of gelatin, a light-refracting interface (referred to by Nolte as 'the tube') and a gelatinous outer coating. However, in *Corynoneura* and *Thienemanniella* the outer coating is missing and the central matrix is surrounded by a hyaline skin. In other species, such as some *Cricotopus* and *Rheocricotopus*, the light-refracting interface is absent and the eggs are simply embedded in gelatin.

Generalizations concerning the 'bale-shaped' masses of other Orthocladiinae are difficult to make. The eggs, which are spirally coiled within the mass, may or may not be surrounded by a tube. In some species, a broad gelatinous ribbon, supporting the eggs, is wound into a spiral and a thin refractive layer separates this from a thick outer gelatinous layer.

The minute egg-masses of some terrestrial Orthocladiinae comprise an inner layer of gelatin, a light-refracting interface and an outer gelatinous layer. The outer layer is a thin sheath and the inner layer is watery. Both types of gelatin may become dehydrated so that sometimes the eggs may be tightly packed and at other times loosely dispersed within the mass (Thienemann and Strenzke, 1940a, 1940b).

The masses of Chironomini characteristically have a gelatinous stalk and central fibres, which do not occur in the stalked masses of Tanypodinae. The fibres consist of a pair of light-refracting filaments extending throughout the length of the mass. In some species, such as *Chironomus*, the fibres are conspicuous but in others, such as *Dicrotendipes* (Silina, 1959) and *Microtendipes* (Baz, 1939), they are difficult to see or may be absent.

5.4 EGG PREDATION

Drifting egg-masses are likely to be eaten by fish although no instances of this appear to have been recorded. The mucilaginous coating around chironomid eggs is believed to confer some protection against predators but they were the principal food of the water mite *Hydrodroma despiciens* in lakes in Berkshire, England (Wiles, 1982). Egg-masses were also favoured sites for the deposition of spermatophores of the mite. According to Wiles, the palps of *H. despiciens* are well adapted for probing the mucilaginous sheath surrounding the eggs and are used to cut and break up the mucilage, allowing the rostrum to penetrate the chorion. Not all chironomid eggs were equally vulnerable. While most presented no problem to the mites, eggs of *Parachironomus vitiosus* Goetghebuer, which are encased in particularly sticky mucilage, proved difficult to deal with. Although some eggs of this species were eaten, the mites spent considerable amounts of time extricating themselves from the sticky mass. Other species of mite did not feed on chironomid eggs. Whereas predation on chironomid larvae by water mites is not uncommon (e.g. ten Winkel, 1987) this appears to be the only confirmed instance of predation on the eggs.

5.5 DEVELOPMENT DURING THE EGG STAGE

5.5.1 DEVELOPMENTAL PERIOD

The time taken for eggs to develop to the point of hatching was stated by Thienemann (1954) to lie between 2.5 and 6 days. However, this is a simplification since developmental rates are strongly influenced by temperature. The time taken for eggs of *Tanytarsus barbitarsis*, from salt lakes in South Australia, to develop (Kokkinn, 1990) was described by the relationship: $D(t) = 8712.32t^{-2.7}$, where D(t) is the time, in days, taken for 50% of the eggs to hatch and t is mean temperature. This relationship explained 99% of the variation in hatching time. In contrast, rates of larval development were strongly influenced by salinity as well as temperature. Rate of embryonic development was, therefore, a poor predictor of overall development rate. Unlike larvae, embryonic chironomids have near ideal ionic and trophic conditions, since for most of their development they are isolated from surrounding fluids and are supplied with their own food source (Kokkinn and Williams, 1988).

Thienemanniella vittata eggs collected in southern England (Williams, 1981) hatched in a minimum of 4 days at 20 °C, 6 days at 15 °C, 13 days at 10 °C and 31 days at 5 °C. However, these were eggs at the outside of composite masses; eggs further within the masses, and possibly subjected to low oxygen concentrations (Pinder, 1992), took much longer to develop. At the highest temperature (20 °C) some eggs took 44 days to hatch; at 5 °C eggs were still hatching after 119 days. *Chironomus tentans* eggs developed fully in 17.5 days at 8.8 °C and 3 days at 22.1 °C (Sadler, 1935) and these times are similar to the minimum developmental periods for *T. vittata* eggs at comparable temperatures.

Mortality among *T. vittata* eggs was also strongly influenced by temperature (Williams, 1981). Survival rates were highest, with 90% hatching successfully, at 15 °C, which is close to the temperature of the river at the time the eggs were collected. At 20 °C, however, only 28% of eggs hatched. Predictably, eggs of *Chironomus pulcher* that Wiedemann collected from the tropical Lake Chad were adapted to much higher temperatures. When reared at 25–30 °C, they hatched after only 48 hours with a mere 1% mortality (Dejoux, 1971).

The preceding examples are all derived from experimental studies of egg development. Although field data are scarce, the indications are that developmental rates under field conditions are generally rapid. In Germany, *Zavreliella marmorata* (van der Wulp) [as *clavaticrus*] eggs developed fully in 90 hours during August and September, but in October their development required an additional 60 or 70 hours to reach completion (Zavřel, 1926). *Bryophaenocladius virgo* eggs hatched after 4–5 days in the field (Thienemann and Strenzke, 1940a). Thienemann (1954) quoted similar periods of development for both *B. virgo* and *P. holsata* at temperatures between 17 and 24 °C.

5.5.2 ECLOSION OF THE FIRST INSTAR

During the early stages of development the embryo of *Tanytarsus neoflavellus* Malloch [as *Calopsectra neoflavellus*] occupies only about two-thirds of the space within the egg but close to the time of hatching it fills virtually the whole of the space within the chorion (Davis, 1966). Some species, such as *Chironomus pulcher*, become folded more or less double prior to eclosion (Figure 5.4a) (Dejoux, 1971), whereas others become spirally coiled or twisted (Figure 5.4b) (Thienemann, 1954). The larva of *T. neoflavellus* lies straight within the egg, even when fully developed (Figure 5.4c). The sequence of events leading to eclosion in this species was described by Davis (1966). About 12 hours before hatching the prolarva begins to move, first by stretching and later by rolling repeatedly, but as the time of eclosion draws closer it becomes too large to be able to do more than simply squirm within the now tightly fitting chorion. Around 40 minutes before hatching the prolarva begins to drink, swallowing every 3 to 5 seconds, causing it to swell further, until the pressure causes the chorion to rupture, very suddenly.

Almost invariably in this species, the rupture occurs on the concave side of the egg, posterior to the anterior parapod. In consequence, the parapod and about five body segments are immediately expelled (Figure 5.4d). Subsequent twisting movements result in the larva becoming completely free after a further minute or so. In other species the larvula may emerge tail first, as in the case of *Paratanytarsus grimmii* (Langton *et al.*, 1988), or head first (Thienemann, 1954).

In the case of an unidentified species, Davis (1966) noted that the split in the chorion always occurred dorsally, towards the back of the head and over a small dark point, which evidently served as an egg-burster, allowing the head to emerge first from the shell (Davis,1966). Larvae were also described as emerging head first by Branch (1923) and Thienemann (1954). Thienemann believed that the split in the chorion was achieved using the mandibles, which he observed to be very active just prior to eclosion. However, it is possible that these mandibular movements are not intended to damage the chorion but accompany drinking, as in the case of *T. neoflavellus*.

Neither Thienemann nor Branch mentions the presence of an egg-burster on the head of the prolarva or newly hatched larvula, although such structures do occur in some other Nematocera such as *Simulium* larvae, which also eclose by imbibing fluid until the egg-burster forces a split in the chorion. In spite of her observations on the eggs and larvulae of very many

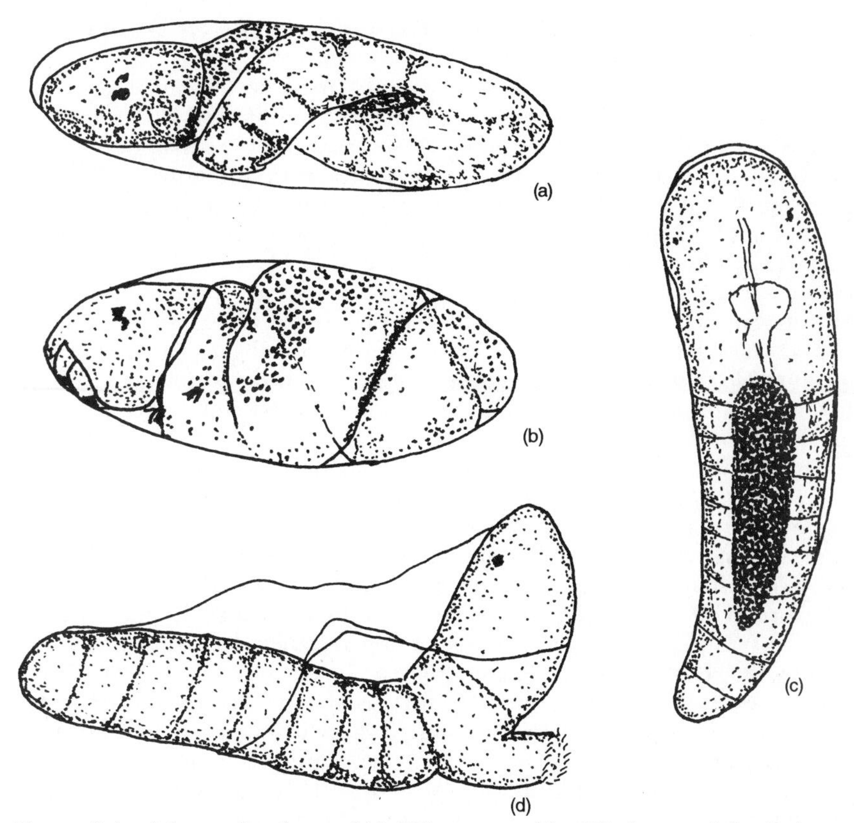

Figure 5.4 Advanced embryo of (a) *Chironomus pulcher* Wiedemann (after Dejoux, 1971); (b) *Pseudosmittia gracilis* (Goetghebuer) (after Thienemann and Strenzke, 1940); (c) *Tanytarsus neoflavellus* (Malloch) (after Davis, 1966). (d) Hatching larvula of *T. neoflavellus* (after Davis, 1966).

species of chironomid, both temperate and tropical, Nolte (personal communication) has seen 'egg-bursters', of the kind described by Davis (1966), only in Diamesinae. In the case of *Diamesa incallida*, she describes this structure as 'a tiny, dark knot, located between the eye-spots and close to the dorsal occipital margin of the head'.

5.6 MORPHOLOGY OF FIRST-INSTAR LARVAE

At the point of eclosion the larva of *Chironomus pulcher* is about 0.35 mm in length (Dejoux, 1971). Like *T. neoflavellus* (Figure 5.4a), the large head capsule of the newly hatched first instar makes up about a quarter of the total length of the larva. Freshly hatched larvae, or larvulae, are virtually transparent, but the head soon becomes sclerotized and reduced in volume to more usual proportions (Dejoux, 1971).

Changes in morphology associated with larval development were reviewed by Kalugina (1959), who commented that in almost all cases the morphology of the first instar, and in some cases also the behaviour, differs markedly from that of later instars. In some cases, she remarks that the differences in morphology are such that the first and last instars could be keyed out as belonging to different genera. Morphological changes associated with larval development are also described, and discussed in a phylogenetic context, by Mozley (1979). Some changes in the morphology of taxonomically important characters occur with each larval moult, but the changes accompanying the transition to third and fourth instar are principally quantitative (e.g. differences in antennal ratio). Those associated with the first larval moult are often much more substantial and frequently also qualitative.

The mouthparts of first and later instars of six species of Chironomini were compared, intraspecifically and interspecifically, by Olafsson (1992). The shape of the mandible of first-instar larvae was similar to that of older larvae, but features such as the seta interna and pecten mandibularis were lacking in the first instar. The labrum and premandibles also had the same basic structure in all instars but labral setae were absent or weakly developed in first-instar larvae and the premandibular brush was also lacking at this stage.

Surprisingly, Olafsson (1992) did not refer to the mentum, which is often conspicuously different in form in the first instar. The mentum of first-instar *Endochironomus impar* (Walker), for example, has a trifid median tooth (Figure 5.5a), of which the central part is longest, while the second instar has a bifid central tooth that is distinctly shorter than the first lateral teeth (Figure 5.5b) (Kalugina, 1959). Similar changes in the menta of certain *Eukiefferiella* and *Tvetenia* species were noted by Storey (1986). In each case he observed that the ventromentum of the first instar was clearly differentiated from the dorsomentum which was not the case in later stages. The ventromentum of the first instar of *E. ilkleyensis*, for example, has three transparent, similarly-sized median teeth (Figure 5.5c), while the mentum of later instars is strongly sclerotized, with a single, broad median tooth (Figure 5.5d). Among the Tanypodinae it is common for the ligula of first instar larvae to possess fewer teeth than later instars (Kalugina, 1959).

Frequently, the first moult also results in striking changes in the antennae, especially in respect of the relative length of the basal segment. In Chironominae and Orthocladiinae this is exceptionally short and squat in the first instar with a disproportionately long distal blade (e.g. Figure 5.5e).

Olafsson (1992) pointed out the differentiation in the shape of the head capsule between the carnivorous Chironomini, such as *Parachironomus* and *Cryptochironomus* species, and other non-carnivorous genera of the same tribe. Carnivorous genera have an oval head capsule narrowing towards the anterior. In detritivorous larvae the head capsule is more globular or in some species may even be wider anteriorly. This distinction is not evident in the first-instar larvae, all of which have a head capsule shaped like that described for carnivorous larvae of later instars. The morphology of the mouthparts of

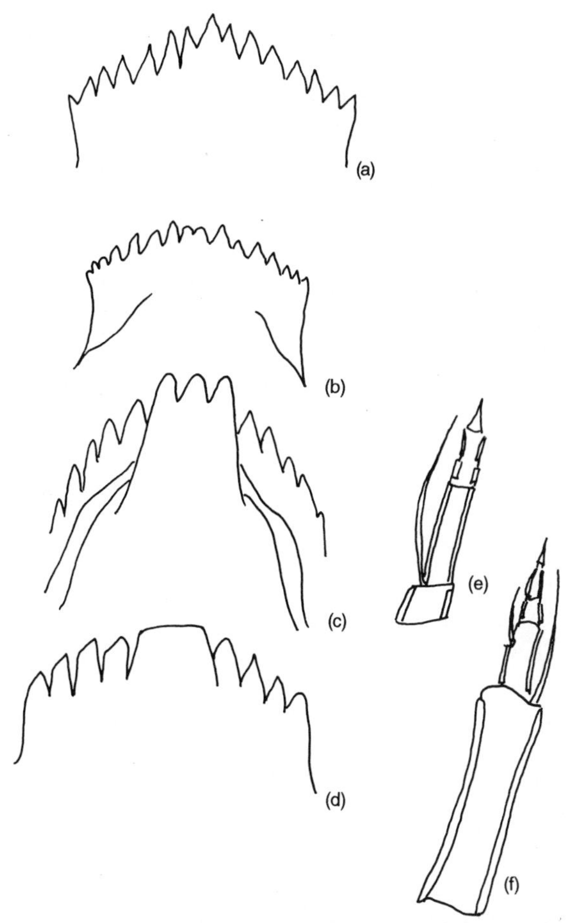

Figure 5.5 Menta of *Endochironomus impar* (Walker) (after Kalugina, 1959): (a) first instar, (b) second instar. *Eukiefferiella ilkleyensis* (Edwards): (c) mentum of first instar; (d) mentum of fourth instar; (e) antenna of first instar; (f) antenna of fourth instar.

the first-instar larvae also resembles in many ways that found within the later stages of carnivorous species – for example, in the morphometry of the mandibles and the lack or poor development of the setae. Olafsson (1992) wondered whether these similarities might be interpreted as adaptations to the ingestion of relatively large food particles.

Changes in the morphology of the thorax and abdomen are usually less striking, although first-instar larvae generally possess a few long, lateral setae, which are not present in older larvae of most species. Early, first-instar larvae of *Stenochironomus gibbus* (Fabricius) have a body shape similar to that of most other chironomid larvae, and are free-living, but before moulting they adopt the wood-mining behaviour that is characteristic of later instars. This change in behaviour is accompanied by an alteration in body shape as

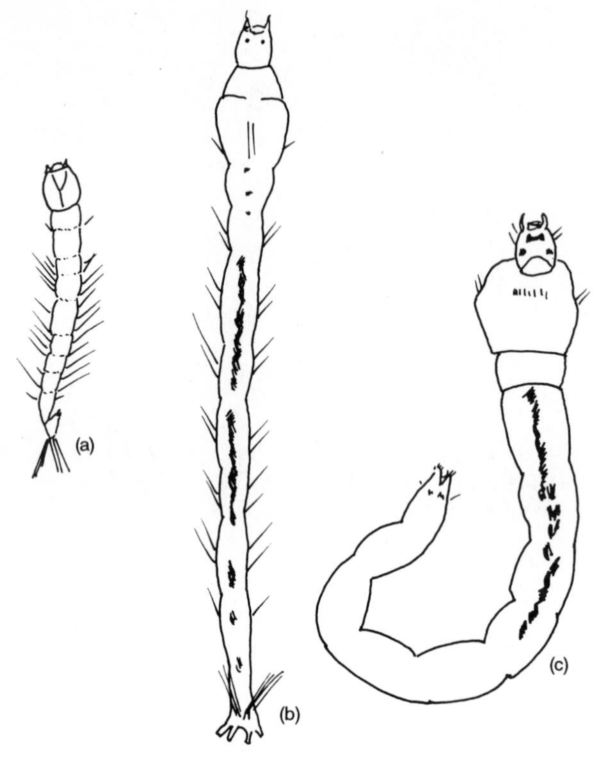

Figure 5.6 *Stenochironomus gibbus* (Fabricius). (a) Newly hatched first instar; (b) later first instar after adoption of wood-mining habit; (c) later instar, not to scale. (After Kalugina, 1959.)

they acquire the exceptionally elongate abdomen and expanded thorax that are typical of later instars of this species (Kalugina, 1959) (Figure 5.6).

5.7 DRIFT OF EGGS AND YOUNG LARVAE

5.7.1 DRIFT OF EGG-MASSES

(a) Rivers

Many egg-masses are attached to a fixed object by the female at the time of oviposition, but egg-masses are also often found firmly attached to submerged objects and it is evident that attachment has occurred some time after the eggs were laid. Newly deposited egg masses of the majority of chironomid species have a specific gravity slightly greater than 1: they sink slowly when dropped on to the surface of a water-body, while the gelatinous coating takes up water. Hence, in rivers and streams, egg-masses that are deposited freely will tend to drift for a while before encountering a firm substratum, such as a plant or stone. Egg-masses of several taxa, including Pentaneurini, *Rheocricotopus*, *Synorthocladius*, *Polypedilum convictum* and *Tanytarsus* sp., as well as other Chironomini and Tanytarsini, were found by Williams (1982) adhering to submerged stones and he deduced that they would have drifted for some distance before becoming so attached. The drift of *Potthastia gaedii* eggs before adhesion to submerged plants has already been described. Williams (1982) also found that some spherical egg-masses, which reached the bed of the river, became covered in silt and continued to be transported along the bed of the stream.

On one occasion Williams estimated that, over a 5-hour period, around 50 000 egg-masses passed a sampling point on a section of river with a cross-sectional area of 2.25 m^2. Similar observations were made on two other occasions and in each case a pronounced peak occurred immediately after dark, although a low-level of drift continued throughout the day. Large numbers of egg-masses were found to drift in a Florida stream, but Soponis and Russell (1984) did not provide information on diel periodicity of their occurrence. Spherical egg masses of *Polypedilum*, and elongate masses of an unspecified Orthocladiinae were both common in this situation.

The extremely sticky nature of the mucilage when first extruded probably ensures that the majority of egg masses drift for only a short distance before becoming attached, at least in shallow or weedy streams and rivers. However, Williams (1982) described an unusual feature of the egg masses of *P. convictum* that prolongs the period during which they remain in the drift. Immediately after their collection he noticed a number of egg-masses floating at the water surface. The club-shaped masses all contained a characteristically spiralled line of eggs, with between 1.5 and 6 helical turns enclosing a gas vacuole, which gradually shrank over a period of 8–11 hours. As the gas vacuole (which was presumably entrapped air) became smaller, the masses gradually sank, until after 2.5–4 hours all had reached the bottom of the sample bottles.

Clearly the numbers of drifting egg-masses in streams may be substantial but the significance of this drift, in terms of dispersal, is more difficult to assess. Downstream drift was considered by Williams (1982) to be an elegant means of dispersal. This conclusion may be justified in the case of *P. convictum*, since larvae of this species are characteristic of habitats that are rich in fine organic detritus. The progress downstream of drifting *P. convictum* egg-masses will be slower in pools than in faster flowing sections. Statistically, therefore, they are more likely to come to rest in slowly flowing, depositional zones of streams or rivers, where conditions are suited for larval feeding.

However, it is also perfectly possible that the presence of trapped air in the masses observed by Williams (1982) is accidental, and that such masses represent only a small proportion of the total. Nolte (in press) favours this explanation, having observed a similar phenomenon in *Apedilum elachistus* Townes living in rock pools in the Mato Grosso. Around one in a hundred of the masses produced by this species contained a bubble of air; the remainder, lacking this feature, sank immediately. The sticky nature of the mucilage and

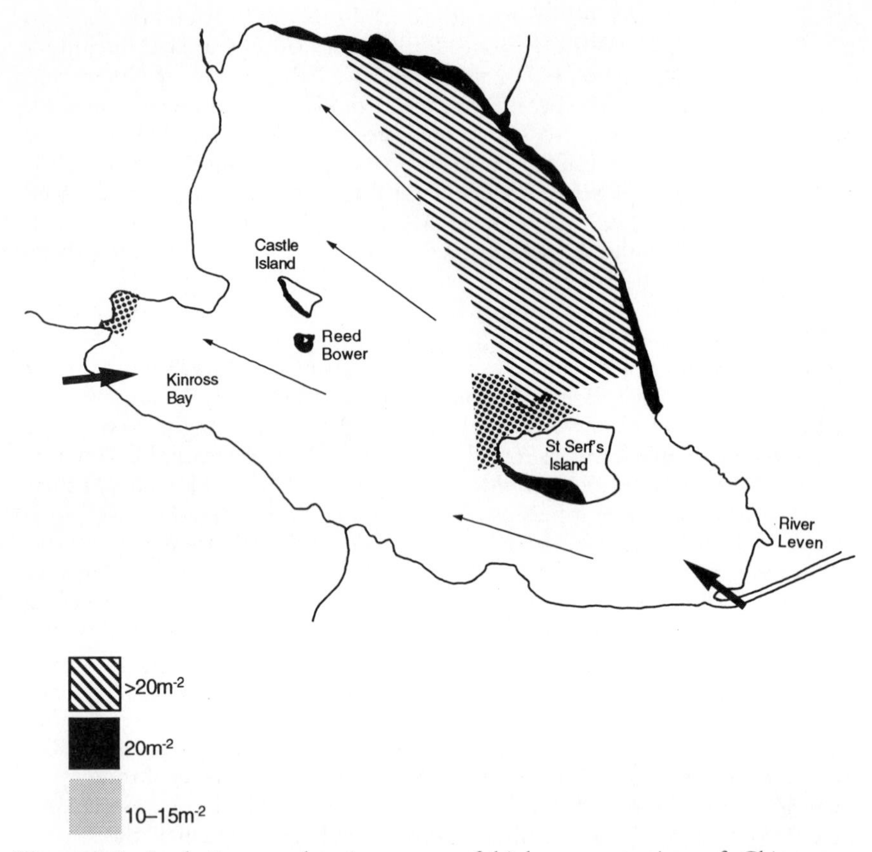

Figure 5.7 Loch Leven, showing areas of high concentration of *Chironomus anthracinus* Zetterstedt eggs on 6th June 1971. Arrow indicate predominant wind directions. (After Davies, 1976.)

the specific gravity of the masses of most species are more likely to cause them to become attached to an object above the stream-bed, assuring that the eggs develop in well-oxygenated situations. First-instar larvae are generally better fitted to a dispersive role than egg-masses, since they can actively re-enter the drift if they encounter unsuitable conditions.

(b) Lakes

Some species of lacustrine Chironomidae oviposit over the whole surface of a lake (e.g. Rempel, 1936) but most lay their eggs along the shoreline (e.g. Sadler, 1935; MacDonald, 1956) and this is the case in respect of *Chironomus anthracinus* (Zetterstedt) in Loch Leven, Scotland (Davies, 1976a). However, oviposition site was of little importance to the eventual distribution of egg-masses in sediments since they were redistributed by wind-induced currents in the loch. Following a mass emergence of *C. anthracinus*, Davies

(1976a) observed large numbers of egg-masses floating near the water surface and being driven across the loch by a south-easterly wind of around 20 km h^{-1}. High densities of egg-masses were associated with wind-induced 'Langmuir spirals'. Most egg-masses were transported in a direction roughly 30° to the east of the prevailing wind direction. This resulted in an accumulation of very large numbers of egg-masses, primarily along the north and north-east shores of the loch (Figure 5.7).

5.7.2 DISPERSAL OF FIRST-INSTAR LARVAE

Recently eclosed larvae of *Chironomus* sp. contain a quantity of embryonal yolk within their guts (Miall and Hammond, 1900). The literature gives no indication of whether or not this is the usual situation in Chironomidae but the presence of yolk may well be important in sustaining the larvae during the immediate post-hatching period. Having escaped from the egg the larvae must excavate their way clear of the surrounding jelly. Larvae reared in the laboratory often spend several hours within the mass, during which time they feed on the gelatin (*pers. obs.*, L.C.V. Pinder, U. Nolte). Hinton (1980) considered the gelatin to consist mainly of carbohydrate and Nolte (personal communication) believes that this source of nutrition is important in helping to sustain the young larvae during the dispersal phase. In addition, larvae often consume detrital particles and diatoms from the outer surface of the remains of the egg-mass.

Newly hatched larvae of *Apedilum elachistus*, living in rock pools in Brazil, were observed by Nolte (in press) to move away from light but this is unusual, since young larvae of the majority of freshwater species show a strong phototactic response. Kalugina (1959) described first-instar larvae, reared in petri dishes, as beating against the wall of the container, on the side turned towards the light. Many other authors have noted strong phototaxis (e.g. Davies, 1973; Luferov, 1971), which occurs in members of all of the major subfamilies of Chironomidae and appears to be characteristic of the young larvae of most freshwater species. It has also been noted among Orthocladiinae and Chironominae inhabiting tidal rock-pools (Stuart, 1942). During this phase of their behaviour, larvae swim vigorously using the typical figure-of-eight motion and in a lake or river they may be carried well away from the site of hatching before regaining contact with the substratum.

The swimming ability of first-instar larvae is no doubt enhanced by their long abdominal setae but they are not capable of directional swimming and are at the mercy of the currents as long as they remain suspended in the water column. During this period they are apparently capable of feeding by ingesting small, suspended particles of detritus (Morduchai-Boltovskoy and Shilova, 1955). The ability of larvae to ingest small particles while swimming in the water column was demonstrated by Kalugina (1959), who eliminated the possibility that they could have been scraping material from the floor or sides of the experimental container.

The phototactic behaviour of first-instar larvae of freshwater chironomids, together with their ability to feed in the water column, makes them well suited to a dispersive role. Many authors have referred to the planktonic activity of first-instar chironomid larvae in lakes (e.g. Lellak, 1968; Davies, 1974, 1976b; Hamilton, 1965). The pelagic behaviour of first-instar larvae of *Chironomus transvaalensis* Kieffer allowed it to colonize the newly created shoreline of Lake Kariba extremely quickly (McLachlan, 1970). A tendency for first-instar larvae to be planktonic has also been observed among some marine species (Tokunaga, 1935).

From time to time larvae cease their vigorous swimming and may then sink slowly to the bottom (Kalugina, 1959). Whether or not the photo-positive swimming behaviour is resumed depends very largely on whether they encounter a suitable substratum. When they do so, they no longer respond positively to light and they settle to a mode of existence similar to that of the older larvae. The initial phototactic behaviour may be wholly suppressed if the substrate on which larvae hatch is suitable for their further development. Under these conditions, according to Kalugina (1959), newly emerged larvae of *Chironomus anthracinus* will immediately begin to build tubes and remain on the bottom of experimental containers.

6

The habitats of chironomid larvae

L.C.V. Pinder

6.1 INTRODUCTION

Anyone who has taken even a casual interest in freshwater biology will be familiar with the larvae of Chironomidae. There are species that thrive in almost every conceivable freshwater environment (Edward, 1986; Oliver, 1971; Pinder, 1986). Ice-cold glacial trickles, hot springs, thin films, minute containers of water in the leaf axils of plants and the depths of great lakes all have their characteristic species or communities. There are semi-aquatic species, living in moist soil or vegetation and others that are truly terrestrial. Some tolerate brackish water while others thrive in intertidal pools and, unusually among the insects, a few are truly marine.

In spite of this enormous diversity of habitat, few species show conspicuous morphological adaptations to their mode of life. *Metriocnemus*, a morphologically homogeneous genus, is a good example. It exhibits one of the broadest spectra of larval habitat of any dipteran group (Cranston and Judd, 1987), inhabiting the margins of running and standing waters, madicolous zones (thin films of water over rock or plant surfaces), phytotelmata (plant-held waters), coastal rock-pools and fully terrestrial biotopes.

6.2 LOTIC HABITATS

The term **lotic** includes all running waters, from the smallest of streams to the largest of rivers. Clearly this encompasses a wide range of habitats, varying dramatically in size, water chemistry, temperature regime, current velocity and substrate type. Even within a relatively short reach of river or stream there may be an alternation of riffles and pools, offering very different environmental conditions. The modern tendency is to regard the downstream changes in rivers as progressive, that is as a continuum, rather than zonal (e.g. Vannote *et al.*, 1980). However, the **river continuum concept** was developed to describe the situation in relatively undisturbed North

The Chironomidae: Biology and ecology of non-biting midges
Edited by P.D. Armitage, P.S. Cranston and L.C.V. Pinder
Published in 1995 by Chapman & Hall. ISBN 0 412 45260 X

American river systems, flowing through natural forested catchments. The concept continues to evolve but it remains unclear to what extent it can be applied to rivers in different biomes (Winterbourn *et al.*, 1981; Ward, 1992). Some Australian running waters, for example, in which seasonality affects flow and allochthonous input may increase in summer (e.g. Lake, 1982) rather than being restricted to winter, exhibit variation in community structure, both spatially and temporally, without showing consistent trends (Lake *et al.*, 1986).

Earlier attempts, largely of European origin, to take account of downstream changes in rivers adopted a zonal concept (Illies, 1961b; Illies and Botosaneanu, 1963). This concept (Chapter 15), which is described in some detail by Ward (1992), provides a convenient framework in which to begin discussing patterns of distribution and community structure among riverine chironomids. However, by no means all lotic systems fit into such a scheme. Some rivers, arising from springs in lowland regions, may have a relatively steep gradient and coarse substratum and have a fauna rather similar to that of an upland stream. A slow-flowing, lowland river may also include swiftly flowing shallows, where it encounters more resistant geological formations, so that a combination of **rhithral** (fast flow and coarse substratum) and **potamal** (low current speed and sandy or muddy substrata) characteristics may result. Anthropogenic factors, such as construction of dams and other means of flow-regulation, also have a major impact on many river systems.

The rhithral zone (Figure 6.1) is characterized by having a steep gradient and high current velocity, with high levels of dissolved oxygen and a predominantly coarse substratum. In the potamal region the current is slower, the water more turbid and the river bed is typically composed of sand and silt. Oxygen depletion may occur in this region. A **crenal**, or spring, zone is present in rivers that are fed by springs, while in high mountains yet another zone, the **kryon**, comprising very cold brooks fed by water from melting glaciers, may also occur (Figure 6.1).

6.2.1 GLACIER-BROOKS: CHIRONOMIDS OF THE KRYON

The upper reaches of mountain streams, fed by melting ice, have a fauna that is distinct from that of spring-fed mountain streams. Most studies of the kryon in Europe (Thienemann, 1954; Sæther, 1968; Steffan, 1971), North America (Elgmork and Sæther, 1970), the Himalaya (Kohshima, 1984; Sæther and Willassen, 1987) and tropical Africa (Harrison, 1965; Willassen and Cranston, 1986) have shown the community to be dominated by the chironomid genus *Diamesa*. According to Steffan (1971) a true kryon or glacier-brook fauna persists for only a short distance downstream from the glacier and both Steffan (1971) and Kownacka and Kownacki (1975) agreed that temperature is the dominant controlling factor. Where summer water temperatures exceed 2 °C a more diverse fauna begins to develop and the species characteristic of glacier brooks decline in relative abundance. In glacier-brooks of northern Scandinavia, *D. lindrothi* Goetghebuer is the most abundant species in the uppermost zone, where the temperature amplitude is

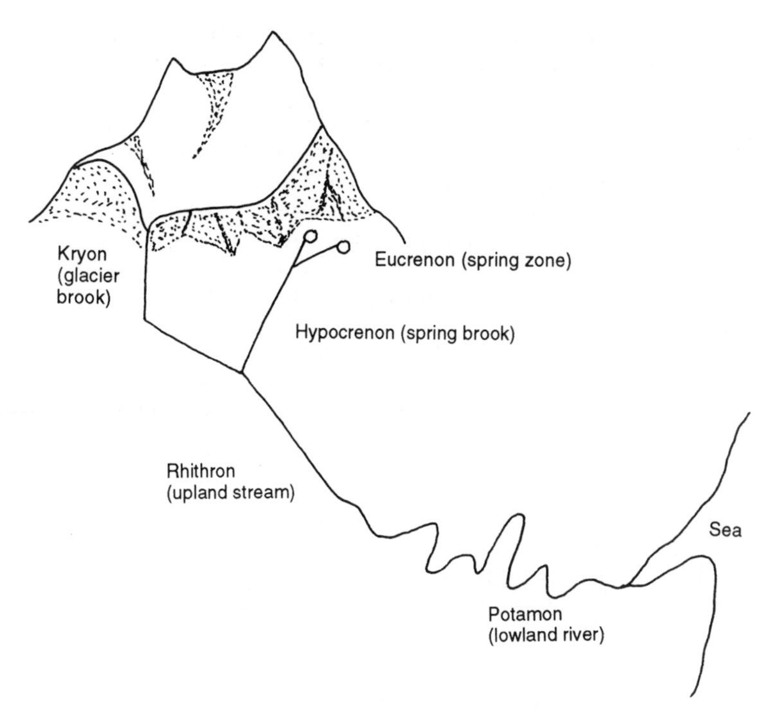

Figure 6.1 Simplified longitudinal zonation scheme for rivers.

very low (Steffan, 1971). Further downstream, with temperatures rising to about 3.5 °C, Orthocladiinae larvae, especially species of *Eukiefferiella* and *Tvetenia*, occur together with *Diamesa*.

Characteristic kryal species are, by definition, well adapted physiologically to low temperatures. They may also have to contend with high current velocities. Mature larvae of *D. lindrothi* are only found in small cavities in rocks, where they spin wide-meshed nets behind which they can pupate without danger of being washed away or crushed by the shifting of the substratum (Steffan, 1971).

Steffan believed that kryal chironomids derive their nutrition from wind-blown organic particles that are released from the ice as it melts. Primary production is low at such cold temperatures but blue-green algae (Cyanophyceae) were found to be growing and providing nutrition for a species of *Diamesa* that lives in tunnels draining melt water in Himalayan glaciers (Kohshima, 1984).

6.2.2 SPRINGS AND SPRING-BROOKS: CHIRONOMIDS OF THE CRENON

Spring sources (the **eucrenon**) are of three broad types. Seepages that form marshy areas, whether small or extensive in area, are sometimes known as

helocrenes. More substantial outflows of ground-water may form ponds or small lakes and such habitats are referred to as **limnocrenes**, while the term **rheocrene** is used to indicate situations where the spring forms a stream or spring-brook. In spite of the range of physical conditions that may be encountered in these situations, a distinctive crenal fauna has been widely recognized (e.g. Illies and Botosaneanu, 1963).

Water temperature at the point of emergence of a spring-brook is almost constant throughout the year and approximates closely to the annual mean air temperature. Temperate springs and spring-brooks are thus warm in winter and rather cool in summer. Spring-fed streams also tend to have a fairly constant discharge and hence a stable substratum that encourages the growth of macrophytes. In well-oxygenated rheocrenes, invertebrates may be abundant but diversity is generally low (Minshall, 1968; Ward and Dufford, 1979). Four categories of fauna may be distinguished (after Hynes, 1970):

- Groundwater species that occur as accidental inhabitants of the surface water, very close to the point of issue.
- Semi-terrestrial species that normally inhabit moist soil or vegetation may find suitable conditions for a fully aquatic life style in springs and in the upper reaches of spring-fed streams.
- Typical crenal species, restricted to this kind of habitat.
- More widespread faunal elements that find suitable conditions in spring zones.

Animals included within the first of these categories are primarily non-insectan. However, a number of chironomids fall into the second group and, according to Hudson (1987), the banks of a water-body together with its floodplain and upland habitats form a continuum, extending from the shore into the terrestrial zone. Thirteen genera, typical of the fauna of banks, were listed by Hudson (1987) (Table 6.1). For this purpose, banks were defined as

Table 6.1 Chironomid genera characteristic of bankside habitats (including springs) (after Hudson, 1987)

Boreochlus
Brundiniella
Chaetocladius
Diplocladius
Metriocnemus
Micropsectra
Paraboreochlus
Parachaetocladius
Parametriocnemus
Prodiamesa
Pseudosmittia
Psilometriocnemus
Trichotanypus

'the moist slope immediately bordering a stream course (including springs)'. Of these genera, *Micropsectra* and, more especially, *Metriocnemus* include species that occupy a wide range of habitats, while *Parachaetocladius*, *Parametriocnemus* and *Prodiamesa* are regarded as accidental inhabitants of the bank zone, normally being wholly aquatic. The remaining genera listed in Table 6.1, however, are characteristic elements of the crenal fauna. Overwhelmingly, the chironomid fauna of these cool habitats is dominated by Orthocladiinae, but one taxon deserving particular mention is a tanytarsine, *Lithotanytarsus emarginatus* Goetghebuer. This species is confined to limestone precipitating springs and spring-brooks in the western Palaearctic. Why the larvae should be restricted to this type of situation is not clear but their tubes become encrusted with limestone, creating in time a characteristic type of tufa, well illustrated by Thienemann (1954).

6.2.3 MADICOLOUS BIOTOPES

The term **madicolous** is used to describe the fauna of films of water less than 2 mm thick (Vaillant, 1956). Such habitats occur, for example, where a thin trickle of water runs over a rock face. Usually the substratum is partly or entirely covered with a film of algae or with a thin layer of fine detritus. Madicolous biotopes are often associated with seepages of ground-water that provide a constant flow. An alternative term that has often been used to describe this biotope, especially in the European literature, is **hygropetric** (e.g. Illies, 1978). However, to a lesser but significant extent the biotope also occurs along stream margins, where the substrate is often mosses or higher plants rather than rock, so that the broader term madicolous is preferable. Oliver and Sinclair (1989) list 47 species that have been recorded from madicolous biotopes, including one Podonominae, four Tanypodinae, seven Diamesinae, 20 Orthocladiinae and six Tanytarsini. Prodiamesinae and Chironomini have not been found in this type of habitat.

Few of the species that have been recorded as madicoles are restricted to this biotope. Of the species listed from Holarctic madicolous habitats by Oliver and Sinclair (1989), only four were regarded as obligate madicoles. These were *Hudsonimyia karelana* Roback (Tanypodinae), *Limnophyes fumosus* Johannsen (Orthocladiinae), both from the Nearctic, and *Syndiamesa hygropetricus* (Kieffer) (Diamesinae) and *Orthocladius fuscimanus* (Kieffer) (Orthocladiinae) from the Palaearctic. *Metriocnemus obscuripes* (Holmgren), known earlier as *hygropetricus* (Kieffer), one of the best known of all madicoles, also occurs in springs and small streams (Thienemann, 1954), while Cranston (1982b) noted its occurrence in sewage lagoons and in percolating filter beds of sewage works. These may be regarded as the ecological equivalents of natural hygropetric biotopes and a number of other typically madicolous species, including *Orthocladius* (*Eudactylocladius*) *fuscimanus* Kieffer, *Metriocnemus hirticollis* (Kieffer) and *Limnophyes minimus* (Meigen), have also been found in sewage filters (Cranston, 1984). *Metriocnemus obscuripes*, along with *M. hirticollis* (which may be only a variety of *M. obscuripes*) is also well known in southern England as a pest of water-cress, where it lives in films of water on the aerial parts of the plant (Cranston and Judd, 1987).

6.2.4 UPLAND STREAMS: CHIRONOMIDS OF THE RHITHRAL ZONE

Typically, the rhithral stream has clear water, a high gradient and consequently a high current velocity, high concentration of dissolved oxygen and a predominantly coarse substratum. Illies and Botosaneanu (1963) intended the distinction between the rhithron and potamon to apply to tropical as well as temperate systems. Different temperature criteria for separating the zones were therefore defined for each of the two geographical regions: for temperate regions the rhithron was described as having an annual range of monthly mean water temperature of less than 20 °C, while in the tropics the limit was set at 25 °C.

Chironominae are relatively scarce in the faunas of such streams, which are dominated by cold stenotherms, predominantly Orthocladiinae but often including Diamesinae. The subfamily Podonominae is also primarily rheophilic and cold-adapted but has far more species in the southern than in the northern hemisphere. The small subfamily Aphroteniinae, confined to South America, southern Africa and Australia, was considered to be another cold-adapted subfamily, characteristic of austral southern or upland catchments (Brundin, 1966). More recently, however, Cranston and Edward (1992) have shown that this is not necessarily the case in Australia and that the genus *Aphroteniella* is much more widespread, occurring in warmer streams and standing waters even in subtropical and tropical parts of the continent.

Lake *et al.* (1986) argued that Australian lotic systems show longitudinal zonation less consistently than elsewhere, with marked temporal and spatial variation sometimes occurring within the same stream order. Edward (1986), in his discussion of several studies that have been carried out within the subcontinent, showed that the fauna of some western Australian streams undergo major changes associated with the markedly (mediterranean) seasonal rainfall; Lake *et al.* (1985) made compatible observations in eastern Australia. The lack of seasonal change in the fauna is considered to be the consequence of the unpredictable nature of the climate in the eastern part of the country (Edward, 1986). Chironomids comprised 32% of the fauna in the upper catchment of the La Trobe River, in Victoria, and 40% of the fauna in streams of the northern jarrah forest, in the Darling Range of Western Australia (Edward, 1986, referring to the work of Marchant *et al.*, 1984; Metzeling *et al.*, 1984; Bunn *et al.*, 1986). A high proportion of species from the upper catchment of the La Trobe River was cold-adapted while few cold stenotherms were found in western Australian streams, where the temperature rarely falls below 10 °C. Only nine Orthocladiinae and no Podonominae or Diamesinae (Table 6.2) were recorded by Bunn *et al.* (1986); subsequently one species of Podonominae and two of Aphroteniinae were found in temperate western Australian streams (Cranston and Edward, 1992). However, the high montane catchment of Devils Creek, in the South Island

Table 6.2 Chironomid communities in streams in Australia and New Zealand (after Edward, 1986)

	Number of taxa			
	AUSTRALIA		NEW ZEALAND	
Subfamily	La Trobe River Victoria	Jarrah Forest W. Australia	Waitakere River N. Island	Devils Creek S. Island
Tanypodinae	7	5	4	7
Chironominae	22	17	4	4
Orthocladiinae	17	9	6	14
Podonominae	2	–	–	7
Aphroteniinae	1	1	–	–
Diamesinae	–	–	2	1
Total	49	32	16	33

of New Zealand, was dominated by Orthocladiinae, along with seven species of Podonominae and one Diamesinae (Cowie, 1980).

Podonominae and Diamesinae are also characteristic elements of the faunas of mountain and high mountain streams in tropical South America. The lower limit of their distribution, according to Brundin (1966), is about 1700–1800 m above sea level, roughly corresponding with the lower limit of the cloud forests.

6.2.5 LOWLAND RIVERS: CHIRONOMIDS OF THE POTAMAL ZONE

Whereas the chironomid fauna of upland streams is generally dominated by Orthocladiinae and other taxa adapted to cool, well-oxygenated situations, the dominant chironomid subfamily in potamal communities is Chironominae. In the main, species of this subfamily are adapted to living in soft sediments and are better able to tolerate the higher temperatures and sometimes lower oxygen concentrations that occur in large lowland rivers. However, where there are faster flowing riffles with a more stable substratum, or where there is aquatic vegetation, Orthocladiinae and other taxa more characteristic of the rhithron may thrive.

The chironomid fauna of the lower River Inn was studied by Reiss and Kohmann (1982), mainly using data derived from light-trapping. Not all of the species they listed necessarily came from the river (for example, the list includes *Camptocladius stercorarius* (Degeer), an inhabitant of cattle dung) but most of them probably did. As expected, Chironominae was the dominant subfamily with 72 species (51% of the total) but there were also 43 species (31%) of Orthocladiinae and seven Diamesinae, as well as 14 Tanypodinae and four Prodiamesinae. These proportions accord well with Thienemann's (1954) generalization that, in temperate regions, Chironominae make up about 55% of the total number of species in lowland rivers, with about 30% being Orthocladiinae.

Unstable sandy substrata, which are commonly found in potamal sections of large rivers, are avoided by many species but notable exceptions are to be found in the orthoclad genus *Rheosmittia* and among the so-called *Harnischia* group of genera. All species in this group are predominantly predatory, with oligochaetes being a major component of the food of many species. A number of genera within the *Harnischia* group are characteristic of sandy substrata, some being also restricted to very large rivers. These include, for example, the genera *Beckidia*, *Chernovskiia*, *Cyphomella* and *Robackia*.

The distinction between the rhithral and potamal zones is usually clear in rivers that arise in the mountains and extend into the lowlands but this is not always the case. Hynes and Williams (1962) noted the tendency for the rhithron to be short, relative to the potamon, in streams of the Amazon basin, where there is a tendency for typically potamal species to extend to high altitudes. The dominance of Chironominae in streams of the central Amazonian rain-forest was remarked upon by Fittkau (1964). The temperature in these streams is remarkably constant, fluctuating diurnally by only 1 °C or 2 °C. They are fed by helocrenes with a constant temperature of 24.5 °C and, although the upper reaches have a strong gradient, the degree of erosion is mostly weak because the watercourses are repeatedly dammed by fallen trees. In consequence the bed, even in the upper sections, is mainly composed of sand. This, together with the high water temperatures, renders these habitats largely unsuitable for most Orthocladiinae, which contribute only 10% of the total number of chironomid species, with Tanypodinae making up a further 10% and the remainder being Chironominae. Highest densities of invertebrates occur in the tangles of exposed roots, where velocity is swift, and *Cricotopus* spp. (Orthocladiinae) are to be found in such situations (Fittkau, 1964).

In contrast, Kownacki and Zosidze (1980) investigated several rivers and streams flowing into the Black Sea from the western Caucasus. In only one of these, the River Coloki, was a potamon distinguishable. The other rivers were entirely rhithral in character and faunal differences were greater between rivers than between upstream and downstream sections. In the Adzariskali, the largest of the rivers that they studied, Kownacki and Zosidze (1980) distinguished three types of chironomid community. At altitudes in excess of 940 m above sea level (a.s.l.), *Diamesa* spp. (mainly *Diamesa thienemanni* Kieffer) were dominant, along with *Cricotopus* and *Orthocladius* spp. Downstream of this, but above 640 m, the dominant species was *Orthocladius rivicola* Kieffer. *Cricotopus* spp. and *Orthocladius* spp. were also abundant in this zone and *Eukiefferiella ilkleyensis* (Edwards) and *Cardiocladius* sp. occurred at lower densities. A third type of community was found in the next downstream section, extending to about 275 m a.s.l., and this was dominated by *Cricotopus* and *Orthocladius* spp., with *O. rivicola* present but no longer one of the dominant species.

The principal factors that determine the community that develops in these streams are the nature of the substratum, water chemistry and annual fluctuations in discharge (Kownacki and Zosidze, 1980). Altitude, and therefore temperature, was relatively unimportant, as may be illustrated by reference to another river, the Coroch, where the dominant chironomids between 30 and 44 m a.s.l. were of the same taxa that characterized the fauna of the Adzariskali at 650–940 m a.s.l.

6.3 LENTIC HABITATS

Lentic or standing-water habitats are extremely varied, ranging in size from large lakes to tiny pools. The chironomid fauna of lakes, especially the profundal or deepwater fauna, has been extensively used in systems of lake classification. These are largely based on the distribution of indicator species and communities in relation to water quality and, more particularly, the trophic status of the lakes. This topic is therefore more appropriately dealt with in Chapter 15. In the present chapter lake faunas will be described briefly, using a few contrasting examples, but the broader emphasis will be on providing an overview of the chironomids of stillwater habitats.

6.3.1 CHIRONOMIDAE OF LAKES

The beds of large lakes can broadly be divided into three ecological zones: the littoral, the sublittoral and the profundal. The **littoral** zone is the shallow lake margin, characterized by the presence of aquatic macrophytes. The **profundal** zone is that part of the lake below which there is insufficient light to support photosynthesis; the **sublittoral** is intermediate between the littoral and the profundal. The littoral zone of the exposed shores of lakes, where wave action maintains a relatively coarse, stable substratum, supports a chironomid fauna that is rather similar to that of the rhithral zone of rivers. Typically, this is dominated by Orthocladiinae. On the other hand, the muddy benthos of the profundal, where oxygen may sometimes be lacking, is typically dominated by Chironominae. An extensive overview of the chironomid faunas of lakes in Europe, North America, Japan and the tropics during the first half of this century was provided by Thienemann (1954).

In this chapter the chironomid fauna of a large Icelandic lake, Thingvallavatn (Lindegaard, 1992), is described briefly as an example of a more recent study. The situation in Thingvallavatn is then compared with that in Lake Cromwell, a small dimictic lake in Canada (Harper and Cloutier, 1986); the Vorderer Finstertaler See, a high alpine lake (Bretschko, 1974); and the tropical Lake Chad (Dejoux, 1968). The annually inundated forests of the Amazon and lentic waters in arid regions, such as much of Australia, are special situations and are described separately.

Lindegaard (1992) points out that spatial variations in zoobenthic diversity and density have been encountered in nearly all studies of lacustrine benthos. A number of factors responsible for this variation were identified by Jónasson (1978) and were summarized by Lindegaard (1992):

- Physical factors (wind-induced currents, responsible for food and oxygen circulation).

- Chemical factors, such as oxygen concentration and hydrogen sulphide concentration.
- Physiological adaptations of species to cope with different physical and chemical conditions.
- Food quality and quantity and the ability of individual species to utilize the available food.
- The type of substratum.
- Intra- and interspecific competition.

The influence of substratum type on patterns of chironomid distribution is discussed in section 6.4. Others factors, such as water quality (Chapter 16), food (Chapter 7) and species interactions (Chapter 12) are discussed elsewhere.

(a) Thingvallavatn, a large Icelandic lake

Chironomids and Oligochaeta dominated the fauna of Thingvallavatn, with 24 species and 16 species respectively (Lindegaard, 1992). The distribution of the Chironomidae, in relation to depth, is shown in Table 6.3. Six taxa, including several species of *Diamesa* and five Orthocladiinae, were typical of the extreme margin of the lake, the surf zone. An additional 11 taxa occurred in the littoral zone, of which four extended into the sublittoral. One of these, *Pseudodiamesa nivosa* Goetghebuer, contributed 12% of the annual production of the lake zoobenthos. Only three species, all Chironominae, were confined to the profundal zone. Two Orthocladiinae, *Heterotrissocladius grimshawi* Edwards and *Orthocladius (Pogonocladius) consobrinus* Holmgren, extended into this zone, although the latter only occurred in very small numbers in the profundal and attained its maximum abundance in the littoral. *H. grimshawi* was also most abundant in the littoral but was relatively common to depths of 110 m. *Chironomus islandicus* Kieffer and *Paracladopelma nigritula* Kieffer were also characteristic of the '*Nitella* zone'. *Nitella* (Charophyceaea) was found growing in regions of the sublittoral where a more gentle slope allowed sand and silt to accumulate.

Mean number of chironomids was very low in the profundal zone of Thingvallavatn. The average density of the most abundant species, *C. islandicus*, was only 34 larvae m^{-2}, compared with 275 larvae m^{-2} in the *Nitella* zone. The second most abundant profundal species was *T. gracilentus* (Holmgren) but this situation was reversed in another Icelandic lake, Lake Myvatn, which has a maximum depth of only about 4 m. *T. gracilentus* dominated, being responsible for 67.1% of the total annual benthic production, with *C. islandicus* subdominant, contributing 23.6% of the annual production (Lindegaard and Jónasson, 1979).

The fauna of Thingvallavatn resembles that of arctic freshwater communities where Chironomidae become increasingly dominant the colder the climate becomes (Danks, 1981). However, Thingvallavatn is not an arctic lake and is at a similar latitude to many Scandinavian lakes that have a more varied fauna and two to three times as many chironomid species. The low number of species and dominance of Chironomidae in Thingvallavatn was attributed by Lindegaard (1992) to the geographical isolation of Iceland, rather than to its climate.

Table 6.3 Distribution of Chironomidae in Thingvallavatn, Iceland (data derived from Lindegaard, 1992)

Zone	Depth	Chironomid taxa
Surf zone	Lake margin	*Diamesa* spp. *Eukiefferiella minor* *Orthocladius frigidus* *Orthocladius oblidens* *Rheocricotopus effusus* *Thienemanniella* sp.
Littoral	0–10 metres	*Arctopelopia griseipennis* *Macropelopia nebulosa* *Diamesa* spp. *Pseudodiamesa nivosa* *Corynoneura* sp. *Cricotopus* spp. *Heterotrissocladius grimshawi* *Orthocladius (Pog.) consobrinus* *Psectrocladius barbimanus* *Psectrocladius limbatellus* (as *edwardsi*) *Micropsectra* sp.
Sublittoral	10–20 metres	*Pseudodiamesa nivosa* *Corynoneura* sp. *Heterotrissocladius grimshawi* *Orthocladius (Pog.) consobrinus*
Profundal	> 20 metres	*Heterotrissocladius grimshawi* *Orthocladius (Pog.) consobrinus* *Chironomus islandicus* *Paracladopelma nigritula* *Tanytarsus gracilentus*

(b) Lake Cromwell, a small dimictic lake in Canada

In Lake Cromwell, Quebec, five types of site were identified and called **nuclei** by the authors of this study (Harper and Cloutier, 1986). Nucleus A sites were in very shallow water, equivalent to the surf zone of Lindegaard (1992). Nucleus B sites were littoral stations in somewhat deeper water, where there was floating or submerged vegetation, and nucleus C sites were also in shallow water but in areas where vegetation was scant. Nucleus D sites were in 'the deeper littoral', often below the vegetation fringe. This corresponds with the sublittoral zone, as defined previously. Nucleus E sites were in the deeper parts of the lake.

The highest number of individuals and species of Chironomidae were found in the most shallow sites and the lowest number of both occurred in the deepest sites. This is similar to the situation in Thingvallavatn. The dominant species in each of the five zones are listed in Table 6.4, which shows that, in

Table 6.4 Common chironomid species, in order of dominance, in the five habitat categories (nuclei) identified in Lake Cromwell (Harper and Cloutier, 1986)

Nucleus	Description	Species
A	Extreme marginal shallows	*Paratanytarsus* sp. A *Procladius bellus* *Polypedilum albulum* *Labrundinia pilosella* *Phaenopsectra flavipes* *Microtendipes pedellus* *Pseudochironomous banksii* *Dicrotendipes modestus*
B	Vegetated littoral zone	*Procladius bellus* *Paratanytarsus* sp. A *Procladius denticulatus* *Guttipelopia guttipennis* *Paratanytarsus recens* *Parachironomus potamogeti* *Cladopelma amachaera* *Polypedilum albulum* *Tanytarsus nemorosus*
C	Largely unvegetated littoral zone	*Paratanytarsus* cf. *laetipes* *Procladius bellus* *Paratanytarsus* sp. A *Guttipelopia guttipennis* *Parachironomus potamogeti* *Procladius denticulatus* *Psectrocladius elatus* *Polypedilum albulum*
D	Sublittoral	*Procladius bellus* *Parachironomus potamogeti* *Guttipelopia guttipennis* *Procladius denticulatus* *Paratanytarsus* sp. A *Cladopelma amachaera* *Microtendipes pedellus* *Dicrotendipes nervosus*
E	Profundal	*Parachironomus potamogeti* *Paratanytarsus recens* *Procladius bellus* *Procladius denticulatus* *Psectrocladius elatus* *Paratanytarsus* cf. *varelus* *Cricotopus trifasciatus* *Paratanytarsus* sp. A

this shallow lake, there is no clear distinction between the littoral and profundal faunas.

(c) Vorderer Finstertaler See, a high mountain lake

Under harsh climatic conditions, as in high mountains and in the arctic, the chironomid fauna of lakes is poor in species. This is illustrated by Bretschko's (1974) study of Vorderer Finstertaler See, situated in Austria, 2237 m a.s.l. Ice cover lasts from November to June but in spite of this there is no oxygen limitation. The littoral consists of bare gravel. The lake bottom is flat and composed of a thick layer of gyttja, while the intermediate slope is characterized by gravel and boulders covered with a layer of fine silt.

In this situation Orthocladiinae and Diamesinae are dominant. Orthocladiinae are represented by *Corynoneura scutellata* Winnertz, *Cricotopus alpicola* , *Heterotrissocladius grimshawi* and *H. marcidus* Walker, with two Diamesinae, *Protanypus forcipatus* (Eggert) and *Pseudodiamesa branickii* Nowick. Chironominae are represented only by three species of the tribe Tanytarsini: *Micropsectra radialis* Goetghebuer (as *coracina* Kieffer), *M. contracta* Reiss and *Paratanytarsus austriacus* Kieffer. The only other chironomids are a tanypodine, *Zavrelimyia melanura* (Meigen), and *Prodiamesa olivacea* Meigen (Prodiamesinae). According to Bretschko (1974) the community is typical of similar alpine or arctic lakes but less diverse than the fauna of comparable lakes in the Pyrenees, arctic Sweden and Canada.

(d) Lake Chad, a tropical African lake

Lake Chad provides a marked contrast to the northern lakes already described. Chad is a large, shallow lake, situated between 12° and 14° north of the equator. Dejoux (1968) used various types of light trap to investigate the fauna of the eastern part of the lake. While this type of sampling may be selective, in that some species may be more likely to be trapped than others, the results nevertheless demonstrate the dearth of Orthocladiinae in tropical lakes. Of 93 species that were sampled by Dejoux, 76 belonged to the Chironominae, 16 to the Tanypodinae and only one, *Cricotopus albitibia* Walker, to the Orthocladiinae.

6.3.2 AMAZONIAN LENTIC WATER-BODIES

Fittkau (1971) observed that lakes, in the true sense, do not occur in Amazonia but the periodically flooded valleys of the Amazon and its tributaries form many different and unstable biotopes. Many 'lakes' are created during the six months of high water levels and these fragment into many smaller water-bodies as the flood waters recede. The waters of the varzea, which are fed from rivers that flow through the rich Andean foothills, are rich in nutrients and often covered with floating vegetation. Strong depletion of oxygen, even in shallow waters, is commonplace and the floating vegetation forms the most important substratum for chironomids. Most of the genera of chironomids that utilize this habitat also occur in temperate lentic waters. They include *Chironomus*, *Cryptochironomus*, *Parachironomus*, *Dicrotendipes*, *Xenochironomus*, *Polypedilum* and *Tanytarsus*, as well as the characteristic neotropical genus, *Goeldichironomus*. *Goeldichironomus natans* Reiss is the most prolific colonizer of the floating vegetation, but this species has a permanent littoral population from which the floating meadows can be recolonized as they reform following a period of low water level (Reiss, 1977). Orthocladiinae, apart from *Cricotopus*, do not occur in this region.

The rivers of central Amazonia also have broad valleys which periodically flood, but their waters are very poor in nutrients and this is reflected in the paucity of the chironomid fauna of the forest lakes that form in this region.

6.3.3 AUSTRALIAN LENTIC WATERS

Information on the chironomid fauna of lentic water bodies in Australia is summarized by Edward (1986). He lists 14 eurythermic, and mainly multivoltine, species that are common throughout Australia in such situations. The majority are Chironominae (*Chironomus tepperi* Skuse, *C. cloacalis* Atchley and Martin, *Kiefferulus intertinctus* (Skuse), *K. martini* (Freeman), *Cryptochironomus griseidorsum* (Kieffer), *Cladopelma curtivalva* (Kieffer), *Polypedilum nubifer*, *Paratanytarsus grimmii* (Schneider) [as *P. parthenogenetica*], *Tanytarsus fuscithorax* Skuse and *T. barbitarsus* Freeman), but the list also includes two species of Tanypodinae (*Procladius paludicola* Skuse and *P. villosimanus* Kieffer) as well as the Orthocladiinae *Cricotopus albitibia* (in actuality a complex of species, none of which is *albitibia* – P.S. Cranston, *pers. obs.*) and *Paralimnophyes* [as *Limnophyes*] *pullulus* (Skuse).

Endorheic lakes, and other standing waters created by saline ground-water infiltration, and pools that form in the large rivers during low flow are important habitats for lentic species in Australia. Salinities increase as these pools dry out – until the rains flush out the systems their only chironomid inhabitants are the halobiontic species *T. barbitarsis* and the euryhaline *Procladius paludicola*. All of the other species listed are considered to be principally freshwater forms, but in western Australia, though less often in the east, some have been recorded from saline waters (Edward, 1983).

Although many lentic habitats in Australia have high conductivities, through groundwater salinity and high evaporation rates, there are many low-conductivity lentic waters. These include the billabongs (seasonal and permanent freshwater lagoons) of the post-monsoonal streams and rivers of the north, the perched lakes on coastal sand dunes, and the alpine and subalpine lakes of the montane south-east and Tasmania. The chironomid faunas of some of these habitats have been investigated (with varying degrees of taxonomic thoroughness). In monsoonal Northern Territory, lentic chironomid diversity is dominated by species of *Conochironomus*, *Procladius*, *Polypedilum*, *Dicrotendipes*, notably *D. lindae* Epler and *Cryptochironomus griseidorsum* (Marchant, 1982; Outridge, 1988; Cranston, 1991). Australian east-coast dilute perched lakes may support *Aphroteniella* Cranston and Edward (1992), many apparently endemic *Harnischia*-complex and Tanytarsini species, with *Kiefferulus martini* present in low pH conditions. High altitude and latitude dilute lakes such as those of Tasmania may be dominated by *Riethia* and *Procladius* (Timms, 1978), with species of *Tanytarsus*, *Parakiefferiella* and several undescribed genera of Orthocladiinae present (P.S. Cranston, *pers. obs.*).

6.4 SUBSTRATUM AND CHIRONOMID DISTRIBUTION IN LAKES AND RIVERS

Many studies have identified the nature of the substratum as an important factor influencing patterns of chironomid distribution and community structure (e.g. Lindegaard-Petersen, 1971; Mackey, 1976a,b, 1977a; McLachlan, 1969) but, although specific preferences are usually apparent, many species are capable of utilizing a variety of substrata (Pinder, 1980). Substrata utilized by freshwater Chironomidae were broadly categorized by Pinder (1986) as hard rock, soft sediment (including sand and mud) submerged wood and aquatic plants. Hard rock is intended in this sense to include boulders, stones and pebbles as well as bed rock.

6.4.1 THE HYPORHEIC ZONE

In coarse, loosely packed substrata of stones or coarse gravel invertebrates may penetrate a considerable distance vertically and laterally, giving rise to a hyporheic fauna. In a riffle area of Salem Creek, Ontario, Godbout and Hynes (1982) concluded that the densities of invertebrates in the hyporheos were at least three times as great as in the surface zone. Orthocladiinae dominated the hyporheos in early summer and Tanytarsini in early autumn. Larvae of certain otherwise abundant species are notoriously difficult to find and it is possible that they are confined to the hyporheos. This was argued to be likely in the case of *Krenosmittia* spp. by Ferrington (1984). It is also tempting to postulate this for *Buchonomyia thienemanni* Fittkau, the larval habitat of which remains unknown in spite of efforts to locate larvae in rivers where pupal exuviae can be found (Murray and Ashe, 1981).

6.4.2 MACROPHYTES

The distribution of chironomid larvae in relation to substratum type was studied by Pinder (1980) in the River Frome, a lowland English river, and its tributary, the Tadnoll Brook. Of 36 species collected at a single site, 30 were found in gravel, 17 in soft sediments (a mixture of sand and silt) and 23 were associated with the dominant aquatic plant, *Ranunculus penicillatus* var. *calcareus* (R.W. Butcher) C.D.K. Cook. Of the species on *R. calcareus*, roughly 75% were Orthocladiinae, while this subfamily made up about 65% of the taxa from gravel but only about 32% in soft sediment. Most of the species living on *R. calcareus* also occurred on gravel. Notable exceptions were *Cricotopus trifasciatus* (Panzer) and *Cricotopus* sp. A, both of which eat living plant material in their third and fourth instars (Williams, 1981).

Eukiefferiella ilkleyensis also eats *R. calcareus* in the later instars and although this species occurred in small numbers on the gravel it was very abundant in samples taken from the macrophyte. For the majority of species, however, the submerged plants and gravel apparently fulfil similar ecological functions – providing a firm substratum in a well-oxygenated situation and allowing diatoms, a major dietary item for most of the Orthocladiinae, to grow in abundance.

Orthocladiinae also dominated the fauna of Linding Å, a Danish lowland river. Lindegaard-Petersen (1972) attributed this to the relatively low summer temperatures but Pinder (1980) considered that the similar situation in the River Frome was attributable to the abundance of macrophytes and Tokeshi and Pinder (1985) also felt that the presence of macrophytes is an important factor, facilitating the colonization of middle and lower reaches of rivers by Orthocladiinae.

The presence of vegetation in aquatic ecosystems often results in a very substantial increase in the area available for colonization. In an Alberta stream, Boerger *et al.* (1982) showed that the surface area of macrophytes was between 5 and 10 times that of the river bed, but estimated that plants harboured only 30–40% of the total chironomids. Most studies, however, have demonstrated a positive relationship between the presence of macrophytes and the abundance and diversity of Chironomidae (e.g. Barber and Kevern, 1973; Driver, 1977; Moore, 1980). Various patterns of microdistribution have been described for species of chironomid and other invertebrates on macrophytes (Drake, 1982; Tokeshi and Pinder, 1985). Tokeshi and Pinder (1985) showed that patterns of microdistribution were strongly influenced by plant morphology and Cyr and Downing (1988) found that more species of invertebrate were found on *Myriophyllum* spp. than on plants with broad leaves. Abundance, contrary to Krecker's (1939) model, was not, however, related to the degree of dissection of the plants.

A number of species of Orthocladiinae and Chironominae mine in plant tissues. Larvae of five species were found mining in different parts of the yellow water lily, *Nuphar lutea* (Linnaeus) Smith, by Van der Velde and Hiddink (1987) who divided them into three categories: **obligate phytophages**, which feed mainly on living plant tissue; **seston eaters**, which filter feed in mines and can also consume plant tissue; and **facultative phytophages**, which use the plant tissue only for making mines. Some *Cricotopus* spp., such as *C. trifasciatus* (Panzer), which have been recorded as mining in plants (e.g. Van der Velde and Hiddink, 1987) are also capable of living on the outer surface of finely divided plants, such as *Ranunculus calcareus,* where the later instars feed mainly on plant tissues (Pinder, 1992) (see also Chapter 7).

6.4.3 SUBMERGED WOOD

Prior to human intervention, submerged and partially immersed wood ('snags') formed a dominant aquatic habitat. Although management of riparian vegetation and the unsnagging and canalization of waterways has reduced its significance, submerged wood still constitutes an important substratum in streams in many parts of the world. The chironomid fauna of this substratum has been investigated extensively by Anderson and colleagues (e.g. Anderson *et al.*, 1984) and the biology of xylophagous (wood-eating) chironomids (section 7.3.4) was reviewed by Cranston and Oliver (1988b). Fifty-six taxa of invertebrates were listed by Dudley and Anderson (1982) as being 'closely associated' with submerged wood while a further 129 were recorded as facultative colonizers of the wood-surface. Many of these graze the aufwuchs that collects on the wood but, among those that burrow and tunnel into the wood, Diptera, especially Chironomidae and Tipulidae, are dominant in both abundance and diversity (Anderson *et al.*, 1984). Several chironomid taxa have only been found mining in wood and with guts containing wood fibres. Within the subfamily Chironominae these are *Stenochironomus, Harrisius* and *Xestochironomus*; among the Orthocladiinae they are *Chaetocladius ligni* Cranston and Oliver, *Orthocladius* (*Symposiocladius*) *lignicola* Kieffer and *Xylotopus par* (Coquillet) (Cranston and Oliver, 1988b).

6.4.4 SOFT SEDIMENTS

Low oxygen concentration is rarely a problem for invertebrates living on clean stones or submerged macrophytes but may be for species that live in soft sediments with a high content of organic matter. Oxygen consumption within the mud of deep lakes can create a sharp gradient of reducing oxygen concentration in the few millimetres or centimetres above the sediment surface (Alsterberg, 1922). Larvae may react to low oxygen concentration in various ways (Konstantinov, 1971a). Tube-building Chironomidae may raise the openings of their tubes above the level of the sediment in which they are living. In the case of *Chironomus plumosus* Linnaeus, Konstantinov (1971a) observed that the lower the oxygen concentration, the further the tubes were extended. In experiments in aquaria the height of the tube entrance ranged from 1 mm to 20 mm, maintaining a concentration of 0.7–0.9 mg $O_2\,l^{-1}$ at the mouth of the tube.

In contrast to Konstantinov's observations on *C. plumosus*, Jónasson and Kristiansen (1967) observed that *Chironomus anthracinus* Zetterstedt larvae in Lake Esrom constructed tubes with 'chimneys', extending 2–3 cm above the mud but only in oxygen-rich water. In unaerated water, tubes were built entirely within the substratum and mutual connections beneath the mud surface were extensive. These anastomoses may well allow for a degree of mutualism, whereby a proportion of larvae are free to feed while others irrigate the network of interlinked tubes.

Sediment depth has a major influence on population density and tube-shape in *C. plumosus*. In a laboratory study, McLachlan and Cantrell (1976) showed that, in sediment less than 10 mm thick, larvae constructed only horizontal tubes, whereas in deeper sediments U-shaped and J-shaped vertical tubes were made. High population densities only occurred in sediments that were sufficiently deep to permit construction of the vertical tubes.

Undulations of the body drive fresh water through the tubes, replenishing oxygen and flushing out metabolites and carbon dioxide. According to Konstantinov (1971), frequency of undulation increases with falling oxygen concentration until a point is reached below which all activity ceases. For Chironominae this point was determined by Konstantinov (1971a) to be between 0.4 and 0.6 mg l^{-1}. However, more recently, Smit *et al.* (1993) have demonstrated clear interspecific differences in behaviour among Chironomini, in response to low oxygen concentrations. They studied three species: *Lipiniella araenicola* Shilova, *Chironomus muratensis* Ryser, Scholl and Wülker and *C. nudiventris* Ryser, Scholl and Wülker. The first two were exposed to experimental conditions, ranging from air saturation to virtual anoxia, for periods of 60 to 90 minutes. In the case of *L. araenicola* the proportion of time spent in ventilating the tube steadily increased with falling oxygen concentration. Under anoxic conditions, 85% of the time was spent in this type of activity and the time spent resting approached zero. *C. muratensis* behaved quite differently. Falling oxygen concentration had no effect on activity, except under anoxic conditions when larvae became totally inactive. *C. nudiventris* was exposed only to water saturated with oxygen at 3 mg l^{-1}. No difference in behaviour was noted over a 90-minute period, but exposure to low concentrations of oxygen for 16 hours resulted in a marked increase in ventilatory activity.

Differences between the responses of *Chironomus anthracinus* and *Stictochironomus sticticus* Fabricius [as *S. histrio* Fabricius] were shown by Heinis and Crommentuijn (1989) to be compatible with their distribution in the field. In air-saturated water both species spent about 10% of their time ventilating. For the remaining time, *S. sticticus* continued to be mainly active but *C. anthracinus* spent almost half the time at rest. Reducing oxygen concentration to 3 mg l^{-1} had no effect on the behaviour of *C. anthracinus* but *S. sticticus* spent progressively more time ventilating as the concentration of oxygen was reduced. At a concentration of 3 mg l^{-1}, around 70% of the time was spent ventilating, leaving proportionately less for other activities such as feeding. Further reduction in oxygen concentration, to 1 mg l^{-1}, resulted in the death of *S. sticticus* larvae within 48 hours. *C. anthracinus* became totally inactive but survived for some time.

In the field, *S. sticticus* was usually found in shallow water and sandy sediments, where oxygen levels normally were high. Within the normal range of oxygen concentrations encountered in this type of situation, variation in the frequency of ventilation of the tubes is likely to be sufficient to maintain an adequate oxygen supply. On the other hand, *C. anthracinus* larvae inhabit hypolimnetic sediments and are likely to encounter periods of hypoxia or even anoxia. In this case, a strategy which enables them to function more or less normally, or at least to survive, under such conditions is advantageous (Heinis and Crommentuijn, 1989).

The research by Smit *et al.* (1993), together with that of Heinis and various co-workers (Heinis and Crommentuijn, 1989; Heinis and Davids, 1993; Heinis *et al.*, 1989), has added considerably to understanding of the relationship between patterns of chironomid distribution and oxygen concentration in lakes. The distribution of Chironomidae in three lakes in the Netherlands was compared by Heinis and Davids (1993). Maarsseveen I is an oligo-mesotrophic lake which stratifies in summer but in which hypoxic or anoxic conditions only occur towards the end of the period of stratification. A clear zonation of the chironomid fauna is evident here, with the littoral dominated by *Cladotanytarsus* sp. and *Stictochironomus sticticus*, the sublittoral by *Tanytarsus bathophilus* Kieffer and the profundal by *Chironomus anthracinus*. Temporal changes in the distribution of *T. bathophilus* (Heinis *et al.*, 1989) are of particular interest. The species is found at all depths from July to September, but disappears from the region below the oxycline during the summer stratification. A concurrent increase in density in the lower epilimnetic zone, where oxygen concentrations are generally high, suggests that the disappearance from the hypolimnion is the result of migration, not mortality.

The nearby lake Maarsseveen II is eutrophic and oxygen depletion in the hypolimnion begins as soon as the lake stratifies, in June, and continues until the autumn turnover. The principal chironomid species here are *Cladotanytarsus* sp. and *Stictochironomus sticticus*, both of which are confined to a narrow littoral band. There are no chironomids at all in the deeper regions of Maarsseveen II. Lake Gijster is also eutrophic but is artificially destratified. The profundal of this lake is dominated by *Tanytarsus bathophilus*, with *Chironomus* spp. virtually absent (Heinis and Davids, 1993). This absence of *C. anthracinus*, in spite of its behavioural and physiological adaptations to low oxygen concentrations, is explained by the long period of anoxic conditions in Maarsseveen II, extending each year from June to November. Under experimental conditions, exposure to anoxia for 75 days resulted in 20% larval mortality among *C. anthracinus*, but after exposure for 150 days all were dead (Nagell and Landahl, 1978).

The importance of oxygen and behavioural adaptations that permit some species to tolerate, or avoid, oxygen depletion, as factors governing the distribution of Chironomidae in soft sediments will be clear from the foregoing. However, other factors also may be important. Particle size, for example, is a major factor influencing the distribution of some species. Larvae of *Kiefferulus* [as *Nilodorum*] *brevibucca* Freeman avoid coarse sand and show a clear preference for silt rather than fine sand or mixtures of different particle sizes (McLachlan, 1969). *Glyptotendipes paripes* Edwards was confined to a narrow zone along the margin of a humic lake, largely as a result of a preference for particles between 1100 and 1700 μm in diameter (McLachlan, 1976b).

Most studies suggest that chironomids do not usually penetrate more than a few centimetres into fine sediments (e.g. Ford, 1962) but Berg (1938) detected larvae of *Chironomus anthracinus* to a depth of 40 cm in the mud of Lake Esrom. The activities of chironomid larvae in sediment have important implications for sediment and water chemistry. The release of phosphorus from aerobic sediment, taken from Lake Mendota, Wisconsin, increased as a result of the activity of both *Chironomus riparius* Meigen and *Chironomus tentans* (Fabricius) (Gallepp, 1979). The rate of release was related to larval density and biomass as well as temperature and Gallepp (1979) concluded that chironomid larvae can have a significant effect on the seasonal availability of phosphorus that is released from aerobic lake-sediments.

Oxygen conditions within sediment are also influenced to some extent by water movements caused by the undulatory activity of larvae within their tubes. *Micropsectra* larvae kept at 4 °C showed an activity pattern consisting of 10–20 minutes spent irrigating their tubes, interspersed with 20–40 minutes resting. Oxygen concentration in front of the tubes showed large fluctuations (Frenzel, 1990), indicating periodic outflow of water with a low oxygen concentration. Within the sediment, fluctuations could also be detected to a distance of 2.5 mm from the tube. On a time-scale of some tens of minutes, conditions close to the tube could change from oxic to anoxic and back again. Where high densities of larvae occur, oxygen conditions in the surface layers of sediment may be dominated by the activities of chironomids. The redox potential of activated sewage sludge was also shown by Edwards (1958) to be influenced strongly by the larvae of *Chironomus riparius*, with consequent effects on other aspects of sediment chemistry, including pH, and levels of ammonia and oxidized nitrogen.

6.5 THERMAL SPRINGS

Thermal springs have a temperature that is higher than the mean ambient air temperature. In extreme cases the issuing water may be close to boiling. The waters of thermal springs are also very rich in dissolved minerals and, even when saturated with oxygen, hold less oxygen than cooler water bodies. Such factors undoubtedly have an influence on the fauna that develops but Lamberti and Resh (1983) concluded that temperature effects were of predominant importance. There are no insects that, in an active state, can survive temperatures greater than 50 °C and very few that can tolerate temperatures above 40 °C (Pennak, 1978). Hot springs are distributed throughout the world and at temperatures above 40 °C a similar and characteristic 'hot spring' community occurs everywhere, consisting mainly of air-breathing insects. Only three species of chironomid have been recorded at such high temperatures: *Cricotopus sylvestris* (Fabricius) from Iceland (Tuxen, 1944); a species of *Chironomus*, as *Camptochironomus*, stated to be 'near *tentans*', from Yellowstone Park (Brues, 1924); and *Chironomus cylindricus* from New Zealand (Winterbourn, 1969). These records are particularly remarkable because, unlike any of the other insects from such hot springs (with the exception of one Odonata from Yellowstone Park), they utilize dissolved rather than atmospheric oxygen.

6.6 EPHEMERAL WATER-BODIES

6.6.1 TEMPORARY STREAMS

Ephemeral or intermittent streams occur throughout the world and in some regions are the dominant type of lotic habitat (Ward, 1992). In spite of this,

Boulton and Suter (1986) concluded that very little is known about the fauna of temporary streams in Australia and that the situation in other parts of the world is little better.

Hynes (1975) considered that oviposition by flying insects, originating from permanent waters, is the most important means of recolonizing tropical, ephemeral streams in southern Ghana and, like Harrison (1966), who studied an intermittent stream in Zimbabwe (Rhodesia), he found no evidence of drought-resistant eggs or of diapausing larvae.

There are many temporary streams, known as winterbournes, arising from chalk aquifers in southern England. These spring-fed headwater streams usually become dry during the summer, although the extent of drying varies from year to year. Some sections become dry annually but others dry out only occasionally, following a prolonged period of drought. The fauna of one of the latter type was investigated, before and after a drought that caused many headwaters to dry out to a greater extent than usual. Twenty-three chironomid taxa were recorded here, by Ladle and Bass (1981), all of which were among the usual components of the fauna of permanent headwater streams in the region. None were observed to have specific behavioural or other adaptations to inhabiting temporary streams. Most were much less abundant in the period following restoration of flow than during the previous wet period and recolonization seems to have been mainly, or entirely, through oviposition. However, cocoons were found in the dry bed of another English winterbourne. These contained larvae of *Endochironomus* sp., folded double and inactive, (M.T. Furse and L.C.V. Pinder, *pers. obs.*). Grodhaus (1976) also found dormant, drought-resistant larvae of two species of *Phaenopsectra* (Chironomini) in Kansas, within specially constructed cocoons, and Edward (1968) noted the presence of larval *Paraborniella tonnoiri* Freeman and *Allotrissocladius amphibius* Freeman in cocoons during summer drought in Australia.

Cocoons (section 10.3.3) are also sometimes associated with overwintering larvae and as such have been recorded for a variety of species, including *Endochironomus tendens* (Fabricius) (Sæther, 1962), *Eukiefferiella claripennis* (Lundbeck) (Madder *et al.*, 1977) and by Danks and Jones (1978) for *Cryptocladopelma edwardsi* (Krusemann), *Polypedilum* sp., *Endochironomus nigricans* (Johannsen) and *Dicrotendipes modestus* (Say). The primary function of cocoons, under freezing conditions, is believed to be in conferring a degree of protection from mechanical damage (Danks, 1971a), but in dry stream beds they also probably play a part in maintaining high humidity around the dormant larva.

6.6.2 CHIRONOMIDS OF TEMPORARY RAIN-POOLS

In many parts of the world rainwater gathers in depressions in rock surfaces and, depending on their situation and size and rainfall patterns, these temporary habitats may persist for a few days or for months before drying out. Two distinct strategies enable insects, including chironomids, to exploit these unpredictable habitats. They may either be drought resisting and able to survive at least some degree of desiccation or they may be opportunists with persistent populations in permanently wet or moist habitats.

A most spectacular example of the first strategy is provided by *Polypedilum vanderplanki* Hinton, discovered by Hinton (1951) living in pools on granite in Africa. The larvae are capable of becoming almost totally desiccated but quickly rehydrate and resume activity and development when the pool in which they are living is filled by the next rainfall. The larvae are able to survive at least ten cycles of dehydration and rehydration (Hinton, 1951). In the dehydrated condition they are able to withstand remarkable extremes of temperature, from −270 °C to 102 °C (Hinton, 1960a, b). On completion of his investigation of *P. vanderplanki* Hinton sealed some larvae and dry sediment in glass tubes containing silica gel which was still blue (i.e. dry) when the tubes were reopened 17 years later (Adams, 1983). Adams first transferred some of the dry soil to a saturated atmosphere and after four hours immersed it in water at room temperature. The account of what followed is best told in his own words.

> I stared down the binocular microscope, scanning the mud slowly at low and high power and was beginning to wonder whether so many years of desiccation would prove to be too long for the larvae to survive, when a slight movement within the mud persisted, a deep red finger-like projection appeared and an insect, almost as old as myself wriggled into the daylight for the first time in seventeen years.

The chironomids and other animals living in temporary pools in Africa have been extensively investigated by McLachlan and various colleagues (e.g., Cantrell and McLachlan, 1982; McLachlan, 1981b, 1985b). Pools that last for one week or less are inhabited by *P. vanderplanki* or by the ceratopogonid midge, *Dasyhelea thompsoni* de Meillon, both of which have physiological adaptations to withstand desiccation. *D. thompsoni* favours pools that are contaminated with the faeces of carnivores, such as civets, but such places are avoided by *P.vanderplanki*.

Chironomus imicola Kieffer has no marked resistance to dehydration and only inhabits pools that persist for several weeks. It is rarely found in pools with *P. vanderplanki* (Cantrell and McLachlan, 1982) and introductions of *P. vanderplanki* to pools inhabited by *C.imicola* generally fail (McLachlan, 1985b) unless the pool is sufficiently shallow for metabolites produced by *C. imicola* to be flushed out by the rains. *P. vanderplanki* only persists, therefore, in short-lived pools which *C. imicola* must recolonize after each dry period. When direct competition does occur, body size appears to dictate the outcome. In short-lived pools *P. vanderplanki* has the advantage in this respect but in more persistent pools the bigger growing *C. imicola* generally prevails.

Polypedilum vanderplanki remains the only chironomid in which cryptobiosis has been observed, but a number of other species are able to withstand a degree of desiccation and survive for some time in pools with little moisture. Adams (1983) mentions another species, *P. dewulfi*, which lives in rock pools in Nigeria and is able to withstand a substantial degree of dehydration, and *Phaenopsectra pilosella* from temporary pools in California. Like a number of other species, the larva of *P. pilosella* aestivates in a cocoon during the summer dry period and, according to Adams (1983) larvae have recovered after 4 years in dry soil.

In Australia drought-resistant adaptations have been recorded in two genera, *Paraborniella* and *Allotrissocladius* (Edward, 1986). *Paraborniella tonnoiri* has a wide distribution but in Western Australia it is confined to the more inland granite outcrops, where summer thunderstorms are more likely to occur (Edward, 1986). When the pools are filled in winter the larvae are quiescent but they do not diapause during the dry period and they feed actively following a thunderstorm (Jones, 1974).

Allotrissocladius is known only from Western Australia and also inhabits pools on granite outcrops (Edward, 1986; Jones, 1974). In contrast to *P. tonnoiri*, larvae of *Allotrissocladius* are active when the pools are filled in winter and diapause in cocoons during the summer dry period.

6.7 PHYTOTELMATA

Phytotelmata are small aquatic habitats associated with living plants. Many examples were provided by Thienemann (1954) and the more recent literature is reviewed by Cranston and Judd (1987). Phytotelmata range in size from tiny droplets of water associated with leaf axils to more substantial quantities that may accumulate in tree-holes. Two species of *Metriocnemus*, *M. scirpi* Kieffer and *M. inopinatus* Strenzke, appear to be restricted to the very small phytotelmata in the leaf axils of *Scirpus sylvaticus* Linnaeus and were found in Germany by Strenzke (1950a); and *Metriocnemus* larvae have also been found in leaf axils of teasels (*Dipsacus* spp.) in the Soviet Union (Borob'ev, 1960) and in the USA (Baumgartner, 1986). Other records of phytotelmata containing chironomid larvae include the axils of wild bananas (*Metriocnemus wittei* Freeman) and the axils of *Lobelia sattimae* R.E. Fries and T.C.E.Fries (*Metriocnemus lobeliae* Freeman) (Cranston and Judd, 1987). The latter also report two *Metriocnemus* spp. that inhabit the mucilaginous liquid in the leaf-axils of *Senecio brassicae* R.E. Fries from high altitudes in Africa. These are able to tolerate high daytime temperatures alternating with freezing to –12 °C at night.

World-wide, the types of phytotelmata that are most commonly colonized by Chironomidae are hollows in trees and water-retaining plant structures such as those of pitcher-plants, bromeliads and pandanus. Tree-hole phytotelmata form where water collects in situations such as the forks of main branches or at the bases of bracket fungi (Williams and Feltmate, 1992), and most often in the hollows formed where lost branches have rotted. In Europe, *Metriocnemus martinii* Thienemann larvae seem to be restricted to water-filled tree-holes and are particularly associated with beech, *Fagus sylvaticus* Linnaeus. The biology of *M. martinii* was studied by Kitching (1972). Densities of larvae were higher in holes in the canopy layer than in the field layer of the woodland, perhaps because of differences in the quality of the detritus, as food, contained in them (Kitching, 1972).

Tree-hole faunas have also been investigated in North America, Indonesia and Australia but tree-hole *Metriocnemus* seem to be restricted to the palaearctic (Cranston and Judd, 1987). Larvae of both *Polypedilum* and *Limnophyes*, but not *Metriocnemus*, occur in tree-holes in California (Grodhaus and Rotramel, 1980). The principal phytotelmata that are occupied by chironomids in temperate North America, according to Cranston and Judd (1987), are pitcher-plants belonging to the genus *Sarracenia*. The best known of these associations is that between *Sarracenia purpurea* Linnaeus, which grows on bogs in the eastern part of the USA, and *Metriconemus knabi* Coquillett. Insects are attracted and trapped by the plant; the larval *M. knabi* feed on the insect detritus and enhance the uptake of nutrients by the plant (Bradshaw and Creelman, 1984). In California and southern Oregon another member of the Sarraceniaceae, *Darlingtonia californica* Tilley, occurs in place of *Sarracenia* and *Metriocnemus edwardsi* Jones takes the place of *M. knabi* (Cranston and Judd, 1987).

There are no chironomids in the liquid of the Australian endemic family of pitcher-plants, the Cephalotaceae. A third family of pitcher-plants, Nepenthaceae, which is unrelated to the Sarraceniaceae, occurs in south-east Asia, Sri Lanka and Madagascar. Although an extensive invertebrate community has long been known to be associated with the pitchers of this family, no chironomids were reported prior to 1976 when Cranston (reported in Cranston and Judd, 1987) discovered larvae of *Metriocnemus* and *Polypedilum* in a *Nepenthes* pitcher on Mount Kinabalu, Sabah. Subsequent studies of *Nepenthes* by Kitching and colleagues have revealed a frequent presence of *Polypedilum*. This has been identified as *P. convexum* Johannsen, an acidophilic species not restricted to pitchers, occurring widely in the Old World from southern Africa (as *P. anale* Freeman) to northern Australia (P.S. Cranston, *pers. obs.*). Kitching's studies encompass many other south-east Asian and Australasian phytotelms and have elucidated a diverse chironomid fauna. These, and other unpublished observations may be summarized:

- *Polypedilum*, predominantly *P. convexum* but including at least one undescribed species, are the most abundant and widespread inhabitants, occupying bamboo internodes, axillary water of many plants, and rot holes in trees.
- *Chironomus* and *Kiefferulus* species have been found in Papua New Guinean tree holes.
- The orthoclad *Compterosmittia* is widespread but apparently is never abundant. Other orthoclads include *Bryophaenocladius* but *Metriocnemus* is absent.
- Phytotelmata with more complex foodwebs frequently contain Tanypodinae: these include south-east Asian and Australasian *Paramerina divisa* Johannsen and Australian *'Anatopynia' pennipes* Freeman that appear to be restricted to phytotelmata.
- Podonominae have been found in axillary water of the Tasmanian giant heath, *Richea pandaniformis* Hook.f. (Cranston and Kitching, in press).

In tropical America many bromeliads (Bromeliaceae) retain water in their leaf axils. The larger, longer lasting of these phytotelmata may develop a

Table 6.5 Chironomidae recorded from brackish waters in the Netherlands (Parma and Krebs, 1977)

Chironomus salinarius	*C. (Camptochironomus) pallidivittatus*
C. halophilus	*Glyptotendipes barbipes*
C. plumosus	*Procladius choreus*
C. riparius (thummi)	*P. breviatus*
C. riparius (piger)	*Psectrocladius varius*
Microtendipes deribae	*Cricotopus sylvestris*

substantial fauna (Cranston and Judd, 1987). *Metriocnemus* and *Polypedilum* larvae have been recorded from bromeliads in Costa Rica and the same two genera, together with tanypodine larvae, occurred in many Bromeliaceae in the Virgin Islands (Cranston and Judd, 1987).

6.8. SALINE HABITATS

6.8.1 BRACKISH WATERS

A number of chironomid species are tolerant of a wide range of salinities (e.g. Rawson and Moore, 1944) and chironomids may be a major component of the fauna of brackish waters. A study of brackish ditches in the Netherlands by Parma and Krebs (1977) produced a list of 13 species (Table 6.5) most of which are also found in non-saline habitats. The most characteristic species in this situation were *Chironomus salinarius* Kieffer, *C. halophilus* Kieffer and *Microchironomus deribae* (Kieffer). These three may be regarded as typical of a wide range of saline conditions. For example, *M. deribae* was recorded by Laville and Tourenq (1975) at a salinity of 42% and it is a frequent inhabitant of waters subject to infiltration by sea water in Europe. *C. salinarius* has also been recorded from waters of similarly high salinity and is tolerant of extreme changes of salinity in pools subject to evaporation (Neumann, 1961).

In parts of Australia, where salinities increase as pools dry out during summer, a number of species normally regarded as freshwater forms have developed a marked tolerance to changes in salinity (Edward, 1986). In Western Australia some of these have the widest range of tolerance to salinity of any Australian chironomids, whereas populations in the less arid eastern part of Australia do not exhibit such tolerance. This leads to the conclusion that the western populations have been subjected to this type of habitat condition for a very long time (Edward, 1986; Halse, 1981). *Kiefferulus longilobus* (Kieffer) is a species characteristic of highly saline conditions such as salt-production lagoons and prawn and sea-fish farms; it is widespread throughout the Indopacific region. Under certain conditions *K. longilobus* populations may be sufficiently large for the adults to constitute a considerable nuisance (Cranston *et al.*, 1990) – a problem that is growing in subtropical Australia in association with marine canal-side housing developments (P.S. Cranston, *pers. obs.*).

6.8.2 INTERTIDAL AND MARINE CHIRONOMIDS

Animals inhabiting the intertidal region are subjected to substantial fluctuations in their physical, chemical and biotic environment. Within the intertidal region Neumann (1976) identified four zones which merge into each other rather than being strictly delimited.

The supra-littoral is the 'splash zone', which is only inundated under the influence of exceptionally high tides or during storms. The supra-littoral fringe of the mid-littoral zone is submerged by high spring tides, once every two weeks, while the lower part is submersed for most of the time, being exposed only at low spring tides every two weeks. The remainder of the mid-littoral zone is covered and uncovered twice each day by the ebb and flow of the tide. Intertidal organisms generally are adapted to live only in a relatively narrow range of the intertidal region. The nature of the substratum is a further important factor influencing the distribution of species and the structure of the intertidal community (Neumann, 1976). Most intertidal chironomids inhabit rocky shores but *C. salinarius* often attains high densities on muddy substrata on protected shores, where oxygen concentrations may be low.

Species of Chironomidae adapted to life in the intertidal zone have been recorded from coasts all over the world. Taxonomically they are diverse, including representatives of three subfamilies (Table 6.6). Telmatogetoninae is primarily an intertidal subfamily although, according to Cranston (1989b), some species are associated with coastal freshwater seepages, especially of sewage polluted waters, or with harbour or river mouths where there is reduced salinity. Among the Orthocladiinae, *Clunio*, *Thalassosmittia*, *Semiocladius* and the monospecific genus *Tethymyia* appear to be exclusively intertidal. *Clunio* spp. occur most commonly on rocky shores and are frequently associated with *Ostrea* and *Mytilus* beds (Cranston *et al.*, 1989). There is some doubt as to whether any *Eretmoptera* are truly intertidal. *E. murphyi* Schaeffer has terrestrial larvae (Cranston, 1985) and the larvae of *E. browni* Kellogg are not known, although adults were found associated with the foreshore in California (Cranston *et al.*, 1989). The association of *Cricotopus* and *Halocladius* with the intertidal zone is relatively weak. *Halocladius* and a few *Cricotopus* spp. have a wide range of salinity tolerance, but are more characteristic of coastal brackish waters and may also occur in inland brackish habitats. A similar comment applies to the intertidal *Chironomus* and *Tanytarsus* but, as will be seen, the other intertidal genus of this subfamily, *Pontomyia*, includes at least one species which is among the very few insects to have adapted to a truly marine existence.

Anal tubules, which are sites of active salt absorption (Strenzke and Neumann, 1960) are usually reduced or lacking in intertidal species, which must be capable of hypo-osmotic regulation of their blood concentration. This ability was demonstrated for *Halocladius* by Sutcliffe (1960) and for *Clunio marinus* Haliday by Neumann (1961). *Clunio* has been shown to be capable of completing its life cycle at salinities between 4% and 40% (Palmén and Lindeberg, 1959) and a similar ability was demonstrated for *C. salinarius*

Table 6.6 Chironomidae recorded from intertidal habitats

Subfamily	Genus/species
Telmatogetoninae	*Telmatogeton*
	Thalassomya
Orthocladiinae	*Clunio*
	Thalassosmittia
	Eretmoptera
	Semiocladius
	Tethymyia
	Cricotopus (some)
	Halocladius
Chironominae	*Chironomus* (some)
	Tanytarsus (some)
	Pontomyia

at salinities ranging from 1% to 50% (Neumann, 1961). However, species with an ability to tolerate lesser salinities, of up to about 8%, may thrive in high levels of the supra-littoral (Neumann, 1976).

The genus *Pontomyia* is widespread in the Pacific, where larvae of several species have been found associated with various algae and in sediment below beds of seaweed (Tokunaga, 1932; Hashimoto, 1959). In discussing the marine chironomids of Australia, Edward (1986) mentions that *Pontomyia cottoni* inhabits sand between tide marks, apparently in areas where there is a freshwater influence. Hashimoto (1976) recorded the occurrence of *P.natans* Edwards below the extreme low water mark, but always close to the shore. He concluded that in spite of being well adapted to a marine existence the genus had not succeeded in colonizing the open sea. However, Bretschko (1982) discovered a population, almost certainly parthenogenetic, of *P. natans* or a very similar species, in the Caribbean Sea. Larvae were found in the lagoon and on the outer reef at Carrie Bow Cay, Belize. Larvae occurred in association with *Thalassia* and in muddy sand within the lagoon as well as in sandy substrata and coral rubble on the reef and on the outer slope of the fore-reef to a depth of at least 30 m. *Pontomyia* is thus the only genus of chironomid, and one of the very few insects, that can be regarded as truly marine.

6.9 TERRESTRIAL CHIRONOMIDS

The only subfamily of Chironomidae with terrestrial representatives is Orthocladiinae. Those genera that are wholly terrestrial, or which include terrestrial representatives (based on Cranston *et al.*, 1989), are listed in Table 6.7. Early studies on terrestrial chironomids, largely by Strenzke (e.g. Strenzke, 1950a, 1950b) were reviewed by Thienemann (1954).

When larvae occur in permanently wet soil or damp vegetation the distinction between terrestrial and aquatic or semi-aquatic habitats is often

Table 6.7 Genera of Orthocladiinae that are wholly or partly terrestrial or semi-terrestrial

Genus	Ecology
Antillocladius	Semi-aquatic/semi-terrestrial
Bryophaenocladius	Mainly terrestrial and semi-terrestrial
Camptocladius	Terrestrial, in cow dung
Chaetocladius	A few are semi-aquatic
Chasmatonotus	Probably terrestrial
Doithrix	Semi-aquatic
Eretmoptera	At least one terrestrial species
Gymnometriocnemus	Terrestrial
Limnophyes	Partly terrestrial and semi-terrestrial
Lipurometriocnemus	Probably semi-aquatic or semi-terrestrial
Paraphaenocladius	Mainly terrestrial
Parasmittia	Terrestrial, in humic soil
Pseudorthocladius	Semi-aquatic
Psilometriocnemus	Semi-aquatic
Smittia	Almost all are terrestrial

difficult to make. This point is well illustrated by a study of the terrestrial chironomids of Spitzbergen, by Sendstad *et al.* (1977). They identified five different communities, associated with a range of soil, moisture and vegetational characteristics. The number of trapped adult chironomids in the driest of the habitats they investigated was only 26 m^{-2}. The fauna comprised a species of *Smittia*, designated as sp. A, and unidentified female Chironomidae. These same taxa, along with *Metriocnemus ursinus* Holmgren, were also recorded from a dry lichen heath, also with continuous plant cover, including *Saxifraga oppositifolia* Linnaeus and *Cetraria delisei* (Bory ex Schaerer) Nyl. The number of individuals taken from a metre square at this site averaged only 14. Much higher densities were recorded from two other sites, described respectively as fairly wet and very wet. The first of these was a moss tundra, with 269 adults m^{-2}; the other was dominated by *Deschampsia alpina* (L.) Roem. and Schult., with mosses between the clumps of grass. This site produced 411 individuals m^{-2}. Not only were densities higher at these wetter sites but diversity was also greater. Four taxa were identified from the moss tundra and five from the *Deschampsia alpina* habitat. *Paraphaenocladius impensus* Walker occurred in both, as did *Metriocnemus ursinus* and *Limnophyes* sp. *Smittia* sp. A occurred in the less wet of the two sites, while additional taxa in the wetter site were *Chaetocladius perennis* (Meigen) and unidentified female chironomids.

The fifth site examined by Sendstad *et al.* (1977) was a beach area, described as intermediate in wetness, between the dry lichen heath and the very wet *Deschampsia* heath, but influenced by sea-spray. Densities were very low here, perhaps as a result of the saline influence. Only 11 adults m^{-2} were captured. In addition to unidentified females the recorded taxa were *Smittia brevipennis* (Boheman) and *Pseudosmittia* sp.

The densities recorded by Sendstad *et al.* (1977), since they refer to numbers of emerging adults, are probably not a true reflection of larval densities. Densities of *Eretmoptera murphyi* larvae in damp moss and peat on Signy Island in the antarctic averaged 1500 m^{-2}, according to Cranston (1985).

The ability to continue development at low temperatures has been demonstrated for several terrestrial species. *Limnophyes pusillus* Eaton living on the Kerguelen Isles continued to grow slowly at temperatures as low as 2 °C and emerged when soil temperatures were between 5 °C and 7 °C (Delettre, 1978). Larvae of *Parasmittia* sp. living in *Armoricum* moorland completed all of their development at low temperatures during winter, when the moor was saturated with water (Delettre and Baillot, 1977). In summer, when the moor was dry, the fourth instar larvae were dormant in deeply placed cocoons. Adults emerged and laid eggs in the autumn. Several other terrestrial species are also known to resist drought by migrating deep into the soils and lying dormant in tubes or cocoons (Delettre, 1988b) but *Limnophyes minimus* Meigen, according to Delettre (1986), does not do this. He found this species to be abundant in heathland soils during winter, but no larvae were present during summer, following emergence of the adult insects. In this case Delettre believed that recolonization must occur each autumn from neighbouring, wetter habitats.

7

Larval food and feeding behaviour

M.B. Berg

7.1 INTRODUCTION

Interest in the food and feeding behaviour of larval chironomids has expanded in recent years. This is primarily because of:

- improvement of taxonomic keys to allow their identification;
- attempts to control pestiferous emergences from lentic habitats (Hilsenhoff, 1966; Ali, 1990);
- recognition of the energetic importance of chironomids in freshwater ecosystems (Benke *et al.*, 1984; Soluk, 1985; Berg and Hellenthal, 1991, 1993).

The latter issues have stimulated studies addressing the trophic basis of chironomid production and the factors controlling chironomid abundance. In the past, a major hindrance to conducting these studies has been the difficulty associated with larval taxonomic identification. The situation has improved over the last decade with the publication of generic keys covering broad geographic areas (e.g., Wiederholm, 1983, 1986, 1989; Coffman and Ferrington, 1984). These publications have enabled researchers who are not chironomid specialists to begin including chironomids in aquatic ecological studies.

This chapter is restricted to the literature published since Thienemann's (1954) survey of chironomid food and feeding. Much of the information published on chironomids in the past four decades has indicated that a diversity of feeding modes, types of food and feeding behaviours exist within subfamilies, tribes, genera and even species of chironomids. The extent of this diversity was recognized by Monakov (1972) when he stated, 'In no other group is there so much variation in the methods and mechanisms of feeding in morphologically close species.' This diversity makes strictly taxonomic generalizations difficult, if not confusing or misleading. Therefore, this chapter will begin by examining the general classes of feeding modes displayed by larval Chironomidae. The second section will discuss our current knowledge of the types of food materials consumed by chironomids with a discussion of food selectivity. The third section will

The Chironomidae: Biology and ecology of non-biting midges
Edited by P.D. Armitage, P.S. Cranston and L.C.V. Pinder
Published in 1995 by Chapman & Hall. ISBN 0 412 45260 X

examine specialized feeding-related behaviours and discuss the effect of chironomids on their food supply. Finally, suggestions for future studies of chironomid feeding will be presented.

7.2 MODES OF FEEDING

One widely used approach for the ecological classification of chironomids and other aquatic insects is based on functional feeding groups (Cummins, 1973; Wiggins and Mackay, 1978; Cummins and Klug, 1979; Lamberti and Moore, 1984; Merritt and Cummins, 1984; Merritt *et al.*, 1992). This approach classifies organisms based upon their mode of feeding, i.e. the morpho-behavioural adaptations for food acquisition. Examining feeding behaviour from a functional feeding perspective allows for generalizations in a taxonomically and ecologically diverse group such as chironomids (Anderson and Sedell, 1979). Although a functional group approach has contributed substantially to our understanding of the dynamics of energy flow in aquatic ecosystems (Vannote *et al.*, 1980; Benke *et al.*, 1984), it also has been misused by researchers who mistakenly equate a functional group (e.g. collector-gatherers) with the ingestion of a single food type (e.g. detritus) (Anderson and Sedell, 1979). In fact, aquatic insects within a functional feeding group often ingest a variety of items from different food categories (Cummins, 1973; Cummins and Klug, 1979). This has led to the view that, in general, many aquatic insects are opportunistic omnivores (Berrie, 1976; Anderson and Sedell, 1979). It will be apparent throughout this chapter that the widespread occurrence of omnivory applies equally well to chironomids.

Based on larval feeding modes, chironomids can be grouped into six general categories: collector-gatherers, collector-filterers, scrapers, shredders, engulfers and piercers. It is important to recognize that most chironomids are not restricted to a single mode of feeding. Even closely related taxa may exhibit different feeding modes. For example, *Cryptochironomus* has been reported to be an obligate predator (Armitage, 1968; Monakov, 1972), whereas other studies have shown that some *Cryptochironomus* species, such as *C. burganadzeae* Chernovskij, possess mouthparts adapted for collecting sediment (Shilova, 1960; as cited in Monakov, 1972), although this species may actually belong to *Harnischia*. The major food item for a species of *Cryptochironomus* in a Canadian stream was detritus (Ward and Williams, 1986). Species of *Chironomus* are also known to use different modes of feeding. Although *Chironomus plumosus* is typically considered to filter food in the profundal zone of lakes (McLachlan, 1977b), *C. anthracinus* feeds primarily on recently deposited sediments (Jónasson, 1972; Johnson, 1985). These examples demonstrate that even though functional group designations are based partially in morphology, there is considerable flexibility in the mode of feeding within the Chironomidae. A variety of factors including larval size, food quality and sediment composition can influence the feeding behaviour of larval midges (Armitage, 1968; McLachlan, 1977b; Baker and McLachlan, 1979; Hodkinson and Williams, 1980).

Differences in mode of feeding can be an important factor in reducing competition between chironomids. Differences in feeding modes associated with variations in tube morphology of *Chironomus plumosus* in Lady Burn Lough were suggested by McLachlan (1977b) to reduce intraspecific competition for food. Similarly, Rasmussen (1984a, 1985) hypothesized that different feeding modes allowed the collector-filterer *Glyptotendipes paripes* and the collector-gatherer *Chironomus riparius* Meigen to coexist at high densities.

7.2.1 COLLECTOR-GATHERERS

Collector-gatherers, also termed deposit-feeders, feed on material sedimented or deposited on submerged substrata. This functional feeding group predominates where fine particulate organic matter (FPOM) accumulates, such as in lakes and in areas of reduced current velocities in rivers and streams. Collecting-gathering is the most common feeding mode exhibited by chironomids and is represented within all subfamilies. Most chironomids are collector-gatherers at some time during larval development. Many taxa are deposit-feeders for most of the larval lifespan (Coffman and Ferrington, 1984), whereas other taxa, such as many Tanypodinae, are deposit-feeders only as early instars (Oliver, 1971; Baker and McLachlan, 1979).

Both tube-dwelling and free-living chironomids exhibit collector-gatherer feeding behaviour. Most tubicolous collector-gatherers, such as species of *Paratendipes*, *Paralauterborniella*, and *Chironomus*, among others, feed by extending the head and anterior part of the body outside the tube while using the posterior prolegs to maintain contact with the inner surface of the tube (Darby, 1962; Hilsenhoff, 1966; Jónasson and Kristiansen 1967; Edgar and Meadows, 1969; Jónasson, 1972; McLachlan, 1977b, Rasmussen, 1984a, b; Johnson, 1985; Leuchs and Neumann, 1990). Foraging areas, therefore, are restricted to a region immediately surrounding the tube and are distinguished by differences in colour between the light brown, recently foraged areas (oxidized sediments) and the darker, underlying anoxic sediments (McLachlan, 1977a; Rasmussen, 1984a). A variation of this typical collecting-gathering behaviour is exhibited by *Microtendipes pedellus* (De Geer). A larva feeds by extending the anterior part of the body from the tube and, with the use of the anterior prolegs, spreads a sheet-like net of sticky salivary secretions over the surface of the sediments. The larva then withdraws into the tube, dragging in the salivary sheet and attached food (Oliver, 1971). Non-tubicolous collector-gatherers such as *Tanypus*, *Thienemanniella* and *Corynoneura* typically forage along surficial sediments ingesting deposited material (Darby, 1962; Roback, 1969b; Tarwid, 1969; Coffman and Ferrington, 1984).

7.2.2 COLLECTOR-FILTERERS

Collector-filterers are suspension feeders that filter food materials from the water column. The size of particles ingested by filter-feeding chironomids generally is smaller than that ingested by collector-gatherers (McLachlan, 1977b). Filter-feeding by taxa within the subfamily Chironominae is the most widely studied chironomid feeding behaviour (Jónasson and Kristiansen, 1967; Konstantinov, 1971b; McLachlan and Cantrell, 1976; McLachlan, 1977b; Brennan, 1981; Hodkinson and Williams, 1980; Johnson, 1985, 1987). Using salivary silk secretions, larvae construct conical catchnets across the lumen of their tubes or retreats (Wallace and Merritt, 1980; Nilsson, 1984). Net construction for some larvae can take as little as 40 seconds (Konstantinov, 1971b). Tubes with catchnets are associated with larvae inhabiting a diverse range of habitats including sediments (e.g. *Chironomus* and *Glyptotendipes*) (Rasmussen, 1984a), submerged substrata including wood (e.g. *Rheotanytarsus*) (Benke *et al.*, 1984), or vascular plant tissues (e.g. *Polypedilum* (*Polypedilum*), *Polypedilum* (*Pentapedilum*) and *Endochironomus*) (Oliver, 1971; Kondo *et al.*, 1989). Larvae within lotic habitats use the action of the current to carry food particles into the net. In lentic habitats and depositional areas of streams and rivers, larvae use body undulations to induce a feeding current through the tubes (Darby, 1962; Konstantinov, 1971b; Brennan, 1981; Nilsson, 1984; Ali *et al.*, 1987; Ferrington, 1992).

Filter-feeding chironomids ingest food particles captured by catchnets in two basic ways. Some species of *Glyptotendipes*, *Parachironomus*, *Polypedilum* (*Pentapedilum*), *Endochironomus*, and *Chironomus* ingest the entire net and its contents and then reconstruct the net (Provost and Branch, 1959; Darby, 1962; Shilova, 1965: as cited in Monakov, 1972; Konstantinov, 1971b; Nilsson, 1984). Other taxa, such as *Rheotanytarus*, remove and ingest sections of the net, and then repair the missing section (Oliver, 1971; Nilsson, 1984).

Unique filter-feeding mechanisms have been described for *Kiefferulus martini* Freeman and *Odontomesa fulva* (Kieffer). *K. martini* larvae possess 'feeding brushes' that extend from the head capsule in a manner similar to those of Simuliidae. These brushes are presumably used to filter food from the water (Martin, 1963; Oliver, 1971). Larvae of *O. fulva* feed by swallowing water and then forcing it back out with strong contractions of the gut. Food particles contained in the expelled water are passively filtered by numerous labral setae (Shilova, 1966: as cited in Monakov, 1972).

7.2.3 SCRAPERS

Scrapers usually possess well-developed mandibles that are used to shear food material from the surfaces of rocks, sediments, wood and other submersed objects. Many Diamesinae, Orthocladiinae and a few Chironominae feed primarily as scrapers. Although scrapers are most abundant in flowing waters, some taxa, primarily the Chironominae, also occur in lentic habitats. Scrapers may be free-living, tubicolous or portable-case builders.

The Diamesinae are free-living and gather food by moving along the substratum and scraping material with the mandibles and, in some cases, well-developed labral lamellae (M. Berg, *pers. obs.*). Orthocladiinae that feed as scrapers, such as many *Cricotopus* species, generally are tube-builders and feed either while extending outside the tube or during the process of tube construction (LeSage and Harrison, 1980a; Hershey, 1987; Wiley and Warren, 1992). The portable case-building tanytarsine *Constempellina* has developed an elaborate morpho-behavioural mechanism that combines locomotion and feeding (Ferrington, in press). As the larva extends anteriorly out of the case, lateral movements of the head and thorax combined with brushing movements of the anterior prolegs scrape food materials from the substratum. In the process of pulling the case forward, the larva bends the head ventrally and uses the labral setae and other pharyngeal structures to scrape materials from the surface of the substratum into the pharyngeal region. As the larva continues to flex the head, the mandibles begin to close thereby pushing the food further into the pharyngeal region.

7.2.4 SHREDDERS

Shredders use either chewing, mining, gouging or grating (rasping) to acquire food. Shredders are associated with coarse particulate organic matter (CPOM) such as living vascular plants (*Cricotopus myriophylli* Oliver and some other *Cricotopus* species, and species of *Endochironomus*, *Hyporhygma*, *Polypedilum* and *Stenochironomus* (*Petalopholeus*)), submerged wood (*Brillia flavifrons* Johannsen, *Xylotopus*, *Stenochironomus* (*Stenochironomus*), *Stictochironomus*, *Orthocladius (Symposiocladius) lignicola* Kieffer), macro- or colonial algae (*Cricotopus nostocicola* Wirth, *Acamptocladius* and *Rheocricotopus*) or leaf litter (*B. flavifrons* and some species of *Chironomus*) (Brock, 1960; Anderson *et al.*, 1978; Anderson and Cummins, 1979; Anderson and Sedell, 1979; Cranston, 1982a; Pereira *et al.*, 1982; Kangasniemi and Oliver, 1983; Borkent, 1984; Coffman and Ferrington, 1984; Oliver, 1984; Stout and Taft, 1985; Ward *et al.*, 1985; Noda *et al.*, 1986; Anderson, 1989).

Darby (1962) described an interesting shredding mechanism exhibited by *Cricotopus bicinctus* (Meigen) in Californian rice fields. Larvae feed on the green alga *Spirogyra* by orienting the filaments lengthwise using the mandibles and structures of the epipharynx. Large algal filaments are forced over the mental teeth in a grating-like motion that causes the cell wall to be broken, thereby releasing the cellular contents.

7.2.5 ENGULFERS AND PIERCERS

The last two functional feeding groups, engulfers and predatory piercers, attack and ingest all or part of the prey (engulfers) or pierce the tissues and withdraw the fluids of the prey (piercers). Engulfing and piercing most commonly are associated with the Tanypodinae but species of some Orthocladiinae (*Cricotopus*, *Nanocladius*, *Metriocnemus*, *Rheocricotopus*) and Chironominae (*Micropsectra*, *Rheotanytarsus*, *Endochironomus*, *Chironomus*, *Glyptotendipes*, *Cryptochironomus*, *Parachironomus*, *Polypedilum*, and *Xenochironomus*) also may display engulfing or piercing feeding behaviour (Darby, 1962; Oliver, 1971; Monakov, 1972; Dendy, 1973; Loden, 1974; Coffman and Ferrington, 1984). Engulfing or piercing may sometimes be used in conjunction with another mode of feeding (Cummins, 1974; Cummins and Klug, 1979). For example, a filter-feeder such as *Rheotanytarsus* may engulf prey that have been caught in the catchnet (Coffman, 1967; Cummins, 1973). The collection of food and actual ingestion of food may involve more than one feeding mode.

7.2.6 FLEXIBILITY OF FEEDING MODE

Although some taxa are classified easily into one of the above functional feeding groups, others exhibit considerable flexibility in their mode of feeding. The ability of some species to switch from deposit-feeding to another mode, and vice-versa, is often associated with changes in sediment composition. For example, Hodkinson and Williams (1980) suggested that *Chironomus plumosus* (Linnaeus) exhibited high flexibility in its mode of feeding in response to allochthonous leaf inputs into an experimental woodland pond. In an unshaded control pond with little leaf input, *C. plumosus* inhabited the sediments and fed primarily by filter-feeding. In the presence of leaf litter accumulations in the experimental pond, however, larvae fed by deposit-feeding because their occurrence on the underside of leaves prevented them from filter-feeding.

7.3 FOOD SOURCES

Given the diverse feeding modes exhibited by chironomids, it is not surprising that these insects also ingest a wide variety of foods. Although some may be selective in their choice of food (see below), diets generally are quite flexible (Tarwid, 1969; Roback, 1969a; Alfred, 1974; Baker and McLachlan, 1979; Mackey, 1979; Dusoge, 1980; Kajak, 1980; Sephton, 1987; Johnson *et al.*, 1989). Diets may change as larvae mature or because of seasonal changes in food availability (Armitage, 1968; Brennan *et al.*, 1978; McLachlan *et al.*, 1978; Baker and McLachlan, 1979; Williams, 1981; Tokeshi, 1986a; Ward and Williams, 1986; Sephton, 1987). Chironomids can be considered to ingest five categories of food: algae, detritus and associated microorganisms, macrophytes, woody debris and invertebrates.

There is conflicting literature on the relative importance of different foods in chironomid diets. Some of these discrepancies may result from the use of different criteria to assess importance (e.g. assimilation efficiency, gut fullness etc.) and comparing diets from field-collected larvae with those from laboratory feeding experiments. However, an equally plausible explanation is that these discrepancies are merely a manifestation of the flexibility of chironomid feeding and their ability to adapt to local environmental conditions. Because of the wide variety of food items ingested by even a single chironomid, attempts to discuss chironomid diets in general terms would be forbidding. Therefore, the following sections will discuss examples in which each food category represents the major component of chironomid diets.

Several criteria have been used to denote the importance of different food sources. Commonly used criteria are gut contents (e.g. Armitage, 1968; Kawecka and Kownacki, 1974; Mason and Bryant, 1975: Kawecka *et al.*, 1978; Johannsson, 1980), direct observations of larval feeding (Abul-Nasr *et al.*, 1970a; Paterson and Cameron, 1982; Borkent, 1984; Van der Velde and Hiddink, 1987) and correlations between food availability and larval growth, population densities, or secondary production (e.g. Marker, 1976; Pinder, 1977; Cattaneo, 1983; Rasmussen, 1984b; Storey, 1986; Tokeshi, 1986b, Johnson and Pejler, 1987). Energetic criteria such as assimilation or absorption efficiencies and the trophic basis of production are increasing in popularity and use (e.g. Johannsson, 1980; Johannsson and Beaver, 1983; Rasmussen, 1984a; Johnson, 1985: Marker *et al.*, 1986; Smith and Smock, 1992). In this chapter when reports of gut fullness contradict energetic contributions of food items, the energetic criterion will be used as the indicator of importance because the presence of material in the alimentary tract is not necessarily an indicator of nutritional importance (Cummins, 1973). For example, Johannsson and Beaver (1983) used a criterion of gut fullness and found that the diatoms *Stephanodiscus* and *Melosira* were important in the diet of *Chironomus plumosus* f. *semireductus*. When assimilation efficiencies were examined, however, algae were determined to contribute little to chironomid energetics. Thus, diatoms would not be considered the major food source for this species.

Laboratory studies of chironomid feeding are valuable in examining the ability of taxa to ingest and assimilate different foods (Cummins, 1973). They have been intentionally omitted from this discussion, however, because of possible artifacts from feeding in an unnatural environment with limited choices of food. Such studies can yield results that are inconsistent with those from field data. For example, Baker and McLachlan (1979) found that *Procladius choreus* was primarily a detritivore–herbivore in the field, but laboratory studies revealed a diet of primarily animal material. Although the incidence of predation could have been underestimated due to field-collected larvae using a piercing-sucking feeding mode, their study does point out the difficulties in comparing laboratory and field diets. Inconsistencies between laboratory and field studies of chironomid diets should not be unexpected, given the broad range of foods that can be ingested by chironomids.

7.3.1. ALGAE

Studies of chironomid feeding have reported conflicting results about the importance of algae in chironomid diets. Some studies have found that algae, particularly diatoms, are unimportant as a food source for chironomids (Davies, 1975: Moore, 1979a,b, 1980; Johannsson and Beaver, 1983), whereas others have reported that algae are a major portion of the diet (Coffman, 1967; Jónasson and Kristiansen, 1967; Kajak and Warda, 1968; Armitage, 1971; Alfred, 1974; Mason and Bryant, 1975: Kajak, 1977; Johannsson, 1980; Johnson *et al.*, 1989). These contradictions may be due to the use of different criteria to assess dietary importance.

The frequently cited importance of algae, particularly diatoms, is not surprising given that the primitive chironomid habitat is thought to have been rich in diatoms (Brundin, 1966). Chironomid densities, standing stock biomass and secondary production are often closely synchronized with the availability and ingestion of algae (Jónasson and Kristiansen, 1967; Jónasson, 1972; Pinder, 1977; Brennan *et al.*, 1978; Johannsson, 1980; Cattaneo, 1983; Rasmussen, 1984b; Ali, 1989; Berg and Hellenthal, 1992). Three general groups of algae are ingested by chironomids: benthic, suspended (phytoplankton) and epiphytic. For the purposes of this chapter, recently sedimented algae will be considered with phytoplankton because of their strong association with phytoplankton production and to distinguish them from truly benthic forms.

(a) Benthic algae

Most studies reporting benthic algae as the primary food source for chironomids have noted that diatoms usually constitute most of the diet (Table 7.1) (Armitage, 1968; Kajak and Warda, 1968; Roback, 1969b; Tarwid, 1969; Titmus and Badcock, 1981; Williams, 1981; Soluk, 1985; Marker *et al.*, 1986; Hambrook and Sheath, 1987; Smit *et al.*, 1991). Green and blue-green algae also have been shown to be important for some chironomids (Brook, 1954; Brock, 1960; Darby, 1962; Ward *et al.*, 1985). Kajak and Warda (1968) examined the relationship between the feeding of non-predatory chironomids (Chironominae) and food availability in three Polish lakes. They concluded that these chironomids fed primarily on algae in surficial sediments (66–72% of gut contents) and that among diatoms, blue-green and green algae, only diatoms were readily assimilated. Marker *et al.* (1986) also reported high digestion rates in chironomids feeding on the diatoms *Achnanthes minutissima* Kütz (85–95%) and *Meridion circulare* Ag. (50%).

The dietary importance of diatoms is not restricted to taxa typically considered to be herbivorous. Armitage (1968) found that small species of Tanypodinae fed primarily on diatoms and the desmid *Closterium*. Other studies also have reported that diatoms can be the dominant food in the alimentary tracts of Tanypodinae, especially *Tanypus* (Roback, 1969b; Tarwid, 1969; Titmus and Badcock, 1981). These studies emphasize the need to examine chironomid diets directly rather than to assume that all species within a particular subfamily or genus consume one type of food.

Benthic diatoms have been reported to be important in the diets of epipsammic chironomids. Soluk (1985) found that alimentary tracts of *Rheosmittia* from an unstable sand habitat were tightly packed with diatoms. Similarly, Smit *et al.* (1991) found that benthic diatoms were an important food source for *Lipiniella araenicola* Shilova in a sand flats area of the Rhine–Meuse estuary.

In addition to the predominance of diatoms in the diets of chironomids, benthic green algae also have been reported to be a major food source. Brook (1954) observed that chironomid larvae on a sand-filter bed ingested only filamentous green algae and filamentous diatoms. In a study of the

Table 7.1 Studies reporting benthic algae to be the dominant food source for chironomids. The dominance of benthic algae was based upon the following criteria of importance: GC = gut contents, IN = inferred, EN = energetic criteria

Taxa	Location	Dominant Food	Criterion of Importance	Reference
Tanypodinae				
Ablabesmyia monilis Linnaeus	L. Kuusijärvi, Finland	*Closterium* Diatoms	GC	Armitage, 1968
Procladius spp.	L. Kuusijärvi, Finland	*Closterium* Diatoms	GC	Armitage, 1968
Tanypus carinatus Sublette	several locales	Diatoms	GC	Roback, 1969b
T. neopunctipennis Sublette	several locales	Diatoms	GC	Roback, 1969b
T. nubifer Coquillett	several locales	Diatoms	GC	Roback, 1969b
T. punctipennis (Meigen)	L. Sniardwy, Poland	Diatoms	GC	Tarwid, 1969
T. stellatus Coquillett	several locales	Diatoms	GC	Tarwid, 1969
Tanypus sp.	Black Horse L. England	*Synedra* *Navicula*	GC, IN	Titmus and Badcock, 1981
Diamesinae				
Diamesa latitarsis (Goetghebuer)	Mnichowy Stream, Poland	*Chamaesiphon polonicus*	GC, IN	Kawecka *et al.*, 1978
D. steinboecki Goetghebuer	Mnichowy Stream, Poland	*Chamaesiphon polonicus*	GC, IN	Kawecka *et al.*, 1978
Orthocladiinae				
Cricotopus bicinctus (Meigen)	rice fields, California	*Spirogyra*	GC	Darby, 1962
C. nostocicola Edwards	Sagehen Creek, California	*Nostoc* (trichomes)	GC	Brock, 1960
C. nostocicola	Nostoc Creek, Oregon	*Nostoc*	GC	Ward *et al.*, 1985
Cricotopus spp.	Big Vermillion R., Illinois	Diatoms Desmids	GC	Wiley and Warren, 1992
Rheosmittia sp.	Sand R., Alberta	Diatoms	GC	Soluk, 1985
Chironominae				
Lipiniella araenicola Shilova	Haringvliet, Netherlands	Diatoms	GC	Smit *et al.*, 1991
Pagastiella orophila (Edwards)	L. Kuusijärvi, Finland	Diatoms	GC	Armitage, 1968
'non-predatory chironomids'	three lakes, Poland	Diatoms	EN	Kajak and Warda, 1968

chironomids inhabiting rice fields, Darby (1962) noted the importance of *Spirogyra* in the diet of *Cricotopus bicinctus*.

Blue-green algae have been reported to be unimportant as a food source for chironomids because they are either relatively undigestible or not ingested at all (Kajak and Warda, 1968; Konstantinov, 1971b; Davies, 1975: Johnson *et al.*, 1989). This generalization does not apply to all blue-green algae and some may be suitable as food for aquatic insects (Gregory, 1983). Kawecka *et al.* (1978) found that two *Diamesa* spp. fed primarily on the blue-green alga *Chamaesiphon polonicus* (Rostaf.) Hansg. In addition, one of the few cases of an obligate mutualistic plant–herbivore interaction in streams is between the orthoclad *Cricotopus nostocicola* and the benthic blue-green alga *Nostoc parmelioides* Kutzing (Brock, 1960; Gregory, 1983; Ward *et al.*, 1985; Dodds and Marra, 1989). *C. nostocicola* inhabits the gelatinous matrix of *N. parmeliodes* and feeds primarily on *Nostoc* trichomes (Brock, 1960). In addition to serving as a food resource, the alga also provides the chironomid with shelter and protection against predation and desiccation. The presence of the chironomid induces a morphological change in the alga indirectly resulting in an increase in the alga's primary productivity (Ward *et al.*, 1985).

(b) Suspended algae

The ingestion of phytoplankton and its role as a dominant food source for chironomids is most commonly associated with filter-feeders, but collector-gatherers also may consume recently sedimented algae. As with the other algal categories, chironomids that ingest phytoplankton as a major food source consume algae from several taxonomic groups (Table 7.2).

Diatoms are frequently reported as the major food source for chironomids feeding on phytoplankton. Johnson *et al.* (1989) found that algal carbon, especially from the diatom *Melosira*, was a nutritionally important food source for *Chironomus plumosus*, contributing up to 84% of the midge's average daily carbon requirement. Higher chironomid growth rates often coincide with increased availability and ingestion of planktonic diatoms (Jónasson and Kristiansen, 1967; Rasmussen, 1984a,b; Johnson and Pejler, 1987) but diatoms may contribute little to gut contents when viewed on an annual basis (Johnson, 1987).

Suspended blue-green algae are also reported to be a major component of chironomid diets. Ali (1990) found that blue-green algae, primarily *Anacystis*, *Gloeocapsa*, *Lyngbya* and *Merismopedia*, represented a major portion (68% of gut contents) of the diet of filter-feeding *Chironomus crassicaudatus* Malloch. Similarly, Provost and Branch (1959) reported that *Glyptotendipes paripes* (Edwards) primarily fed on phytoplankton (98.7% of diet) with blue-green algae accounting for 60.7% of the total phytoplankton. This latter study also showed a high contribution of phytoplankton to *Chironomus decorus* Johannsen diets (91.7%), but green algae replaced blue-green algae as the dominant food ingested (51% of total phytoplankton). In a study examining the interactions between *Chironomus imicola* Kieffer and tadpoles in tropical rock pools, McLachlan (1981b) demonstrated the importance of sedimented algae in the diets of the chironomid. In the absence of tadpoles, *C. imicola* had a low growth rate and fed primarily on epipelic blue-green algae. In the presence of tadpoles that graze on epineustic green algae, chironomid diets shifted to

Table 7.2 Studies reporting suspended algae to be the dominant food source for chironomids. The dominance of suspended algae was based upon the following criteria of importance: GC = gut contents, IN = inferred, EN = energetic criteria

Taxa	Location	Dominant Food	Criterion of Importance	Reference
Chironomini				
Chironomus anthracinus Zetterstedt	L. Erken, Sweden	Diatoms	IN	Johnson and Pejler, 1987
C. anthracinus	L. Esrom, Denmark	Diatoms	GC, IN	Jónasson and Kristiansen, 1967
C. costatus Johannsen	temporary pond, India	Dinoflagellates	GC	Alfred, 1974
C. crassicaudatus Malloch	L. Monroe, Florida	Blue-greens	GC	Ali, 1990
C. decorus Johannsen	Polk County Lks, Florida	Greens	GC	Provost and Branch, 1959
C. imicola Kieffer	rock pools, Africa	Greens	GC	McLachlan, 1981
C. plumosus Linnaeus	L. Erken, Sweden	Diatoms	IN	Johnson and Pejler, 1987
C. plumosus	L. Vallentunasjön, Sweden	Diatoms	EN	Johnson *et al.*, 1989
Glyptotendipes paripes (Edwards)	Polk County Lks, Florida	Blue-greens	GC	Provost and Branch, 1959
G. paripes	Stephenson Pond, Alberta	Diatoms	GC, IN, EN	Rasmussen, 1984a, b

sedimenting epineustic algae that were associated with tadpole faeces. The ingestion of this nutrient-rich food source resulted in higher growth rates of *C. imicola*.

(c) Epiphytic algae

Relatively few studies have examined the relationship between epiphytic algae and associated chironomids in either lakes (Cattaneo, 1983) or streams (see review of chalk stream work by Pinder (1992)). The studies that have been conducted, however, have reported the importance of epiphytic algae as a food source for larval chironomids. Diatoms are frequently considered to be the major algae consumed by epiphytic chironomids (Darby, 1962; Mason and Bryant, 1975: Williams, 1981; Cattaneo, 1983; Tokeshi, 1986a), but green algae also have been reported as a major dietary constituent (Table 7.3) (Botts and Colwell, 1992).

In a study examining the relationship between assemblages of epiphytic

Table 7.3 Studies reporting epiphytic algae to be the dominant food source for chironomids. The dominance of epiphytic algae was based upon the following criteria of importance: GC = gut contents

Taxa	Location	Dominant Food	Criterion of Importance	Reference
Diamesinae				
Potthastia gaedii (Meigen)	Raba River, Poland	Diatoms	GC	Kawecka and Kownacki, 1974
Orthocladiinae				
Cricotopus bicinctus Mg (as *dizonias*)	Alderfen Broad, England	Diatoms *Cladophora*	GC	Mason and Bryant, 1975
Cricotopus sp.	Raba River, Poland	Diatoms	GC	Kawecka and Kownacki, 1974
Psectrocladius sp.	L. Padgett, Florida	*Cosmarium*	GC	Botts and Colwell, 1992
Thienemaniella cf. *fusca*	L. Padgett, Florida	*Cosmarium*	GC	Botts and Colwell, 1992
Chironominae				
Glyptotendipes glaucus (Meigen)	Alderfen Broad, England	Diatoms *Cladophora*	GC	Mason and Bryant, 1975
Polypedilum gr. *pedestre*	Raba River, Poland	Diatoms	GC	Kawecka and Kownacki, 1974
Pseudochironomus richardsoni Malloch	Eel River, California	*Epithemia* *Cocconeis*	GC	Power, 1991

chironomids and epiphytic biomass in Lake Memphremagog, Canada, Cattaneo (1983) found that grazing chironomids fed extensively on diatoms, resulting in a positive correlation between epiphytic diatoms and chironomid biomass. Epiphytic diatoms associated with filamentous green algae have also been reported to be an important food source for chironomids. Power (1991) and Kawecka and Kownacki (1974) reported that chironomids associated with *Cladophora* were feeding mainly on epiphytic diatoms.

Mason and Bryant (1975) examined the grazing of *Glyptotendipes glaucus* (Meigen) and *Cricotopus bicinctus* (as *dizonias*) on periphyton associated with *Typha* stems and found that both species primarily fed on diatoms and filamentous green algae (*Cladophora*). In contrast, Botts and Cowell (1992) reported that *Psectrocladius* and *Thienemanniella* cf. *fusca* on *Typha* stems mostly consumed green algae (*Cosmarium*).

7.3.2 DETRITUS

The most commonly reported food ingested by chironomids and identified from gut contents is detritus (Table 7.4) (Tarwid, 1969; Monakov, 1972;

Table 7.4 Studies reporting detritus to be the dominant food source for chironomids. The dominance of detritus was based upon the following criteria of importance: GC = gut contents, IN = inferred, OB = observational, EN = energetic criteria. NR = source of detritus not reported

Taxa	*Location*	*Source of detritus*	*Criterion of importance*	*Reference*
Tanypodinae				
Ablabesmyia monilis	R. Thames, England	NR	GC	Mackey, 1979
Macropelopia adaucta Kieffer (as *Anatopynia goetghebeuri*)	3 sites northern UK	NR	GC	Baker and McLachlan, 1979
Procladius bellus (Loew)	Laurel Creek Res., Ontario	NR	GC	Sephton, 1987
P. choreus Mg.	R. Thames, England	NR	GC	Mackey, 1979
P. choreus	3 sites northern UK	NR	GC	Baker and McLachlan, 1979
P. culiciformis (Linnaeus)	3 sites northern UK	NR	GC	Baker and McLachlan, 1979
P. denticulatus Sublette	Great Slave L., Canada	NR	GC	Moore, 1979b; 1980
Diamesinae				
Diamesa nowickiana Kownacki & Kownacka	Mnichowy Str., Poland	NR	GC	Kawecka *et al.*, 1978
Prodiamesinae				
Prodiamesa olivacea (Meigen)	Mattma Str., Germany	*Sphaerotilus natans*	GC, EN	Jankovic, 1974
P. olivacea	R. Thames, England	NR	GC	Mackey, 1979
Orthocladiinae				
Corynoneura coronata Edwards	R. Thames, England	NR	GC	Mackey, 1979
Cricotopus bicinctus	R. Thames, England	NR	GC	Mackey, 1979
C. sylvestris Fabricius	R. Thames, England	NR	GC	Mackey, 1979
Tvetenia (as *Eukiefferiella*) *bavarica* gp.	Olszowy Str., Poland	Beech leaves	GC	Kawecka and Kownacki, 1974
Heleniella ornaticollis Edwards	Olszowy Str., Poland	Beech leaves	GC	Kawecka and Kownacki, 1974
Heterotrissocladius changi Sæther	Great Slave L., Canada	NR	GC	Moore, 1979b, 1980

Table 7.4 Continued

Taxa	*Location*	*Source of detritus*	*Criterion of importance*	*Reference*
Metriocnemus knabi (Coquillett)	Pitcher-plants	Invertebrates	OB	Paterson and Cameron, 1982
Orthocladius sp.	Rouge R., Ontario	NR	GC	Ward and Williams, 1986
Synorthocladius semivirens (Kieffer)	R. Thames, England	NR	GC	Mackey, 1979
Chironominae				
Chironomus anthracinus	Uchinskoye Res., USSR	*Dreissena polymorpha* pseudofaeces	GC, IN	Izvekova and Lvova-Katchanova, 1972
C. anthracinus	L. Erken, Sweden	NR	GC	Johnson, 1987
C. cingulatus Meigen	Black Horse L., England	NR	GC	Titmus and Badcock, 1981
C. decorus	Great Slave L., Canada	NR	GC	Moore, 1979b, 1980
C. lugubris Zetterstedt	Blaxter Lough, England	Peat	IN	McLachlan *et al.*, 1979
C. plumosus	L. Erken, Sweden	NR	GC	Johnson, 1987
C. plumosus	Rybinsk Reservoir, Russia	NR	GC	Margolina, 1971
C. plumosus	Great Slave L., Canada	NR	GC	Moore, 1979b, 1980
C. plumosus	Duck Pits, England	Invertebrate faeces	IN	Hodkinson and Williams, 1980
C. plumosus cf. *semireductus*	Bay of Quinte, L. Ontario, Canada	Bacteria	IN	Johannsson and Beaver, 1983
C. riparius	Stephenson Pond, Canada	NR	GC, EN	Rasmussen, 1984a, 1985
Cladopelma (as *Cryptochironomus*) *virescens* (Mg.)	R. Thames, England	NR	GC	Mackey, 1979
Cladotanytarsus cf. *iucundus* Hirvenoja	L. Kuusijärvi, Finland	Plant	GC	Armitage, 1968
C. cf. *molestus* Hirvenoja	L. Kuusijärvi, Finland	Plant	GC	Armitage, 1968
C. nigrovittatus (Goetghebuer)	L. Kuusijärvi, Finland	Plant	GC	Armitage, 1968
Cryptochironomus digitatus (Malloch)	Great Slave L., Canada	NR	GC	Moore, 1980
Cryptochironomus sp.	Rouge R., Ontario	NR	GC	Ward and Williams, 1986
Dicrotendipes modestus (Say)	Great Slave L., Canada	NR	GC	Moore, 1979b, 1980

Table 7.4 Continued

Taxa	*Location*	*Source of detritus*	*Criterion of importance*	*Reference*
D. modestus (as *Limnochironomus pulsus*)	L. Kuusijärvi, Finland	Plant	GC	Armitage, 1968
D. modestus (as *Limnochironomus pulsus*)	R. Thames, England	Plant	GC	Mackey, 1979
D. nervosa (Staeger)	R. Thames, England	Plant	GC	Moore, 1980
Endochironomus albipennis (Meigen)	Uchinskoye Res., U.S.S.R.	*Dreissena polymorpha* pseudofaeces	GC, IN	Izvekova and Lvova-Katchanova, 1972
E. albipennis	L. Piaseczno and L. Glebokie, Poland	NR	GC	Kornijów, 1992
Glyptotendipes pallens (Meigen)	R. Thames, England	NR	GC	Mackey, 1979
Micropsectra atrofasciata (Kieffer)	Olszowy Str., Poland	Beech leaves	GC	Kawecka and Kownacki, 1974
Micropsectra sp.	Carp Creek, Michigan	NR	GC	Pringle, 1985
Micropsectra spp.	R. North Tyne, England	Allochthonous	GC	McLachlan *et al.*, 1978 Brennan *et al.*, 1978
Phaenopsectra flavipes (Meigen)	R. Thames, England	NR	GC	Mackey, 1979
Polypedilum nubeculosum (Meigen)	Black Horse L., England	NR	GC	Titmus and Badcock, 1981
Polypedilum sp.	Rouge R., Ontario	NR	GC	Ward and Williams, 1986
Stempellinella (as *Stempellina*) *minor*	L. Kuusijärvi, Finland	Plant	GC	Armitage, 1968
Tanytarsus debilis (Mg.) (as *samboni*)	L. Kuusijärvi, Finland	Plant	GC	Armitage, 1968
Tanytarsus sp. *gregarius* gr.	L. Kuusijärvi, Finland	Plant	GC	Armitage, 1968
Tanytarsus sp.	Great Slave L., Canada	NR	GC	Moore, 1979b, 1980
Tanytarsus spp.	R. North Tyne, England	Allochthonous	GC	McLachlan *et al.*, 1978 Brennan *et al.*, 1978

McLachlan, 1977b; Baker and McLachlan, 1979; Johannsson, 1980; Pereira *et al.*, 1982; Pinder, 1986). In this chapter, detritus includes all non-living particulate organic matter and associated non-photosynthetic microorganisms (Cummins, 1973; Boling *et al.*, 1975; Anderson and Sedell, 1979). The ingestion of woody debris is often included in discussions of detritus processing (Anderson and Sedell, 1979) but xylophagy will be considered separately here because of the exclusive association of some chironomids with woody debris.

It is not uncommon for detritus to account for 50–70% of gut contents of chironomids (Armitage, 1968; Coffman *et al.*, 1971; Margolina, 1971; Kawecka and Kownacki, 1974; McLachlan, 1977b; Kawecka *et al.*, 1978; McLachlan *et al.*, 1978; Mackey, 1979; Moore, 1980; Titmus and Badcock, 1981; Ward and Williams, 1986; Sephton, 1987) although values >90% have been reported (Johnson, 1987; Kornijów, 1992). Even taxa usually considered to be predaceous, such as Tanypodinae, have been reported to ingest large quantities of detritus (Armitage, 1968; Baker and McLachlan, 1979; Hildrew *et al.*, 1985; Smith and Smock, 1992).

Although many studies report the importance of detritus in the diets of chironomids, the nutritional value of the non-living component of detritus has been questioned. Because much of the material ingested by detritivorous invertebrates is undigestible (cellulose, lignin and ash) and rapidly passes through the gut, the microbial component accounts for much of the nutritional value (Rodina, 1971; Bjärnov, 1972; Cummins, 1974; Berrie, 1976; Baker and McLachlan, 1979; Cummins and Klug, 1979; Mackey, 1979; Moore, 1979b; Ward and Cummins, 1979; Lamberti and Moore, 1984) and can be important in synthesizing essential vitamins that can be used by chironomids (Rodina, 1971; Ladle, 1982).

Attached microorganisms such as bacteria, fungi and protozoans can serve as a direct food source for chironomids (Rodina, 1971; Jankovič, 1974; Baker and Bradnam, 1976) or they can play a role in transforming relatively refractory detritus into a form that is more nutritious and readily digestible (McLachlan *et al.*, 1979; Ward and Cummins, 1979; Johannsson and Beaver, 1983; Ward and Williams, 1986). Detritus from more labile inputs (e.g. algal cells, plant reproductive parts) represents a higher quality food source than more refractory inputs (e.g. wood, leaf litter) because of its higher microbial biomass and relatively rapid rates of decomposition.

It is important to distinguish between the quality and quantity of detritus when discussing its role as a food source because the nutritional quality of detritus can vary considerably. Baker and Bradnam (1976) found that although the alimentary tracts of *Chironomus riparius* (as *thummi* Kieffer) contained large numbers of bacteria and that bacteria alone could be an adequate food source (50% of bacteria ingested are digested), they were not as quantitatively important as other detrital components. Rodina (1971) also showed that *Chironomus plumosus* could complete larval development on a diet solely of bacteria. Similarly, *Paratanytarsus grimmii* (Schneider) in an enclosed water distribution system was found to complete its parthenogenetic life cycle on a diet almost entirely of bacteria (M. Berg, *pers. obs.*).

Detrital quality has been shown to affect chironomid population densities (Hodkinson and Williams, 1980; Toscano and McLachlan, 1980; Walentowicz and McLachlan, 1980) and life history patterns (Ward and Cummins,

1979). It has been suggested that the poor food quality of peat (high in organics but low in total organic nitrogen) limits chironomid densities in erosional areas of the River North Tyne, England (Toscano and McLachlan, 1980; Walentowicz and McLachlan, 1980). Other studies have also shown a strong positive relationship between detrital nitrogen content and both larval biomass (Muthukrishnan and Palavesam, 1992) and adult fecundity (Palavesam and Muthukrishnan, 1992). Ward and Cummins (1979) suggested that seasonal differences in detrital quality could account for the failure of *Paratendipes albimanus* (Meigen) to complete more than one generation in a Michigan stream, whereas lentic populations feeding on higher quality detritus have been reported to exhibit a bivoltine life cycle. They found an inverse relationship between detrital standing crop in the stream and *P. albimanus* growth, and attributed poor growth to low detrital microbial activity (indicated by respiration). Conversely, highest growth rates coincided with the period of maximal detrital microbial respiration. Although *P. albimanus* consumed primarily detritus throughout the year, larval growth was related to the seasonal variations in microbial biomass, or quality, of the detritus. Izvekova and Lvova-Katchanova (1972) reported that *Endochironomus albipennis* (Meigen) and *Chironomus anthracinus* Zetterstedt feeding on zebra mussel (*Dreissena polymorpha*) pseudofaeces (material not entering the alimentary tract of the mussel, but rather passing through the mantle cavity and ejected with a coating of mucus) had higher assimilation and growth rates than those that fed on mussel faecal matter. They suggested that the increased nutritive value of the pseudofaeces was due to high numbers of bacteria associated with the mucopolysaccharide coating.

The role of bacteria in chironomid nutrition may differ in lakes of different trophic states. Goedkoop and Johnson (1993) found that in oligotrophic lakes, bacterial carbon can account for up to 47% of the carbon needs of chironomids, whereas in eutrophic and hypereutrophic lakes, bacterial carbon accounts for only 2% of the carbon demand. Because of the rapid degradation of phytoplankton in the water column of oligotrophic lakes, bacterial carbon plays an important role in the diet of profundal chironomids.

7.3.3 MACROPHYTES

Few studies have reported direct chironomid feeding on aquatic macrophytes (Table 7.5). Of the 36 species of chironomids recorded from rice fields in California, Darby (1962) noted that only one fed on aquatic macrophytes. These results are consistent with the broader view that macrophytes are unimportant as a food source for aquatic invertebrates (Hutchinson, 1975; Wetzel, 1975; Gregory, 1983; Lamberti and Moore, 1984). Recent reviews, however, have provided compelling evidence that challenge this generalization (Lodge, 1991; Newman, 1991). In his review of herbivory and detritivory on freshwater macrophytes, Newman (1991) found that the incidence of invertebrate feeding on macrophytes is more common than generally acknowledged. It also has been suggested that, evolutionarily, it is unlikely that a food resource as abundant as macrophytes should be

Table 7.5 Studies reporting macrophytes to be the dominant food source for chironomids. The dominance of macrophytes was based upon the following criteria of importance: GC = gut contents, IN = inferred, OB = observational

Taxa	*Location*	*Dominant food*	*Criterion of importance*	*Reference*
Orthocladiinae				
Cricotopus myriophylli Oliver	Okanagan Valley Lks, British Columbia	*Myriophyllum spicatum*	GC	Kangasniemi and Oliver, 1983
C. sylvestris	California	*Potamogeton*	GC	Darby, 1962
C. sylvestris	California	Germinating rice seeds and leaves	GC	Darby, 1962
C. trifasciatus (Meigen)	Japan	Rice leaves	IN, OB	Noda *et al.*, 1986
C. trifasciatus	R. Frome, England	*Ranunculus calcareus* leaves	GC	Williams, 1981
Eukiefferiella ilkleyensis (Edwards)	R. Frome, England	*Ranunculus calcareus* leaves	GC	Williams, 1981
Chironominae				
Chironomus tepperi Skuse	N.S.W., Australia	Rice seedling roots	OB	Treverrow, 1985
Chironomus. sp.	Egypt	Rice rootlets and germinating kernels	OB	Abul-Nasr *et al.*, 1970a
Chironomus. sp.	Texas	Rice seeds and seedlings	IN	Way and Wallace, 1989
Endochironomus lepidus (Meigen)	Netherlands	*Nuphar lutea*	OB	Van der Velde and Hiddink, 1987
Endochironomus subtendens (Townes)	Okanagan Valley Lks, British Columbia	*Myriophyllum spicatum*	GC	Kangasniemi and Oliver, 1983
Polypedilum (Pentapedilum) tigrinum (Hashimoto)	Japan	*Nymphoides indica*	GC	Kondo *et al.*, 1989
Stenochironomus (Petalopholeus) spp.	Several localities	Live and dead leaves	GC, OB	Borkent, 1984
Tanytarsus formosana Kieffer	Japan	Rice seedling leaves	OB	Hiroaki and Miyazaki, 1986
Tanytarsus sp.	Texas	Rice seeds and seedlings	IN	Way and Wallace, 1989

unexploited (Lodge, 1991). Perpetuation of the view that macrophytes are unimportant as a food source is more a result of *a priori* assumptions of researchers and a paucity of studies rather than experimental demonstration of their unimportance (Lodge, 1991). Explanations for the lack of grazing on aquatic macrophytes usually include the presence of tough cell walls, heavily lignified structures, low nitrogen content and the presence of allelopathic substances (Gregory, 1983; Otto, 1983; Lamberti and Moore, 1984; Ostrofsky and Zettler, 1986). Recently, some of these assumptions have been challenged (Sand-Jensen and Madsen, 1989; Lodge, 1991; Newman, 1991).

Van der Velde and Hiddink (1987) described three groups of chironomid–macrophyte associations: obligate phytophages, seston eaters and facultative phytophages. **Obligate phytophages** feed primarily on living plant tissue, whereas **seston eaters** mine in the leaves or stems of vascular plants and either filter-feed seston or consume living plant tissue. **Facultative phytophages** feed on seston and use the plant solely as a substratum for mining and filter-feeding. They listed only *Endochironomus lepidus* (Meigen) as an obligate phytophage on *Nuphar lutea* leaves and peduncles. Williams (1981) reported that the leaves of *Ranunculus photophilus* var. *calcareus* (R.W. Butcher) C.D.K.Cook became a major food source for fourth instar larvae of two *Cricotopus* species and *Eukiefferiella ilkleyensis* (Edwards). Kondo *et al.* (1989) found *Polypedilum* (*Pentapedilum*) *tigrinum* (Hashimoto) associated with floating leaves of *Nymphoides indica* (L.) O. Kuntze, but it was concluded that they were primarily seston feeders consuming leaf material only in the process of excavating channels in spongy mesophyll tissues.

One of the most frequently cited interactions between chironomids and macrophytes is the damage associated with chironomid feeding activities on the leaves, seeds, seedlings and roots of rice plants (Noda *et al.*, 1986; Way and Wallace, 1989; Ferrarese, 1993). The genera most commonly reported to feed on rice plants are *Tanytarsus*, *Paralauterborniella*, *Chironomus*, and *Cricotopus* (Darby, 1962; Abul-Nasr *et al.*, 1970a; Clement *et al.*, 1977; Noda and Miyazaki, 1986; Treverrow, 1985; Way and Wallace, 1989). Although Darby (1962) found a diverse assemblage of chironomids in California rice fields, most taxa were believed to have used macrophytes primarily as substrata for attachment or as a source of periphyton rather than as a direct food source. One of the few exceptions was *Cricotopus sylvestris* (Fabricius), which was found to feed on pondweed (*Potamogeton*) and burhead (*Echinodorus cordifolius* Griseb.). Ferrarese (1993), in an extensive listing of rice-field chironomids from Italy, reported *Cricotopus sylvestris* and *Chironomus* sp. as dominant.

Another commonly cited example of a chironomid–macrophyte interaction involves *Cricotopus (Isocladius) myriophylli* and Eurasian water milfoil, *Myriophyllum spicatum* L. (Kangasniemi and Oliver, 1983; Oliver, 1984). All larval activities, including case construction and pupation, occur on the macrophyte. Larvae feed exclusively on apical buds and the presence of even a single larva can affect shoot growth and flower emergence resulting in reductions in *M. spicatum* populations (Kangasniemi and Oliver, 1983; Macrae *et al.*, 1989; Macrae *et al.*, 1990). Macrae *et al.* (1989) tested the effect of *C. myriophylli* on twelve macrophytes (not including *M. spicatum*) and found that only *Myriophyllum exalbescens* (Fern.) Jeps. was fed upon. This suggested that *C. myriophylli* might serve as a relatively specific biocontrol for *M. spicatum*. Further experiments on non-target macrophytes are needed, especially given that *C. sylvestris*, a known feeder on rice plants, is closely related to *C. myriophylli*. *Endochironomus subtendens* has also been documented to feed on *M. spicatum* but damage to macrophytes was localized and had little effect on plant growth (Kangasniemi and Oliver, 1983).

7.3.4 WOODY DEBRIS

Studies examining the ingestion of woody material by chironomids and other aquatic invertebrates have focused on lotic systems because wood represents a large proportion of the total organic material in streams. Wood also plays an important role in affecting stream biota through its influence on energy flow, nutrient dynamics and stream morphology (Anderson *et al.*, 1978; Anderson and Sedell, 1979). Although many lake littoral zones also possess large quantities of woody debris, these habitats have been little studied.

When discussing the association between larval chironomids and woody debris, it is important to distinguish organisms that feed directly on wood (true xylophages) from those that use wood primarily as a substratum for attachment or for grazing attached periphyton (surface associates) (Pereira *et al.*, 1982; Cranston and Oliver, 1988b; Anderson, 1989). Many of the studies cited in this section discuss both aspects of chironomid–wood associations; however, the following discussion will focus on the actual ingestion of wood and its role as a food source for chironomids.

Most studies of wood-associated chironomids support the general conclusion that relatively few midges can be considered to be true xylophages (Table 7.6) (Dudley and Anderson, 1982; Kaufman and King, 1987). The low incidence of true xylophagy reflects the poor nutritional value of wood (Anderson *et al.*, 1978; Anderson and Sedell, 1979; Cummins and Klug, 1979). Woody material is highly refractory (80% lignin and cellulose) with very high C:N ratios of 300–1300:1 (Anderson *et al.*, 1978; Anderson and Cummins, 1979; Anderson, 1982; Pereira *et al.*, 1982). It might be expected that microorganisms play an important nutritional role for xylophages; however, as with detritivorous chironomids, it often is difficult to distinguish whether nutrition is derived from wood or from associated microorganisms (Pereira *et al.*, 1982).

Because of the low nutritional value of wood, assimilation efficiencies of aquatic xylophages are believed to be low. It has been suggested that a symbiotic gut flora might be present to enhance nitrogen fixation and cellulose digestion (Anderson and Sedell, 1979; Cranston, 1982a), but studies demonstrating the presence of gut symbionts in chironomids are lacking. Kaufman *et al.* (1986) reported a localized area of ectoperitrophic bacteria associated with the midgut of *Xylotopus par* Coquillet but, because of its location outside the peritrophic membrane, these bacteria were not believed to play a role in the digestion of wood.

In addition to the problem of distinguishing true xylophages from surface associates, the lack of taxonomic studies and the general practice of grouping all detritus into a single category has obscured the prevalence and importance

Table 7.6 Studies reporting woody debris to be the dominant food source for chironomids. The dominance of woody debris was based upon the following criteria of importance: GC = gut contents, IN = inferred, OB = observational. NR = type of wood not reported

Taxa	*Location*	*Dominant food*	*Criterion of importance*	*Reference*
Orthocladiinae				
Brillia sp.	Coniferous forests, Oregon	Douglas fir	IN	Anderson *et al.*, 1978
Brillia sp.	Western US streams	NR	GC	Dudley and Anderson, 1982
Brillia sp.	W. Oregon streams	NR	GC	Pereira *et al.*, 1982
Eukiefferiella sp.	Western US streams	NR	GC	Dudley and Anderson, 1982
Paraphaenocladius sp.	Western US streams	NR	GC	Dudley and Anderson, 1982
Chaetocladius ligni Cranston & Oliver (as n. gen. nr. *Limnophyes* or *Tokunagaia*)	Western US streams	NR	GC	Dudley and Anderson, 1982
Orthocladius lignicola as *acutilabis* K. Kieffer	Western US streams	NR	GC	Dudley and Anderson, 1982
Chironominae				
Glyptotendipes. sp.	Western US streams	NR	GC	Dudley and Anderson, 1982
Harrisius pallidus Freeman	New Zealand streams	NR	GC	Anderson, 1982
Phaenopsectra sp.	Western US streams	NR	GC	Dudley and Anderson, 1982
Polypedilum (*Pentapedilum*) sp.	Western US streams	NR	GC	Dudley and Anderson, 1982
Polypedilum (*Polypedilum*) sp.	Western US streams	NR	GC	Dudley and Anderson, 1982
Stenochironomus (*Stenochironomus*) sp.	Widespread	Angiosperms	GC, OB	Borkent, 1984
Stenochironomus sp.	Western US streams	Angiosperms	GC, OB	Dudley and Anderson, 1982
Stenochironomus sp.	W. Orgeon streams	Angiosperms	GC, OB	Pereira *et al.*, 1982

of chironomid–wood associations (Cranston and Oliver, 1988b). Pereira *et al.* (1982) conducted one of the first studies that distinguished between wood, fungal components, and amorphous detritus in invertebrate alimentary tracts. They found that of the 27 chironomids examined, only the wood-boring or mining *Brillia*, *Stenochironomus* and two undescribed Orthocladiinae were true xylophages. Similarly, Dudley and Anderson (1982) also reported that *Brillia* and *Stenochironomus* ingested wood in western USA streams. In one of the few studies examining wood-associated invertebrates outside North America, Anderson (1982) reported two xylophagous species of chironomids from New Zealand streams: *Harrisius pallidus* Freeman and an undescribed Orthocladiinae. Most of the dominant xylophagous taxa reported by Anderson et al. (1978) in Oregon have functional analogues within the New Zealand fauna. For example, *H. pallidus* and the unidentified Orthocladiinae were considered analogous to *Brillia* and *Stenochironomus*.

Cranston and Oliver (1988b) identified 10 genera as being either true xylophagous taxa or having a distinct association with wood. Other than the association of taxa within the *Stenochironomus* complex, they could find no phylogenetic relationships between genera. They did note, however, that wood-mining taxa shared some morphological characteristics that included a relatively flimsy body cuticle, elongate and heavily sclerotized ventromental teeth (within the Orthocladiinae), broader and shorter mandibles and a narrowed anterior labrum that folds beneath the frontoclypeus. In addition, Anderson (1982) observed morphological similarities between *Harrisius* and *Stenochironomus*, with both possessing haemoglobin and a flattened thorax. It has also been suggested that aquatic insects that exploit wood as a food resource possess life history characteristics adapted to the poor nutritional quality of wood (Anderson and Cummins, 1979). These characteristics include:

- Long life cycles with low metabolic activity.
- High ingestion rates to obtain sufficient amounts of digestible carbon and nitrogen from the wood and associated microorganisms.
- Dietary supplements of higher quality food such as periphyton.
- Some combination of the above characteristics.

Because *Xylotopus par*, an obligate xylophage, lacks some of these characteristics, morphology and phenology may not be accurate indicators of a xylophagous habit (Kaufman and King, 1987; Cranston and Oliver, 1988b).

Borkent (1984) provided one of the few detailed descriptions of the feeding mechanism of a xylophagous chironomid. The feeding activity of *Stenochironomus unictus* Townes begins with the mandibles opening (abducting) so that the inner surface of each mandible is parallel. Through a series of head extensions, labiohypopharynx withdrawal and labrum extensions, the labral structures expand outwards. As the mandibles adduct, the labiohypopharynx moves anteriorly and the labrum posteriorly. The wood particle is passed rapidly from the oral cavity to a region just posterior to the anterior prolegs. These feeding actions ensure that only small pieces of wood are ingested.

7.3.5 ANIMAL MATTER

Larval chironomids feed on a variety of invertebrates and most commonly ingest other aquatic insects (including other chironomids), oligochaetes, and zooplankton (Table 7.7). The incidence of predation by chironomids,

Table 7.7 Studies reporting animal matter to be the dominant food source for chironomids. The dominance of animal matter was based upon the following criteria of importance: GC = gut contents, OB = observational, EN = energetic criteria

Taxa	*Location*	*Dominant food*	*Criterion of importance*	*Reference*
Tanypodinae				
Ablabesmyia sp.	unnamed pond and streams	Oligochaetes	GC	Loden, 1974
'*Anatopynia*' sp.	unnamed pond	Oligochaetes	GC	Loden, 1974
Conchapelopia pallidula (Meigen)	Raba R., Poland	Orthoclads	GC	Kawecka and Kownacki, 1974
Conchapelopia sp.	unnamed streams	Oligochaetes	GC	Loden, 1974
Labrundinia sp.	unnamed pond	Oligochaetes	GC	Loden, 1974
'*Pentaneura*' sp.	unnamed stream	Oligochaetes	GC	Loden, 1974
Procladius imicola Kieffer (as *nigriventris*)	L. Kuusijärvi, Finland	faunal remains	GC	Armitage, 1968
P, signatus (Zetterstedt)	L. Kuusijärvi, Finland	faunal remains	GC	Armitage, 1968
Procladius. sp.	unnamed pond	Oligochaetes	GC	Loden, 1974
Thienemannimyia festiva (Meigen)	Lough Neagh, N. Ireland	Chironomids	GC	Tokeshi, 1991a
Thienemannimyia spp.	Buzzards Branch, Virginia	Chironomids	EN	Smith and Smock, 1992
Chironominae				
Chironomus attenuatus Walker	unnamed stream	Oligochaetes	GC	Loden, 1974
Cryptochironomus pararostratus gr.	L. Kuusijärvi, Finland	Oligochaetes	GC	Armitage, 1968
Cryptochironomous sp. cf. *redekei*	L. Kuusijärvi, Finland	Oligochaetes	GC	Armitage, 1968
Cryptochironomus sp.	unnamed stream	Oligochaetes	GC	Loden, 1974
Endochironomus sp.	unnamed pond	Oligochaetes	GC	Loden, 1974
Glyptotendipes sp.	unnamed pond	Oligochaetes	GC	Loden, 1974
Polypedilum sp.	unnamed pond and stream	Oligochaetes	GC	Loden, 1974
Tribelos sp.	unnamed pond	Oligochaetes	GC	Loden, 1974

especially Tanypodinae, can be substantially underestimated if the presence of animal material in the gut contents is used as the sole indicator of a carnivorous diet. This is primarily due to the ability of many tanypods to pierce and withdraw body fluids from their prey (Konstantinov, 1971b). The diet of a predator using this mode of feeding would not be evident from gut analyses (Armitage, 1968; Roback, 1977a; Baker and McLachlan, 1979).

Despite the vast amount of literature documenting the ingestion of invertebrates by chironomids, few midges can be considered obligate carnivores (Oliver, 1971; Pinder, 1986).

Most reports of chironomids feeding on animal matter have come from studies of Tanypodinae, and more specifically *Procladius*. The ingestion of crustacean zooplankton is frequently an important component of *Procladius* diets (Kajak *et al.*, 1968; Roback, 1969a; Tarwid, 1969; Kajak and Dusoge, 1970; Goulden, 1971; Baker and McLachlan, 1979; Dusoge, 1980; Kajak, 1980; Titmus and Badcock, 1981; Vodopich and Cowell, 1984); however, chironomids can also make a significant contribution (Kajak and Dusoge, 1970; Monakov, 1972; Roback, 1976; Vodopich and Cowell, 1984). These results are in contrast to other studies that have reported the importance of oligochaetes in the diets of Tanypodinae (Konstantinov, 1971b; Loden, 1974; Roback, 1976). Some of the conflicting reports of tanypod diets result from using numbers of prey items in gut contents as an indicator of importance. Tarwid (1969) found that, on a numerical basis, *Procladius* primarily preyed upon cladocerans, but on a biomass basis, chironomids were as important as cladocerans. In their study of macroinvertebrate predators in a south-eastern USA blackwater stream, Smith and Smock (1992) found that gut contents (on an areal basis) of *Thienemannimyia* spp. were 67–82% detritus and 18–33% animal material; however, only 27% of the secondary production was attributed to detritus and 73% was attributed to animal matter.

Results from other studies further support the observation that Tanypodinae diets generally consist of animal matter. Tokeshi (1991a) found that gut contents of *Thienemannimyia festiva* (Meigen) in Lough Neagh, Northern Ireland, were almost entirely of animal origin and that 91% of larval alimentary tracts contained chironomid prey, 14.4% contained harpacticoid copepods, 7.7% contained oligochaetes (primarily naidids) and <7% contained other crustaceans, caddisflies and ceratopogonids. On a numerical basis, 81.5% of prey items were chironomids. Kawecka and Kownacki (1974) reported that *Thienemannimyia geijskesi* (Goetghebuer) in the Olszowy Stream, Poland, exhibited a predatory diet that included chironomids (mainly *Eukiefferiella*), tardigrades, mayflies and rotifers. Similarly, they found that Orthocladiinae were a major dietary component for *Conchapelopia pallidula* (Meigen) in the Raba River, Poland.

Despite studies demonstrating the importance of animal material in the diets of *Procladius* and other Tanypodinae, most studies have reported an omnivorous diet consisting of aquatic invertebrates, detritus and algae (Armitage, 1968; Roback, 1969a; Izvekova, 1971; Mackey, 1977b; Baker and McLachlan, 1979; Mackey, 1979; Dusoge, 1980; Vodopich and Cowell, 1984; Hershey, 1986). Kajak (1980) also reported the omnivorous feeding habits of *Procladius* and other Tanypodinae and suggested that this flexibility in feeding may explain why this group often dominates areas with poor environmental conditions.

Although an omnivorous diet for *Procladius* and other tanypods may be the rule rather than the exception, the higher quality of food from animal sources may be important for larval growth (Konstantinov, 1971b). Vodopich and Cowell (1984) reported that *Procladius culiciformis* (Linnaeus) did not moult beyond the third instar when fed only algae and detritus. Similarly, growth of *Ablabesmyia monilis* (Linnaeus) decreased when denied animal food (Mackey, 1977b).

Although most instances of predation are attributed to Tanypodinae, taxa in other groups also are known to ingest animal material either occasionally or as a major component of the diet. *Cryptochironomus* is frequently reported to feed primarily on animal material (Armitage, 1968; Titmus and Badcock, 1981). In a study of the chironomid fauna of a shallow woodland lake in south Finland, Armitage (1968) found that only *Cryptochironomus* could be considered an obligate predator. The gut contents of *Cryptochironomus* contained oligochaete setae and chironomid remains, but much of the predation on chironomids was suggested to have occurred within the sample before sorting. When samples of *Cryptochironomus* gut contents were examined immediately after collection, only oligochaete remains were found.

Predation by Orthocladiinae is relatively uncommon but *Cardiocladius* larvae have been reported to feed on larval Simuliidae (Cranston *et al.*, 1983) and pupal Hydropsychidae (Parker and Voshell, 1979). Dendy (1973) described the predation of *Nanocladius alternantherae* Dendy and Sublette on eggs and larvae of other chironomids. The feeding mechanism involved turning the eggs lengthwise and swallowing the entire egg, or using the mandibles to break the egg and drawing out the contents. *N. alternantherae* fed on larvae by swallowing the entire prey body, by decapitating the prey and eating only the body, or by consuming the prey from the posterior end and discarding the head capsule. *N. alternantherae* was not, however, considered an obligate predator because it also was observed to ingest aufwuchs.

7.4 FOOD SELECTIVITY

Chironomids can ingest food either non-selectively on the basis of availability or selectively on the basis of size, nutritive value or food type. The occurrence of selective or non-selective feeding can vary considerably within a single species. The observation that conspecific taxa within the same stream feed selectively at some sites and non-selectively at others (Ward and Williams, 1986) supports the view that chironomid feeding preferences and selectivity are influenced greatly by local biotic and abiotic conditions (Vodopich and Cowell, 1984). Larvae also may exhibit ontogenetic changes from selective to non-selective feeding as they mature. Williams (1981) found that first instar larvae of several taxa in the River Frome selectively ingested diatoms to the virtual exclusion of detritus. As larvae matured, however, they exhibited less selectivity and consumed diatoms and detritus in proportion to their availability.

A requirement for using food availability as a criterion for non-selective feeding is an assessment of available food materials in the environment (Provost and Branch, 1959; Davies, 1975: Ward and Williams, 1986; Ali, 1990; Tokeshi, 1991a; Botts and Cowell, 1992); chironomids that exhibit a distinct preference for a particular food (based on gut content analyses) may not be selectively feeding if the preferred food is the most abundant item available. Many studies have shown that gut contents of chironomids merely reflect food availability in the environment (Provost and Branch, 1959; Kajak and Pieczynski, 1966; Goulden, 1971; Biever, 1971; Monakov, 1972; Alfred, 1974; McLachlan and Cantrell, 1976; McLachlan, 1977b; Mackey, 1979; Moore, 1979a; Dusoge, 1980; Ali, 1989, 1990). For example, diet variability of *Procladius choreus* depended on the accessibility of prey items (Kajak and Dusoge, 1970). Dusoge (1980) reported that, during the summer, *Procladius* fed primarily on numerically dominant entomostracans even though chironomids were the preferred prey (Luferov, 1958, as cited in Kajak and Dusoge, 1970); *Procladius* diets merely reflected prey availability.

It also is important to distinguish between 'actual' and 'realized' food availability. Actual availability is a measure of the presence of the food item in the environment and is usually measured by sampling the habitat. Realized availability is a measure of availability from the perspective or scale of the chironomid, not the researcher. This distinction is necessary to avoid erroneous conclusions of selective feeding in instances where the most commonly ingested food is not the most abundant item in sample collections. Hershey (1986) found that *Procladius* consumed a higher proportion of Orthocladiinae and a lower proportion of Chironomini than occurred in the habitat. Such an observation would indicate selective feeding. However, because orthoclads grazed outside their tubes, they were more susceptible and probably more apparent to *Procladius* than tube-dwelling, filter-feeding Chironomini that remained inside their retreats. The realized availability of Orthocladiinae prey resulted in increased consumption, not an active selection of Orthocladiinae over Chironomini. Similarly, *Procladius* preyed more effectively on collector-gatherer Chironomini (first instar *Chironomus*) than on filter-feeder Tanytarsini of the same size (second instar *Paratanytarsus*). These results demonstrate that the feeding mode of prey can be important in influencing predator behaviour (Hershey and Dodson, 1985; Hershey, 1986).

Other aspects of prey behaviour also affect susceptibility to chironomid predation (Loden, 1974). Baker and McLachlan (1979) suggested that the preferential ingestion of tubificids over chironomids by a tanypod predator was due to the lack of resistance to attack. Crawling activities of ostracods resulted in increased predation compared with less susceptible, swimming cladocerans.

Selection of food on the basis of size is primarily a function of larval body size. Large chironomids generally ingest larger food items than small larvae (Jónasson and Kristiansen, 1967; McLachlan *et al.*, 1978; Williams, 1981). Thus, size-selective feeding is more of a constraint on small larvae and explains the common observation that large, late-instar chironomids often exhibit more flexible and less restrictive diets than smaller, early instars (Hilsenhoff, 1966; Armitage, 1968; Shilova, 1968, as cited in Monakov, 1972; Coffman *et al.*, 1971; Brennan *et al.*, 1978; Baker and McLachlan, 1979).

Size-selective feeding is not exclusively associated with one group of chironomids or with a particular feeding mode (Jónasson and Kristiansen,

1967; McLachlan *et al.*, 1978; Baker and McLachlan, 1979; Moore, 1979a; Tokeshi, 1991a; Botts and Cowell, 1992). Non-predaceous chironomids can appear to feed size-selectively because of limitations imposed by mesh sizes of catchnets (Provost and Branch, 1959) or by the size of feeding structures such as the mentum (McLachlan *et al.*, 1978; McLachlan, 1981a). In the River Frome, Williams (1981) observed that early instars of epiphytic chironomids preferred small diatoms (e.g. *Cocconeis*) to longer, chain-forming diatoms (e.g. *Diatoma* and *Melosira*), but Marker *et al.* (1986) found that chironomid larvae in experimental channels selected for large cells of *Synedra ulna* (Nitzsch) Ehr. although they were not particularly abundant. In contrast to the latter study, larval chironomids have been observed to consume *Synedra* either non-selectively (Williams, 1981) or not at all (Tokeshi, 1986a).

Despite the common view that filter-feeding chironomids are non-selective feeders, Konstantinov (1971b) observed that behavioural actions of larvae allow rejection of food particles of inappropriate type or size. Botts and Cowell (1992) showed that two epiphytic chironomids, *Psectrocladius* sp. and *Thienemanniella* cf. *fusca*, could exploit whatever material was available, but some algae (especially *Cosmarium*) were preferentially ingested because of their larger cell size. Even Tanypodinae that use a piercing/sucking feeding mode can display size-selective feeding because of difficulties in capturing and in resisting the attacks of large prey (Baker and McLachlan, 1979).

Among the few studies conducted, there is little consensus on the ability of chironomids to select food on the basis of nutritive value. Larvae have been reported to select sediments high in organic matter (Davies, 1975) or foods that elicit maximal growth rates (Konstantinov, 1971b; Baker and McLachlan, 1979). Kajak and Warda (1968) reported that Chironominae had higher proportions of algae and lower proportions of detritus in their gut contents than in the sediments. Similarly, *Chironomus plumosus* can selectively feed on bacteria and algae from surficial sediments (Johnson *et al.*, 1989). Small Tanypodinae were found by Armitage (1968) to select algae from the sediments to the exclusion of other food materials. In contrast, other studies found that larvae did not distinguish between foods of different nutritive value such as detritus and charcoal (McLachlan *et al.*, 1978) or detritus and sand (Mattingly *et al.*, 1981).

7.5 FEEDING-RELATED BEHAVIOURS

7.5.1 WOOD-MINING

Wood texture is an important determinant of the ability of xylophagous taxa to exploit wood (Dudley and Anderson, 1982; Cranston and Oliver, 1988b). Quality is influenced by the species of wood, extent of waterlogging and extent of decay (soundness) (Anderson and Sedell, 1979). Chironomid mining is generally restricted to the wood of angiosperms (Borkent, 1984; Anderson, 1989). Kaufman and King (1987) found that all larval stages of *Xylotopus par* in the Chippewa River, Michigan, were associated with a variety of wood including ash (*Fraxinus pennsylvanica* Marsh.), basswood

(*Tilia americana* L.), elm (*Ulmus*), birch (*Betula*), cottonwood (*Populus*) and occasionally white cedar (*Thuja*). Anderson (1989) found that densities of *Orthocladius lignicola* and *Chaetocladius ligni* Cranston and Oliver were higher on alder (*Alnus*) (maximum density 1749 m^{-2}) than on conifers (25 m^{-2}) and suggested that a similar pattern would be found for other riparian species such as maple (*Acer*), ash or cottonwood. Cranston and Oliver (1988b) noted that maple was a preferred substratum for many wood-mining taxa.

The ability of xylophagous chironomids to construct tunnels or galleries in wood is largely dependent on structural characteristics of the submerged wood (Anderson *et al.*, 1984). The depth of penetration into wood is restricted by oxygen gradients, which explains the predominance of galleries just under the wood surface (Dudley and Anderson, 1982). Many burrowing taxa such as *Stenochironomus* possess haemoglobin that enables them to bore deeper into environments with lower oxygen concentrations. Tunnelling activities of wood-mining larvae result in an extension of the oxygenated zone into the anoxic subsurface layers (Kaufman and King, 1987).

Colonization of wood by *X. par*, *Polypedilum*, *Glyptotendipes* and *Pagastia* is influenced by terrestrial fungal decay before submergence (Anderson *et al.*, 1984; Kaufman and King, 1987; Anderson, 1989). Other taxa, such as *O. lignicola*, *C. ligni* and *Brillia* are found just below the surface in sound wood where microbial processes play a major role in providing suitable habitat for larvae (Cranston, 1982a; Dudley and Anderson, 1982, 1989).

Most wood-mining *Stenochironomus* inhabit permanently submerged wood in lentic and lotic habitats (Borkent, 1984). Wood must be firm, free of surficial deposits, and securely anchored so that it is not susceptible to movement. In his monograph on the *Stenochironomus* complex, Borkent (1984) detailed the feeding-related behaviours associated with wood-mining. Larvae build extensive chambers (2–3× the larval body length) parallel to and just beneath the surface of the wood. Chambers are constructed with small openings to allow water circulation and frass removal. Larvae mine in a U-shaped posture with the posterior prolegs just posterolateral to the head capsule. The thorax is wedged between the upper and lower portions of the chamber and feeding occurs with the ventral surface of the larva toward the outer surface of the wood. This enables larvae to exploit the nutrient-rich cambium layer of the wood.

7.5.2 TUBE-BUILDING

One widely studied feeding-related behaviour of chironomids is tube-building. The ability of chironomids to construct tubes decreases the risk of predation by vertebrates and invertebrates (Walde and Davies, 1984; Dillon, 1985; Hershey, 1985a, 1987; Macchiusi and Baker, 1991) and may minimize their dislodgement by currents (Brennan, 1976).

Stream-dwelling chironomids use salivary secretions to bind detritus, diatoms, filamentous algae, moss, leaves or sand grains into relatively cylindrical retreats. Because the scouring of substrata during spates can remove accumulated material from rock surfaces in streams, flexibility in the use of tube-building materials is highly advantageous (Brennan, 1976). Thus, although larvae may be more efficient in handling some particles than others, the material used in tube construction is primarily determined by particle availability (Brennan and McLachlan, 1979).

Tube-building is initiated by first (Danks, 1971b; McLachlan, 1977a) or second (Hilsenhoff, 1966; Brennan, 1981) instar larvae with the mechanism of tube construction varying considerably depending on species. *Chironomus riparius* construct algal tubes using the mouthparts to gather bundles of algae (Edgar and Meadows, 1969). By applying salivary secretions along the edge of each piece, the larva can attach two pieces together. Once several large bundles have been constructed and transferred to the anterior prolegs, the larva crawls through the bundle to form a tunnel. This process continues for one to three hours until the tube is complete. In a similar manner, *Cricotopus bicinctus* also gathers tube materials with the mouthparts and accumulates the material around the posterior prolegs (LeSage and Harrison, 1980a). Larvae enter the accumulated material when the tube reaches the length of the larval body. Additional materials are then added for enlargement and support. Tubes of 1–2 cm in length can be constructed in 30 minutes, whereas a 3–5 cm tube requires 24 hours.

In addition to constructing their own tubes, larval *Cricotopus* have been reported to reuse abandoned tubes and associated territories, a behaviour termed **territory recycling** (Wiley and Warren, 1992). Using *in situ* time-lapse cinematography, Wiley and Warren (1992) found that territory recycling occurred almost twice as frequently as new tube construction and that as many as six larvae recycled a single tube and territory in less than a 6-hour period.

Rheotanytarsus muscicola Thienemann exhibits an elaborate mechanism of tube and net construction using three types of silk (Kullberg, 1988). The first type of silk, which is used in constructing the tube, is finely textured and is produced by actions of the ventromental and maxillary lamellae. Larvae construct an open-ended tube of detritus, plant fragments and mineral particles that is oriented parallel to the current. Material for tube construction can be captured in the catchnet or gathered from the surrounding substrate. A single, arm-like projection is added to the anterior end of the new tube with more arms being added as the larva grows and the tube is extended. In the process of tube lengthening, portions of older arms may be incorporated into the tube to form ridges that provide structural support. The anterior end of the tube is coated with silk strands and the arms are constructed from mineral particles covered by a smooth layer of silk. The inner surface of the tube is overlain with a sheet of salivary secretions applied with the prolegs.

The second type of silk produced by *R. muscicola* is composed of thicker strands and is generated by ejecting silk through the oral cavity. This silk forms the framework for the capture net. The final type of silk is used to form the net and is produced by forcing salivary secretions through grooves on the anterior edge of the ventromental plates. These strands are suspended between the arm-like extensions to capture materials used for food and tube construction.

Although most epilithic chironomids require the presence of tube-building materials, some lotic taxa can construct tubes without the use of particles. Brennan and McLachlan (1979) reported four taxa from the River North Tyne (*Orthocladius rivulorum* Kieffer, *O. thienemanni* Kieffer, *Synorthocladius semivirens* (Kieffer) and *Eukiefferiella clypeata* (Kieffer)) that could construct tubes entirely of salivary secretions. In a subsequent study, Brennan and McLachlan (1980) examined the differences in tube construction of several *Eukiefferiella/Tvetenia* species and described the variations in tube-building. In areas of reduced current velocities where detrital particles were available, *Tvetenia verralli* (Edwards) and *T. calvescens* (Edwards) (both cited as *Eukiefferiella*) constructed loose tubes that incorporated detritus. In contrast, *E. clypeata* was found under solid sheets of salivary secretions constructed over rock depressions or fissures. Because this species was found primarily on stone surfaces in rapid currents, there was little material available for tube construction.

Lentic chironomids can also construct a variety of tubes depending on the presence of leaf litter accumulations (Hodkinson and Williams, 1980), seasonal changes in food availability (Jónasson, 1972) and sediment depth (McLachlan and Cantrell, 1976). Three basic types of tubes have been recognized (McLachlan and Cantrell, 1976). In shallow sediments (3–10 mm) around lake margins, larvae construct open-ended horizontal tubes. In areas where sediment depth is greater (11–30 mm), such as in the centre of lakes, J-shaped and open-ended U-shaped tubes predominate. J-shaped tubes are sealed at the distal end and are considered to be young U-shaped tubes in the process of construction.

Different types of tubes impose a characteristic feeding mode (McLachlan, 1977b; Hodkinson and Williams, 1980). Horizontal and J-shaped tube-builders are deposit-feeders, although the former may also filter-feed (McLachlan, 1977b; Hodkinson and Williams, 1980). Larvae constructing U-shaped tubes feed primarily as filter-feeders. It has been suggested that the failure of some J-shaped tubes to progress to the U-shaped stage reduces competition for food because of differences in the food material ingested (McLachlan, 1977b; Rasmussen, 1984a,b).

7.6 EFFECTS OF FEEDING ON FOOD SUPPLY

Algal grazers are the most commonly reported group of chironomids causing reductions in their food resources. This effect is largely due to high larval densities, rapid growth rates and high feeding rates of grazers (Botts and Cowell, 1992). Brook (1954) was the first to note that grazing chironomids could have a marked effect on periphyton standing crop, a finding supported by subsequent studies of periphyton–chironomid interactions (Kajak and Warda, 1968; Mason and Bryant, 1975; Eichenberger and Schlatter, 1978; Kesler, 1981; Cattaneo, 1983; Hann, 1991; Wiley and Warren, 1992).

Overgrazing of periphyton by chironomids has been reported from both lotic and lentic habitats. Eichenberger and Schlatter (1978) used artificial channels to examine the effects of grazing by chironomids on benthic algal assemblages. At low grazer densities, chironomids had a positive effect on algae, but when grazing pressure was higher, overgrazing resulted in a reduction in algal standing crop. Similarly, Cattaneo (1983) observed a decline in epiphyte biomass (as chlorophyll a) that coincided with periods of maximum grazer biomass (chironomids and oligochaetes) and concluded that grazing was the major cause of the algal decline. The effects of chironomid–oligochaete assemblages on periphytic algae were examined also by Hann (1991), who found that high invertebrate densities resulted in overgrazing and a reduction in algal biomass, but algal biomass increased after chironomid densities were reduced.

Wiley and Warren (1992) reported a rapid depletion of periphyton resources within *Cricotopus* territories (i.e. the area within one larval body length of the tube) that resulted in the establishment of what they termed a **territory cycle**. This cycle is an alternating sequence of territory establishment (reusing abandoned tubes or new tube construction) and territory/tube abandonment events. Wiley and Warren (1992) observed that the average time from territory establishment to resource depletion and territory abandonment was 2.4 h (range 0.2–7.1 h). Because *Cricotopus* consumption rates exceeded periphyton renewal rates by a mean factor of 1.75 (range 1.02–3.76), it was suggested that the primary factor leading to territory abandonment was overgrazing.

To examine the potential impact of grazing by chironomids on algal biomass in chalk streams, Welton *et al.* (1991) determined gut retention times for chironomids in recirculating stream channels (32–69 minutes) and combined this information with chironomid density and algal standing stock data from studies of similar chalk streams (Marker, 1976; Pinder, 1980). Results of this analysis indicated that chironomids could consume up to 55.6% of available algal biomass per day. Based on this high maximum rate of daily consumption, Welton *et al.* (1991) suggested that chironomids may be responsible for the decline of the spring diatom bloom in chalk streams.

In addition to the effects on algal standing crop, the presence of grazing chironomids also can influence the taxonomic composition of epiphytic communities. In grazer enclosure experiments conducted in Lake Memphremagog, Cattaneo (1983) found that in treatment controls (i.e. no grazer manipulations) diatoms initially dominated the epiphytic community and were later replaced by blue-green algae. In grazer exclosures, however, diatoms remained dominant for a longer period relative to controls. In grazer enclosures (i.e. grazer additions), grazer biomass was 3× the biomass in the controls, resulting in the destruction of epiphyte cover, which was abundant in the controls. These results are consistent with those of Eichenberger and Schlatter (1978). In chironomid exclusion channels, they noted that assemblages of blue-green algae were well-developed after seven weeks. In channels with chironomids, however, grazing pressure prevented the establishment of blue-green algae.

Intense grazing pressure by chironomids, by affecting food availability, can indirectly affect larval densities. Batzer and Resh (1991) found that grazing chironomids in a seasonal wetland could reduce periphyton biomass to the point where chironomid populations were suppressed. Mason and Bryant (1975), however, found that periphyton reduction had little effect on chironomids because larvae were moving into the sediments in preparation for winter at the time that periphyton abundance was low.

Like grazing chironomids, the ability for predaceous chironomids to control prey populations has also been reported (Kajak and Dusoge, 1970; Naeem, 1988a). Loden (1974) observed an inverse relationship between densities of *Chironomus attenuatus* Walker and *Limnodrilus hoffmeisteri* Claparode. Gut content analyses revealed a large number of oligochaete setae, suggesting that chironomids were preying on *L. hoffmeisteri*. This is consistent with other studies that have noted the absence of tubificids at sites where *Chironomus* were present (Brinkhurst and Kennedy, 1965). Although the latter study suggested that this was likely due to competition, Loden (1974) concluded that the elimination of *L. hoffmeisteri* by *C. attenuatus* was due to predation.

The ability for chironomids to affect zooplankton populations has received little attention. However, Goulden (1971) presented several lines of evidence supporting his view that predation by Tanypodinae (*Larsia* spp., *Ablabesmyia* spp., *Procladius bellus* (Loew), and *Arctopelopia* sp.) could explain the decrease in chydorid populations in Lake Lacawac, Pennsylvania. Tanypod and chydorid densities displayed an oscillating predator–prey cycle with a decline in chydorid populations following high tanypod densities and an increase in chydorid populations following low tanypod densities. The numbers of chydorids in tanypod alimentary tracts were positively related to chydorid densities and reflected prey proportions in the field.

7.7 CONCLUSIONS AND RECOMMENDATIONS FOR FUTURE STUDIES

Much of the research conducted on chironomid feeding since the publication of Thienemann's monograph (1954) has demonstrated that most chironomids are omnivorous. Broad generalizations and *a priori* assumptions that closely related taxa feed in the same manner and on similar foods, however, must be used with caution. Such assumptions can result in obscuring the importance of chironomids and their role in energy flow in aquatic ecosystems. For example, assuming that all Tanypodinae are predaceous would substantially underestimate the energetic importance of other foods, such as algae or detritus. Also, observations that conspecific taxa can feed on different foods in response to local environmental conditions points to the necessity to examine material ingested by chironomids rather than rely on literature accounts.

One area that is still in need of research is the importance of detritus and associated microorganisms as food sources for chironomids. The use of staining techniques that allow for the distinction of different detrital components (e.g. DAPI; Merritt *et al.*, 1990) can provide information on the composition of ingested detritus. Because all detritus is not of the same nutritional value, it is also important to begin examining what detrital sources contribute most to chironomid diets. Because chironomids are important energetically in aquatic ecosystems (Benke *et al.*, 1984; Berg and

Hellenthal, 1993), defining their major sources of energy will improve our understanding of energy flow in these systems. A technique that has great potential for examining the source of detritus is the use of high performance liquid chromatography (HPLC). HPLC can be very effective at separating allochthonous and autochthonous fossil pigments of plants from lake sediments (Leavitt *et al.*, 1989) and may therefore provide a method to determine the source of detritus in chironomid diets rather than grouping all amorphous material as detritus.

Many studies of plant–herbivore interactions have neglected the role of chironomids as grazers. This is especially apparent in studies examining grazing on epiphytes and macrophytes. Because of the high densities often attained by chironomids, it is not surprising that some studies have shown that chironomids can substantially reduce plant and animal food resources. Further studies are needed to examine how commonly chironomids can reduce their own food supply to the point where it suppresses chironomid populations. Recent reviews have also suggested that grazing on macrophytes may be more common than the literature indicates. Except for a few studies on economically important macrophytes, chironomid–macrophyte interactions have been unexplored.

The availability of taxonomic keys to identify larval chironomids has resulted in the increased use of midges in addressing basic ecological questions. It has become apparent that chironomids can provide valuable insight into the functioning of aquatic ecosystems. As more information is gathered on the food and feeding-related behaviours of chironomids, the importance of chironomids as consumers in aquatic ecosystems will emerge.

8

The pupa and events leading to eclosion

P.H. Langton

8.1 INTRODUCTION

The pupal stage of the chironomid life cycle has attracted increasing attention since the beginning of the century, stimulated by comprehensive rearing programmes, predominantly by Thienemann and his students, notably Potthast and Bause. The systematic value of the pupal stage was appreciated early as an aid to species identification and, later, in providing additional or alternative characters to larva and adult in estimating phylogenetic relationships. As identification keys have become more comprehensive, pupal exuviae have been used as a simple and rapid way of obtaining ecological information on aquatic habitats, and data on autecology and geographic distribution. However, studies on pupal development, behaviour and physiology are still in their infancy and much of what is deduced is based on chance observations and circumstantial evidence. This is for the greater part due to the short duration of the instar and its rapid development, compounded by the generally opaque silk tube in which most of them live.

8.2 THE ROLE OF THE PUPA IN THE CHIRONOMID LIFE CYCLE

By the time the pupal stage has ecdysed from its larval cuticle, apolysis of the adult cuticle has already occurred. Thus, strictly speaking, it is now a pharate adult, but the term 'pupa' will be used here in its loose sense to refer also to the pupal cuticle containing a developing adult.

The chironomid pupal stage links two active phases of the insect's life – the larva and the adult – both of which are capable of responding to adverse stimuli by avoidance reactions. Whereas the pupal stage in many insects is immobile, most chironomid pupae are active for the greater part of their

The Chironomidae: Biology and ecology of non-biting midges
Edited by P.D. Armitage, P.S. Cranston and L.C.V. Pinder
Published in 1995 by Chapman & Hall. ISBN 0 412 45260 X

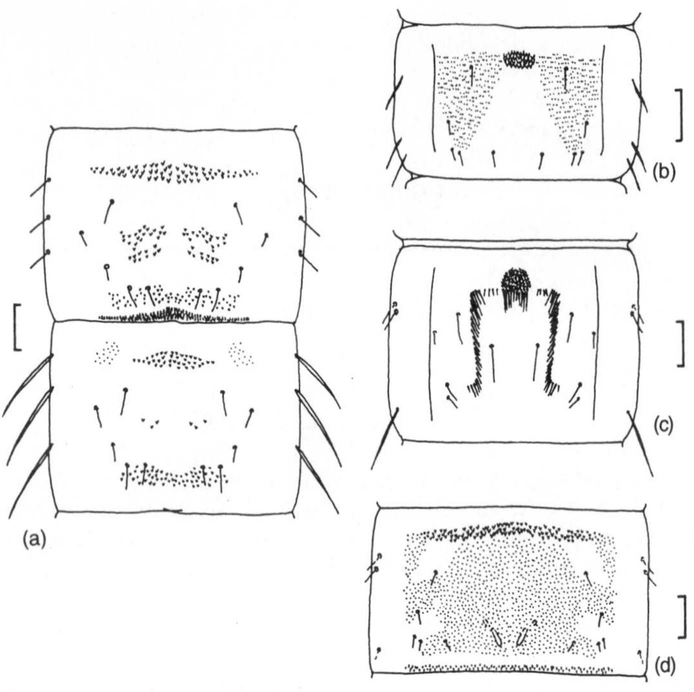

Figure 8.1 Pupal abdominal armament of (a) terga II and III of *Polypedilum (Tripodura) pullum* (Zetterstedt); (b) tergum IV of *Paratanytarsus intricatus* (Goetghebuer); (c) tergum IV of *Paratanytarsus bituberculatus* (Edwards); (d) tergum IV of *Polypedilum pedestre* (Meigen). (Scale line = 0.1mm.)

existence. However, this activity is rarely defensive: free-living tanypod pupae when disturbed at the water surface will swim rapidly to the substratum and there remain immobile for a while, but for most chironomids, their movements are limited to carrying out the three main pupal functions: ecdysis from the larval cuticle; providing sufficient respiratory oxygen; and moving to the water surface for adult eclosion.

8.3 MOULTING FROM THE LARVAL CUTICLE

Shedding the larval cuticle is rarely unsuccessful. The pupal cephalothorax is formed within the larval thoracic segments and its greater bulk already puts a strain on the dorsal suture. Undulatory movements of the abdomen engage the points on the pupal integument with the larval integument, driving the pupa forwards ratchet-like into the larval thorax. This extra pressure causes the suture to split from the back of the larval head to the first abdominal tergum. Continued undulatory movements ease the pupa out. Pupal ecdysis is a straightforward exercise for the free-living Podonominae and Tanypodinae: with the larval cuticle hooked firmly to the substratum by its parapods, the undulatory movements of the pupal abdomen will drive the pupa free of the larval exuviae. However, this is not so with the majority of species which live in the widened end of a long larval tube, or, more usually, within a specially constructed pupal tube, little longer than the pupa itself, with the narrowed openings at each end limiting pupal movement. In these, the larval cuticle has to be eased backwards along the body until it is shed, squashed concertina-fashion in the end of the pupal tube, where it may stay or be discharged. Sometimes the larval cuticle is only partly shed, with the head capsule remaining attached to the pupal abdomen. This does not seem to prevent further development or eclosion, for exuviae with an attached larval cuticle occur.

Structures to effect pupal ecdysis in tube-dwelling forms are primarily concentrated on the abdomen; the wide cephalothorax should slip out easily when pushed from behind, so that, apart from the antero-dorsal area which is immediately exposed when the suture splits, it is embellished only with rounded or flexible mounds, tubercles and/or granules. The abdominal tergites, however, are armed with fields of backwardly-directed, cuticular projections, which may be extensive and amalgamated or reduced to differing degrees. These projections vary in form and size from short squat 'points' to long narrow 'spines'. When these are very small they are collectively referred to as 'shagreen'. Where the larval body is little wider than the pupa within, the armament is of points (Figure 8.1d), whereas, narrow pupae formed within wider larvae require long spines to provide an effective ratchet (Figure 8.lc) (Langton, 1989).

On each tergite there are three primary fields of posteriorly directed cuticular projections: an anterior transverse band, a median quadrate patch and a posterior transverse band. The taxonomic value of variations in the extent and armament of these fields have been described in section 2.4.3. The plethora of patterns so generated defies belief that, in detail, each is a direct response to selection pressure to produce the most efficient ecdysial ratchet for that species.

In addition, on some segments there is an apical transverse band just posterior to the posterior band, the points, hooks or spinules of which are directed forwards (Figure 8.la). Between these two bands is a narrow strip of thin cuticle that allows the apical band to be tucked under the posterior margin of the tergite to prevent its armament from opposing the forward movement generated by the rest. Its function will be discussed in section 8.4.

8.4 STRUCTURES ASSOCIATED WITH MAINTENANCE OF THE DEVELOPING ADULT

Although apolysis of the imaginal cuticle has already taken place by the time the pupa sheds its larval skin, differentiation of internal organs is as yet incomplete. It is rare that the free pupal state lasts more than 72 hours;

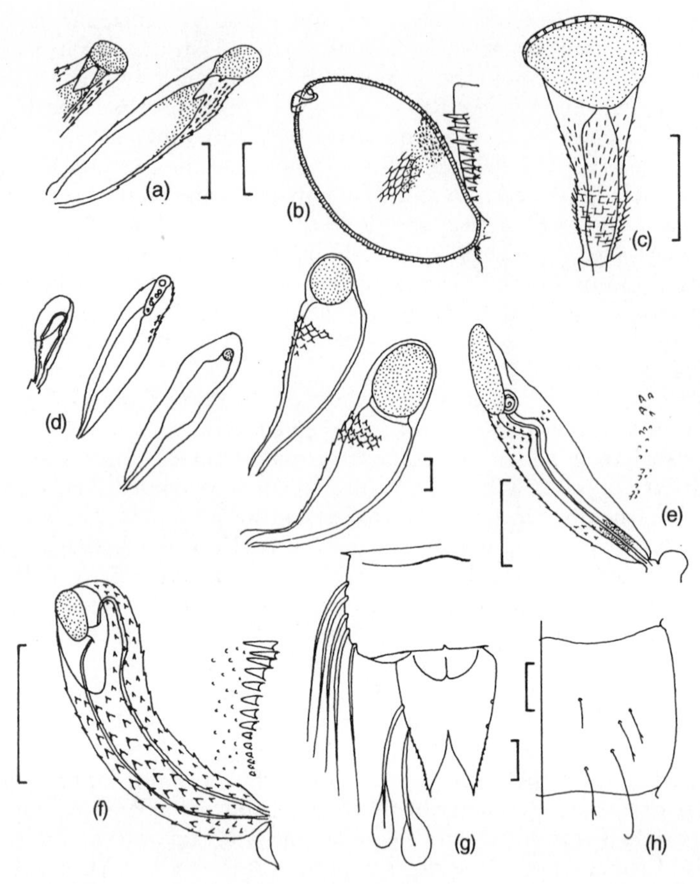

Figure 8.2 Thoracic horns of (a) *Zavrelimyia* sp.; (b) *Ablabesmyia longistyla* Fittkau; (c) *Boreochlus thienemanni* Edwards; (d) *Macropelopia fehlmanni* (Kieffer); (e) *Xenopelopia falcigera* (Kieffer); (f) *Paramerina* sp.; (g) anal segments of *Paramerina* sp.; (h) right half of tergum IV of *Monopelopia tenuicalcar* (Kieffer). (Scale line = 0.1mm.)

development of the adult is rapid and requires a good oxygen supply. As some pupae lack any specialized oxygen absorbing structures, they must be able to obtain the oxygen they need by diffusion through the cuticle, but these are species which live in relatively oxygen-rich habitats. However, most pupae possess structures that enhance the absorption of oxygen, the more universal of which is the thoracic horn.

The most elaborate thoracic horns represent the most primitive condition, seen in the subfamilies Podonominae and Tanypodinae. These free-living pupae have open-ended thoracic horns which are applied to the water/air interface to replenish the oxygen supply directly from the air (Figure 8.2c,f). Some Podonomine pupae are relatively inactive and remain at the water surface, whereas those of the Tanypodinae swim in the manner of mosquito

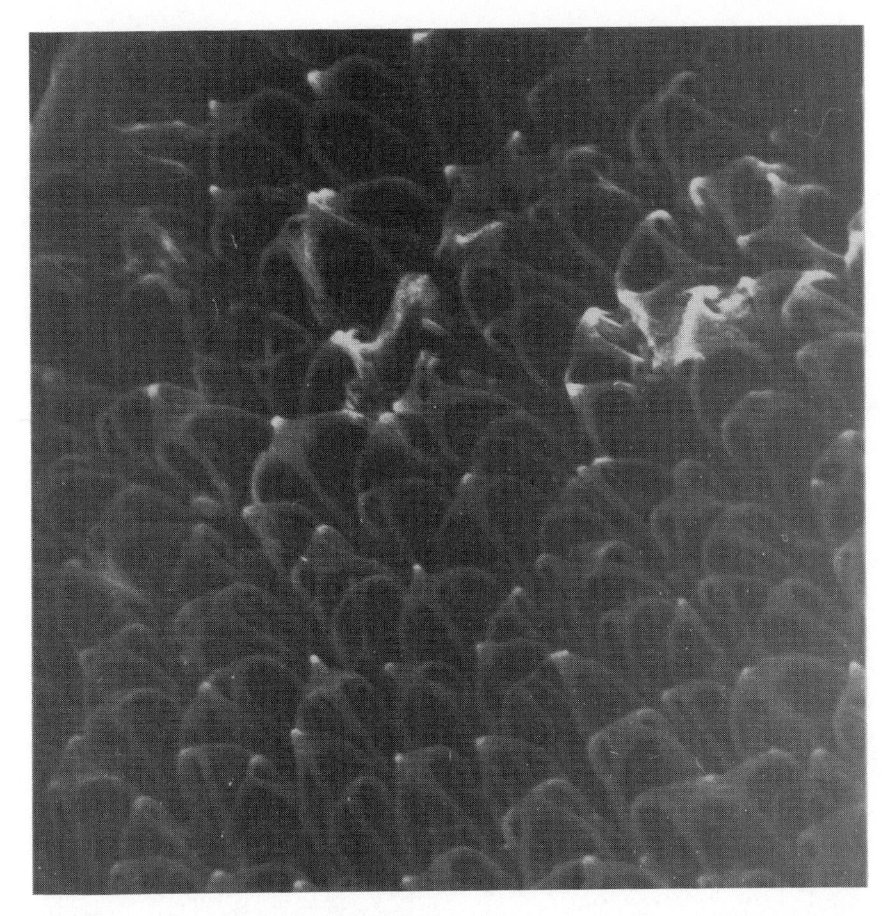

Figure 8.3 Electromicrograph of plastron plate surface of thoracic horn of *Procladius choreus* (Meigen). (Scale line = 1 μm.)

pupae and rise to the surface from time to time. However, in well-oxygenated waters they may never do so, for the plastron plate which fills the opening acts as an efficient physical gill.

The plastron plate is composed of minute hydrophobic cuticular hoops (Figure 8.3) which will break through the water surface to allow continuity of the air within the horn and the atmosphere, or, when submerged, will support a free water surface from which oxygen can diffuse into the horn. However, the plastron plate is frequently reduced and may be missing altogether; in one species, *Macropelopia fehlmanni* Kieffer, variation ranges from a well-developed plastron plate to complete absence (Figure 8.2d) (Langton, 1984). Species without this structure remain submerged until eclosion and oxygen must diffuse in all over the horn surface.

From the plastron plate a tube, the aeropyle, of varying length according to the species, passes to the horn atrium, which has a very thin outer and inner

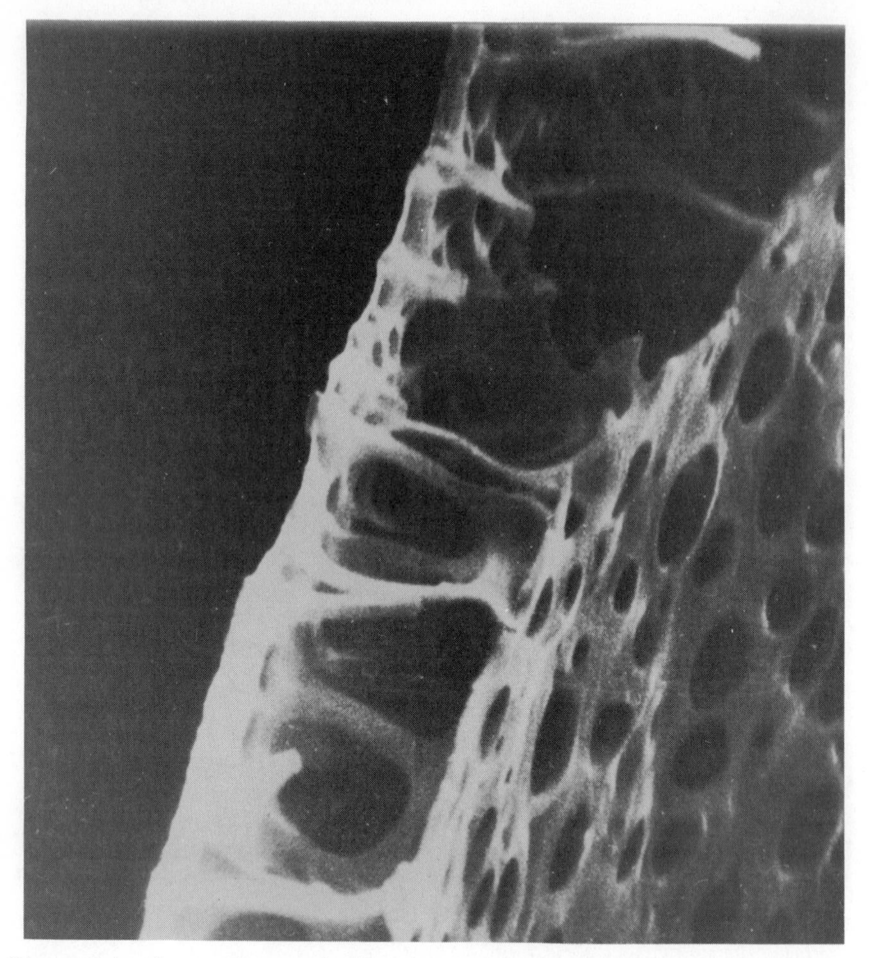

Figure 8.4 Electromicrograph of atrial wall of thoracic horn of *Ablabesmyia monilis* (Linnaeus). (Scale line = 1 μm.)

wall joined by dense simple or branched struts (Figure 8.4). The outer and inner walls are perforated by minute pores. Its plastron-like structure would be exposed if the horn were damaged, presumably thereby enabling it to maintain its respiratory function.

In species where the plastron plate is absent, the horn is frequently almost spherically swollen and the outer wall of the atrium is fused to the wall of the horn (Figure 8.2b), providing additional circumstantial evidence that oxygen is absorbed over the whole surface. Furthermore, the external surface of the horn is extended by means of hollow triangular projections (Figure 8.5). The question then arises: why is it that many tanypodines with a plastron plate have the horn wider, sometimes considerably wider, than the atrium? Buoyancy, required for rising to the surface to breathe, is provided by the

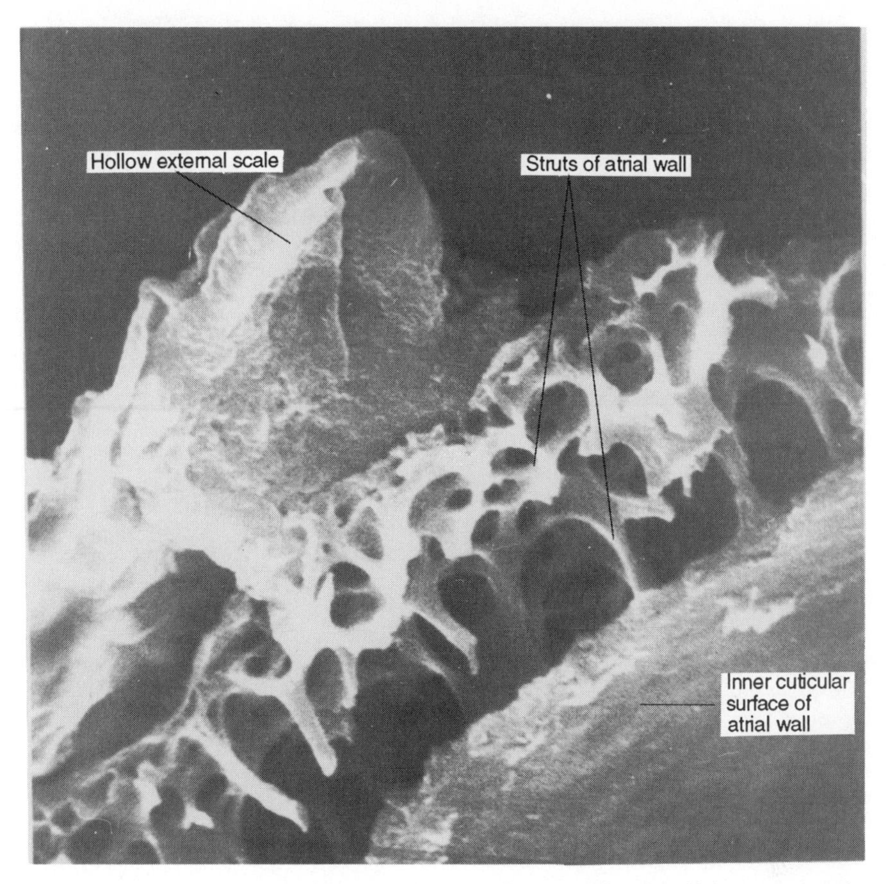

Figure 8.5 Electromicrograph of thoracic horn wall of *Ablabesmyia monilis* (Linnaeus). (Scale line = 1 μm.)

very structures that must contact the surface first, hence the large horn, but the atrium is a physiologically expensive structure to produce and so it is formed narrower than the horn. On the whole, the thicker the atrial wall is, the narrower the atrial diameter (Figure 8.2e).

The buoyancy provided by the horn is also a liability: without expenditure of energy the pupa will tend to rise to the surface, where it is easily seen and invites predation from above and below. In just over half the Holarctic genera of the Pentaneurini, the anal macrosetae are coated with a jelly-like glue of unknown origin (Figure 8.2g). This provides sufficient adhesion to tether the pupa to the substratum, but will come adrift in response to a sudden abdominal flexion. *Monopelopia* is a genus of the Pentaneurini which does not possess adhesive macrosetae, but unusually for the Pentaneurini (and in common with the Anatopyniini and Macropelopiini) it possesses hooked setae on the tergites which serve to anchor the pupa in algae or the divided

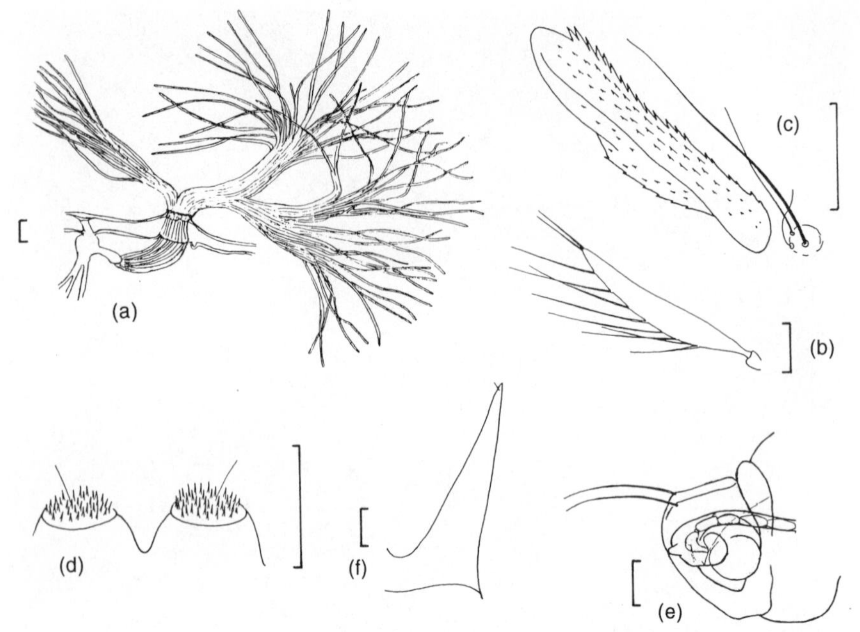

Figure 8.6 (a) Thoracic horn and tracheolar connection to imaginal tracheal system of *Chironomus annularis* (De Geer); (b) thoracic horn of *Cladotanytarsus nigrovittatus* Goetghebuer; (c) thoracic horn of *Zalutschia humphresiae* Dowling; (d) cephalic tubercles and frontal setae of *Phaenopsectra* sp.; (e) head and frontal setae of *Cricotopus (Isocladius) sylvestrix* (Fabricius); (f) cephalic tubercles and frontal setae of *Cryptochironomus redekei* (Kruseman) (Scale line = 0.1mm.)

leaves of submerged hydrophytes (Figure 8.2h). In *Guttipelopia*, a pentaneurine with neither of these adaptations, there is a strong thorax comb, a character it shares with many other genera of its tribe. This is a series of conical or finger-shaped 'teeth' that form a row from one thoracic horn to the other over the top of the thorax. These teeth are generally longest dorsally and shortest next to the horns (Figure 8.2f). The comb may serve to hold the pupa in submerged vegetation. *Clinotanypus*, of the Coelotanypodini, has hooked setae in its anal fringe. When the thoracic horns are closed and narrow, as in *Arctopelopia*, *Rheopelopia* and *Thienemannimyia* of the Pentaneurini, none of these adaptations occur, so providing circumstantial confirmation that pupae with buoyant horns benefit from having anchoring devices below the surface.

Many chironomid pupae are enclosed in a silken tube open at both ends. These pupae under normal circumstances only leave the tube to rise to the surface for eclosion. All the oxygen they require is obtained from the water. Their thoracic horns are simple or branched haemolymph-filled tubules, which in the Chironomini have tracheoles passing into and along them.

These tracheoles are secondary invasions of the thoracic horn, but nevertheless are in continuity with the pharate adult thoracic tracheal system (Figure 8.6a) (Miall and Hammond, 1900; Coffman, 1979). There can be little doubt that they all function as gills, whether they contain tracheoles or not. Additional circumstantial evidence for this function is derived from the habitat and associated thoracic horn. The Chironomini, particularly the genus *Chironomus*, contains species which are known to be most tolerant of anoxic conditions and these are forms with highly plumose thoracic horns (Figure 8.6a). The Orthocladiinae and Tanytarsini are to be found on vegetation or near the water surface; they have unbranched thoracic horns (Figure 8.6b,c). In those species with no direct connection to the pharate adult tracheal system, it is possible that the oxygen diffuses from the base of the horn to the spiracle via the moulting fluid between the two cuticles.

Respiratory movements of the pupa drive water through the tube. These are more sustained when the oxygen level drops below that of the preferred habitat; species which normally develop in oxygen-rich habitats show increased activity at a higher oxygen concentration than do those which normally live in hypoxic conditions. This behaviour mirrors that observed by Heinis and Crommentuijn (1992) concerning larval respiratory movements. Species living in fast-flowing waters construct their tubes in line with the current so that the water flows straight through – their only activity is to escape the tube for eclosion. Those that live in stagnant water or less oxygenated, flowing water perform undulatory movements of the abdomen to drive water through the tube. To enable this activity to be performed without expelling the pupa from its tube, the body is suspended from the tube by a pad, band or row of anteriorly directed hooks in the apical band of tergite II (Figure 8.7) (Lenz, 1951). As the abdomen is wafted up and down, the pupa rocks on this pivot. At the same level of the body, in many tube-dwelling chironomid pupae conical or bulbous lateral swellings protrude, the pedes spurii. Humphries (1937) first suggested that these were used as pivots for respiratory movements; this is supported by their extreme development in species with reduced hook rows, e.g. *Parametriocnemus stylatus* (Kieffer) (Figure 8.8a). In some species the apical bands of more than one segment have enlarged hooks; for instance, in the genus *Eukiefferiella* there may be hook rows on tergites III–V and in *Clunio* they are to be found on tergites III–VII or III–VIII. Attachment by a series of segments would hamper the respiratory movement, so there must be another function here. *Eukiefferiella* is characteristic of oxygen-rich, fast-flowing water and *Clunio* of marine intertidal habitats, where a firm attachment is necessary to prevent eviction by forceful water movements. Similarly, in those species with pedes spurii B on more than one segment (as in another coastal species, *Halocladius fucicola* (Edwards), where they occur on segments I–III and are covered with sharp points apically) the function is most probably to prevent the pupa from being dislodged from its tube.

Whereas in the active tanypodine pupae expanded anal lobes and fringed abdomen are adaptations for swimming, this cannot be true for the fringed anal lobes of tube-dwelling species. Their only excursion from the tube is to rise to the surface for eclosion, which is primarily effected by increased

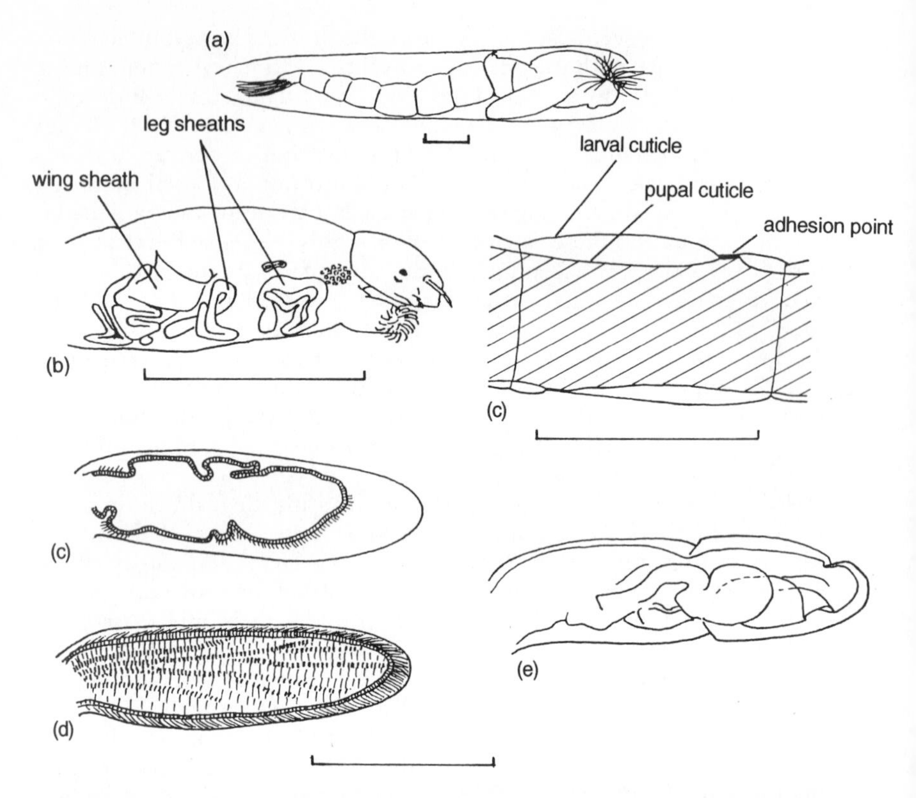

Figure 8.7 (a) Pupa of *Chironomus* in pupal tube; (b) anterior part of late fourth instar *Chironomus* larva showing position of developing pupal wing and leg sheaths. Pupal wing sheaths of *Chironomus*: (c) early development of imaginal wing bud; (d) fully developed imaginal wing; (e) expanded wing prior to eclosion. (Scale line = 0.1mm.)

buoyancy at this time. The extent of the fringe of broad, flattened, hollow setae and the length of these setae reflects the habitats in which the species develops. The lower the oxygen concentration in the habitat, the more extensive is the fringe. The fringe in these forms has developed to make the respiratory movements more effective in driving water through the tube. In some genera, e.g. *Thienemanniella* and *Corynoneura*, the long fringe appears to compensate for the lost thoracic horn.

The greatest development of divided thoracic horn and anal fringe occurs in the mud-dwelling *Chironomus* species (Figure 8.8b). In addition, they usually possess two pairs of tubular projections on sternite VIII, similar to those found on many larvae in that genus. Here the pupal tube is formed within the surface of the mud. If the oxygen level falls below that capable of sustaining the pupa at maximum respiratory activity, the cephalothorax is

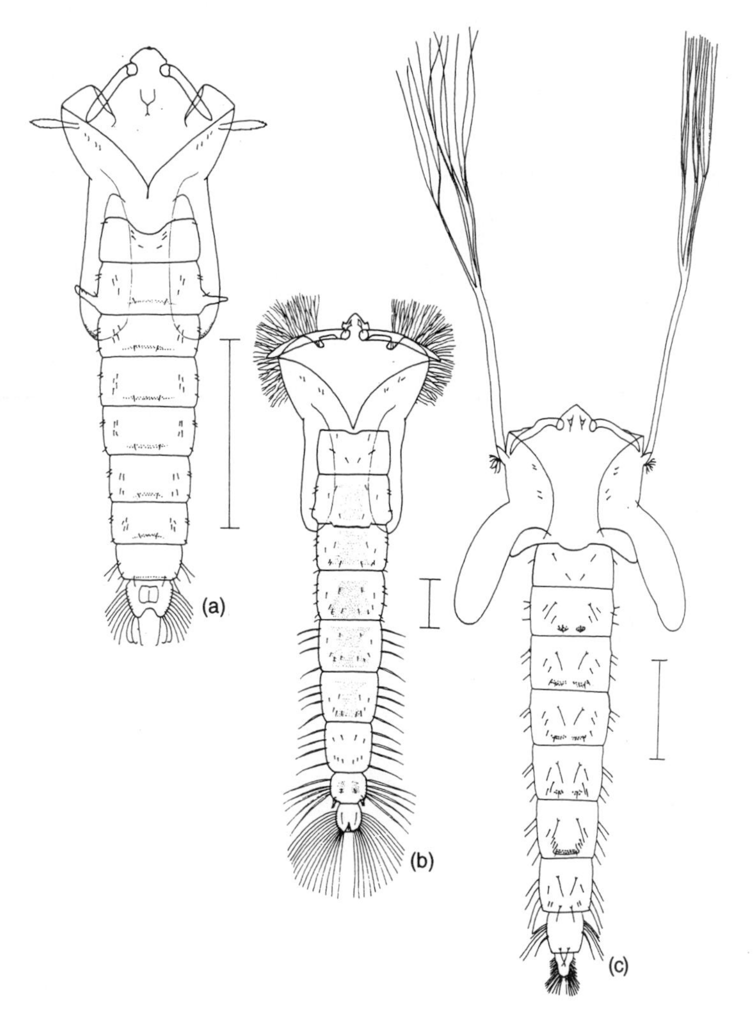

Figure 8.8 Pupal exuviae of (a) *Parametriocnemus stylatus* (Kieffer); (b) *Chironomus* sp.; (c) *Cryptotendipes holsatus* Lenz. (Scale line = 0.1mm.)

pushed out of the tube and the hook row is engaged near its mouth. Undulatory movements of the abdomen now also wag the cephalothorax with its bushy thoracic horns in the water. In extreme oxygen depletion the pupae may vacate their tubes completely and lie on one side on the mud surface, undulating the body freely in the water.

The sand-dwelling pupae of *Cryptotendipes* are interesting for their reduced anal lobe fringes and exceedingly long anterior branch of the thoracic horns – over half the total length of the pupa, rigid and usually unbranched for half their length (Figure 8.8c). *Cryptotendipes* pupae lie in tubes constructed vertically in the substratum; abdominal movements cause the projecting

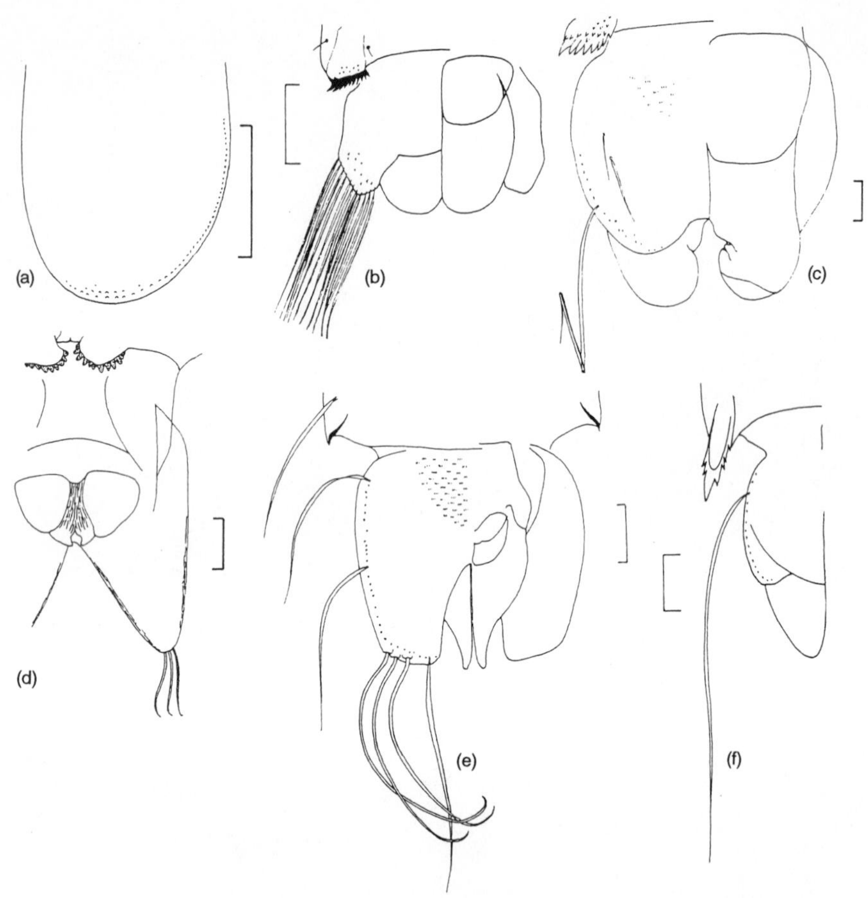

Figure 8.9 (a) Apex of wing sheath showing pearl row of *Corynoneura edwardsi* Brundin. Anal segments of: (b) *Neozavrelia* sp.; (c) *Tanytarsus niger* Anderson; (d) *Euryhapsis* sp.; (e) *Zalutschia humphresiae* Dowling; (f) *Polypedilum arundinetum* Goetghebuer. (Scale line = 1 μm.)

anterior branches of their thoracic horns to waft about in the water (F. Reiss, *pers. comm.*).

It is very rare that tube-dwelling species attached to objects above the mud surface leave their tubes: anoxic conditions reduce predation, but free pupae at higher oxygen levels are easy prey. These pupae become progressively more active as the oxygen level falls, intermittently unhooking themselves from the tube and turning round in order to reverse the flow of water. Despite the apparent difficulty of this manoeuvre, it is effected with remarkable speed. The hook row is on the apical part of the tergite and so can be disengaged by being tucked under the posterior edge of the tergite simply by drawing the next segment forwards. The apical bands of the next two or three segments, as mentioned previously, are composed of anteriorly directed spinules or points, which oppose the posterior points of the tergite. They can be used to grip the tube temporarily as the hook row is disengaged. Undulatory movements now of a higher pitch will engage various toothed structures on the postero-lateral corners of the abdominal segments to effect locomotion: the spurs of *Chironomus* and many *Polypedilum* species (Figure 8.9f), the postero-lateral 'comb' on segment VIII of most Tanytarsini (Figure 8.9b,c) and many Chironomini, the strengthened postero-lateral corners of the posterior segments of the orthocladiine genera *Zalutschia* (Figure 8.9e) and *Heterotrissocladius*, the ventro-lateral pedes spurii A (swellings armed with spines or points), and the strongly armed posterior margin of sternite VIII in some Orthocladiinae (e.g. *Euryhapsis* (Figure 8.9d) and *Heterotrissocladius*). It is possible also that the pearl row on the apices of wing sheaths, a row or band of granular projections, that occurs in some Orthocladiinae and Tanytarsini has the same function (Figure 8.9a).

The pupa receives information about its position in the larval cuticle, its position in the tube and the relative positions of the segments to each other through sensory setae. In most species all setae, except those of the fringe, are sensory. Each is set in a boss of cuticle and from its base passes a neural connection to the pharate adult nervous system. The severed neural connection can be seen in pupal exuviae attached to the inside of the boss. According to W.P.Coffman (*pers. comm.*) this is because the nerve fibre is surrounded by a thin cuticular sheath, which remains after ecdysis. As soon as these connections are severed, the characteristic undulatory movements of the pupal stage cease and the adult sensory system takes over.

The length of the sensory setae of tergites and sternites reflects in general the length of the cuticular armament. For instance, long spines on tanytarsine tergites are accompanied by long setae (Figure 8.1c), whereas, in the same tribe, tergites armed with short points are usually provided with shorter setae (Figure 8.1b). Flattened sensory setae may detect the flow of water past the pupa. The minute 'O' setae at the anterior edge of the tergites and sternites or on the intersegmental membranes, as suggested by Coffman (1979), are probably concerned with providing information of the position of one segment relative to another. The sensory needs of the pupa are minimal compared with those of the larva and adult, for the pupa is produced within a tube little larger than itself where it will remain until the final moments before eclosion. It has a limited number of simple behaviour patterns; a few tactile setae is all that it requires to provide the information that it needs.

The cephalic tubercles are hollow, usually surmounted by a sensory seta and located to sense signals arriving in front of the pupa. It is significant that species with long cephalic tubercles have short apical setae, e.g. *Cryptochironomus redekei* (Kruseman) (Figure 8.6f), whereas long frontal setae are associated with the absence of cephalic tubercles, e.g. some *Cricotopus* species (Figure 8.6e). Frequently the tubercles are armed with spinules, most highly developed in *Sergentia* and *Phaenopsectra* (Figure 8.6d). It is possible that these, along with spinous setae on tanytarsine thoracic horns, deter access to the tube by other organisms.

8.5 DEVELOPMENT OF THE PUPA AND IMAGO

The form of the pupa reflects much more closely that of the adult than the larva. The few places within the pupa that the pharate adult does not fill are the cephalic tubercles, frontal warts, thoracic horns, cephalothoracic mounds, pedes spurii B and anal lobes. However, the development of the external features of the imago begins in the fourth instar larva and has been described in detail for *Chironomus* by Miall and Hammond (1900). The antennae develop in imaginal grooves which run down the dorsal surface of the head epithelium from the larval antennae. The imaginal head, being larger than the larval head, arises within the larval prothorax through the folding inwards and back of the larval head hypodermis from the back of the head. The legs and wings arise within hypodermal pockets on their respective thoracic segments. These structures are usually referred to as 'imaginal folds', 'imaginal discs' or 'wing buds' and 'leg buds'. However, the reference to imaginal structure overlooks the intervening pupal stage. These structures are more accurately pupal folds or discs, wing sheath buds or leg sheath buds.

Miall and Hammond ignored the development of the pupal cuticle in their account. The following is an extension of their observations to include the formation of the pupa in *Chironomus*. Apolysis of the larval cuticle is gradual and patchy. It begins at the points where invagination of the hypodermis occurs to form the pouches in which projecting pupal structures arise. The leg sheath and wing sheath rudiments grow out from the bottom of their pouches and, as they do, more of the hypodermis comes adrift from the larval thoracic cuticle. However, their growth in length is greater than the space so provided and they become characteristically twisted and folded (Figure 8.7b). Only when enough cells have been generated to produce the wing and leg sheaths is the pupal cuticle laid down, which in the case of the wing sheath buds is when they are about three-fifths the length of the fully extended wing sheaths. The pupal cuticle of these structures is closely corrugated transversely and less closely longitudinally to accommodate the sheaths in their cramped conditions. Apolysis follows immediately and the marginal cells of the wing and leg rudiments differentiate: some (legs) or nearly all (wings) begin to develop long external filaments which will form the setae/macrotrichia when the imaginal cuticle is laid down.

Apolysis of the abdominal segments occurs with the exception of several adhesion points on each tergite and sternite. When these adhesion points separate from the larval cuticle, their position on the pupal cuticle is recorded by the smooth adhesion marks, previously referred to as 'muscle marks' or 'Fensterflecken', (Langton, 1994) (Figure 8.7c). The first set of these to separate from the larval cuticle are the lateral longitudinal connections of tergites and sternites: contraction of the dorsiventral muscles attached beneath these connections has the double effect of pulling away from the larval cuticle and flattening the pupal abdomen, the tergites and sternites of which are considerably thicker than that of the larva. The remaining anterior mediolateral and posterior mediolateral connections are only pulled away upon ecdysis. The function of these connections is to enable the organism during these changes to move about: movements of the pupal cuticle are transmitted directly through the larval cuticle to the substratum.

Upon ecdysis, pressure generated within the cephalothorax inflates the sheaths to their full extent; hardening is immediate. The wing and leg rudiments rapidly differentiate and grow to accommodate much of the space within their sheaths (Figure 8.7d). The imaginal cuticle is soon laid down and in the case of the wing and leg rudiments is corrugated in the same way as the pre-ecdysis pupal sheaths. The wing rudiments at this stage are about three-fifths the length of the extended adult wing.

8.6 ECLOSION

As pharate adult development nears completion, the wings begin to expand within their sheaths. Because their surface area now exceeds the space to accommodate them, they buckle (Figure 8.7e). Moulting fluid between the cuticles is absorbed and replaced by air, the hook row is disengaged and strong undulatory movements of the abdomen drive the pupa out of the tube. Although the pupa rises naturally to the water surface due to its increased buoyancy, its 'swimming' movements continue. It is now that the pupa is at its most vulnerable and the species suffers greatly from predation. Under experimental conditions in a tank with fish, few of the pupae of *Chironomus nuditarsis* Keyl reached the surface during daylight hours, although eclosion occurred throughout the day and night (Fischer and Rosin, 1968). Presumably the erratic movements performed by pupae as they rise increases their chances of survival, albeit marginally. However, these movements have another function: the spines on the adult abdomen are now sufficiently stiff to engage the pupal cuticle and rack the abdomen forwards towards the pupal thorax, inflating the adult thorax with haemolymph. Furthermore, the separation of the abdominal segments from the pupal cuticle severs the former neural connections and a different form of abdominal movement is initiated. This is the 'abdominal peristalsis' of various authors, perhaps universal in eclosing insects (Hughes, 1980), where the longitudinal muscles of the abdomen are contracted in successive segments from behind forwards, followed by extension. The effect is more powerful in driving the abdomen forwards as it engages the adult setae dorsally and ventrally with the pupal cuticle. At this point eclosion may be thwarted by the penetration of the pupal cuticle by the adult spines. The effect of the increased pressure within the thorax is to split the suture along the mid-dorsal line.

The adult dorsum may already be exposed when the pupa reaches the water surface; or, especially when the distance to the water surface has been short, the pupae may swim around at the surface before the suture splits. Eclosion begins with the adult thorax rising out through the parted suture. Then a Y-shaped cephalic suture splits to release the head. The upper part of the pupal thoracic cuticle snaps outwards along the pleura (which are relatively unsclerotized), taking with it the upper half of the eye sheaths and flattening the frontal apotome forwards, forming a raft on the water surface.

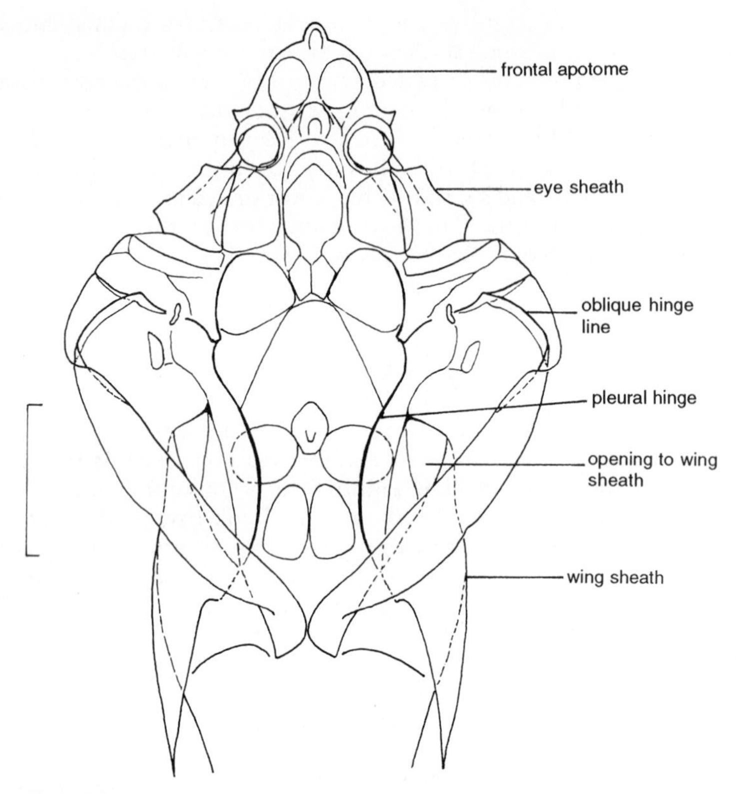

Figure 8.10 Cephalothorax of *Chironomus* pupal exuviae, dorsal view. (Scale line = 1mm.)

A weak hinge line passing from beneath the antepronotum obliquely backwards to the dorsal suture kinks inwards to allow the initially concave scutum to flatten out (Figure 8.10).

The inner openings of the wing sheaths are now nearly horizontal instead of inwardly directed. This releases the compressed wings which extend, simultaneously further inflated to full size, pushing the insect upwards and forwards out of the exuviae. This is sufficient to withdraw the antennae, and the legs from the sinuous part of the leg sheaths. The legs are withdrawn, extended and immediately hardened, so that they may support the animal on the water surface. The body bends forwards, withdrawing the wing tips, and the tarsi are brought down to lie on the water surface. The wings are tanned rapidly. The tip of the abdomen is withdrawn, leaving the imago perched above its now empty exuviae. The whole process of eclosion takes less than a minute; 20 seconds has been recorded (Langton, 1980). The imago usually leaves immediately, flying to marginal vegetation or other solid substratum to roost.

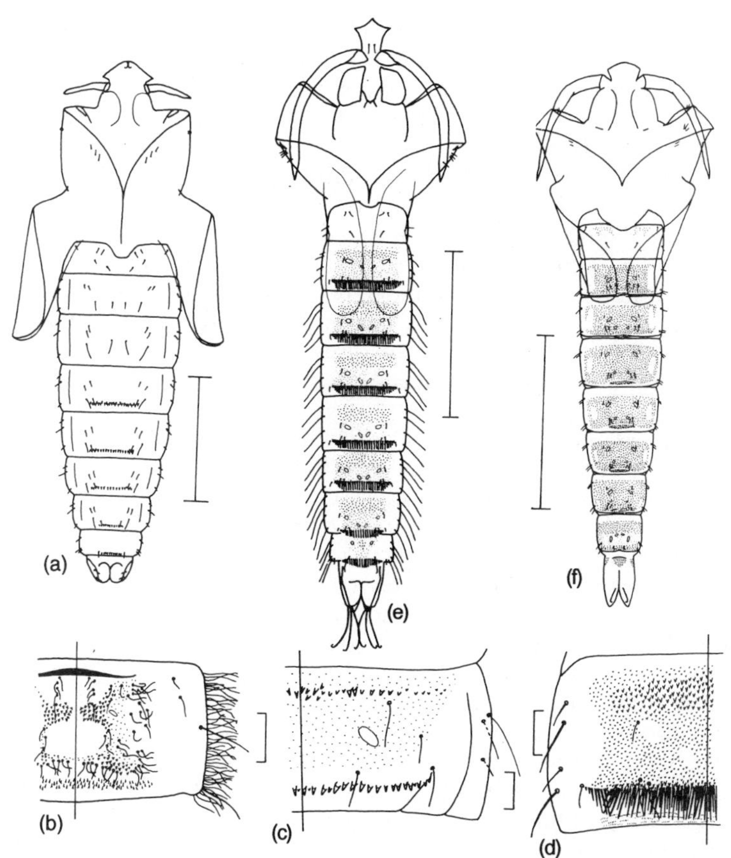

Figure 8.11 (a) Pupal exuviae of *Orthocladius (Euorthocladius) ashei* Soponis; (b) abdominal segment IV (dorsal) of *Epoicocladius ephemerae* Kieffer; (c) abdominal segment II (dorsal) of *Pseudorthocladius filiformis* (Kieffer); (d) abdominal segment II (dorsal) and (e) pupal exuviae of *Limnophyes edwardsi* Sæther; (f) pupal exuviae of *Pseudosmittia recta* (Edwards). (Scale line = 0.1mm.)

8.7 SPECIALIZED ADAPTATIONS RELATED TO HABITAT

Variations of the general patterns of pupal structure have been discussed previously, but the chironomids that have invaded habitats outside the normal range for the family have characteristic features worthy of elaboration. Perhaps the least extreme of these habitats is torrential water as there is a continuity of colonization of a stream from source to mouth. Maintaining station is not quite the problem it might seem, for the water flowing over rocks creates regions of reduced flow, particularly in the boundary layer immediately over the rock surface (Ambühl, 1959). However, the high oxygen levels remove the necessity to develop respiratory structures and economy is a strong selection force. Here are found pupae with greatly reduced thoracic horns, small anal lobes and absence of hook row II (Figure 8.11a). The cuticular armament of the abdominal tergites is frequently reduced in extent (e.g *Orthocladius* (*Euorthocladius*) species) or size (e.g. *Paratrichocladius skirwithensis* (Edwards)), though it usually extends to segment VIII, whereas in most chironomids it is reduced on segment VI and absent from segments VII and VIII except for small patches of fine shagreen.

Chironomid species which have invaded habitats that are not part of a body of stagnant or running water are referred to as 'terrestrial', though the microhabitat in which the larvae and pupae develop may be wet to the extent that they are fully submerged in a film or pocket of water. As with species inhabiting torrents, respiratory structures are greatly reduced and for the same reason. Terrestrial pupae fall into two groups according to their structural features.

The first group contains heavily armed pupae with long spines or robust teeth on the posterior margins of most of the abdominal segments, often both dorsally and ventrally. This is because the surface of water is much more difficult to break through when curved than when flat. When these species eclose they crawl up through the wet vegetation or over the surface of the wet substratum until the film of water over their cephalothorax is broken through. They have been observed to crawl 2 cm above the water surface in smooth plastic rearing containers before eclosion, surface tension of the water film being drawn up with them providing sufficient adhesion. Many hygropetric and madicolous species show these adaptations, as do moss-dwelling forms (Figure 8.11c–e).

Pupae of the second group are characterized by a covering of small points over most of the abdominal terga and sterna (Figure 8.11f). These species develop in damp, rather than wet, habitats, where problems with breaking through a meniscus do not occur. The species of *Smittia* and *Pseudosmittia* are frequently encountered in soil or damp peat, where they form their pupal tubes with the cephalic end at the substratum surface. When these eclose, they push only the cephalothorax through the tube opening, leaving all, or at least much, of the abdomen within, in the manner of tipulids.

Up to this point in this account of chironomid pupae all the modifications recorded are embellishments of a very similar structure. This is not surprising. The general structure of chironomid imagines is very similar and as the pupa acts as an economical container for the developing adult, the pupal gross morphology is going to be similar. However, in the intertidal species of the subfamily Telmatogetoninae, there is an astonishing modification of the posterior abdominal segments – astonishing because neither the larva nor the adult is sufficiently different caudally from other members of the family to expect it. Here we find a 'terminal disc' which is formed from both segments VIII and IX. Its main development is dorsal, where the tergites of the two segments produce a longitudinally oval disc, continuously crenulate around the margins, though weaker posteriorly, and on tergite IX fringed with fine simple or branched hairs. It is tipped upwards at the anterior end to produce a sloping shield at an angle of about 45° to the longitudinal axis. The whole structure is concave, causing the crenulate margin to be nearly vertical to disc

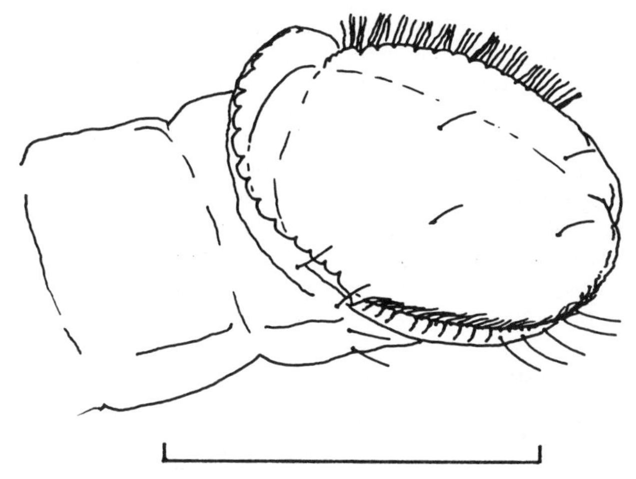

Figure 8.12 Anal segments of pupa of *Telmatogeton japonicus* Tokunaga. (Scale line = 0.1mm.)

anteriorly. It is strongly sclerotized and studded with tubercles and the sensory setae are conspicuously strong. The ventral surfaces of these segments are little modified, except for the presence of sensory setae and a pair of strong anteriorly directed hooks towards the posterior margin of segment IX (Figure 8.12). The pupal tube is to be found on vertical rock faces amongst seaweeds in the intertidal zone. In the absence of direct observation it would appear that the disc acts as an operculum to the tube, preventing an inrush of water from beneath as a wave dashes against the rock, the hooks holding the structure firm to the tube, while allowing drainage of water out between waves to provide a respiratory stream (Wirth, 1947; Leonard, 1972). Perhaps, also, water may be held around the pupa when the tide is out by using the disc to block the exit completely. A different function for telmatogetonine pupal structure is suggested by Robles (1984), who observed that the larvae of *Telmatogeton alaskensis* Coquillett and *T. trilobatus* (Kieffer) are extremely aggressive and attempt to evict other larvae from their tubes if they occur within the feeding area. Robles proposes that the highly cuticularized anterior thorax and the operculum-like posterior end of the pupa, which block both openings to the tube when the tide is out, are adaptations to resist eviction by conspecific larvae. Doubtless, the adaptations allow such aggression amongst the larvae, but were the pupae not 'preadapted' to resist eviction, the species could not have survived.

Similar agonistic behaviour has been observed in larval *Telmatogeton japonicus* Tokunaga (P.S.Cranston, *pers. comm.*). Fourth instar larvae and pupae formed a transverse band across a vertical sea wall covered in *Enteromorpha* in which the *T. japonicus* tunnelled. Above and below this band the proportion of earlier instars was much higher and it appears possible, therefore, that larval aggression in this species ensures that pupation occurs in a zone optimal for pupal survival and/or adult eclosion.

Phoretic and parasitic species reflect in their pupal structure the habitat of their hosts: *Epoicocladius ephemerae* (Kieffer) lives in streams phoretically on the sand-burrowing nymphs of the mayfly *Ephemera danica* (Müller). Unusually for the Orthocladiinae, its pupal abdomen has a lateral fringe of sensory setae and an increased number of these both dorsally and ventrally (Figure 8.llb), which reflect its need to swim out of cul-de-sacs in the sand as it makes its way to the surface for eclosion. *Symbiocladius rhithrogenae* (Zavřel), on the other hand, is reputedly parasitic on another mayfly larva (*Rhithrogena* sp.), which lives flat on the surface of stones and rocks in fast-flowing streams. The *Symbiocladius* pupa has the characteristic modifications of torrenticolous species: absent thoracic horn, lack of hook rows, reduced anal lobes and the abdominal tergite armament extended to segments VII and VIII.

8.8 PUPAL VARIATION AND FREQUENCY OF ABNORMALITIES

The foregoing account demonstrates that, despite the overall similarity of chironomid pupae, in detail pupal structure varies greatly, such that the exuviae can be used to identify most species. The differences of tergite armament have been generated through variations in the extent of the fields and the strength of the individual cuticular projections. The patterns that these processes achieve are reflected across the subfamilies; for instance, a posterior transverse row of strong teeth occurs in the Tanypodinae (*Nilotanypus*), Buchonomyiinae (*Buchonomyia*), Diamesinae (*Diamesa, Syndiamesa and Pseudokiefferiella*), Orthocladiinae (*Krenosmittia, Metriocnemus, Heleniella* and others) and Chironominae (*Cryptochironomus* and *Harnischia* in the Chironomini, and *Constempellina* in the Tanytarsini). Paired anterior point patches formed from the median field occur in the Orthocladiinae (*Acricotopus*, some *Psectrocladius, Orthocladius* (*Pogonocladius*) and others), in the Chironomini (*Omisus* and *Lauterborniella*) and in the Tanytarsini (many genera). These are but two examples of such correlations. It has been suggested (Langton, 1989) that a common gene pool for cuticular modification exists throughout the family and that, even if phenotypic structures have been lost through selection, the genes are maintained in an inactive form, enabling the structure to reappear, contrary to Dollo's Law. The reason advanced for such a facility is that, unlike the larva and imago where selection will be effected through competition and survival of the fittest, the pupa has to be able to generate variants spontaneously which will allow larvae to evolve in habit and habitat without jeopardizing completion of the life cycle.

Circumstantial evidence was derived from the frequent occurrence of gross aberrations in natural populations. For the tergite armament these mainly take the form of patches or bands being repeated on tergites that normally do not possess them (Figure 8.13a,b). They may occur on one side only, rendering them easy to recognize as aberrations, but, though less commonly, they also appear as identical patterns in form and extent to the segment before or behind. Also, the absence or reduction of a patch or band

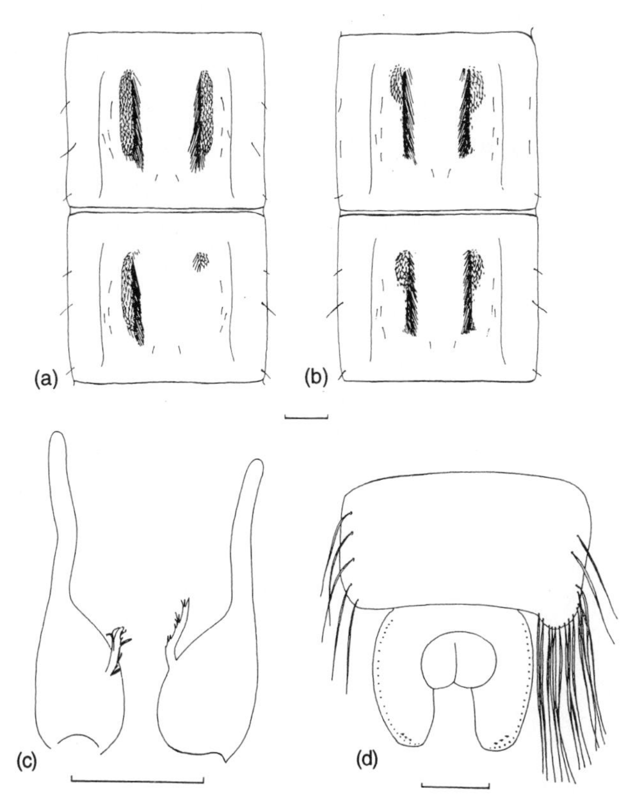

Figure 8.13 (a) (b) Abdominal segments IV and V of aberrant *Tanytarsus gregarius* Kieffer; (c) thoracic horn aberration of *Tvetenia discoloripes* (Goetghebuer); (d) lateral setation aberration, segment VIII of *Rheocricotopus fuscipes* (Kieffer). (Scale line = 0.1mm.)

occurs, but is much more rarely seen in exuviae, perhaps because pupal survival to eclosion is reduced.

These aberrations are not restricted to abdominal cuticular armament. They have been found in the lateral setation of the abdomen (Figure 8.13d) and the structure of the thoracic horn. In an example of *Tvetenia discoloripes* (Goetghebuer) (Orthocladiinae), both thoracic horns have the apical filamentous spine greatly reduced and possess a small toothed branch anteriorly (Figure 8.13c); here the toothed branch may represent the reappearance of the unmodified orthocladiine horn. Specimens of a *Zavrelimyia* species (Tanypodinae) from one of the most northerly lakes in Scotland show a reduction of the plastron plate and a progressive reduction of the neck of the atrium (Figure 8.2a). This condition has not been encountered in *Zavrelimyia* before, but, as previously described, in *Macropelopia fehlmanni* such variation is normal (Figure 8.2d).

Some populations show greater than normal variation, as if genetic suppression has been removed. Such variation was seen in a population of

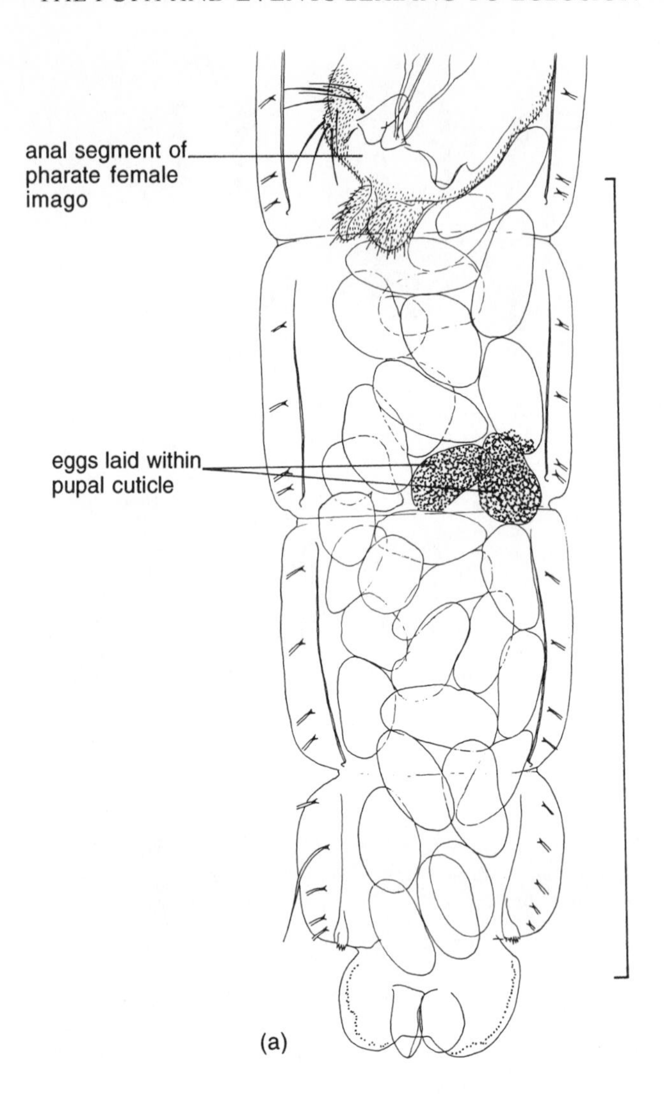

Tanytarsus gregarius Kieffer from a small loch in Scotland. Spine bands normally restricted to segment IV were repeated on segment V, commonly on one side, but with symmetrical examples present (Figure 8.13b). Most specimens were aberrant, but the exuviae of all the other species collected at the same time had normal, restricted variation. The hypothesis was put forward that the conserved bank of pupal genes is tested relatively frequently, perhaps by transposons, and those patterns which confer greater success on adult eclosion will become the normal form for the pupa as the less successful forms are eliminated. Such a process would generate not only the welter of forms to be found in chironomid pupae, but also high incidence of homoplasy, the recurrence of similar patterns in distantly related taxa.

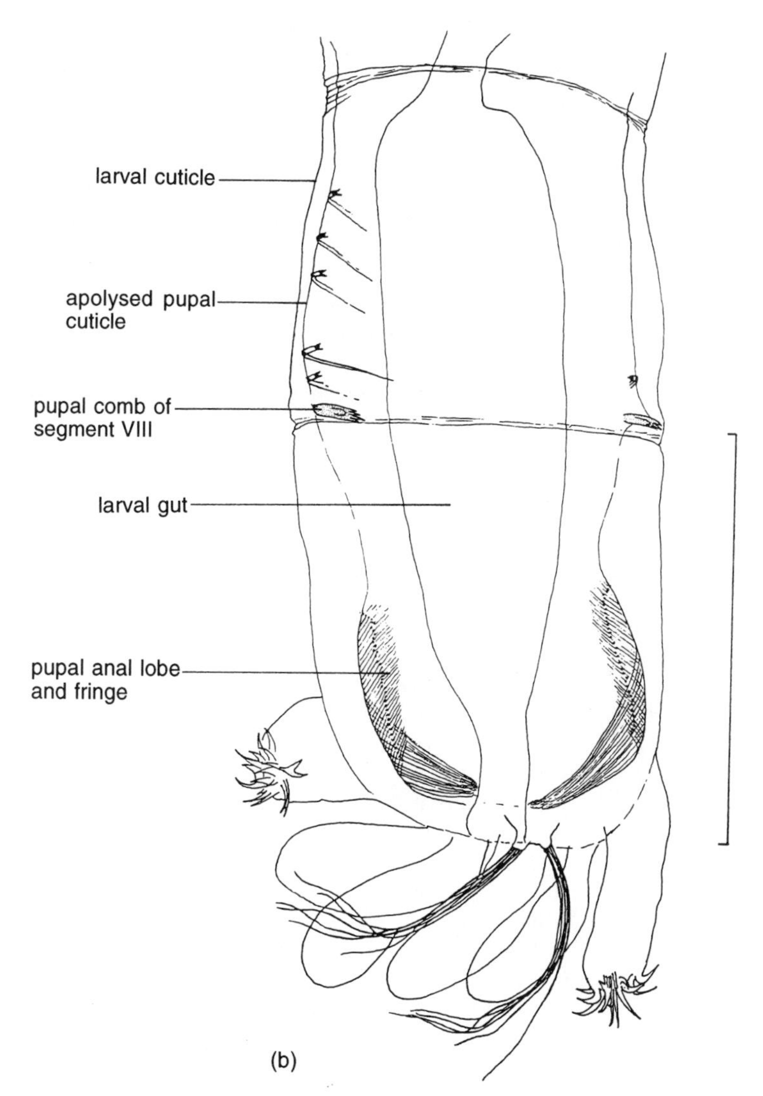

Figure 8.14 Posterior abdomens of (a) pharate adult female of *Paratanytarsus grimmii* (Schneider), pupal cuticle containing discharged eggs; (b) fourth instar *Chironomus* larva containing nearly fully apolysed pupa. (Scale line = 1mm.)

8.9 PRE-ECLOSION EGG-LAYING AND PARTHENOGENESIS

Parthenogenesis is most easily detected in collections of pupal exuviae where a distinct form is represented only by female specimens. As parthenogenesis has enabled a species to perpetuate itself through many generations without eclosing, it is appropriate to discuss the phenomenon here. Parthenogenetic species were first detected by the chance rearings of females which proceeded to lay eggs that hatched quite normally into larvae. This was the case, for example, with *Paratanytarsus grimmii* (Schneider) (Grimm, 1870) and *Pseudosmittia baueri* Strenzke (Strenzke, 1960c). However, incorrect conclusions can be derived from such observations, even if the clone is carried through a number of generations. For instance, in the case of *Paratanytarsus laccophilus* (Edwards), both parthenogenetic and bisexual populations occur (Lindeberg, 1958). Rearing of the parthenogenetic form has so far not yielded a single male, but that may be due to the lack of appropriate environmental switch. Lindeberg (1971) has shown that in northern Finland all specimens of *Tanytarsus gregarius* that eclose late in the season are female, prompting him to postulate the presence of seasonal parthenogenesis in this species.

The occurrence of parthenogenesis in the northern hemisphere appears to be most common in the far north, but this may reflect the distribution of scientists interested in the phenomenon. Exuvial collections from Ellesmere Island in the Canadian Arctic reveal a parthenogenetic member of the *Tanytarsus lestagei* group (a complex of sibling species hitherto known only to be bisexual), and a *Micropsectra* close to *nigripila* (Johannsen), a bisexual species found further south.

Paratanytarsus grimmii is always parthenogenetic. It has achieved notoriety through its capacity to breed in water distribution systems, and much time and expense has been incurred in eradicating the midge: naturally, customers object to midge larvae in their tap water. It is world-wide in distribution (Langton *et al.*, 1988) and a multiplicity of clones has been generated, presumably through random mutations. Some of these are capable of laying their eggs within the pupal cuticle and, as they do not eclose, they do not need an air space above the water in which they live. It is possible that in nature the species is hyporheic in habit. Sufficient fine allochthonous material enters water distribution systems to support a large population that spills over into domestic supplies (Williams, 1974). The problem has been exacerbated by the replacement of old filter beds with more efficient types, followed by a population explosion in the pipes believed to be due to the filtering off of predators which were previously entering the systems but unable to breed therein (Langton, 1974).

Although the submerged pharate adult performs eclosion movements and the pupal cephalothorax splits along the suture, ecdysis ceases at this point. Shortly afterwards an egg-mass is laid (Figure 8.14a). This absorbs water and swells, tearing the abdominal cuticle of the pupa and allowing the larvae to escape on hatching. The ability to reproduce in this way was enabled by a 'pre-adaptation': a reduction in the time between eclosion and egg-laying. In bisexual species, egg-laying has to be delayed until the eggs are fertilized. The sequence of events – roosting after eclosion until the time of day when the males swarm, flying to the swarming sites, mating and subsequent flight to an oviposition site – means that the first egg-mass is laid 24 hours or more after eclosion. However, parthenogenesis enables this period to be greatly reduced and in eclosing *P. grimmii* the first egg-string is laid shortly after eclosion, often while the adult is still resting on the surface above its pupal exuviae. This ensures perpetuation of the species in a suitable habitat. A second, smaller egg-string is laid about 24 hours later, allowing time for dispersal and the colonization of new habitats.

8.10 THE CHIRONOMID PUPA AS AN INDEPENDENT ORGANISM

In *Chironomus*, and probably most other Chironomidae, the late fourth instar larva contains a pupa: the leg sheaths and wing sheaths are fully formed though not yet extended; the thoracic horn is fully developed; the anal lobes complete with a fringe of filaments lie in segment IX; and the postero-lateral toothed spurs of the pupa are to be seen in segment VIII (Figure 8.14b). From the larval neck to the bases of its parapods, anal tubules, procercus and anus, it is a pupa lying within the larval cuticle. However, apolysis of the larval head so far has been restricted to the infoldings of the hypodermis for the development of the various pupal head structures. This compound organism is under larval control. It continues to feed and digest food in a hitherto unaltered gut and to egest undigested waste. Its responses are larval, curling up when disturbed, foraging for food from the opening to its tube and extending its tube from the still conspicuous salivary glands. The final period to pupal ecdysis is extremely rapid. Apolysis of the cuticle of the parapods, anal gills and procercus occurs without the laying down of new cuticle. The muscles of the head degenerate and apolysis takes place followed by the laying down of the cuticle of the pupal head. The contents of the parapods, anal gills and procercus are withdrawn and ecdysis immediately begins. Momentarily it is a pharate pupa; then it is free.

Already apolysis has occurred within the wing and leg sheaths, and differentiation of the imaginal structures has begun, but it is an organism in its own right, performing activities it shares with neither the larva nor the adult. A substantial part of the midge genome is invested in it to allow for a rapid response to larval and adult evolution. Its flattened abdomen, so necessary for propelling the respiratory current, imposes on the imago a similarly flattened body (though this may be obscured when swollen with eggs). It can also be the final free-living stage in the life cycle, in which the trapped female imago functions only as its ovary, generating eggs to perpetuate its kind. This view of the chironomid pupa is a far cry from that expressed by Miall and Hammond (1900) in their otherwise excellent account of the structure and function of the 'Harlequin Fly':

> The pupa of *Chironomus* is hardly more than the fly enclosed in a temporary skin, and details of its structure cannot be understood without constant reference to the structure of the fly.

9

Behaviour and ecology of adults

P.D. Armitage

9.1 INTRODUCTION

Adult chironomids are short-lived and their behaviour is concerned largely with reproduction. Chironomid swarms are familiar and field and laboratory observations of reproduction are numerous. For ephemeral adults, mating requires spatial and temporal synchronization of emergence and subsequent adult location.

For the purposes of this chapter, commencement of adult life is defined by the emergence of the free adult midge from the pupal skin. Emergence is rapid and the adult is able to fly almost immediately. The eclosion of most studied chironomids is synchronized through photoperiod with some modification by temperature (Kureck, 1979, 1980; Danks, 1978). Exceptionally the males of some species are protandrous, emerging before females, and some even assist the female in emerging from the pupa (section 9.3.3).

Post-emergence behaviour ranges from almost instant readiness for mating to a refractory period of shelter followed by movement to the swarming site. Generally those species which emerge in harsh conditions do not move far from the emergence site and mating and oviposition take place within minutes of emergence; this is particularly so in marine chironomids (Neumann, 1976; Cheng and Collins, 1980).

The relationship between timing of emergence and environmental variables has been the subject of many studies (e.g. Brundin, 1949; Palmén, 1955; Mundie, 1959; Darby, 1962; Kureck, 1966; Coffman, 1974; Neumann, 1976; Wartinbee, 1979; LeSage and Harrison, 1980a; Learner *et al.*, 1990; Pinder *et al.*, 1993). The seasonality of emergence relates to the duration of the larval period, which varies with temperature, oxygen concentration, photoperiod, water level and food resources (Chapter 10).

In this chapter, all aspects of emergence are considered, together with details on swarming, mating and dispersal.

The Chironomidae: Biology and ecology of non-biting midges
Edited by P.D. Armitage, P.S. Cranston and L.C.V. Pinder
Published in 1995 by Chapman & Hall. ISBN 0 412 45260 X

9.2 EMERGENCE – SEASONAL/ GEOGRAPHICAL ASPECTS

9.2.1 ARCTIC

Latitude, or its climatological equivalent such as elevation, has an overwhelming influence on emergence period and synchronicity (Corbet, 1964). Synchronous emergence appears to be a characteristic feature of Arctic chironomids. At Ellesmere Island in the Canadian Arctic, Oliver (1968) found that 75% of the total seasonal emergence from a small tarn took place within 7 days. Welch (1973) reports an equivalent value of 10–30 days for Arctic Char Lake.

Water temperature remains close to freezing for 8–9 months of the year in the Arctic and the season available for larval growth is short. Synchronicity increases the chances of sexual encounters and hence reproductive success. At the Ellesmere sites, there are marked differences in emergence period from pond to pond due to depth/temperature variations. The time at which emergence of 50% of the annual total is reached for a given species may be 3 weeks apart in different pools (Danks and Oliver, 1972a). Thus adults of one species are present throughout the Arctic summer, increasing the chance of reproductive success.

Physiological mechanisms for synchronization must be especially effective because cohorts may have several years to diverge in growth and development (Butler, 1980). High pond species (Danks and Oliver, 1972a) are considered to be 'absolute spring species' – that is they overwinter as prepupae that pupate without feeding following a temperature cue in the spring and emerge synchronously. The emergence of chironomids from the larger and deeper Char Lake is less synchronous than that of pond species and there is no winter diapause. In tundra ponds in Alaska all dominant species exhibited highly synchronized emergence (Butler, 1980); however, species' emergence periods were staggered within the ponds and it is suggested that the pond environments were homogeneous, preventing the broad seasonal overlap of species observed in Ellesmere Island. The distinct emergence times may reproductively isolate closely related species. However, Butler (1982a,b) reports an example of two species of *Chironomus* in which emergence periods never overlap within any one pond, but differences in the onset of emergence from individual ponds gives rise to a period of regional coexistence of adults of the two species. Alternative mechanisms for reproductive isolation, notably swarming behaviour and site, must be significant.

The highly synchronized emergence behaviour of high latitude chironomids is probably controlled by a heat-sum threshold. In contrast to the high degree of synchrony observed in Arctic lakes, stream studies suggest that emergence is not synchronized by diapause and that emergence does not cease after all overwintered mature larvae have completed development (Hayes and Murray, 1987).

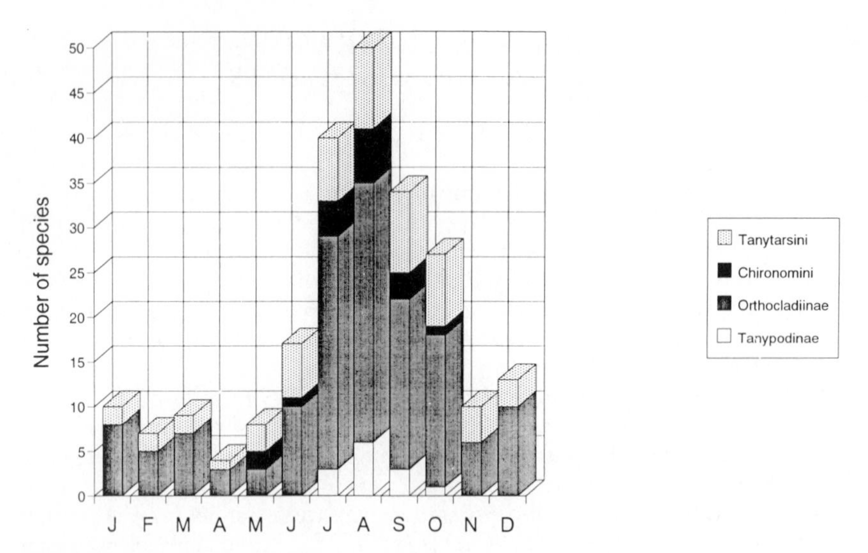

Figure 9.1 The seasonal pattern of emergence of species in a temperate zone stream. (Based on data in Pinder, 1974.)

9.2.2 THE TEMPERATE ZONE

In temperate areas, synchronicity of emergence is less evident than in Arctic regions. In northern temperate zones, emergence may take place in 5–10 or more months of the year in both standing (Humphries, 1938; Brundin, 1949; Mundie, 1957; Reiss, 1968; Sandberg, 1969; Potter and Learner, 1974; Koskenniemi and Paasivirta, 1987) and running waters (Illies, 1971; Coffman, 1973; Ringe, 1974; Pinder, 1974; Boerger, 1981; Caspers, 1983b; Rempel and Harrison, 1987). In the southern hemisphere, year-round emergence of chironomids occurs in New Zealand running waters (Boothroyd, 1988) and in Australia's River Murray (Cranston and Hillman, *unpubl. obs.*) and streams of southern Queensland (Cranston, Hillman, Arthington and Pusey *unpubl. obs.* and in preparation).

The overall pattern of chironomid emergence derives from the individual emergence patterns of component species. Species often emerge at particular times in the year and generalizations may apply to higher taxonomic groupings, even to tribal and subfamily levels (Figure 9.1) where phylogenetic constraints may operate (Learner *et al.*, 1990).

In the northern hemisphere, as a generalization, Diamesinae and some Orthocladiinae emerge in winter, Orthocladiinae dominate the spring and autumn emergence, and Chironominae and Tanypodinae most commonly emerge during the summer months (Coffman, 1973; Rempel and Harrison, 1987) with Tanypodinae often dominant in autumn emergence. In the southern hemisphere, the few Diamesinae, most Podonominae and all Aphroteniinae are spring emergers (Cranston and Edward, 1992) as are many other southern Gondwanan taxa (Edward, 1986; D.H.D. Edward, *pers. comm.*). In contrast, taxa with northern Gondwanan or world-wide distributions (e.g. many Tanytarsini, *Cricotopus*, *Polypedilum*, *Chironomus*) emerge predominantly in warmer months.

The temporal cues and controls for emergence are most likely to be temperature-related, with rising spring temperatures associated with the onset of emergence. Overwintering chironomids complete their development when the cumulative degree days exceed their species-specific threshold (Mundie, 1957). Accordingly, the timing of emergence varies between years according to seasonal variations in temperature (Titmus, 1979; Butler, 1980). Modifications to the natural temperature cycle may alter emergence patterns (Coler and Kondratieff, 1989) and artificial elevation of temperature in an experimental channel by 10 °C caused the peak emergence periods of some chironomids to be advanced by 1 to 4 weeks (Nordlie and Arthur, 1981). Altitudinal factors, by acting on the thermal regime, may also determine voltinism and emergence patterns (Huryn, 1990).

Coffman (1973) found onset of the emergence of lotic Orthocladiinae under natural conditions took place when water temperature first increased, whereas the peak of Chironomini and Tanypodinae coincided with the attainment of maximum water temperature for the year. Tanytarsini started to emerge at about the time that maximal diel fluctuations in water temperature occurred. Goddeeris (1990) found the early spring emergence of *Tanytarsus sylvaticus* (van der Wulp) was synchronized by a diapause in the fourth instar larvae in the winter. The autumn emergence of *Tokunagayusurika akamusi* (Tokunaga) may be triggered by reduction in bottom water temperature from about 18 to 10 °C – a further drop below 10 °C is thought to terminate pupation (Iwakuma *et al.*, 1989).

Photoperiod can also determine the timing of larval diapause (Shilova and Zelentsov, 1972; Ineichen *et al.*, 1979) and, in conjunction with temperature, can control diapause and influence the timing of emergence (Neumann and Kruger, 1985). However, phenology is not determined completely by temperature and photoperiod: significant relationships between food supply (in the form of phytoplankton production) and emergence have been demonstrated in Canadian lakes (Welch, 1976; Davies, 1980; Welch *et al.*, 1988). Larval biotic interactions, in the form of predation and competition for food, modify production and may affect the timing of adult emergence (Johnson and Pejler, 1987; van der Bund, 1992; Johnson and Goedkoop, 1993). In streams in the Laurentides (Quebec, Canada), Tanypodinae showed two emergence patterns – short and synchronous (four species) and extended over many weeks (eight species); the latter was most affected by weather conditions (Cloutier and Harper, 1978). Two species, *Meropelopia flavifrons* (Johannsen) and *A. americana* (Fittkau) [both as *Arctopelopia*], had different emergence patterns depending upon whether they coexisted in the same section of stream or occurred alone. This phenomenon tends to ensure genetic isolation and minimizes interspecific competition.

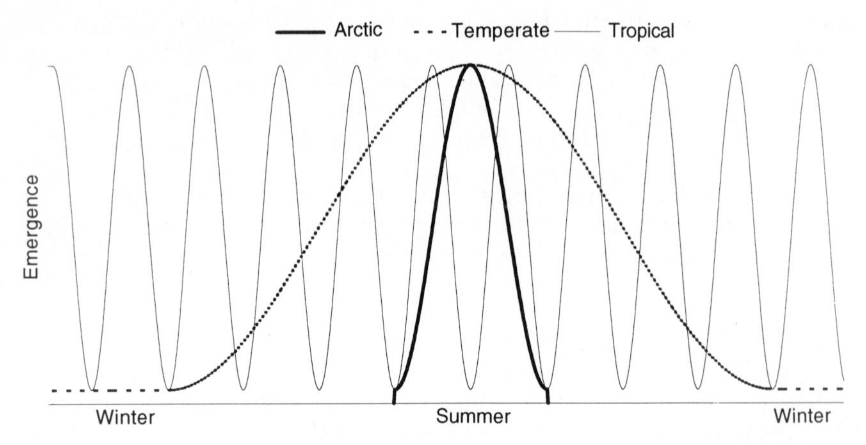

Figure 9.2 Stylized representation of patterns of emergence in latitudinal zones. Note the fluctuations in the tropics which are linked to the lunar cycle.

9.2.3 TROPICS, SUBTROPICS AND MEDITERRANEAN ENVIRONMENTS

In year-round warm environments, chironomid emergence takes place throughout the year (Corbet, 1964; MacDonald, 1956). However, there is seasonal variation with respect to climatic variables other than temperature. In Lake Chad (Africa), chironomid adults were most abundant in the cold dry season and at a minimum in the rainy season (Dejoux, 1969). In the basin of the River Nile in Sudan, mass outbreaks of chironomids, including the pest species *Cladotanytarsus lewisi* (Freeman), occur in the cooler winter months (Rzóska, 1964), but this probably relates more to the timing and severity of rainfall in the Nile catchment causing elevated nutrient levels that encourage algal production (Cranston *et al.*, 1981).

In contrast, in Lake Bangweulu (Zambia) south of the equator, the main chironomid emergence period was from June to November in the period of increasing temperature that precede seasonal rains (Fryer, 1959). In Lake George (Uganda), which straddles the equator, no seasonality of emergence was noted for *Procladius brevipetiolatus* Goetghebuer but a clear lunar periodicity was observed with greatest abundances in the first quarter of the lunar month and smallest in the third quarter (McGowan, 1975). Similar patterns were observed for *Conochironomus acutistilus* (Freeman) in Jinja (Uganda) (Tjönneland, 1962).

Where temperature changes are slight, other environmental fluctuations may be used as cues for hatching. In subtropical Transvaal (South Africa), which has a long dry season, no close association was found between emergence and the onset of the wet season (Frank, 1965). However, for one species, *Dicrotendipes* [as *Chironomus*] *chambiensis* Goetghebuer, a clear emergence peak was associated with the period April/May, just before the onset of the dry season.

In higher latitudes, often with a more mediterranean climate, the seasonal periodicity of emergence typical of temperate regions becomes more evident. In Israel there is emergence throughout the year but with peaks in autumn (October) and spring (April/May) (Kugler and Wool, 1968). In subtropical southern California (Frommer and Sublette, 1971) and Louisiana (Buckley and Sublette, 1964) there is a broad period of emergence with a summer peak and low densities in the winter (November to February). In the mediterranean climate of Perth (Western Australia), emergence of dominant species of nuisance chironomids is highest following periods of warm water temperatures and is reduced in late autumn and winter (Pinder *et al.*, 1993).

Latitudinal shifts in patterns of emergence are illustrated diagrammatically in Figure 9.2.

9.3 DIEL EMERGENCE

9.3.1 PATTERNS AND CONTROLS

Within the seasonal patterns of emergence observed for many aquatic insects, there are also daily rhythms. These have been studied for Chironomidae in the laboratory (Phillipp, 1938; Remmert, 1955; Fischer and Rosin, 1968; Kureck, 1979, 1980); in lakes and ponds (Miller, 1941; Scott and Opdyke, 1941; Morgan and Waddell, 1961; Danks and Oliver, 1972b; Wrubleski and Ross, 1989; Learner *et al.*, 1990; Pinder *et al.*, 1993); in running waters (Sprules, 1947; Learner and Edwards, 1966; Kureck, 1966; Coffman, 1973; Wartinbee, 1979; Singh and Harrison, 1982; Vilchez-Quero and Lavandier, 1986; Boothroyd, 1988); and in brackish and marine environments (Caspers, 1951; Palmén, 1955; Neumann, 1966, 1976).

Few chironomids emerge evenly throughout the day; most have been noted to emerge at particular times within the 24-hour period. There is considerable interspecific variation both temporally, in diel emergence (Figure 9.3) and seasonally, in the northern (Coffman, 1974; Cobo and Gonzáles, 1991) and southern (P.S. Cranston and T.D. Hillman, *pers. obs.*) hemispheres. The view of Morgan and Waddell (1961) that species must be treated separately and observations should be at least monthly through the emergence season is amply confirmed.

Environmental cues for the timing of emergence have been attributed to changes in light intensity and/or water temperatures. In a shallow lake in Scotland, the period of maximum emergence related to change in the time of sunset (Morgan and Waddell, 1961). For species with a June generation and another in August, the time of maximum emergence shifted between generations in step with change in time of sunset. One species, *Psectrocladius psilopterus* (Kieffer), peaked in both evening and morning, suggesting that changing light intensity cued emergence in this species.

Seasonal differences in the timing of emergence peaks have also been observed in running waters (Coffman, 1974). In spring and mid-autumn the maximum daily emergence took place during the mid–late afternoon,

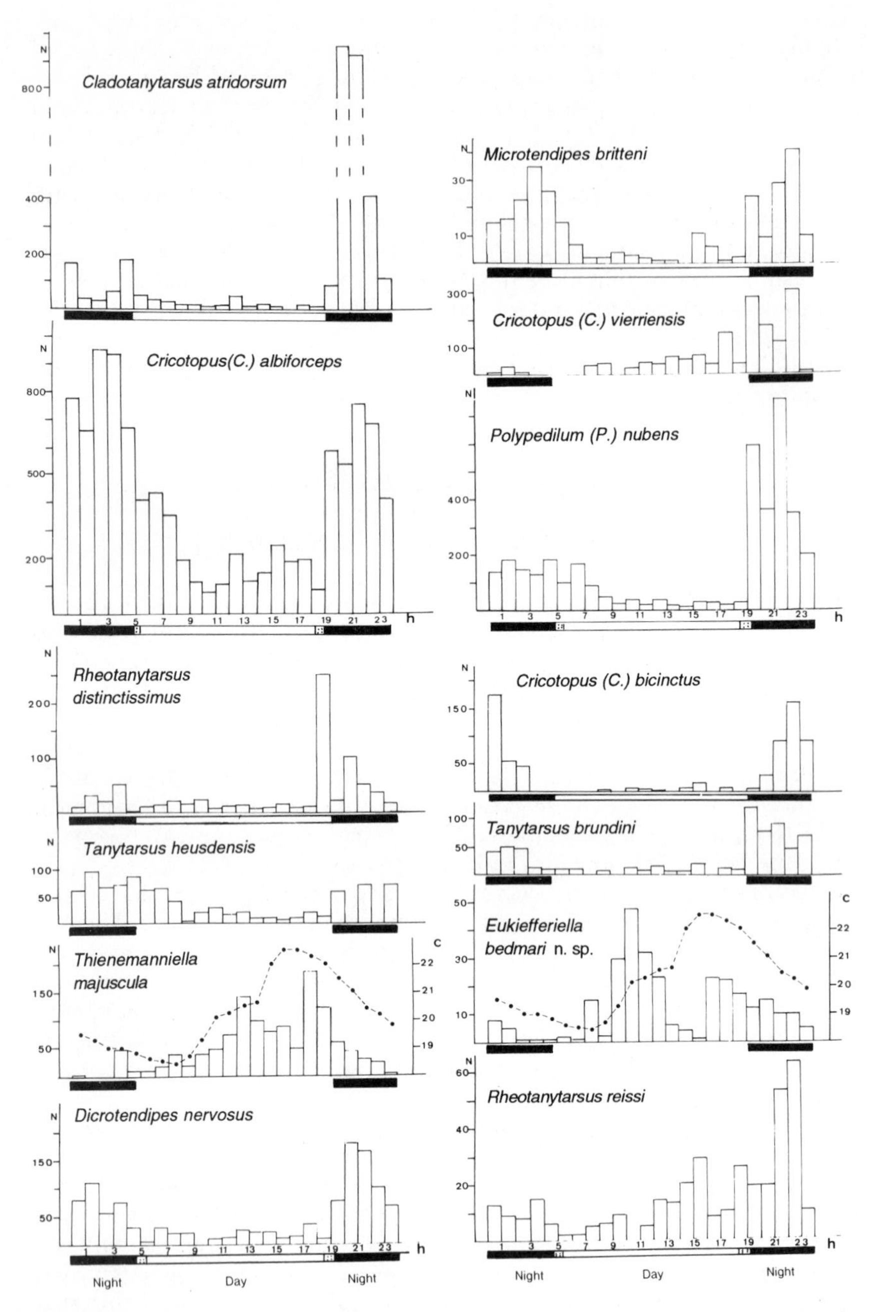

Figure 9.3 Interspecific variation in diel patterns of emergence as illustrated by the drift of pupal exuviae. (Based on Vilchez-Quero and Lavandier, 1986. Reproduced by permission of *Annales de Limnologie*.)

contrasting with the situation in summer and early autumn, when two maxima were observed. One afternoon maximum consisted mainly of small-sized orthoclads belonging to *Corynoneura* and *Thienemanniella*, and the second in late evening was composed mainly of Chironominae. Coffman (1974) suggested that the afternoon peak was cued by maximum light intensity.

Separating the roles of light intensity and water temperature in controlling eclosion is difficult. Even in the Arctic of Spitzbergen, which has 24-hour summer illumination, air and water temperatures cycle in correlation with diel changes in light intensity (Remmert, 1965). Diel periodicity of Spitzbergen chironomid emergence is related to these congruent cycles. In the Canadian high Arctic, Danks and Oliver (1972a,b) found that the diel patterns of pond emergence of many species related to increasing or peak water temperatures. In contrast to Remmert's (1965) finding of emergence in Spitzbergen during the afternoon, Danks and Oliver found all species to emerge maximally during the middle of the day. The influence of temperature on emergence is particularly important in the high arctic where short-term temperature changes may inhibit adult activity. At high latitudes (81° 47′ N), Danks and Oliver believed light to be an unreliable measure of temperatures, which lie near critical thresholds.

Working on channels in Linesville Creek (Pennsylvania, USA), Wartinbee (1979) postulated complex interactions between light and temperature effects to account for observed diel patterns of emergence. A species of *Polypedilum (Tripodura)* began to emerge in late evening and continued into the early daylight hours. Emergence in this species was suggested to be a response to falling temperature rather than emergence during a period with a particular water temperature. *Rheotanytarsus* cf. *exiguus* (Johannsen) and *Sublettea* [as *Tanytarsus*] *coffmani* (Roback) had major emergence peaks immediately after darkness, with minor pulses in the morning. These patterns were attributed to responses to changes in light, with temperature controlling the number of individuals able to respond. As chironomid pupae prepare for emergence, they inflate themselves with a gas bubble which separates the pupal skin from the pharate adult. Wartinbee suggests that temperature determines the rate at which the pupal gas bubble expands, thereby affecting the numbers of individuals ready to emerge.

Daily maximum–minimum water temperatures at either bank of Duffin Creek, Ontario, showed some correlation with chironomid emergence (Williams, 1982). Immediately before and during the early summer peak in emergence on the east bank, minimum water temperatures were 2–3 °C higher than on the west bank. This may have stimulated the earlier emergence on the eastern side. Immediately before the midsummer emergence peak on the west bank, maximum temperatures were generally 1–4 °C higher than on the east bank. Williams postulates that this may have attracted insects in a pre-emergent state to the western side and suggests that in contrast to control of emergence by light, control by water temperature may be more finely tuned.

In a study in southern Spain, most emergence from the Guadalquivir took place at dusk and in the 4 hours following, with a reduction in light intensity proposed as the cue for emergence in most species (Vilchez-Quero and Lavandier, 1986) (Figure 9.3). In contrast, *Microtendipes britteni* (Edwards) reacted to both increasing and falling light intensity. Temperature, which varied between about 19 and 22 °C, appeared unrelated to the pattern or numbers of emerging chironomids. The same conclusions were made by Pinder *et al.* (1993) regarding the primacy of light intensity in controlling the immediate post-dusk emergence of *Polypedilum nubifer* (Skuse), *Tanytarsus bispinosus* Freeman and *Cryptochironomus griseidorsum* (Kieffer) from lakes in Perth (Western Australia).

Diel patterns of emergence from a hyporhithral site on the River Aude (Eastern Pyrenees) appeared similarly unaffected by the small variations in temperature (=1 °C) over 24 hours (Gendron and Laville, 1993). In spring, the majority of species had a bimodal emergence pattern with peaks just after dawn (7.00–9.00 h) and around sunset (18.00 h). In the autumn the species had a rather unimodal pattern of emergence, some at noon only and others at night. Of the species with small exuviae (<3 mm), 70% emerged during daylight hours and 63% of species with exuviae larger than 5 mm emerged during darkness.

In contrast to the interspecific variations seen amongst members of a single aquatic community above, there is one exceptional example of nearly complete congruence in diel eclosion pattern amongst all chironomid species in a single habitat. Jackson (1988) observed emergence restricted to the dusk and immediate post-dusk period amongst the pool of species that emerged from a stream in the Sonoran Desert (Arizona, USA). The extreme daytime air temperatures, low humidity and diurnal predators were argued to have selected for synchronized nocturnal emergence in all species.

Laboratory-based evidence concerning the relationship between photoperiod and temperature derives from Kureck (1979; 1980), who demonstrated experimentally that *Chironomus riparius* [as *thummi*] had a biphasic response, emerging at midday from cold (9–12 °C) water and mainly after dusk from warm (16–25 °C) water. At medium temperatures no intermediate emergence peak was observed. These results allow the interpretation of earlier findings concerning *C. riparius* of daytime eclosion (Phillipp, 1938) and biphasic, but mainly at dusk (Strenzke, 1960b); they refute the suggestion of Remmert (1962) that different species were involved.

9.3.2 LUNAR INFLUENCES

The moon controls tidal movements and usually provides some nocturnal illumination. Both properties, either singly or together, influence the emergence periodicity of chironomids in tropical and marine environments (Caspers, 1951; Fryer, 1959).

(a) The tropical environment

In the apparently homogeneous conditions encountered in tropical lakes, the sequence of emergence, mating and egg-laying might be expected to take place continuously, resulting in non-discrete cohorts of larval instars. However, MacDonald (1956), working on Lake Victoria (Uganda), noted that the life cycles of the two common species, *Chironomus* sp. and *Tanypus guttatipennis* (Goetghebuer), were related to lunar periodicity. Larvae of both populations pupated at the same lunar phase, just before the new moon. Although these species dominated collections, evidence was found that at least ten other species emerged in a similar pattern.

Lunar periodicity of emergence of the species *Chironomus brevibucca* (Kieffer) from Lake Bangweulu (Zambia) was observed by Fryer (1959). This species appeared in large numbers shortly after the appearance of the full moon. Similarly, a lunar periodicity was reported for two chaoborids and *Procladius brevipetiolatus* Goetghebuer at Lake George, Uganda, by McGowan (1975). Abundances were greatest in the first quarter of the lunar month and least in the third quarter. *Conochironomus acutistilus* (Freeman) was attracted to a light trap beside Lake Victoria at Jinja, Uganda in a periodicity that Tjönneland (1962) interpreted as a biphasic lunar pattern (of emergence) with peaks at around moon ages of 7 and 21 days. At higher latitudes, in eastern Transvaal, Frank (1965) found that *P. brevipetiolatus* emerged during the dark phase of the moon and suggests that moonlight may inhibit emergence of this species.

Although lunar periodicities clearly do occur in several tropical chironomids, there have been no published investigations of the phenomenon and the mechanisms controlling them have not been elucidated. This contrasts with the situation in marine areas, where the phenomenon has been well studied.

(b) The marine environment

The link between emergence and lunar/tidal cycles was first described by Chevrel (1894) for a species in the genus *Clunio*. Since then the phenomenon has been observed in several studies (Tokunaga, 1935; Caspers, 1951; Oka and Hashimoto, 1959; Neumann, 1966) and the subject has been comprehensively reviewed by Neumann (1976) in his account of adaptations of chironomids to intertidal environments.

Marine chironomids face a variety of problems in order to complete their development, namely physiological stress, exposure to wave action and the peculiar problems associated with lunar-driven tidal cycles. Life history stages must be synchronized to the environmental conditions pertaining at a site.

Where there is appreciable tidal movement, emergence and reproduction must take place during the period of ebb tide unless it is to take place on the moving water surface. The appearance of adults is affected by diurnal, fortnightly and sometimes annual changes in tide. The emergence of *Clunio* species varies locally according to variations in the tide. Caspers (1951) found that a Heligoland population of *Clunio marinus* Haliday emerged twice daily at the ebb-tide. In reviewing marine chironomids, Hashimoto (1976) noted that *Clunio tsushimensis* Tokunaga also emerges at spring tide but the diurnal emergence time changes with the seasonal variations in the height of the tide.

On the Pacific coast of Japan, the lowest tide occurs in the morning during the period June to September and at night during the period December to February. In the summer, *C. tsushimensis* emerges in the morning and in winter, at night. In transitional periods of March–May and October–November, when the difference between the two low tides is small, emergence takes place bimodally, in the morning and at night (Oka and Hashimoto, 1959).

The mechanisms controlling the periodicity of emergence of *Clunio* have been investigated by Neumann (see Neumann, 1976, for a review; and Neumann and Kruger, 1985). In a series of experimental studies, *Clunio* was shown to have two physiological time clocks. The first uses moonlight to synchronize the time of pupation. The second uses the day–night cycle to synchronize the time of eclosion of the adult. Tidal factors have no direct influence on emergence. The two systems work together to allow reliable advanced programming of insect development which is synchronized to tidal movements.

In southern Japan, *Clunio takahashii* Tokunaga lives in the midlittoral zone and emerges only when the habitat is exposed at low tide (Hashimoto, 1965). Such conditions occur twice a day but only for a short period of about 30 minutes. A similar pattern was observed for *Clunio marinus* in the Arctic (Neumann and Honegger, 1969). The tidal timing mechanism corresponds with the 'principles of an hourglass timer' which inhibits emergence for about 11–13 h during the final phase of pupation, when started by a temperature rise of a degree or more. In the Arctic this start takes place every 12.4 h during exposure of the midlittoral habitat in midsummer (Pflüger and Neumann, 1971; Pflüger, 1973).

In marine habitats with little tidal movement or where chironomids occupy the upper levels of the tidal zone, timing of emergence is probably achieved through the circadian clock mechanisms which control temporal programming in aquatic and terrestrial insects. In the open-sea race of *C. marinus* in the Baltic Sea, in which larvae live at a depth of 4–10 m, diurnal emergence occurs between midnight and morning and is controlled by the day/night cycle (Palmén and Lindeberg, 1959). In species of *Pontomyia* living in the Pacific Ocean, tidal cycles apparently are not crucial for synchronized emergence, which takes place at dusk or at night (Cheng and Collins, 1980). There is little tidal influence along the Black Sea coast, but the littoral habitat of *Clunio* is regularly exposed in the morning as a result of offshore winds which cause a slight fall in the water level. Emergence takes place early in the morning to coincide with the exposure of the habitat (Caspers, 1951).

Latitudinal and temporal shifts in tidal movements and amplitudes and the requirement for synchronization of emergence of *C. marinus* has led to the development of a variety of local races. Neumann (1976) has collated the information for geographic groups along the European coast, each of which is genetically adapted to local conditions. Although five races are recognized between the north of Spain and the Arctic, these are not considered as separate species on account of their morphological similarity and the fact that no sympatric populations have been observed.

9.3.3 PROTANDRY AND SEX RATIOS

Protandry is the term used to describe the phenomenon whereby males emerge slightly before females. The phenomenon may assist in outbreeding or in enhancing mating success. This pattern is standard for solitary bees and wasps but occurs frequently in many other groups of insects (Thornhill and Alcock, 1983). It has been observed in Chironomidae in arctic (Danks and Oliver, 1972a, b; Butler, 1980), northern temperate (Palmén, 1962; Boerger, 1981), temperate (Danks, 1978), subtropical (Wool and Kugler, 1969), tropical (McLachlan, 1986a) and marine habitats (Caspers, 1951; Hashimoto, 1962). The subject is reviewed by Wiklund and Fagerström (1977) and discussed in Thornhill and Alcock (1983).

Danks and Oliver (1972a) considered that protandry generally is partially controlled by female development requiring greater amounts of, for example, temperature (Haufe and Burgess, 1956). This is reflected in a sexual difference in growth rate (Fischer and Rosin, 1969) that Danks (1978) located between instar IV and emergence for *Chironomus decorus* Johannsen and *Endochironomus nigricans* (Johannsen). Butler (1980) also suggests that the slight difference in the value of the heat sum threshold for males and females may be the mechanism behind protandrous emergence.

In species with short-lived adults, advantages would be derived from an abundance of ready-to-mate males awaiting female emergence, thereby increasing the chances of mating success. Some species depend on the earlier emergence of the males: *Clunio marinus* males emerge within 2 minutes of arrival of the pupa at the surface but females take longer to emerge (Caspers, 1951). In this and in other species within the genus, female emergence may fail in the absence of males (Hashimoto, 1957; Oka and Hashimoto, 1959; Neumann, 1966). In the Baltic population of *Clunio marinus*, the female rarely emerges from the pupa without male assistance (Olander and Palmén, 1968). The activity of the protandrous males ensures that the exuviae are stripped from the females.

Protandry is not universal in chironomids and LeSage and Harrison (1980a) found no evidence for either daily or seasonal protandry in their study of Canadian *Cricotopus* species. The numerical bias towards females that they observed was attributed to parasitism by mermithids, which differentially damaged males (LeSage and Harrison, 1980b). In arctic ponds, one *Tanytarsus* species regularly showed 65–70% females at emergence, but the sex ratio of a sympatric congener did not differ significantly from 1:1 (Butler, 1984). Other examples of unbalanced sex ratios have been reported for six species of Tanytarsini which Lindeberg (1971) attributed partially to the occurrence of sporadic parthenogenesis.

In contrast to protandry, which may have some adaptive value, there is no evidence that deviations from a sex ratio of 1:1 have any value for aquatic insects (Williams, 1979).

9.3.4 DIEL EMERGENCE AND CHIRONOMID GROUPS

Learner *et al.* (1990) collated and analysed data on diel emergence from a range of studies in temperate and polar regions of the Northern Hemisphere. Their analysis showed a strong association between the diel emergence pattern displayed by a particular species and the tribe to which it belongs, although there is more variability in some tribes than others. For the taxa for which information was available the authors found:

- Chironomini species exhibit a generally crepuscular or nocturnal pattern of emergence;
- Macropelopiini species are characteristically diurnal with peak emergence in the morning;
- Metriocnemini species are also diurnal but with an afternoon peak of emergence;
- Tanytarsini species in general emerged mainly at dusk but with less abundant emergence at dawn or in the early morning;
- Orthocladiini appear to emerge usually during daylight but with peaks in early morning or late afternoon/early evening.

If advantage accrues to a species in emerging at some specified time of day, then Learner *et al.* (1990) assume that the association of a distinctive emergence pattern with a particular major taxonomic group reveals ancient patterns with little stimulus for subsequent change. Although changes in diel patterns can be achieved in the laboratory, Learner *et al.* state that these studies fail to show how such different patterns become established and persist in the field. However, the studies of geographical races of *Clunio* species in Europe (see Neumann, 1976) cited above seem to show how such patterns can be established in the genetic codes of species.

9.4 SWARMS AND SWARMING

9.4.1 THE INCIDENCE OF SWARMING

The aggregation of individuals prior to mating as seen in the aerial swarm of many insects is a typical mating behaviour of aquatic insects (Wiley and Kohler, 1984). It predominates in the archaic order Ephemeroptera, is found in the Trichoptera and occurs sporadically in Odonata. This behaviour reaches its highest development in the complex mating dances of the males of certain nematoceran Diptera, particularly the Tipulidae, Culicidae and Chironomidae. There are many references to the phenomenon in the literature (see Downes, 1969, and Sullivan, 1980, for reviews). These range from descriptive accounts (Miall and Hammond, 1900) to more detailed behavioural studies (e.g. Gibson, 1945; Syrjämäki, 1964; Kon, 1989).

The characteristic swarming of Chironomidae brings them to the notice of the public as a visually striking cloud of insects or even as a potential nuisance (Chapter 13) despite their inability to bite humans. Dense columnar swarms of midges are sometimes seen extending upwards from tree-tops, roofs or towers where they have been mistaken for plumes of smoke (Thienemann, 1954) and for residents around lakes excessive numbers of chironomids can cause problems during periods of maximum emergence (Mulla, 1974; Beattie, 1981; Chapter 13).

9.4.2 THE OCCURRENCE AND LOCATION OF SWARMS

Swarms are most frequently seen in calm weather (wind speeds less than 11 km h^{-1} when the air temperature exceeds 10 °C. Generally no swarming occurs at wind speeds greater than 20 km h^{-1} (Beattie, 1981) and swarms observed by Gibson (1945) during wind gusting at 32 km h^{-1} at 5 m above the ground, formed only in the most sheltered places. Actual wind speed in excess of 15 km h^{-1} experienced by the swarms prevented swarming.

Lindeberg (1964) stresses the difficulty of providing an exact description of the swarming sites of a species because, although most species have characteristic swarming sites and specific times of flying, changing weather (wind, light intensity and humidity) and diversity of the terrain may modify these. Nevertheless, remarkable constancy of swarming sites for certain species have been observed. Year after year and during each generation, the individuals of a species swarm at exactly the same spot in favourable weather conditions.

Lindeberg (1964) categorizes swarming sites as: above the water, at the shoreline, in an open wood, in the shade at the edge of a wood or on open land such as meadows or moorland. The site may be further subdivided according to altitude into four main strata:

- Very near the ground (often at the shore) or at 0.5–2.0 m (in the wood, in open places).
- 5–15 m above high shrubs and between the tree-tops.
- Above the tops of the trees.
- Very high up in the open air.

In urban areas swarms may be associated with buildings (Gibson, 1945; Thienemann, 1954; Mulla, 1974). Some authors (Harnisch in Gruhl, 1924; Edwards, 1929; Gibson, 1945) suggest that the altitude of the swarm is proportional to the size of the fly. However, this is not supported by Lindeberg (1964), who found small species swarming above tree-tops and larger species closer to the ground, apparently because of a preference for shade.

9.4.3 COMPOSITION OF SWARMS

The numbers of adults in swarms range from very low, as for example in *Smittia extrema* (Holmgren) from Spitzbergen where swarms ranged in size from one or two to about 15 individuals (Syrjämäki, 1968a), to hundreds of millions of individuals, as in the extensive swarms of *Limnophyes minimus* (Meigen) reported by Gibson (1945) from a sewage treatment works in England and the zonal swarms of *Tanytarsus gracilentus* (Holmgren) and *Chironomus islandicus* (Kieffer) observed around Lake Myvatn, Iceland (Lindegaard and Jónasson, 1979).

Swarms may contain both sexes (Nielsen and Nielsen, 1962), males of several species (Darby, 1962) or more usually males of a single species (Lindeberg, 1964). Only 1% of swarms containing >50 individuals of *Stictochironomus crassiforceps* (Kieffer) were females (Syrjämäki, 1964) but in smaller swarms of up to 10 individuals, one or two females were present.

9.4.4 SWARM MARKERS

The position and boundary of the swarm is related to a discrete marker that is visually recognized by the swarming adults. Downes (1969) notes that swarm markers, though taking many forms, are usually objects that human beings would also regard as useful landmarks – large objects or those that contrast against ground or sky, or with sharp boundaries or conspicuous angles. The marker may present a range of characteristics to the adult fly depending on the angle at which it is approached and the height at which it is viewed. This means that a single marker can be used by more than one species simultaneously (Downes, 1958), a point supported by Lindeberg (1964) in his observations of 'composite swarms' of chironomids.

In terms of the response of swarms to different coloured markers, the quantity of light (intensity) appears to be more important than the quality (colour) but midges are attracted to both dark and light markers (Gibson, 1945). The flies respond to the contrast between the marker and the background. Experimentally, swarms have been produced artificially by placing a light-coloured marker on the ground (Syrjämäki, 1964). The response of swarming chironomids to bright objects and their tendency to accompany a moving object can be a source of annoyance, for example at a boat club on a small reservoir in south Wales bald-headed individuals have been followed by swarms while out in their boats. At Cow Green Reservoir (England), the author was followed along the shore by large swarms of *Chironomus anthracinus* Zetterstedt which were apparently attracted by an orange life-jacket. When this was removed and placed on the ground, the swarm was displaced and formed above it.

Swarm size and shape may be determined by the size of a marker, with large markers frequently attracting large swarms. Increases in the numbers of males joining a swarm may cause the shape to change from spherical to columnar (Syrjämäki, 1964; Sullivan, 1980).

9.4.5 CLIMATIC MODIFICATION OF SWARMING

The orientation relative to a marker of a swarm may be modified through the effects of wind, temperature, light and humidity. The general effects of wind on formation of swarms has been discussed above but wind also influences the orientation and behaviour of swarms. The dances of chironomids within a swarm vary and include oblique cruising across the wind current, vertical movement and direct flight against the wind with retirement whenever the wind speed increases, followed by advances as a result of more rapid flight (Gruhl, 1924; Gibson, 1945). All individuals within the swarm face into the wind but this uniformity is lost in still air. In calm conditions the direction of the body axis of *S. crassiforceps* is perpendicular to the shore line with the heads pointing shorewards (Syrjämäki, 1964). It is self-evident that swarming individuals should face the wind: for by flying against the air current and adjusting their own velocity to that of the wind they are able to maintain station with respect to the ground below and gain lift using the air currents. Orientation in flight is achieved only by visual inspection of the apparent direction of movement of the landscape (Kennedy, 1940; Haskell, 1966).

The intensity of swarming appears to correlate with temperature. The number and size of swarms of *S. crassiforceps* are positively correlated with rising temperature (Syrjämäki, 1964). However it is difficult to separate the influence of temperature from that of light as a controller of swarming. It has been demonstrated that light intensity plays an important role in the onset and cessation of swarming (Gibson, 1945; Syrjämäki, 1964). Several authors have also noted that the light intensity under which swarming occurs is inversely correlated with temperature (Syrjämäki, 1966; Römer and Rosin, 1971; Reisen *et al.*, 1983; Kon, 1984). *Chironomus yoshimatsui* Martin and Sublette formed swarms after sunset in summer but before sunset in spring and autumn (Kon, 1984), a pattern repeated in *Chironomus samoensis* Edwards [as *C. flaviplumus*] (Kon, 1989). An interesting feature which emerges from these observations is the low light at which some species can maintain swarm structure. Gibson (1945) found that *L. minimus* was able to form swarms at a light intensity of 0.2 lux, and Kon (1984) records swarming at 1 lux for *C. yoshimatsui*. Swarms of *Phaenopsectra* [as *Lenzia*] *flavipes* were observed by Syrjämäki (1964) when the light intensity was so low that it could not be measured. Orientation on a ground marker would be difficult, but it is possible that the weak contrast between the sky and an object in relief would provide sufficient visual clues.

The physical effects of rain are known to inhibit swarming, and humidity has been considered to affect it in some way, but three separate studies (Gibson, 1945; Nielsen and Greve, 1950; Syrjämäki, 1964) have failed to show any relationship between relative humidity and the number and size of swarms. However, Syrjämäki (1964) found that sexual activity was clearly reduced at a relative humidity of 0%. In addition, 'resting' adult midges appear to be distributed along a moisture gradient. Syrjämäki (1960, 1963) found that older adults showed a preference for more humid locations than freshly emerged specimens, and Wilson (1969) observed resting *Chironomus* sp. at their greatest densities within 10 cm of the edge of a lake.

9.4.6 DIEL PERIODICITY OF SWARMING

Swarming is linked with the diurnal change of illumination and/or temperature and Syrjämäki (1964) has divided chironomids into daylight and twilight swarmers. The latter group may be subdivided according to whether dusk or dawn is preferred. Most species studied swarm only at dusk (*Chironomus pseudothummi* Strenzke: Syrjämäki, 1966) but there are examples of species which swarm at both dawn and dusk (*Chironomus strenzkei* Fittkau: Syrjämäki, 1965) and those that swarm only at dawn (*Glyptotendipes paripes* (Edwards): Nielsen and Nielsen, 1962). In the Arctic, in conditions of continuous daylight, a diel periodicity has been observed in *S. extrema* (Syrjämäki, 1968a). No swarming was observed between about 22.00 and 02.00 h, irrespective of the weather, but it is unclear whether this periodicity is under exogenous or endogenous control. *C. yoshimatsui* has diel activity under endogenous control (Kon, 1985) but later work has shown that this may be modified by temperature (Kon, 1986). The questions are complex and more work is required to determine the relative importance of the factors involved in maintaining the rhythmic activity of chironomid adults.

9.4.7 SEASONAL ASPECTS OF SWARMING

Seasonal aspects of swarming have been studied in multi-voltine species. In general the features investigated include all those dealt with above. The work of Kon (1984; 1989) clearly shows the relationship between swarm occurrence and duration between spring and autumn. Swarming periods for *C. yoshimatsui* were longer (2–3 h) in the spring and autumn than in the summer (<1 h). A possible explanation is suggested by the work of Chiba *et al.* (1982) on mosquitoes. As temperature becomes lower the evening and morning peaks occur earlier and later, respectively. This results in greater activity in lighter times of the day at low temperatures than at higher. Conversely in some mosquitoes the proximate factor which determines the timing of swarming may be light intensity. The response of particular species to specific light intensity will vary with temperature (Nielsen and Nielsen, 1962; Chiba, 1967), so that the timing of the swarming activity appears to change with temperature. The controlling mechanisms in the Chironomidae are as yet unresolved but, as indicated by the work of Kon (1984, 1989), they are likely to be similar to those outlined here for mosquitoes.

9.5 MATING AND OVIPOSITION

9.5.1 THE FUNCTION OF SWARMING

The commonly held view is that the swarm, which with few exceptions (section 9.4.3) is composed largely of males, ensures the meeting of the two sexes for mating purposes (Downes, 1969). Swarms may provide the female with the most efficient means of becoming inseminated. Swarming may allow sexual selection, since large male size, fast flight and complex manoeuverability may indicate genetic quality (Sullivan, 1980). Conversely, McLachlan (1986b) equates small size and aerobatic ability of males with reproductive success (section 9.6.3).

The aerial swarm is predominantly monospecific. Although Lindeberg (1964) noted that more than one species may be obtained by random collections through a swarm, close observation of the swarm shows that each species forms its own assembly, usually at different altitudes. Despite overlapping ranges, the swarm centres differ and the populations are actually isolated.

Although swarms evidently have a central role in the mating procedure of many species of chironomid, it is not a universal behaviour pattern. Some species, such as *C. pseudothummi* and *C. yoshimatsui*, copulate only in the air (Syrjämäki, 1966; Kon, 1985), whereas others such as *S. crassiforceps*, *C. strenzkei* and *T. akamusi* do so both in swarms and also on the ground (Syrjämäki, 1964, 1965; Otsuka *et al.*, 1986). Yet other species dispense with the swarming habit altogether and they are the subject of section 9.5.4. This variety in mating locations and associated behaviour has resulted in a diversity of mating positions – face-to-back, end-to-end, face-to-face (Syrjämäki, 1964) – which have led to structural changes in the hypopygium (Reiss, 1966, 1971; Fittkau, 1971b). In *S. crassiforceps*, the male mounts from above (end-to-face position) and the switch to end-to-end position is accompanied by a rotation between the 7th and 8th abdominal segments which allows the male to remain upright (Syrjämäki, 1964). In the marine genera *Pontomyia* and *Clunio*, the hypopygium rotates 180° before copulation (Hashimoto, 1957).

9.5.2 MATING IN THE SWARM

Mating in the aerial swarm has been described in a number of studies in addition to those mentioned above (*L.* [as *Spaniotoma*] *minimus*, *Metriocnemus obscuripes* Kieffer [as *longitarsus*], *Chironomus dorsalis* (auctt. nec Meigen): Gibson, 1945; *Chironomus salinarius* Kieffer: Koskinen, 1969; *Chironomus plumosus* (Linnaeus): Hilsenhoff, 1966; Römer and Rosin, 1969; *C. riparius* Meigen: Caspary and Downe, 1971 [and as *thummi*] and *Chironomus piger* Strenzke: Miehlbradt and Neumann, 1976).

A recent account is given by Kon (1984) for *C. yoshimatsui*. Males gradually leave the resting area to form a swarm in which they fly in a zigzag manner over a fixed marker. The flying males never contact one another in the swarm. Soon after the onset of swarming the females gradually enter the swarm. Females were easily distinguished from males on account of their thicker silhouette and a flight pattern which consisted of slow movement in a straight line, rather like the 'offering flights' described for *S. crassiforceps* (Syrjämäki, 1964). At a distance of slightly less than 2 cm, the male reached for and grasped the female, which then flew carrying the male in the end-to-end position in which the male hung motionless. Copulation lasted for 6–7 seconds in this position in the air. After copulation the male sometimes returned to the swarm.

Doubt has sometimes been expressed that copulation during swarming actually results in ejaculation of sperm (Syrjämäki, 1964 on *S. crassiforceps*) but Kon (1986) found evidence for insemination in *C. yoshimatsui* and suggests that in this species swarming has an epigamic function and is obligatory for mating. Caspary and Downe (1971) found that in matings of 2–3 seconds in *C. riparius* sperm was not found in the female ducts, whereas matings of 5–10 seconds regularly produced sperm. When swarming was suppressed, no matings occurred.

Sex recognition is based on an auditory response. The large plumose antennae of many male nematocerans are able to detect the female flight sound in the air (Wishart and Riordan, 1959; Belton, 1974; Ikeshoji, 1981). Males recognize females by their flight sound, which is generally lower than that of the males (Säwedal and Hall, 1979), This was noted by Syrjämäki (1966) who, while humming a Finnish folk song in the vicinity of a swarm of *C. pseudothummi*, found that the whole swarm descended and the nearest swarmers flew very vigorously near his mouth. In a later study he was able to use this feature to initiate swarming of *S. extrema* (Syrjämäki, 1968b) using a tone of about 300Hz. Römer and Rosin (1969) confirmed that *C. plumosus* swarms respond to flight tone. The tonal frequency was found to be linearly related to temperature by Säwedal and Hall (1979). In the Japanese *Rheotanytarsus kyotoensis* (Tokunaga), Ogawa (1992) found the most attractive frequency changed by about 11Hz $°C^{-1}$ between 8.8 and 18.2°C. Working on *Smittia* sp., Delettre (1984) suggests a link between the flight tone, temperature, size of swarm and number of couplings. Syrjämäki (1965) observed a density-dependent behavioural factor in the swarming of *C. strenzkei* where flight speed rose with increased size of the swarm. Olfactory signals may also be important in some species but have not been investigated.

An additional density-dependent feature of swarming has been observed by Neems *et al.* (1992). Large swarms of *C. plumosus* attracted more females than small ones, but despite this the probability of mating was found to be greatest in the smallest swarms. This was offset by predation risk being higher in small swarms.

9.5.3 ISOLATION AND BIOLOGICAL FUNCTION

The separation of swarms of two or more species within a small area or over the same marker (Lindeberg, 1964) is clearly demonstrated in the study of LeSage and Harrison (1980a). These authors found that *Cricotopus* species from a stream could be approximately grouped on the height of swarming. One group swarmed over grass and less than 1 m above the ground (*Cricotopus annulator* Goetghebuer, *C. festivellus* (Kieffer), *C. infuscatus* (Malloch)) and the other group swarmed about 2–3 m above the ground (*C. bicinctus* (Meigen), *C. luciae* LeSage, *C. trifascia* Edwards and *C. varipes* Coquillett). Not only were swarms separated by height, but species were clearly differentiated by the horizontal distance of the swarm from the stream (Figure 9. 4).

In addition to these spatial factors, circadian rhythms of flight activity and mechanisms that shift the flight activity of multivoltine species seasonally serve also to bring sexes of the same species together for mating (Kon, 1987). The frequency of female flight sound which releases male mating behaviour

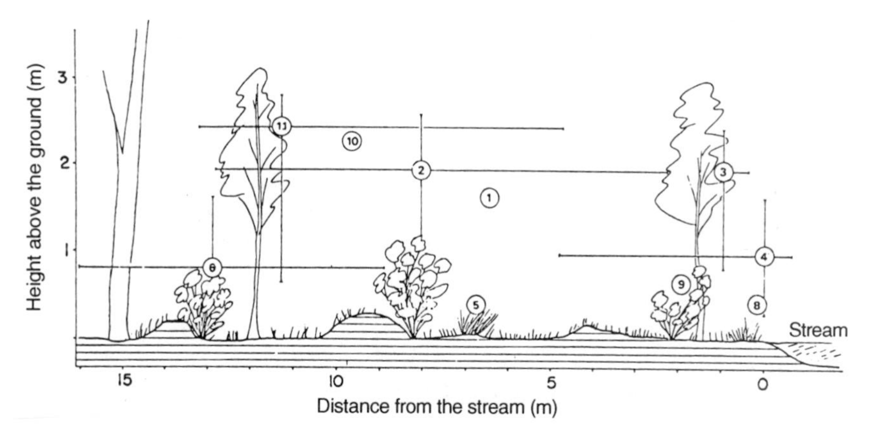

Figure 9.4 Location and height of *Cricotopus* swarms along Salem Creek (Ontario, Canada); numbers of swarms collected in parentheses. 1, *C. sylvestris* (1); 2, *C. bicinctus* (74); 3, *C. varipes* (9); 4, *C. triannulatus* (10); 5, *C. infuscatus* (3); 6, *C. festivellus* (7); 8, *C. annulator* (2); 10, *C. luciae* (1); 11, *C. trifascia* (11). (Reproduced from LeSage and Harrison, 1980a, by permission of *Archiv für Hydrobiologie*.)

in the swarm (Kon, 1989) has a similar effect. Spatial, temporal and ethological factors operate on both sexes to isolate populations and this isolating function of swarms avoids interspecific copulation, with its waste of time, effort and gametes (Wiley and Kohler, 1984).

In his review of mating systems in chironomid midges, Kon (1987) notes that swarming tactics may secure high mating rates of females when population densities are low. If little effort is required for the female to travel to the swarm, then swarming tactics alone will secure a consistently high mating rate for an individual female. In contrast, for an individual male a combination of swarming with searching methods may be cost-effective. For example, *T. akamusi* uses both these tactics and can copulate with virgin females which are located by searching before swarming (Kon *et al.*, 1986). This technique appears to maximize the chances of a successful mating and it is somewhat surprising that some species are apparently restricted to mating in swarms (*C. riparius*: Caspary and Downe, 1971; *C. yoshimatsui*: Kon, 1986).

The biological function of swarming is still a complex issue. Syrjämäki (1964) cites many examples where swarming is not considered to be part of the sexual activity of the species. Nielsen and Greve (1950) and Nielsen and Haeger (1960) suggest that swarming in mosquitoes is of no real importance as an occasion for mating. Swarming is said to play no part in the meeting of sexes of *S. crassiforceps* because emergence is at midnight when there is no swarming, and they immediately fly to the nearest shore, where they remain within a restricted area (Syrjämäki, 1964). This of course does not prevent sexual encounter in a swarm, and Downes (1969) finds it implausible that retained swarming is a functionless ritual, since the habit is costly both of time and effort and is likely to be retained as an integral part of the species' life history. It seems more likely that the truth, as usual, lies somewhere between these two views. For example, in the case of *S. crassiforceps*, although most matings do occur on the ground, some were reported within the swarm. Swarming may have been an integral part of this species' behaviour but its evolution is moving away from swarming towards 'walking type' mating behaviour (Syrjämäki, 1964). Thus, in this species, swarming bears no biological significance and represents only a 'phylogenetic contraint'. The issues are further discussed and elaborated upon in Sullivan (1980) and McLachlan and Neems (in press).

9.5.4 NON-SWARMING MATING ACTIVITY

'Swarming' and 'searching' are treated as two distinct tactics and mating systems are grouped into three types by Kon (1987). In the first, mating takes place only in the air by swarming; in the second, mating occurs in the aerial swarm or on the ground by searching; and in the third the aerial phase is omitted and mating takes place only on the ground (or water) by searching. Swarming aspects are covered in some detail above: in this section the searching technique is reviewed.

Non-swarming mating behaviour can be considered as a response to particular environmental conditions. In cold, wet and windy locations the opportunities for swarming are restricted and swarming might actually be disadvantageous – selection for mating behaviour that dispenses with swarming is high. Examples are seen in marine (Hashimoto, 1976) and temperate and arctic chironomids (Oliver, 1968; Hågvar and Østbye, 1973; Ferrington and Sæther, 1987).

Before considering more extreme departures from the swarming habit, it is useful to examine species which retain some aspects of swarming. In *S. crassiforceps*, aerial swarming is reduced in importance and most copulation occurs on the ground (Syrjämäki, 1964). There are no specific courtship patterns. The male is the active partner and the female remains passive throughout the mating process apart from attempts to escape the attentions of the male. The male responds sexually to a wide range of objects including living and dead males and females of the same species, and females of different species. Responses varied greatly and copulation was stimulated most by general activity and particularly to the movements of the female's legs. Complex courtship behaviour appears to be unnecessary where high densities of both sexes occur together at a site.

Mating in *T. akamusi* is also in a swarm and/or on the ground, with males preferring freshly emerged specimens of either sex. This is attributed to the selective advantage of copulating with a virgin female since the first copulation can provide the dominant contribution to fertilization (Otsuka *et al.*, 1986).

Multiple copulation by female chironomids is reported rarely and most species mate once or twice (Martin and Porter, 1977). However, there is indirect evidence of multiple insemination in *Chironomus oppositus* Walker

(Martin and Lee, 1989). In contrast, male chironomids may copulate frequently (Kon, 1986).

Pelagic swarming has been adopted in some species that have abandoned aerial swarming as in *Abiskomyia virgo* Edwards, a species with both parthenogenetic and sexual strains. The adults never leave the surface of the water and mate on this substrate. Males in particular are structurally modified with a reduced antennal plume, relatively stout legs and terminalia rotated up to 90° (Lindeberg, 1974).

Mating without aerial swarms in freshwater environments is seen in the genus *Diamesa*. A Himalayan wingless glacier midge (*Diamesa* sp.) mates on the snow around the glacial meltwater drainage channel (Kohshima, 1985). Males appear to search for females by walking around the emergence site. Kon (1987) observed male *Diamesa japonica* Tokunaga gliding on the surface of a fast-flowing stream and mating with newly emerged females. A similar mating strategy is described for the winter-emerging temperate orthoclad *Oliveridia hugginsi* Ferrington and Sæther (Ferrington and Sæther, 1987). Males glide actively across the water surface and search for females. Copulation lasts for 3–7 seconds and *O. hugginsi* is reluctant to fly and could only be induced into making very short 1–2-second jumping flights, never above 15 cm. In the orthoclad *Belgica antarctica* Jacobs, large aggregations of males and females occur together at one time in space and frenzied sexual activity can be correlated with temperature maxima near 21 °C (Peckham, 1971). Surface mating provides the most selective advantage for winter-emerging temperate species (Hågvar and Østbye, 1973; Ferrington and Sæther, 1987).

Non-aerial swarming and corresponding morphological adaptations are evident in marine midges where lunar/tidal diel emergence rhythms (Neumann, 1976) influence adult reproductive behaviour (section 9.3.2). *Pontomyia* are small flightless midges which occur on many tropical and subtropical oceans shores and reefs (section 6.8). The males have reduced antennal plumes and possess oar-like wings which propel them over the water surface (Cheng and Collins, 1980) in a motion which was mistakenly called swimming by early authors. The females, which are wingless and have very short legs, float passively on the water surface until found by the males. Males appeared to be strongly attracted to freshly emerged females and mating usually took place within 60 seconds of female emergence, with copulation lasting for a 'few minutes'.

The marine littoral genus *Clunio* has been the subject of numerous studies (e.g. Hashimoto, 1957; Olander and Palmén, 1968; Neumann and Honegger, 1969; Hashimoto, 1976). Female *Clunio* are vermiform, being flightless and virtually legless, whereas males skate over intertidal pools, exposed rocks, seaweed and sand. Copulation takes place on shore exposed at low tide (Caspers, 1951) or on the water surface (Olander and Palmén, 1968). Copulation begins immediately after the male has stripped the pupal exuviae off the female (which takes some 15 seconds). The female in the end-to-end position is dragged along by the flying male until copulation is complete (Figure 9.5).

Other common denizens of the intertidal zone belong to the subfamily

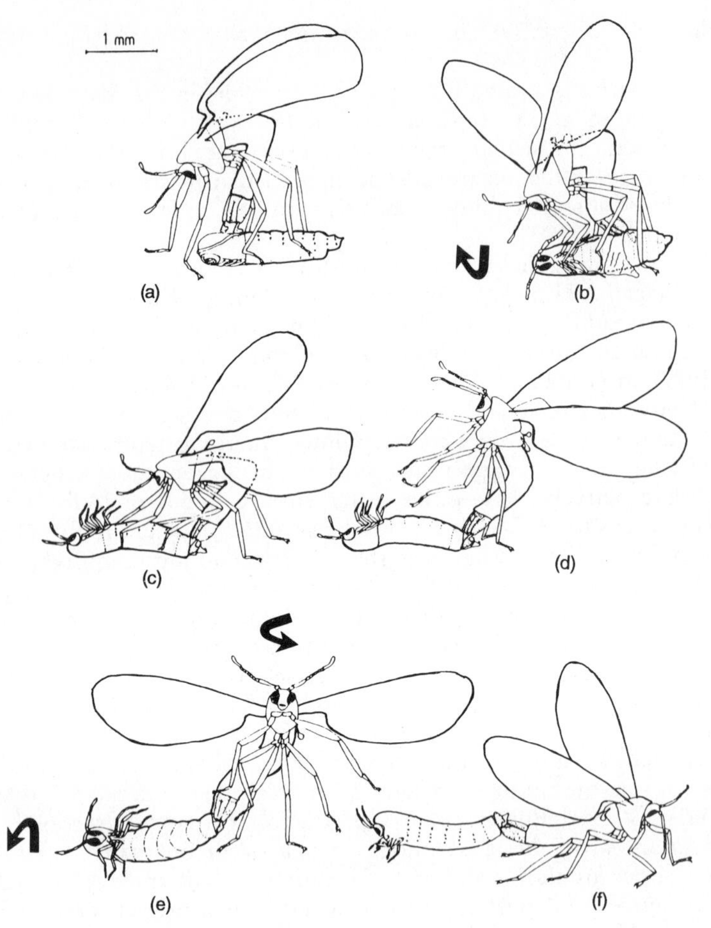

Figure 9.5 The positions in copulation of *Clunio marinus*. (Reproduced from Dordel, 1971, by permission of The Entomological Society of Canada.)

Telmatogetoninae. Most members of *Telmatogeton* live in the littoral zone amongst seaweed such as *Enteromorpha* but some species live in the splash zone of waterfalls in Hawaii (Wirth, 1947). Adults of most species can fly but normally run rapidly over the surface of rocks in the splash zone near the high water mark. Copulation takes place on exposed rocks and is accompanied by male hypopygial rotation.

9.5.5 SEARCHING VERSUS SWARMING

Successful swarming is unlikely to be attained in exposed locations and the 'searching' strategy is most common in harsh environments where the adaptations described above aid successful copulation. Furthermore, some species living in harsh locations that do swarm can also mate on the ground, for example *Thalassosmittia*. Since such species maximize their chances of mating, it is appropriate to ask why swarming alone is so common.

It is puzzling that males swarm despite severe competition for mates and females are attracted to a location (the swarm site) which includes no resource for them (Kon, 1987). Kon (1987) questions how selection for swarming took place if there is no resource for the female, now or in the past. On the other hand, swarming need not have current adaptive significance, but may be no more than an historic burden derived from the origin of flight in aquatic insects. The precise location of swarm sites allows effective mate finding compared with searching, and this is particularly beneficial if individuals are rare, or dispersed and at low density. Under these conditions, mate location in a swarm ensures continuation of the behaviour.

9.5.6 PARTHENOGENESIS

Parthenogenesis is widespread amongst insects (Suomalainen *et al.*, 1976) although not particularly common in the Chironomidae. It is found mainly in the Tanytarsini and Orthocladiinae (Lindeberg, 1958; Edward and Colless, 1968; Delettre and Cancela de Fonseca, 1979; Cranston, 1987).

The adaptive significance of parthenogenesis is discussed by Thornhill and Alcock (1983). Asexual forms arising in a population with males and females may acquire a 'huge reproductive advantage' by producing only daughters under certain conditions. If a sexually reproducing female invests her resources equally in sons and daughters and it is assumed that parthenogenetic females secure as many resources for offspring production as sexual females, then there will be twice as many female offspring produced by the asexual as the sexual female. If survival of the parthenogenetic females is equivalent to the sexual ones, then the proportion of asexual forms double from generation to generation, leading eventually to the elimination of sexual females (and by association sexual males) by asexual parthenogenetic females (Maynard-Smith, 1971).

For chironomids in certain environments, parthenogenesis may confer a considerable reproductive advantage. It removes not only the necessity for the quasi-synchronous emergence of the two sexes but also the need for the adults to congregate either in swarms or on the land surface (Delettre and Cancela de Fonseca, 1979). Some species which have adopted parthenogenesis include the pest species *Paratanytarsus grimmii* (Schnieder) which is able to reproduce in water mains (Langton *et al.*, 1988); *Bryophaenocladius furcatus* Kieffer, a pest of greenhouse crops (Cranston, 1987); and *Eretmoptera murphyi* Schaeffer, an inhabitant of damp moss in the sub-Antarctic (Cranston, 1985). In all these species, parthenogenesis has imparted a reproductive advantage in environments in which normal sexual reproduction involving swarm formation would be difficult. In the sub-Antarctic Kerguelen Isles, *Limnophyes pusillus* (Eaton), which reproduces sexually in Europe though suspected of facultative parthenogenesis (Goetghebuer, 1932), is parthenogenetic, presumably as an adaptation to the harsh fluctuating environment (Delettre and Cancela de Fonseca, 1979). Crafford (1986) noted similar adaptations in nearby Marion Island. More puzzling is the occurrence of both parthenogenetic and normal populations of *A. virgo* in rather similar lacustrine environments within the Arctic Circle (Lindeberg, 1974).

Sporadic parthenogenesis has been reported in some species. Beermann (1955) observed that the embryonic development of eggs of *Chironomus* (*Camptochironomus*) began without fertilization but no larvae hatched. Grodhaus (1971) noted that some egg batches of *Chironomus atrella* (Townes), *C. 'attenuatus'* Walker and *C. stigmaterus* Say showed evidence of parthenogenesis and some hatched larvae reached the fourth instar stage. Lindeberg (1971) suggested that the occurrence of sporadic parthenogenesis may account for the unbalanced sex ratios observed in some northern chironomid populations, but this is untested and the frequency of the phenomenon and incidence of complete and successful development is unknown.

9.5.7 OVIPOSITION

Records of the observation of oviposition are infrequent, but females of aquatic chironomids deposit their eggs at or close to the surface of the water at dusk or dawn (Oliver, 1971). Syrjämäki (1965) found that most egg masses of *C. strenzkei* appeared during the night. A sudden start to oviposition by *Cricotopus* species at dusk in a Canadian stream was noted by LeSage and Harrison (1980a). This was apparently triggered by falling light and stopped by oncoming darkness but there was also minor oviposition at dawn. LeSage and Harrison also observed a generic succession of ovipositing females, with Chironominae depositing first followed by *Cricotopus* and then *Orthocladius*. Statistical analyses indicated that a similar succession took place within the genus *Cricotopus*.

The potential reproductive capacity seen within Chironomidae, as measured by the number of female gametes (fecundity), can range from single eggs to egg-masses containing up to about 3000 eggs (Nolte, 1993). Generally speaking large species (with size measured by wing length) carry more eggs than do small species. Thus a species of *Corynoneura*, with a wing length of about 1 mm, produces about 70 eggs per female whereas *Polypedilum convictum* (Walker) (wing length about 2 mm) produces about 300 eggs per female (Svensson, 1979). Most species lay a single egg-mass but some species (*Chironomus tepperi* Skuse) lay up to six egg batches, particularly if the females are able to feed. The numbers of eggs per mass and their fertility are reduced in successive batches (Martin and Porter, 1977).

(a) Oviposition sites

Most aquatic chironomids lay their eggs at or near the water surface. Credland (1973a) used a floating string on which females oviposited and from which the eggs were removed for experimental purposes.

Egg-masses are usually deposited individually but in some cases mass depositions are observed, in lakes (Wesenburg-Lund, 1913), streams (Terry, 1913), rivers (Dinulesco, 1932; LeSage and Harrison (1980a) and the Baltic Sea (Remmert, 1960). The subject is reviewed in detail in Nolte (1993). Occasionally the masses may be so large as to be visible at a distance – for example, the ribbon 10–15 cm wide of *Cardiocladius capucinus* Zetterstedt [as *leoni* Goetghebuer] egg-masses along the rocky shore of the River Danube (Romania) (Dinelesco, 1932). The mass deposition of an unidentified chironomid on wet road surfaces near the River Main in Germany (Noll, cited in Thienemann, 1954) comprised densities of about 90 egg-masses per m^{-2}. Considering the area covered and the number of eggs per mass (c. 300) it was estimated that 600×10^6 eggs had been deposited.

Occasionally oviposition may cause financial loss, as in South Africa when egg-masses of *Kiefferulus* [as *Chironomus (Nilodorum)*] *brevibucca* (Kieffer) were deposited on car roofs. Very few egg-masses were laid on light cars or those with dull black vinyl tops; the shiny roofs of dark-coloured cars were preferred, presumably mistaken for bodies of water (Theron, 1972). The combination of dew, gelatinous egg-mass and sunlight caused the paintwork to discolour, crack and lift and 180 new cars had to be resprayed as a result of this damage.

For chironomid species living in marine littoral environments, oviposition must be synchronized with tidal exposure. This derives from the synchronization of emergence (Neumann, 1976) and oviposition may take place within 5–20 minutes of emergence (Oka and Hashimoto, 1959; Morley and Ring, 1972; Cheng and Collins, 1980). In *Telmatogeton* (including former *Paraclunio*), eggs are laid singly without a gelatinous coat (Saunders, 1928; Tokunaga, 1935) but in marine orthoclads including *Clunio*, *Thalassosmittia*, *Halocladius* and Chironominae, eggs are laid in gelatinous masses. Oviposition in the tanytarsine *Pontomyia cottoni* takes place within 0.5 to 5 minutes of mating.

Oviposition sites may be extremely specific. For example, in *Metriocnemus knabi* Coquillett, an inhabitant of water held in the modified leaves ('pitchers') of the pitcher-plant, *Sarracenia purpurea* (Sarraceniaceae), oviposition occurs into overwintered or newly opened leaves (Paterson and Cameron, 1982). There is also evidence that ovipositing females may be able to discriminate between solutions of pollutants. Females of *C. riparius* showed some ability to discriminate between different cadmium (Cd) solutions, although they were rather insensitive, and avoided water which was toxic to first instar larvae. On the other hand females avoided ovipositing into solutions which were acutely toxic to eggs (Williams *et al.*, 1987).

9.6 DISPERSAL AND FLIGHT

Chironomids colonize new habitats predominantly through passive dispersal of fertilized females. Although their flight is generally weak, adults have been found at high elevation (Glick, 1960) and over oceans and seas (Holzapfel and Perkins, 1969; Cheng and Birch, 1977). Hoffrichter (1973) has cited a particular instance of vehicular traffic as a dispersal agent for chironomids and viable adults are amongst the insects intercepted by quarantine of international air traffic into Australia. In an unusual case of anthropogenic dispersal, Block *et al.* (1984) deduced the accidental introduction of *E. murphyi* to maritime Antarctic Signy Island from sub-Antarctic South Georgia with experimentally transferred plant material. General high fecundity also ensures that the arrival of one or a few females may have the potential to colonize a habitat.

Most studies of colonization have followed the development of larval communities in newly created water bodies (Armitage, 1977; Street and Titmus, 1979; Caspers, 1983a; Nolte, 1988; Matěna, 1990; Layton and Voshell, 1991) and little attention has been paid to the 'pool' of potential colonizing adults in the aerial flow. An exception is the work of Delettre (1993) who related the distribution of larvae to the movements of adult chironomids.

Chironomidae are likely to be dispersed unintentionally during their regular movement. Three main periods of movement in adult chironomids can be distinguished (Oliver, 1971):

- Initial dispersal from the eclosion site to a resting area and subsequent adjustments to prevailing weather conditions.
- The swarming flight.
- The oviposition flight of females.

9.6.1 INITIAL DISPERSAL

Movement from the eclosion site, particularly in lacustrine chironomids, is influenced strongly by the wind (Nielsen and Nielsen, 1962; Davies, 1976a) but light may also focus the direction of flying adults (Beattie, 1981; Kokkinn and Williams, 1989; section 13.4.1). Attraction to light causes nuisance to humans: massed emergences may be attracted to artificial lights and enter houses, hospitals and business premises (Ali, 1980a) where they may cause considerable damage and inconvenience (Chapter 13). The eventual resting place of species emerging in more natural environments (where attraction to light is not a problem) is determined by a combination of temperature and wind (Konstantinov, 1961). These two factors control the ambient humidity, which is an important factor for the adult must avoid water loss.

Wilson (1969) observed shoreline aggregations of summer-emerging adult chironomids (*Chironomus plumosus/staegeri*?) which appeared to be related to humidity gradients. These adults showed a diel activity pattern involving resting in the woods bordering the lake during the day from 10.30 to 18.00 h, at which time midges slowly streamed out of the woods until the sun sank below the lake horizon at 20.40 h. At this time the air became full of swarming midges until at 21.15 h 'hordes' descended and settled on the sand of the lakeshore. The population on the shoreline reached its peak abundance at about midnight. When the sun brightened the morning sky at 05.00 h, midges started to leave the beach. The rate of departure increased when the sun shone directly on the shoreline and by 10.30 h almost the entire population had sought shelter in the woods. This example illustrates an extrinsic control of diel periodicity.

Such humidity reactions may be related to the observed diel emergence of some summer species which emerge after sunset (Palmén, 1955; Syrjämäki, 1966). The responses to humidity may not be as marked in spring-emerging species because of the cooler temperatures (Oliver, 1971), but this supposition is speculative.

9.6.2 THE SWARM IN DISPERSAL

The swarm is not a dispersal phase but the action of wind on large aggregations of individuals may disperse populations away. Beattie (1981) reports the removal of most of the adults away from a lake as a result of unsuitable weather conditions. Wind in general inhibits swarming and induces the males to land (Gibson, 1945; Syrjämäki, 1964; Paasivirta, 1972). Delettre (1984) found that species of terrestrial chironomids (*Smittia* 'sp. 1' and *S. pratorum* Goetghebuer) avoided downwind transport through inhibitory effects of the wind. This behaviour, together with larval resistance to drought, may account for the persistence of populations and their permanent occupation of suitable habitats (Rainey, 1976; Delettre, 1988b).

In contrast, Delettre (1988a) reported two other species of terrestrial chironomids, *Pseudosmittia longicrus* (Kieffer) and *L. minimus* which continue to fly when upward air transport is at a maximum. He suggests that this frees them from the 'boundary layer' (Taylor, 1974), a layer of air near the ground in which insects can control their flight. Thus, despite limited flight ability, wide-range dispersal becomes possible.

9.6.3 FLIGHT AND WING LENGTH

The relation between wing length (a postulated indicator of dispersal ability) and predictability of habitat duration has been examined by McLachlan (1985a). His hypothesis that wing length was negatively related to habitat duration or predictability has been criticized by Vepsäläinen (1986), who argued that extrapolation of concepts on wing length and dispersal ability from the intraspecific to the interspecific level are invalid. Delettre (1988a) failed to reveal any differences in habitat duration or predictability when wing length of four terrestrial species were used as predictors. Delettre (1988a) cites large variation in wing length, and the fact that ranking of species according to wing length may group species which display very different escape strategies, as reasons for the failure of the hypothesis.

McLachlan (1983a), studying rain-pool chironomids in Africa, noted a positive relationship between adult female size and fecundity and dispersal ability. Since the breeding pools are ephemeral, dispersal is necessary to provide habitat for the next generation. The mean dispersal distance in the two species, *Chironomus imicola* Kieffer and *Polypedilum vanderplanki* Hinton, ranged from about 200 to 400 m.

Flight or aerobatic ability of male chironomids has been considered to be advantageous to insects mating on the wing (McLachlan, 1986a) and this theme has been developed in a number of studies (McLachlan, 1986b; Neems *et al.*, 1990; Neems *et al.*, 1992; McLachlan and Neems, in press). It is argued that small males are more likely to acquire females than large males (McLachlan and Allen, 1987), a view which contrasts with the convention that large size is a universal determinant of mating success. These theories require testing over a much larger group of chironomid species because there are many cases where wing size is either the same in both sexes or where males have larger wings.

9.6.4 FOOD AND FEEDING

It has been the common view that chironomids do not feed as adults but an increasing body of evidence (Downes and Colless, 1967; Goff, 1972; Downes, 1974; Schlee, 1977; Burtt *et al.*, 1986; Cranston, 1988) clearly demonstrates that feeding occurs over a wide range of species in the group. The natural foods reported include fresh fly dropping, nectar, pollen and honeydew (aphid excretion) and in the laboratory adults have taken sucrose solutions.

Although feeding takes place, the question remains as to whether it affects the life of the individual. Burtt *et al.* (1986) found that starved females of *C. plumosus* flew on average about twice as long as starved males, implying that females emerge with larger reserves of energy than do males. When individuals were allowed to feed, female flight time increased by 52% as against males whose flight time increased by 160%. The authors suggest that the extra energy obtained through feeding is used by the sexes in different ways. Males increase their duration of flight whereas females use the energy to extend their life-span which was found to be slightly longer (3.1 days) than that of males (2.6 days). These responses are thought to improve reproductive success by increasing the chances of successful mating and they may do in certain species. However, mating in chironomids often takes place almost immediately after eclosion (Olander and Palmén, 1968; Otsuka *et al.*, 1986) and feeding will have no direct effect on these species, unless further matings take place in longer-lived individuals. Thus Martin and Porter (1977) note that feeding affects maturation of the second egg-mass of *Chironomus tepperi*. Food therefore may increase the flight time and longevity of some species which in turn may increase the chances of random elements such as wind dispersal, but the effect will depend on the species reaction to wind (see Delettre, 1988a).

Chironomids attracted to flowers to seek nectar and/or pollen may pollinate some plants, or at least increase the chances of fertilization. In the primitive angiosperm *Pseudowintera colorata* (Winteraceae), species of *Smittia* are attracted to stigmatic exudate and the midge is cited as a major contributor to successful pollination (Lloyd and Wells, 1992). Other chironomids are said to be involved in the sex lives of plants, notably at high elevations and latitudes, but these are frequently anecdotal observations in connection with the feeding of midges on nectar and pollen (see above).

9.6.5 OVIPOSITION FLIGHT

After mating the female may fly away from the swarming site to deposit her eggs. There are a number of reports of directional flight of the female prior to oviposition but the site selection process is not understood (Oliver, 1971). The importance of site selection is questioned by Davies (1976a, b) who suggests that movements of first instar larvae determine distribution in standing water-bodies.

The egg-laying flight of *S. crassiforceps* off the south coast of Finland was observed by Syrjämäki (1964). When the sea was calm, females were observed flying towards the open sea and back to the shore at 5–20 cm above the water. It is suggested that these females were going to lay their eggs and returning from egg-laying. Paasivirta (1972) noted that *T. gracilentus* flew low over the lake on calm evenings touching the water surface with her abdomen every 0.5–1.0 m. Oviposition flights may be quite long and McLachlan (1983a) records flights of up to 850 m for the rain-pool species *C. imicola*.

Females of *Cricotopus* species from a Canadian stream were observed to move from their resting places amongst riparian vegetation and concentrate over riffles (LeSage and Harrison, 1980a). The flight pattern consisted of a zigzag motion either across or parallel to the direction of flow for a period of 1 to 2 minutes. This activity was followed by a drift downstream of 5–15 m followed by ascent and zigzag flight. This behaviour was repeated constantly throughout the ovipositing period which reached a peak at 20.00 h.

In the less specialized marine chironomids such as *Halocladius*, adults swarm over land 100–200 m away from the tidal zone in sheltered areas, and the females return to the intertidal habitat during low water to deposit their eggs (Neumann, 1976).

Parthenogenetic forms show little dispersive activity in the oviposition flight. In the harsh climate of the Kerguelen Isles, *L. pusillus* lays its eggs very soon after emergence (Delettre, 1984).

9.6.6 BRACHYPTERY AND APTERY

In many marine species where life cycles are synchronized to tidal movements, females are brachypterous or apterous and oviposition takes place almost immediately after mating (Olander and Palmén, 1968; Cheng and Collins, 1980). In non-discrete contiguous habitat such as the seashore, dispersal is the result of an expansion of the core population laterally along the coastline but this may be enhanced by currents acting on unattached life-history stages such as rising pupae, sinking egg-masses, swimming larvae and emerged adults on the water surface. In some species, for example *Clunio takahashii*, these susceptible stages are reduced because the male mates with the morphologically degenerate female which does not leave the 'nest-tube'. The female lays the egg-mass in the tube, thereby reducing chances for dispersal (Hashimoto, 1965).

In the Antarctic, the apterous adults of *B. antarctica* disperse by walking a few metres. However, passive modes offer a greater chance of dispersal. Edwards and Baust (1981) observed dispersal of adults to the sea via meltwater streams. Survival on the sea surface may be several days and it is suggested that short-range transport by surface rafting is possible. In addition adults may be swept off the ground and carried by the wind, but Peckham (1971) notes that the adults 'perform a holding behaviour' when confronted with a sudden wind and if this is prolonged will seek shelter in lower moister strata of moss. Fragments of substrate containing larvae and adults may be blown about, which will disperse the species. Birds may also pick up clumps of plant material containing *B. antarctica* and transport them over larger distances.

9.6.7 DISPERSAL – A SUMMARY

Most dispersal of Chironomidae is passive and modified by factors such as wind, species reaction to wind speed, location of swarms and general activity of the species which may make them more susceptible to be dispersed. Flight ability and the taking of food may enhance the chances of dispersal. Flightless forms may also be dispersed passively by wind and water currents.

10

Life cycles and population dynamics

M. Tokeshi

10.1 INTRODUCTION

As insects belonging to the Endopterygota (or Holometabola), the life cycle of Chironomidae is divided into four distinct stages, i.e. egg, larva, pupa and imago. Notwithstanding the large number of species within the family, chironomids share one conspicuous life history characteristic in that the last two stages are generally very short in duration, while the lengths of egg and larval stages vary substantially between and within species. This seems to be related to the fact that, as in some other aquatic insect orders such as the Ephemeroptera and Plecoptera, chironomid adults mostly rely upon the energy stored during the larval stage to accomplish reproduction, the single most important task assigned to them. Though feeding is known to occur in adults (Chapter 9), this constitutes a tiny proportion of energy acquisition and does not contribute significantly to overall reproductive output, a situation which is clearly different from that of many other related dipterans such as blood-sucking blackflies and mosquitoes. Clear demarcation of energy acquisition and reproduction along the phases of metamorphosis implies that the adult, reproductive phase is heavily dependent on the preceding feeding phase. If a fixed amount of energy is to be allocated to the maintenance of the adult body and gamete production, the highest reproductive output would be achieved by minimizing the length of the adult stage, the strategy that seems to have been adopted by the vast majority of chironomid species. It then follows that in terms of determining the overall pattern of life cycle, the larval stage assumes a disproportionate importance in the Chironomidae.

This chapter deals with life-cycle phenomena such as voltinism, growth and development with particular reference to the larval stage and examines their variation within and between species, taking into account various factors affecting life cycles. These are then linked to population dynamics and abundance, which together constitute an important area of population ecology of the Chironomidae.

The Chironomidae: Biology and ecology of non-biting midges
Edited by P.D. Armitage, P.S. Cranston and L.C.V. Pinder
Published in 1995 by Chapman & Hall. ISBN 0 412 45260 X

10.2 LIFE CYCLE PATTERNS

10.2.1 VOLTINISM

(a) Problems of interpreting voltinism

Voltinism refers to the number of generations that a population passes in a year, which has generally been studied through observation of adult emergence and/or regular sampling of a larval population. However, the observation of adult emergence patterns alone is insufficient to establish the voltinism of a particular species. This is because the number of emergence peaks observed does not necessarily correspond to voltinism if there are different cohorts within a population, or if individuals from a single cohort emerge over an extended period, during which there may be two or more peaks. Thus it is always desirable to collect information on the larval population to determine voltinism. On the other hand, it should be noted that recording changes in larval population density alone is insufficient, because population fluctuations do not necessarily correspond with voltinism where successive generations are partially overlapping.

What is needed is growth/instar analysis of individuals at a reasonably frequent interval (adjusted to the growth rate of the species concerned), so that the progression of a cohort or population can be followed closely throughout the year. In reality, however, this principle is not followed stringently enough in many chironomid studies for a variety of reasons and the information thus obtained always contains some degree of ambiguity or uncertainty. In particular, a sampling interval of one month or longer produces data which are often too coarse for life history analyses of multivoltine species, easily leading to speculation rather than to concrete knowledge. Similarly, the use of a sieve for sample separation introduces bias to the analysis of instar composition, especially when mesh size of 200 μm or more is used with medium-sized to small species of chironomid. It is therefore necessary to exercise much caution in interpreting data on voltinism.

(b) Interspecific variation in voltinism

The number of generations per year is known to vary from 1/7 (i.e. one generation taking 7 years to complete; Butler, 1982b) to 5 (LeSage and Harrison, 1980a; Drake, 1982; Tokeshi, 1986b), but some species that show overlapping, continuous emergence of adults may have more than 5 generations in a year (e.g. 7 generations estimated for *Chironomus riparius* Kieffer in a polluted English river; Learner and Edwards, 1966). Voltinism of less than 1 (each generation taking more than one year to complete) has been reported mainly for species inhabiting cold Holarctic regions. For example, Welch (1976), investigating the general ecology of Chironomidae in an arctic lake of northern Canada, found that the life cycle spanned 2–3 years for six common taxa including *Pseudodiamesa arctica* (Malloch), *Heterotrissocladius oliveri* Sæther, *Trissocladius* sp., *Orthocladius* spp., *Lauterbornia* sp. and

Paracladius quadrinodosus Hirvenoja. Similarly, *Sergentia coracina* (Zetterstedt) from Lake Langvatn in Norway (Aagaard, 1982) and *Chironomus islandicus* (Kieffer) from Lake Myvatn in Iceland (Lindegaard and Jónasson, 1979) were estimated to take 2–3 years to complete a life cycle. In oligotrophic Lake 95, Greenland, *Chironomus* spp. appear to have 2–3 year life cycles and *Arctopelopia melanosoma* (Goetghebuer) a 2-year life cycle (Lindegaard and Mæhl, 1993). Jónsson (1985) suggested that *Procladius* spp. and *Cryptochironomus* from the littoral of Lake Esrom have life cycles of 1 or 2 years. The longest life cycle recorded to date refers to two sibling species of the genus *Chironomus* in arctic tundra ponds in Alaska, where an annual open-water season lasts about 90 days (Butler, 1982b). In this case the constant presence of seven larval cohorts during a 3-year study period suggested that larvae take 7 years from hatching to emergence as adults.

The majority of chironomid species so far studied, however, seem to demonstrate voltinism of 1 or 2, with the latter being most dominant, though this may reflect nothing more than the traditional research bias towards cold temperate regions. Notwithstanding this bias and inevitable inaccuracies in available data, the distribution of voltinism among a total of 125 chironomid taxa (species were amalgamated in some cases) is as follows: one generation per year (**univoltine**), 33%; two generations (**bivoltine**), 44%; three or more generations (**multivoltine**), 18%. To date, four or more generations per year has been suggested mainly for lotic species, including *Paratanytarsus natvigi* (Goetghebuer) (5 gen.) and *Cricotopus fuscus* (Kieffer) (5 gen.) from River Kennet, southern England (Drake, 1982); *Tvetenia calvescens* (Edwards) (5 gen.) and *Thienemanniella majuscula* (Edwards) (5 gen.) from River Tud, eastern England (Tokeshi, 1986b); *Corynoneura lobata* Edwards (4 gen.), *Rheocricotopus fuscipes* (Kieffer) (4 gen.) and *Orthocladius frigidus* (Zetterstedt) (>4 gen.) from Rabis Baek, Denmark (Lindegaard and Mortensen, 1988); and *Eukiefferiella brevinervis* (Malloch) from Juday Creek, Indiana, USA (Berg and Hellenthal, 1991, 1992). The suggestion of seven annual generations for a lentic species, *Tanytarsus barbitarsis* Freeman from a saline lake in Australia (Paterson and Walker, 1974b), requires further confirmation, since monthly sampling is considered inadequate for determining voltinism in such a case. On the other hand, it is likely that the frequency of species with four or more generations per year is substantially under-reported, because of difficulties of analysing multivoltine life cycles with interval sampling, and because multivoltine species are more abundant in warmer regions of the earth, where research is sparse.

In terms of differences in higher taxa, the Orthocladiinae appear to include more species with multivoltine life cycles than the Chironominae, as attested by the fact that species with four or more generations per year mentioned above all belong to the former. Conversely, the Chironominae seem to encompass a higher proportion of low-voltine species, including those with less than one generation per year. This is an interesting observation, since the Orthocladiinae are in general considered to be cold-adapted while the Chironominae are warm-adapted (Oliver, 1971), which may suggest that species of Orthocladiinae are capable of rapid development and hence shorter life cycles under the temperate climate. In this respect further investigations

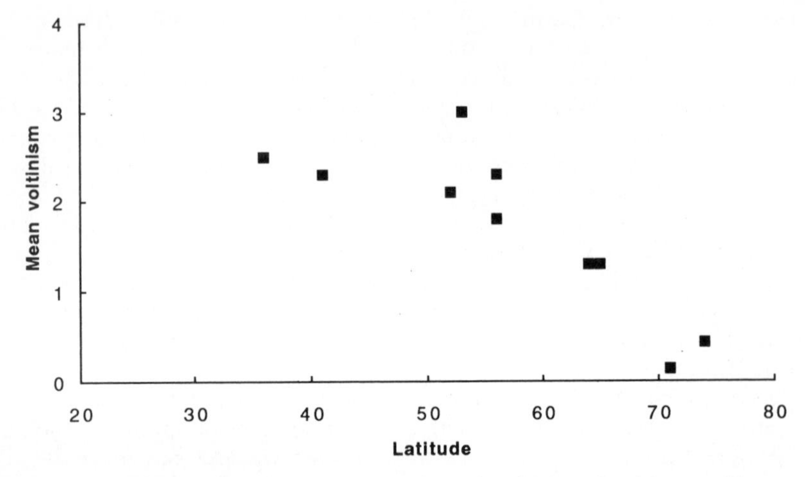

Figure 10.1 Relationship between latitude and voltinism in chironomid assemblages.

and comparison with subtropical and tropical species, which appear to be dominated by the Chironominae, would be highly interesting.

In contrast to the Orthocladiinae and Chironominae, species of Tanypodinae studied to date have a narrower range of voltinism, from one to three generations per year.

(c) Latitude and voltinism

Latitude affects photoperiod and temperature, two major factors known to control insect development. Species in lower latitudes can in general grow more rapidly and pass more generations in a year, though drought and high temperatures may constitute negative factors in tropical/subtropical regions (see section 10.3.3(b) concerning drought-tolerance and summer aestivation). For a set of ten studies which documented the voltinism of major species within a habitat ranging in latitude from 36° to 74° N, voltinism (mean number of annual generations per species in a habitat) is negatively correlated with latitude ($p^2 = 0.69$, $P < 0.01$, Figure 10.1). Similarly, the range in voltinism within a habitat (i.e. difference between maximum and minimum values of voltinism among species) tends to decline with latitude.

(d) Within- and between-habitat variation in voltinism

A lake or a stream habitat contains many chironomid species with differing voltinism. Drake (1982) and Tokeshi (1986b), both working on the faunas of English rivers, reported voltinism ranging from one to five generations, while Potter and Learner (1974) reported a range of one to three generations, the majority of them two, from a reservoir in south Wales. It may appear from these that in the same latitude lotic chironomid assemblages are more diverse in voltinism and at the same time include species with higher voltinism (more generations per year) than lentic ones. A parallel case is found in Denmark, where the littoral chironomid fauna of Lake Esrom has voltinism of 1/2 to three generations per year (Jónsson, 1985), while the lotic chironomid fauna of the Rabis Baek demonstrates one to four generations (Lindegaard and Mortensen, 1988).

(e) Intraspecific variation in voltinism

Populations of a chironomid species inhabiting different habitats may show variation in voltinism. *Chironomus plumosus* (Linnaeus), one of the most ubiquitous species in the northern Holarctic, is generally reported to have two generations per year in central Europe, western Russia and North America (Potter and Learner, 1974; Terek and Losos, 1979; Frank, 1982; Hilsenhoff, 1967; Jankovič 1971), while three annual generations occur in shallow, eutrophic Lake Kasumigaura in central Japan (Iwakuma, 1986a) and only one generation in high-altitude (2055 m) Big Bear Lake in California (Siegfried, 1984) as well as in Lake Erken, Sweden (Johnson and Pejler, 1987). Furthermore, Laville (1972) reported a 2-year life cycle for this species inhabiting a high-altitude lake (2285 m) in the Pyrenees. Another interesting example concerns *Paratendipes albimanus* (Meigen) which is also widely distributed in the Holarctic and possesses eurytopic habitat choice, including pools, ponds and lakes as well as upland and lowland rivers (Thienemann, 1954). In standing waters the species is either univoltine (Brundin, 1949; Iovino and Miner, 1970; Learner and Potter, 1974) or bivoltine (Townes, 1938; Thienemann, 1950; Reiss, 1968), but in running waters only univoltine populations have so far been recorded (Dittmar, 1955; Lehman, 1971; Ward and Cummins, 1978). On a smaller spatial scale, populations of *Chironomus pseudothummi* Strenzke living in two small ponds in England seemed to differ in voltinism, one univoltine and the other bivoltine (Smith and Young, 1973).

The fact that temperature is a major factor controlling growth in chironomids (cf. section 10.3.2(b)) suggests that inter-annual fluctuations in temperature can induce variation in voltinism even within the same body of water. For example, a small part of the population of *Chironomus anthracinus* Zetterstedt in Lough Neagh, Northern Ireland, may take 2 years to reach maturity if temperature is low and food supply scarce during the growth period, while the majority complete the life cycle in one year (Carter, 1980; M. Tokeshi, *pers. obs.*). Similarly, the majority of *Tokunagayusurika akamusi* (Tokunaga) in Lake Kasumigaura emerge in one year, but some may take 2 or even 3 years to emerge (Iwakuma, 1986a). In the case of *C. anthracinus* larvae in Lake Esrom, Denmark, some proportion may emerge as adults in one year, while most of them take 2 years (Jónasson, 1961, 1972). Furthermore, the *C. anthracinus* population in Lake Esrom seems to demonstrate a spatial variation in voltinism such that those inhabiting the shallower part (less than 17 m deep) of the lake emerge after one year whereas those in the profundal take two years to complete a life cycle (Jónasson, 1965). Similarly, *C. plumosus* may be multivoltine in the shallow littoral of some lakes but univoltine at deeper sites (Maitland *et al.*, 1972; Terek and Losos, 1979). A population of *Pseudodiamesa branickii* (Nowicki) in an upland stream in Germany appeared to encompass a mixed voltinism, i.e. one group bivoltine and the other trivoltine (Nolte and Hoffmann, 1992). In this case the generation time was affected by the photoperiod which eggs and larvulae experienced. Inter-annual variation in voltinism among multivoltine species, in particular those with three or more generations in a year, must be a fairly common phenomenon in the Chironomidae, but to date there are virtually no reliable data which can unequivocally demonstrate its existence.

(f) Body size and voltinism

Studies on the Chironomidae published to date appear to indicate that species with small body size tend to pass through more generations in a year than those with large body, other things being equal. Indeed, small to medium-sized species of *Corynoneura*, *Thienemanniella*, *Nanocladius*, *Eukiefferiella*, *Tvetenia* and *Tanytarsus* are often observed to possess tightly overlapping generations which make the estimation of voltinism rather difficult, while large-bodied species such as *Chironomus* spp. generally demonstrate two generations or less per year in cold temperate regions.

In a study of lotic chironomid faunas in the Appalachian mountains, USA, Huryn (1990) derived an empirical model which described the influence of thermal regime on growth and voltinism, using multiple regression analysis. Differences in voltinism across species were predicted to depend upon the terminal body size of larvae (i.e. maximum body size in the fourth instar) as well as on temperature regime. Specifically, the model predicted that *Corynoneura*, which reached a terminal size of 1.5–2.5 mm in the Appalachian streams, has up to seven generations in a year at the study site, corresponding well with the continuous emergence of adults observed through most of the year. In contrast, *Micropsectra* (5.5 mm) and *Pseudorthocladius* (c. 7 mm) were predicted to have bivoltine and univoltine life cycles, respectively, which also agreed with the observation.

The dependence of voltinism on body size, however, may not be absolute and there are perhaps a good number of exceptions. Among species inhabiting temporary rain-pools in Africa, large-sized *Chironomus imicola* has a substantially shorter generation time than small-sized *Polypedilum vanderplanki*, the minimum time from egg to adult being 12 days for the former and 35 days for the latter (McLachlan and Cantrell, 1980; McLachlan, 1983a).

10.2.2 EMERGENCE PHENOLOGY

In conjunction with voltinism, emergence phenology relates to an important facet of life cycle, i.e. temporal scheduling of reproductive activities. Voltinism specifies the frequency with which reproduction occurs on a yearly basis, whereas emergence phenology refers to the actual seasonal timing of such activity. Thus, species with the same voltinism do not necessarily have the same pattern of emergence phenology. The Chironomidae as a group demonstrate a wide variation with respect to emergence phenology, with emergence being observed virtually all through the year (e.g. Pinder, 1974). Here, attention is drawn to different patterns of emergence phenology seen amongst the Chironomidae.

(a) Emergence occurring once a year

Univoltine species as well as those with a generation time of more than one year (**merovoltine** species; Pritchard, 1983) show a single emergence in a year. In the case of species occurring in arctic/subarctic habitats, adult emergence is often restricted to a short, ice-free period in summer (Welch, 1976; Moore, 1979b; Butler, 1982b). In Char Lake, north-western Canada, chironomids which have passed 2–3 years as larvae emerge as adults between mid-July and September, with mating usually occurring on the ice or at the shore edge (Welch, 1976). Similarly, June–August is the major period of adult emergence for univoltine chironomid species including *Arctopelopia griseipennis* (van der Wulp), *Heterotrissocladius grimshawi* (Edwards), *Psectrocladius limbatellus* (Holmgren) (as *edwardsi*), *Rheocricotopus effusus* (Walker) and *Paracladopelma nigritula* (Goetghebuer) from subarctic Lake Thingvallavatn, Iceland, which is ice-bound from December till April (Lindegaard, 1992). *Procladius signatus* (Zetterstedt) and *P. cinereus* Goetghebuer from Lake Malsjøen, Norway, also emerge in mid-summer (July–August) (Aagaard, 1978). On the other hand, summer emergence is relatively uncommon among univoltine species inhabiting ice-free habitats.

In addition to summer emergence of adults, emergence in spring is equally (or more) common among univoltine species. For example, in another subarctic lake in north-western Canada, Moore (1979b) observed major adult emergence 2–3 weeks after the ice-melt in May/June. This closely resembled the situation in mesotrophic Lake Masjøen in Norway where the majority of univoltine species (*Procladius barbatus* Brundin, *Heterotrissocladius marcidus* (Walker), *Heterotanytarsus apicalis* (Kieffer), *Parakiefferiella bathophila* (Kieffer), *Polypedilum bicrenatum* Kieffer, *Micropsectra insignilobus* Kieffer and *Tanytarsus chinyensis* Kieffer) were observed to emerge in late May–early June, i.e. 2–3 weeks after the ice period (early November–early May) ends (Aagaard, 1978). Two univoltine species in the littoral of Lake Esrom, *Psilotanypus lugens* Kieffer and *Stictochironomus sticticus* (Fabricius), also emerge 1–2 months after ice-melt, in May–June (Jónsson, 1985).

It is worth noting, however, that some species of Chironomidae do not require completely ice-free conditions to achieve emergence. For example, Herrmann *et al.* (1987) documented the emergence of *Diamesa leona* Roback in mid-winter, mainly between December and February, from the high-altitude Arkansas River in central Colorado, USA. Brachypterous (i.e. wing reduced, not permitting flight) adults of this species were often observed to crawl on the submerged ice shelf and aggregate to mate in air-filled ice pockets in the ice shelf, being protected from the wave action of the river. A similar emergence pattern was observed for an Antarctic chironomid, *Belgica antarctica* Jacobs (Sugg *et al.*, 1983) and some temperate chironomids including species of *Metriocnemus* may be univoltine with the adults emerging in winter (M.Tokeshi, *pers. obs.*). Mid-winter emergence may serve as a mechanism for reducing predation mortality of adults, since many potential predators are either inactive or migrate to warmer localities in winter.

In habitats without an extended period of ice cover, spring emergence, as distinct from summer emergence, is frequently observed among univoltine species. In Lough Neagh, Northern Ireland, where ice does not form in winter, two major univoltine species, *Chironomus anthracinus* and *Stictochironomus sticticus* (Fabricius), normally emerge in early April and May, respectively (Carter, 1973; M.Tokeshi, *pers. obs.*). In a weekly sampling of emerging adults from a stream in southern Ontario, a single emergence in April/May was observed for *Thienemanniella xena* (Roback), *Eukiefferiella* cf. *coerulescens*, *E.* cf. *devonica*, *Tvetenia* cf. *duodenaria*, *Chaetocladius* spp., *Parametriocnemus lundbeckii* (Johannsen), *Heleniella curtistila* Sæther and *Paraphaenocladius* sp., whereas no univoltine species emerged in summer (Singh and Harrison, 1984).

Some univoltine species are known to emerge very early in spring. In Lake Thingvallavatn, mentioned earlier, univoltine *Pseudodiamesa nivosa* (Goetghebuer) emerges immediately after ice-melt, in late April–early May (Lindegaard, 1992). *Hydrobaenus pilipes* (Malloch) [as *Trissocladius grandis*], which occurs commonly on the submerged macrophytes in northern European lakes and rivers, is usually abundant in swarms in the first half of April, but is only rarely seen thereafter (Thienemann, 1954; Mozley, 1970). Similarly, Tokeshi (1986b) observed that *Orthocladius obumbratus* Johannsen [= *O.* (*O.*) Pe 9 of Langton, 1984] attained rapid growth in March and emerged in late March–early April in eastern England. Some other species of *Orthocladius* and *Hydrobaenus* [as *Trissocladius*] also appear to be univoltine with early spring emergence (often lasting for a very short period), the fact that tends to lead to their being overlooked in regular sampling schemes concentrated on the spring–summer period.

While spring/summer emergence is observed most commonly among univoltine (and multivoltine) species, there are examples of univoltine species emerging in autumn. Data from Starnberger See in southern Germany suggested that the main emergence of *Tanytarsus lugens* Kieffer and of *Micropsectra radialis* Goetghebuer (as *coracina*), both profundal species, occurred in September–October and late October–early November, respectively, while for the latter species sporadic emergence was seen throughout spring and summer (Gerstmeier, 1989b). In Lake Kasumigaura, central Japan, large and abundant *Tokunagayusurika akamusi* emerge as adults between late October and early December, the main emergence taking place in early November (Iwakuma, 1986a). Autumn emergence (late August–early September) was also observed in two profundal species, *Sergentia coracina* and *Stictochironomus rosenschoeldi* (Zetterstedt), from Lake Malsjøen which was ice-covered from November onwards (Aagaard, 1978).

Emergence in univoltine species is often of short duration, lasting no more than one month and intensive emergence occurring within a period of 1–2 weeks. Such well-synchronized emergence is considered to maximize the chance of successful reproduction, an aspect which is all the more important for species with a single period of reproductive activity in a year. On the other hand, there may be some cases where emergence occurs for an extended period among (supposedly) univoltine species (cf. Cloutier and Harper, 1978).

(b) Emergence occurring twice or more per year

By far the most common pattern of seasonal emergence in bivoltine species is spring and summer/early autumn emergence. In contrast, emergence in early spring, late autumn or winter, as has been documented for univoltine species, is rare among bivoltine species. In general, overall seasonal timing of emergence appears to be more or less synchronized among bivoltine species in any one locality. For example, in the littoral of Lake Esrom, Denmark, bivoltine species including *Microtendipes* spp., *Dicrotendipes modestus* (Say) (as *pulsus*), *Polypedilum nubeculosum* (Meigen), *Endochironomus albipennis* (Meigen) and *Tanytarsus usmaensis* Pagast all had spring emergence in May/June and summer emergence in August, while the second generation of *Tanytarsus bathophilus* Kieffer emerged in September/October (Jónsson, 1985). Similarly, *Cricotopus infuscatus* (Malloch), *C. festivellus* (Kieffer), *C. annulator* Goetghebuer, *C. varipes* Coquillett, *C. slossonae* Malloch, *C. triannulatus* (Macquart), *C. trifascia* Edwards and an unnamed species of *Cricotopus* all had two emergence periods, first in May/June and second in September/October, in Salem Creek in southern Ontario (LeSage and Harrison, 1980a). Furthermore, the majority of species emerging from a eutrophic reservoir in south Wales, including *Procladius choreus* (Meigen), *Chironomus (Camptochironomus) tentans* Fabricius, *Chironomus plumosus*, *Endochironomus albipennis*, *Glyptotendipes paripes* (Edwards), *Microchironomus tener* (Kieffer), *Microtendipes pedellus* (De Geer), *Parachironomus cinctellus* (Goetghebuer), *P. monochromus* (van der Wulp), *P. vitiosus* (Goetghebuer), *Cladotanytarsus mancus* (Walker), *Tanytarsus mendax* Kieffer (as *holochlorus*) and *Paratanytarsus inopertus* (Walker), had two major peaks of adult emergence in May/June and September/October (Potter and Learner, 1974). In a Danish stream, Rabis Baek, *Prodiamesa olivacea* (Meigen) emerged in May/June and August/September, while *Brillia modesta* (Meigen) emerged in April/May and September/October (Lindegaard and Mortensen, 1988). *Sergentia coracina* in Starnberger See, southern Germany, had a similar pattern of emergence in April/May and September/October (Gerstmeier, 1989b).

Species with three or more generations in a year generally emerge between spring and autumn, wherein emergence peaks are separated to a greater or lesser extent. In a weekly sampling of emerging adults from a stream in southern Ontario, three distinct emergence periods were observed for *Nilotanypus fimbriatus* (Walker), *Eukiefferiella* cf. *ilkleyensis*, *E.* cf. *brevicalcar*, *E.* cf. *claripennis*, *Parametriocnemus lundbeckii* and *Micropsectra* spp. (Singh and Harrison, 1984). *Tanytarsus lugens* Kieffer from the Eglwys Nunydd Reservoir in south Wales also had three separate emergence periods between February and October (Potter and Learner, 1974). In other cases, emergence may be continuous with overlapping generations.

It is generally the case among multivoltine species that the first, spring emergence is stronger than later emergence(s) (e.g. chironomid species studied by Jónsson (1985) and Singh and Harrison (1984, cited above). This may relate to the fact that larvae of the spring generation are better synchronized in terms of growth and development due to overwintering and distinct environmental cues such as rising temperature and increasing photoperiod in spring.

(c) Intraspecific variation in emergence pattern

Patterns of emergence may vary temporally and spatially, depending on various environmental factors. For example, *Chironomus anthracinus* generally has a single spring emergence (April/May) in cold Holarctic (e.g. Lake Esrom: Jónsson, 1985; Lough Neagh: Carter, 1980), but the population in Lake Erken, Sweden, emerges in summer (Johnson and Pejler, 1987). On a smaller spatial scale, intraspecific variations in emergence pattern have been observed for species of Tanypodinae occurring in different stream sites within the single catchment area of the Laurentides, Canada (Cloutier and Harper, 1978). Furthermore, two species of genus *Meropelopia*, *M. flavifrons* (Johannsen) and *M. americana* (Fittkau), demonstrated different emergence patterns in sympatry and allopatry; in sympatric sites the two species had short, non-overlapping emergence periods, while in allopatric sites emergence periods were more extended.

10.3 GROWTH AND DEVELOPMENT

Life cycle patterns such as voltinism and emergence phenology are ultimately determined by how growth and development are achieved in the natural environment. Here, **growth** refers to increase in body size, which is in general restricted to larvae, particularly second to fourth instars in the Chironomidae (the first instars often represent the planktonic/dispersive phase, lasting for a very short period), while **development** refers to the progress towards reproductive maturation from the egg stage onwards. To elucidate growth and development, it is useful to resort to both field and laboratory observations. Here, aspects of growth and development which bear upon life-cycle patterns in the Chironomidae are reviewed briefly.

10.3.1 PATTERNS OF GROWTH

Absolute and relative duration of larval instars and their seasonal timing constitute an important aspect of growth in the Chironomidae. Duration of different instars varies from one species to another, with the last, fourth instars often being longer than earlier instars. In particular, species which take one year or more to complete a generation spend the longest period of larval life in the fourth instars. For example, the population of *Chironomus anthracinus* in Lough Neagh, Northern Ireland, passes through the first to third instars between May and July and spends the remaining 9 months as

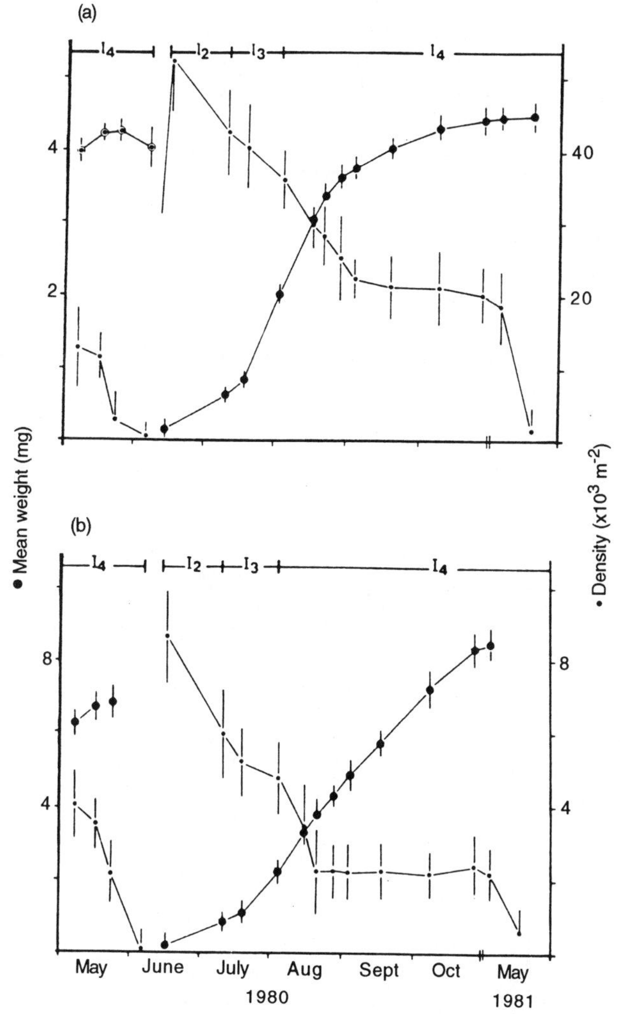

Figure 10.2 Larval growth pattern in (a) *Chironomus riparius* Meigen and (b) *Glyptotendipes paripes* (Edwards). (After Rasmussen, 1984b.)

fourth instars (Carter, 1980). The same pattern of growth is seen for this species in Lake Erken, Sweden, the only difference being a 2-month delay in the seasonal cycle (Johnson and Pejler, 1987). A similar pattern occurs with univoltine *Chironomus riparius* and *Glyptotendipes paripes* in a prairie pond in Canada, where larvae are in the fourth instars from mid August till emergence in the following May, while earlier instars are seen only between June and early August (Figure 10.2; Rasmussen, 1984b). A terrestrial chironomid *Limnophyes pusillus* Eaton which had a univoltine life cycle in the Kerguelen Isles, French Antarctic, showed a similar pattern of growth

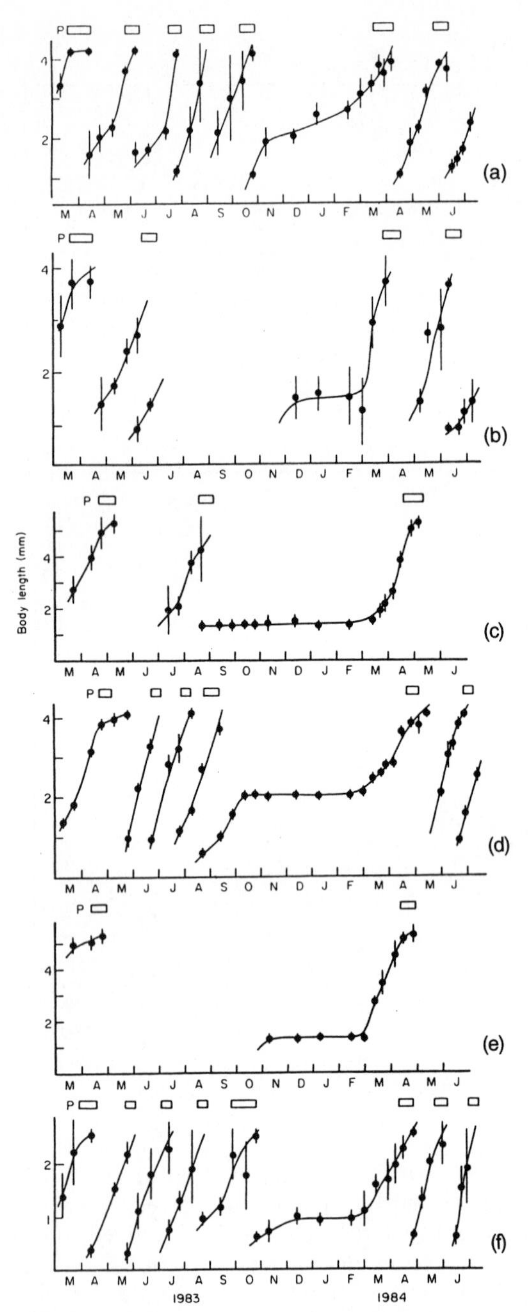

Figure 10.3 Larval growth pattern (mean body length ± 95% CL) in (a) *Tvetenia calvescens* (Edwards); (b) *Eukiefferiella ilkleyensis* (Edwards); (c) *Rheocricotopus chalybeatus* (Edwards); (d) *Rheotanytarsus curtistylus* (Goetghebuer); (e) *Cricotopus (C.) annulator* Goetghebuer; (f) *Thienemanniella majuscula* (Edwards). (After Tokeshi, 1986b).

(Delettre, 1978). Larvae grew rapidly during austral summer and spent the rest of the year as third or fourth instars until emergence in spring. In all these cases fourth-instar larvae are present virtually all through the year, while earlier instars are restricted to a few months. Fourth instars of *Cryptochironomus redekei* Kruseman and *Polypedilum nubeculosum* also occurred all through the year in Hjarbæk Fjord, Denmark (Lindegaard and Jónsson, 1987). Similarly, a commensal chironomid *Epoicocladius flavens* Malloch, which has a univoltine life cycle in eastern England, occurred as first and second instars only between June and August, while third and fourth instars coexisted in other months, with the latter being always predominant (Tokeshi, 1986b).

Where overwintering occurs as second or third instars in uni- and multivoltine species, however, the period of fourth instars can be shorter than that of third or second instars and may constitute less than 20% of the entire life cycle. For example, the majority of individuals of *Fleuria lacustris* Kieffer, *Polypedilum bicrenatum* and *Cladotanytarsus* spp. in Hjarbæk Fjord were in the third instars from late summer till next spring (Lindegaard and Jónsson, 1987), and the second instars were longest in duration in the overwintering generations of *Cricotopus (C.) annulator*, *Rheocricotopus chalybeatus* (Edwards) and *Rheotanytarsus curtistylus* (Goetghebuer) in the River Tud, eastern England (Tokeshi, 1986b). Similarly, the length of the fourth instar stage amounts to less than 20% of the entire larval life in *Paratendipes albimanus* in a headwater stream in Michigan (Ward and Cummins, 1978).

Irrespective of the relative duration of different instars and their seasonal timing, almost invariably the largest amount of growth in both absolute and relative terms occurs during the fourth-instar stage. This is mainly due to the overall geometric increase in body size from one instar to another within a species: fourth instars are on average 5–8 times larger in mass than third instars, which in turn are 5–8 times larger than second instars. Thus, where overwintering occurs as second or third instars, growth curves tend to show a sharp upturn towards the end of larval life, coinciding with rapid growth as fourth instars in spring/summer when conditions are favourable (e.g. Figure 10.3). In non-overwintering generations, such a rapid growth prevails to a greater or lesser extent through the whole of larval life, resulting in a short generation time.

Many chironomid species do not grow at all under low winter temperatures (Figure 10.3b–f), but some species do keep growing, albeit at a slower rate. The overwintering generation of *Tvetenia calvescens* (Edwards) maintains a reduced but steady growth throughout winter (Figure 10.3a), in contrast to other species coexisting in an eastern English river. Similarly, a continuous growth over winter was observed in *Eukiefferiella brevinervis* and *C. mancus* group sp. A in a midwestern US stream (Berg and Hellenthal, 1992) and in the majority of chironomid species inhabiting Char Lake in arctic Canada (Welch, 1976). But the most intriguing case in this respect refers to *Tokunagayusurika akamusi*, a large-sized species indigenous to lakes in Japan. Its growth and development from eggs to fourth instars occurs in winter between December and March, and thereafter the fourth instars do not seem to grow significantly until their emergence in November (Figure 10.4, Iwakuma, 1986a). Thus its pattern of life cycle is completely different

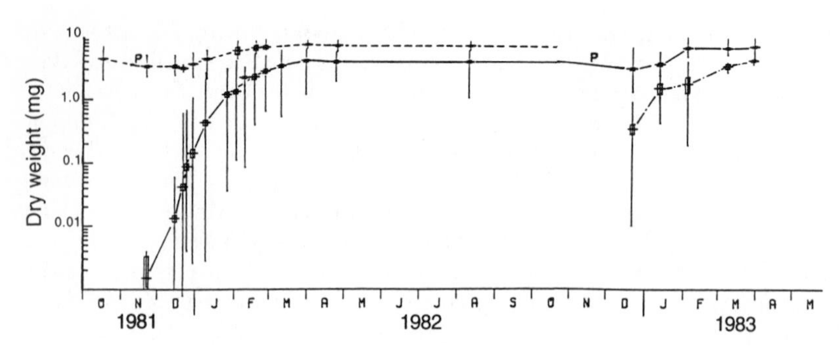

Figure 10.4 Larval growth pattern (mean dry weight ± 95% CL) in *Tokunagayusurika akamusi* (Tokunaga). (Adapted from Iwakuma, 1986a.)

from most other species, particularly *Chironomus plumosus* which is another major species with large body size inhabiting central Japanese lakes. *Hydrobaenus pilipes* [= *Trissocladius grandis*], a northern European species, also appears to grow mainly in winter with emergence taking place in late winter–early spring (Mozley, 1970). Some species of Podonominae and Aphroteniinae in Australia also seem to complete all their growth in winter.

10.3.2 FACTORS AFFECTING GROWTH AND DEVELOPMENT

(a) Egg development

Chironomid eggs in general seem to hatch in a relatively short time, within a few days to one month (Hilsenhoff, 1966; Danks, 1971a; Frank, 1982; Iwakuma, 1986a). Even for univoltine or merovoltine species it seems frequently to be the case that the egg stage lasts no more than 2 weeks. On the other hand, it is rather difficult to prove egg dormancy or delayed hatching under field conditions, even where it is suspected (cf. some species of *Orthocladius*: Tokeshi (1986b). Young larvulae which may remain dormant in the sediment are as difficult to observe as scattered eggs in the field.

Very few studies have examined the factors controlling egg development in the Chironomidae, in contrast to studies on other groups of aquatic invertebrates. In *Chironomus plumosus* rate of egg development was positively related (or hatching time was negatively related) to temperature within the range of 5–30 °C (Figure 10.5a). A similar but less pronounced positive relationship between egg development rate and temperature was found for *Tokunagayusurika akamusi* within the range of 4.5–25 °C (Figure 10.5b). Interestingly, in both of these species eggs failed to hatch at higher temperatures, i.e. 35 °C for *C. plumosus* and 30 °C for *T. akamusi* (Iwakuma, 1986a). This clearly demonstrates that very high temperatures are detrimental to egg development.

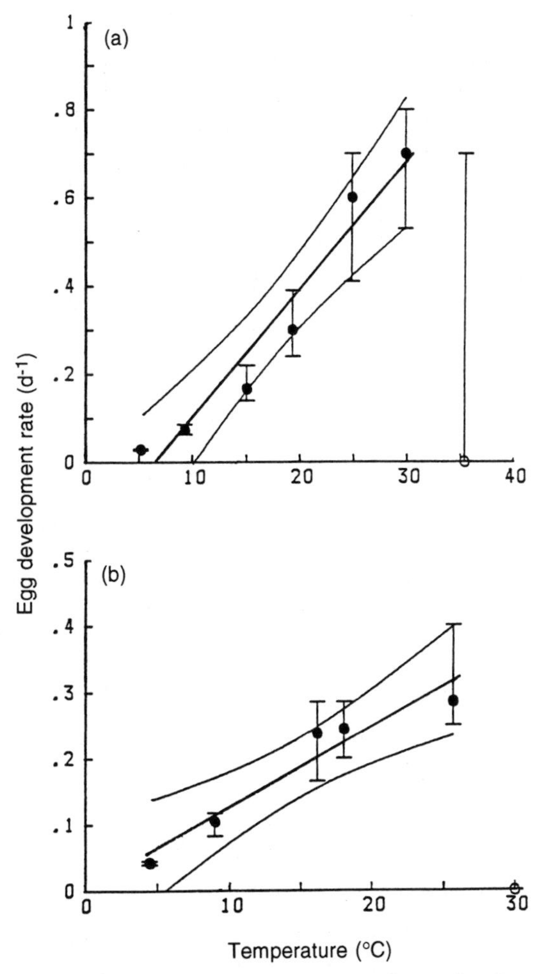

Figure 10.5 Relationship between temperature and egg development rate in (a) *Chironomus plumosus* (Linnaeus) and (b) *Tokunagayusurika akamusi* (Tokunaga). (Adapted from Iwakuma, 1986a.)

(b) Temperature and larval growth

As in egg development, temperature constitutes a major controlling factor in larval growth. A number of studies have demonstrated that growth rate of larvae is increased at higher temperatures. For example, Menzie (1981) reported that in a laboratory rearing experiment *Cricotopus sylvestris* completed larval development in 28 days at 15 °C and in 10 days at 22 and 29 °C, while Konstantinov (1958b) documented for the same species the developmental time of 21 days at 18 °C and 14 days at 22 °C. Comparison of growth in *Paratendipes albimanus* at three different temperatures (10, 15 and 20 °C) also

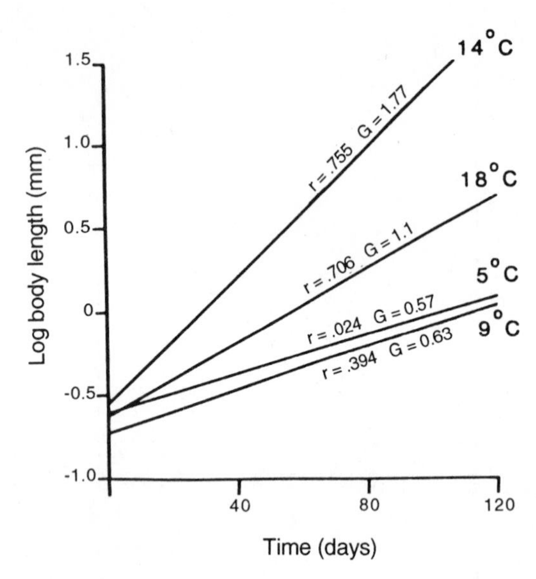

Figure 10.6 Relationship between time and body size at different temperatures in *Eukiefferiella ilkleyensis* (Edwards). (After Storey, 1987.)

showed that larval growth was enhanced at higher temperatures (Ward and Cummins, 1979). Graham and Burns (1983) and Ladle *et al.* (1984) demonstrated that growth in *Chironomus zealandicus* Hudson and *Micropsectra aristata* Pinder, respectively, was positively related to temperature. Similarly, the length of life cycle was reduced at higher temperatures in *Dicrotendipes conjuncus* Walker and *Polypedilum nubifer* (Skuse) in a laboratory experiment (Edward, 1986).

These cases notwithstanding, probably a more general picture relating larval growth to temperature is the one with the highest growth rate being associated with certain (intermediate) temperatures and decreasing rates at higher or lower temperatures. The existence of such temperature maxima for growth is naturally expected, since increasing maintenance costs at higher temperatures will ultimately cancel out the benefits of higher metabolism. When larvae of *Tokunagayusurika akamusi* were grown at five different temperatures, growth rate increased from the lowest value at 1.7 °C to the maximum at 12.3 °C, and again declined at 16.5 °C. A similar pattern was observed for *Eukiefferiella ilkleyensis* (Edwards) grown at 5–18 °C with 'winter' diet: the highest growth rate was achieved at 14 °C (Storey, 1987; Figure 10.6). *Cricotopus bicinctus* (Meigen) also grew faster at 15 °C than at 10 or 20 °C (Mackey, 1977a). In this case it is notable that development rate, as opposed to growth rate, was highest at 20 °C, meaning that larvae matured and pupated in a shorter time than at 10 or 15 °C. A slightly different approach employing experimental chambers located at a field site in Georgia, USA, revealed that the relationship between estimated daily growth rate and temperature was best described as a second-order polynomial (Figure 10.7),

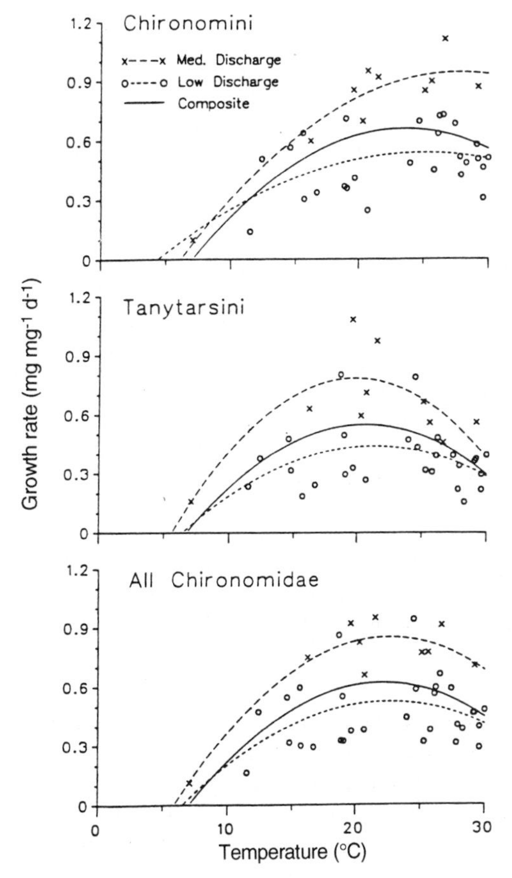

Figure 10.7 Relationship between growth rate and temperature in (a) total Chironominae; (b) total Tanytarsini; (c) total Chironomidae. (After Hauer and Benke, 1991.)

suggesting growth maxima at 21–24 °C for total Chironominae, total Tanytarsini, and the Chironomidae as a whole (Hauer and Benke, 1991).

(c) Food and larval growth

As in any animal species, the quality and quantity of food has a direct influence on growth in the Chironomidae. Growth of a shredder *Brillia flavifrons* (Johannsen) in Michigan streams was higher on the diet of fresh tag alder (*Alnus rugosa*) leaves than on the diet of autumn-shed leaves (Figure 10.8; Stout and Taft, 1985). Similarly, fourth instar larvae of a collector-gatherer *Stictochironomus annulicrus* (Townes) achieved higher growth rate when detritus and faeces of *Tipula* were mixed in increasing proportions with the sand substrate (Mattingly *et al.*, 1981). The addition of nitrogen-rich

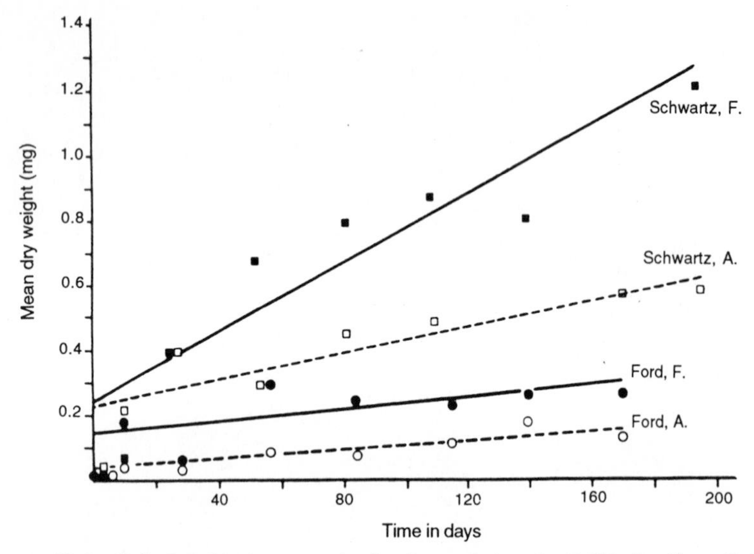

Figure 10.8 Relationship between body size and time in *Brillia flavifrons* (Johannsen) supplied with different food resources. F = fresh leaves; A = autumn-shed leaves. Schwartz and Ford refer to two separate trials. (After Stout and Taft, 1985.)

macrophyte tissue (*Ranunculus*) to the diet of diatoms and detritus also resulted in a higher growth rate in *Eukiefferiella ilkleyensis* grown from the first instars at different temperatures (Storey, 1987). Nitrogen content, however, may not necessarily be the sole dietary factor. *Paratendipes albimanus* grown on four different food resources seemed to respond directly to microbial abundances as expressed by the level of adenosine triphosphate (ATP), rather than to nitrogen content (Ward and Cummins, 1979). In this case, the substrate based on natural detritus, which had relatively high nitrogen content but a low ATP level, led to the lowest rate of growth, whereas the substrate enriched by pignut hickory (*Carya glabra*) leaves, high in both nitrogen and ATP, led to the highest growth rate at all temperatures (Figure 10.9).

As shown by these experiments with *Eukiefferiella* and *Paratendipes*, it is most likely that temperature and food combine to influence growth of chironomid larvae in natural environments. Indeed, field experiments using an in situ enclosure in Lake Ontario (Johannsson, 1980) revealed that seasonal growth in *Chironomus plumosus* f. *semireductus* fluctuated under the twin influence of temperature and food quality. No growth was observed when temperature dropped to 10 °C and the quality of diatoms as food (e.g. thickness of frustules and perhaps nitrogen content) as well as the occurrence of blue-green algal blooms seem to have affected assimilation efficiency and hence growth rate. Furthermore, it is likely that different species respond differently to the same conditions of temperature and food; substrate enrichment led to a substantial increase in growth rate of *Glyptotendipes paripes*, but not of *Chironomus riparius* (Rasmussen, 1985).

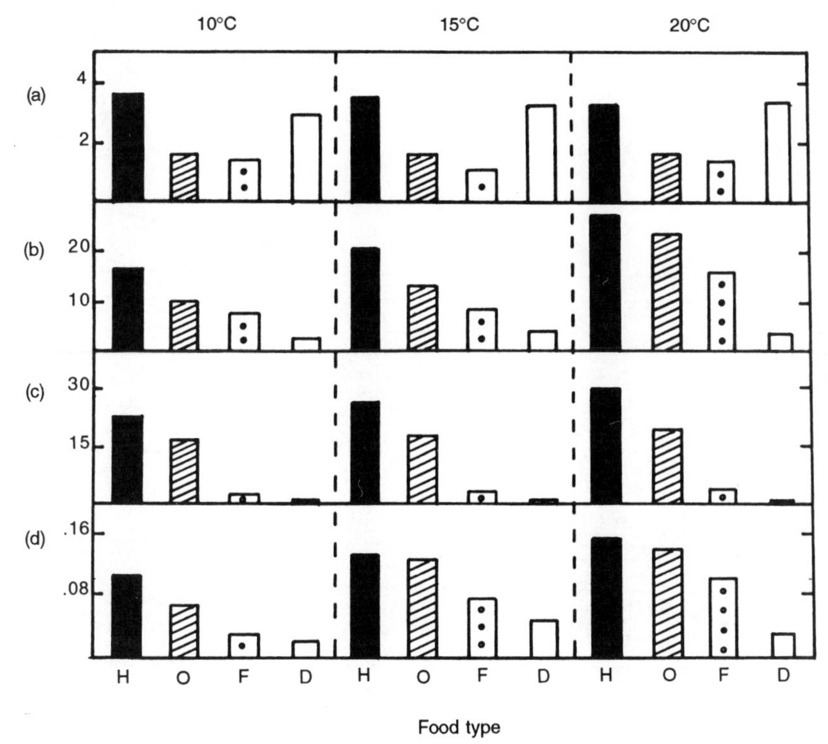

Figure 10.9 Comparison of (a) substrate nitrogen (% total N), (b) bacterial respiration (μl O_2 mg AFDW^{-1} h^{-1}), (c) substrate ATP (nM g AFDW^{-1}) and (d) daily growth rate (mg mg^{-1} d^{-1}) of *Paratendipes albimanus* (Meigen) under four different substrate (food) conditions at three different temperatures. H = pignut hickory leaves; O = white oak leaves; F = insect faeces; D = natural detritus. (After Ward and Cummins, 1979.)

(d) Biotic interactions and larval growth

Intra- as well as interspecific competition may affect larval growth and development in the Chironomidae, mainly (but not always) through the exploitation of food resources. Behavioural interference could also reduce the opportunity for feeding, thus negatively affecting growth and development.

Kajak (1963) examined the effects of artificially increasing the abundance of *Chironomus plumosus* larvae on the growth, development and survival of this and other species of chironomid using experimental cages placed on the bottom of a shallow Polish lake. Increased density resulted in poor growth and development of *C. plumosus*, particularly of young larvae; larvae were smaller in body size and a smaller proportion of them reached pupal stage than in the control. Similar effects of intraspecific competition were seen in *Tokunagayusurika akamusi* (Iwakuma, 1986a), where the body size of larvae at the end of growing season was negatively correlated with their density at the

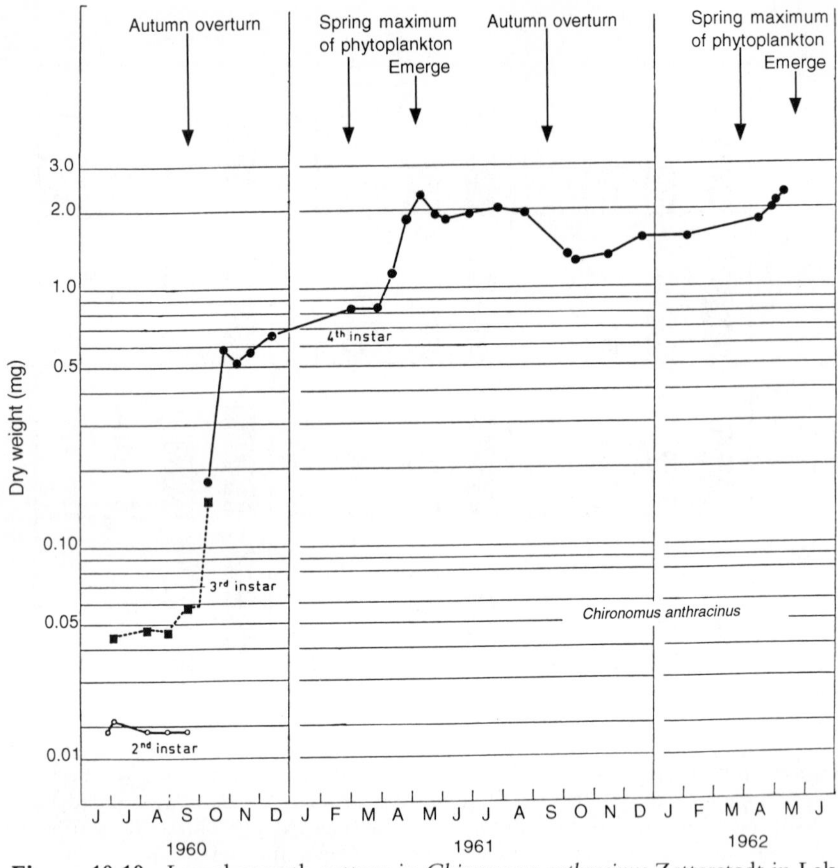

Figure 10.10 Larval growth pattern in *Chironomus anthracinus* Zetterstedt in Lake Esrom. (After Jónasson and Kristiansen, 1967.)

beginning of growing season. Similarly, the increased density of *C. plumosus* seemed to affect negatively the growth and abundance of *C. anthracinus* (Kajak, 1963) and *T. akamusi* (Iwakuma, 1986a) in Polish and Japanese lakes, respectively. In the case of *Chironomus riparius* and *Glyptotendipes paripes* which were abundant in a small prairie pond in Alberta, Canada (Rasmussen, 1985), density-dependent intraspecific competition was implicated as a mechanism to reduce growth rates, but interspecific effects were weak or undetectable.

(e) Other factors affecting larval growth

In addition to temperature, food and biotic interactions, there is a host of factors which may influence growth in the Chironomidae, including oxygen content of water, pH, toxic substances, discharge, photoperiod etc. There are, however, relatively few studies that have investigated in detail the effects of these on larval growth. In a comparison of chironomid communities from three different substrate types (snag, backwater and sand), mean growth rates of all the species were highest in the snag community (Stites and Benke, 1989), presumably due to a better feeding opportunity. In the case of *Chironomus anthracinus* inhabiting the profundal of Lake Esrom, larval growth is restricted to two short periods in a year, i.e. immediately after an autumnal overturn which raises oxygen content as well as temperature near the bottom, and during the spring maxima of phytoplankton (Figure 10.10; Jónasson and Kristiansen, 1967). On the other hand, extremely slow growth and development seen in arctic *Chironomus* species could not definitely be related to either temperature or food (Butler, 1982b). Rather than the absolute level of temperature, large diurnal and seasonal fluctuations of arctic habitats (Oliver and Corbet, 1966; Corbet, 1967) may exert a strong negative influence on growth and development. The general picture to emerge from these observations is that growth in the Chironomidae is influenced by combinations of factors which may vary at different times of the year.

Apart from external environmental factors as mentioned above, there is an intrinsic factor – body size, which is known to affect the instantaneous growth rate of larvae. Growth rate decreased with increasing body size (length/biomass) in *Orthocladius calvus* Pinder inhabiting an artificial stream channel in England (Ladle *et al.*, 1985) as well as in *Corynoneura taris* Roback, Tanytarsini and *Polypedilum* spp. from the 6th-order river in Georgia, USA (Hauer and Benke, 1991). For an assemblage of chironomid species inhabiting Appalachian mountain streams (Huryn, 1990), growth rate (G, % d^{-1}) was described as a function of body size (L, length in mm) and of average daily temperature T (°C), thus:

$$G = 0.051 - 0.068 \ln L + 0.006\ T$$

which suggests that growth rate is higher for smaller larvae and at higher temperatures (Figure 10.11).

10.3.3 DIAPAUSE AND AESTIVATION

One extremity on the spectrum of growth and development is the temporary arrest of such physical change termed **diapause**, **dormancy** or **aestivation**. This is a phenomenon which is widely seen in poikilotherms such as insects and amphibians to survive through unfavourable environmental conditions. It should be noted that, in the Chironomidae, diapause/aestivation is restricted to the egg, larval and pupal stages, whereas many insects and other animal species undergo this phenomenon as both adults and immatures. This is mainly the reflection of a reduced adult life span in the Chironomidae compared with other species. Thus, diapause/aestivation in the Chironomidae not only implies a period of inactivity but truly involves a growth/developmental arrest, because the normal course of physiological progress as eggs, larvae or pupae is temporarily halted. Such developmental arrest occurs widely amongst chironomid species and therefore constitutes an important aspect in their life cycles.

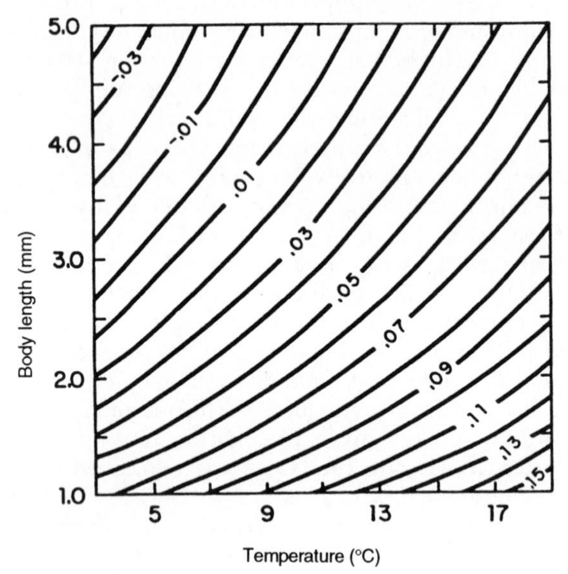

Figure 10.11 Model relating chironomid growth rate (contours, % wt d^{-1}) to body size and temperature. (After Huryn, 1990.)

(a) Overwintering and diapause

In temperate and arctic regions, low temperatures in winter constitute a particularly unfavourable condition for growth and development in insects, including the Chironomidae. Nevertheless, the Chironomidae often represent the most abundant group of macroinvertebrate in arctic habitats (Oliver, 1968). It is known that various chironomid species pass through winter at different developmental stages (e.g. Danks, 1971a; Welch, 1976; Pinder, 1983; Goddeeris, 1987; see previous section on patterns of growth) and even within a species the instar composition of overwintering larvae may vary substantially from year to year according to climatic conditions. In general, however, broad differences exist among taxa with respect to overwintering stages which closely link to the overall patterns of phenology and life history.

Cessation of feeding under a decreasing temperature regime in late autumn–winter marks the arrest of larval growth, and larvae of some species were often observed without food in the gut during the coldest months of winter (e.g. *Thienemanniella majuscula* Edwards: Tokeshi, 1986b). At the same time physiological change such as reduction in bodily water content may occur (Danks, 1971a). *Polypedilum halterale* (Coquillett) (as *simulans*) and *Einfeldia pagana* (Meigen) (as *synchrona*) larvae collected in winter had c. 5% less water content than those collected in summer (Danks, 1971a). Similarly, dehydration seems to occur fairly commonly among cold-temperate and arctic species exposed to the freezing of the surrounding substrate (Olsson, 1981).

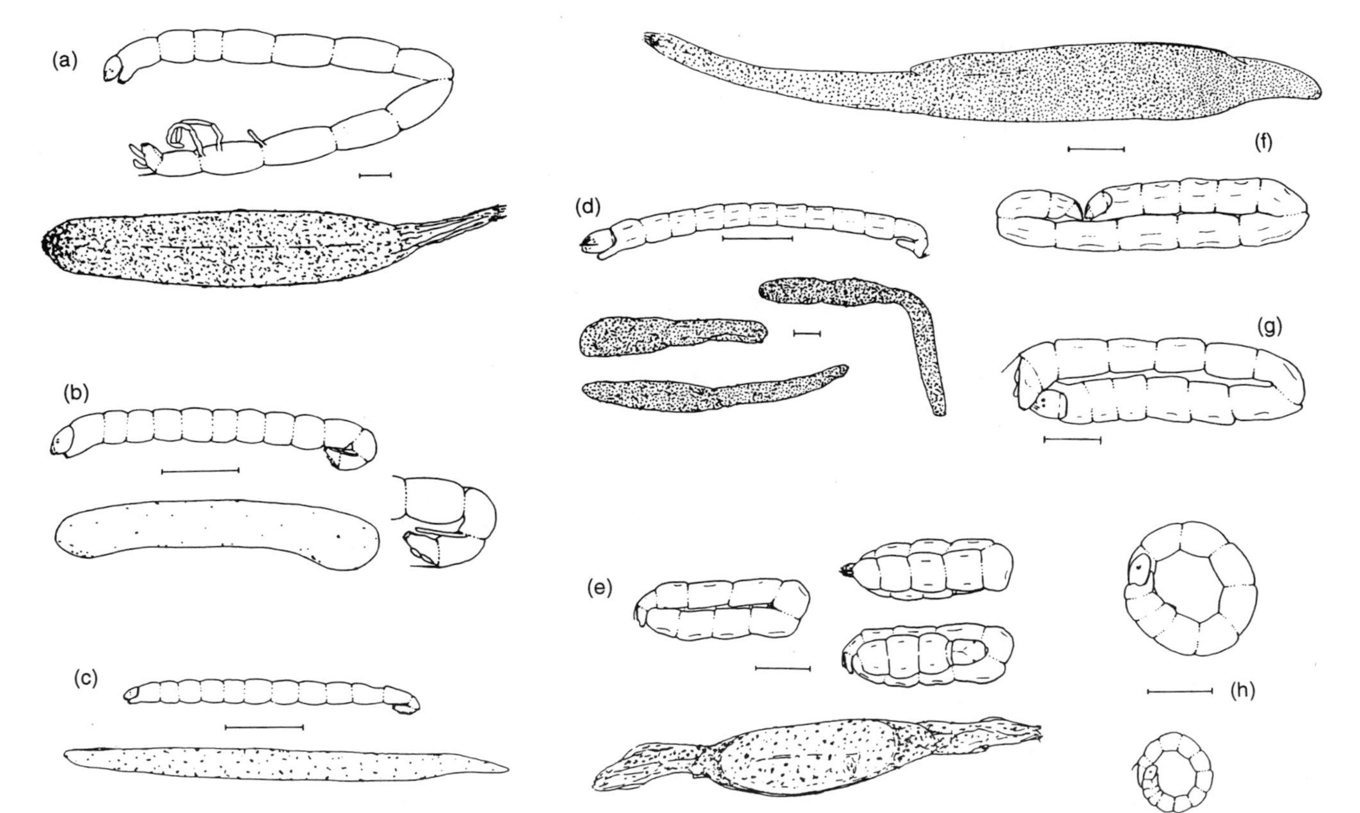

Figure 10.12 Winter cocoons and the posture of a larva inside the cocoon. (a) *Chironomus tentans*, fourth instar; (b) *Einfeldia pagana*, third instar; (c) *Polypedilum simulans*, third instar; (d) *Tanytarsus norvegicus*, fourth instar; (e) *Psectrocladius* sp., fourth instar; (f) *Stictochironomus unguiculatus*, fourth instar; (g) *Chironomus* sp., fourth instar; (h) *Procladius culiciformis*, second and third instars. (Scale lines = 1mm.) (After Danks, 1971a.)

Some species inhabiting cold-temperate and arctic regions are known to produce cocoons in which larvae overwinter. Cocoons are made either by sealing off an existing tube (e.g. some species of Tanytarsini: Armitage, 1970; *Stictochironomus* and *Tanytarsus*: Danks, 1971a) or by building a new structure with smooth, more or less transparent walls (*Endochironomus*: Thienemann, 1921; Sæther, 1962; *Einfeldia*: Danks, 1971b). Inside the cocoons, larvae take up a tightly folded posture which appears to be largely species-specific (Figure 10.12). In shallow ponds in southern Canada, Danks (1971a) found that most species examined in detail built cocoons in winter and suggested that winter cocoons occur in all subfamilies of the Chironomidae, the only exception being the Tanypodinae which have never been observed with cocoons. The occurrence of winter cocoons, however, rarely reaches 100% within a population and some larvae overwinter without cocoons, suggesting that cocoon formation is not an obligatory physiological pathway for survival in winter. Nevertheless, larvae of *Einfeldia pagana* and *Polypedilum halterale* within cocoons better survived experimental freezing than those without cocoons (Danks, 1971a). In a Canadian pond the proportion of larvae in winter cocoons varied from only 3.5% in *Cladopelma edwardsi* (Kruseman) second instars to 93.7% in *Dicrotendipes modestus* second instars (Danks and Jones, 1978). At the same time, cocoon formation tends to be restricted to the coldest period in winter when bottom temperatures are near freezing (c. 1 °C), thus occurring well after ice cover (Danks, 1971a). This indicates that temperature strongly controls cocoon-forming behaviours. Perhaps the greatest benefit of winter cocoons relates to the protection of larvae from the mechanical stress of freezing and thawing. In conjunction with dormancy and cocoon formation, *Endochironomus* larvae seem to migrate from submerged plant habitats to the bottom substrate at the beginning of winter (Jónsson, 1985; Kornijów, 1992), which presumably reduces the chance of being trapped by the surface ice cover. Penetration into the sediment during the period of diapause was also observed in *Chironomus plumosus* (Hilsenhoff, 1966) and in *Einfeldia pagana* (Danks, 1971b).

Dormancy in winter is closely related to freezing tolerance in some temperate and arctic species. Although the Chironomidae as a whole are considered to be cold-adapted (Oliver, 1971), the ability to withstand freezing conditions may vary amongst species. In a freezing experiment at –4 °C for 24 h, well over 80% of larvae of *Stictochironomus, Chironomus* and *Procladius* were alive after three days, while c. 67% of *Tanytarsus* survived (Danks, 1971a). When larvae collected in frozen mud were thawed, all these four taxa showed over 90% survival; many were also alive after 12 weeks of extended freezing at –4 °C. Similarly, Olsson (1981) observed that 96% of larval Chironomidae from frozen sediment of a northern Swedish river were alive on thawing and some larvae survived 5-month experimental freezing at –4 °C. Thus it appears that freezing tolerance is ubiquitous among species which live in habitats customarily exposed to such conditions.

(b) Drought tolerance and summer aestivation

Environmental conditions unfavourable for chironomid growth and development are not restricted to low temperatures and freezing in winter.

Gradual diminution of water body due to evaporation represents another type of unfavourable condition, commonly seen in temporary waters of tropical/subtropical Africa, Australia and other arid regions. Survival through the period of drought is a prerequisite for species living in such habitats. Drought tolerance is also an important factor for terrestrial chironomids which are frequently exposed to reduced soil humidity.

In northern California temporary pools exist for 3–6 months during winter–spring, drying up at other times of year. Grodhaus (1980) sampled dry soil of such habitats in summer and identified eight species of chironomid in six genera, *Hydrobaenus*, *Paratanytarsus*, '*Calopsectra*' (= *Tanytarsus*), *Tribelos*, *Phaenopsectra* and *Wirthiella*. All of these were second or third instar larvae encased in cocoons, capable of swimming or crawling within 48 h of being returned to wet conditions. Grodhaus (1980) also noted that *Tribelos atrum* (Townes) and *Phaenopsectra pilicellata* Grodhaus survived an extended drought of 32 months. Larvae of *Paraborniella* and *Allotrissocladius* species in Australia survive summer dry periods in a partially dehydrated state (Jones, 1975; Edward, 1986). A well-known case of exceptional drought-resistance refers to an African species *Polypedilum vanderplanki* Hinton (Chapter 6), which remained alive in dry material for as long as 10 years (Figure 10.13; Hinton, 1960b). The same species also tolerated an experimental regime of repeated dehydrations and heat treatment (56–57 °C for 2 h) (Hinton, 1968). If species possess such a high degree of drought resistance, their habitats are considered to be effectively permanent, though it is still true that growth is arrested for as long as a habitat remains dry.

Summer diapause or aestivation may also occur without the habitat being dried up. In such cases unfavourable conditions do not refer to desiccation but to elevated temperatures (above the optimum for the species concerned) and/or reduced level of dissolved oxygen. Larvae of *Tokunagayusurika akamusi* in central Japanese lakes aestivate deep in the sediment (>40 cm) in summer when surface temperatures exceed 25 °C; they reappear in the

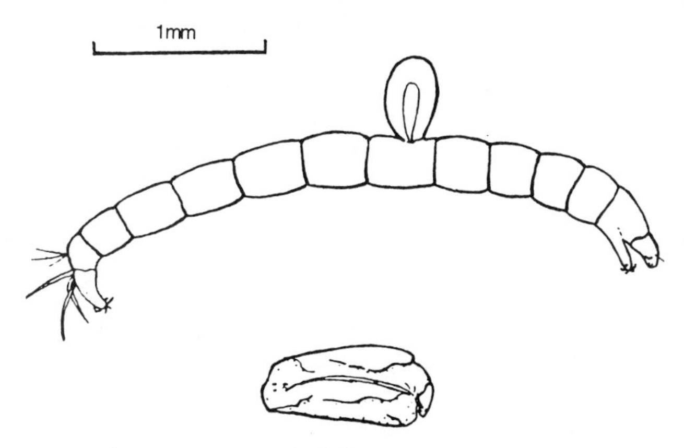

Figure 10.13 *Polypedilum vanderplanki* Hinton in a drought-resistant, dehydrated state and after immersion in water. (After Hinton, 1960.)

shallower layers of the sediment in autumn with adult emergence occurring in late autumn/early winter (Yamagishi and Fukuhara, 1972; Iwakuma, 1986a). *Hydrobaenus pilipes*, which occurs in various aquatic habitats in northern Europe, might aestivate as eggs or larvae (Mozley, 1970), and larvae of the same species collected from vernal pools in California were already in cocoons before the pools had become dry (Grodhaus, 1980). Aestivation in this species terminated when temperature and photoperiod were both reduced (from 23 to 3°C and 15.5 to 10.5 h, respectively) (Hudson, 1971).

In the present state of knowledge, terrestrial chironomids appear to possess basically the same mechanisms of coping with drought conditions as their aquatic counterparts. Delettre (1988b) examined the possibility of ecophysiologically-based drought resistance in *Smittia* and *Pseudosmittia* inhabiting heathlands and concluded that they were incapable of controlling the rate of evapotranspiration through the body wall. Rather, vertical migration in the substrate and formation of cocoons/tubes, as seen in aquatic species, seem to be the strategy adopted under drought conditions. More research is apparently needed to clarify exact mechanisms of drought resistance in many other terrestrial chironomid species.

(c) Factors controlling diapause/aestivation

As in overall growth and development (cf. previous section), many factors could potentially influence the commencement and termination of diapause/aestivation in the Chironomidae, including temperature, photoperiod, light intensity, humidity, oxygen concentration, food abundance etc. However, the first two of these – temperature and photoperiod – seem to be of particular importance in controlling diapause/aestivation. For example, photoperiod controls diapause in *Metriocnemus knabi* (Coquillett) which inhabits the pitcher-plants in North America (Paris and Jenner, 1959). Fischer (1974) observed that *Chironomus plumosus* from a stratified Swiss lake stopped development under short-day conditions at three temperatures (10, 15 and 20°C). Interestingly, another species of *Chironomus* from the same lake, *C. nuditarsis* Keyl, entered diapause only when short days were accompanied by lower temperatures. Similarly, diapause in a marine chironomid *Clunio marinus* Haliday occurred under combined conditions of short day (LD 8:16) and low temperatures (7–10°C) (Neumann and Krüger, 1985).

As noted earlier, formation of winter cocoons in cold-temperate and arctic species is strongly dependent upon ambient temperatures. These cocoons seem to occur in significant numbers only when temperatures drop below 2°C, more usually at near-freezing temperatures. At the opposite end, summer aestivation also appears to be temperature-controlled, though relevant data are scarce. In the case of *Tokunagayusurika akamusi* the temperature in the sediment below 40 cm where larvae aestivate in summer does not exceed 25°C, whereas the sediment surface temperature may reach 30°C in mid-summer (Iwakuma, 1986a). In addition to lower temperatures and reduced temperature variations, other factors such as oxygen depletion near the sediment surface and predation might also have been involved here.

(d) Significance of diapause/aestivation in the life cycle

Apart from the immediate benefit of survival through unfavourable conditions and perhaps the better synchronization of adult emergence, the ability to enter into temporal dormancy must have some importance in terms of life history strategies and the exploitation of a wide range of habitats. From an evolutionary perspective, colonization of habitats which are temporarily unsuitable for growth and development, such as arctic freshwaters and seasonal water bodies in the arid regions, could not have evolved if the benefit of doing so (e.g. high productivity during a short period of favourable conditions, significantly reduced predation/competition pressure from other taxa, etc.) did not exceed the cost of perfecting the capacity for tolerance and diapause. The fact that a large number of chironomid species have indeed successfully invaded such habitats may suggest that those seemingly unfavourable conditions have not constituted insuperable difficulties for Chironomidae as a whole, and that the benefits were greater than the costs. Thus, it is possible that the ability to enter into diapause/aestivation played a significant role in modifying chironomids' life cycles and expanding their habitats through evolutionary time.

10.4 POPULATION DYNAMICS

Whilst voltinism, emergence phenology, growth and development are mainly concerned with qualitative, phenomenological characteristics of the life history, **population dynamics** are concerned with the quantitative aspect of abundance and its variation through life cycles. This section considers the seasonal dynamics of chironomid populations, including the analysis of life tables and the long-term (inter-annual) population fluctuations.

10.4.1 PROBLEMS OF STUDYING SEASONAL DYNAMICS IN CHIRONOMIDAE

Although seasonal dynamics of larval chironomid populations have frequently been investigated, the extent of ambiguity and uncertainty which available data encompass is such that it is extremely difficult to draw a general conclusion or to make useful comparisons among different studies. It is therefore important to consider what the data on chironomid population dynamics may imply in broad terms. Seasonal population fluctuations as demonstrated by a particular data set may result from:

- Demographic trends from newly-hatched larvae to pupation and emergence as adult, including density-dependent mortality.
- Patterns of successive recruitment into the larval stage of different cohorts/parts of a cohort resulting from an extended oviposition period or extended/delayed hatching of eggs.

- Environmental stochasticity including density-independent mortality.
- Various sampling errors.

Combination of these four aspects in different proportions makes the interpretation of data on numerical fluctuations problematical, if not totally impossible.

Amongst the four potential contributors to numerical fluctuations, sampling errors are often a serious problem in chironomid research. Errors can be attributed to:

- limitation of the apparatus used;
- sample sorting methods;
- mesh size of sieves;
- inadequacy in terms of number of sample units taken on each occasion;
- sampling intervals being too wide with regard to a species' generation time.

For example, the Ekman–Birge sampler which is widely used for sampling profundal benthos in lakes may not penetrate deep enough into the sediment, thus resulting in a substantial underestimation (by over 50%) of deep-burrowing chironomid species such as *Chironomus* (Shiozawa and Barnes, 1977; Frank, 1982; Iwakuma, 1986a). Furthermore, grab samplers which cannot be sealed completely may allow a variable proportion of larvae, in particular small forms, to escape while being pulled back to the water surface. In rivers with sandy, stony or gravelly bottoms no conventional sampling technique can sufficiently sample deeper layers and interstitial habitats, necessitating elaborate techniques such as freeze-cores (cf. Soluk, 1985; Schmid, 1992a, 1993).

Sample sorting is particularly tedious with chironomid larvae and the use of flotation techniques with sucrose or manganese phosphate may not be efficient (e.g. Welch, 1976), especially with some tube-dwelling species. Subsampling, which is frequently employed to reduce the overall amount of samples to be sorted, may introduce additional errors in population estimation. Mesh sizes chosen for sieving represent another ubiquitous problem. Use of large mesh sizes allows efficient sorting of samples, but results in loss of small-sized larvae. A mesh size of 200 μm would seriously underestimate the abundance of young instars and larvae of small species such as *Corynoneura*, *Thienemanniella* and *Tanytarsus*. Depending on the species, even relatively small mesh sizes such as 100 μm may allow the majority of first instars and an unknown (perhaps large) proportion of second instars to be lost. It is therefore important to realize that numerical results thus obtained are mesh-size dependent and the accuracy of estimation may vary at different times of year as size compositions of larval populations change through time. Furthermore, inadequacy in sample sizes and sampling intervals, which often relates to the time-consuming nature of sorting and enumerating chironomid samples, introduces error variations in population estimation and may mask real population trends. As in the case of analysing voltinism (section 10.2.1), sampling intervals of one month or longer are often inappropriate for

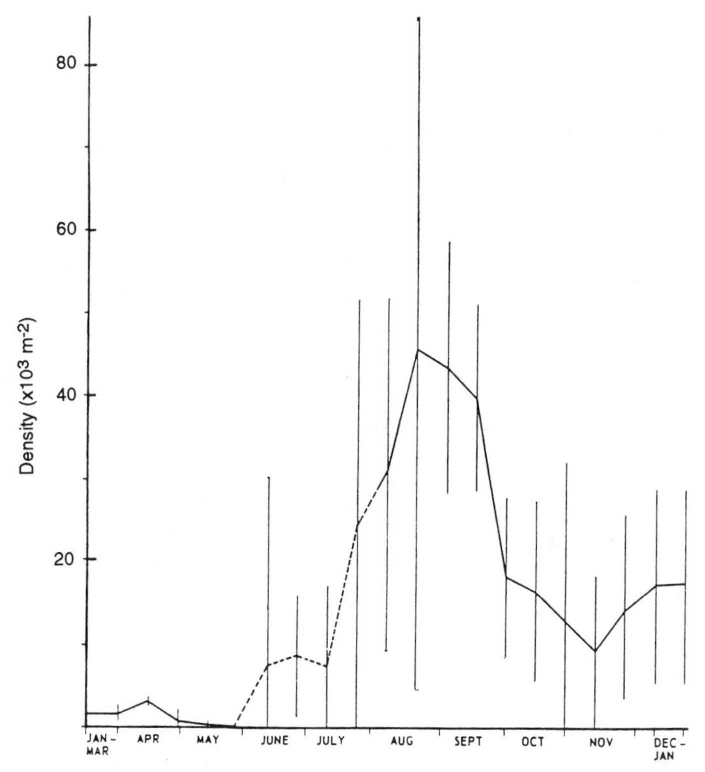

Figure 10.14 Population fluctuations (mean ± 95% CL) in *Einfeldia pagana* (Meigen) (as *synchrona*). (After Danks, 1971b.)

analysing seasonal population trends of multivoltine species, particularly during spring/summer when changes occur rapidly.

These points should always be taken into consideration whenever population studies on the Chironomidae are planned and data are interpreted. Some examples of chironomid population data are presented in the next section to illustrate these points.

10.4.2 PATTERNS OF SEASONAL DYNAMICS

Danks (1971b) investigated the life history of *Einfeldia pagana* (as *synchrona*) by taking a sample of two or three units at weekly intervals using a corer. Samples were sieved through a 420 μm screen and sorted by sugar flotation. As shown in Figure 10.14, 95% CLs given for numerical values are extremely wide, resulting mainly from a small number of sample units taken on each occasion and also a fairly large mesh size used. Nevertheless, some population trends may be discernible at this coarse resolution, owing in large part to the non-overlapping, univoltine nature of life cycle. Adults emerged

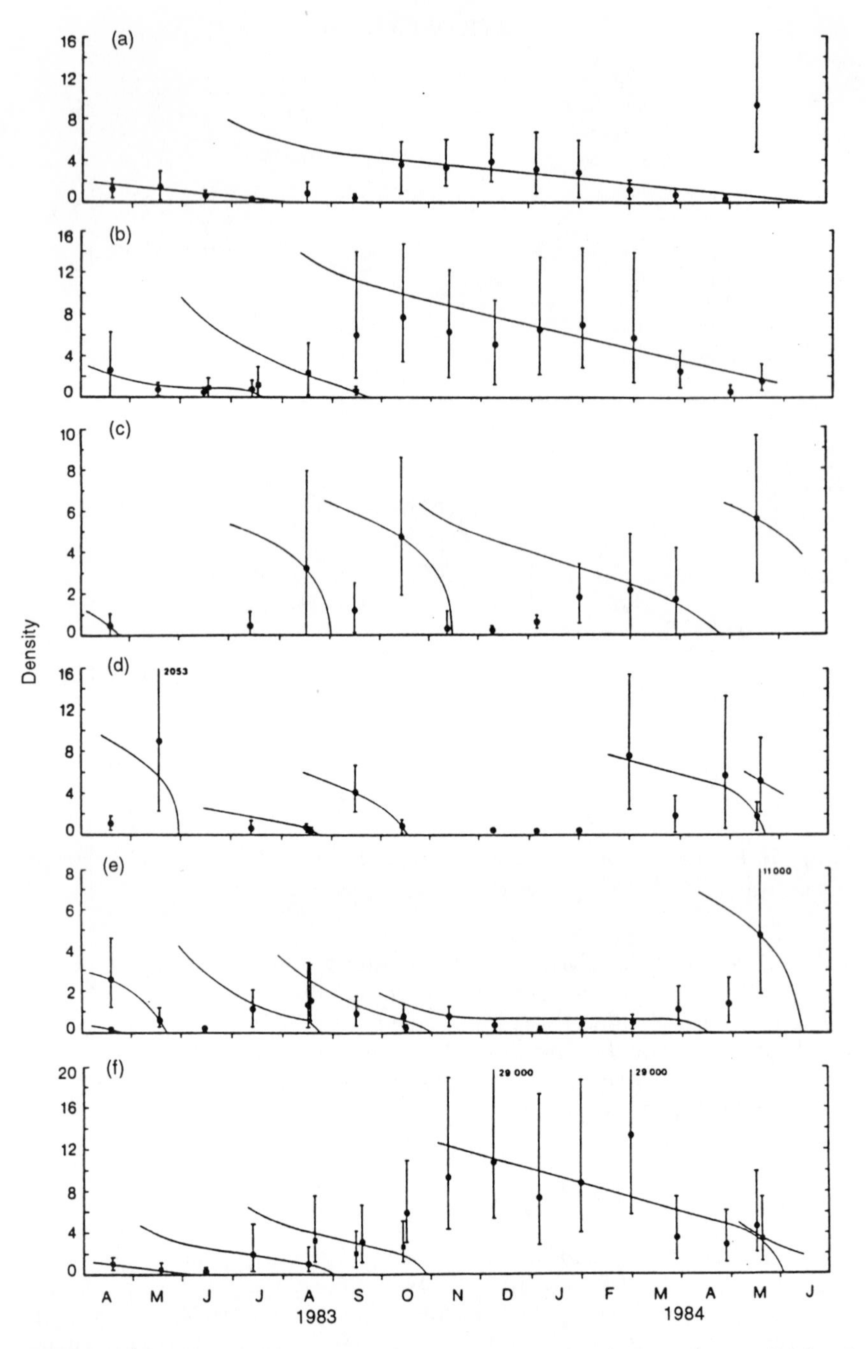

Figure 10.15 Population fluctuations in (a) *Conchapelopia melanops* (Meigen); (b) *Prodiamesa olivacea* (Meigen); (c) *Corynoneura lobata* Edwards; (d) *Eukiefferiella brevicalcar* (Kieffer); (e) *Rheocricotopus fuscipes* (Kieffer); (f) *Tvetenia verralli* (Edwards). Density scales (a)–(d), no. $\times 10^2$ m^{-2}; (e) (f) no. $\times 10^3$. (Adapted from Lindegaard and Mortensen, 1988.)

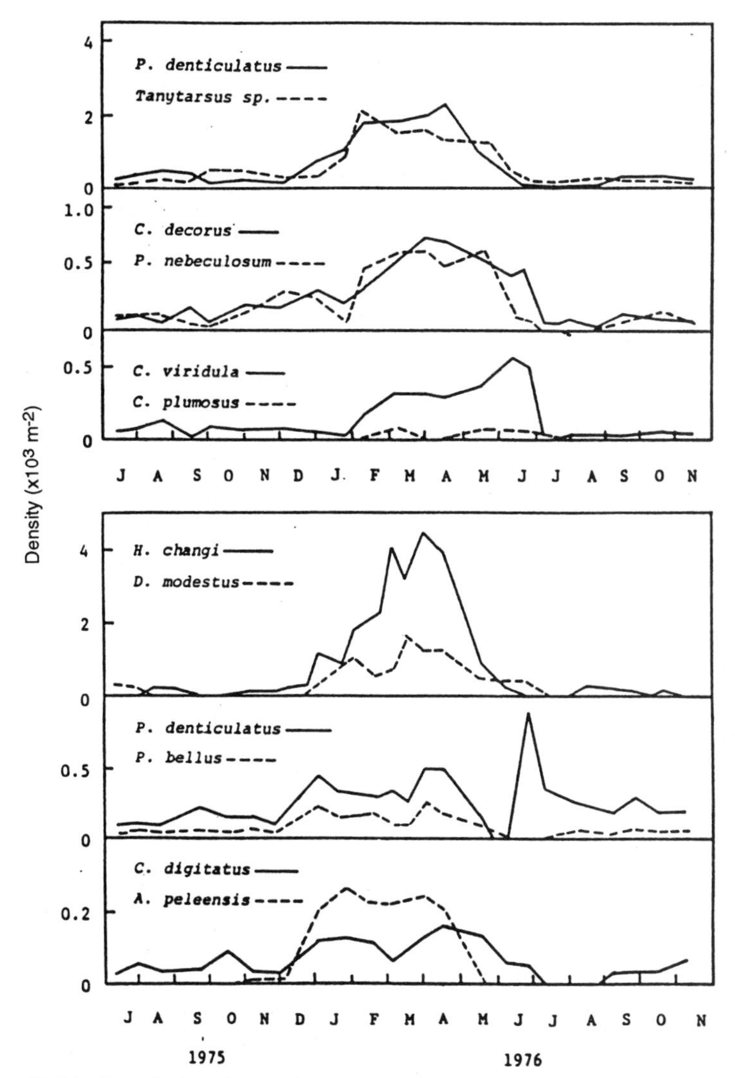

Figure 10.16 Population fluctuations in some arctic chironomid species. Fourth instars only. (Adapted from Moore, 1979b.)

in May and the larval population built up towards summer, though exact numerical values must have been severely underestimated during this period due to the use of a large mesh size with predominantly small-sized larvae. The population declined towards autumn and thereafter appeared to remain stable.

Lindegaard and Mortensen (1988), working in a lowland stream in Denmark, took monthly samples of 21–25 cores which were sieved through a 210 μm mesh. Despite a relatively large number of sample units, 95% CLs

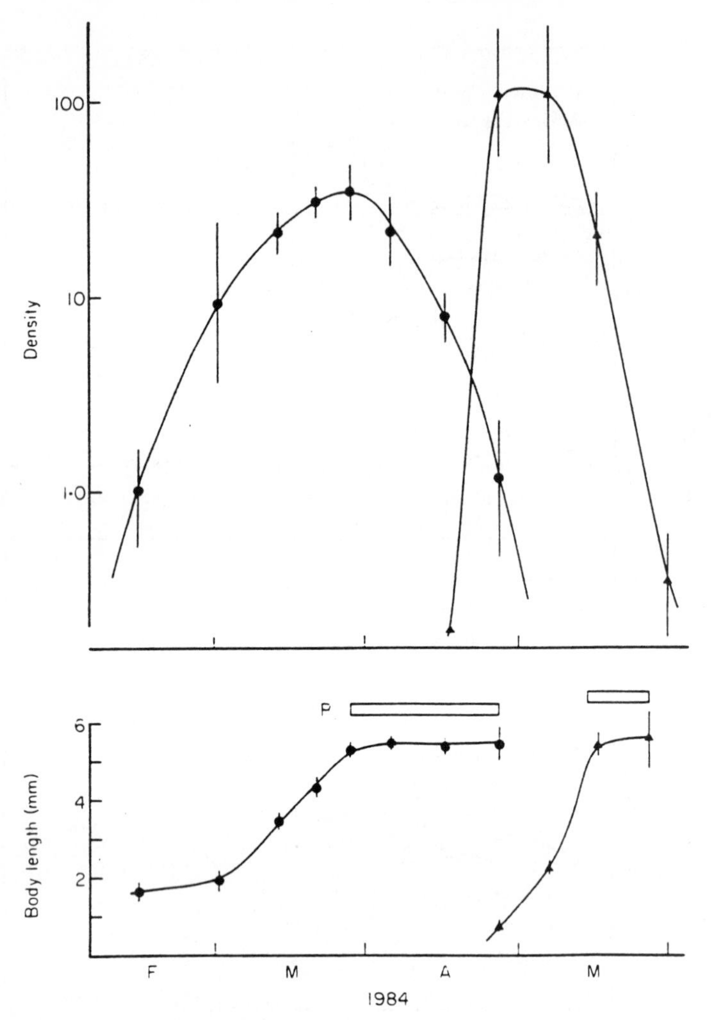

Figure 10.17 Population fluctuations (upper graph, mean density ± 95% CL) and larval growth (lower graph, mean dry weight ± 95% CL) in *Orthocladius (O.) obumbratus* Johannsen (●) and *O. (O.) oblidens* (Walker) (▲) in River Tud, eastern England. (After Tokeshi, 1986b.)

were still large (Figure 10.15), reflecting highly patchy distribution of larvae at the spatial scale of a corer (50.3 cm^2). At the same time, the mesh size used was still large enough to allow most of the first and second instars to pass through, and therefore density values are considered to refer to later instars only. In both univoltine *Conchapelopia melanops* (Meigen) (Figure 10.15a) and bivoltine *Prodiamesa olivacea* (Figure 10.15b), the density of the overwintering generation remained at a more or less stable level through autumn to winter and thereafter gradually declined towards adult emergence in spring/summer. Population trends for the summer generation of *P. olivacea* were harder to recognize. Data on other small to medium-sized Orthocladiinae species with multivoltine life cycles – *Corynoneura lobata*, *Eukiefferiella brevicalcar* (Kieffer), *Rheocricotopus fuscipes* and *Tvetenia verralli* – were more difficult to interpret, despite their common occurrence; the problems of mesh size and sampling interval must compound the complicated population trends here. In *Tvetenia verralli* (Edwards), the density (of late instars) appeared to be higher during winter than at other times of year.

Figure 10.16. shows population trends of fourth instar larvae of 11 chironomid species from a subarctic lake in north-western Canada (Moore, 1979b). Four sampling units were taken at fortnightly intervals and sorted with a 250 μm screen. Moore (1979b) restricted population estimation to the fourth instars only, considering that the abundances of other instars could not reliably be determined by this mesh size. In all the 11 species, the populations appeared to remain at an elevated level (though confidence limits or SD were not given) all through winter (ice cover between December and May).

Jónsson (1985) sampled chironomid populations from the littoral of Lake Esrom, Denmark, for over a year. A total of 15 sampling units was taken at irregular intervals using an Ekman sampler and sorted with a mesh size of 200 or 100 μm. His data generally demonstrated steady population levels, particularly during autumn/winter. However, the use of 200 μm mesh before October, 1979 (thereafter changed to 100 μm) is thought to have underestimated abundance values on earlier sampling occasions.

Tokeshi (1986b) used apical sections of the spiked water-milfoil *Myriophyllum spicatum* as a unit for sampling epiphytic chironomid populations. A total of 20–50 units was sampled at variable intervals of 5–7 days (spring/summer) to one month (mid winter) and no sieving/subsampling was employed, thus eliminating the possibility of instar bias. Different species demonstrated characteristic patterns of population fluctuations. *Orthocladius obumbratus* (Figure 10.17) sharply built up its population in early spring and reached its peak density in late March, thereafter showing a sharp decline in abundance as adult emergence occurred throughout April. A similar pattern, but shifted a month later, was observed in another species of *Orthocladius*, *O. oblidens* (Walker) (sp. A of Pinder) (Figure 10.17). *Cricotopus (C.) annulator* (Figure 10.18a), with a univoltine life cycle, appeared in large numbers in autumn, followed by a slow decline in density through winter and sudden disappearance in spring as they emerged as adults. In contrast, the overwintering generation of bivoltine *Rheocricotopus chalybeatus* gradually increased its population density through autumn–winter, probably due to delayed hatching, until its disappearance as emerging adults in April (Figure 10.18b). In this case the summer generation with lower density maxima was also discernible. The overwintering generation of multivoltine *Tvetenia calvescens* (Figure 10.18c) also appeared to increase its population density in winter, though patterns at other times of year were more variable. Population trends of *Thienemanniella majuscula* (Figure 10.18d) and *Eukiefferiella ilkleyensis* (Figure 10.18e) were difficult to interpret; demographic and environmental stochasticity coupled with low overall densities may have

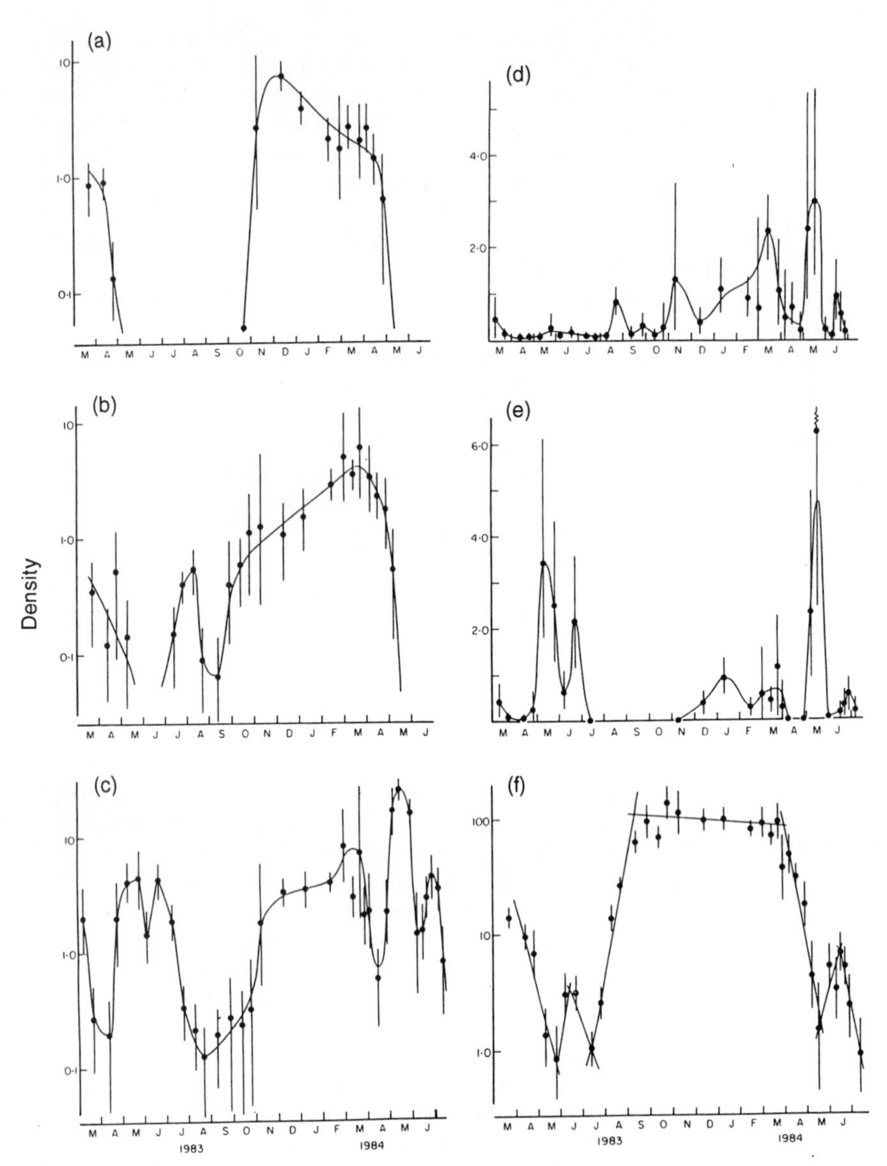

Figure 10.18 Population fluctuations (geometric mean ± 95% CL) in (a) *Cricotopus (C.) annulator* Goetghebuer; (b) *Rheocricotopus chalybeatus* (Edwards); (c) *Tvetenia calvescens* (Edwards); (d) *Thienemanniella majuscula* (Edwards); (e) *Eukiefferiella ilkleyensis* (Edwards); (f) *Rheotanytarsus curtistylus* (Goetghebuer). (After Tokeshi, 1986b.)

obscured the patterns. Finally, *Rheotanytarsus curtistylus* showed another unique pattern (Figure 10.18f). Its population built up strongly in summer (July–September) but thereafter remained almost stationary until emergence

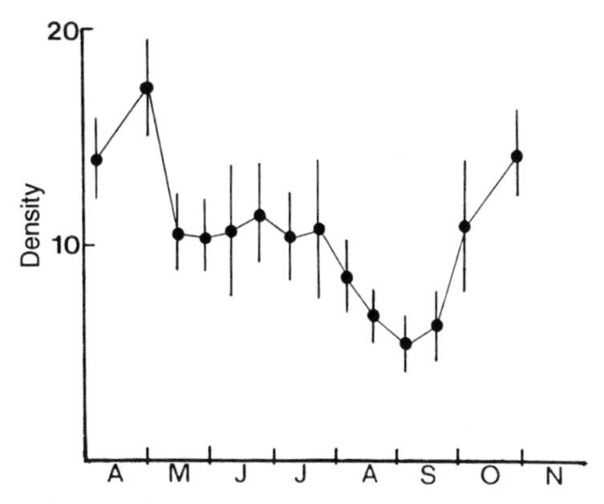

Figure 10.19 Population fluctuations (mean ± 95% CL) in *Metriocnemus knabi* (Coquillett). (After Paterson and Cameron, 1982.)

in spring. Interestingly, apart from the overwintering generation, three other generations were overlapping to some extent and did not correspond with population fluctuations.

Mackey (1977b) took weekly samples of *Nuphar lutea* (L.) leaves in the River Thames during April–October to investigate the epiphytic chironomid fauna. At a site where sampling was not disrupted by flood the two most abundant species, *Cricotopus sylvestris* and *C. bicinctus*, showed well-separated peaks of abundance, perhaps corresponding to different cohorts. At other sites, however, the patterns were less clear.

Small, discrete habitats such as phytotelmata provide an ideal opportunity for a complete census of populations without the need for sieving or subsampling, whilst facilitating adequate sample replications. Seasonal dynamics of *Metriocnemus knabi* larvae in leaves of the pitcher-plant *Sarracenia purpurea* in New Brunswick, Canada (Paterson and Cameron, 1982), are shown in Figure 10.19. The chironomid species colonize the pitcher habitat during April–October when the leaf fluid is not frozen. Larval density declined in May and August when adult emergence occurred.

In a backwater area of the Austrian River Danube, fluctuations in larval density as sampled by a multi-corer reflected life history patterns of different species (Figure 10.20, Schmid, 1992a). A decline in larval density, particularly of fourth instars, corresponded to adult emergence. Four species – *Prodiamesa olivacea*, *Harnischia curtilamellata* (Malloch), *Polypedilum scalaenum* (Schrank) and *P. laetum* (Meigen) – had two generations in a year; *Cryptochironomus defectus* (Kieffer) had three. Accurate information on population fluctuations of chironomid species inhabiting rivers with gravel substrate is particularly scarce, due to sampling difficulties. One such stream in Austria was investigated with the freeze-coring method that enabled sampling to a depth of 70 cm (Schmid, 1993a). Abundances of three young instars combined and of the fourth instars of four species, *Nilotanypus dubius* (Meigen), *Orthocladius (Euorthocladius) frigidus* (Zetterstedt), *O. (E.) rivulorum* (Kieffer) and *Heleniella ornaticollis* (Edwards), showed relatively small

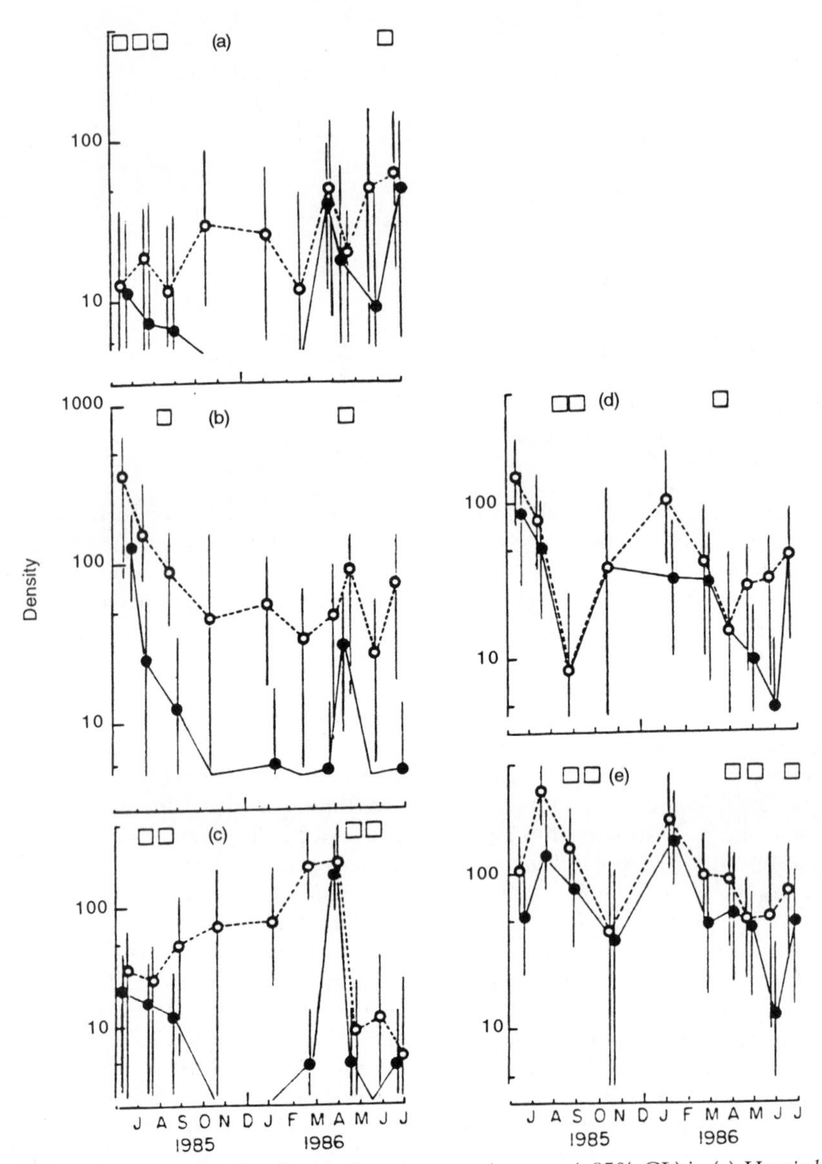

Figure 10.20 Population fluctuations (geometric mean ± 95% CL) in (a) *Harnischia curtilamellata* (Malloch); (b) *Polypedilum scalaenum* (Schrank); (c) *P. laetum* (Meigen); (d) *Prodiamesa olivacea* (Meigen); (e) *Cryptochironomus defectus* Kieffer. Solid line = fourth instar; broken line = all instars; squares indicate adult emergence. (Adapted from Schmid, 1992.)

fluctuations, with an increasing number of fourth instars in early summer indicating summer emergence of adults (Figure 10.21). An interesting case of very stable population density through the year refers to a commensal species

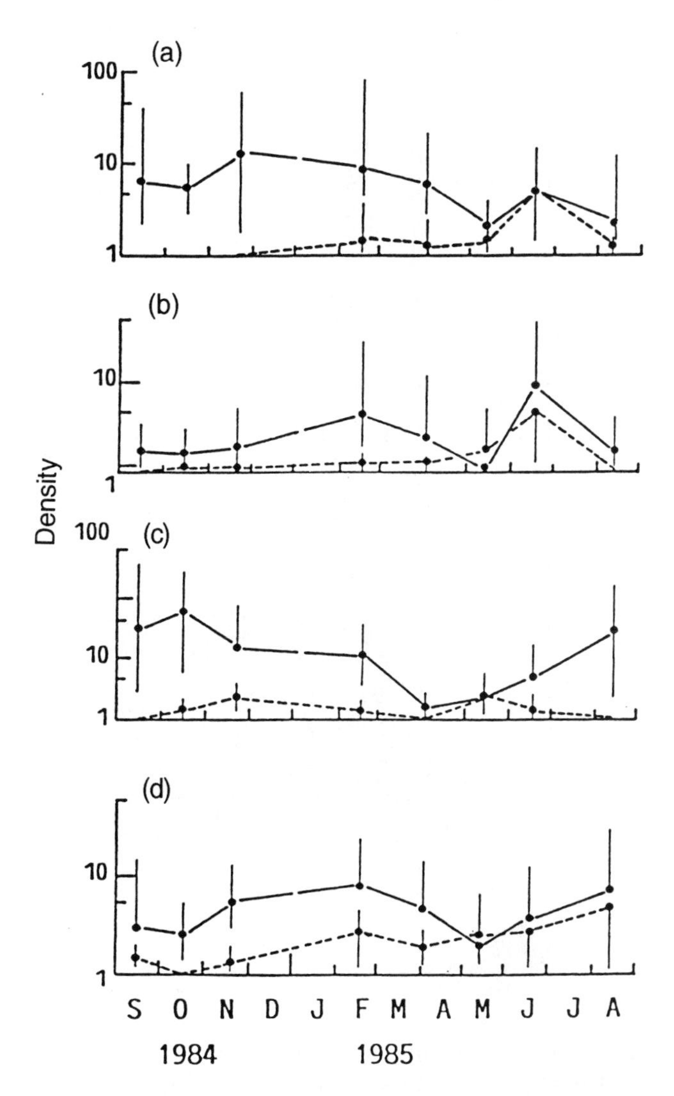

Figure 10.21 Population fluctuations (geometric mean ± 95% CL) in a mountain stream in Austria in (a) *Nilotanypus dubius* (Meigen); (b) *Orthocladius (E.) frigidus* (Zetterstedt); (c) *O. (E.) rivulorum* (Kieffer); (d) *Heleniella ornaticollis* (Edwards). Solid line = first, second and third instars combined; broken line = fourth instar. (After Schmid, 1993.)

Epoicocladius flavens (Malloch) on its host, the burrowing mayfly *Ephemera danica* (Tokeshi, 1986a; Chapter 12; Figure 12.9). Whether such a stable population trend is widespread among commensal chironomids remains to be investigated.

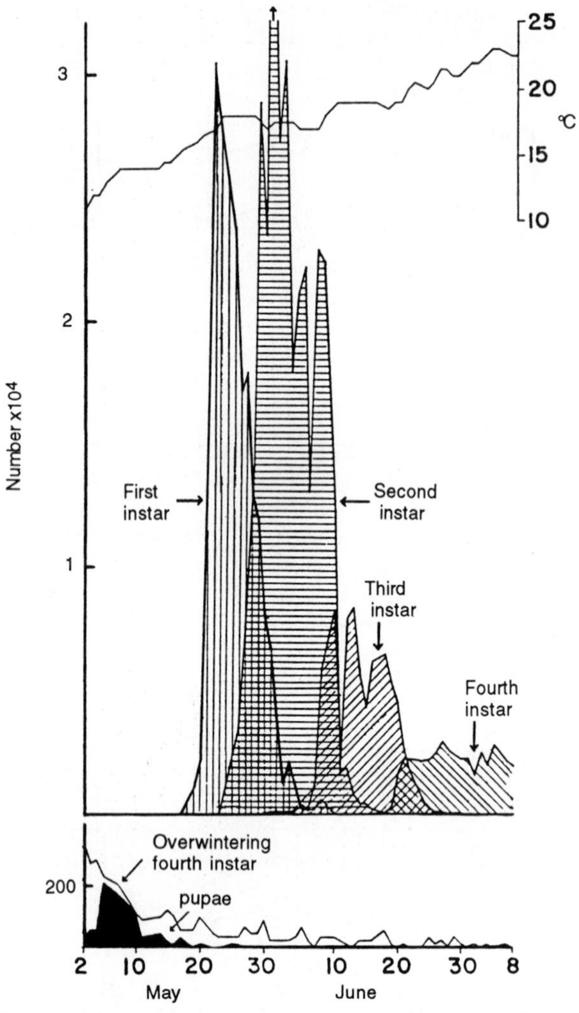

Figure 10.22 Abundances of different instars of *Chironomus plumosus* (Linnaeus), based on daily sampling in Lake Winnebago, Wisconsin. (After Hilsenhoff, 1966.)

10.4.3 DEMOGRAPHIC PATTERNS AND LIFE TABLE ANALYSIS

The foregoing section shows that seasonal dynamics encompass a variety of patterns, being influenced by a number of factors. If the main objective of investigation is to clarify the demographic changes in abundance, a more straightforward approach is to follow the fate of a single cohort through an intensive sampling of a field population. Naturally, this is more easily achieved with populations with relatively long life cycles.

In one of the most detailed demographic studies on chironomids, Hilsenhoff (1966) collected samples of *Chironomus plumosus* using four Ekman dredges at daily intervals from the beginning of May till July in Lake Winnebago, Wisconsin. The cohort clearly showed a sharp decline in abundance from the second to the third instars, and a less pronounced decline from the third to the fourth (Figure 10.22). The abundance of first instars must have been substantially underestimated by the mesh size used (>400 μm).

A more systematic approach to the analysis of demographic trends involves the construction of a life table (cf. Southwood, 1966). The approach has been widely applied to terrestrial insect populations but rarely to freshwater insects, including the Chironomidae. In a detailed study on *Chironomus anthracinus* in Lake Esrom, Jónasson (1972) presented a life table based on field sampling of larvae. Despite inefficient sampling of first and early second instars, the survivorship curves derived from the table clearly showed that summer stagnation, autumn overturn and fish predation together exerted a significant influence on the life cycle of *C. anthracinus*. Similarly, Iwakuma (1986a) produced a stage-specific life table of *Tokunagayusurika akamusi* (Table 10.1) based on a combination of field and laboratory investigations. A high mortality of 98% occurred from egg to first instars and fish predation removed a large proportion of pupae. Larvae were susceptible to fungal infection until early fourth instars. A detailed analysis such as these needs to be conducted with other species of Chironomidae.

Table 10.1 Stage-specific life table of *Tokunagayusurika akamusi* (1981 year-class) sampled in Takahamairi Bay, Lake Kasumigaura, Japan; numbers are expressed as no. of individuals m^{-2} (modified from Iwakuma, 1986a)

Stage	Number surviving at start of interval x	Number dying within interval x	Mortality within interval x (%)	Mortality factor
Eggs (expected) – Instar I	522,000	511,300	98	Unknown (failures in oviposition/hatching, predation by gobid fish, etc.)
Instar II	10,700	2,370	20	Unknown (predation by gobid fish)
Instar III	8,330	1,250	15	Disease (fungi)
Instar IV (feeding)	7,080	1,687	24	Disease and predation
Instar IV (non-feeding)	5,393	1,095	20	Aestivation, predation and emergence failure
Pupae	4,284	3,361	78	Predation by fish
Adults	923 (females 518)			(Adult mortality not accounted for)
Expected no. of eggs	693,200			

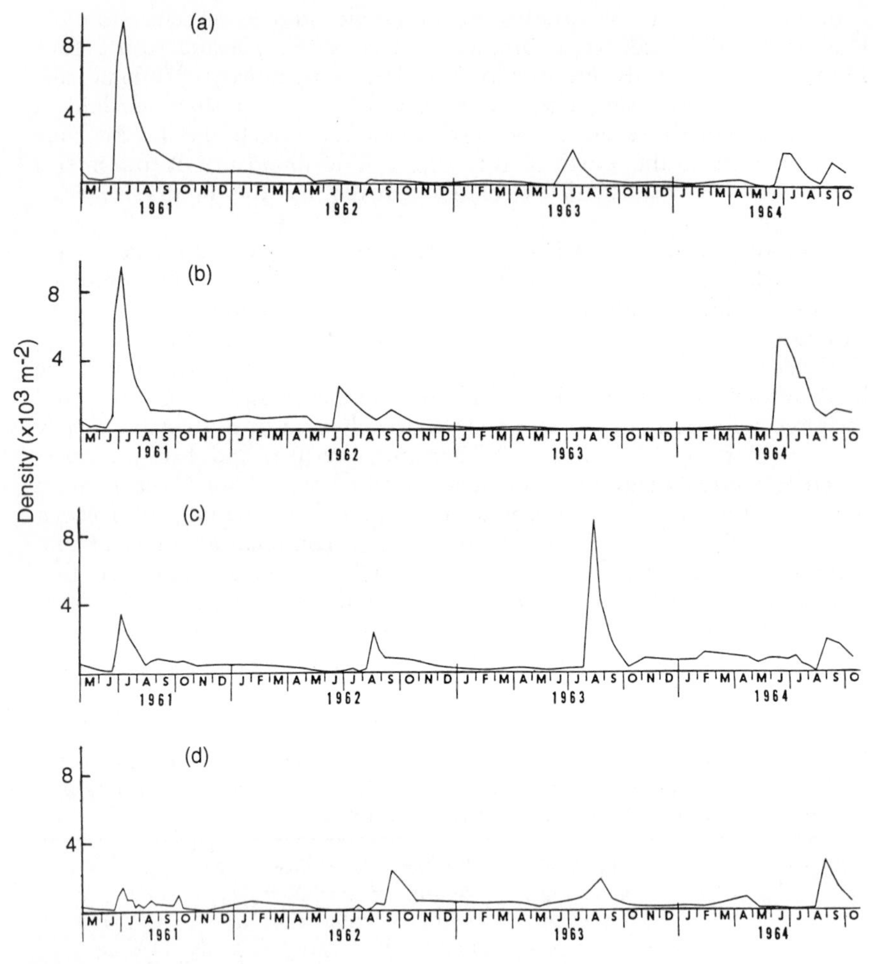

Figure 10.23 Population fluctuations in the fourth-instar *Chironomus plumosus* (Linnaeus) at four different sites in Lake Winnebago, Wisconsin. (a) North Buoy; (b) Central Buoy; (c) West Buoy; (d) South Buoy. (After Hilsenhoff, 1967.)

10.4.4 LONG-TERM (INTER-ANNUAL) DYNAMICS

Apart from seasonal variation, inter-annual variation in abundance constitutes an important aspect of population dynamics. Long-term population studies on the Chironomidae have been confined virtually to lake profundal species, most notably of the genus *Chironomus* (Jónasson, 1965; Hilsenhoff, 1967; Yamagishi and Fukuhara, 1971; Lindegaard and Jónasson, 1979; Frank, 1982; Iwakuma, 1986a).

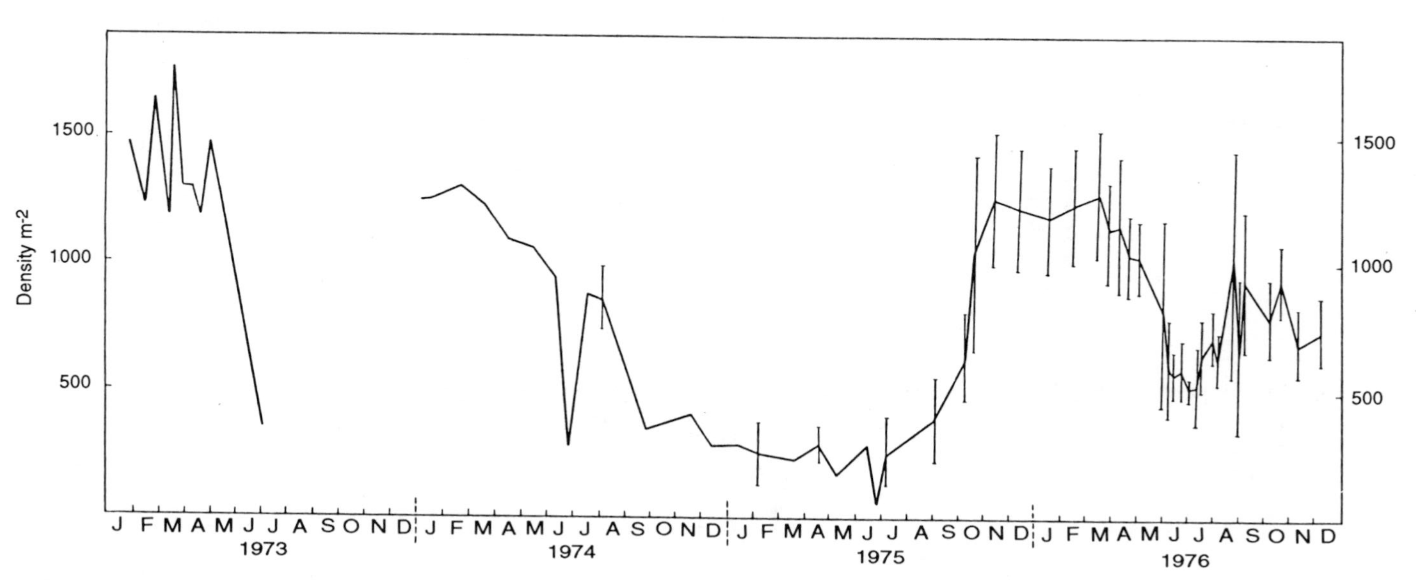

Figure 10.24 Population fluctuations of *Chironomus plumosus* (Linnaeus) in Federsee, Germany. (After Frank, 1982.)

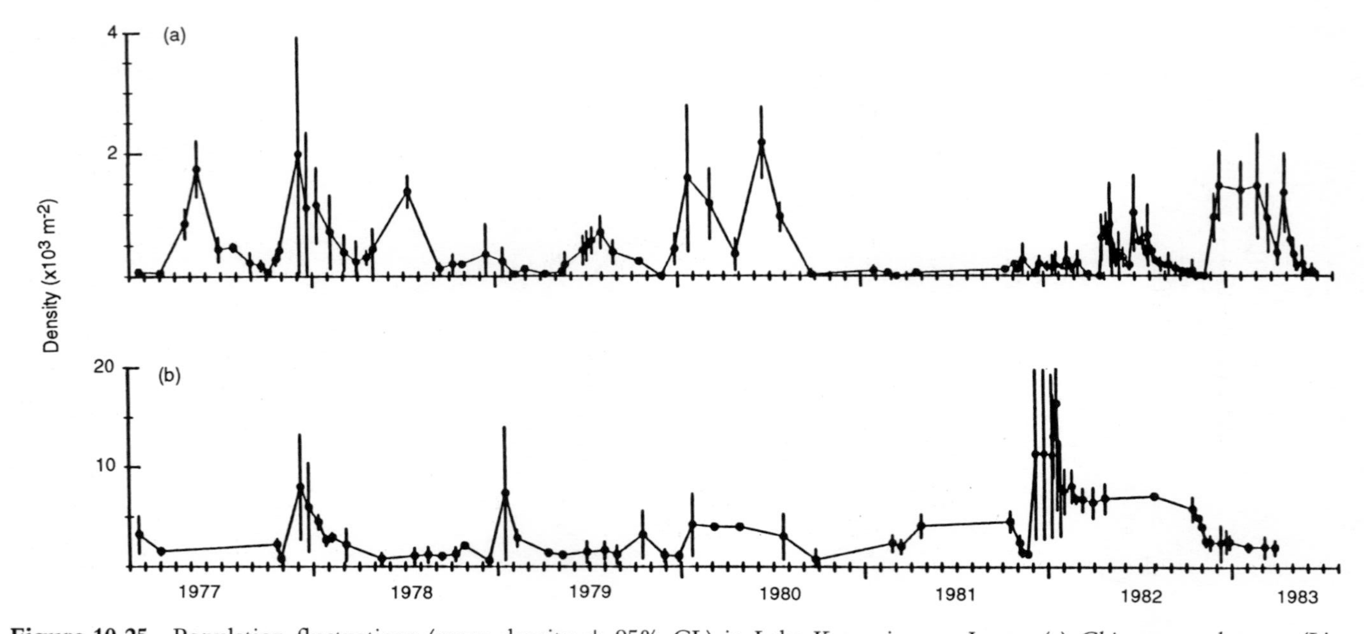

Figure 10.25 Population fluctuations (mean density ± 95% CL) in Lake Kasumigaura, Japan. (a) *Chironomus plumosus* (Linnaeus); (b) *Tokunagayusurika akamusi* (Tokunaga). (After Iwakuma, 1986.)

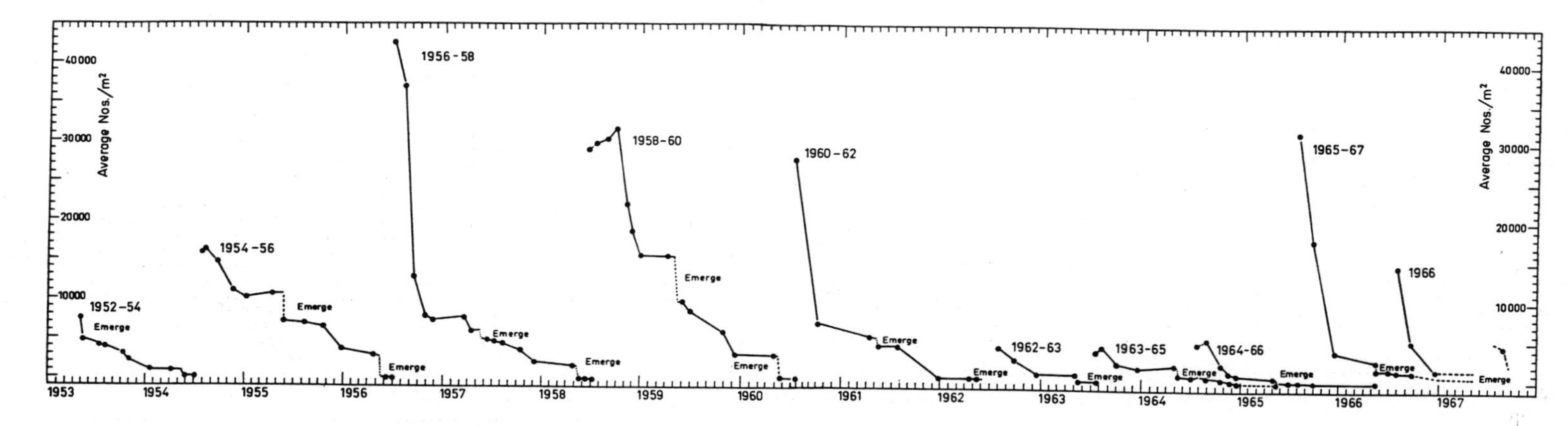

Figure 10.26 Population fluctuations of *Chironomus anthracinus* (Zetterstedt) in Lake Esrom, Denmark. (After Jónasson, 1972.)

Hilsenhoff's (1967) 4-year study of *C. plumosus* at four different sites in Lake Winnebago, Wisconsin, showed single yearly peaks in abundance of fourth instar larvae (Figure 10.23), though peaks did not occur at all sites in all the years. This, however, may have been partly due to sampling artifact, i.e. unreplicated samples reflecting spatial heterogeneity in distribution rather than true population trends. Frank's (1982) 4-year study on the same species in central Europe did not reveal such yearly peaks (Figure 10.24); rather the population seems to undergo a longer cycle of fluctuation. Iwakuma (1986a) reported on the population dynamics of *C. plumosus* over 8 years in Lake Kasumigaura (Figure 10.25a). In this case the population showed two or three peaks in a year, reflecting the occurrence of two/three generations. Despite these fluctuations, the population demonstrated notable stability in overall abundance over the years; it is interesting that the low abundance in 1981 was followed by a swift recovery to the previous level. Iwakuma's (1986a) other long-term data on *Tokunagayusurika akamusi* (Figure 10.25b) also showed remarkable stability in population density over 8 years, with yearly peaks in winter representing recruitment of young larvae. The population density of fourth instar *C. islandicus* larvae in Lake Myvatn, Iceland (Lindegaard and Jónasson, 1979), remained at nearly the same level for 5 years, apart from somewhat higher values observed in the first year. Fourth-instar *Tanytarsus gracilentus* from the same lake reached comparable annual peak abundances for 4 years, though the peak in the last year was lower. Interestingly, no young larvae of both *C. islandicus* and *T. gracilentus* seemed to be recruited in the last year of investigation.

In one of the longest studies, Jónasson (1972) documented the population dynamics of *C. anthracinus* over 14 years (Figure 10.26). Between 1952 and 1962, 2-year life cycles were predominant, beginning in even numbered years (1952, 1954, 1956, 1958 and 1960), except the 1962 year class, which had a one-year life cycle. Although a proportion of a cohort emerged in one year, their progeny did not normally establish themselves in odd numbered years (except 1963 and 1965), presumably due to predation/interference competition from mature larvae present in the habitat.

10.4.5 FACTORS AFFECTING CHIRONOMID POPULATION DYNAMICS

As the above examples of seasonal and inter-annual dynamics imply, chironomid populations are influenced by a combination of abiotic factors and biotic factors. Environmental variations (temperature, photoperiod, oxygen level, water level and movement, food availability etc.) and biotic interactions such as predation, competition, parasitism and diseases operate on intrinsic life-cycle patterns, resulting in complex population dynamics. Despite this general picture, there are conspicuous gaps in our knowledge, e.g. the role of diseases/parasitism in chironomid population dynamics. Above all, studies aiming at a finer resolution of different factors are necessary, along with an effort to improve population estimations by reducing sampling errors. In this respect, more quantitative approaches to life cycles need to be pursued in chironomid research.

11

Production ecology

M. Tokeshi

11.1 INTRODUCTION

Production ecology represents one of the major areas where much effort has been directed within chironomid research. Indeed, chironomids are perhaps the most frequently treated group among macroinvertebrates in freshwater productivity studies. This is attributable to the fact that larval chironomids often dominate lake benthos which have been subject to numerous productivity studies, particularly at the instigation of the International Biological Programme since the mid 1960s. In most lentic systems, immature Chironomidae form an important link between primary producers, such as phytoplankton and benthic algae, and secondary consumers, most notably fish. Because the latter often assume an economic importance in many lake systems world-wide, productivity of chironomids as food for fish has been considered of immediate relevance to fish productivity. In addition, increasing problems of lake eutrophication (section 15.2) focused attention on the Chironomidae as a potential agent of processing and removing excess nutrients, particularly phosphorus, from freshwater systems. Three general characteristics of lake chironomids (their abundance in the benthos, sediment detritus feeding and emergence as adults with a large proportion perishing on the land) make them more convenient as removers of phosphorus than artificial removal of fish or marginal aquatic plants. Thus, studies of chironomid productivity can contribute to understanding ecosystem dynamics in freshwaters, with important applied implications.

Conceptually, productivity estimations allow comparison of various components of an ecosystem on a common ground and also enable inter-ecosystem comparisons. Traditionally, however, productivity studies in the Chironomidae tended to remain merely an exercise of calculation with no more than a cursory comparison of published productivity values. While some useful insights have been obtained, conceptual progress has been particularly slow in this area and a more integrated approach is needed in future.

In principle, the study of productivity is dependent upon the knowledge of

The Chironomidae: Biology and ecology of non-biting midges
Edited by P.D. Armitage, P.S. Cranston and L.C.V. Pinder
Published in 1995 by Chapman & Hall. ISBN 0 412 45260 X

life cycles, seasonal population dynamics and larval growth. Therefore the kind of information presented in the previous chapter naturally forms a base and has a close bearing on estimation of productivity. Consequently, errors and uncertainties associated with all these aspects, as discussed in the previous chapter, could also be transferred to productivity estimates to a greater or lesser extent.

This chapter reviews the ecophysiological basis of productivity, methodological aspects, and patterns of productivity in the Chironomidae, focusing on both lentic and lotic habitats. Factors affecting productivity are examined and the relationships between chironomid production and freshwater ecosystem functioning are considered.

11.2 BACKGROUND OF PRODUCTIVITY ESTIMATION

11.2.1 ENERGETIC BASIS OF PRODUCTIVITY

Productivity (or alternatively, **production**) is defined technically as the rate at which new biomass is accumulated per unit area. From an energetics point of view production relates to the rates of ingestion, assimilation, excretion and respiration:

Ingestion (I)	→	assimilation (A)	→	production (P)
	└	excretion (E)	└	respiration (R)

Thus, production essentially amounts to the total energy ingested minus the energy lost through excretion and respiration ($P = I - E - R$), summed over individuals in an area through time. Though in practice estimation of chironomid productivity generally proceeds without recourse to these related components, they are nevertheless important as an energetic basis of production and therefore available information to date is summarised here.

(a) Ingestion

Ingestion or feeding rate refers to the amount of food intake per unit time. Direct observation and quantification of ingestion is difficult, though not impossible, with chironomid larvae, especially if they feed on detritus or algae which do not form a discrete food package. In the case of algae, however, it is possible to adopt a radiometrical technique to quantify the rate of ingestion. Essentially, the method involves labelling algae with ^{14}C and allowing chironomid larvae to feed on them in a chamber. The time course of ^{14}C accumulation in the larval body is then observed by periodically sacrificing larvae and determining the amount of ^{14}C by liquid scintillation counting. Johannsson (1980) and Kesler (1981) used this technique to study the ingestion rate of *Chironomus plumosus* (Linnaeus) f. *semireductus* from Lake Ontario and of *Corynoneura scutellata* Winnertz from a pond in Rhode Island, respectively. In the former, measurements were taken at ambient temperature at 2–3-week intervals from May to November and ingestion rate was

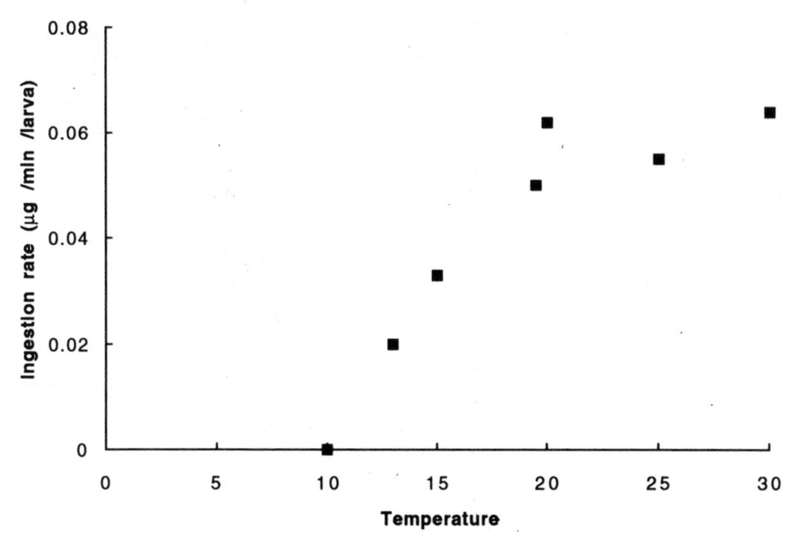

Figure 11.1 Relationship between ingestion rate (as measured by bodily accumulation of ^{14}C) in *Corynoneura scutellata* Winnertz and temperature. (After Kestler, 1981.)

expressed as 'gut-filling time', i.e. the time required to reach the point of levelling off of radioactive uptake. Within the temperature range of 10–22 °C, gut-filling time was inversely related to temperature, while between 22 and 24.5 °C it remained at a minimum value of 0.5 h and below 10 °C it was either very long or larvae did not feed at all. Thus, temperature controlled ingestion rate in this species, but larval size and time of day had no effect. Total amount of ingestion per day was estimated to be nearly five times as much as body weight. In the case of *Corynoneura*, ingestion rate was obtained as the slope of a line fitted to the relationship between ^{14}C in larval body and time. This ingestion rate increased from 0 at 10 °C to a maximum at 30 °C (Figure 11.1), which is similar to the situation in *Chironomus*. In addition to the radiometric method, Kesler (1981) also employed a gravimetric method whereby faecal material and other non-adhering cells were removed from experimental algal substrate after it was offered to chironomid larvae and the difference between this and control substrate was determined. The value thus obtained seems to fit in with the general trend based on the radiometric method.

(b) Assimilation

Assimilation refers to the proportion of total energy ingested which is actually incorporated into the organism's body. Since assimilation is placed between ingestion and production (see above) there are two ways of expressing it: $A = I - E$; or $A = P + R$.

Accordingly, two different approaches exist for estimating assimilation. Kesler's (1981) gravimetric method mentioned above provides rough estimates of ingestion and excretion, though the latter may be underestimated if soluble excretion is ignored. A more direct method of estimating assimilation was pioneered by Johannsson (1980), who effectively used the anterior and posterior sections of the midgut content as ingested and excreted matter, respectively. The method involves cutting the alimentary tract into 5–10 sections and determining the organic content of the most anterior and posterior sections, the difference being regarded as assimilation. Note, however, that unless gut passage time is also estimated and the value is converted to a 'per unit time' basis, the method simply produces assimilation 'efficiency' rather than assimilation 'rate'. Assimilation efficiency in *C. plumosus* f. *semireductus* was found to vary seasonally, with high values corresponding with the latter part of diatom (*Melosira* and *Stephanodiscus*) blooms and low values with diatom scarcity and blue-green algal blooms in late summer (Johannsson, 1980; Johannsson and Beaver, 1983). When the same technique was applied to *Glyptotendipes paripes* (Edwards) and *Chironomus riparius* Meigen coexisting in a pairie pond in Canada, the former had a significantly higher assimilation efficiency than the latter (11.9% and 5.9%) (Rasmussen, 1984a). Similarly, Johnson (1985) used this technique to estimate carbon and nitrogen absorption by two *Chironomus* species, *C. plumosus* and *C. anthracinus* Zetterstedt in Lake Erken, Sweden. The two species had comparable mean assimilation efficiencies, with 26.8% C and 29.3% N in *C. plumosus*, and 24.6% C and 28.1% N in *C. anthracinus*.

As shown above, assimilation can also be obtained as a sum of production and respiration (e.g. Benson *et al.*, 1980; Iwakuma, 1986a). However, because production is the focus of attention here, this 'reversal' approach to estimating assimilation is of minor importance in the present context.

(c) Excretion

The overall magnitude of excretion by chironomid populations is virtually unknown, though some investigated the release of key nutrients, in particular nitrogen and phosphorus, into the environment (e.g. Fukuhara and Sakamoto, 1988). Gardner *et al.* (1983) examined the patterns of nitrogen excretion in *Chironomus* spp. from Lake Michigan by monitoring *o*-phthalaldehyde reactive nitrogen (ammonium plus primary amines) in an incubation flow cell (cf. Gardner and Scavia, 1981). Nitrogen excretion by chironomid larvae occurred in clear spurts several times per hour rather than continuously and the weight-specific ammonium release rate ranged between 3 and 15 nmol NH_4 (mg AFDH)$^{-1}$ h^{-1}. In another experiment (Gardner *et al.*, 1981) phosphorus excretion was also measured, yielding values of 0.27–0.57 nmol P (mg AFDW)$^{-1}$ h^{-1} at 5 °C and 0.68–0.81 nmol P (mg AFDW)$^{-1}$ h^{-1} at 20 °C. Fukuhara and Yasuda (1985) made a more detailed measurement of phosphorus excretion by *Chironomus plumosus* and *Tokunagayusurika akamusi* (Tokunaga) from Lake Suwa, Japan, and found excretion rates to be positively related to temperature within the range of 5–30 °C and negatively related to body weight. At 15 °C, weight-specific phosphorus excretion rates were 0.144 and 0.073 PO_4-P μg (mg dry wt)$^{-1}$ day^{-1} for *C. plumosus* and *T. akamusi*, respectively. Similar patterns were

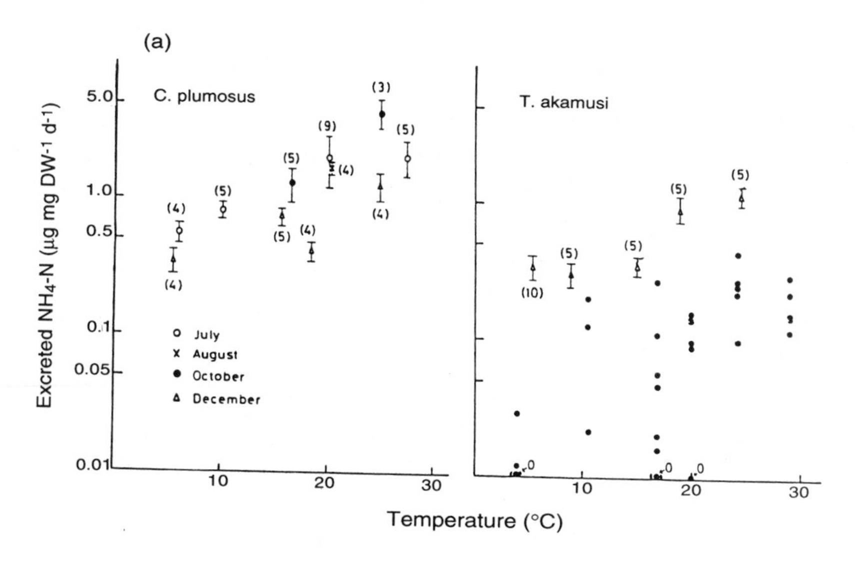

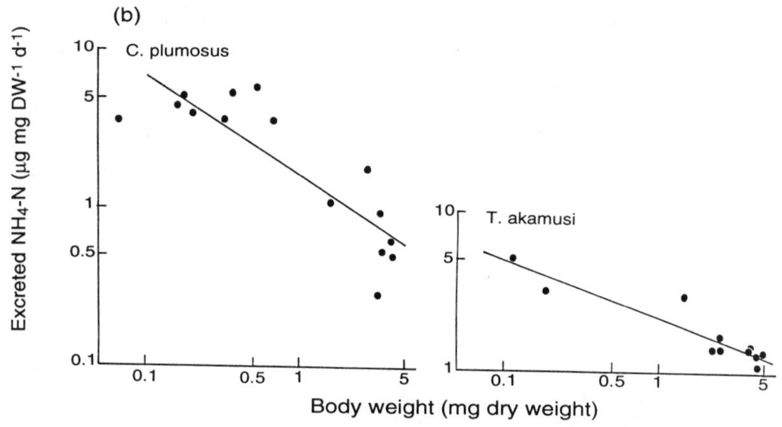

Figure 11.2 Weight-specific ammonia excretion in *Chironomus plumosus* (Linnaeus) and *Tokunagayusurika akamusi* (Tokunaga) in relation to (a) temperature; (b) body weight. (After Fukuhara and Yasuda, 1989.)

observed for the rate of ammonium excretion in these species, positivecorrelation with temperature and negative correlation with body weight (Figure 11.2a,b; Fukuhara and Yasuda, 1989).

(d) Respiration

Respiration rate can be measured directly by monitoring oxygen consumption of organisms in a closed chamber. A wide range of aquatic organisms have been subjected to such measurements, but a relatively small number of studies dealt with Chironomidae. Respiration rate is dependent upon both temperature and body size, which has been demonstrated for a number of chironomid taxa, particularly large-bodied lentic species such as *Chironomus* spp. (Johnson and Brinkhurst, 1971; Konstantinov, 1971a; Ripley, 1980; Iwakuma, 1986a; Hamburger and Dall, 1990). In general, weight-specific respiration rate scales positively with temperature and negatively with body size. Johnson and Brinkhurst (1971) fitted a general model which relates respiration to temperature and body size:

$$\ln R = a + bT + c \ln W$$

where R is respiration rate (μg O_2 (mg AFDH)$^{-1}$day^{-1}), T is temperature (°C) and W is ash-free dry weight (mg). Parameter values for *Chironomus* sp. and *Procladius* spp. from Lake Ontario were a=1.62, b=0.115 and c=–0.186 for the former and a=2.02, b=0.115 and c=–0.340 for the latter. In addition to temperature and body size, Iwakuma (1986a) incorporated dissolved oxygen concentration U (g O_2 m^{-3}) into an integrated model of respiration rate R thus:

$$R = R_d f(W)\, g(T)\, h(U)$$

where R_d is the weight-specific respiration rate of a larva of weight W_d at water temperature T_d under dissolved oxygen concentration U_d, and $f(W)$, $g(T)$ and $h(U)$ are dimensionless functions relating to weight, temperature and dissolved oxygen concentration, respectively. For *Tokunagayusurika akamusi*, the equation was:

$$R = 5.96 \times 10^{-4}\, W^{-0.359} e^{0.0941 T}\, U/(2.00 + U)$$

For *Chironomus plumosus*,

$$R = 6.73 \times 10^{-3}\, W^{-0.245}\, T\, U/(2.28 + U) \quad (T<20.4\,°C)$$

$$R = 0.0128\, W^{-0.245}\, U/(2.00 + U) \quad (T>20.4\,°C)$$

whereby the second equation indicates that respiration rate does not vary with temperature above 20.4 °C in this species.

11.2.2 METHODOLOGICAL BASIS OF PRODUCTIVITY STUDIES ON THE CHIRONOMIDAE

Five methods of estimating production are commonly used in chironomid studies:

- Size–frequency method (Hynes and Coleman, 1968; Hamilton, 1969; Benke, 1979; Hynes, 1980; Menzie, 1981; Waters and Hockenstrom, 1980).
- Instantaneous growth method (Johnson and Brinkhurst, 1971; Waters and Crawford, 1973).
- Removal–summation method (Winberg *et al.*, 1971).

- Increment–summation method (Winberg *et al.*, 1971).
- Allen curve method (Allen, 1951).

Less common methods include:

- Mean population biomass multiplied by an average P/B ratio (Waters, 1969; Kajak and Ryback, 1966; Kimerle and Anderson, 1971).
- Estimation through measurements of respiration (McNeill and Lawton, 1970; Humphreys, 1979; McLusky and McFarlane, 1974).
- Annual biomass of emerging adults multiplied by a factor of 2.8 (Iwakuma, 1986b).

These 'indirect' methods, however, should be regarded as a very rough approximation (there is no way of knowing the accuracy of values thus obtained in particular cases) and used only where others cannot be applied. Although theoretically all these methods should produce broadly similar values, in practice calculation from the same data set may lead to a substantial discrepancy among different methods. This is to a large extent inevitable, as different methods rely on different assumptions regarding productivity calculations. It is therefore important to choose a method which is considered to introduce the least amount of errors in estimation under given circumstances. This involves consideration of population patterns of organisms and sampling strategies adopted. For a broad review of estimating productivity, see reviews in Downing and Rigler (1984) and Plante and Downing (1990).

Density and biomass constitute two major components of production estimation. Problems of estimating density are treated in section 10.4. Obtaining accurate estimations of density is especially problematic with small species and young instars, and reliance on sieving for sample processing exacerbates the situation. There are two ways in which inaccurate estimation of the density of small larvae could affect overall productivity estimation. First, where mean body weight and mean population density are combined in the calculation of productivity (e.g. increment-summation method), inaccuracy in density estimation will seriously affect the values of mean body weight and, consequently, productivity. Second, small larvae are easily underestimated, directly resulting in underestimation of productivity. This is considered a serious problem particularly with small species such as Tanytarsini and *Corynoneura*. In large species such as *Chironomus* the contribution of first and to some extent second instars to overall larval production may be small in proportion, and their underestimation may not drastically change final estimation. Considering the general pattern of an insect life cycle, in which the vast majority die as eggs or first instars without significant growth and geometric increase in larval body size, production by early instars is likely to amount to no more than 10% (or much lower) of the total in most cases. This, however, need to be verified.

In terms of underestimating productivity values, perhaps a more significant error occurs with underestimation of the abundance of late instar larvae which contribute most to overall production. Fourth instars of some large species are capable of burrowing deep into the sediment, resulting in a low sampling efficiency of the Ekman–Birge dredge. This factor alone could underestimate population biomass and production by 50% or more (e.g. Iwakuma and Yasuno, 1981). Late instars of some highly mobile species such as *Thienemanniella* may present a similar problem of underestimation, if disturbance by sampling device induces escape response.

The accuracy with which growth and population trends of a cohort are traced ultimately determines the reliability of productivity values obtained. Accordingly, the number of sampling units taken on each occasion and sampling frequency are both very important in production studies. In this context, the problem is that it is possible to make calculations with sparse samples and a small number of sampling units, leading to a seemingly 'quantitative' figure which in fact encompasses an unknown degree of uncertainty. Although Winberg *et al.* (1971) recommended the use of smoothed survivorship curves superimposed on population data, unless the latter are based on frequent sampling coupled with detailed analysis of growth, this practice would entail largely subjective judgement and hence an unknown amount of error. Variations due to sampling errors represent perhaps the most serious but often ignored aspect in production studies on the Chironomidae; comparison of values derived from such precarious estimation is fraught with dangers. This problem is more pronounced with some multivoltine species with strongly overlapping cohorts. At the same time, the existence of different sources of error, some of which are mentioned in this and the next sections, makes the reporting of three or more significant digits for production values largely meaningless for the majority of cases. Therefore, all the values considered here are rounded to the first two significant digits only.

11.2.3 BIOMASS ESTIMATION AND CONVERSION

Since 'production' in ecological terms refers to production of biomass, estimating current biomass (standing crop) is an essential part of production studies. In the case of Chironomidae, large numbers of (often small) individuals encountered and variation in body size combine to make biomass estimation not an easy task.

One way to facilitate biomass estimation is to obtain length–weight relationships, which are best done separately for different species. Body length L (anterior to posterior margin) is in general related to (dry) weight W thus:

$$W = a L^{b}$$

or

$$\ln W = \ln a + b \ln L$$

where a and b are parameters. Values of $\ln a$ and b for some chironomid taxa are given in Mackey (1977a) and Nolte (1990). The latter examined the relationships between these measures and the body shape of larvae. If only approximate values of biomass are to be estimated, for example in fish diet studies, average values of ln $a = -5.3$ and $b = 2.3$ (Smock, 1980) may be applied to all the individuals, irrespective of species.

Apart from measuring length and weight individually to derive the above relationship, another approach which is widely practised, particularly in connection with the size–frequency method of estimating production, is to obtain biomass for each of a number of size classes (or instars) separately. In this case, weight can be measured for a group of individuals of similar size, thus avoiding problems of weighing small individuals separately (cf. Tokeshi, 1986a; Berg and Hellenthal, 1991).

It is sometimes considered desirable to use ash-free dry weight (AFDW), calorific values or carbon weight (C. mg) rather than simple dry weight to express biomass and production. Ash-free dry weight can be obtained by subtracting weight after ignition (e.g. 550 °C, 1 h) from dry weight (60 °C, 24 h) (cf. Lindegaard and Mæhl, 1993) or, alternatively, using a conversion factor (1 g dry wt = 0.9 g AFDW). Apart from directly measuring calorific content by devices such as bomb calorimeter, dry weight is converted to calorific values by 1 g dry wt = 5.5 kcal = 23.0 kJ, and to carbon weight by 1 g dry wt = 0.79 g C for chironomid species in general (Cummins and Wuycheck, 1971). Similarly, it is sometimes more practicable to measure wet weight rather than dry weight, though error variation may be larger with the former than the latter. Wet weight can roughly be converted to dry weight by 6 g wet wt = 1 g dry wt (Waters, 1977). **For ease of comparison all the productivity values reported in original literature are converted, where necessary, to dry weight in this chapter.**

When samples preserved in alcohol or formalin are used, biomass estimation is subject to another source of error. Because preservatives replace body fluid gradually over a period, the longer the period of preservation the larger is the weight loss and its variation. Lindegaard and Jónsson (1987) and Benson *et al.* (1980) used a loss (multiplication) factor of 1.67 and 1.25, respectively, to estimate unpreserved dry weight, but the application of a single such value to all preserved samples under different conditions may require caution. In addition to the kind of preservative used and length of preservation, body size of specimens may affect weight loss through differing surface area/mass ratios. When specimens are preserved for different lengths of time, it is theoretically possible to derive a relationship between weight loss and time (e.g. Berg and Hellenthal, 1991), though predictive power of such relationship may not be high for individuals of different species and sizes (cf. Dermott and Paterson, 1974). Where use of preserved material is unavoidable, perhaps the least problematic approach is to obtain length–weight relations with fresh specimens and measure the lengths of preserved specimens to calculate their biomass.

11.3 PATTERNS OF PRODUCTIVITY

11.3.1 TOTAL CHIRONOMID PRODUCTION

(a) Lentic habitats

Annual production by the Chironomidae as a whole in a given lentic habitat varies from less than 1.0 g dry wt m^{-2} yr^{-1} to nearly 100 g dry wt m^{-2} yr^{-1}. Overall, total chironomid production in lentic habitats seems to have two peaks in frequency distribution on a logarithmic scale of production (Figure 11.3a): one representing those less than 2 g dry wt m^{-2} yr^{-1} and the other between 16 and 32 g dry wt m^{-2} yr^{-1}, while values between 2 and 16 g dry wt m^{-2} yr^{-1} are moderately represented. On this scale (i.e. with chironomid production being taken as a major criterion), it is reasonable to classify values less than 2 g dry wt m^{-2} yr^{-1} as low productivity (**oligotrophy**), 2–8 g dry wt m^{-2} yr^{-1} as intermediate productivity (**mesotrophy**) and 8–32 g dry wt m^{-2} yr^{-1} as high productivity (**eutrophy**), while those over 32 g dry wt m^{-2} yr^{-1} are considered as extremely productive situations (**hyper-eutrophy**).

It is important to note, however, that productivity may vary substantially at different sites in a lake (e.g. littoral versus profundal) and therefore a balanced view needs to be sought on the basis of an extensive sampling if a lake's trophic status is to be assessed. For example, chironomid production ranged between c. 0.3 and 18 g dry wt m^{-2} yr^{-1} in different areas of Lake Ontario (Johnson and Brinkhurst, 1971). Proportion of chironomid production among the entire macro-zoobenthos production varies from c. 20% to well over 90% (Lindegaard, 1989), with profundal zones of lakes in subarctic and cold temperate regions frequently dominated by chironomid production.

One of the lowest values recorded to date comes from an arctic Canadian lake (Char Lake, 74° 42′ N) where total annual production by seven major species of chironomid (*Pseudodiamesa arctica* (Malloch), *Heterotrissocladius oliveri* Sæther, *Trissocladius* sp., *Orthocladius lapponicus* Goetghebuer, *Orthocladius* sp., *Lauterbornia* sp. and *Paracladius quadrinodosus* Hirvenoja) amounted to 0.51–0.71 g dry wt m^{-2} yr^{-1} for two consecutive years (Welch, 1976). Similar low values of total annual production (less than 1 g dry wt m^{-2} yr^{-1}) were obtained from a number of lakes in Russia (Lakes Krivoe and Krugloe: Alimov *et al.*, 1972; Lakes Zelewtskoye and Myastro: Winberg *et al.*, 1972, 1973) and a few sites in Lake Ontario, Canada (Johnson and Brinkhurst, 1971). Most of these lakes may be considered to represent arctic/subarctic situations. Studies in a number of other subarctic and cold-temperate lakes (e.g. Lakes Lisunie and Plosek: Kajak and Ryback, 1966; Lake Erken, 59° 51′ N: Johnson and Pejler, 1987; Lake 95, 60° 52′ N: Lindegaard and Mæhl, 1993; Alderfen Broad, 53° N: Mason, 1977; Thingvallavatn, 64° 10′ N: Lindegaard, 1992) reported values between 1.0 and 2.0 g dry wt m^{-2} yr^{-1} of annual chironomid production.

Annual production of 2–8 g dry wt m^{-2} yr^{-1} has been recorded from temperate lakes such as Upton Broad, 53° N (Mason, 1977) and a gravel pit, 52° N (Titmus and Badcock, 1980) in England, Lake Sniardwy in Poland

(Kajak and Ryback, 1966) and Lake Batorin, 54° N (Winberg *et al.*, 1972) in Russia as well as subarctic Red Lake, 64° N (Andronikowa *et al.*, 1972). Similarly, production by *Chironomus* spp. which dominate the chironomid fauna in an arctic Alaskan pond amounted to 3.4–5.4 g dry wt m^{-2} yr^{-1} in 3 years (Butler, 1982c).

Intermediate to high chironomid productivity (8–16 g dry wt m^{-2} yr^{-1}) has been observed in a number of cold/warm temperate lakes including Lake Esrom (Jónsson, 1985), Lake Ontario (Bay of Quinte at Big Bay, Johnson and Brinkhurst, 1971), Lake Norman (Wilda, 1984) and Lakes Taltowisko and Mikolajskie (Kajak and Ryback, 1966). High productivity of 16–32 g dry wt m^{-2} yr^{-1} has been reported from what are generally regarded as eutrophic temperate lakes such as the Eglwys Nunydd reservoir in Wales (11 species totalling 20 g dry wt m^{-2} yr^{-1}, Potter and Learner, 1974), Lake Kasumigaura in Japan (two dominant species (*Tokunagayusurika akamusi* and *C. plumosus*) totalling 31 g dry wt m^{-2} yr^{-1}, Iwakuma, 1986a), the shallow waters (2 m depth) of Lake Esrom (>11 species, totalling 19–21 g dry wt m^{-2} yr^{-1}, Jónsson, 1985) and Lake Hayes in New Zealand (*Chironomus* spp. totalling 29 g dry wt m^{-2} yr^{-1}, Graham and Burns, 1983). Four species of chironomid (*Chironomus anthracinus*, *Glyptotendipes paripes*, *Polypedilum nubeculosum* (Meigen) and *Dicrotendipes modestus* (Say) (as *pulsus*) in the mud habitats of Loch Leven together reached 29 g dry wt m^{-2} yr^{-1} of annual production (Charles *et al.*, 1974), whereas in the sand habitats *G. paripes*, *D. modestus* and *Stictochironomus* spp. produced 16–45 g dry wt m^{-2} yr^{-1} in 2 years (Maitland and Hudspith, 1974).

High chironomid productivity, however, may not be restricted to temperate lakes. In subarctic Lake Myvatn (65° 35′ N), Iceland, 11 species of chironomid (of which two, *Tanytarsus gracilentus* (Holmgren) and *Chironomus islandicus* (Kieffer), were predominant) had production values of 13–38 (mean = 30) g dry wt m^{-2} yr^{-1} in 3 years of investigation (Lindegaard and Jónasson, 1979), despite the presence of ice cover for half of a year.

Extremely high production values (on average >32 g dry wt m^{-2} yr^{-1}) recorded to date generally come from special habitats where, often, a single species dominates the fauna. *Tanytarsus barbitarsis* Freeman was the only abundant benthic invertebrate in Lake Werowrap in Australia, a small, shallow, highly alkaline lake with fluctuating salinities (c.36–56 g/l), resulting in an annual production estimate of 66 g dry wt m^{-2} yr^{-1} (Paterson and Walker, 1974a). Hjarbaek Fjord in Denmark, after 15 years of being cut off from the sea by a dam which led to a decrease in salinity and an increase in nutrients and phytoplankton production, had 95% of its benthic fauna dominated by the Chironomidae (Lindegaard and Jónsson, 1987). Their total annual production reached 49 g dry wt m^{-2} yr^{-1}, of which *Fleuria lacustris* Kieffer accounted for 42%, *C. plumosus* 23%, *Cladotanytarsus* spp. 13%, *Cryptochironomus redekei* (Kruseman) 11%, *Polypedilum bicrenatum* Kieffer 4% and *P. nubeculosum* 4%.

The highest values of production were obtained for a population of *Glyptotendipes barbipes* (Staeger) inhabiting sewage lagoons in Oregon (Kimerle and Anderson, 1971). In the secondary lagoon annual production was estimated to be 92 g dry wt m^{-2} yr^{-1} in one year but declined to 7.4 g dry wt m^{-2} yr^{-1} in the following year, whereas in the primary lagoon values of 33 and 3.6 g dry wt m^{-2} yr^{-1} were obtained for the two consecutive years. These values, however, were derived from multiplying mean biomass by a turnover ratio (see previous section) and should be treated with caution. In a narrow nearshore band of the secondary lagoon containing 90% of chironomid biomass, annual production was estimated to be c. 160 g dry wt m^{-2} yr^{-1}, by far the largest value reported for a chironomid population.

Within lentic systems shallower areas may often support higher density and biomass of chironomids and hence lead to higher productivity. In a hyper-oligotrophic lake in Greenland chironomid production in the 0–4 m depth zone was 5.7 times that of the 12–20 m zone and 1.7 times that of the 8–12 m zone (Lindegaard and Mæhl, 1993). Similarly, annual production by two species of *Chironomus*, *C. zealandicus* Hudson and *C.* sp. a, in Lake Hayes, New Zealand, amounted to well above 60 g dry wt m^{-2} yr^{-1} at up to 10 m depths but ranged from 28 to 43 g dry wt m^{-2} yr^{-1} at 16 m depths and only 3.7–4.8 g dry wt m^{-2} yr^{-1} at 25 m depths (Graham and Burns, 1983). Indeed, in an analysis of productivity in 37 lentic systems (Iwakuma, 1986b) it has been found that overall chironomid production is negatively correlated with mean depth of water bodies.

(b) Lotic habitats

In comparison with lakes and other static waters, lotic habitats may at first sight appear to support a relatively small chironomid production. Proportion of lotic habitats with low productivity (less than 2 g dry wt m^{-2} yr^{-1}) is indeed higher than that of lentic habitats (35% v. 27%, cf. Figure 11.3a.b). However, as far as information available to date is concerned, it is perhaps more notable that chironomid production in lotic habitats (Figure 11.3b) and in lentic habitats (Figure 11.3a) spans a similar range of values. Furthermore, as with lentic data, two peaks in the frequency distribution of production values can be recognized, one below 2 g dry wt m^{-2} yr^{-1} and the other between 8 and 32 g dry wt m^{-2} yr^{-1} (Figure 11.3b). Thus, the productivity classification scheme with reference to the Chironomidae in lentic habitats (i.e. oligotrophy, <2 g dry wt m^{-2} yr^{-1}; mesotrophy, 2–8 g dry wt m^{-2} yr^{-1}; eutrophy, 8–32 g dry wt m^{-2} yr^{-1}; and hyper-eutrophy, >32 g dry wt m^{-2} yr^{-1}) can be applied to lotic habitats without modification. Oligotrophy and eutrophy appear to be more strongly demarcated in lotic habitats than in lentic habitats.

Very low chironomid production of less than 1 g dry wt m^{-2} yr^{-1} have been recorded for temperate rivers including Factory Brook in Massachusetts, USA (Neves, 1979), Rold Kilde in Denmark (Iversen, 1988), Sand River in Alberta, Canada (Soluk, 1985), Rivers Caribou and Blackhoof in Minnesota, USA (Krueger and Waters, 1983) and River Warta in Poland (Grzybkowska *et al.*, 1990). Values between 1 and 2 g dry wt m^{-2} yr^{-1} have been obtained for Horowiki River in New Zealand (Hopkins, 1976), Bisballe Baek in Denmark (Mortensen and Simonsen, 1983) and Hunt Creek in Michigan, USA (Wiley, 1978). The fact that these studies encompass a variety of substrate types may indicate that low chironomid productivity is

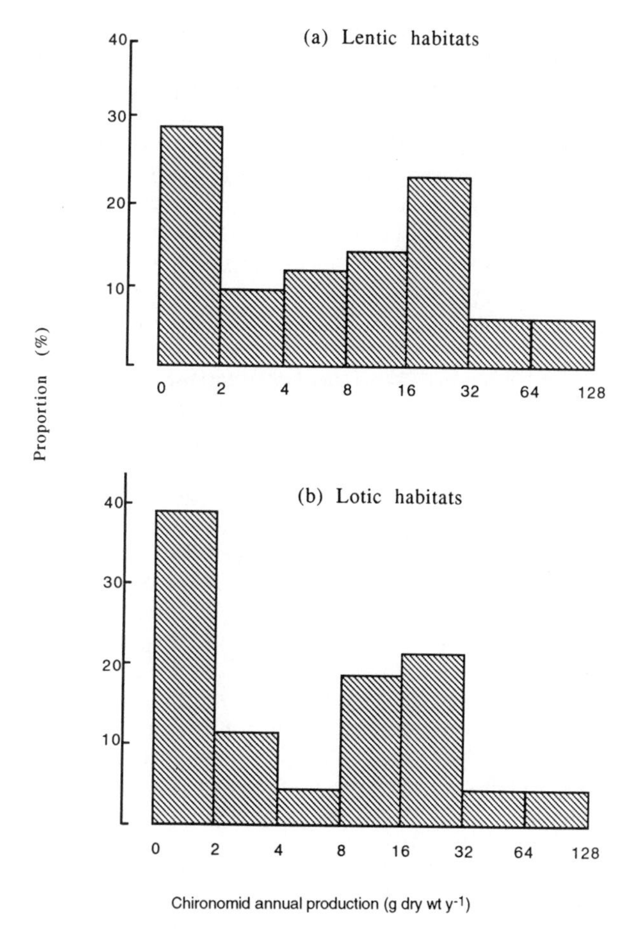

Figure 11.3 Total chironomid production in (a) lentic and (b) lotic habitats.

not a direct function of substrates. Examples of intermediate productivity (2–8 g dry wt m^{-2} yr^{-1}) include the flint zone of River Thames in England (Mackey, 1977b), Saale in Germany (Flössner, 1976) and North Branch Creek in Minnesota, USA (Krueger and Waters, 1983).

Moderate to high production in the range of 8–32 g dry wt m^{-2} yr^{-1} occurred in a number of temperate rivers. Annual chironomid production in Rabis Bæk, a lowland stream in Denmark, was estimated as 13 g dry wt m^{-2} yr^{-1}, of which *Micropsectra* spp. made up 47%, *Tvetenia verralli* (Edwards) 24%, *Prodiamesa olivacea* (Meigen) 14% and *Rheocricotopus fuscipes* (Kieffer) 8% (Lindegaard and Mortensen, 1988). Similar values were reported for two adjacent lowland rivers in central Poland, the Widawka and the Grabia (Grzybkowska, 1989). Both rivers had chironomid production of 12 g dry wt m^{-2} yr^{-1}, but with significant difference in species composition. In the Widawka production was dominated by *Chironomus* sp. and *Prodiamesa olivacea* which together comprised 74% of total production, whereas in the Grabia *Eukiefferiella* sp. and *Cricotopus* (*Orthocladius*?) sp. together accounted for 55% of total production with only 13% contributed by *Chironomus* and *Prodiamesa*.

Benke *et al.* (1984) made a detailed study of invertebrate production in a blackwater river in the coastal plain of Georgia, USA, with special reference to three different types of habitat – snag, sandy and muddy. Annual chironomid production was in the moderate to high category: 9.5–9.7 g dry wt m^{-2} yr^{-1} for the snag habitat, 10–24 g dry wt m^{-2} yr^{-1} for the sandy habitat and 3.7–15 g dry wt m^{-2} yr^{-1} for the muddy habitat. Production values of similar magnitude were obtained for a Sonoran desert stream (18 g dry wt m^{-2} yr^{-1}, Fisher and Gray, 1983) and Hudson River, New York (19 g dry wt m^{-2} yr^{-1}, Menzie, 1981). Higher values were recorded for chironomid production in the *Nuphar* and *Acorus* zones of River Thames (27 and 30 g dry wt m^{-2} yr^{-1}, respectively) (Mackey, 1977b) and in Juday Creek, Indiana (30 g dry wt m^{-2} yr^{-1}, Berg and Hellenthal, 1991). In the latter case five species together accounted for 80% of total chironomid production: *Diamesa nivoriunda* (Fitch), 34%; *Cricotopus bicinctus* (Meigen), 17%; *Pagastia* sp., 10.2%; *C. trifascia* Edwards, 9.7%; *Orthocladius obumbratus* Johannsen, 9.6%.

Extremely high production by the Chironomidae of over 32 g dry wt m^{-2} yr^{-1} has been documented for a small number of lotic habitats. The highest value estimated to date (77 g dry wt m^{-2} yr^{-1}) refers to one site with sandy/muddy substrate in Bityska River, Czechoslovakia (Zelinka *et al.*, 1977). Interestingly, other sites with stony substrate in the same river had very low chironomid production (c. 1 g dry wt m^{-2} yr^{-1}), suggesting that the high value mentioned above may relate to a very local condition and should be treated with caution. Jackson and Fisher (1986) estimated annual chironomid production of 58 g dry wt m^{-2} yr^{-1} for a Sonoran desert stream, in contrast to earlier value (18 g dry wt m^{-2} yr^{-1}) given by Fisher and Gray (1983). Similar high production values were estimated in an artificial recirculating stream in England, with three species (*Chaetocladius melaleucus* (Meigen), *Micropsectra aristata* Pinder and *Synorthocladius semivirens* (Kieffer)) together reaching 35–56 g dry wt m^{-2} yr^{-1}.

11.3.2 FACTORS AFFECTING OVERALL CHIRONOMID PRODUCTIVITY

Since the calculation of production entails combining variation in individual growth and that in population density, production etimates are subject to a whole range of factors affecting growth and development of individuals as well as the dynamics of populations. Thus temperature, dissolved oxygen level, food availability, predation, competition, disturbance events etc. all affect productivity and it is often very difficult to tease apart the contributions of different factors. Nevertheless, some factors, most notably food availability, may have an overriding effect on the production of chironomid populations. For example, in some lentic systems benthic chironomid production has been found to be strongly influenced by phytoplankton

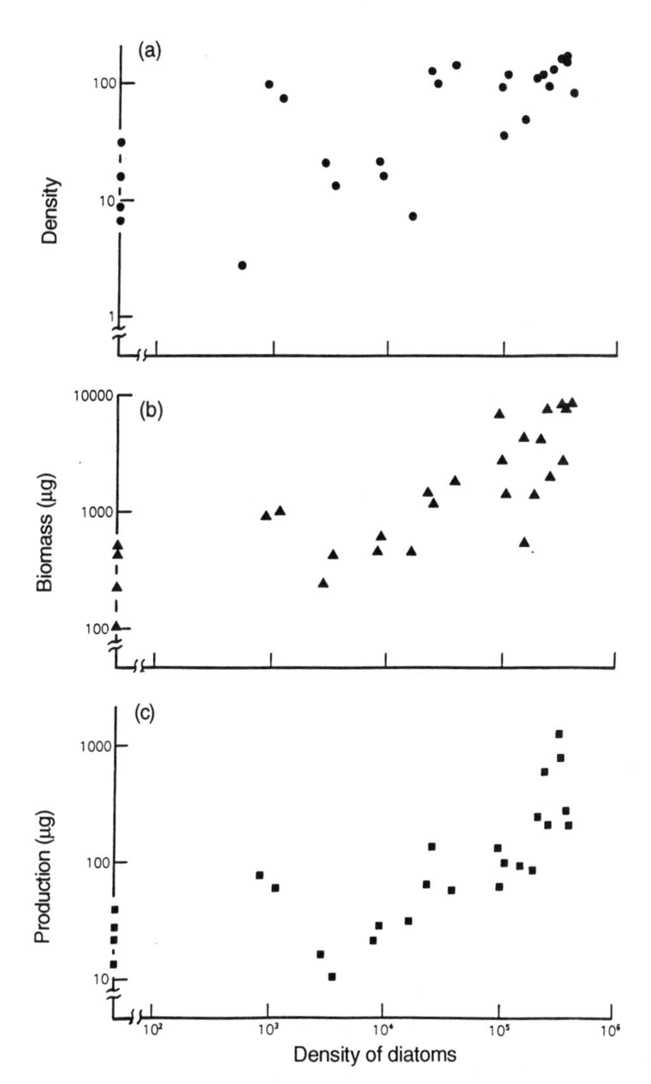

Figure 11.4 Relationship between the density of diatoms and (a) chironomid density, (b) chironomid biomass, and (c) chironomid production in an epiphytic chironomid assemblage. (After Tokeshi, 1986a.)

production (e.g. Loch Leven, Scotland: Bindloss, 1974; Charles *et al.*, 1974; Lake Memphremagog, Canada: Ross and Kalff, 1975; Dermott *et al.*, 1977; a prairie pond, Canada: Rasmussen, 1984b); and Welch *et al.* (1988) suggested that chironomid production is roughly predictable from total phytoplankton production throughout the latitudinal range of the small Canadian lakes. In an analysis of production data for lentic chironomid species, annual production values decreased with increasing mean water depth (Iwakuma, 1986b), presumably due to a decreasing proportion of the euphotic zone

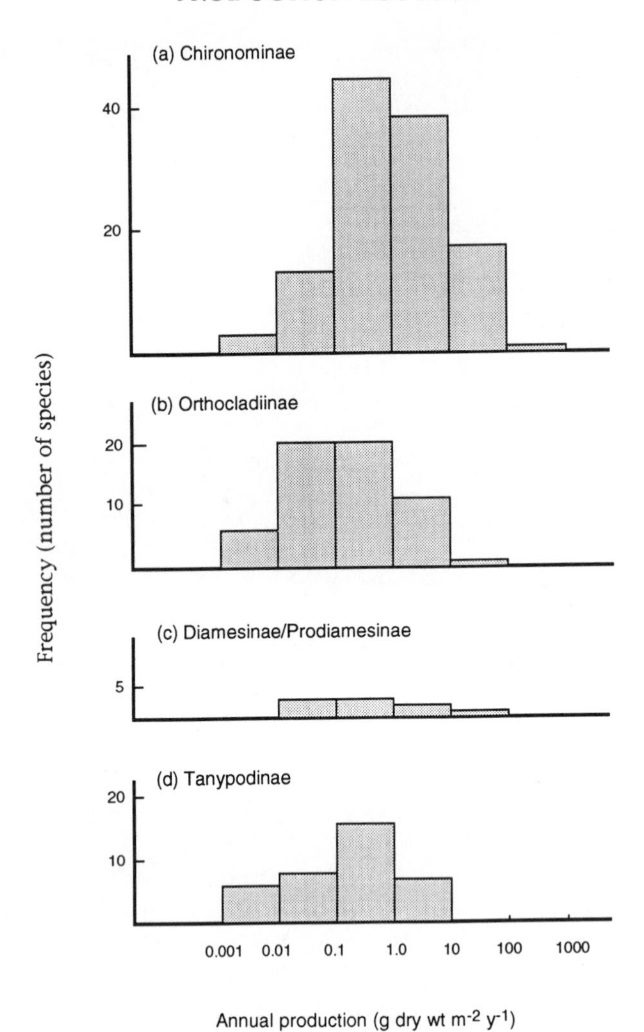

Figure 11.5 Annual production in different subfamilies of the Chironomidae.

where phytoplankton production occurs. In lotic systems diatom production is also highly seasonal and known to affect chironomid production (e.g. Figure 11.4; Tokeshi, 1986a). Similarly, the availability of allochthonous matter may affect the abundance of stream-dwelling chironomids (e.g. Richardson, 1991).

11.4 INTERSPECIFIC VARIATIONS IN PRODUCTIVITY

Variations in productivity among different chironomid taxa (subfamilies to species) are examined with respect to three related measures of productivity:

annual production, annual mean biomass and production to biomass (P:B) ratios.

11.4.1 ANNUAL PRODUCTION

Annual production of the Chironomidae on a species-by-species basis varies by 5–6 orders of magnitude, in the range of 1 mg–100 g dry wt m^{-2} yr^{-1} (Figure 11.5). Overall, production values of 0.1–1.0 g are predominant, but with significant variation among different subfamilies. It is apparent that species of the subfamily Chironominae have higher productivity than other subfamilies, with 85% of production values exceeding 0.1 g dry wt m^{-2} yr^{-1}; corresponding proportions are 55% in the Orthocladiinae, 67% in the Diamesinae plus Prodiamesinae and 62% in the Tanypodinae. At the same time, high production values over 10 g dry wt m^{-2} yr^{-1} comprised a significant proportion (16%) among the Chironominae, in contrast to other subfamilies where annual production of >10 g rarely occurred. Within the Chironominae, species inhabiting lentic habitats are more productive than those in lotic habitats (peak frequency 1–10 g dry wt m^{-2} yr^{-1} in the former and 0.1–1.0 g dry wt m^{-2} yr^{-1} in the latter).

Annual production in the Orthocladiinae was mostly below 10 g dry wt m^{-2} yr^{-1} with values between 0.01 and 1.0 g predominating. Data on the Diamesinae/Prodiamesinae are perhaps too sparse to make any definitive comment, apart from the fact that production values ranged widely between 0.01 and 100 g dry wt m^{-2} yr^{-1}. The peak frequency of annual production in the Tanypodinae occurred in the 0.1–1.0 g range, other values being more or less evenly spread between 0.001 and 10 g dry wt m^{-2} yr^{-1}. Species are treated below according to their subfamily placement.

(a) Chironominae

Because of their preponderance in lentic habitats where many production studies have been undertaken, species of Chironominae command a particularly important role in energy dynamics of lentic systems. Large-sized Chironominae species such as *Chironomus* and *Glyptotendipes* are often associated with high values of annual production, generally in excess of 1 g dry wt m^{-2} yr^{-1} and often over 10 g dry wt m^{-2} yr^{-1}. In the case of the genus *Chironomus*, for example, 25% of estimated production values (= 28) were over 10 g dry wt m^{-2} yr^{-1} and 71% over 1 g dry wt m^{-2} yr^{-1}. Values for *C. plumosus* range from 1.7 g dry wt m^{-2} yr^{-1} in Vallentunasjorn, Sweden, to 22 g dry wt m^{-2} yr^{-1} in Lake Sevan, Russia (Ostrovskii, 1982). Similarly, annual production by *C. anthracinus* ranges from 1.4 g dry wt m^{-2} yr^{-1} in Lago Varese, Italy (Bonomi, 1962) to 26 g dry wt m^{-2} yr^{-1} in the muddy substrate of Loch Leven, Scotland (Charles *et al.*, 1974). In Lake Hayes, New Zealand, two species of *Chironomus*, *C. zealandicus* and *C.* sp., had annual production of 29 g dry wt m^{-2} yr^{-1}. On the other hand, a low value of annual production (0.63 g dry wt m^{-2} yr^{-1}) was recorded for *C. commutatus* Keyl inhabiting a high-altitude lake (2285 m) in the Pyrenees (Laville, 1975). Apart from the genus *Chironomus*, *Glyptotendipes* species were reported with high production values, such as 92 g dry wt m^{-2} yr^{-1} for *G. barbipes* in a sewage lagoon (Kimerle and Anderson, 1971) and 41 g dry wt m^{-2} yr^{-1} for *G. paripes* in the sandy substrate of Loch Leven (Maitland and Hudspith, 1974). High annual production was also documented for *Microtendipes chloris* (Meigen) in Lake Mikolajskie, Poland (24 g, Kajak and Rybak, 1966), *Micropsectra aristata* in an artificial stream (22–29 g, Ladle *et al.*, 1984) and *Microchironomus* (= *Leptochironomus*) *deribae* (Freeman) in Lake Nakuru, Kenya (17 g, Vareschi and Jacobs, 1984).

Among the subfamily Chironominae, species of tribe Tanytarsini, despite their small body size, are sometimes associated with high annual production. Apart from *Tanytarsus barbitarsis* in Lake Werowrap mentioned earlier, *T. gracilentus* in Lake Myvatn, Iceland, was estimated to have annual production of 21 g dry wt m^{-2} yr^{-1} and a number of cases falls in the range of 1–10 g dry wt m^{-2} yr^{-1}, including *T. lugens* (1.6 g) and *Paratanytarsus* (as *Tanytarsus*) *inopertus* (Walker) (2.4 g) from a Welsh reservoir (Potter and Learner, 1974), *T.* sp. (2.1 g) from Lake Esrom (Jónsson, 1985).

(b) Orthocladiinae

As in the Chironominae, annual production by species of Orthocladiinae ranges between 0.001 and 100 g dry wt m^{-2} yr^{-1}, but the proportion of relatively low production values (less than 0.1 g dry wt m^{-2} yr^{-1}) is higher among the Orthocladiinae than among the Chironominae (45% and 25%, respectively). Very low annual production of less than 0.01 g dry wt m^{-2} yr^{-1} has been recorded for species inhabiting cold-temperate/subarctic freshwater systems studied by Lindegaard and his colleagues, such as *Brillia longifurca* Kieffer and *Thienemanniella* cf. *vittata* (Edwards) from Rabis Bæk, Denmark (Lindegaard and Mortensen, 1988), *Cricotopus tibialis* (Meigen) from Lake Myvatn, Iceland (Lindegaard and Jónasson, 1979) and *Psectrocladius barbimanus* Kieffer, *Rheocricotopus effusus* (Walker) and *Thienemanniella* sp. cf. *acuticornis* (as *morosa*) from another lake (Thingvallavatn) in Iceland (Lindegaard, 1992). High production values exceeding 1 g dry wt m^{-2} yr^{-1} were documented mainly for lotic species, including *Tvetenia verralli* in Rabis Baek (3.8 g, Lindegaard and Mortensen, 1988), *Cricotopus bicinctus* (5.0 g, *C. triannulatus* Macquart (1.3 g), *C. trifascia* (2.9 g), *Eukiefferiella brevinervis* (Malloch) (1.4 g) and *Orthocladius obumbratus* (2.9 g) from Juday Creek, Indiana (Berg and Hellenthal, 1991), *Cricotopus sylvestris* (Fabricius) (5.8 g) from Hudson River, New York (Menzie, 1981) and *C. sylvestris* (2.9 g) and *Eukiefferiella gracei* (Edwards) (3.5 g) from River Grabia, Poland (Grzybkowska, 1989). The highest values of annual production reported to date for orthoclads refer to *Chaetocladius melaleucus* in an artificial recirculating stream in England (11–25 g dry wt m^{-2} yr^{-1}, Ladle *et al.*, 1984) and *Tokunagayusurika akamusi* in Lake Kasumigaura, Japan (18 g, Iwakuma, 1986a).

(c) Diamesinae and Prodiamesinae

Amongst a small number of production estimates concerning the Diamesinae and Prodiamesinae, the highest value recorded (c. 10 g dry wt m^{-2} yr^{-1}) was for *Diamesa nivoriunda* from Juday Creek, Indiana (Berg and Hellenthal, 1991) and the lowest (1.1 mg) for *Diamesa* sp. from Lake Thingvallavatn, Iceland (Lindegaard, 1992). *Prodiamesa olivacea*, one of the most common species in Europe, had an annual production of 0.30 and 4.8 g dry wt m^{-2} yr^{-1} in two adjacent rivers in central Poland (Grzybkowska, 1989) and 1.8 g in a Danish lowland stream (Lindegaard and Mortensen, 1988). In the latter site, *Diamesa insignipes* Kieffer and *Odontomesa fulva* (Kieffer) also occurred, but with substantially lower annual production than in the case of *P. olivacea* (0.095 and 0.074 g dry wt m^{-2} yr^{-1} for *D. insignipes* and *O. fulva*, respectively).

(d) Tanypodinae

Annual production in the Tanypodinae ranges from 0.001 to 10 g dry wt m^{-2} yr^{-1}, with values between 0.1 and 1.0 g predominating (Figure 11.5d). Amongst Tanypodinae, species of *Procladius* are most frequently studied, though it is often very difficult to separate the several species of this genus that often coexist in the same habitat. Nevertheless, the majority of annual production values in *Procladius* species fall in the range of 0.1–1.0 g dry wt m^{-2} yr^{-1}, e.g. *Procladius* spp. in Hjarbaek Fjord (0.56 g, Lindegaard and Jónsson, 1987), Big Vihorlate Lake (0.80 g, Terek and Losos, 1979), Lake Sniardwy (0.20 g, Kajak and Dusoge, 1976), Deer Lake (0.53 g, Uutala, 1981) and the littoral of Lake Esrom (0.73 g, Jónsson, 1985); *P. islandicus* (Goetghebuer) in Lake Myvatn (0.51 g, Lindegaard and Jónasson, 1979), *P. pectinatus* Kieffer in Lake Esrom (0.52 g, Jónasson, 1972), *P. denticulatus* Sublette in Lake Memphremagog (0.14–0.86 g, Dermott *et al.*, 1977), *Procladius* sp. in River Gravia (0.76 g, Grzybkowska, 1989) and *P. bellus* (Loew) in Laurel Creek Reservoir (0.11–0.18 g, Sephton and Paterson, 1986). Production values exceeding 1.0 g dry wt m^{-2} yr^{-1} include 2.4 g for *Procladius* sp. in a Texas pond (Benson *et al.*, 1980), 2.9 g for *P. bellus* in Big Bear Lake (Siegfried, 1984), 1.8 g for *P. choreus* (Meigen) in the River Thames (Mackey, 1977a) and the highest value, 3.4 g, was recorded for *P. choreus* in a eutrophic Welsh reservoir (Potter and Learner, 1974). Apart from a production value of 2.0 g estimated for *Psilotanypus rufovittatus* (van der Wulp) in the last mentioned study, species of Tanypodinae other than *Procladius* seem to have relatively low annual production, generally less than 1 g and more frequently below 0.1 g dry wt m^{-2} yr^{-1}.

11.4.2 ANNUAL MEAN BIOMASS

Calculated annual mean population biomass in different species of Chironomidae ranges from less than 1.0 mg to over 10 g dry wt m^{-2}, with values between 0.01 and 1.0 g most frequently encountered (Figure 11.6). As in annual production, annual mean biomass varies among subfamilies. Species of Chironominae have higher biomass than other subfamilies, the peak frequency being in the 0.1–1.0 g range and values over 1.0 g constituting a significant proportion (24% of total). The Orthocladiinae, Diamesinae,

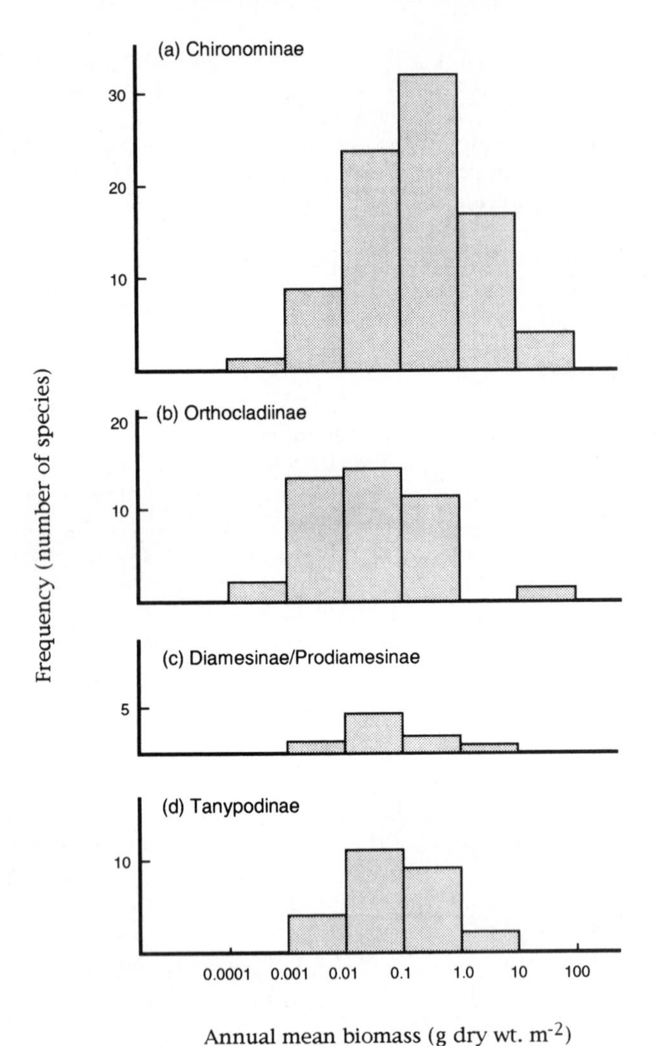

Figure 11.6 Annual mean biomass in different subfamilies of the Chironomidae.

Prodiamesinae and Tanypodinae are most frequently observed with biomass in the range of 0.001–1.0 g and rarely above 10 g.

(a) Chironominae

The majority of mean biomass values in the Chironominae are in the range of 0.01–10 g (Figure 11.6a). Very high biomass of c. 10 g or over was reported for *Glyptotendipes barbipes* in a waste stabilizing lagoon (11 g, Kimerle and Anderson, 1971), *Chironomus anthracinus* in Lake Esrom (10 g, Jónasson, 1972) and unidentified species of *Chironomus* in a carp pond in Bohemia (19 g,

Ruzickova, 1987). In general, large-bodied species such as *Chironomus* and *Glyptotendipes* tend to have high mean population biomass, often exceeding 1.0 g, which in turn contributes to high production values. On the other hand, small species of Chironominae such as *Tanytarsus* could in some circumstances demonstrate a high population biomass. For example, *Tanytarsus gracilentus* in Lake Myvatn, Iceland, attained mean biomass of 2.8 g and annual production of 21 g, the highest values among 11 species in this habitat. Overall, however, it is often the case that low biomass is associated with small rather than large species. Mean biomass of less than 0.01 g has been observed for small species including *Cladotanytarsus mancus* gr. A and *Tanytarsus* gr. A in Juday Creek (Berg and Hellenthal, 1991) and *Cladotanytarsus vanderwulpi* (Edwards) in River Widawka (Grzybkowska, 1989).

(b) Orthocladiinae

Mean biomass in the Orthocladiinae ranges from <1 mg to 10 g dry wt m^{-2} (Figure 11.6b). Low biomass of <0.01 g was reported for the majority of Orthocladiinae species in Rabis Bæk (Lindegaard and Mortensen, 1988) as well as for *Eukiefferiella brevicalcar* (Kieffer) in River Widawka (Grzybkowska, 1989). Intermediate values (0.1<B<1.0 g) were estimated for species in Juday Creek (Berg and Hellenthal, 1991), including *Cricotopus bicinctus* (0.73 g), *C. trifascia* (0.46 g), C. *triannulatus* (0.12 g), *Eukiefferiella brevinervis* (0.15 g), *Orthocladius obumbratus* (0.37 g) and *Parametriocnemus lundbeckii* (Johannsen) (0.13 g) and for *C. sylvestris* (0.28 g) in Hudson River (Menzie, 1981). Mean biomass of 0.01 to 1.0 g seems to occur widely among species inhabiting different habitats. By far the highest value recorded for an orthoclad was for a large, univoltine species *Tokunagayusurika akamusi* in Lake Kasumigaura, Japan (11 g, Iwakuma, 1986a).

(c) Diamesinae and Prodiamesinae

Mean biomass in the Diamesinae and Prodiamesinae reported to date ranges from 9.4 mg for *Odontomesa fulva* in Rabis Bæk (Lindegaard and Mortensen, 1988) to 2.1 g for *Diamesa nivoriunda* in Juday Creek (Berg and Hellenthal, 1991). *Prodiamesa olivacea* had mean biomass of 0.27 g and 0.014 g in two adjacent rivers in central Poland (Grzybkowska, 1989). There is need for more information on these taxa.

(d) Tanypodinae

Species of Tanypodinae have mean population biomass within the range of 0.001–10 g dry wt m^{-2}, with values between 0.01 and 1.0 g more frequently encountered (Figure 13.6d). High biomass values observed include 3.4 g for *Procladius pectinatus* in Lake Esrom (Jónasson, 1972) and 1.2 g for *P. bellus* in Big Bear Lake (Siegfried, 1984). Low values include 6 mg for *P. culiciformis* (Linneaus) in Lake Kasumigaura (Iwakuma, 1987), 4.3 mg for *Apsectrotanypus trifascipennis* (Zetterstedt) and 1.2 mg for *Zavrelimyia barbatipes* (Kieffer) in Rabis Bæk (Lindegaard and Mortensen, 1988) and 8.4–13 mg for *Procladius bellus* in Laurel Creek (Sephton and Paterson, 1986). As with Diamesinae and Prodiamesinae, more information is necessary to derive a general pattern of population biomass in this group.

11.4.3 PRODUCTION:BIOMASS (P:B) RATIOS

Annual production (P) to annual mean biomass (B) ratios express the rates of biomass turnover and have been frequently quoted as an important measure of productivity. Because P/B ratios crucially depend on the length of larval life, which is particularly variable among species of Chironomidae, estimated values of P/B ratios encompass a wider range in the Chironomidae than in any other aquatic invertebrate taxa. At the same time P/B ratios can vary intraspecifically, since populations inhabiting different geographical localities or even microhabitats may show variation in voltinism and other aspects of life cycle (cf. Chapter 10). As with annual production and annual mean biomass, P/B ratios appear to vary among different subfamilies of Chironomidae (Figure 11.7), though more data may be needed to verify such patterns.

(a) Chironominae

On the available evidence, the subfamily Chironominae seems to cover a wide range of P/B values. P/B ratios of less than 10 are more common, but values above 10 occur frequently (30% of all, Figure 11.7a). There appear to exist two weak modes in the frequency distribution of P/B ratios, one between 1.0 and 3.0 and the other between 5.0 and 7.0, the latter mostly encompassing multivoltine species. Large species with relatively long larval life such as *Chironomus* spp. tend to have low P/B values. For example, *Chironomus* species inhabiting an arctic pond in Alaska were estimated to have a larval life of 7 years and a P/B ratio of 0.49 (Butler, 1982b). *C. anthracinus*, which usually has a life cycle of 2 years in Lake Esrom, had a P/B ratio of 1.3 (Jónasson, 1972), while the same species with a univoltine life cycle in Lake Erken had a P/B ratio of 2.3 (Johnson and Pejler, 1987). One of the most widely distributed species, *C. plumosus*, was estimated to have a P/B ratio of 1.6 in Alderfen Broad, U.K. (Mason, 1977), 1.7 in Federsee, Germany (Frank, 1982) and 2.4 in Lake Beloie, Russia (Borutski, 1939; Borutski *et al.*, 1971). All these cases of low P/B ratios are nevertheless associated with relatively high values of annual production, suggesting that high production is largely attributable to high mean biomass of the species involved.

At the other end of the spectrum, some Chironominae species were estimated to have very high P/B ratios. One of the highest values obtained to date for Chironominae refers to species inhabiting the snag substrate of subtropical Satilla River, USA: 147–166 for *Polypedilum* spp., 176–184 for species of Tanytarsini, 80 for Chironomini sp., 65 for *Stenochironomus* spp. and for *Tribelos*/ *Xenochironomus* spp. (Benke *et al.*, 1984). In Lake Norman, USA, *Tanytarsus* spp., *Stempellina* spp. and *Cryptochironomus* spp. had P/B ratios of 135, 109 and 100, respectively (Wilda, 1984). In central Polish rivers

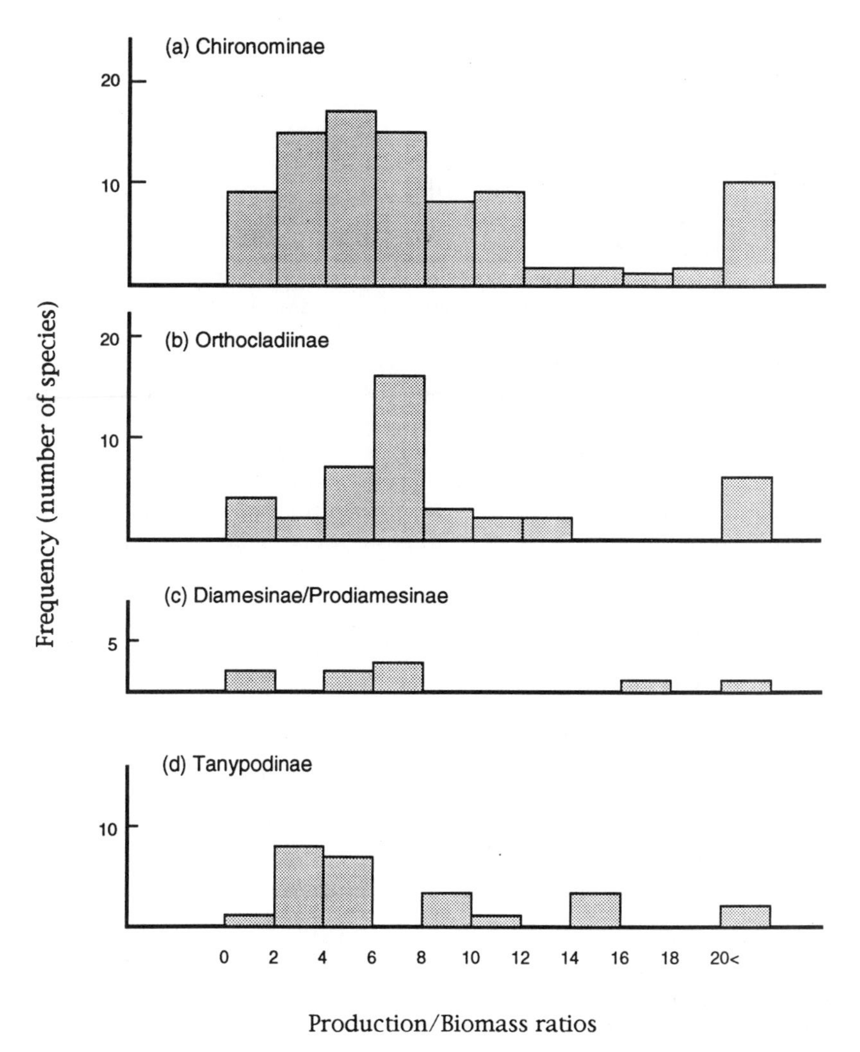

Figure 11.7 Production to biomass ratios (P/B) in different subfamilies of the Chironomidae.

Cladotanytarsus vanderwulpi and *Polypedilum convictum* (Walker) were estimated to have P/B ratios of 45–46 and 32–34, respectively (Grzybkowska, 1989), while *P. convictum* in Juday Creek had a value of 22 (Berg and Hellenthal, 1991). *Microchironomus* (as *Leptochironomus*) *deribae* in tropical Lake Nakuru, Kenya, had a P/B ratio of 33 and an annual production of 17 g dry wt m^{-2} yr^{-1}; in this case high production was due to a high turnover of biomass.

(b) Orthocladiinae

In the Orthocladiinae, P/B ratios vary as widely as in the Chironominae, though values appear to be more strongly concentrated in the range 6.0–8.0 with no secondary peak in lower ranges (Figure 11.7b). This may reflect the fact that the majority of species of Orthocladiinae are multivoltine. Indeed, those orthoclad species reported with low P/B ratios tend to be inhabitants of high latitudes where larval life takes one year or longer to complete, such as *Heterotrissocladius oliveri* (P/B ratio of 0.8–1.1), *Trissocladius* sp. (1.9–2.1) and *Orthocladius* spp. (1.4–2.3) in an arctic lake (Welch, 1976). Another univoltine species, *Tokunagayusurika akamusi* inhabiting eutrophic temperate lakes in Japan, had a P/B ratio of 1.5 (Iwakuma, 1986b). High P/B values of over 20 were observed for *Eukiefferiella devonica* gr. A in Juday Creek (21, Berg and Hellenthal, 1991) and for species inhabiting two adjacent rivers in central Poland (Grzybkowska, 1989): *Cricotopus sylvestris*, 23–26; *Eukiefferiella brevicalcar*, 73; *E. gracei*, 82. By far the highest values, however, are again from the snag and sandy substrate of Satilla River mentioned earlier (Benke *et al.*, 1984): *Cricotopus*, 99–118; *Corynoneura*/*Thienemanniella*, 200–355; *Parakiefferiella*, 306–337. These extremely high P/B values result from estimates of very short larval life and it is yet to be confirmed whether separate species consistently repeat very short life cycles throughout a year under field conditions.

(c) Diamesinae and Prodiamesinae

Though limited in number of studies, the Diamesinae/Prodiamesinae seem to show a substantial variation in P/B ratios, ranging from 1.1 to 1.5 for *Pseudodiamesa arctica* (Malloch) in Char Lake (Welch, 1976) to 21 for *Prodiamesa olivacea* in River Grabia (Grzybkowska, 1989). *Diamesa insignipes*, *Odontomesa fulva* and *Prodiamesa olivacea* coexisting in Rabis Bæk had similar values of P/B ratio (7.1, 7.1 and 7.0, respectively).

(d) Tanypodinae

In the Tanypodinae, P/B ratios span as wide a range of values as in other subfamilies, but the peak occurrence is in the range of 3.0–5.0, which is somewhat lower than in the Orthocladiinae and Chironominae (Figure 11.7d). Whether this is related to the carnivorous nature of this taxon is an interesting aspect to explore. On the other hand, high values were recorded for some species, e.g. 113–120 for *Procladius* spp. and 45–49 for *Coelotanypus* spp. in the muddy substrate of the Satilla River (Benke *et al.*, 1984) and 13 for *P. bellus* in Laurel Creek (Sephton and Paterson, 1986).

11.5 TEMPORAL VARIATION IN PRODUCTIVITY

11.5.1 SEASONAL VARIATIONS

Production does not normally remain at the same level through the year; rather, seasonal variations in production reflect temporal patterns of resource utilization and of life cycles in a given population. As in biomass, the largest proportion of increase in production occurs in the fourth instar stage, implying that monthly population production will be higher in months when larger proportions of larvae are in the fourth instars. Where the period of fourth instars is extended as in the case of uni/mero-voltine species, temporal fluctuations in the abundance of food resources and/or temperature conditions and other factors may also affect seasonal production. Among lotic chironomid species inhabiting submerged *Myriophyllum* (Tokeshi, 1986a), an average of 85.1% of the total annual production was attained during three months between 1 March and 31 May: 100% in *Orthocladius obumbratus* [= *O.* (*O.*) Pe 9 of Langton, 1984], 95% in *O.* (*O.*) sp. A, 94% in *Eukiefferiella ilkleyensis*, 91% in *Cricotopus annulator*, 89% in *C. bicinctus*, 85% in *Rheocricotopus chalybeatus* (Edwards) and *Thienemanniella majuscula*, 73% in *Tvetenia calvescens* and 54% in *Rheotanytarsus curtistylus* (Goetghebuer). All these species were dependent on diatoms as food, which were most abundant during spring–summer (Tokeshi, 1986a). In the profundal of Lake Ngapouri, North Island, New Zealand, *Chironomus zealandicus* had total annual production of 20 g dry wt m^{-2} but its monthly production varied from 0.13 to 4.3 g dry wt m^{-2}, while in the littoral of the same lake another species of *Chironomus* (*C.* sp. a) had a lower annual production (5.2 g dry wt m^{-2}) and smaller variation in monthly production (0.2 to 0.75 g dry wt m^{-2}, Forsyth, 1986; Figure 11.8). Similarly, *Polypedilum pavidum* Hutton in Lake Okaro, North Island, New Zealand, had monthly production fluctuating from 0.27 to 12 g dry wt m^{-2} (Forsyth and James, 1988). In all these cases low production values were due to the disappearance of fourth instars through pupation and adult emergence.

Weekly estimates of production in *Glyptotendipes barbipes* in a sewage lagoon showed a gradual increase from c. 1 g dry wt m^{-2} in May to 18 g dry wt m^{-2} in August, thereafter declining again to less than 1 g dry wt m^{-2} in November (Kimerle and Anderson, 1971). Similarly, Mackey (1977a) estimated chironomid production between April and October in the *Acorus* and *Nuphar* zones of the Rivers Thames and Kennet and found that total daily production was highest in the mid season, i.e. June–July, with differences of 2–3 orders of magnitude between the highest and the lowest values recorded.

Not all species achieve high production in spring–summer. In Lake Kasumigaura, central Japan, production by *Tokunagayusurika akamusi* reached its highest level in winter months when food (deposited phytoplankton) was still available; the species appears to be well adapted for growth under low temperature conditions (Iwakuma, 1986a).

In multivoltine species cohorts occurring in different seasons may show a large difference in total cohort production. For example, in an Indiana stream (Berg and Hellenthal, 1991) differences of more than one order of magnitude were observed among different cohorts of *Pagastia* sp. A, *Cricotopus bicinctus*, *C. trifascia*, *Eukiefferiella brevinervis* gr., sp. A, and *Orthocladius obumbratus*. On the other hand, other species demonstrated little difference in cohort production, e.g. *Cladotanytarsus mancus* gr., sp. A, *Rheotanytarsus exiguus* gr., sp. A and *Cricotopus triannulatus*.

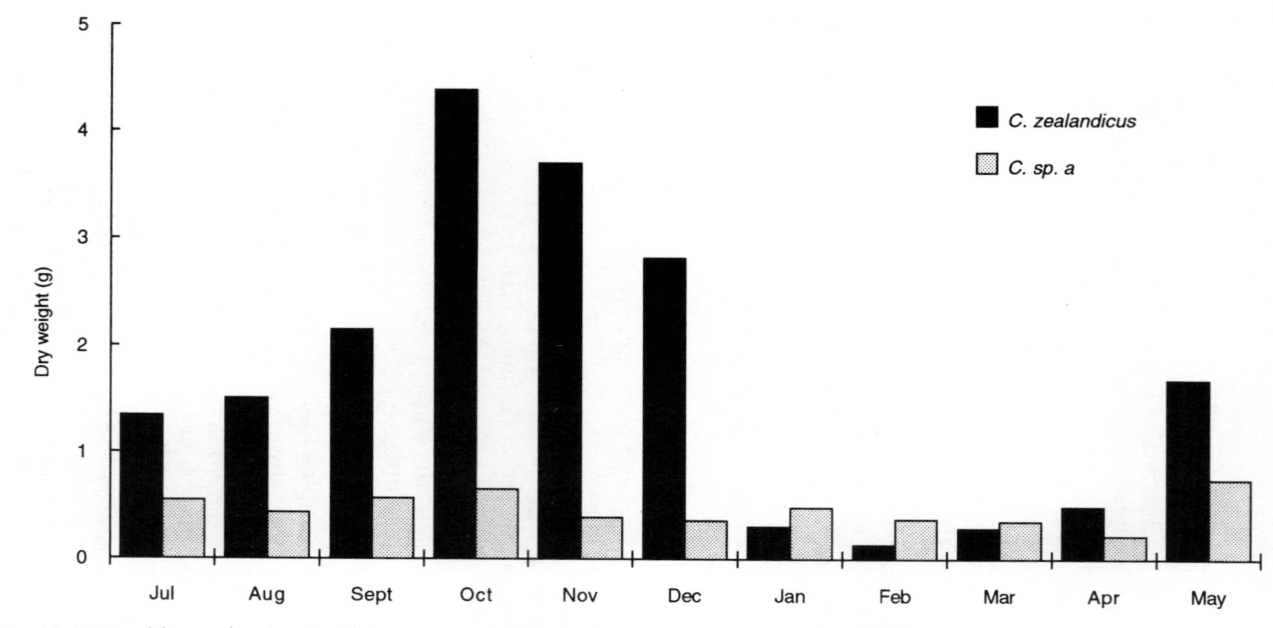

Figure 11.8 Monthly production in *Chironomus zealandicus* and *C.* sp. a. (Data from Forsyth, 1986.)

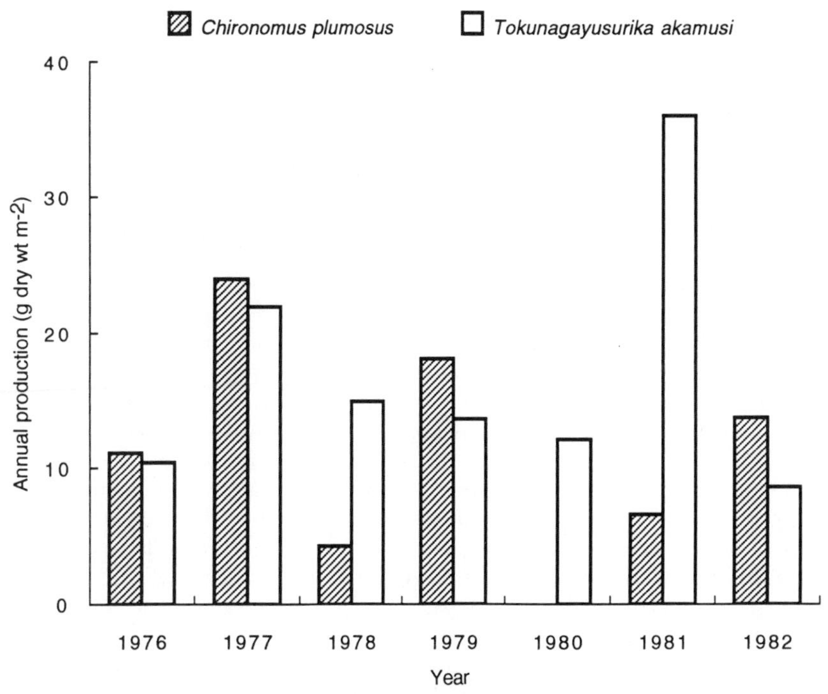

Figure 11.9 Variation in the annual production of *Tokunagayusurika akamusi* (Tokunaga) and *Chironomus plumosus* (Linnaeus) in Lake Kasumigaura, Japan. (After Iwakuma, 1986a.)

11.5.2 INTER-ANNUAL VARIATIONS

Observations on the yearly variation in productivity of benthic chironomid species are often made difficult by the amount of effort required to gain reliable estimates of population density and biomass for an extended period. This is particularly true for species undergoing rapid growth and turnover of generations in a complex habitat. It is therefore understandable that long-term data on chironomid production have been restricted to lake-dwelling species with relatively slow growth.

Lindegaard and Jónasson (1979) reported on the production of benthic chironomid species over a 3-year period in Lake Myvatn, Iceland. Annual production by the most abundant species, *Tanytarsus gracilentus*, was estimated to be 35 g dry wt m^{-2} in the first two consecutive years, but declined to 12 g dry wt m^{-2} in the third year. A similar but less pronounced change in annual production was observed for *Chironomus islandicus*, the second most abundant species: 1972, 8.6 g dry wt m^{-2}; 1973, 8.3 g dry wt m^{-2}; 1974, 5.2 g dry wt m^{-2}. These two species together comprised the bulk of benthic production in this lake, reaching 95%, 92% and 76% of the total annual production in the three years studied. On the other hand, no clear reason could be offered for the conspicuous decline in the production of these species in the third year. Variation in the pattern of annual production was less clear for other, less abundant species; overall, values remained well within one order of magnitude.

In the shallow, eutrophic Federsee in south-west Germany, *Chironomus plumosus* had a relatively stable level of annual production during a four-year period: 11, 6.8, 9.6 and 8.6 g dry wt m^{-2}, respectively, in 1973–1976 (Frank, 1982). Interestingly, variations in biomass and P/B ratios were slightly larger: biomass – 8.5, 5.1, 3.0 and 5.1 g dry wt m^{-2}; P/B – 1.3, 1.3, 3.2 and 1.7. Similarly, two closely-related species of *Chironomus* in an arctic pond (Butler, 1982b,c) showed low temporal variation in total annual production (5.4, 3.6 and 3.4 g dry wt m^{-2} in 1975–1977) and in biomass (12, 7.7 and 6.7 g dry wt m^{-2}).

Amongst long-term studies on chironomid production, Iwakuma's (1986a) data on *Chironomus plumosus* and *Tokunagayusurika akamusi* extend to 7 years and, though different sampling devices were employed in the course of the study, provide useful information on the overall variability in annual production of these species. Four to five-fold differences in annual production were observed for both *C. plumosus* and *T. akamusi* between the minimum and maximum years of production, but patterns of fluctuations were different between the two (Figure 11.9). At the same time, temporal variation in production followed that in annual mean biomass more closely in *C. plumosus* (mean of annual production = 13 g dry wt m^{-2}, annual mean biomass = 2.3 g dry wt m^{-2}) than in *T. akamusi* (17.6 and 11.4 g dry wt m^{-2}, respectively).

12

Species interactions and community structure

M. Tokeshi

12.1 INTRODUCTION

The fact that the Chironomidae are extremely species-rich and occur in a wide variety of habitats points to their varied roles in ecosystems and the possibilities of interaction with different kinds of organism. This opens up many opportunities for community-orientated studies involving the Chironomidae but ecological research on chironomids has largely been dominated by autoecological approaches with a conspicuous lack of community perspectives, apart from typological studies where chironomid faunas were associated with some environmental characteristics. Studies on species interactions and community structure involving chironomids are limited in number and often lack analytical rigour, though the situation is steadily improving. This chapter reviews community aspects of the Chironomidae including predator–prey relations, host–parasite/commensal relations, competitive/non-competitive relations, the roles of disturbance and stochasticity, and community assembly.

12.2 PREDATOR–PREY RELATIONS

12.2.1 CHIRONOMIDAE AS PREDATORS

Species belonging to the subfamily Tanypodinae are generally known as predators. However, the extent of carnivory varies from one species to another and many are better described as omnivores feeding not only on animals but also on detritus, algae and other organic matter. Amongst the Tanypodinae, species of the genus *Procladius* have most frequently been subjected to dietary analyses (Armitage, 1968; Roback, 1969a; Mackey, 1979; Baker and McLachlan, 1979; Dusoge, 1980; Vodopich and Cowell, 1984),

The Chironomidae: Biology and ecology of non-biting midges
Edited by P.D. Armitage, P.S. Cranston and L.C.V. Pinder
Published in 1995 by Chapman & Hall. ISBN 0 412 45260 X

though sample sizes were often small and/or laboratory feeding trials rather than field-derived samples were the basis of information.

In subtropical Lake Thonotosassa, Florida, larvae of *Procladius culiciformis* (Linnaeus) showed positive feeding selection for chironomids, ostracods and cladocerans, and negative selection for rotifers (Vodopich and Cowell, 1984). At the same time their survival and growth were high on the diet of oligochaetes and slightly lower on chironomids and zooplankton; a detritus/ algae diet failed to allow development into the fourth instar. It appears that, despite omnivorous characteristics often revealed by diet analyses, *Procladius* species are crucially dependent upon animal food to complete larval development (cf. Baker and McLachlan, 1979). Therefore the availability of prey must be an important factor for these predatory chironomids.

In a laboratory feeding experiment, *Procladius* larvae fed more readily on first-instar *Chironomus* than on second-instar *Paratanytarsus* (Hershey, 1986); while in the field, species of Orthocladiinae were more frequently represented in the guts of *Procladius* predators than species belonging to other subfamilies. This is attributed to differences in foraging behaviour; free-living larvae and tube-dwelling larvae which frequently come out of their tubes to feed are more susceptible to predation than larvae which remain in the tube for feeding (cf. filter-feeding *Paratanytarsus*). Thus predaceous chironomids feed more heavily on certain types of prey, though they are in principle generalist predators capable of handling a variety of prey.

Another important aspect of prey selection concerns the selection of prey sizes, which is little studied among predaceous chironomids. Kajak and Dusoge (1970) noted that *Procladius choreus* (Meigen) consumed small non-predatory chironomid larvae of not longer than 5 mm. Diet analysis of different instars of three species of tanypods, *Trissopelopia longimana* (Staeger), *Zavrelimyia barbatipes* (Kieffer) and *Macropelopia adaucta* Kieffer [as *goetghebueri*], in an acid and iron-rich stream in southern England revealed that the range of prey sizes ingested scales with predator size, such that larger predators feed more on large prey items (Hildrew *et al.* 1985). Similarly, the fourth instar larvae of *Thienemannimyia festiva* (Meigen) in the littoral of Lough Neagh, Northern Ireland, consumed a larger number of prey items as well as larger prey than the third instars (Tokeshi, 1991a, Figure 12.1). This seems to be a general pattern among gape-limited insect predators such as the Odonata, Plecoptera, Ephemeroptera and Megaloptera in freshwater habitats.

One of the mechanisms whereby a predator species can efficiently utilize prey populations is aggregative responses by predators to prey patches. This was the case with *Procladius culiciformis* (Vodopich and Cowell, 1984), the distribution of which was significantly related to densities of total organisms, total chironomids, *Glyptotendipes* and *Cryptochironomus*. Interestingly, however, the density of oligochaetes was negatively correlated with *Procladius* density, despite the fact that the highest growth rate of the latter was achieved by oligochaete prey in a laboratory experiment.

The impact of chironomid predators on their prey communities has rarely been a subject of study, partly because the prey involved are typically small species/individuals that are hard to enumerate accurately, particularly under natural and semi-natural conditions. Another reason for the paucity of

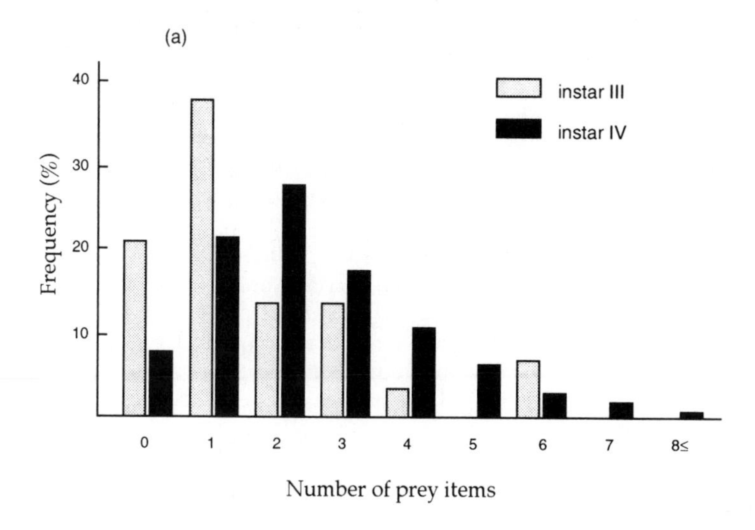

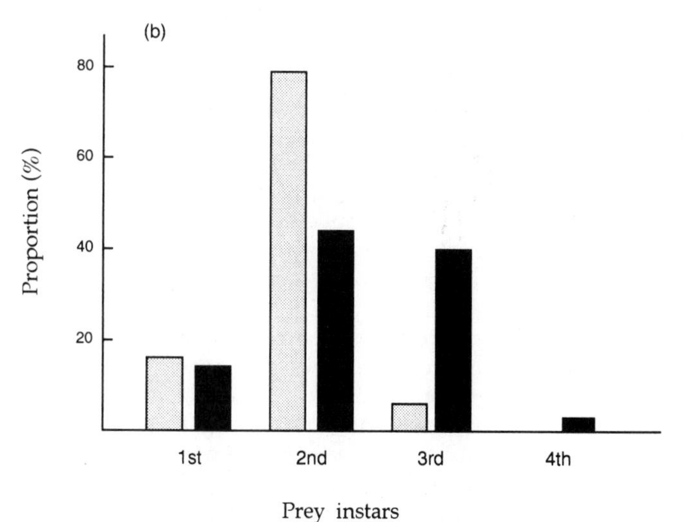

Figure 12.1 (a) Frequency distribution of number of prey items and (b) proportions of different instars of chironomid prey, in the guts of third and fourth instar *Thienemannimyia festiva* (Meigen). (After Tokeshi, 1991a).

studies relates to the fact that chironomid predators themselves are often prey for larger invertebrate and vertebrate predators such as Odonata larvae and fish, thus necessitating a more integrated approach geared towards food web research in order to unravel multi-level trophic systems. In this respect it is notable that the role of chironomid predators has primarily been assessed in lake bottom systems which possess relatively simple trophic structures (e.g. Kajak, 1980). Seasonal fluctuations in the density of *Procladius* in a eutrophic Polish lake demonstrated a delayed cycle in relation to the fluctuations in relative prey biomass (i.e. the ratio of potential prey biomass to predator biomass), thus indicating interactions between predators and prey (Dusoge, 1980). However, the precise causal relationships, whether the predator controls the prey abundance or vice versa, cannot be determined from this observation and prey biomass *per se* did not seem to correlate with *Procladius* abundance. In this study the daily food requirement of *Procladius* was estimated to be 20–60% of their body weight, the figure that appeared to exceed the estimate of benthic production.

In an eastern English river, total density, diversity, species richness and equitability in an epiphytic chironomid community were neither positively nor negatively related to the presence/absence of the tanypod *Zavrelimyia* sp. (Tokeshi, 1993a). In this case the overall density of predators may have been too low to affect the prey community, in addition to the patchy habitat structure which is considered to reduce the effectiveness of predation (Tokeshi, 1994).

Small water-bodies such as plant-held waters (phytotelmata, section 6.7) can provide an ideal set-up for experimental analyses of predator–prey relations and community structure in general. In an arthropod community of the fluid-filled bracts of *Heliconia imbricata* (Kuntze) in Costa Rica (Naeem, 1988a) a predatory chironomid *Pentaneura* sp. kept the density of a resident mosquito *Wyeomyia pseudopecten* Dyar and Knab low while completely excluding the non-resident *Trichoprosopon digitatum* (Rondani). In this case *Pentaneura* sp. can maintain its population by feeding on a harpacticoid copepod as an alternative prey through the period of low abundance of the two mosquito species.

One interesting and yet neglected aspect of the Chironomidae as predators relates to their feeding on microbial communities. Microorganisms are believed to constitute an important food resource for many detritivorous chironomids and other aquatic insects (cf. Cummins, 1974), being nutritiously more valuable than detrital matter. In a eutrophic Swedish lake the concentration of bacteria was found to be significantly higher in the gut of *Chironomus plumosus* Linnaeus than in the sediment, thus indicating selective feeding by *Chironomus* on bacterial prey (Johnson *et al.*, 1989). In laboratory experiments the feeding activity of *Chironomus* larvae decreased the abundance of bacteria but increased the cell-specific production (Johnson *et al.*, 1989). It is likely that the direct effect of predation on bacteria is confounded by other processes such as sediment mixing by burrowing larvae and excretion of organic matter, which may lead to an increase in bacterial abundance (e.g. Kajak *et al.*, 1968).

12.2.2 CHIRONOMIDAE AS PREY

In the context of overall food web structures of aquatic ecosystems, the role of chironomids as prey to many predatory organisms is more conspicuous and significant than their role as predators mentioned above. Indeed, chironomids fall prey to virtually all types of predators, both invertebrates and vertebrates, in aquatic habitats. Some predators are known to undergo ontogenetic shift in diet: small nymphs of a stonefly *Acroneuria (Calineuria) californica* Banks feed mainly on chironomids and other dipterans, whereas intermediate and large nymphs feed on mayfly nymphs and caddis larvae, respectively (Sheldon, 1969). When the gut contents of predators are compared with the relative abundances of different prey types in the environment, it has often been observed that there is a positive selection for chironomid prey (Siegfried and Knight, 1976; Tompkins and Gee, 1983; Vodopich and Cowell, 1984). Notwithstanding some technical problems associated with this kind of analysis (cf. Peckarsky, 1984), the overall picture to emerge is that the Chironomidae as a group constitute a vital nexus in the overall trophic structures of many freshwater ecosystems.

Size-selective predation on chironomids by invertebrate predators often involves progressively larger prey being taken as predator size/instar increases (Hayashi, 1988; Tokeshi, 1991a). In some cases, however, chironomid prey size was not correlated with predator size (Devonport and Winterbourn, 1976). On the other hand, because of relatively small sizes of many freshwater invertebrate predators, younger instars of chironomids tend to constitute more important prey than the larger, late instars (cf. Kajak, 1980; Tokeshi, 1991a). The small predaceous oligochaete *Chaetogaster diaphanus* was observed to feed mainly on the first instars and never on the third/fourth instars (Tokeshi, 1993a). This must be related to the difficulty of handling a large prey in comparison with a predator's own body size, particularly the gape size.

Amongst fish predators the age 0+ and 1+ individuals of the brook stickleback *Culaea inconstans* Kirtland in central Canadian rivers consumed disproportionately small chironomid larvae in comparison with their abundance in the environment (Tompkins and Gee, 1983). Similarly, the bluegill sunfish *Lepomis macrochirus* Rafinesque (Werner *et al.*, 1983) and the sculpin *Cottus cognatus* Richardson in an Alaskan lake (Hershey, 1985a) seemed to feed more on small chironomids. It is likely that susceptibility to predation is higher in small chironomids than in large ones due mainly to behavioural characteristics: larger chironomid larvae tend to build more secure tubes and/or burrow deeper down into the sediment, thus effectively avoiding predation. This makes a contrast to fish predation on zooplankton prey whereby the largest prey is in general selected (Brooks and Dodson, 1965; Hall *et al.*, 1970; Werner *et al.*, 1983).

Amongst the Chironomidae, free-ranging species such as *Procladius* and *Thienemanniella* seem to be most susceptible to predation, whilst in the case of tube-dwelling species those which spend more time foraging or engaged in other activities outside the tube are more susceptible (cf. Brown *et al.*, 1980; Hershey and Dodson, 1985; Hershey, 1987). In laboratory feeding experiments involving a damselfly predator *Ischnura verticalis* Say and four species of chironomid prey, the predator consistently selected the prey which spent more time out of the tube, the ranking being *Pentaneura inconspicua* (Malloch) (non-tube builder) > *Cricotopus sylvestris* (Fabricius) > *Cricotopus bicinctus* (Meigen) > *Paratanytarsus inopertus* (Walker) (Figure 12.2, Hershey, 1987). Apparently, susceptibility to predation depends on the foraging behaviour of predators, relative sizes of predators and prey, and behavioural characteristics

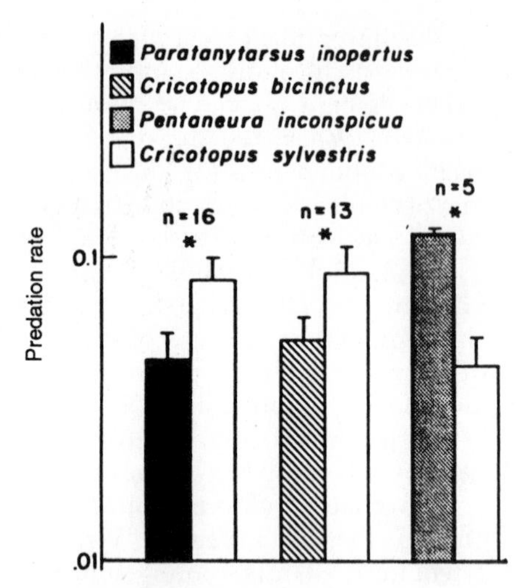

Figure 12.2 Susceptibility to predation by a damselfly *Ischnura verticalis* Say. Densities of pairs of chironomid species were compared. * indicates significant difference ($P<0.05$). (After Hersey, 1987.)

of prey – all of which vary among different species of predators/prey and among different age classes within a predator/prey species. Given the range of possible interactions between these factors, more detailed analyses of predator–prey relations involving different species of predators and chironomid prey are necessary before any generalization is attempted.

In a few cases where predator distribution was analysed in relation to chironomid prey distribution, predators showed aggregative responses (Vodopich and Cowell, 1984; Hildrew and Townsend, 1976). The distributions of a predaceous caddis *Plectrocnemia conspersa* and, to a lesser extent, a megaloptera *Sialis fuliginosa* were more closely related to the combined density of stonefly and chironomid prey than to the abiotic factors in a southern English stream (Hildrew and Townsend, 1976). Predaceous *Rhyacophila dorsalis* showed an aggregated distribution in relation to chironomid and simuliid prey on a submersed macrophyte *Potamogeton* x *zizii* (Tokeshi and Pinder, 1985). On the other hand it is notable that the aggregative distribution was not evident on another macrophyte *Potamogeton pectinatus*, with finely divided leaves, indicating that habitat complexity/ patchiness can affect predator responses to prey (cf. Tokeshi, 1994).

Migrant birds such as swallows (*Hirundo rustica*), swifts (*Apus apus*), house martins (*Delichon urbica*) and sand martins (*Riparia riparia*) are known to feed intensively on adult chironomids when they emerge in spring, but little research has been done to elucidate their predator–prey relations (e.g. Tait-Bowman, 1980). Similarly, various species of waterfowl which frequent standing waters feed on macroinvertebrates, including the Chironomidae (e.g. Bengtson, 1975). Studies in Lough Neagh, Northern Ireland, which constitutes an internationally important site for overwintering waterfowl, showed that chironomid larvae, particularly *Chironomus anthracinus* which is the dominant species in the extensive profundal zone of this lake, were a major food item in tufted duck (*Aythya fuligula* (L.)), goldeneye (*Bucephala clangula* (L.)), scaup (*Aythya marila* (L.)) and pochard (*Aythya ferina* (L.)), with the last species exclusively feeding on chironomid prey (Winfield, 1991). Considering the substantial feeding capacity of waterfowl, their role as predators of the Chironomidae should warrant more attention.

12.2.3 PREDATION IMPACT ON CHIRONOMIDAE

Studies assessing the impact of predation in freshwater habitats have often included the Chironomidae as a major prey category. This applies to both invertebrate and vertebrate predators of the Chironomidae. In a third-order stream in New York, chironomids were the most important prey in terms of numbers for two stonefly predators, *Acroneuria carolinensis* and *Agnetina capitata* (Pictet) (Peckarsky, 1985). When colonization experiments lasting 7 days were conducted using cages with and without stonefly predators, it was found that chironomid density was significantly reduced in cages with predators (Peckarsky, 1985). Interestingly, there was no detectable reduction in chironomid density when predators were confined in a small container within the cage, which suggests that prey reduction was the result of direct predation and/or contact escape by prey. Furthermore, when similar experiments spanning 3 days were performed with three stonefly predators, *Megarcys signata* (Hagen), *Pteronarcella badia* (Hagen) and *Kogotus modestus* (Banks), in a high-elevation Colorado stream, there was no significant effect of predation on the densities of chironomids and other prey taxa (Peckarsky, 1985). Another experiment involving a stonefly predator *Doroneuria baumanni* Stark and Gaufen over a 25-day period in British Columbia (Lancaster, 1990) showed a clear reduction in the density of chironomids through direct predation (Figure 12.3). In the littoral of Lake Maarsseveen, Holland,

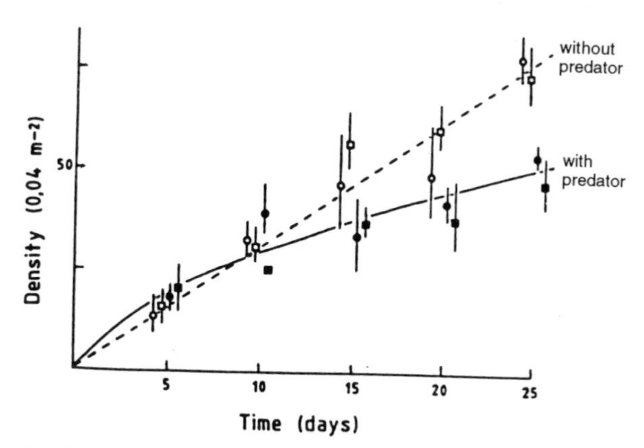

Figure 12.3 Variation in the density of chironomids with and without a stonefly predator *Doroneuria baumanni* Stark and Gaufen. (After Lancaster, 1990.)

predation by water mites of the genus *Hygrobates* is thought to account for a large decline in abundance of *Stictochironomus sticticus* larvae (ten Winkel and Davids, 1987).

The impact of fish predation on chironomids and other macroinvertebrates has been intensively studied, particularly since the 1980s. In a relatively large-scale experiment in a stream in the Colorado mountains, Allan (1982) reduced the standing stock of trout (mainly *Salvelinus fontinalis*) to 10–25% of the initial level within a 1220 m exclusion section over a 4-year period. Subsequently, comparison of densities of chironomid and other invertebrate taxa between the trout-exclusion and the upstream/downstream control sections revealed no consistent difference. Similar results showing no significant effect of brook trout predation were obtained from two third-order streams in northern Quebec, Canada (Reice and Edwards, 1986). Furthermore, 3-month enclosure experiments with juvenile creek chubs (*Semotilus atromaculatus* Mitchill) in a second-order stream in New York (Gilliam *et al.*, 1989) showed no significant predation effect on the density of Chironomidae and Sphaeriidae, though the Oligichaeta and Isopoda, the two dominant taxa in the benthic fauna, were negatively affected by predation.

In contrast, 21-day experiments in a second-order Appalachian stream (Flecker, 1984) resulted in a significant increase in abundance of the Chironomidae and a stonefly *Leuctra* in the fish-exclusion cages, with chironomid density negatively related to the density of sculpins, *Cottus bairdi* Girard and *C. girardi* Robins. In a similar vein, small artificial pools without the mosquitofish *Gambusia affinis* Holbrook supported a high density of the Chironomidae and other insect taxa, while pools with the predator were virtually devoid of insects within two months of fish introduction (Hurlbert *et al.*, 1972). In eutrophic Lake Kasumigaura, Japan, predation by benthic fish (mainly the goby *Tridentiger obscurus* (Temmink and Schlegel)) and the prawn *Macrobrachium nipponense* (de Haan) is estimated to account for 73–86% and 24% of mortality during the emergence period of *Tokunagayusurika akamusi* (Tokunaga) and *Chironomus plumosus*, respectively (Iwakuma, 1986a).

The fact that different taxa/size classes of Chironomidae have different susceptibility to predators, as revealed in some prey selection studies (cf. section 12.2.2), points to the possible importance of treating different chironomid taxa separately rather than lumping all as chironomids, when assessing the impact of predation. In 10-day field experiments involving predator enclosure/exclosure in a first-order stream in Alberta, Canada, the densities of *Thienemanniella* and other Orthocladiinae were found to be negatively affected by a stonefly *Kogotus nonus* Needham and Claassen, while no such effect was detected with respect to the densities of *Corynoneura* and *Stempellinella* (Figure 12.4; Walde and Davies, 1984). In this case a significant reduction in prey density occurred only when prey density was relatively high. In an Alaskan lake mentioned earlier (Hershey, 1985a) fish exclusion (particularly the sculpin *Cottus cognatus*) over a period of 3 weeks resulted in a higher total density of chironomids in bare sediments but not in macrophyte-covered sediments. When different genera of chironomid were analysed separately, no single taxon was found to be significantly affected by fish predation, but it was revealed that reduced total chironomid density in bare

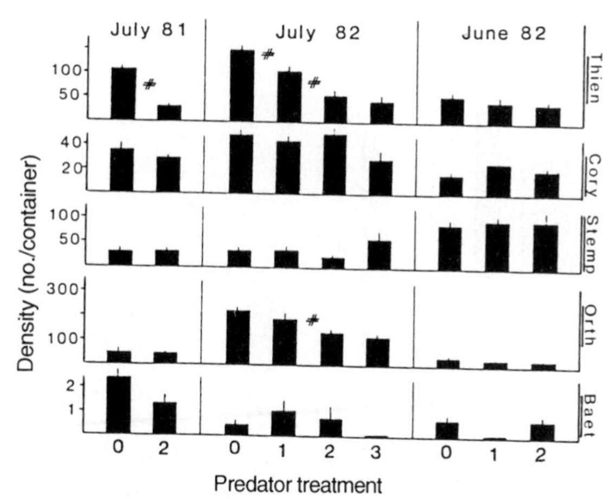

Figure 12.4 Densities of *Thienemanniella*, *Corynoneura*, *Stempellinella*, other Orthocladiinae and *Baetis* under different levels of predation pressure from a stonefly *Kogotus nonus*. # indicates a significant difference (P<0.05). (After Walde and Davies, 1984.)

sediments was due to reduction in density of small (<7 mm) chironomids, irrespective of species. Thus fish predation seems to be directed mainly to small chironomids, the impact of which is apparent only in the exposed habitats without macrophyte cover.

In a set of 3-month experiments spanning a year in the littoral of a reservoir, Thorp and Bergey (1981b) analysed the effects of predation by fish and turtles on chironomids which were classified into four functional groups (predators, collectors, scrapers and shredders) according to Merritt and Cummins (1978). While densities of total chironomids and the two numerically dominant functional groups, predators and collectors, were not significantly related to the presence/absence of vertebrate predators, densities of less abundant scrapers and shredders were significantly lower in the predator exclusion cages (Figure 12.5). In this case there was no evidence to suggest that predators affected the size distribution of chironomids. Neither the total number of genera nor the number of genera in each functional group seemed to be affected by predation.

Another example of the complex nature of chironomid response to fish predation is seen in a year-long caging experiment in the littoral of a mesotrophic farm pond in North Carolina (Gilinsky, 1984). In replicated treatments the density of the bluegill sunfish *Lepomis macrochirus* was set at three different levels, high (10 fish m^{-2}), low (3.3 fish m^{-2}) and no fish, and benthic invertebrates were sampled at regular intervals. Amongst predaceous tanypods, the most abundant species *Clinotanypus pinguis* (Loew) showed a marked effect of fish predation, with its density significantly reduced in high predation treatment in all seasons (Figure 12.6a). In stark contrast, another tanypod *Procladius* spp. showed a completely different pattern of response to fish predation, with higher density being achieved in low/high predation

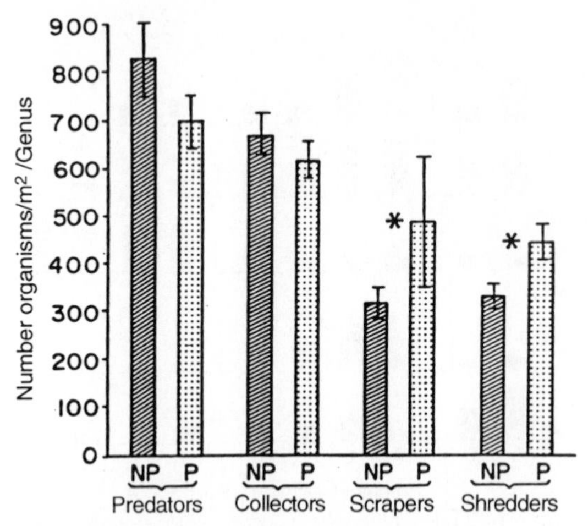

Figure 12.5 Densities of four functional groups of chironomid, (P) with and (NP) without vertebrate predators (fish and turtles). (After Thorp and Bergey, 1981b.)

treatment than in no predation treatment in autumn–winter (Figure 12.6b). Another group of chironomids, herbivorous benthic species including *Tanytarsus* spp., *Polypedilum tritum* (Walker), *Cladotanytarsus* sp. and *Pagastiella* cf. *ostansa* (Webb), had patterns of response similar to *Procladius*: increased density in high predation treatment in autumn and winter and no difference in spring and summer (Figure 12.6c). It is also notable that bluegills had no effect on the abundance of other macroinvertebrate species. A detailed analysis such as this is clearly more useful in unravelling predator–prey relations involving an ecologically diverse group such as the Chironomidae.

Rather than taking a direct form of predation, predators can affect certain groups of organisms in an indirect manner through an intermediary of other predators in the food web. In a 3-month experiment with the bluegill sunfish (*Lepomis macrochirus*) in artificial ponds, the presence of fish predators led to reduced abundance of large invertebrate predators (the Libellulidae and Coenagrionidae: Odonata) and herbivores (*Hyalella*), which in turn resulted in an increase in abundance of smaller predators (the Tanypodinae) and herbivorous chironomids (Crowder and Cooper, 1982). In contrast, such cascading effects of predation were not observed in an 8-day experiment with the brown trout *Salmo trutta* in a southern English stream (Schofield *et al.*, 1988). Although the density of a predatory caddis *Plectrocnemia conspersa* was significantly reduced in the fish enclosure cages, there was no corresponding increase in density of six chironomid taxa (*Macropelopia* spp., the Pentaneurini, *Brillia* spp., *Heterotrissocladius marcidus* (Walker), *Prodiamesa olivacea* (Meigen) and the Tanytarsini) nor of other invertebrate species. The two predators, the brown trout and the caddis, had broadly similar diets and might have complemented each other in terms of overall predation pressure on benthic prey, thus leading to no detectable change in the prey community.

These experimental studies have demonstrated that diverse patterns can

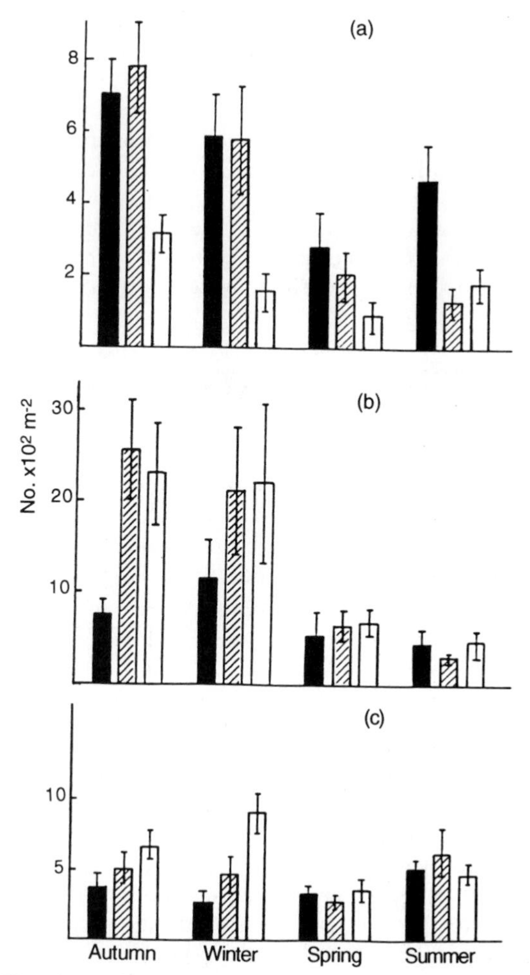

Figure 12.6 Densities of (a) *Clinotanypus pinguis* (Loew), (b) *Procladius* spp. and (c) herbivorous chironomids under three different levels of predation pressure from the bluegill sunfish. ■, No pressure; ▨, low pressure; □, high pressure. (After Gilinsky, 1984.)

occur as a response by chironomid communities to various forms of predation. However, it is worth noting that, highly instructive as they are, experimental studies employing predator exclusion/enclosure cages have generally been restricted in terms of spatial and temporal scales and that generalization on the basis of a study carried out at a single spatio-temporal scale can be misleading. In addition, a particular mesh size used for caging seriously influences the outcome of experiments. Spatial scale may play a crucial role in the organization of many aquatic communities (cf. Tokeshi, 1994), including the Chironomidae, and further research is needed in this direction.

12.3 HOST–PARASITE/COMMENSAL RELATIONS

Parasitic and commensal life involves direct and sustained bodily contact between organisms, the effects of which need to be analysed from the points of view of both hosts and parasites/commensals. The family Chironomidae makes a contrast to other aquatic insect taxa in that species with parasitic/commensal modes of life occur widely (Table 12.1). Broadly, three aspects are examined here: the Chironomidae as hosts to other organisms, as parasites and as commensals, with considerations of the evolution of commensalism.

12.3.1 CHIRONOMIDAE AS HOSTS TO OTHER ORGANISMS

Chironomids are known to be parasitized by a variety of taxa including mermithid nematodes, water mites, fungi and microsporidians (Hilsenhoff and Lovett, 1966; Hunter, 1968). Disease agents such as bacteria and viruses may also be considered as parasites, but there is virtually no ecological information on these. A selection of cases is reviewed here.

(a) Mermithid nematodes

Nematodes of the family Mermithidae parasitize a wide range of insect taxa, including the Chironomidae (Nickle, 1972). Despite their common occurrence, very few studies have investigated in detail the effects that nematodes are likely to have on chironomid populations. Nematode worms generally occur as endoparasites in the last larval instar, pupae and adults of the Chironomidae (Hilsenhoff, 1967; Golini and Sherry, 1979; LeSage and Harrison, 1980b). It is possible that earlier instars are parasitized but nematodes are too small to be readily recognized, since the initial contact between parasite and host is thought to be established when a chironomid larva, while grazing algae and detritus, ingests the nematode eggs (and juveniles?) resting on the substrate surface. The fact that a nematode worm is often seen occupying the entire body cavity of the last instar larva of medium-sized chironomid species such as *Cricotopus* spp. (M. Tokeshi, *pers. obs.*) seems to suggest that small, earlier instars do not constitute appropriate food resources for parasitizing mermithids.

Among mermithids there appear to be two different modes of parasitization. In the first case nematodes spend their entire life cycle in water, reaching maturity and leaving the chironomid host before or soon after pupation, whereby the host is invariably killed. In the second case nematodes remain in the host until after emergence as adult, escaping from the host only when it returns to water for mating/oviposition. In the latter case nematodes are known to cause varying degrees of morphological abnormalities in the adult characteristics, including empty thorax/abdomen and intersex morphologies (Wülker, 1958, 1960, 1964; LeSage and Harrison, 1980b). In adult

Table 12.1 Parasitic/commensal chironomids and their hosts

Chironomid species	Host	Locality	Reference
(Parasitic species)			
Parachironomus varus Goetghebuer	*Physa fontinalis* (L.) (Gastropoda:Lymnaeidae)	Germany	Meier, 1987
	Radix ovata (Drap.) (Gastropoda:Lymnaeidae)	Germany	Meier, 1987
Cryptochironomus ex. gr. *pararostratus*	*Lymnaea* spp. (Gastropoda:Lymnaeidae)	Poland	Czeczuga *et al.*, 1968
Baeoctenus bicolor Sæther	*Anodonta cataracta* Say (Bivalvia:Unionidae)	Canada	Gordon *et al.*, 1978
	Anodonta implicata Say (Bivalvia:Unionidae)	Canada	Gordon *et al.*, 1978
Xenochironomus xenolabis Kieffer	Porifera: Spongillidae	Canada	Steffan, 1968
Epoicocladius sp. #1	*Ephemera guttulata* Pictet	USA	Jacobsen, *pers. comm.*
Cardiocladius sp.	*Hydropsyche incommoda* Hagen	USA	Parker & Voshell, 1979
	Hydropsyche venularis Banks	USA	Parker & Voshell, 1979
	Cheumatopsyche spp.	USA	Parker & Voshell, 1979
Eurycnemus sp.	*Goera japonica* Banks	Japan	Kobayashi, in press
Symbiocladius rhithrogenae (Zavřel)	*Heptagenia lateralis* (Curtis)	Europe	Codreanu, 1939
	Rhithrogena spp.	Europe	Codreanu, 1939; Soldán, 1978
	Epeorus spp.	Europe	Soldán, 1978
Symbiocladius equitans (Claassen)	*Heptagenia* spp.	Canada	Wiens *et al.*, 1975
	Rhithrogena spp.	Canada	Wiens *et al.*, 1975
	Epeorus spp.	USA	Claassen, 1922
Symbiocladius aurifodinae Hynes	Leptophlebiidae spp.	Australia	Hynes, 1976
Symbiocladius sp. A*	*Eperous latifolium* Uéno	Siberia	Matena & Soldán, 1982
	Rhithrogena cf. *tianshanica*	Siberia	Matena & Soldán, 1982
Symbiocladius sp. B*	Heptageniidae spp.	Japan	Ueno, 1930; Tokeshi, *pers. obs.*
Symbiocladius (*Acletius*) *wygodzinskyi* Roback	*Thraulodes** (Ephemeroptera:Leptophlebiidae)	S. America	Roback, 1965

Table 12.1 Continued

Chironomid species	Host	Locality	Reference
(Commensal species)			
Epoicocladius flavens (Malloch)	*Ephemera dancia* Müller (Ephemeroptera:Ephemeridae)	Sweden	Svensson, 1976, 1978, 1980
		England	Tokeshi, 1986c, 1988
		Czechoslovakia	Soldán, 1988
	Ephemera vulgata L., *Ephemera lineata* (Eaton)	Czechoslovakia	Soldán, 1988
Epoicocladius sp. #3	*Ephemera guttulata* Pictet[1]	N. America	Jacobsen, 1992
Epoicocladius sp. A*	*Ephemera orientalis* MacLachlan	Korea	Matena & Soldán, 1986
Epoicocladius sp. B*	*Ephemera* sp.	Vietnam	Matena & Soldán, 1986
Epoicocladius sp. C*	*Ephemera strigata* Eaton	Japan	Takemon, *pers. comm.*
*Eukiefferiella** sp.	*Brachycentrus occidentalis* (Trichoptera:Brachycentridae)	USA	Gallepp, 1974
Eukiefferiella sp.	*Cryphocricos peruvianus* De Carlo (Hemiptera:Naucoridae)	Cent./S. America	Roback, 1977b
Eukiefferiella ancyla Svensson	*Ancylus fluviatilis* Müller (Gastropoda:Ancylidae)	Sweden	Svensson, 1986
Dratnalia potamophylaxi (Fittkau & Lellak)	*Potamophylax cingulatus* Stephens (Trichoptera:Limnephilidae)	Cent./N. Europe	Dratnal, 1979
Nanocladius (Plecopteracoluthus) downesi (Steffan)	*Acroneuria abnormis* (Newman) (Plecoptera:Perlidae)	N. America	Steffan, 1965; Bottorff & Knight, 1987
	Acroneuria lycorias (Newman) (Plecoptera:Perlidae)	Canada	Steffan, 1965
	Paragnetina media (Walker) (Plecoptera:Perlidae)	Canada	Steffan, 1965
	Paragnetina immarginata (Say) (Plecoptera: Perlidae)	Canada	Steffan, 1965
	Nigronia serricornis Say (Megaloptera:Corydalidae	USA	Hilsenhoff, 1968
	Corydalus cornutus (L.) (Megaloptera:Corydalidae	USA	Tracy & Hazelwood, 1983
	Chauliodes pectinicornis (Say) (Megaloptera: Corydalidae)	USA	Benedict & Fisher, 1972
Nanocladius (P.) branchicolus Sæther	*Acroneuria lycorias* (Newman)	Canada	Dosdall & Mason, 1981
	Pteronarcys dorsata (Say) (Plecoptera:Pteronarcyidae)	Canada	Dosdall *et al.*, 1986
Nanocladius (P.) bubrachiatus Epler	*Traverella* sp. (Ephemeroptera:Leptophlebiidae)	Honduras	Epler, 1986

Table 12.1 Continued

Chironomid species	Host	Locality	Reference
Nanocladius (s.s.) *rectinervis* (Kieffer)	*Nigronia serricornis* Say	Canada	Gotceitas & Mackay, 1980
Nanocladius cf. *rectinervis* (Kieffer)	*Pteronarcys dorsata* (Say)	Canada	Dosdall *et al.*, 1986
Nanocladius cf. *spiniplenus* Sæther	*Pteronarcys dorsata* (Say)	Canada	Dosdall *et al.*, 1986
*Dactylocladius** *commensalis* Tonnoir	*Neocrupira hudsoni* Lamb (Diptera:Blepharoceridae)	New Zealand	Tonnoir, 1923
Cricotopus sp. nr. *tremulus* (L.)	*Pteronarcys dorsata* (Say)	Canada	Dosdall *et al.*, 1986
Tvetenia vitracies (Sæther)	*Pteronarcys dorsata* (Say)	Canada	Dosdall *et al.*, 1986
Paratanytarsus confusus Palmen	*Pteronarcys dorsata* (Say)	Canada	Dosdall *et al.*, 1986
Polypedilum convictum (Walker)	*Pteronarcys dorsata* (Say)	Canada	Dosdall *et al.*, 1986
Polypedilum sp.	*Corydalus cornutus* (L.)	USA	Furnish *et al.*, 1981
Rheotanytarsus exiguus Johannsen	*Macromia georgina* (Sel.) (Odonata:Macromiidae)	USA	White & Fox, 1979
Rheotanytarsus spp.	*Pteronarcys dorsata* (Say)	Canada	Dosdall *et al.*, 1986
	Corydalus cornutus (L.)	USA	Furnish *et al.*, 1981
	Boyeria vinosa (Say) (Odonata)	USA	White *et al.*, 1980
	Macromia spp. (Odonata)	USA	White *et al.*, 1980
	Calopteryx maculata (Beauvois) (Odonata)	USA	White *et al.*, 1980
	Austroaeschna atrata Martin (Odonata)	Australia	Hawking & Watson, 1990
	Nectopsyche exquisita (Walker) (Trichoptera:Leptoceridae)	USA	White *et al.*, 1980
	Stenonema smithae Traver (Ephemeroptera:Heptageniidae)	USA	White *et al.*, 1980
	Tricorythodes sp. (Ephemeroptera:Tricorythidae)	USA	Wilda, 1987
	Oxytrema (*Elimia*) *carinifera* (Lamark) (Gastropoda:Pleuroceridae)	USA	Vinikour, 1982
	Goniobasis sp. (Gastropoda)	USA	Mancini, 1979
Ablabesmyia sp.	*Anodonta* spp. (Bivalvia: Unionidae)	USA	Roback *et al.*, 1979
	Fusconaia sp., *Quadrula* sp. (Bivalvia:Unionidae)	USA	Roback *et al.*, 1979

Table 12.1 Continued

Chironomid species	Host	Locality	Reference
Xenochironomus canterburyensis (Freeman)	*Hyridella menziesi* (Gray) (Bivalvia:Lamellibranchia)	New Zealand	Forsyth & McCallum, 1978
Ichthyocladius spp.	Astroblepidae (Pisces)	Amazonia	Freihofer & Neil, 1967
	Loricariidae (Pisces)	Amazonia	Fittkau, 1974

* Tentative designations – confirmation necessary. [1]Another unidentified chironomid species was also found on this mayfly; parasitic nature suspected but requires confirmation.

females of some *Cricotopus* species, reproductive organs (the spermathecae and ovaries) are partially or completely destroyed (Wülker, 1970; LeSage and Harrison, 1980b).

Mermithid parasitization of chironomid larvae seems to be heaviest in spring–summer, when larvae are at the most active phase of feeding. Similarly, emerging adults of *Cricotopus triannulatus* Macquart were most heavily parasitized in late spring–early summer (Figure 12.7; LeSage and Harrison, 1980).

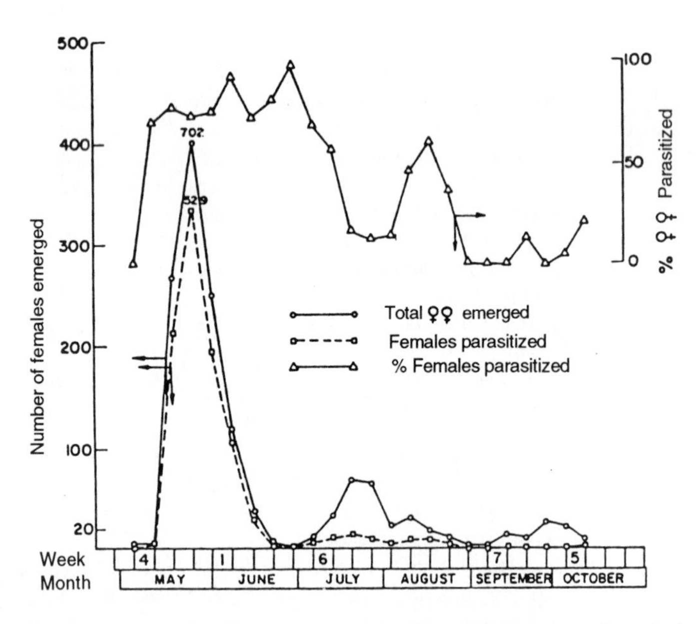

Figure 12.7 Emergence of *Cricotopus triannulatus* Macquart and variation in the proportion of females parasitized by mermithid nemadodes. (After LeSage and Harrison, 1980b.)

Susceptibility to parasitism must be related to the seasonal occurrence and feeding habits of larval hosts; some species, such as *Cricotopus* spp., are more frequently parasitized than coexisting larvae of other taxa (M.Tokeshi, *pers. obs.*). Furthermore, the frequency of parasitism can vary from year to year, as has been revealed for *Cricotopus triannulatus* and *C. infuscatus* (Malloch) (66.9% of total females in one year to 25.7% in the following year, and 93.7% to 37.2%, respectively), while other species may show little variation in parasitism (e.g. *C. bicinctus*, 5.0% to 2.6%, LeSage and Harrison, 1980b).

It is likely that heavy parasitization by mermithid nematodes significantly depresses the chironomid populations, possibly in a density-dependent manner (cf. LeSage and Harrison, 1980b). Unfortunately, the lack of detailed population studies, particularly involving larval dynamics and their incidence of nematode parasitization, precludes further speculations on this important aspect.

(b) Water mites

Water-mites (Hydrachnellae: Acari) possess a complex life cycle with a parasitic larval stage (cf. Böttger, 1976; Smith and Oliver, 1976). Water-mite larvae parasitize a variety of insect taxa including the Odonata, Plecoptera, Trichoptera, Hemiptera, Coleoptera and Diptera, with the Chironomidae constituting a particularly important host taxon. Mite larvae locate chironomid pupae in the substrate and cling to them with their legs. On emergence of the chironomid host as winged adults, mite larvae are carried out of the water and begin a truly parasitic life, albeit short, by sucking the body fluid of the host. When the host returns to the water, mites leave the host and re-enter the aquatic habitat where they metamorphose and reach maturity. Adult water mites are known as predators, consuming in particular chironomid eggs and hatched larvae (e.g. ten Winkel *et al.*, 1989).

Amongst a total of 34 species of chironomid that emerged from a small reservoir in Wales, 13 species were found to be parasitized by water-mite larvae of the genera *Arrenurus*, *Hygrobates*, *Piona* and *Unionicola*, with *Arrenurus* accounting for >90% of all mites (Booth and Learner, 1978). The incidence of parasitism varied greatly among chironomid species, with a maximum of 42% (of total individuals caught) in *Procladius choreus* to less than 1% in *Cricotopus bicinctus* [as *dizonias*], *Chironomus anthracinus* Zetterstedt, *Dicrotendipes modestus* Say [as *Limnochironomus pulsus*], *Paratanytarsus inopertus* and *Tanytarsus lugens* (Kieffer). The abundance of *Arrenurus* appears to coincide with the emergence phenology of its principal host, *P. choreus* (Figure 12.8).

The incidence of water-mite parasitization in 11 species of *Cricotopus* from an alkaline stream in Ontario, Canada, varied from 47.4% in *C. infuscatus* to 8.4% in *C. trifascia* with an overall parasitization of 24.9% of all *Cricotopus* adults (LeSage and Harrison, 1980b). Seasonally, the percentage of hosts parasitized by mite larvae increased towards July/August and declined in September.

The fact that water-mite larvae attach themselves as ectoparasites to chironomid hosts for a relatively short period and that hosts are not killed by

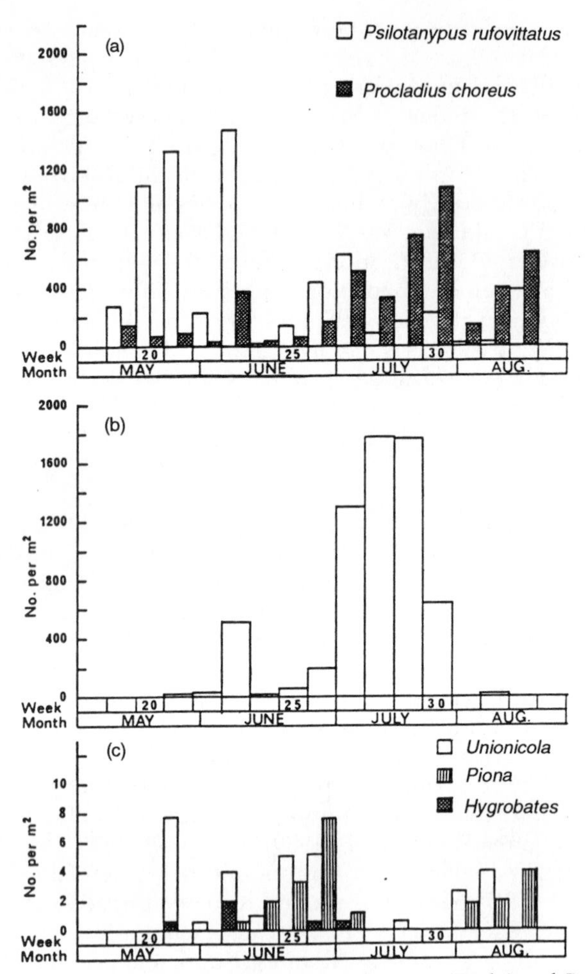

Figure 12.8 (a) Emergence of *Psilotanypus rufovittatus* (Wulp) and *Procladius choreus* (Meigen) and the abundances of water-mites: (b) *Arrenurus*; (c) *Unionicola*, *Piona* and *Hygrobates*. (After Booth and Learner, 1978.)

these parasites may indicate that parasitism by water mites has a negligible effect on chironomid populations, despite its ubiquitous occurrence. It may also be argued that water mites are a successful taxon in freshwater habitats because they have evolved to satisfy themselves with a small supply of nutrients provided by the host, without impairing the host's capacity to reproduce. There is an interesting contrast between ectoparasitic water-mite larvae and endoparasitic mermithid nematodes mentioned in the previous section, in that the two groups seem to be adopting different strategies of parasitism on the chironomid host.

(c) Fungi of *Coelomomyces*

Fungi of the family Coelomomycetaceae are known to parasitize larval mosquitoes as well as chironomids, but their reports are scarce (Laird, 1961; Weiser and Vávra, 1964). In a shallow lake in British Columbia, Canada, the sporangia of two species of *Coelomomyces*, *C. beirnei* Weiser and McCauley and *C. chironomi* var *canadense* Weiser and McCauley, were found in three species of Tanytarsini and one species of *Psectrocladius*, respectively, amongst 25 species of chironomid examined (Weiser and McCauley, 1971). Parasitization by *C. beirnei* appeared more frequent in winter months, but the maximum incidence was only 7.6% of Tanytarsini sp. larvae being parasitized in March. There seemed to be no restriction as to the instars of chironomid parasitized. If the incidence of parasitism is as low as this study suggests, perhaps the impact of *Coelomomyces* parasitization on chironomid populations is minimal. In the absence of more research, however, the relationships between fungal parasites and the Chironomidae remain unresolved.

12.3.2 CHIRONOMIDAE AS PARASITES

Apart from chironomids of the genus *Symbiocladius*, which have long been recognized as ectoparasites of mayfly larvae (Claassen, 1922), *Baeoctenus bicolor* Sæther (Gordon *et al.*, 1978) and *Cryptochironomus* spp. (Czeczuga and Bobiatynska-Ksók, 1968) were observed to lead a parasitic life in freshwater bivalves and gastropods, respectively. Larvae of *Cardiocladius* [*albiplumus* Sæther, G.A. Halvorsen, *pers. comm.*] are reported as ectoparasitic on the pupae of Hydropsychidae (Parker and Voshell, 1979). In addition, one species of *Epoicocladius* is considered as a parasite on *Ephemera* larvae (R. Jacobsen, *pers. comm.*) and *Eurycnemus* sp. on a case-bearing caddis *Goera japonica* Banks (Kobayashi, in press). Table 12.1 lists the names of hosts and chironomid parasites/commensals.

(a) *Symbiocladius* spp.

Symbiocladius has a world-wide distribution, being recorded from North and South America, Eurasia and Australia (Table 12.1). According to Codreanu (1939), only the larvae of early first instars of the European species *S. rhithrogenae* (Zavřel) are free-living and seek mayfly hosts of the genera *Ecdyonurus*, *Heptagenia* and *Rhithrogena*. In France, overwintering first instars grow rapidly in spring and pupate in mid May to early June; two further generations follow by the end of summer, thus producing a total of three generations in a year (Codreanu, 1939). In a brown water river in the Northwest Territories, Canada, *S. equitans* Claasen is thought to have two generations, one overwintering and one summer generation, with apparent alternation of hosts: *Rhithrogena* in winter and *Heptagenia* in summer (Wiens *et al.*, 1975). Parasitizing larvae attach themselves firmly under the wing pads of the host and feed on its haemolymph, causing disruption of moulting and preventing the normal development of reproductive organs (Soldán, 1979). After pupation of the parasite, younger mayfly larvae generally die whilst older ones may reach the adult stage, but with high rates of sterility and/or mortality. The incidence of parasitism was 28–64% on *Rhithrogena* hosts and 18–69% on *Heptagenia* hosts in Romania (Codreanu, 1939), while in Canada it was 25–65% on *Rhithrogena* and 5–45% on *Heptagenia* (Wiens *et al.*, 1975). Thus, species of *Symbiocladius* appear to have a high potential of significantly reducing the population size of ephemeropteran hosts, whilst in the long term the parasite and the host populations must persist in a dynamic equilibrium. In this context more detailed studies on the population dynamics of both the parasite and the host are required to understand this interactive system.

(b) *Baeoctenus bicolor*

Larvae (third and fourth instars) and pupae of *Baeoctenus bicolor* were found in the mantle cavity of unionid bivalves *Anodonta cataracta* Say and *A. implicata* Say in a mesotrophic reservoir in New Brunswick, Canada (Gordon *et al.*, 1978). *Baeoctenus* larvae construct tubes of particulate organic matter which are normally attached to the anterio-dorsal surface of gills near the labial palps of *A. cataracta*. The gills around each larval case were extensively damaged, suggesting that larvae actively fed on gill tissue. The incidence of parasitism fluctuated between 4.2 and 70% from late November to late June, though these figures need to be interpreted with caution because of small sample sizes, whilst no larvae were found between July and mid November. Despite the apparently parasitic nature of the larvae, it is unknown how seriously the host is affected by this parasitism and how widely the chironomid species occurs.

(c) '*Cryptochironomus*' spp.

'*Cryptochironomus*' spp. have been found in the bodies of various species of freshwater gastropods, in particular *Lymnaea*, *Radix* and *Physa* (Jutting, 1938; Guibé, 1943; Czeczuga and Bobiatynska-Ksók, 1968; Young, 1973; Meier, 1987). Chironomid larvae inside a snail's body are in direct contact with the digestive glands and the gonads, and the infestation by all four larval instars and the relatively high content of haemoglobin and glycogen in the body of chironomid larvae suggest that they are truly parasitic (Czeczuga and Bobiatynska-Ksók, 1968). The proportion of infested snail individuals varied seasonally: up to 37% in a population of *Radix auricularia* in Poland (Czeczuga and Bobiatynska-Ksók, 1968); 70% in *Lymnaea peregra* and 34% in *Physa fontinalis* in England (Young, 1973); and 58% in *Radix ovata* in Germany (Meier, 1987). Meier (1987) reported that infested snails died soon after chironomid larvae left them for pupation. An annual life cycle was apparent in the population studied in England (Young, 1973), with very low infestation in the summer months corresponding to adult emergence and a steady rate of infestation between autumn and spring. In contrast, two generations, one in May–June and the other in late summer, might have occurred in *Parachironomus varus* (Goetghebuer) infesting *Radix ovata* Drapamaud (Meier, 1987).

(d) *Cardiocladius albiplumus* Sæther

Larvae of *Cardiocladius albiplumus* (as *Cardiocladius* sp.) were observed to enter the enclosed pupal cases of net-spinning caddisflies, *Hydropsyche* and *Cheumatopsyche*, in a North American river (Parker and Voshell, 1979). Usually only one chironomid larva was found in a pupal case, with a small minority containing two. Caddis pupae showed various degrees of damage, from one or two holes in the abdomen to virtually complete loss of the body. Therefore, host mortality upon parasitization is considered to be very high and this situation is more akin to predation or 'parasitoidism' than parasitism in the strict sense.

12.3.3 CHIRONOMIDAE AS COMMENSALS

Chironomid larvae occur as commensals on other aquatic organisms such as the Ephemeroptera, Plecoptera, Trichoptera, Hemiptera, Megaloptera and even fish (Table 12.1; Tokeshi, 1993b). Furthermore, mutualistic relationships are suspected between some chironomid larvae (particularly *Cricotopus*) and the blue-green alga *Nostoc* (e.g. Brock, 1960; Wirth, 1975; Ward *et al.*, 1985; Dodds and Marra, 1989). The list of potential commensals is extensive and will undoubtedly increase in the future, but to date detailed analyses of host–commensal relationships have been restricted to *Epoicocladius* species on *Ephemera* hosts. Accordingly, the following section deals with this relationship. For a recent suggestion on the nomenclature of species in this genus, see Jacobsen (1993).

(a) *Epoicocladius flavens* (Malloch)

The European species *Epoicocladius flavens* is a commensal of the mayfly genus *Ephemera*, in particular *E. danica*, and occurs widely in conjunction with the hosts' distributions in Europe. The life cycle varies within its geographical range: one generation a year in eastern England (Tokeshi, 1986c), partially two generations in southern Sweden (Svensson, 1979) and two generations in Czechoslovakia (Soldán, 1988). Unlike *Symbiocladius* spp. parasitic on Heptageniid mayflies which have a long overwintering period as first instars (Codreanu, 1939; Wiens *et al.*, 1975), *E. flavens* larvae hatching in summer rapidly reach the third and fourth instars in autumn and the fourth instars predominate the population through most of the year in an English river (Figure 12.9c, Tokeshi, 1986c). The discrepancy may relate to the basic difference between commensal and parasitic modes of life. Larger larvae of *Epoicocladius* have better capacity for attaching to the host than small ones, making it more advantageous to reach late instars as quickly as possible. In contrast, in the case of *Symbiocladius* it is more important to coincide its

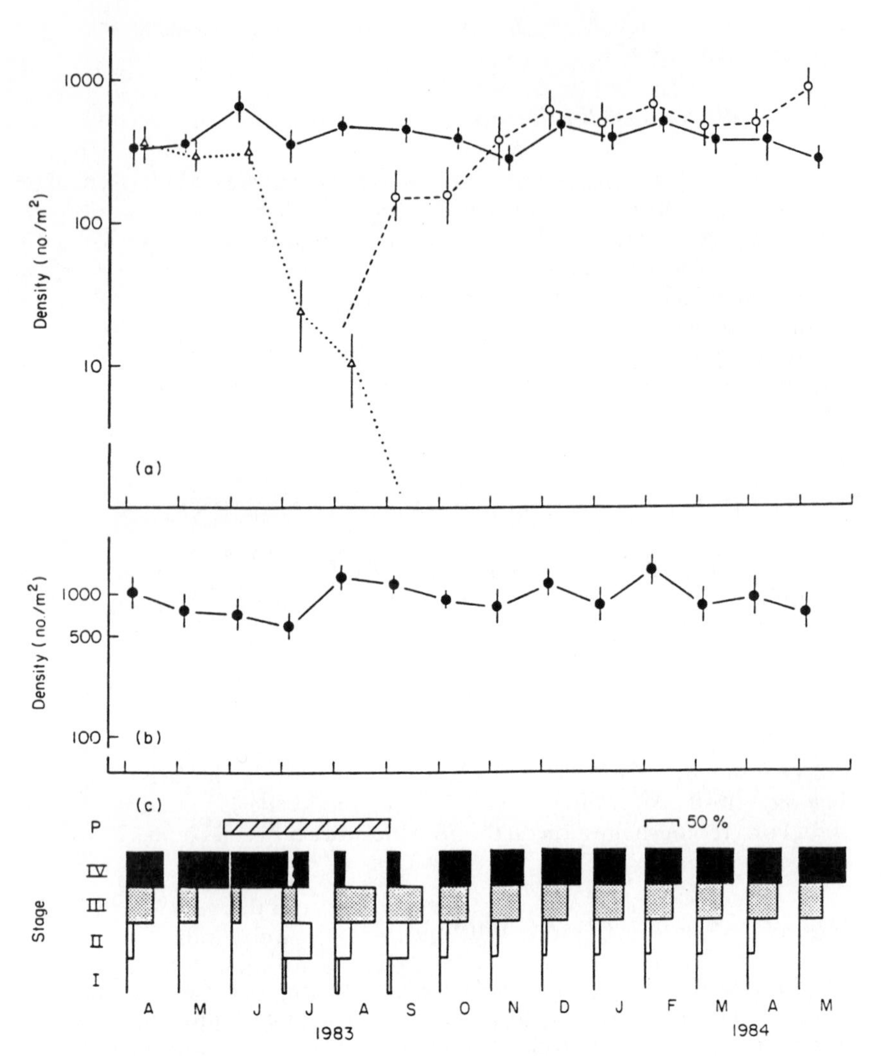

Figure 12.9 Variation in the density (geometric mean ± 95% CL) of (a) the mayfly host *Ephemera danica* Maller (dotted line 1981 year-class, continuous line 1982 year-class and broken line 1983 year-class); and (b) *Epoicocladius flavens* (Malloch). (c) Instar composition of *E. flavens*. (After Tokeshi, 1986c.)

development with that of the host when the latter is growing rapidly in spring, but not before when the host is still small.

The most striking feature of the commensal–host relationship between *Epoicocladius* and *Ephemera* observed in eastern England is the stable population density throughout the year of *E. flavens* (Figure 12.9b) which is coupled to the density of large individuals (>10 mm in body length) of *Ephemera danica* Müller (Figure 12.9a), indicating that mortality of *E. flavens*, once attached to a suitable host, is very low. At this locality the host *E. danica* generally has a 2-year life cycle (Tokeshi, 1985), resulting in the constant presence of relatively large mayfly larvae which serve as main hosts to chironomid commensals. A similar pattern occurred in Czechoslovakia (Soldán, 1988).

The fact that *Epoicocladius* spp. seem to be widespread and common in the northern hemisphere where *Ephemera* mayflies occur indicates that this association is ancient in origin and has been successful, particularly on the part of *Epoicocladius*. That pupation and adult emergence of *Epoicocladius* are well synchronized and occur just before the emergence of the host may also attest to a long evolutionary history. *Epoicocladius* larvae appear to feed on organic particles which settle on the body surface of the host, while relying on the superior mobility of *Ephemera* larvae to be transported in the sandy or silty substrate. The commensal relation does not seem to involve any advantage or disadvantage for the *Ephemera* host.

12.3.4 EVOLUTIONARY SIGNIFICANCE OF COMMENSALISM IN CHIRONOMIDAE

Existence of many chironomid species living in association with a variety of aquatic organisms points to some kind of advantage accrued to the commensal mode of life among the Chironomidae. There are four interrelated factors which are considered to favour commensalism in the Chironomidae (Tokeshi, 1993b). First, a commensal chironomid can rely upon better feeding opportunities afforded by a constant supply of detritus or algae collected on the host's body, which is enhanced by the host's morphology and behaviour. Second, increased mobility with minimal energy expenditure on the part of chironomid commensals may constitute a significant factor, to avoid inhospitable microhabitats and to seek optimum living conditions in terms of food supply, oxygen concentration etc. In this respect the 'living' substrate is better suited to cater for the needs of chironomids than the 'dead' one. Third, commensal life may confer better protection from disturbances which in particular characterize running-water environments. Hosts possess a firmer grip on the substrate and better locomotive ability to escape from local disturbances, and their large bodies may serve as protective shields against objects of disturbance such as rolling stones on the stream bed. Fourth, commensal life may reduce predation risk for chironomid larvae. There are three aspects bearing on this point: physical camouflage on the host which itself is less palatable than the chironomid; escape and defence capability of the host; and the size effect of predation. The last point relates to the fact that many predators, particularly invertebrates, are gape-limited (cf. section 12.2) and therefore incapable of feeding on larger prey. A chironomid larva, which is otherwise a convenient prey for many predators, can increase its effective size by adhering to a larger organism, thereby decreasing the potential number of predators capable of feeding on it. The combination of these factors is likely to have influenced the evolution of commensalism in the Chironomidae.

Transition from commensalism to parasitism is considered slight as long as parasitism involves a minor effect on the host's fitness. The evolution of parasitic habit of *Baeoctenus bicolor* Sæther in *Anodonta* hosts (section 12.3.2(b); Gordon *et al.*, 1978) must have arisen from a seemingly harmless commensal life as exhibited by *Ablabesmyia* sp. in *Anodonta* and other bivalve hosts (Roback *et al.*, 1979). A parallel example may be seen between parasitic and commensal *Epoicocladius* spp. (R. Jacobsen, *pers. comm.*). When the parasite begins to significantly affect the fitness of the host, however, the dynamics of the two populations become more closely coupled and the commensal will no longer enjoy a 'free ride'. In this context comparative studies on the ecology of *Symbiocladius* and commensal species such as *Epoicocladius*, as well as species with an ordinary mode of life, will be highly useful.

12.4 COMPETITIVE/NON-COMPETITIVE RELATIONS

12.4.1 COMPETITION AND COEXISTENCE

Despite the advances made, particularly since the 1970s, with respect to the analysis of **interspecific competition** in various ecological communities, only a very small proportion of studies on the Chironomidae has dealt with this aspect in great detail. Furthermore, few chironomid studies adopted rigorous experimental or analytical approaches to the issue of competition (see the next section).

Profundal chironomid communities of lakes, with a relatively small number of constituent species, can be convenient systems in which to investigate the often complicated phenomena of interspecific competition. Kajak *et al.* (1968) conducted a 7-day laboratory experiments involving the addition of *Chironomus plumosus* larvae to cylinders containing the benthic fauna from a Polish lake, and observed a reduction in abundance of herbivorous chironomids, particularly *Tanytarsus gregarius* and *Cladotanytarsus mancus*. In similar vein, *C. plumosus* may have influenced the seasonal emergence pattern of *C. anthracinus* in Lake Erken, Sweden (Johnson and Pejler, 1987), in that the peak emergence of the latter species occurred in summer rather than in spring as has been observed elsewhere, presumably resulting in the lessening of competitive interactions over food resources.

In eutrophic Japanese lakes *C. plumosus* often coexists with another large species, *Tokunagayusurika akamusi* (Yamagishi and Fukuhara, 1971; Iwakuma, 1986a). Larvae of the latter species burrow deep into the sediment and aestivate in summer months, which effectively results in the avoidance of outright competition for food with *C. plumosus* and (less significantly) other benthic species. In autumn *T. akamusi* larvae come back to the surface layer and emerge as adults, and larvae of the next generation appear in early winter. Subsequently, new larvae of *T. akamusi*, being capable of growth under low temperature conditions, can exploit the winter deposit of phytoplankton and detritus without serious competition from *C. plumosus*. When data covering

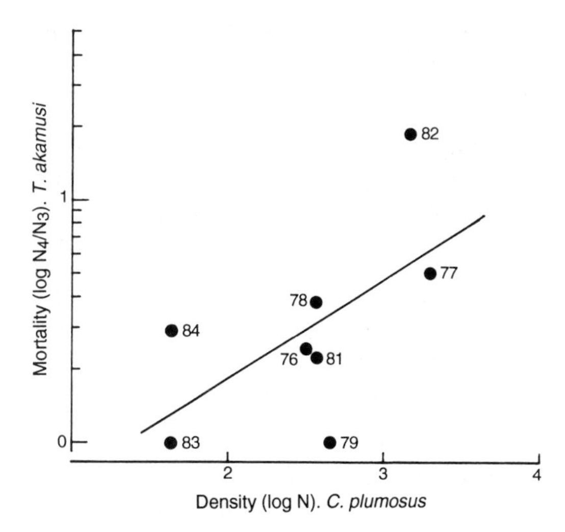

Figure 12.10 Mortality of *Tokunagayusurika akamusi* (Tokunaga) from third to fourth instars (expressed as log density ratios) in relation to the density of *Chironomus plumosus* (Linnaeus) over 8 years. (After Iwakuma, 1986a.)

8 years were analysed, mortality between the third and fourth instars of *T. akamusi* appeared to increase with the increasing density of *C. plumosus* (Figure 12.10; Iwakuma, 1986a), thus reinforcing the view that the two species may indeed exist in a competitive relationship.

Competition need not be restricted to interactions between species of the same taxonomic group but can involve taxonomically distant species. In the pitcher-plant *Darlingtonia californica* densities of two coexisting scavengers, a chironomid *Metriocnemus edwardsi* Jones and a slime mite *Sarraceniopus darlingtoniae* Fashing and O'Connor, are negatively correlated, with the former being capable of consuming resources at a faster rate than the latter (Naeem, 1988b). In the rocky littoral of an arctic Alaskan lake the gastropod *Lymnaea* shares the epilithic algal resource with chironomids and other invertebrate species (Cuker, 1983). The density of tube-dwelling *Paratanytarsus* was significantly reduced in enclosures with a high density of *Lymnaea* snails. Free-ranging species such as *Corynoneura*, *Cricotopus* cf. *sylvestris* and *Zalutschia trigonacies* Sæther were also negatively affected (but to a lesser extent than *Paratanytarsus*) by high snail density, while predaceous *Arctopelopia* and *Ablabesmyia* were only weakly affected. Thus it appears that less mobile chironomid species are more susceptible to the grazing snail on the stone-surface habitat. This makes an interesting contrast to the results of some predation studies (section 12.2) where less mobile, tube-dwelling chironomids were found to be least susceptible to predation. If both of these cases apply to a single community, it then follows that the advantage and disadvantage of being a free-ranger/tube-dweller may shift as relative importance of predation and competition in the community varies through time or in space. For species which tend to suffer more from predation, a less

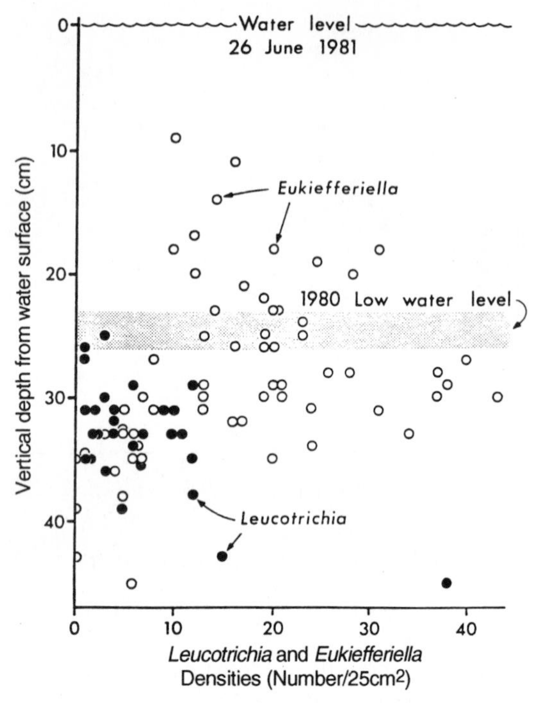

Figure 12.11 Densities of *Leucotrichia* and *Eukiefferiella* in relation to vertical depth. (After McAuliffe, 1984.)

mobile mode of life with a fixed tube may be the best option to adopt, whereas those with more competitive pressure from superior grazers such as snails may better adopt a free-living mode.

Chironomids tend to be competitively inferior to other larger-bodied species such as gastropods and caddis larvae, and may easily be displaced by them if on the same microhabitat. On the other hand, the relatively short generation time of many chironomid species and their capacity to colonize new habitats enable them to utilize ephemeral habitats more efficiently. In a lake-outlet stream in Montana, USA, species of *Eukiefferiella* mainly colonized the upper part of a stone while competitively dominant, territorial caddis *Leucotrichia* occupied the surfaces below the previous August's low water mark (Figure 12.11; McAuliffe, 1984). Thus they partitioned space vertically and the abundance of *Eukiefferiella* larvae on stones was negatively correlated with that of *Leucotrichia* when the latter was experimentally manipulated (Figure 12.12). On a microhabitat scale, *Eukiefferiella* larvae never occurred next to *Leucotrichia* cases containing active larvae and tube-building *Rheotanytarsus* larvae also tended to avoid the territories of *Leucotrichia*.

On a similarly small spatial scale, segregation in space indicative of past/present operation of competitive interactions has been observed among

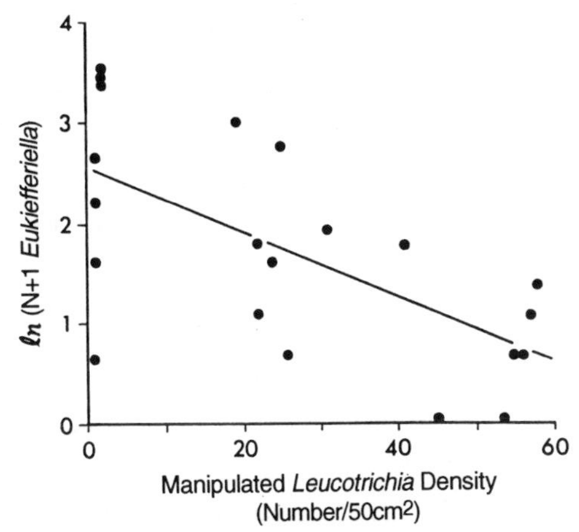

Figure 12.12 Relationship between densities of *Eukiefferiella* and *Leucotrichia* when the latter was experimentally manipulated. (After McAuliffe, 1984.)

parasites/commensals on a host. For example, a tanypod *Ablabesmyia* sp. and water-mites *Unionicola* spp. both live inside the unionid mussels, but were never found between the same pair of demibranchs of a mussel (Roback *et al.*, 1979). Interestingly, the two taxa can apparently share the same side of the mussel foot if *Unionicola* occupies the labial palp whilst *Ablabesmyia* sp. occupies the interbranchial space. Thus, competitive interactions between these taxa, if any, occurs on a very small spatial scale.

In a similar vein, two taxonomically unrelated species, a chironomid *Epoicocladius flavens* and a ciliate protozoan *Scyphidia* sp., share the commensal mode of life on the burrowing mayfly *Ephemera danica* (Tokeshi, 1988). In this case the two species showed clear spatial segregation with respect to the host's body size, with *E. flavens* predominantly infesting mayfly nymphs larger than 8 mm in length and *Scyphidia* sp. infesting those smaller than 8 mm (Figure 12.13). This pattern was consistently observed throughout the year and few *E. danica* hosts were observed with the two commensals attached. On a larger spatial scale, rain-pools on granite exposures in tropical Africa are predictably inhabited by either of the two species of *Chironomus*, *C. imicola* Kieffer and *C. pulcher* Wiedemann (McLachlan, 1988). The former species is capable of dispersing longer distances than the latter after the long dry season in pools of river water. These examples may suggest some form of competitive interactions, but other factors may also be implicated and further investigations are necessary to clarify causal relationships of the patterns observed.

Apart from the negative effects of competition, the presence of certain organisms may be beneficial to others. For example, the density of detritivorous chironomids was significantly higher in the vicinity of bivalve molluscs (*Elliptio complanata* (Lightfoot), *Anodonta cataracta* and *Lampsilis ochracea* Say) in a small reservoir in New Brunswick, Canada (Sephton *et al.*,

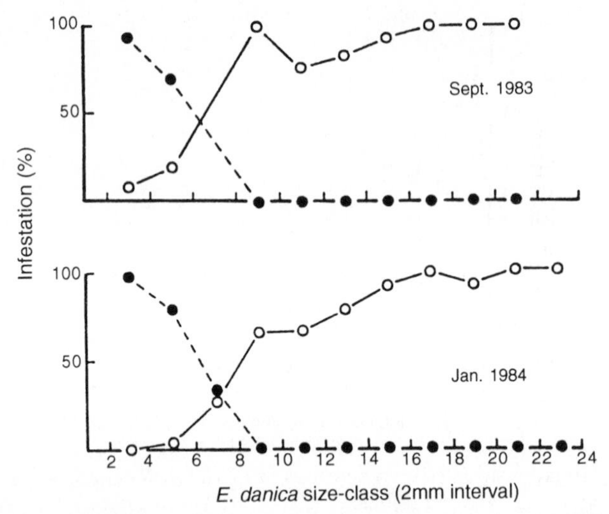

Figure 12.13 Proportions of the mayfly host *Ephemera danica* Müller infected by *Epoicocladius flavens* (Malloch) (open circles) and by *Scyphidia* sp. (closed circles), with respect to different host size classes. (After Tokeshi, 1988.)

1980), due perhaps to faeces and pseudofaeces excreted by bivalves which can serve as nutritive food resource for chironomids. This kind of 'facilitation' may in some cases be as important as competitive interactions in a community.

12.4.2 PATTERNS OF RESOURCE UTILIZATION

The disjunct occurrences of two species are easily observed on a spatio-temporal scale, as has been seen in a number of studies cited in the previous section. This does not, however, necessarily indicate the importance of interspecific competition in a multi-species community context. If the role of competition in community organization is to be understood, patterns of **resource utilization** (which is the hub of competitive interactions) need to be analysed in a more rigorous manner than a mere description, to exclude such possibilities as patterns expected from random processes. Though the importance of such an approach has been recognized in studies of various animal communities (cf. Strong *et al.*, 1984), there are as yet few attempts to analyse a chironomid community from this perspective.

In a small river in eastern England the epiphytic invertebrate community on the spiked water-milfoil *Myriophyllum spicatum* was dominated by the Chironomidae, of which nine species (*Orthocladius oblidens* (Walker) (= *O.* sp. A of Pinder), *O. obumbratus* Johannsen (= *O.* Pe 9 of Langton, 1984), *Thienemanniella majuscula* (Edwards), *Eukiefferiella ilkleyensis* (Edwards), *Tvetenia calvescens* (Edwards), *Rheotanytarsus curtistylus* (Goetghebuer), *Cricotopus annulator* Goetghebuer, *C. bicinctus* (Meigen) and *Rheocricotopus chalybeatus* (Edwards)) accounted for over 95% of total numbers and biomass throughout the year (Tokeshi, 1986a). Consequently, patterns of resource utilization in terms of time, food and space among these nine species were analysed with a set of null models.

Temporal pattern of resource utilization was expressed as population production of each species through a year; production is considered to reflect the level of resource utilization more accurately than either the number of individuals or biomass. Subsequently, overlap in temporal resource utilization was calculated for a total of 36 different pairs of species using nine production curves and mean overlap value was obtained for the whole community. If interspecific competition is important with respect to the temporal patterns of resource utilization in this community, the observed temporal overlap would be significantly smaller than expected by chance. To test this a first null model was constructed by placing nine resource utilization curves randomly on the time axis and a mean overlap value was obtained. This process was replicated 500 times to give a frequency distribution of 500 values of overlap (Figure 12.14a). Because the occurrence of a resource utilization curve on the time axis may be restricted, even without competitive interactions, by the temperature regime prevailing at the study site, a second null model was constructed incorporating the seasonality factor: the peaks of resource utilization curves were assumed to occur randomly but within March–October, avoiding the four winter months (November–February). Comparison of the observed value of temporal overlap with values obtained from these two null models (Figure 12.14) show that the observed value is not significantly smaller than would be expected from either of the models, indicating that there is no competition-induced temporal segregation in this community.

In fact, the observed value was significantly larger than the values obtained in the null models, which is related to the fact that all the chironomid species are heavily dependent upon diatoms as food resource (Chapter 11, Figure 11.4), the availability of which peaks in spring–summer. Indeed, an average of 85.1% of total annual production of chironomid species occurred within 3 months between March and May. Consequently, there was no dietary segregation in this community.

The lack of temporal and dietary segregation naturally led to the analysis of spatial distribution and overlap among species. In this community the leafy, apical section of *Myriophyllum* constitutes a natural habitat unit, within which there is little scope for resource partitioning. Therefore, if interspecific competition is an important organizing force with respect to space utilization, it is surmised that different species would be more or less segregated on different apices of *Myriophyllum*, thereby reducing the extent of spatial overlap and the possibility of outright competition. In the first place, colonization experiments conducted at different times of year revealed that there is no significant difference in capacity to colonize new habitats among chironomid species. On the basis of this knowledge, significance of spatial segregation was tested by constructing two different models of random colonization. In the first model chironomid individuals were assumed to colonize habitat units in a random fashion (Figure 12.15a), while in the

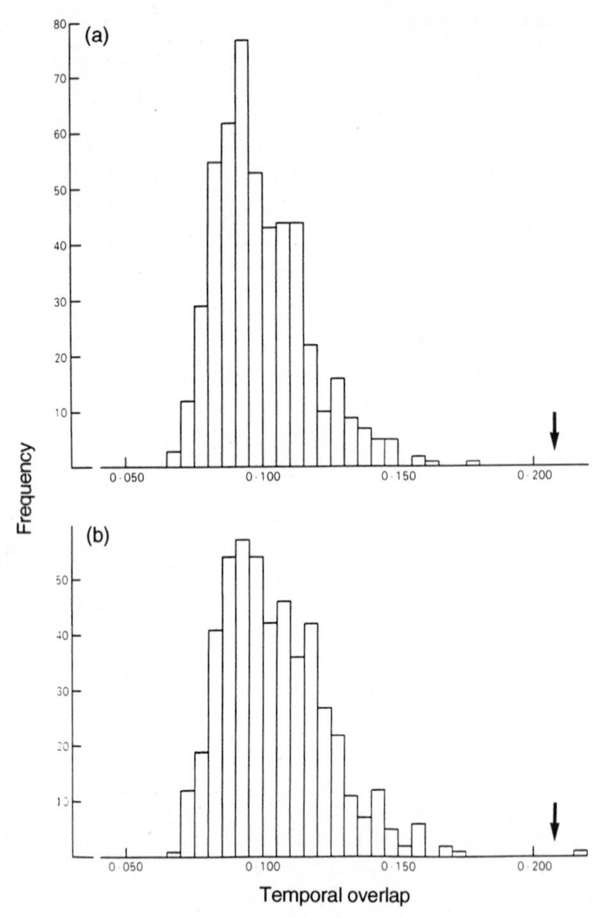

Figure 12.14 Frequency distributions of 500 values of overlap obtained from (a) temporally unrestricted and (b) restricted random resource utilization models, compared with the observed value of temporal overlap (arrow) in an epiphytic chironomid community. (After Tokeshi, 1986a.)

second model groups or 'patches' of individuals were assumed to colonize the habitat randomly (Figure 12.15b). The second model was based on the consideration that colonization is likely to occur as patches of individuals, as a result of batch laying of eggs and/or non-random distribution of larvae in the flowing water. If competition in space is strong enough, observed overlap values will be significantly smaller than would be expected from both of these models. Consequently, 200 replications were made for each model with monthly data.

Results showed that amongst a total of 140 pairs of species observed through the year, 80 pairs had significantly small values of spatial overlap in comparison with values derived from the 'individual'-colonization model.

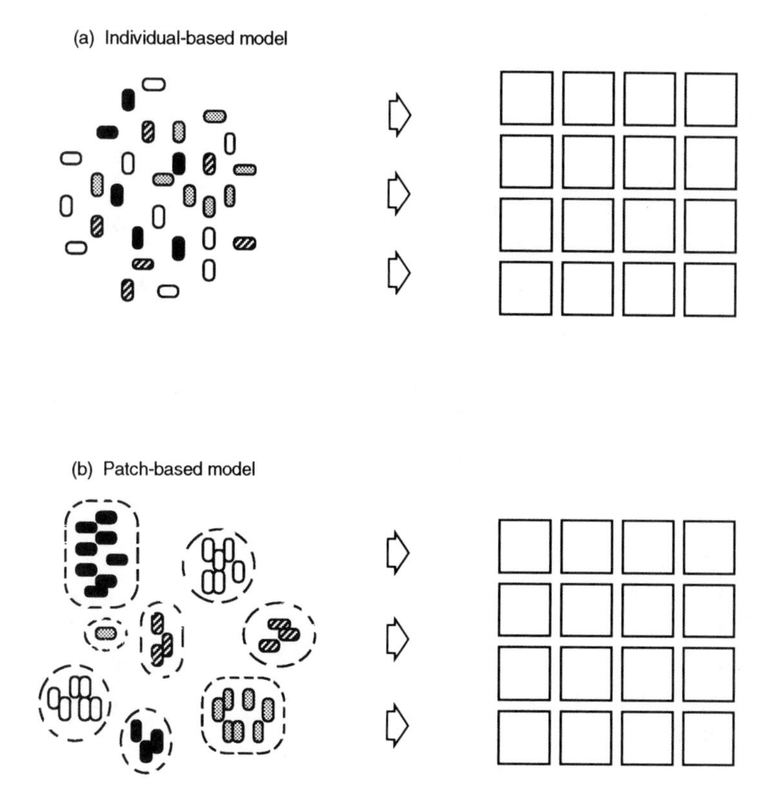

Figure 12.15 Two models of habitat colonization by chironomid larvae, (a) 'individual' random colonization, and (b) 'patch' random colonization. Different symbols represent different species.

This suggests that over half the pairs were segregated spatially in terms of individual occurrences. However, only four out of these 80 pairs had a value significantly smaller than would be expected from the 'patch'-colonization model, indicating that the vast majority of cases of reduced spatial overlap can be attributed to random colonization by patches of individuals (Figure 12.15b). In addition, the fact that 60 pairs of species (42.9% of the total examined) did not show any noticeable spatial segregation suggests that competition with respect to space utilization, if any, is very weak in this epiphytic chironomid community.

The general conclusion to emerge from this analysis is that interspecific competition is not important in the community organization of chironomids on *Myriophyllum* and that 'random patch formation' and the absence of strong competitive interactions enhance the coexistence of species. A similar study was carried out on the benthic chironomid community in the River Danube (Schmid, 1992a) and results were in close agreement with the study on epiphytic species (Figure 12.16), but with a higher proportion of species pairs (28%) showing spatial segregation on the basis of patch randomization.

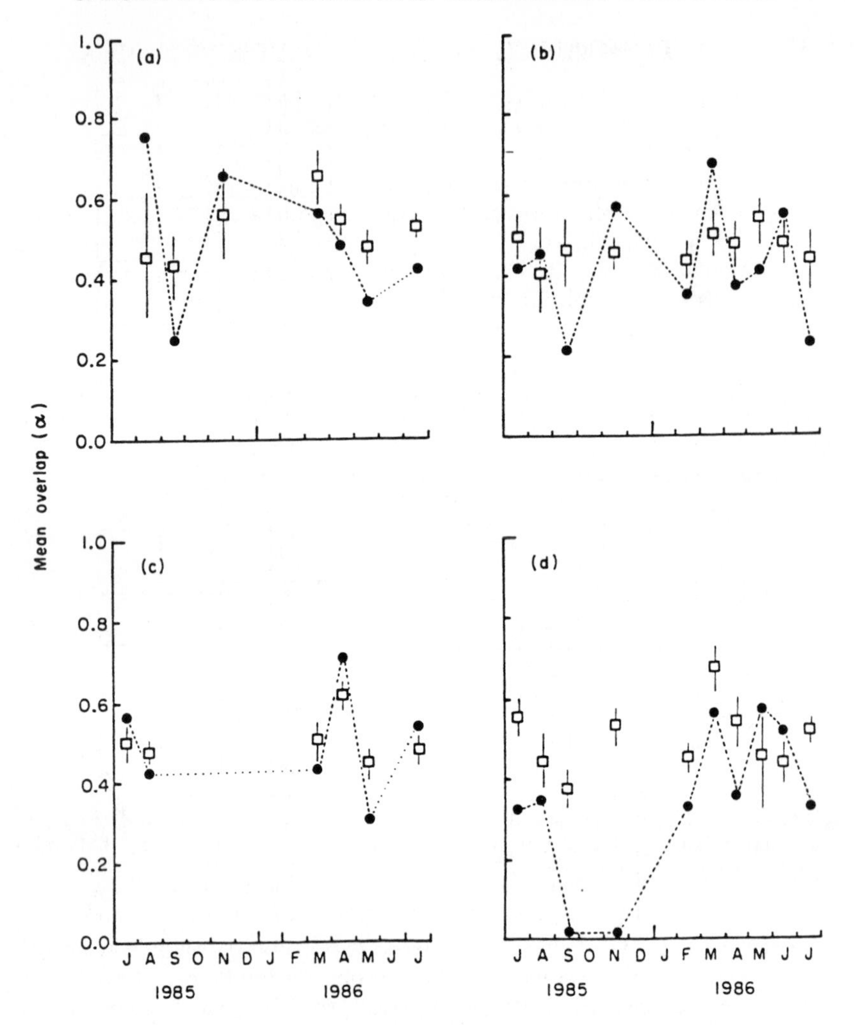

Figure 12.16 Variation in mean spatial overlap in benthic chironomid communities of the River Danube. Closed circles are observed values and squares represent mean values (with 95% CL) from a patch randomization model. (a) 0–5 m depth; (b) 6–10 m; (c) 11–15 m; (d) 16–20 m. (After Schmid, 1992b.)

Thus, random patch formation may not be an uncommon phenomenon in chironomid communities (cf. Schmid, 1993b). Furthermore, predation does not appear to be an important factor in this community, with predators being fairly scarce throughout the year (Tokeshi, 1986a, 1993a). Attention has thus been drawn to the importance of stochastic processes and patchy habitat structure (Tokeshi, 1992) which may have a significant bearing upon these and other communities (cf. 'stochastic patch dynamics model', Tokeshi, 1994).

12.5 DISTURBANCE AND HABITAT STRUCTURE

12.5.1 DISTURBANCE AND CHIRONOMID COMMUNITIES

Disturbance is a common factor in ecological communities and takes on a variety of forms, both biotic and abiotic, natural and artificial. Physical disturbances are particularly prevalent in running-water systems and lotic chironomid communities may be strongly influenced by them (e.g. Schmid, 1992a,b). Drift and recolonization are fundamental aspects of chironomid community organization in running waters (Tokeshi and Pinder, 1986; Tokeshi and Townsend, 1987).

In a 6-week field experiment in a fourth-order stream in North Carolina, disturbance events were simulated by artificially tumbling cobbles within cages at different frequencies and the recovery of macroinvertebrates was monitored (Reice, 1985). Most taxa, including the Chironomidae (which, together with the Simulidae, accounted for more than half of all invertebrates), significantly decreased in number following the disturbance, while no taxon increased with increasing disturbance, but recovery to near-normal population densities was achieved in about 4 weeks. Similarly, Tokeshi and Townsend (1987) observed for epiphytic chironomid species that time to reach 95% of the mean natural density after complete denudation of a *Myriophyllum* habitat was less than 2 weeks (5.6–13.2 days).

Disturbance may either ameliorate or accentuate the effects of predation, depending on how and whether predators and prey are influenced by disturbance events. In the course of 10-day experiments with a stonefly predator *Kogotus nonus*, fine sediment was poured into experimental cages at regular intervals (Walde, 1986). In two of three experiments, high levels of sediment deposition eliminated the predation effect on Orthocladiinae larvae, though their density was nevertheless reduced by the disturbance itself. In contrast, in one experiment high level of sediment deposition led to a significant negative effect of predation on chironomid density, presumably because the fine sediment happened to fill up the interstitial space where chironomid larvae can take refuge from predation.

Disturbance in terms of chemical properties of water, particularly pH, can occur as natural processes, though events of human origin are fast increasing in number and extent. Experimental reduction of pH over 8 days in a lake outflow stream in Ontario, Canada, resulted in increased abundances of species belonging to the subfamily Chironominae, while the densities of Orthocladiinae (*Thienemanniella*, *Eukiefferiella*, *Parametriocnemus* and *Psectrocladius*) were reduced (Hall, 1990). Thus, this form of disturbance can modify the species composition of chironomid communities.

12.5.2 HABITAT STRUCTURE AND CHIRONOMID COMMUNITIES

In general, complex habitats are associated with a larger number of species than simple habitats, though the operational definition of habitat

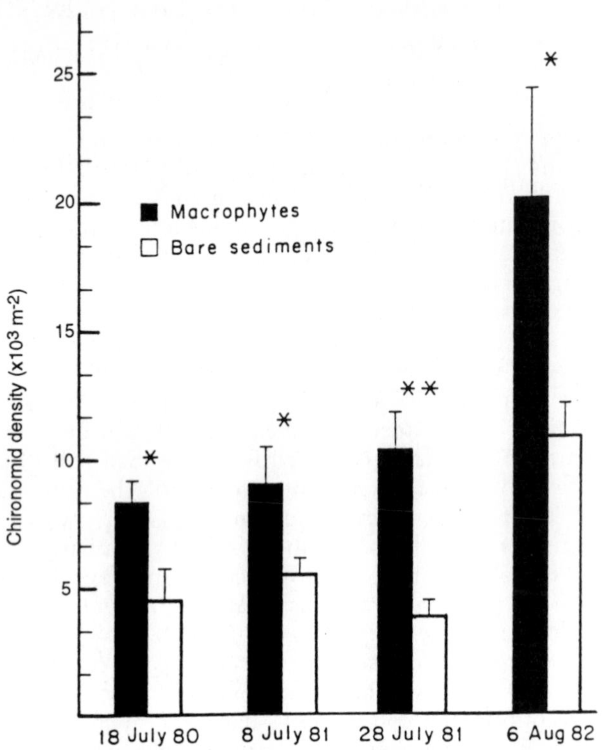

Figure 12.17 Density of chironomid larvae in the macrophyte bed and in bare sediments, as affected by sculpin predation in an arctic lake. ⋆$P<0.05$; ⋆⋆$P<0.01$. (After Hershey, 1985a.)

'complexity' is fraught with problems. In a similar context, finely divided habitats can support more (potentially competing) species than coarsely divided habitats (Tokeshi, 1994). This may at least partially account for a relatively large number of chironomid species found on some submerged macrophytes (i.e. divided, structurally complex habitat) in comparison with bottom sediments (i.e. structurally less complex habitat).

Apart from mediating competitive interactions among species with similar resource requirements, finely divided/complex habitat can seriously curtail predators' foraging efficiency, thereby allowing the maintenance of higher total density of prey species. For example, the sculpin *Cottus cognatus*, which was a major predator of chironomid larvae in an arctic lake, was less efficient in preying on chironomids in the macrophyte habitat than in the bare sediments (Figure 12.17; Hershey, 1985a).

Similarly, predation by the bluegill sunfish *Lepomis macrochirus* had no significant effect on the densities of chironomids and other invertebrates in the macrophyte habitat, while predation effects were evident in the macrophyte-free habitat (Gilinsky, 1984). In a similar but slightly different context, the effect of predation by a chironomid *Allotrissocladius* sp. on another chironomid *Paraborniella tonnoiri* in temporary rain-pools of Western Australia seemed to be influenced by habitat characteristics (Jones, 1974). In deep, vegetated pools ample food supply allows the latter to grow rapidly to reach the third or fourth instars, thus effectively escaping gape-limited predation which falls mainly on the first and possibly second instars. In contrast, *P. tonnoiri* individuals in shallower, non-vegetated pools, where food is less abundant, grow more slowly and are thus exposed to heavy predation pressure.

Where chironomid faunas are surveyed and measurements are taken of different variables of habitat, it is possible to use a number of multivariate analyses to discover possible association and grouping of chironomid species with respect to habitat characteristics (e.g. Rae, 1985). This is considered a useful approach particularly at an early stage of investigation. The elucidation of causal mechanisms, however, requires more detailed analyses and/or experimental approaches. Unfortunately, there are very few studies on the Chironomidae which integrate different types of analyses and approaches.

12.6 PATTERNS OF COMMUNITY ASSEMBLY

12.6.1 ANALYSIS OF FAUNAL ASSEMBLY

Faunal investigations of the Chironomidae have often led to typological classification of chironomid communities associated with different habitats or environmental gradients, either within a single water-body (e.g. Thorp and Chesser, 1983) or among different water-bodies (e.g. Brundin, 1956; Sæther, 1975; Chapter 15). It is generally possible to recognize some species typically associated with a particular habitat type, thus conveniently serving as 'indicator' species. Though undoubtedly useful as a first step towards recognizing faunal characteristics, this approach remains a purely descriptive exercise without physiological investigations and/or detailed analyses of community structure. In this connection, use of multivariate statistical methods such as PCA, DCA, TWINSPAN and DECORANA is helpful in adding objectivity to faunal analyses, especially where data involve a large number of taxa as is typically the case with the Chironomidae. For example, chironomid communities in the littoral of a reservoir in South Carolina were examined in detail using PCA, revealing temperature and plant biomass/diversity as most important factors correlated with chironomid density and diversity (Thorp and Chesser, 1983). In the River Danube, few clear species-groups were identified in an analysis of chironomid faunal association using the unweighted average linkage method (UPGMA) with the percentage similarity index (Schmid, 1992b). Similarly, clusters of species were not evident in an analysis of stream-dwelling chironomid assemblages in Italy using PCA, DCA and non-metric multidimensional scaling (MDSCAL) (Rossaro, 1992).

Where feasible, it is useful to pose a specific testable hypothesis about faunal assembly considering possible biological processes involved. One such hypothesis concerns the relationship between number of species and number of genera in a local fauna (cf. Elton, 1946; Williams, 1964; Grant, 1966). Because congeneric species are thought to have more similar resource requirements than do species of different genera, competition among congeneric species would be more severe and likely to result in a faunal pattern where species are widely spread out among different genera (Tokeshi, 1991b). In other words, given the total number of species, more genera are represented in a local community than would be expected by chance from a larger geographical assemblage of species. Tokeshi (1991b) tested this hypothesis using chironomid faunal data from three rivers in southern England (Pinder, 1974; Wilson, 1977; Drake, 1982). When the number of genera represented at each site was compared with numbers expected as a random sample from a larger geographical pool of species, i.e. Britain as a whole (Pinder, 1978; Langton, 1984), there was no evidence of generic spread (i.e. more genera than expected by chance) in the local faunas. Thus, competition does not seem to be important in the generic structures of local chironomid faunas at the three sites. Apart from purely ecological reasons, the tenuous link between the taxonomic status/uncertainty of different genera within the Chironomidae and the ecology of species remains a serious problem in this kind of analysis (Tokeshi, 1991b), pointing to the importance of detailed ecological knowledge of different species.

In addition to the analysis of generic spread, generic association among the three sites (c. 70–100 km apart from one another) was also analysed for the Chironomidae as a whole and three major subfamilies separately. Generic association among sites was stongest in the Tanypodinae, intermediate in the Orthocladiinae and weakest in the Chironominae, which may suggest inter-subfamily differences in dispersal ability and other ecological characteristics.

12.6.2 FAUNAL COMPOSITION AND SPECIES ABUNDANCE PATTERNS

Chironomid faunal investigations often lead to the examination of taxonomic composition, particularly the distribution of species among different subfamilies (e.g. Coffman, 1973; Drake, 1982; Lenat, 1987). This exercise, however, has virtually been limited to lotic studies, presumably because researchers perceive that lentic habitats are invariably dominated by the subfamily Chironominae, whereas lotic habitats may present a more varied picture and hence merit closer analysis. Nevertheless, most lotic studies in the temperate northern hemisphere have reported that species of Orthocladiinae dominate the fauna, followed by the Chironominae, while other subfamilies represent small proportions. Thus, this is considered a general pattern of taxonomic composition in temperate lotic habitats, though further observations of various types of lotic habitats may generate different patterns.

Where an entire chironomid assemblage is examined quantitatively, it is generally the case that only a small number of species account for a large proportion of total number of individuals or total biomass. For example, Laville (1972) noted that in a high-altitude Pyrenean lake, three species

(*Psectrocladius sordidellus* (Zetterstedt), *Chironomus commutatus* Keyl and *Zavrelimyia melanura* (Meigen)) accounted for 77% of total number and over 84% of total biomass of the chironomid community consisting of 32 species. In an epiphytic chironomid community six out of 19 species accounted for more than 95% of total numbers and biomass (Tokeshi, 1986a).

An analysis of faunal composition can naturally be extended to the analysis of '**species abundance patterns**', one of the fundamental aspects of ecological communities which requires both theoretical and empirical approaches (cf. Tokeshi, 1993c). Despite the importance of the subject and the fact that the earliest model of species abundance pattern, i.e. the Geometric Series, was proposed with reference to the benthic fauna of lakes including the Chironomidae (Motomura, 1932), there have since been few studies on species abundance patterns involving the Chironomidae.

Data on emerging adult chironomids from Lake Erken, southern Sweden (Sandberg, 1969), did not appear to conform to the Broken Stick model of MacArthur (1957). However, there are technical problems of testing the fit of this model, which have often been ignored in empirical studies (Tokeshi, 1990b, 1993c).

With a large, heterogeneous assemblage of species a Log-normal pattern of species abundance is generally expected (May, 1975; Tokeshi, 1993c). Except for the abundance class representing very rare species, species abundance data including a total of 112 species derived from emergence trap sampling along a 150 m stretch of a third-order river in Alberta, Canada, conformed to the Log-normal model (Boerger, 1981); the frequency of rare species in the sample was higher than expected from the model. Similarly, Schmid (1992a) found that the Log-normal fitted some of the benthic chironomid data from the River Danube (Figure 12.18). Nevertheless, statistically-oriented species abundance models such as the Log Series and the Log-normal have practical difficulties in conferring biologically meaningful interpretation to data (cf. Tokeshi, 1993c). In this respect niche-oriented models may be more versatile, especially with regard to relatively small assemblages of closely related species which constitute ecologically more tangible 'communities' than heterogeneous collections of species from larger areas. Tokeshi (1990a) compared data on a chironomid community on *Myriophyllum* mentioned earlier with seven niche-oriented models (the Geometric Series, Dominance Preemption, Random Fraction, MacArthur Fraction, Dominance Decay, Random Assortment and Composite model). For both numerical (i.e. numbers of individuals) and biomass data, the observed species abundance pattern conformed with the Random Assortment model (Figure 12.19), suggesting that niches of different species in this community are largely unrelated. This agrees well with the results of the analyses of resource utilization patterns (see previous section), pointing to the stochastically dynamic nature of this community. In the light of this, apart from the effect of 'statistical' heterogeneity, the fit of Log-normal to data such as Boerger (1981) and Schmid (1992a) may also be interpreted as indicating the importance of stochasticity in these assemblages.

In connection with species abundance patterns, one aspect which has received much attention is the relationship between abundances and body

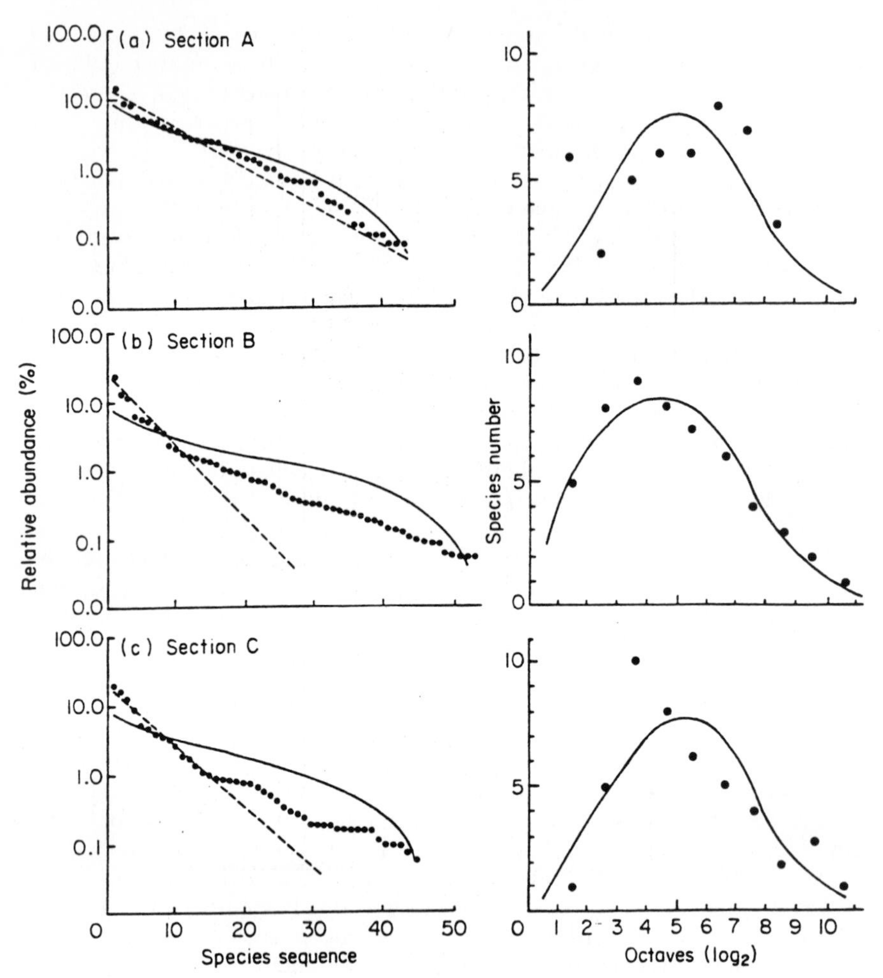

Figure 12.18 Species abundance patterns in benthic chironomid communities of River Danube. Diagrams on the left were fitted with MacArthur's Broken Stick model (continuous line) and the Geometric Series model (broken line), those on the left show a possible fit of the Log-normal model. (After Schmid, 1992a.)

sizes of species. This 'density/body-size allometry' refers to a steady decrease in population density against increasing body size among species, such that total resource use or energy consumption is more or less the same among species of different sizes in a community. If such a relationship exists, it indicates that abundance is a function of body size while species maintain an equitable share of resources within a community. For the epiphytic chironomid community it has clearly been demonstrated that this is not the case (Tokeshi, 1990b), suggesting that abundances of different chironomid species are not controlled by processes associated with body size as expected of an equilibrium community.

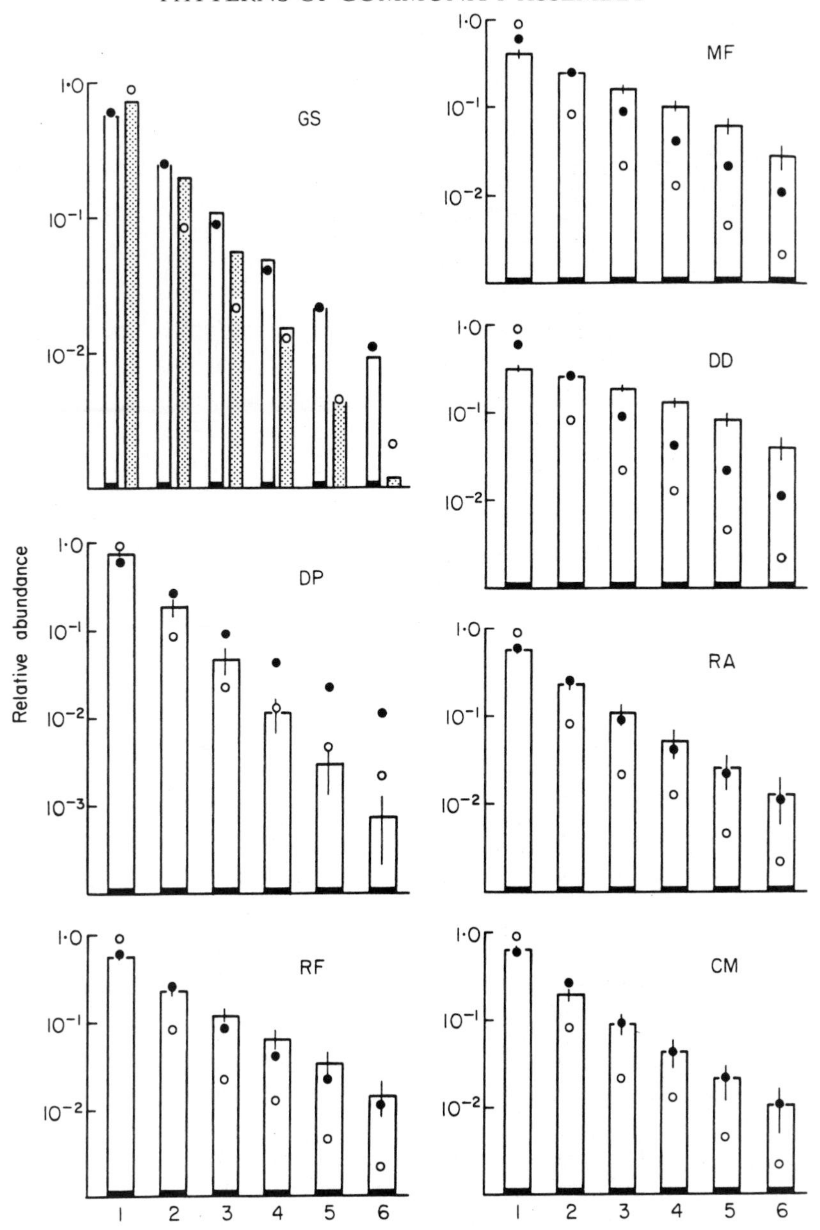

Figure 12.19 Species abundance patterns in a chironomid community on *Myriophyllum*. Two different data sets (spring–summer data = closed circles; autumn–winter data = open circles) were compared with expected values (histograms) from the Geometric Series (GS) model, the Dominance Preemption (DP) model, the Random Fraction (RF) model, the MacArthur Fraction (MF) model, the Dominance Decay (DD) model, the Random Assortment (RA) model and the Composite (CM) model. (After Tokeshi, 1990a.)

Part Three

Interactions with Humans

13

Nuisance, economic impact and possibilities for control

A. Ali

13.1 BACKGROUND

The emergence and swarming of adult chironomids in massive numbers has been observed near certain aquatic habitats for decades. In 1831, John James Audubon noted during his travels on the St Johns River, Florida, USA, that the 'blind mosquitoes' (chironomid midges) were so abundant that they covered almost every object on the boat and even extinguished candles in the cabins. Tales of hordes of midges extinguishing bonfires, signal lights etc. near lakes and rivers were also common in Florida at the turn of the century (Patterson, 1965). Elsewhere in the USA, the midge *Chironomus* [as *Tendipes*] *plumosus* (Linnaeus) has plagued Wisconsin residents living by Lake Winnebago for at least a century (Hilsenhoff, 1962).

More recently, annoyance of the public by adult midges has increased significantly and has been reported from many parts of the world (Ali, 1991a). This is primarily due to the increasing tendency of humans to locate their dwellings closer to lakes and rivers (midge habitats) to enjoy water-related recreations; a continuous increase of new midge-producing habitats, such as holding ponds and effluent discharge channels at sewage treatment facilities, ponds on golf courses, and residential-recreational lakes; and eutrophication of inland waters due to human population growth and rapid industrialization. Generally, water-bodies in urban and suburban areas exposed to intensive human use (residential, recreational, agricultural etc.) and receiving point and non-point discharges are becoming increasingly eutrophic. As a consequence of these stresses, plant and animal diversity in these habitats is gradually decreasing. This leads to an increase in populations of more pollution-tolerant organisms, such as some species of Chironomidae which constitute a major component of the macroinvertebrate faunal community. Larval densities may reach tens of thousands per square metre of the habitat.

The Chironomidae: Biology and ecology of non-biting midges
Edited by P.D. Armitage, P.S. Cranston and L.C.V. Pinder
Published in 1995 by Chapman & Hall. ISBN 0 412 45260 X

The suspected trend of 'global warming' may also have some bearing on the increase in midge populations. In warmer waters, the productivity of phytoplankton and benthic algae, which serve as food for the larvae of many chironomid species, rises due to increased utilization of solar energy in the process of photosynthesis, thus creating favourable conditions for chironomid development.

Eutrophication generally allows high chironomid production and certain species are biological indicators of pollution. However, under heavily polluted conditions chironomid survival may be very limited. For example, until recent decades, chironomid midge emergence from natural and artificial habitats in Japan was unnoticed because of its low and tolerable levels. In the past two decades, however, the intensity and frequency of midge emergence from many such habitats in different parts of Japan have grown to intolerable levels (Inoue and Mihara, 1973; Tabaru, 1975). The main reason for this change is the relative improvement of water quality after enforcement of the Water Pollution Control Act in 1972. Prior to 1972, most rivers and other drainage systems in urban areas were so heavily polluted that they supported very little animal life (Anon., 1975; Inoue, 1976). Subsequently, the improved quality of water has permitted the rapid colonization and breeding of large populations of chironomids (Tabaru, 1986). This is particularly true amid urbanized and industrial growth areas where wastewater treatment plants generate large discharges.

Obviously human activities variously interfere with and disturb the ecological balance in aquatic environments. In many situations, this imposed imbalance results in population explosions of Chironomidae. The adults emerge in phenomenal numbers and cause a variety of nuisance, economic and medical problems for human populations within the dispersal range of these flies.

13.2 NUISANCE AND ECONOMIC PROBLEMS

Chironomid swarms often limit human activity outdoors because the adults can be inhaled or fly into the mouth, eyes, or ears; adult swarms may cause asphyxia in cattle (Grodhaus, 1963). Under hot weather conditions, midges congregate in shady cool areas and deposit meconium or release egg-masses which stain surfaces occupied by the resting adults. These surfaces frequently require washing or repainting. Midges soil automobiles and cover headlights and windshields, and the swarms create risky driving conditions which may lead to traffic accidents (Figure 13.1). Chironomid swarms are hazardous to passengers and crews on trains, buses, cargo vessels and boats in Venice, Italy (Ali *et al.*, 1992). The attraction of adult midges to light causes great discomfort in residential areas. The adults swarm around lighted outdoor electrical fixtures and other objects which serve as swarm markers; they also enter homes where possible – small species in particular can pass through standard screens on doors and windows. Indoors, adult midges stain laundry, deface walls, ceilings, curtains and other furnishings, contaminate food and create distressful conditions for the residents.

Adult midges may cause a considerable economic loss to the hotel and

Figure 13.1 Swarming adult midges causing traffic hazard in Japan. (With permission of *The Sankei Shimbun*.)

tourism industry in many situations worldwide. For example, an economic impact study of midges was undertaken in 1977 by the Greater Sanford Chamber of Commerce, Sanford, Florida, USA (Anon., 1977). This study revealed that midges emerging from Lake Monroe (a part of the St Johns River facing the city of Sanford, Seminole County, Florida) and other nearby habitats inflict a business loss of US$3–4 million annually on Sanford, with just one lakefront motel, the Holiday Inn, spending c. US$50 000 each year on property maintenance and attempts to control the pest. The same study indicated that at least 10 counties in Florida were affected by chironomid problems. Midges can also be a problem for paint, pharmaceutical and food processing industries where hordes of adults may contaminate the final products. There is a considerable loss of working time because midge swarms disrupt night-shift operations through work impairment and/or stoppage of operation.

Accumulations of dead adult midges and the unsightly spider webs spun around resting adults require the frequent cleaning and washing of properties. Adult midges clog air-conditioning wall units and automobile radiators. Body fragments of some chironomid species stick to car paint, causing damage. Dead midges have an offensive odour, like rotting fish, that persists in damp weather for several days after removal of the adults. At times, the dead adults accumulate on roads in such quantities that they make the roads slippery and dangerous for any traffic. Such conditions often prevail in multi-storeyed car-parking lots along waterfronts in Venice, Italy. At Marco Polo Airport near Venice, there is great concern regarding the possibility of airplanes skidding over massive accumulations of dead midges on the runways. Entry of adult chironomids into delicate equipment mounted on airplanes poses danger and additional economic loss to the aviation industry (Barbato *et al*., 1990). The equivalent of US$1 million is spent annually on chironomid control in the Venice area (Ali *et al*., 1992).

Adult midges are not known to be vectors of any disease organism, although when emerging from some polluted habitats they may transport bacteria (Steinhaus and Brinley, 1957) and organic insecticides (Larsson, 1984) to terrestrial environments. They are, however, associated with human allergic reactions including conjunctivitis, dermatitis, rhinitis and asthma (Chapter 14).

The larvae of chironomids also have some notoriety as pests. There are numerous reports of infestations of midge larvae in drinking-water supply systems in various parts of the world (Flentje, 1945b; Williams, 1974; Langton *et al*., 1988). Several midge species are agricultural pests; larvae have been observed to damage rice seedlings and plants (Risbec, 1951; Hely, 1958; Darby, 1962; Cocchi, 1966; Jones, 1968; Treverrow, 1985; Way and Wallace, 1989; Ferrarese, 1993), indoor horticultural crops (Cranston, 1987), roots of Japanese horseradish (Yokogi and Ueno, 1971) and floating leaves of lotus (Tokunaga and Kuroda, 1936).

13.3 NUISANCE SPECIES AND THEIR HABITATS

Of more than 4000 described species of chironomids worldwide (Sublette and Sublette, 1988), nearly 100 species in the subfamilies Orthocladiinae, Tanypodinae and Chironominae have been reported to emerge in large numbers. Pestiferous or allergenic species are listed in Table 13.1. The vast majority of pestiferous species are listed but species of minor importance or those considered to be potentially pestiferous are excluded. In some instances, where a nuisance species has been reported by several authors and occupies more than one habitat type in different geographical locations, only the latest paper is cited in Table 13.1; all types of habitats reported for the species may not be shown.

Pest swarms of midges have been reported to emanate from a great variety of natural and artificial freshwater lentic or lotic habitats as well as some saltwater biotopes. These habitats range in surface area from <1 ha to several thousand hectares and are distributed throughout the world. Freshwater habitats include natural ponds and ditches, sewage filter beds, sewage oxidation ponds, sludge lagoons and other wastewater holding and processing enclosures at treatment plants, aquaculture ponds, eel culturing ponds, ponds on golf courses, phosphate mining pits, reservoirs, swamps, natural and artificial lakes, water-spreading basins for aquifer recharge, drinking-water supply systems, irrigated agriculture (particularly rice paddies), canals, effluent discharge channels, flood control conduits, storm drains and rivers. The saltwater habitats include milkfish ponds, tidal coves, saline lakes, coastal lagoons, intercoastal waterways and brackish water areas and impoundments.

Table 13.1 Species of Chironomidae reported to emerge in massive swarms from a variety of habitats in various countries of the world

Taxon	Habitat(s)	Country	Reference
Tanypodinae			
Conchapelopia cygnus (Kieffer)	Rivers	Sudan	Cranston *et al.*, 1983
Procladius choreus (Meigen)	Reservoirs, rivers, lakes	UK	Cranston, 1988
P. freemani Sublette	Lakes	USA	Ali *et al.*, 1978
P. noctivagus (Kieffer)	Rivers	Sudan	Cranston *et al.*, 1983
P. sublettei Roback	Spreading basins, lakes	USA	Ali *et al.*, 1985a
Tanypus grodhausi Sublette	Sewage ponds	USA	Grodhaus, 1967
T. neopunctipennis Sublette	Impounded salt marshes	USA	Fellton, 1940
T. punctipennis Meigen	Rivers, lakes	UK	Cranston, 1988
T. stellatus Coq.	Lakes	Nicaragua	Palomäki, 1987
Orthocladiinae			
Cricotopus bicinctus Meigen	Polluted rivers	Japan	Tabaru *et al.*, 1987
C. sylvestris (Fabricius)	Storm drains	USA	Ali *et al.*, 1976
Metriocnemus obscuripes (Holmgren) (as *hygropetricus*)	Sewage ponds	UK	Cranston, 1988
Tokunagayusurika akamusi (Tokunaga)	Lakes, ponds	Japan	Tabaru *et al.*, 1987
Chironominae			
Tanytarsini			
Cladotanytarsus lewisi (Freeman)	Rivers	Sudan	Cranston *et al.*, 1981
Cladotanytarsus spp.	Reservoirs	Denmark	Lindegaard and Jónsson, 1987
Micropsectra atrofasciata Kieffer	Drains, canals	UK	Cranston, 1988
Parantanytarsus grimmii (Schneider)	Drinking water supplies	UK	Langton *et al.*, 1988
Tanytarsus barbitarsis Freeman	Saline lakes	Australia	Kokkinn and Williams, 1989
T. gracilentus (Holmgren)	Reservoirs	UK	Cranston, 1988
T. semibarbitarsus Glover	Ponds	Australia	Cranston, 1989
T. sylvaticus (van der Wulp)	Reservoirs	UK	Cranston, 1988
Tanytarsus spp.	Saline water	Japan	Tabaru *et al.*, 1987
Tanytarsus spp.	Lakes, rivers, ponds	USA	Ali, 1980b
Chironomini			
Apedilum elachistus Townes	Lakes	USA	Rodcharoen *et al.*, 1991
A. subcinctum Townes	Storm drains, lakes	USA	Ali *et al.*, 1976
Baeotendipes ovazzai (Freeman)	Saltwater impoundments	Saudi Arabia	(J. Burton, *pers. comm.*)

Table 13.1 continued

Taxon	Habitat(s)	Country	Reference
Chironomus alternans Walker group	Rivers	Australia	Cranston, 1989
C. annularis (Degeer)	Rivers	UK	Cranston, 1988
C. anthracinus Zetterstedt	Reservoirs, canals, lakes	UK	Cranston, 1988
C. atrella (Townes)	Tidal coves	USA	Hitchcock and Anderson, 1968
C. balatonicus Devai, Scholl and Wülker	Lakes	Hungary	Dévai, 1988
C. circumdatus (Kieffer)	Rivers, ponds, ditches	Japan	Tabaru *et al.*, 1987
C. crassicaudatus Malloch	Lakes, reservoirs, ponds	USA	Beck and Beck, 1969
C. decorus Johannsen group	Lakes, reservoirs, ponds	USA	Ali, 1980b
C. dorsalis Meigen	Reservoirs	Japan	Kawai and Konishi, 1988
C. frommeri Atchley and Martin	Reservoirs, lakes	USA	Grodhaus, 1963
C. kiiensis Tokunaga	Eel ponds	Japan	Yasuno *et al.*, 1982
C. maturus Johannsen	Lakes, ponds, ditches,	USA	Sublette and Sublette, 1974
C. nipponensis Tokunaga	Lakes	Japan	Tabaru *et al.*, 1987
C. plumosus (L.)	Lakes, reservoirs, rivers	Denmark	Lindegaard and Jónsson, 1987
C. riparius Meigen	Rivers, sewage channels	UK	Edwards *et al.*, 1964
C. salinarius Kieffer	Saltwater lagoons	France	Sinegre *et al.*, 1990
C. samoensis Edwards	Lakes, reservoirs	Japan	Kawai and Konishi, 1988
C. stigmaterus Say	Lakes, sewage ponds	USA	Beck and Beck, 1969
C. utahensis Malloch	Lakes	USA	Ali, 1980b
C. yoshimatsui Martin and Sublette	Polluted rivers, ponds	Japan	Tabaru *et al.*, 1987
Cryptochironomus fulvus (Joh.)	Lakes	USA	Ali and Fowler, 1983
C. redekei (Kruseman)	Reservoirs	Denmark	Lindegaard and Jónsson, 1987
Cryptotendipes sp.	Impoundments	Netherlands	Beattie, 1981
Dicrotendipes californicus (Johannsen)	Storm drains, spreading basins	USA	Ali, 1980b
D. fusconotatus (Kieffer)	Rivers	Sudan	Cranston *et al.*, 1983
D. lobiger (Kieffer)	Ponds	UK	Cranston, 1988
D. tenuiforceps (Kieffer)	Sewage ponds	India	Hafiz, 1939

Table 13.1 continued

Taxon	Habitat(s)	Country	Reference
Glyptotendipes barbipes (Staeger)	Sewage ponds	USA	Bickley and Ludlam, 1968
G. glaucus (Meigen)	Polluted rivers, ponds	Japan	Tabaru *et al.*, 1987
G. lobiferus (Say)	Lakes	USA	Beck and Beck, 1969
G. pallens (Meigen)	Rivers	UK	Cranston, 1988
G. paripes (Edwards)	Lakes, ponds	USA	Beck and Beck, 1969
G. tokunagai Sasa	Eel ponds, sewage channels	Japan	Sasa, 1979
Goeldichironomus carus (Townes)	Lakes, rivers	USA	Beck and Beck, 1969
G. holoprasinus (Goeldi)	Lakes, rivers	USA	Beck and Beck, 1969
Kiefferulus longilobus Kieffer	Saline waters	Orient	Tang and Chen, 1959
K. longilobus	Saline waters	Australia	Cranston, *et al.*, 1990
K. tainanus (Kieffer)	Eel ponds	Japan	Tabaru *et al.*, 1987
Parachironomus arcuatus (Goet.)	Rivers, ponds	UK	Cranston, 1988
P. frequens (Johannsen)	Rivers	UK	Cranston, 1988
Polypedilum digitifer Townes	Reservoirs	USA	Grodhaus, 1963
P. halterale (Coq.)	Lakes	USA	Ali and Fowler, 1983
P. kyotoense (Tokunaga)	Irrigated agriculture	Japan	Sasa, 1985
P. nubeculosum (Meigen)	Canals, rivers, ponds	UK	Cranston, 1988
P. nubifer (Skuse)	Eel ponds	Japan	Yasuno *et al.*, 1982
P. scalaenum Schrank group	Impoundments	Netherlands	Beattie, 1981

Chironomid species most commonly encountered in heavy midge-producing urban and suburban habitats belong to the genera *Procladius, Tanypus* (Tanypodinae), *Cricotopus, Tokunagayusurika* (Orthocladiinae), *Chironomus, Baeotendipes, Dicrotendipes, Glyptotendipes, Goeldichironomus, Kiefferulus, Apedilum, Polypedilum, Cladotanytarsus,* and *Tanytarsus* (Chironominae). In the USA, *Chironomus crassicaudatus* Malloch, *C. decorus* Johannsen, *C. plumosus, C. riparius* Meigen, *C. utahensis* Malloch and *Glyptotendipes paripes* Edwards are the major pest species. In the UK, *Polypedilum nubeculosum* (Meigen), *G. paripes, Tanypus punctipennis* Meigen and several species of *Procladius* and *Chironomus* are commonly a cause of nuisance complaints. In Australia, *Polypedilum nubifer* (Skuse), *Kiefferulus longilobus* Kieffer, *Tanytarsus barbitarsis* Freeman, *T. semibarbitarsus* Glover and several species of *Chironomus* emerge at nuisance levels. In Japan, *C. yoshimatsui* Martin and Sublette, *C. plumosus,* and *Tokunagayusurika akamusi* (Tokunaga) are the principal pests. In Italy and France, *C. salinarius* Kieffer and *C. plumosus* emerge at nuisance levels. In Hungary, *C. balatonicus* Dévai, Wülker and Scholl and *C. plumosus* predominate. In Sudan, *Cladotanytarsus lewisi* (Freeman) is a serious pest and a cause of allergies. In Tunisia and Saudi Arabia, species of *Baeotendipes* pose pest problems. It is likely that several other species of midge emerge at nuisance levels in many countries, but are presently not reported in the literature.

13.4 THE POSSIBILITIES FOR CONTROL

The nuisance and economic problems posed by chironomids necessitate their control. In the past 50 to 60 years, chemicals, biological control agents and physical and cultural control techniques have been employed in various problem situations. Of these, chemical control has been the most practised. The effectiveness of some predators for the biological control of midges has been studied. Many parasites and pathogens of midges have been identified but their mass culturing and field inoculation for the practical control of midges has not been attempted. Commercially produced bacteria, *Bacillus thuringiensis* serovar *israelensis (B.t.i.)* and *Bacillus sphaericus* have been applied in the field in efforts to reduce midge larvae. Physical and cultural control techniques of chironomids have included mechanical means, habitat management and ecological manipulations, and behavioural manipulations of adult midge populations. The following is a brief review of chironomid control methods that have been employed in a variety of situations worldwide.

13.4.1 PHYSICAL AND CULTURAL CONTROL

(a) Mechanical control

Control of chironomids involving the use of a surface oil to trap emerging adults (Lewis, 1957), mechanical removal of egg masses from sides of rivers and dredging and mixing substrate materials in river beds (Shimizu, 1978) proved ineffective. In concrete-lined storm drains, removal of substrate materials by mechanical means or by natural factors such as rainfall (flushing effect) caused considerable reductions of midge populations (Ali *et al.*, 1976; 1977). In some water treatment and supply systems, where filters and sedimentation tanks were identified as the primary egg-laying sites of chironomids, significant potential midge infestations were prevented by protecting the filters and tanks with water sprays (Flynn and Bolas, 1985). In many midge-affected areas, electrocutor traps are employed. These traps are the most commonly used mechanical means of midge control and provide some relief from adult midges. However, electrocutor traps often malfunction in situations where huge swarms of adults are attracted to them and midge body fragments stick and completely cover the electrocuting grid.

(b) Habitat management and ecological manipulation

Midge control methods through environmental management are poorly developed, primarily because a precise understanding of the ecological bases for midge production in various problem habitats is lacking. Some effective methods used in small semi-permanent habitats include source reduction by alternate operation of sludge lagoons at sewage treatment facilities (Anon., 1963), and rotational flooding and drying of partial areas of midge breeding sources, such as water-spreading basins used for aquifer recharge (Anderson *et al.*, 1964). Increasing depth of a midge habitat where possible (Ali and Mulla, 1976b) and using proper designs for new reservoirs and artificial lakes would be conducive to containing chironomid productivity at low levels (Magy, 1968). Midge populations in some habitats may be manipulated by understanding their interaction with prevailing macrophytes as physical displacers or as producers of toxic chemicals in the midge environment, detrimental to midge larval development and survival (Johnson and Mulla, 1983).

The strategy of midge control through environmental management is highly desirable in midge habitats that spread over thousands of hectares. In such situations, knowledge of the physico-chemical environments of the nuisance species, i.e. the physical and chemical composition of the substrate materials and chemistry of the overlying water in relation to spatial and seasonal abundance and distribution of larvae of the pest species, may provide a clue to their ecological basis of production. Their proliferation could then be discouraged by manipulating nutrients (pollutants) and other factors conducive to their breeding and rapid propagation. For example, Ali (1989) presented environmental data including a large number of physical and chemical parameters and their influence on *G. paripes* and *C. crassicaudatus* populations in Lake Monroe, Florida. In a later study, the feeding behaviour of *C. crassicaudatus* in the lake was reported (Ali, 1990). Based on these studies and earlier observations of Provost and Branch (1959) on the feeding habits of *G. paripes*, it was concluded that the occurrence of high densities of *C. crassicaudatus* and *G. paripes* in Lake Monroe was largely a function of phytoplankton abundance in the water (Ali, 1990). As Lake Monroe is a part of the St Johns River system, it was suggested that the water flow through the lake should be artificially increased at times of midge larval abundance to displace phytoplankton and, in turn, reduce midge densities. Thus it appears that, in some large habitats, midge populations could be reduced by manipulating the larval environment. However, the strategy may have to vary depending upon the habitat's nature and size, plus a multitude of prevailing physical, chemical and biological conditions that would require investigation.

(c) Behavioural manipulations of adult midges

Another possibility of cultural control of chironomids is through the understanding and manipulation of their adult behaviour. For example, studies in the laboratory (Ali *et al.*, 1984) and field (Ali *et al.*, 1986) in Florida on the attraction of *G. paripes*, *C. crassicaudatus* and a few other midge species revealed that among a combination of white (incandescent and fluorescent), yellow, orange, blue, green and red lamps of the same wattage, these midges were attracted mostly to white light. A similar attraction behaviour to artificial light in the field was exhibited by *C. salinarius* inhabiting the lagoon of Venice, Italy (Ali *et al.*, 1994). These observations led to the conclusion that midges respond more to the quantity (power or intensity) than to the quality (colour or wavelength) of light (Ali *et al.*, 1984). However, laboratory studies of Kokkinn and Williams (1989) emphasized the specific attractive spectral components of artificial light to a chironomid, *Tanytarsus barbitarsis*. While adults of this species were reported to be strongly attracted to artificial light, their primary peak of attraction remained in the near-ultraviolet part of the spectrum (370–400 nm), with a secondary peak lying between 490 and 510 nm.

Information on phototaxis of adult chironomids, whether related to light intensity or wavelength, is of practical significance. Many large habitats producing midges at nuisance levels, such as Lake Monroe, Florida, the lagoon of Venice and others, are surrounded by a high density of homes and businesses interspersed with less densely inhabited areas. If dimmer lights could be used in the densely inhabited areas and high intensity (brighter) lights installed in and around the less densely inhabited areas, adult midges would be drawn to the latter areas where suitable control measures could be implemented. This strategy for midge control is particularly desirable in the Venice area where numerous deserted or uninhabited islands exist in the lagoon.

In the case of light spectrum specific attraction, commercially available lamps which emit lights with peaks in those parts of the spectrum may be employed for midge adult diversion, trapping or decoy purposes. Along the lines of adult midge behavioural studies for control purposes, information on adult dispersal (Ali and Fowler, 1983), patterns of abundance (Ali *et al.*, 1983; 1985a), and diel eclosion periodicity (Ali, 1980a) should be useful in reducing the cost of destroying the adults. Applications of insecticides could be synchronized with the emergence of adults, thereby reducing the area to be treated and the amount of material needed.

13.4.2 BIOLOGICAL CONTROL

(a) Pathogens and parasites

Viruses, rickettsiae, fungi, protozoans and nematodes have been found in chironomids. Among viruses, entomopoxviruses (EPV) have been discovered in larvae of many species of Chironomini midges (Ali, 1991a). An EPV significantly reduced field populations of nuisance *C. decorus* midges (Harkrider and Hall, 1978). In the laboratory, Harkrider and Hall (1979) demonstrated the important role of high larval density (crowding) of *C. decorus* populations in the spread of an EPV epizootic. However, a wide range of <1 to 100% larval mortality of midge populations caused by EPV in different field situations has been reported (Weiser, 1948; Huger *et al.*, 1970;

Anthony, 1975; Majori *et al.*, 1986). In addition to the poxviruses, a cytoplasmic polyhedrosis virus (Federici *et al.*, 1973) and a virus closely related to iridoviruses were also found in chironomids (Stoltz *et al.*, 1968).

Fungi of the genus *Coelomomyces* are pathogens of midge larvae (Weiser, 1976). Weiser and McCauley (1971) discovered two *Coelomomyces* infections of Chironomidae in Marion Lake, British Columbia, Canada, and also reported infection of midge larvae by a new fungal species, *Bertramia marionensis* (Weiser and McCauley, 1974).

Microsporidian and ciliophore protozoans parasitize chironomid larvae. *Thelohania* (microsporidia) were observed to infect 0.9–9.5% of field populations of *C. plumosus* in Lake Winnebago, Wisconsin, with no viable adults maturing from the diseased larvae (Hilsenhoff and Lovett, 1966). In another report from California, midge populations were annihilated due to a microsporidian infection in three semi-permanent ponds 3 weeks after initial detection of the diseased larvae (Hunter, 1968). Species of the ciliate *Tetrahymena* parasitize midge larvae (Corliss, 1960); however, no epizootics of ciliates in midge populations have been reported.

Reports on mermithid nematode parasites of chironomids from various parts of the world were summarized by Ali (1991a). In the USA, a monthly infection rate of *Hydromermis contorta* in *C. plumosus* ranged from 0 to 24.8% in a year (Johnson, 1955), while in Germany natural mermithid infections in *Tanytarsus* midges were as high as 100% at certain times of the year (Wülker, 1958; 1961). In another study, the incidence of *H. canopaga* infection in *Tanytarsus* midges in water percolation ponds in California reached 45% in a spring season and it was concluded that the parasite controlled the midge for nearly 2 years after the ponds were first flooded (Chapman and Ecke, 1969). Such incidences, however, are rarely observed in nature.

The bacteria *Bacillus thuringiensis* serovar. *israelensis* (*B.t.i.*) and *B. sphaericus* have been evaluated in the laboratory and field against pestiferous species of midges. In laboratory bioassays, the LC_{90} values of *G. paripes, C. crassicaudatus, C. decorus* and *Tanytarsus* midges ranged from 4.56 to 47.02 ppm for three wettable powders (WP) and one flowable concentrate (FC) of *B.t.i.* (Ali *et al.*, 1981); the potency of these formulations ranged from 1000 to 3500 international units of toxicity (ITU)/mg. In other laboratory studies on *C. salinarius*, two WP and one FC formulation of *B.t.i.* containing 1500 to 3500 ITU/mg were shown to have LC_{90} values ranging between 5.07 and 38.26 ppm (Ali and Majori, 1984; Ali *et al.*, 1985b). However, a more potent WP of *B.t.i.*, containing 6000 ITU/mg, was reported to cause 90% mortality of fourth instar *Chironomus tepperi* at 0.79 ppm in the laboratory (Treverrow, 1985).

In field tests with *B.t.i.*, a WP formulation ABG–6108 containing 1000 ITU/mg, applied at 1, 2, 3, 4 and 10 kg/ha, gave 18–88% larval reductions of Chironomini and Tanytarsini in experimental ponds 0.5 m deep in Florida (Ali, 1981a). The 10 kg/ha (or 2.5 ppm) rate of treatment gave 88% control after 2 weeks of treatment. The same formulation of *B.t.i* applied at 3 kg/ha to a pond 0.6 m deep on a golf course yielded 30–67% larval reductions of Chironomini during 4 weeks post-treatment. Tanypodinae in the pond remained unaffected.

Rodcharoen *et al.* (1991) conducted a number of midge control studies with *B.t.i.* in California. In ponds 0.3 m deep, Vectobac[R] 6AS (aqueous suspension containing 600 ITU/mg) at rates of 11.2 and 22.4 kg/ha yielded a maximum of 37 and 57% reduction, respectively, of populations of *C. decorus*, *Goeldichironomus holoprasinus* Goeldi [as *Chironomus fulvipilus*] and *Apedilum* [as *Paralauterborniella*] *elachistus* Townes during the 2 weeks of observation. Different formulations of *B.t.i.* were evaluated in three separate artificial lakes (Woodward Lake, Woodbridge North Lake and Lake Calabasas), ranging from 8.4 to 21.4 ha and 1.8 to 2.1 m in depth. ABG–6253 (corngrit granules containing 200 ITU/mg) applied at rates of 13.5, 28 and 56 kg/ha to 1.4–2.0 ha separate sections (fingers) of Woodward Lake, yielded 22, 83 and 96% control of *C. decorus*, respectively, at 2 weeks post-treatment. Tanypodinae midges were not affected by these treatments. In Woodbridge North Lake, a technical powder (TP) of *B.t.i.*, ABG–6164 (containing 12 430 ITU/mg) applied at rates of 1.4 and 2.8 kg/ha gave 73 and 87% control, respectively, of *Chironomus* sp. at 2 to 3 weeks post-treatment. In Lake Calabasas, Vectobac[R] TP, containing 5000 ITU/mg, applied at rates of 2.2, 4.5 and 6.7 kg/ha, resulted in a maximum control of 66, 90 and 100%, respectively, of *Chironomus* midges at 2 to 3 weeks post-treatment.

Thus, the *B.t.i.* studies reveal that this biocide is effective mostly against chironomine midges, although rather high rates of treatment (at least 10× the rates established for mosquito larvicidal activity) are required to achieve satisfactory midge control in some situations (Rodcharoen *et al.*, 1991). The use of such elevated rates of *B.t.i.* for chironomid control may be possible in small habitats ranging up to 100 ha, but would not be economical or practical in large habitats spreading over thousands of hectares.

Studies on *B. sphaericus* against chironomids indicate that this bacterium is ineffective against midge larvae at mosquito larvicidal rates (Ali and Nayar, 1986), or even at a rate as high as 22.4 kg/ha (Rodcharoen *et al.*, 1991). Therefore, *B. sphaericus* does not appear to offer any potential for midge control.

(b) Macroinvertebrate predation

Many invertebrates including leeches (Mann, 1957; Elliott, 1973a), dragonfly nymphs (Prichard, 1964), and dytiscid beetles and *Coelotanypus* midges (Bay, 1964) are predaceous on midge larvae and pupae. The subject of midge predation by invertebrates was reviewed by Bay (1974). So far, most accounts in the literature concerning invertebrate predators of chironomids are confined to qualitative observations in the laboratory or field.

The flatworm, *Dugesia dorotocephala*, is the most studied invertebrate predator of chironomid larvae. Ali and Mulla (1983) evaluated *D. dorotocephala* against chironomids in experimental ponds during three consecutive summers. Of the employed rates of 10, 25, 50 and 100 planaria/m^2, the rate of 50 planaria/m^2 reduced midge populations by 32–61% during the 3rd to the 8th week post-introduction of the predator. The rate of 100 planaria/m^2 did not necessarily produce the maximum midge reduction in the ponds because abundance of the flatworm after a certain level may have been regulated by higher predators in the food chain. Since mass rearing of *D. dorotocephala* seems feasible (Tsai and Legner, 1977), this flatworm warrants further development for the biological control of midges.

(c) Predatory fish

Chironomid larvae and pupae comprise a significant part of the diet of a variety of fish (see also Chapter 17). These fish include several sunfish, catfish, desert pupfish, whitefish, young bass, carp, mosquito fish, *Tilapia* and even gilthead seabream (Cook, 1962, 1964; Cook *et al.*, 1964; Bay and Anderson, 1965, 1966; Kimball, 1968; Kugler and Chen, 1968; Legner *et al.*, 1975a; Mezger, 1967; Zur, 1980; Ceretti *et al.*, 1987; Rasmussen, 1990).

The diet of 14 species of fish inhabiting a reservoir in Florida was recently examined (A. Ali, unpublished). Midge larvae and pupae comprised 42–67% by volume and 42–78% by wet weight, of the total food contents of three species of sunfish, *Lepomis macrochirus* (blue gill), *L. megalotis* (longear), and *L. microlophus* (redear), and one species of catfish, *Ictalurus* sp. Hence, increasing predatory pressure by the use of fish at the time of midge nuisance could reduce midge populations in some habitats. In Hjarbæk Fjord, Denmark, chironomids were the principal food item of whitefish (*Coregonus lavaretus*); this fish was considered to have caused some reductions of midge nuisance after its introduction to the fjord (Rasmussen, 1990). However, in some situations of large midge populations, midge nuisance may not be significantly reduced even though bottom-feeding fish, such as catfish, were reported to consume considerable numbers of midge larvae (Hayne and Ball, 1956). In such situations, food supply of fish tends to be continuously replaced as it is consumed.

Introductions of carp, *Cyprinus carpio*, and goldfish, *Carassius auratus*, for midge control in water-spreading basins in the USA was effective only for a short time in reducing midge populations when the fish were stocked at 165–550 kg/ha. Over the long term, factors such as pond silting, growth of filamentous algae and indigenous natural enemies of midges tended to regulate midge populations at the same level provided by the two fish species. Removal of introduced fish from the basins caused only a temporary increase in midge populations until other controls interceded (Bay and Anderson, 1965). The authors concluded that carp were more effective in small sources of chironomid nuisance than in large lakes and temporarily disrupted habitats. Carp have been used for the biological control of chironomids in Japan (Anon., 1974), but no quantitative data on their effectiveness are available. The introduction of fish such as carp for midge control in a habitat requires caution because carp are prolific and aggressive. They can displace the desirable species of fish and become a pest, and could also induce permanent and drastic alterations in a habitat, similar to those produced by the mosquito fish (*Gambusia affinis*) in pond ecosystems (Hurlbert *et al.*, 1972). Introduction of *G. affinis* resulted in increases of phytoplankton populations, change of water quality, and production of heavy algal blooms because the mosquito fish consumed and decimated zooplankton populations which fed upon phytoplankton, causing imbalance in the aquatic ecosystem.

The indiscriminate or preferential feeding of the introduced fish on macroinvertebrate predators of chironomids, which may be more efficient than the fish in reaching protective niches of midge larvae, can also disrupt the trophic structure in a habitat. However, in the presence of predatory fish, such as bass, the number of carp or other midge predatory fish may be regulated, thus diminishing the undesirable effects produced by the latter.

Cook *et al.* (1964) studied the impact of mosquito fish on chironomid populations in Clear Lake, California. They reported that although chironomid larvae and adults comprised the bulk of the summer diet of mosquito fish, these fish feed upon midges when under food stress, concluding that mosquito fish have little value for chironomid control. Similar conclusions were reported by Bay and Anderson (1966) who observed mosquito fish feeding upon chironomid larvae and adults, but causing no appreciable reductions of midge populations even when the predator biomass reached >276 kg/ha.

Exotic fish, such as *Tilapia* spp., could be highly effective for midge control in some situations (Legner and Medved, 1973; Murray, 1976). However, restocking of *Tilapia* would be necessary each summer in areas where winter temperatures decline to 10–12 °C; this would result in fish mortality and the need to remove large numbers of dead fish. As with carp, there are environmental risks involved with the introduction of exotic fish. This problem is of great concern to fish and wildlife biologists and conservation agencies.

In general, effective biological control of chironomids through the use of predatory fish alone could be achieved only in small (<20 ha) and closed habitats. In open and large habitats extending over thousands of hectares, such as Lake Monroe, Florida, and the lagoon of Venice, Italy, midge predatory fish would have to be considered as one component in the overall integrated approach of chironomid management.

13.4.3 CHEMICAL CONTROL

(a) Larvicides

Chemical control of midge larvae by the use of pyrethrum powder was attempted in Germany in the early 1930s (Buchmann, 1932). In the USA, pyrethrins, rotenone, orthodichlorobenzene and trichlorobenzene were the first chemical midge larvicides (Fellton, 1940; 1941). For several years, chlorinated hydrocarbons such as DDT, DDD, Dieldrin and others were used as midge larvicides in a variety of habitats worldwide (Flentje, 1945a; Brown *et al.*, 1961; Anderson *et al.*, 1964; Edwards *et al.*, 1964; Abul-Nasr *et al.*, 1970b) but the problems of biomagnification of organochlorines in the aquatic food-chain, occasional fish mortality and development of resistance in midge larvae necessitated use of organophosphorus (OP) insecticides for midge control. In initial evaluations of several OPs in the laboratory, Dipterex, DDVP, EPN and malathion proved highly toxic to *C. plumosus* larvae at rates of 0.11 kg AI/ha (Hilsenhoff, 1959). In lakes in Wisconsin, granular malathion was successfully used to reduce *C. plumosus* larval populations (Hilsenhoff, 1962). In Florida, more than 100 chemicals were evaluated as potential larvicides for *G. paripes*, but benzene hexachloride and EPN were most effective and were extensively used in the early 1950s to reduce *G. paripes* in natural lakes (Patterson, 1964). The use of BHC and EPN in Florida against chironomids was generally discontinued by 1955 because of the resistance problem (Patterson, 1964; 1965); these chemicals were successfully replaced by fenthion and temephos (Patterson and Wilson, 1966).

In 1969, Mulla and Khasawinah standardized the method of screening midge larvicides in the laboratory. This method utilizes quartz sand to prevent natural larval mortality and cannibalism in test cups and is extremely useful to establish susceptibility levels of pest species of midges to the candidate compounds. Based on laboratory values, the effective field-use rate of an active compound can be determined. Other methods for midge larval bioassays have also been used, including testing the ability of intoxicated larvae to construct nests and tubes using green algae (Sato and Yasuno, 1979) or 0.1 mm glass beads (Tabaru, 1985a). Selected laboratory data on the susceptibility of laboratory colonized or field-collected species of pestiferous midges from a variety of habitats to several OP and experimental pyrethroid insecticides tested in the USA, Japan and Italy are presented in Table 13.2.

It is evident that susceptibility of a midge species to the various OPs or the numbered experimental pyrethroids varies considerably, and different chironomid species or genera occurring in the same habitat may respond differently to a test compound in terms of susceptibility. For example, *C. utahensis* in an artificial lake was highly susceptible to chlorpyrifos (LC_{90} = 0.0038 ppm) but was tolerant to fenthion (LC_{90} = 1.16 ppm) (Ali and Mulla, 1978a). *Chironomus* spp. and *Tanytarsus* spp. in the Santa Ana River spreading grounds, California, were highly susceptible to chlorpyrifos and temephos (LC_{90} <5 ppb), but *Cricotopus* spp. in the same habitat were tolerant to the two insecticides as well as to the other OPs tested (LC_{90} = 0.12–2.1 ppm) (Ali and Mulla, 1976a). Midge fauna of concrete sewage channels and storm drains in California were generally resistant to chlorpyrifos, fenthion, malathion and temephos (Ali and Mulla, 1980).

Laboratory evaluations of the pyrethroids in California, Florida and Venice, Italy, revealed that these compounds were highly toxic to most midge species occurring in a variety of habitats (Table 13.2). The OP-resistant midges in concrete sewage channels and storm drains in California were highly susceptible to decamethrin (FMC–45498) and its chloroanalogue FMC–45497 (LC_{90} = 0.13–12 ppb). In Florida, all the numbered pyrethroids proved highly toxic to *G. paripes, C. crassicaudatus* and *C. decorus* with LC_{90} values ranging from 0.26 to 13 ppb (Ali, 1981b). The range of 0.34 to 4.3 ppb of cypermethrin, permethrin and deltamethrin against *C. salinarius* larvae collected from the lagoon of Venice indicated the superior activity of these pyrethroids against the saltwater midge (Ali *et al.*, 1985b).

Although most pyrethroids are superior in activity to OP compounds against chironomid larvae, in aquatic environments the pyrethroids at midge

Table 13.2 Susceptibility of fourth instar laboratory-colonized and/or field-collected chironomid larvae from a variety of habitats located in different geographical regions of the world to various organophosphorus and pyrethroid insecticides in the laboratory

Insecticide	24 h LC_{90} values in ppm			Larval source	Reference
	Chironomus sp. 51	*Goeldichironomus holoprasinus*	*Tanypus grodhausi*		
Chlorpyrifos	0.0009	0.0019	0.0013	Laboratory colonies (California, USA)	Mulla and Khasawinah, 1969
Fenthion	0.031	0.092	–		
Malathion	0.0054	–	–		
Temephos	0.0013	0.0042	0.0039		
Parathion	0.012	0.0042	0.001		
	Chironomus spp.	*Tanytarsus* spp.	*Cricotopus* spp.		
Chlorpyrifos	0.0015	0.002	0.26	Santa Ana River spreading grounds (California, USA)	Ali and Mulla, 1976a
Fenthion	0.28	0.25	0.21		
Malathion	0.037	0.18	0.24		
Temephos	0.0046	0.0046	0.52		
Phenthoate	0.015	0.09	0.12		
Ethyl parathion	0.064	–	–		
Methyl parathion	0.22	0.084	2.1		
	Chironomus utahensis	*Chironomus decorus*	*Procladius* spp.		
Chlorpyrifos	0.0038	0.019	0.0013	Artificial lakes (California, USA)	Ali and Mulla, 1978a
Fenthion	1.16	0.96	0.014		
Malathion	0.013	0.092	0.014		
Temephos	0.0047	0.07	0.06		
FMC-45497	0.00043	0.0053	0.00026		
FMC-45498	0.00077	0.004	0.00013		
FMC-35171	0.004	0.085	0.0025		
SD-43775	0.0076	0.038	0.015		

Table 13.2 continued

Insecticide	24 h LC_{90} values in ppm			Larval source	Reference
	Dicrotendipes californicus	*C. decorus*	*T. grodhausi*		
Chlorpyrifos	0.16	4.52	0.0039	Concrete sewage channels and storm drains (California, USA)	Ali and Mulla, 1980
Fenthion	0.39	4.72	–		
Malathion	1.42	0.21	–		
Temephos	1.09	42.14	0.31		
FMC-45497	0.012	0.0022	0.00013		
FMC-45498	0.0073	0.00088	0.00024		
FMC-35171	0.05	0.0068	–		
SD-43775	0.16	0.033	–		
	Glyptotendipes paripes		*Chironomus crassicaudatus*		
Chlorpyrifos	0.049	0.30	0.14	Natural lakes and a sewage pond (*C. decorus*) (Florida, USA)	Ali, 1981b
Fenthion	0.052	0.38	0.48		
Malathion	0.0079	0.12	0.16		
Temephos	0.022	0.10	0.19		
FMC-30980	0.0021	0.0097	0.0098		
FMC-33297	0.0053	0.011	–		
FMC-35171	0.0026	0.013	–		
FMC-45497	0.0024	0.0034	0.0027		
FMC-45499	0.00076	0.0026	0.0018		
FMC-52703	0.00026	0.0021	0.0013		
	Chironomus salinarius				
Chlorpyrifos	0.0071			Lagoon of Venice (Italy)	Ali *et al.*, 1985b
Fenitrothion	0.07				

Table 13.2 continued

Insecticide	24 h LC_{90} values in ppm			Larval source	Reference
	Chironomus salinarius			Lagoon of Venice (Italy)	Ali *et al.*, 1985b
Fenthion	0.046				
Temephos	0.027				
Cypermethrin	0.00034				
Deltamethrin	0.0043				
Permethrin	0.0003				
	Chironomus yoshimatsui	*Polypedilum nubifer*	*Psectrocladius* sp.		
Dichlorvos	0.224[1]	0.098	0.076	Laboratory colonies and field collected (Japan)	Sato and Yasuno, 1979
Fenitrothion	4.009[1]	0.0031	0.01		
Fenthion	0.103[1]	0.0012	0.009		
Temephos	0.015[1]	0.00032	0.0042		
Resmethrin	0.202[1]	0.033	0.01		
Bromophos	1.627[1]			Laboratory colony (Japan)	Tabaru, 1985a
Chlorpyrifos	0.0139[1]				
Chlorpyrifos methyl	0.0186				
Dichlorvos	0.199				
Diazinon	1.76				
Fenthion	0.105				
Fenitrothion	8.87				
Temephos	0.397				

[1] LC_{50} values.

larvicidal rates would have a relatively low index of safety to non-target invertebrates and fish (Mulla *et al.*, 1978a,b). Therefore, use of pyrethroids as larvicides would be limited to midge problem habitats such as sewage ponds and wastewater channels where non-target invertebrates and fish would be of minimal concern. Pyrethrins have been successfully used to control midges in public water supply systems (Burfield and Williams, 1975).

Since the early 1970s, many OP insecticides in different formulations have been used to reduce midge larvae in a variety of habitats in California. In artificial residential/recreational lakes, covering <10 to 100 ha and ranging from 1 to 5 m in depth, chlorpyrifos at rates of 0.11 to 0.28 kg AI/ha gave excellent control of chironomine and tanypodine midges for over 1 month (Mulla *et al.*, 1973; 1975; Ali and Mulla, 1977b). In one study, midge control lasted for 4 months in a lake 1–2 m deep receiving chlorpyrifos at 0.22 kg AI/ha (Mulla *et al.*, 1971). Generally, granular formulations of chlorpyrifos gave better control in terms of magnitude and duration than emulsifiable formulations. Chlorpyrifos was also reported to provide satisfactory control of *C. riparius* in a sewage effluent channel in Chicago, Illinois (Polls *et al.*, 1975).

Temephos has also been used in California to reduce chironomid larvae. The rates of temephos applications had ranged from 0.17 to 0.84 kg AI/ha in water percolation basins and residential/recreational lakes. These rates resulted in a wide range of midge control depending upon the nature of the habitat and its midge composition. For example, in water percolation basins, temephos at 0.27–0.38 kg AI/ha gave a maximum of 78% control of *Tanytarsus*, *Chironomus* and *Procladius* midges for 1 week post-treatment (Johnson and Mulla, 1980), while at higher rates of 0.56–0.84 kg AI/ha utilized in residential/recreational lakes, temephos controlled midges for 4–5 weeks (Mulla *et al.*, 1971; 1975). By contrast, much lower rates of 0.17–0.28 kg AI/ha of temephos were needed to achieve satisfactory control of midges in water-spreading basins and in a residential/recreational lake (Ali and Mulla, 1976a; 1977b). Repeated use of temephos in one lake resulted in poor or lack of control of *C. decorus* and *Procladius* spp. even at application rates of 0.33–0.56 kg AI/ha (Johnson and Mulla, 1981). Similar observations on temephos against chironomids were previously reported by Bickley and Ludlam (1968) and Mulla *et al.* (1971). Temephos, in general, yields control of midges for shorter durations than chlorpyrifos even when applied at higher rates than the latter insecticide and in most situations is innocuous to Tanypodinae at field-use rates.

Other OPs, such as fenthion, malathion, phenthoate and methyl parathion, have also been utilized for midge control in artificial lakes in California, but their overall use has been rather limited because of their lower levels of effectiveness compared with chlorpyrifos and temephos (Mulla *et al.*, 1971; 1975). In Japan, temephos, chlorpyrifos methyl, fenthion and fenitrothion have been used as midge larvicides, with temephos employed in most field trials. Temephos at concentrations of up to 2 ppm maintained for 60 minutes gave satisfactory control of *C. yoshimatsui* in polluted rivers (Tabaru 1975; Tabaru *et al.*, 1978; Ohno and Shimizu, 1982). In eel culturing ponds, chironomid populations were significantly reduced by temephos concentrations ranging from 0.05 to 1 ppm (Ohkura and Tabaru, 1975; Yasuno *et al.*, 1982). In wastewater gutters, disinfectant tanks, and discharge channels at sewage treatment facilities, temephos and chlorpyrifos methyl concentrations of 2 ppm maintained for 20–30 minutes gave excellent control of *C. yoshimatsui* (Inoue and Mihara, 1975; Tabaru, 1985b). Temephos at rates of 0.2–0.4 kg AI/ha reduced *C. salinarius* larval populations by 82–92% in the saltwater lagoon of Venice, Italy (Ali *et al.*, 1992). Overall field use of OP insecticides intended to reduce chironomid populations has indicated that temephos, chlorpyrifos, chlorpyrifos methyl, fenthion and a few other OP compounds were effective and suppressed larval populations of nuisance midge species for 2–5 weeks or longer at rates below 0.56 kg AI/ha or <1 to 5 ppm.

At present, temephos is the only chemical registered by the United States Environmental Protection Agency (EPA) for use against chironomid larvae in standing waters in the USA. Temephos has extremely low avian and mammalian toxicity and when used at chironomid control rates causes only temporary reductions of zooplankton and other invertebrates coexisting with chironomid larvae (Ali and Mulla, 1978b). However, repeated and prolonged use of an insecticide such as temephos could result in build-up of resistance in midge larvae as evidenced in California (Pelsue and McFarland, 1971; Ali and Mulla, 1978a; Johnson and Mulla, 1981) and in Japan (Ohno and Okamoto, 1980; Tabaru, 1985c). Therefore, chemical control of midges in a habitat requires a specific strategy, avoiding frequent and indiscriminate use of a chemical and with the rotational use of effective materials where possible.

(b) Insect growth regulators

The discovery and availability of **insect growth regulators** (**IGRs**) in the past two decades have provided additional materials with novel modes of action for chironomid control. These IGRs (having delayed effects) include **juvenile hormone analogues** (**JHAs**) such as methoprene, and **chitin synthesis inhibitors** (**CSIs**) such as diflubenzuron. Mulla *et al.* (1974) evaluated several JHAs (and/or JH mimics) including methoprene (Altosid[R] or ZR–515), RO–20–3600, RO–8–5497 and R–20458 against laboratory-colonized and field-collected populations of *Chironomus* sp. 51 and reported methoprene to be the most effective against this midge (Table 13.3). A 50 ppb concentration of methoprene induced complete inhibition of emergence of *Chironomus stigmaterus* Say and *Tanypus grodhausi* Sublette populations collected from sewage oxidation ponds in California and exposed to the IGR in the laboratory (Mulla *et al.*, 1974). In Japan, methoprene tests in the laboratory resulted in 50% inhibition of adult emergence of *C. yoshimatsui* at 0.65 ppb (Kamei *et al.*, 1982). However, this species was more sensitive to diflubenzuron than to methoprene at comparable concentrations of the two IGRs (Tabaru, 1985d).

Ali and Lord (1980) developed a laboratory bioassay technique for IGRs against midges. This technique was employed to evaluate the IGRs, R–20458, MV–678, SIR–8514, diflubenzuron, UC–62644, UC-84572, UC–75118, UC–75150, UC–76721 and UC–76724 against field-collected early fourth instars of *G. paripes, C. crassicaudatus* and *C. decorus* (Table 13.3). It is

Table 13.3 Activity of juvenile hormone analogue and chitin synthesis inhibitor insect growth regulators (IGRs) against fourth instar laboratory-colonized or field-collected chironomid larvae exposed continuously to the IGRs in the laboratory

IGR	LC_{90} (90% inhibition of adult emergence) values in ppb						Reference
	Chironomus sp. 51	*Chironomus crassicaudatus*	*Chironomus decorus*	*Chironomus yoshimatsui*	*Glyptotendipes paripes*	*Tanypus grodhausi*	
			Juvenile hormone analogue				
Methoprene (ZR-515)	10					10	Mulla *et al.*, 1974
RO-20-3600	250						(California, USA)
RO-8-5497	>1000						
R-20458	>1000						
R-20458			240		700		Ali and Lord, 1980
MV-678			50		69		(Florida, USA)
Methoprene				0.65[1]			Kamei *et al.*, 1982 (Japan)
			Chitin synthesis inhibitor				
Bay SIR-8514			22		7.6		Ali and Lord, 1980
Diflubenzuron		7.4[2]	6		4.1		
UC-62644			5.7		3.1		Ali and Stanley, 1981
UC-84572		2.0[2]			2.4[2]		(Florida, USA)
UC-75118		12.6			22.4		Ali *et al.*, 1987
UC-75150		9.4			10.6		(Florida, USA)
UC-76721		25.3			34.2		
UC-76724		4.5			3.1		

[1] LC_{50} value
[2] Data of Ali and Nayar, 1987 (Florida, USA).

evident that all CSIs tested were generally far superior in activity to the JHAs or their mimics. Among the CSIs, UC–84572, UC–62644, UC–76724 and diflubenzuron, in that order, were the most active, causing 90% inhibition of adult emergence of the exposed midge species at 2.0–7.4 ppb.

Field use of IGRs for midge control has primarily been attempted in the USA (California and Florida) and Japan. In California, Pelsue *et al.* (1974) reported 60–86% control of *C. decorus*, *Dicrotendipes californicus* (Johannsen), *Tanytarsus* sp. and *Procladius culiciformis* (Linnaeus) for 1–3 weeks with methoprene (Altosid®, SR–10) and diflubenzuron (TH–6040, 25% WP) concentrations of 0.1 ppm in water-spreading basins. Mulla and Darwazeh (1975) evaluated different rates of a variety of formulations of the IGRs methoprene, RO–10–3108, R–20458 and diflubenzuron against natural populations of chironomids in experimental ponds and found methoprene and diflubenzuron generally superior in activity to RO–10–3108 and R–20458 at comparable rates of treatment.

Several formulations of methoprene and diflubenzuron were employed for chironomid control in artificial lakes and the latter IGR was also utilized in midge habitats, such as flood control channels and spreading basins used to replenish subsurface water in California. In the lakes, methoprene at 0.11–0.34 kg AI/ha inhibited adult emergence of Chironominae and Tanypodinae midges by 47–100% for 8–19 days (Mulla *et al.*, 1974, 1976), while diflubenzuron at 0.11–0.28 kg AI/ha induced 48–100% midge control for 5–8 weeks (Mulla *et al.*, 1976). In flood control channels and spreading basins, diflubenzuron at 0.11 kg AI/ha gave 88 to 100% control of chironomids for 4 weeks in a channel <1 m deep and 59–64% midge control for 3 weeks in basins 1–2 m deep (Ali and Mulla, 1977a). Diflubenzuron applications at 0.11–0.22 kg AI/ha to artificial lakes provided excellent control of chironomids for 2–5 weeks (Ali and Mulla, 1977b; Ali *et al.*, 1978). In one lake, however, *C. decorus* was controlled for only 1 week post-treatment (Ali and Mulla, 1977b). Johnson and Mulla (1981) also reported the relative ineffectiveness of diflubenzuron as well as SIR–8514 against *C. decorus* in an artificial lake in California.

Evaluations of two formulations (25% WP and 0.5% G) of SIR–8514 at 0.11 and 0.28 kg AI/ha in ponds <1 to 2.5 m deep covering 0.25–2.5 ha on a golf course in California resulted in 50–100% control of *Chironomus* spp. and *Procladius* sp. midges for 5 weeks with the WP formulation, and 70–100% control for 5 weeks with the granular formulation (Johnson and Mulla, 1982). The level and duration of control given by the lower rate of each formulation were almost the same as produced by the higher rate. At present, diflubenzuron is registered for midge control in California under a special local need permission.

In Florida, IGRs have not been used on a large scale for the practical control of midges. However, several field evaluations of methoprene, diflubenzuron and new experimental IGRs in different formulations have been conducted in 6 × 4 m experimental ponds, 45–50 cm deep, supporting natural populations of chironomids. The results indicate that among SIR-8514, MV–678, diflubenzuron, UC–62644, UC–84572, methoprene and a new JH mimic pyriproxyfen, UC–62644 was the most active, inhibiting 94–99% adult

Table 13.4 Effectiveness of insect growth regulators (IGRs) against natural populations of chironomid midges[1] in artificial experimental ponds in Florida, USA

IGR	Formulation	Application rate (g AI/ha)	% Control and duration (weeks)	Reference
SIR-8514	25% WP	28	32–94 (7)	Ali and Lord, 1980
		56	34–98 (7)	
		112	44–99 (7)	
	0.5% G	56	42–99 (7)	
		112	50–100 (7)	
MV-678	EC 4	56	0–30 (3)	
		112	0–70 (3)	
Diflubenzuron	25% WP	28	6–80 (4)	
		56	0–98 (4)	
UC-62644	25% WP	25	94–99 (4)	Ali and Stanley, 1981
		50	98–99 (4)	
		100	98–100 (4)	
UC-84572	10 EC	12	0–67 (4)	Ali and Chaudhuri, 1988
		24	40–96 (4)	
		48	60–100 (4)	
Methoprene	A.L.L.[2]	15	10–19 (2)	Ali, 1991b
		280	74–99 (2)	
	SAN 810 I 1.3 GR[3]	170	61–87 (2)	
	Altosid pellet[4]	220	64–98 (7)	
	Altosid XR briquet[5]	820	38–98 (7)	
Pyriproxyfen	3% Sand G	50	51–100 (10)	Ali *et al.*, 1993
		200	64–100 (11)	
	10 EC	50	32–100 (3)	
		200	47–100 (4)	

[1] Predominantly Chironomini and Tanytarsini species
[2] Altosid® Liquid Larvicide containing 5% S-methoprene
[3] Experimental granules containing 1.3% S-methoprene
[4] Altosid pellets containing 4% S-methoprene
[5] Altosid XR briquet containing 1.8% S-methoprene.

emergence of midges for 4 weeks at a rate as low as 25 g AI/ha (Table 13.4). Other IGRs – SIR–8514, diflubenzuron and UC–84572 – also caused significant reductions of adult midge emergence for several weeks post-treatment at rates ranging from 24 to 112 g AI/ha. Sustained release methoprene, Altosid[R] XR briquet and Altosid[R] pellet formulations proved highly effective, reducing midge emergence by 38–98% for 7 weeks and 64–98% for 7 weeks at rates of 0.82 kg AI/ha (briquet) and 0.22 kg AI/ha (pellets), respectively. A 3% sand granule formulation pyriproxyfen (Nylar[R] or Sumilarv[R]) was also highly effective, giving long-term midge control ranging from 81 to 100% for 9 weeks when applied at 50 g AI/ha to the ponds (Ali *et al.*, 1993). This IGR inhibited >80% emergence of chironomids (predominantly *P. nubifer*) for 3 weeks when applied at 50 g AI/ha to a lake in western Australia (Pinder *et al.*, 1991).

Diflubenzuron and methoprene are commercially available in Japan for midge control purposes. Treatments of two rivers with diflubenzuron at 1 ppm maintained for 60 minutes resulted in satisfactory control of *C. yoshimatsui* for 3 weeks. Methoprene at the same dosage proved less effective in the same river (Tabaru, 1985d). In the gutters and effluent discharge channels of sewage treatment plants, methoprene concentrations of 0.13 and 4 ppm controlled *C. yoshimatsui* only at the higher concentration (Tsumuraya *et al.*, 1982).

The laboratory and field studies on IGRs against midges have shown that most of these compounds are highly effective against a variety of chironomid species at very low concentrations (ppb range). In field situations, some of these compounds induced midge control for several weeks at rates much lower than 0.25 kg AI/ha. The IGRs are desirable in midge control programmes because they inhibit larval or pupal development and unlike OP insecticides have less decimating effects on midge larval biomass, an important component of the aquatic food chain. The IGRs also provide alternate means to control the midge species resistant to OP insecticides.

The use of chemicals, including IGRs, for chironomid control in aquatic environments has been shown to have temporary or chronic effects on non-target biota coexisting with midge larvae (Ali and Mulla, 1978b, 1978c; Ali and Stanley, 1981). Therefore, chemicals in the aquatic environments should be used with great caution and their environmental implications and cost benefits should be evaluated. However, adverse effects of chemicals on non-target biota would be of minimal concern in midge habitats, such as sewage ponds and polluted rivers. In some other situations, only partial areas of a habitat that support large populations of midge larvae could be treated (Ali and Mulla, 1977b). This practice would reduce some midge nuisance and would also be conducive to quicker restoration of the lost non-target organisms from the untreated areas. The toxic effects of the chemical used in such a practice would also diminish sooner due to dilution.

(c) Adult Control

There are limited reports on chironomid adult control. In the 1950s and 1960s, destroying adult midges was the principal means of control in Florida. Malathion and malathion-lethane or naled were used as thermal aerosol fogs. Applications of these materials were made from trucks, boats and airplanes. Low volume aerial sprays of malathion at 0.14–0.27 kg AI/ha were effective and provided control of *G. paripes* within 3 hours and lasting for 4 days when applied at the rate of 0.27 kg AI/ha (Patterson *et al.*, 1966). At present, no specific insecticide is registered in the USA for controlling adult midges. However, some OP compounds, pyrethroids and other insecticides labelled for adult mosquito control in the USA (Rathburn, 1988) probably reduce some adult midge populations when used.

Insecticides against adult midges are used on a very limited basis in Japan (Y. Tabaru, *pers. comm.*). By contrast, midge adulticiding with deltamethrin and malathion is the only means of *C. salinarius* control in and around Venice, Italy (Ali *et al.*, 1992).

13.5 SUMMARY AND CONCLUSIONS

In recent years, massive swarms of adult Chironomidae emanating from a variety of natural and artificial aquatic habitats distributed worldwide have increasingly posed nuisance and economic problems and in some situations medical problems to waterfront residents, workers, visitors and business and industrial establishments. Although less than 100 species of chironomids have been reported to be pestiferous, it is estimated that the annual economic loss attributable to chironomid problems amounts to millions of US dollars on a global basis.

Among chironomid control methods, chemical control is the most studied. A large number of organochlorines, organophosphates, pyrethroids and insect growth regulators (IGRs) has been evaluated in the laboratory and/or in the field against nuisance species of midges. Among OP insecticides, chlorpyrifos and temephos have generally resulted in larval control for 2–5 weeks or longer at application rates below 0.56 kg AI/ha. Frequent use of OP compounds in some habitats has resulted in development of resistance in midge larvae.

Numerous pyrethroids have exhibited far superior activity to that of OP insecticides against chironomid larvae, but most pyrethroids at midge larvicidal rates would have a relatively low index of safety to non-target aquatic invertebrates and fish. Therefore, field use of pyrethroids for midge control would be limited to habitats such as sewage ponds and wastewater channels where non-target invertebrates and fish would be of minimal concern.

The IGRs methoprene, pyriproxyfen and diflubenzuron have shown superior activity, suppressing >90% adult emergence of midges in a variety of habitats at rates below 0.25 kg AI/ha. These IGRs are excellent candidates for controlling midges, particularly the OP-resistant species. However, the scope of chemical control alone directed against midge larvae in large habitats covering hundreds or thousands of hectares would be limited because of the relatively high costs of treating large areas. Chemical displacement, adverse effects on aquatic non-target organisms and other environmental concerns would also severely restrict or prevent the use of chemicals in such habitats.

A number of chironomid larval parasites and pathogens such as viruses, rickettsiae, protozoans, nematodes and fungi have been reported but most of these microbial organisms have not been subjected to mass culturing and inoculating of breeding sources of chironomids for quantitative biological control assessments. *Bacillus thuringiensis* serovar. *israelensis* (*B.t.i.*) is effective against Chironominae midges, but at rates at least 10 times those established for mosquito larvicidal activity; therefore *B.t.i.* has limited scope for midge control. The flatworm *Dugesia dorotocephala* and some fish (e.g. carp, sunfish, catfish, *Tilapia* and others) would be effective in reducing midge larvae in some situations. In large habitats, the use of predatory fish alone may be partially effective but would serve as an important component in the integrated approach to chironomid management. More research is needed on the development and quantitative field assessments of parasites, pathogens and predators of midge larvae, and the environmental implications and impact on trophic structure of these biological control agents in the aquatic midge environments.

Midge habitats that spread over hundreds or thousands of hectares demand exploration of physical and cultural control methods. In such habitats, the physical and chemical composition of substrate materials and chemistry of the overlying water in relation to spatial and seasonal distribution of midge larval populations may provide a clue to the conditions conducive to their proliferation, and the ecological basis of midge production. Their proliferation could then be discouraged by manipulating their environment where possible. Information related to adult dispersal behaviour, patterns of abundance, diel periodicity of eclosion and attraction to light could also be manipulated in some situations to reduce midge nuisance more economically. Better understanding of the larval ecology and adult behaviour of pestiferous Chironomidae is needed.

14

Medical significance

P.S. Cranston

14.1 INTRODUCTION

Many of the most important insect vectors of serious disease to humans and livestock are flies. The mosquitoes (Culicidae), blackflies (Simuliidae) and biting midges (Ceratopogonidae) are familiar vectors belonging to the dipteran suborder Nematocera. This group includes many flies in which females feed on vertebrate and invertebrate hosts. A vertebrate blood meal may be required for egg maturation: during the feeding process parasites may be transferred from fly to host or vice-versa. Complex coevolutionary life history adaptations have taken place in the host and parasite involving the pathogens of diseases such as malaria, dengue, river blindness and blue tongue.

Although the Chironomidae are nematocerans and close relatives of the medically significant families, there is no evidence for chironomid midges biting humans or indeed any other animals. Virtually all Chironomidae lack functional mandibles and are unable to bite. Therefore, it may seem surprising to find a quite substantial literature and a chapter in this book concerning the medical significance of chironomids. Apart from anecdotal reports of asphyxia provoked by some mass emergences of midges discussed in Chapter 13, it has become increasingly clear in the last two decades that allergic disease induced by chironomid flies is widespread, and may be very prevalent.

14.2 ALLERGIC DISEASE, WITH PARTICULAR REFERENCE TO INSECTS

Allergic disease encompasses the symptoms of dermal irritation and/or respiratory allergy and is caused by abnormal stimulation of the immune system. This aberrant immune system response is termed **hypersensitivity**. A rapid (almost immediate) onset of allergic symptoms following triggering of the immune system is termed **Type I hypersensitivity**. This contrasts with a variety of more delayed, or cyclical responses to allergens.

The Chironomidae: Biology and ecology of non-biting midges
Edited by P.D. Armitage, P.S. Cranston and L.C.V. Pinder
Published in 1995 by Chapman & Hall. ISBN 0 412 45260 X

The term **allergen** is used for a macromolecule (usually a protein, glycoprotein, peptide or lipoprotein) which provokes usually an IgE-mediated (immunological) response that results in localized inflammation of the target organ, or systematic reaction such as anaphylaxis. The term is used more loosely to mean the environmental chemicals that cause allergy and the actual source of the proteins (which themselves do not cause the disease). Human reactions to insects (and other arthropods) and their products depend upon the biochemistry of the allergen, its concentration, the means and duration of exposure, and variations in the exposed individual's immune system. The individual's history of prior exposure is important, as is the genetic predisposition (**atopy**) to allergies.

The milder symptoms of allergic disease range from **urticaria** (derived from similarity to the reaction to the stinging nettle genus *Urtica*) and itching of the skin (**pruritis**), to **conjunctivitis** (inflamation of the conjunctiva) to **allergic rhinitis**. The symptoms of allergic rhinitis (vasodilation, nasal congestion and mucous flow) are colloquially referred to in English as hay fever because of the early recognition of the role of grass pollens as provocators of the symptoms. Severe allergic rhinitis leads to permanent cold-like symptoms, loss of smell and taste and sinus problems.

Allergic rhinitis symptoms can overlap with the symptoms of **asthma**, but this latter disease strictly involves some kind of bronchial constriction and an underlying bronchial inflammation. This may range from mild bronchospasm to chronic obstruction of the flow of air. Not all asthma is allergic in aetiology, in the sense that the immune system is stimulated by an allergen – for example, exercise may induce asthma without any obvious allergen to trigger wheezing.

Severe asthma can be a chronic complaint and it can lead to death. Asthma-related mortality arises by problems associated with respiratory failure and **anaphylaxis**. Anaphylaxis is a systemic rather than organ-based reaction caused when the antigen rapidly reaches sensitized mast cells and results in the generalized release of histamine and other vasoactive compounds that result from antigen–antibody reaction. These substances, especially histamine, causes a cascade of reactions including smooth muscle contraction, oedema, fever and shock. Death may result unless adrenaline or anti-histamines are used. Anaphylaxis is the cause of death in those individuals hypersensitive to bee and wasp venoms and several food allergens, most notably legumes such as peanuts. It is also a cause of death amongst some sensitized Sudanese who are exposed to midge swarms, if appropriate therapy is unavailable.

14.2.1 DIAGNOSIS OF ALLERGIC DISEASE

We often talk popularly about hay fever and our allergies. Medically, a causative agent, an allergen, must be established to prove allergic disease. Asthma is often a hidden, misdiagnosed disease, in that the symptoms are assumed to be due to other chronic ailments, such as bronchitis or emphysema, especially when associated with tobacco smoking or viral or

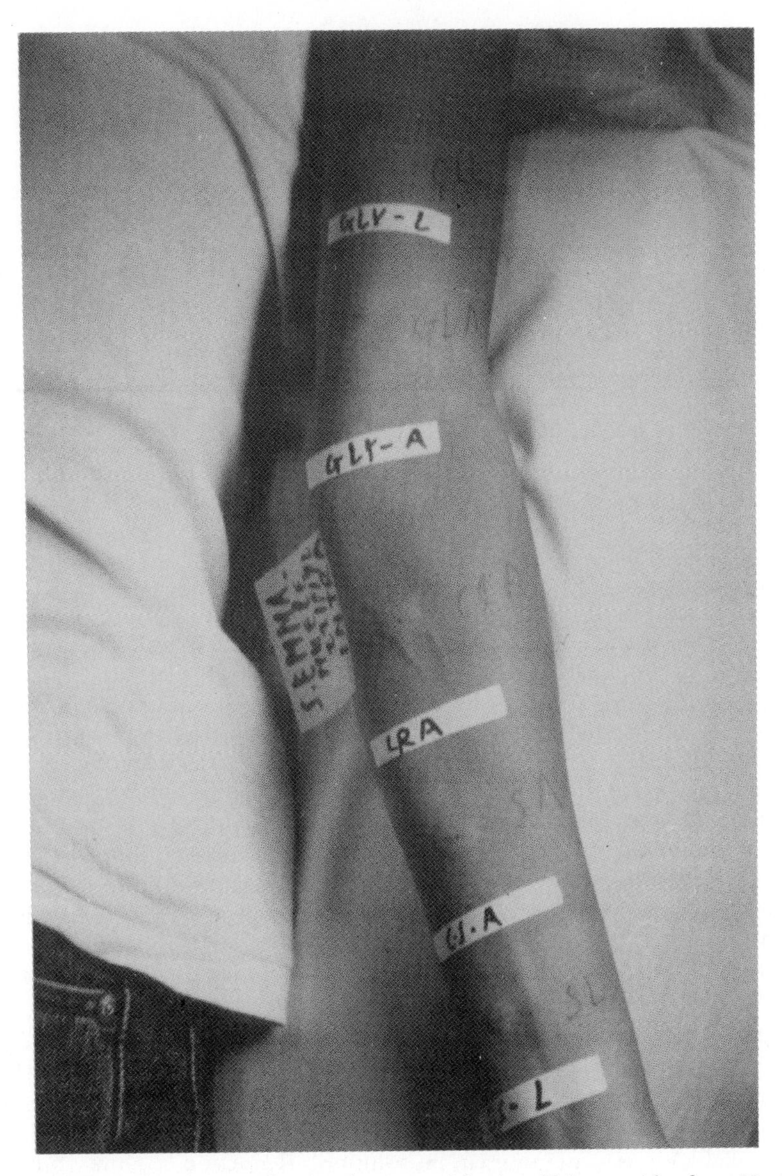

Figure 14.1 Weal responses of patient hypersensitized to Chironomidae from Venice lagoon. Intradermal injection with extracts of: CS-L = *Chironomus salinarius* Kieffer larvae; CS-A = *C. salinarius* adults; CRA = *C. crassicaudatus* Malloch adults; GLY-A = *Glyptotendipes paripes* (Edwards) adults; GLY-L = *G. paripes* larvae. (Courtesy of Dr. Carlo Giacomin.)

bacterial pulmonary infections. The following tests are widely used for clinical diagnosis of asthma.

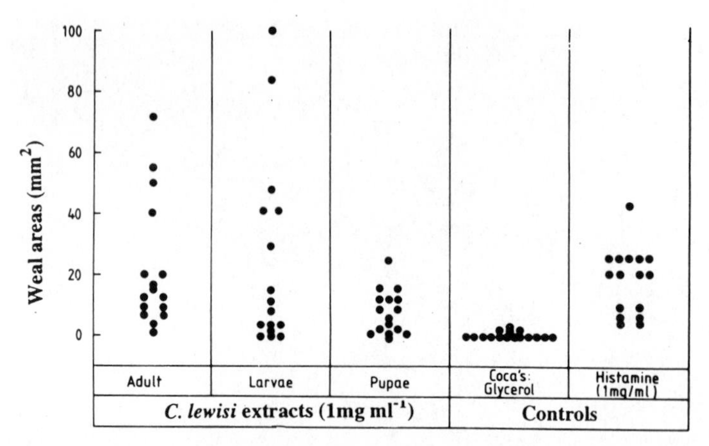

Figure 14.2 Skin-prick test weal sizes in response to extracts of adult, larva and pupa of *Cladotanytarsus lewisi* Freeman. Responses to the negative (Coca's solution) and positive (histamine) controls are shown.

(a) Skin tests

The atopic status of individuals is often detected by a skin test. These include scratch tests, prick tests and intradermal tests in which allergen extracts are applied either superficially or just below the dermis. Immediate hypersensitivity is demonstrated by a weal (wheal) or weal-and-flare response (Figures 14.1, 14.2) relative to histamine positive and inert carrier negative controls. These tests prove that there are circulating antibodies but not necessarily presence of symptoms. Skin tests are the most frequently used means to identify potential causes of seasonal hay fever, house dust mite allergy and allergy to domestic animals. They have been used widely in diagnosis of Sudanese chironomid allergy where immunodiagnostic tests are difficult to perform.

(b) Provocation tests

As long ago as 1873, when grass pollens were found to be important allergens causing allergic rhinitis, pollen was used directly to provoke the nasal or lung membranes. Since the concentration of allergen used is far in excess of natural exposure, nowadays nebulizers are used to deliver more dilute (and more controllable) doses of antigen in solution. Reduction in lung function (reduction in forced expiratory volume) in bronchial provocation is a positive clinical indication for allergenicity of the allergen tested. These tests must be performed under medical guidance, where resuscitation backup is available, preferably in hospital.

(c) The RAST antibody assay

The presence of a specific antibody response when an atopic individual is exposed to an allergen can be detected through increase in the serum of a

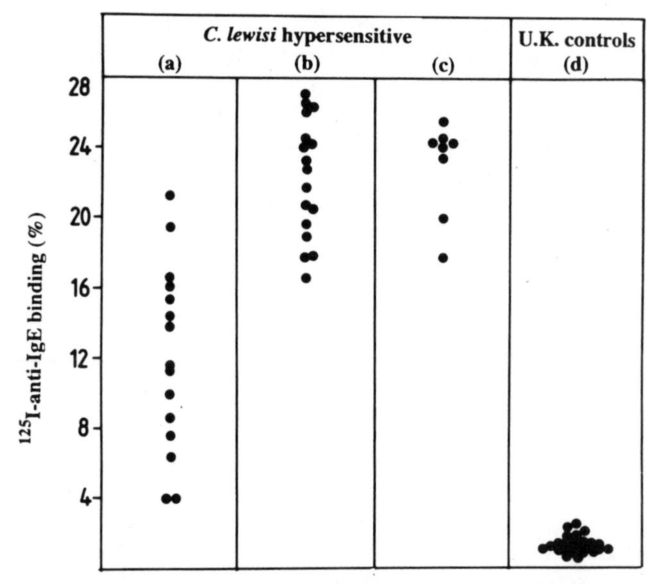

Figure 14.3 Radioallergosorbent (RAST) results (as % ^{125}I-anti-IgE binding) for Sudanese *C. lewisi* sensitized patient's sera grouped by skin test reactions and UK control sera. (a) Weak skin test reaction group; (b) moderate skin test reaction group; (c) strong skin test reaction group; (d) skin test negative UK controls.

particular isotype of immunoglobulin, **IgE**. A commonly used technique, the **radioallergosorbent test** (RAST), binds allergen to an activated paper disc, which is then incubated with the serum to be tested. Any specific IgE antibody from the serum that binds to the allergen on the disc can be demonstrated and quantified by radiolabelled (^{125}I-) anti-IgE absorption, which can be counted by an automatic gamma counter. The results of such a test are portrayed in Figure 14.3. This technique, described in detail by Baur *et al.* (1982) and Tee *et al.* (1984), has been used widely in studies of allergy to Chironomidae.

(d) The ELISA antibody assay

A cheaper and more rapid technique than RAST is available in an **enzyme-linked immunosorbent assay** (ELISA). Of the several ELISA techniques, that described by McHugh *et al.* (1988) involves coating microtitre plates with buffered antigen then incubated with patient's serum. After washing to remove excess reagent, it is further incubated with HRP conjugated goat-derived anti-human IgE antibodies. Following further washing, the microtitre plates may need to be incubated with HRP conjugated rabbit–anti-goat IgG to enhance the sensitivity of the reaction. The final stage anti-mammal antibody, in this case rabbit–anti-goat IgE, is conjugated to an enzyme such as horseradish peroxidase – HRP. Any reaction between serum and antigen is made readable by incubation in the dark with ortho-phenylene diamine with a substrate of hydrogen peroxide, blocked by sulphuric acid or sodium fluoride, and read by absorbance at 410 nm or 492 nm. Results are compared with blanks (no serum), negative controls (human cord serum) and known positive sera. This ELISA was used unsuccesfully by McHugh *et al.* (1988), and a similar one was used with greater success by Japanese workers including Matsuoka *et al.* (1988), who also elaborated an ELISA inhibition test (Matsuoka *et al.* 1990).

14.2.2 HISTORY OF ALLERGY TO INSECTS

Before considering in detail the role of chironomids in allergic disease, it is appropriate to consider the relationship between insects in general and allergic disease. Perhaps one of the earliest records concerned the death of King Menes of Egypt from a wasp or hornet sting (cited in Wirtz, 1984). Restricting this brief review to non-stinging insects, we find references to skin irritation or urtication from lepidopteran caterpillars as far back as early Greek writings. These problems stem from hollow spines containing the secretions of a subcutaneous venom gland, which are released when the spine is broken. Other species have setae (bristles and hairs) containing toxins which cause intense irritation when the setae contact human skin. Urticating caterpillars include the processionary caterpillars that were used in 1870 by the famous naturalist Fabre to sensitize himself, thus developing the skin patch test in the process (cited in Wirtz, 1984). The first report of an occupational disease attributed to an insect was over 250 years ago, in which it was noted that silk carders developed a terrible cough and serious difficulty in respiration from the particles of silkworm remaining in the silk (Bellas, 1990).

Inhalant allergies can be associated with natural populations of insects, as well as to cultured ones such as silkworms. A short historical review of this area is presented by Henson (1966) and Wirtz (1984) presents an excellent review of allergic reactions to various non-stinging arthropods. Mathews (1989) specifically reviews inhalant allergens derived from insects. Amongst aquatic insects, adult mayflies (Ephemeroptera) were cited first in 1913 as a cause of allergy in Detroit, Michigan and Buffalo, New York, where hay fever was induced in a few exposed individuals (Wilson, 1913). Later studies demonstrated that 3.2% of 500 seasonally allergic (asthmatic) patients in Buffalo responded positively to skin tests with mayfly extracts (Parlato, 1938). Adult caddisflies (Trichoptera) may occur in swarms as do the mayflies cited above. Likewise, they have been implicated in allergic disease, once again at Buffalo, as early as 1929 (Parlato, 1929, 1930). The problems were widely distributed around the Great Lakes basin, as adult caddisflies made longer flights than mayflies. Cross-reactivity (sensitivity to more than one insect) was quite common and many people reacted to both mayflies and caddisflies. There is no evidence that chironomid midges were believed to be a problem and there are no reports of testing with their extracts.

However, in those environments that give rise to nuisance numbers of mayflies and caddisflies, it is to be expected that there are also chironomids, perhaps in appreciable numbers but less conspicuous than the mayflies and caddisflies. Historically, these may have been overlooked. The absence of published references should not be taken to imply that there have been no problems. Anecdotally, people that fish for pleasure and are exposed to aquatic insects while engaged in their pastime often complain of 'hay fever' and some are even aware of the irritation of chironomids. For documentation of this association, the individual must see their medical adviser and be referred onwards to a specialist interested enough to treat the occurrence as worthy of further study and documentation in the medical literature. Such a scenario occurs infrequently.

14.2.3 OCCUPATIONAL ALLERGIC DISEASE AND INSECTS

Occupational asthma is induced by allergens inhaled at the work-place. We saw above the early report of allergy amongst silk carders and the list of occupational allergens has grown immensely in the intervening 250 years. Early indications of problems are episodes of respiratory symptoms (breathlessness, wheezing, coughing and chest tightness) that are associated with the work-place, with symptoms abating when away from the work environment. Asthma commonly follows repeated exposure.

In occupational health assessment, it is important to recognize how many individuals who are exposed to the causative agents actually develop occupational asthma. For example, between 5 and 10% of those working with laboratory animals develop occupational asthma, and half must change their employment. Amongst these laboratory animals are mice, rats and rabbits, but many entomological and teaching establishments maintain colonies of insects. As reviewed by Bellas (1990), insects appear to be potentially highly allergenic, with reports of occupational asthma associated with locusts, cockroaches, moths, several granary beetles and fishing baits such as blowfly larvae and mealworms. Freeze-dried larval Chironomidae, especially *Chironomus*, prepared as fish food for use by aquarists are a particularly potent source of allergens for those involved in their preparation, as we shall see in section 14.4.

14.3 HISTORY OF ALLERGY TO CHIRONOMIDAE

In this section, the recognition of the role of chironomids in allergic disease is treated under the heading of the geographical areas in which it occurs. Discussion on the epidemiological findings and allergen determinations is deferred until later sections.

14.3.1 AFRICA, NOTABLY THE SUDAN

Perhaps the earliest report of allergy induced by Chironomidae is due to Carpenter (1920) writing as a naturalist on Lake Victoria. He reported on the biology of swarming gnats, called 'E'sami' by local people, and which he referred to as Chironomidae. The natives compressed the flies into cakes that were used in some unknown manner as food. Carpenter noted the malodour of the dead flies and he stated:

> It is a common saying in Entebbe [Uganda] that the arrival of clouds of these gnats produces an outbreak of nasal catarrh amongst the white inhabitants; possibly this is of the nature of 'hay fever'.

We know from other entomologically and anthropologically skilled observers that the cake is also termed 'kungu' and is prepared from aquatic insects emerging *en masse* from other Ugandan lakes (Chapter 17). Examination of an example of the cake preserved in the Natural History Museum showed that a major component is actually a species of fly belonging to the family Chaoboridae, *Chaoborus (Sayomyia) edulis* Edwards (Cranston, unpubl.). This species undergoes lunar periodic emergence (section 10.2.2) from Lake Victoria, and moves in large numbers towards lakeside lights. As the specific name implies, F.W. Edwards knew of the edibility of the species when he described it. It is uncertain whether the allergic disorders of white residents of Entebbe were due to the chaoborid or the small tanytarsine midges also present.

The best investigated early case, and perhaps the most widespread and prevalent allergic disease associated with chironomids, came from the sub-Saharan African country of the Sudan (Kirk, in Lewis, 1956). The country is bisected by the Rivers Nile, which flow from south and east to the north, where the main Nile enters Egypt downstream of the artificially formed Lake Nasser (Figure 14.4). A substantial part of the Sudanese population lives and works beside the rivers (the White Nile, Blue Nile and the main Nile). At certain times of the year, adult numbers are enormous and the midges fly into the eyes, the nasal passages and the mouths of local inhabitants. The local people call the midges 'nimitti', which is a general name for gnat-like flies, and early publications on the problems refer to 'green nimitti' to differentiate the chironomid midges from, for example, blackflies (Simuliidae) and chloropids (Lewis, 1954). After mass emergence, the dead bodies rest in buildings where they desiccate and fragments become airborne when domestic cleaning takes place. The midges were recognized as causing nuisance as early as 1918, being described as 'troublesome' in 1927. There were indications of asthma in the area and by the 1930s an entomologist was recruited to assess the problem and to suggest control. By the end of the 1930s many cases of asthma were being reported in the north of the country, and asthma camps were established, located some 5 km from the river. The town of Wadi Halfa was particularly plagued and in 1948; 14 of the permanent staff of 67 at the hospital were incapacitated by asthma for the complete chironomid season. In this year it was even suggested that the town be relocated. (It was subsequently; the building of the Aswan Dam caused flooding of the old town of Wadi Halfa by the rising waters of Lake Nasser.)

Initial observations on the medical significance of midges in the Sudan implicated a single species *Cladotanytarsus* [as *Tanytarsus*] *lewisi* (Freeman) (Lewis, 1956). This species attained high densities in the benthos of the

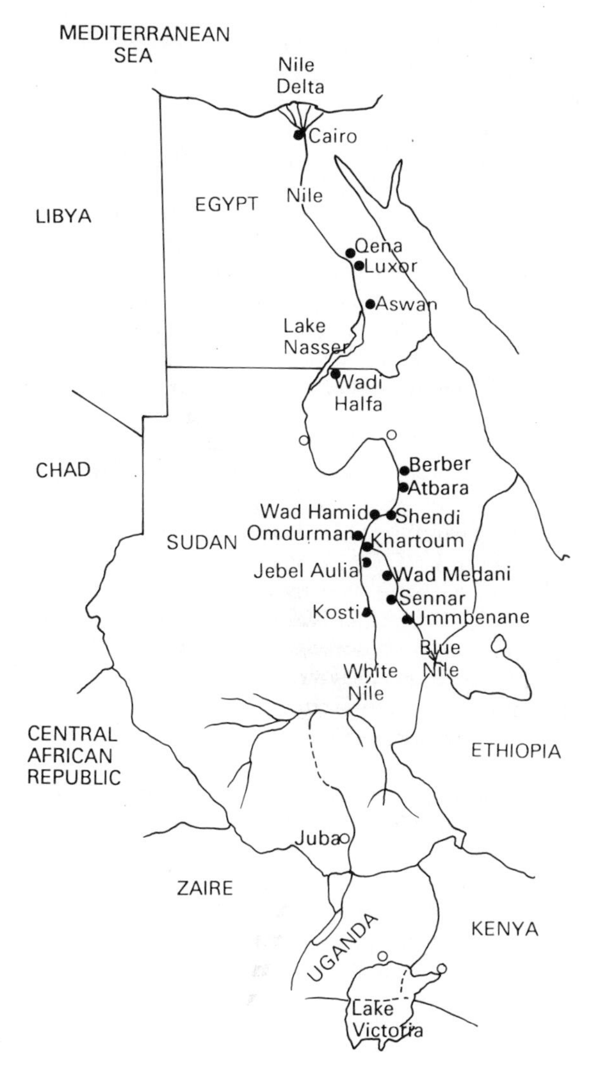

Figure 14.4 Map of the Sudan and adjacent Africa to show localities associated with proved midge allergy.

post-flood Nile and adults emerged in massive numbers during the winter months. Anecdotal stories tell of the midday sky being darkened by the clouds of midges. Electrical lights close to the river attracted such numbers of flies that labourers with buckets and shovels were needed to clear the heaped bodies that accumulated beneath the lights during only one night. Circumstantial evidence related the problems to damming of the Nile, but natural slowing of the flow takes place in the post-flood period providing suitable conditions for midge development (Cranston *et al.*, 1981).

By 1951, Kirk (cited in Lewis 1956) reported that a very high proportion of hospitalized asthmatics in Wadi Halfa gave strongly positive reactions when skin-tested with antigens prepared from *C. lewisi*. The immunological symptoms of the sufferers are predominantly those of inhalant allergic disease, such as rhinitis (cold-like symptoms of nasal obstruction, sneezing, wheezy breathing), conjunctivitis and frequent provocation of asthma.

The application of modern immunological methods to this problem commenced with Kay *et al.* (1978), who established that the Sudanese medical problems derived from increased circulating IgE antibodies, established by the passive sensitization of biopsied lung fragments with Sudanese sera. They subsequently developed a specific radioallergosorbent test (RAST) (Gad El Rab and Kay, 1980) and the results of these and ensuing immunological studies are described in sections 14.5 and 14.6.

14.3.2 UNITED STATES OF AMERICA

Early suggestions of the implication of chironomid midges in allergic disease came from southern USA (Weil, 1938, 1940). In this case an employee at an hydroelectric plant in Alabama developed a seasonal (May to August) hay fever associated with *Tanytarsus* emergence from the dam. Symptoms included cough, wheezing and itching that resulted in 'whelps' (weals) when scratched. Typically for an occupational disease, the symptoms disappeared when he left the dam site.

The worker gave a strong positive response when tested intradermally with extract of dried insects and it was suggested that the midge 'scales' caused the allergy. Weil percipiently suggested that allergy to *Tanytarsus* might prove to be responsible for the symptoms of hay fever in many cases.

Although one may be certain that midge-induced allergy did occur elsewhere in the USA, it was a long time until a second report was published. Kagen and colleagues (1984) in a brief abstract, documented clinical allergy to *Chironomus plumosus* in atopic individuals living close to a eutrophic reservoir in Wisconsin in contrast to matched atopic individuals living further from the source of midges.

14.3.3 JAPAN

In contrast to the allergenic chironomids from America and Africa, many of which originate from waters that are not evidently polluted with organic wastes, the situation in Japan concerns eutrophicated waters. Ironically, in Japan it is improvement in the quality of previously grossly polluted waters that has encouraged some of the increase in adult chironomid nuisance. Concurrent with this 'betterment of the quality of some waters was a deterioration in other previously unpolluted waters by wastewaters produced by rapid urbanization and industrialization' (Tabaru *et al.*, 1987). Adult midges, particularly of *Polypedilum kyotoense* (Tokunaga) occur in enormous numbers. Knowing of the medical problems induced in the Sudan, M. Sasa

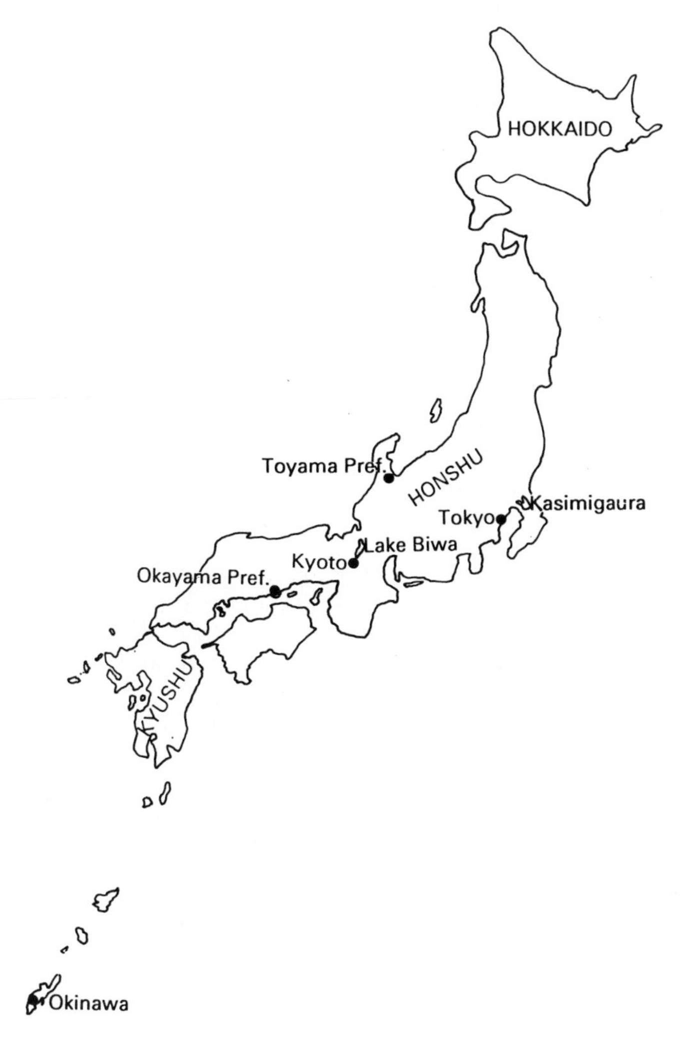

Figure 14.5 Map of Japan to show localities associated with proved midge allergy.

was stimulated to initiate a study of possible allergenicity of Japanese nuisance species (e.g. Sasa, 1985). A group of researchers under the guidance of Sasa extracted allergens and tested for allergic diseases. Three publications (in Japanese) followed shortly thereafter, demonstrating the role of chironomid midges in bronchial asthma in the Toyama region (Igarashi *et al.*, 1985; Ito *et al.*, 1985; Mizukami *et al.*, 1986). The species involved were the Chironominae *P. kyotoense*, *Tanytarsus oyamai* Sasa and *Chironomus salinarius* Kieffer and an Orthocladiinae, *Tokunagayusurika akamusi* (Tokunaga). The problems were not restricted to Toyama; shortly afterwards, researchers in

Tokyo made similar findings and came to similar conclusions (Ito *et al.*, 1985, 1986).

14.3.4 EUROPE

Many European waters are conducive to the development of large populations of nuisance midges (Chapter 13). However, there is a paucity of reported allergy to midges, perhaps because of a belief that the clinician need look no further than the house dust mite (*Dermatophagoides pteronyssinus*), established as an overwhelmingly dominant environmental allergen. It is conceivable that some skin-test or RAST positivity to chironomids may be due to allergen cross-reactivity with *D. pteronyssinus*, but the evidence is equivocal or weak.

The earliest report of a suspected role for chironomids in allergy in Europe came from a survey of allergic disease amongst Swedish farmers from rural Laholm (Eriksson *et al.*, 1985). In contrast to the preconceived ideas concerning the significance of *D. pteronyssinus* and moulds in allergy, skin-prick test and RAST showed a higher frequency of response to 'larval chironomid' and 'Sudanese nimitti' (17 and 15% respectively) than to house dust mite and *Cladosporium* (7% and 2% respectively). This led the authors to suggest that chironomid-associated allergens are of importance in allergic disease in farmers.

Skin tests with chironomid larval allergens showed that 14% of individuals reporting to a dermatology clinic in Britain gave positive responses. In a follow-up investigation of a semi-rural population exposed to midges emanating from a water-storage reservoir, a low incidence and severity of allergic symptoms were found (McHugh *et al.*, 1988). RAST tests for anti-midge antibodies (IgE) correlated with the symptoms revealed by survey, but did not correlate with skin-prick results. This cast doubt on the nature and extent of chironomid allergy in this selected community.

Another community with a lengthy exposure to nuisance midges is Venice, Italy (Ali *et al.*, 1992; section 13.2). The residents exposed to *Chironomus salinarius* have proved to have chironomid-related allergic disease (Figure 14.1), first found in a child (Giacomin and Tassi, 1988) and confirmed for children and adults (Marcer *et al.*, 1990).

14.4 OCCUPATIONAL ALLERGY TO CHIRONOMIDAE

It may come as a surprise to many that there is a commercial trade in chironomids. However, estimates of more than 60 million fish kept as pets in the western part of Germany and their equal popularity in many other affluent countries is a clue as to the nature of the chironomid trade. A common source of fish food is freeze-dried larvae of chironomid midges (Chapter 17), almost exclusively the 'bloodworms' of *Chironomus* species.

These may be sold entire, or blended with other products such as tubificids and brine shrimps.

In the early 1980s, five pet-fish owners reported to the Munich University Chest Clinic that they experienced severe asthma symptoms on exposure to fish-foods. In response to an announcement in two pet-fish journals, 32 individuals identified the symptoms of type I hypersensitivity (wheezing and shortage of breath, rhinitis, conjunctivitis and skin weals) on exposure to dried midge larvae. A further 63 patients were recruited from a production facility for fish-food, as were five co-workers in the research laboratory where haemoglobins were being investigated (Baur, 1982; Baur *et al.*, 1982). From this total of 105 individuals, 48 showed immediate (type I) hypersensitivity reactions. Sera of 100 individuals were available for testing using *in vivo* skin tests and RAST to measure specific IgE to chironomid extracts including haemoglobins.

The skin tests showed characteristic weal-and-flare demonstrating immediate (type I) hypersensitivity; and bronchial challenge with chironomid haemoglobin produced a typical asthma attack. *In vitro* tests showed highly significantly elevated levels of IgE antibodies to chironomid haemoglobins in 33 of 43 symptomatic individuals and in only 1 of 57 asymptomatics.

These early studies on occupational allergy provided the patients, their sera and other materials, including chironomid haemoglobins, for the elucidation of the allergens described in section 14.6.1.

At about the same time, two smaller studies revealed a role for chironomid fish-food in allergy. Also in Germany, Fuchs and Kleinhaus (1982) reported a female patient with a clear hypersensitivity to fish-food containing chironomid larvae. Her one-year long work in a firm that produced aquarium filters involved her in handling chironomid-based fish-food. Symptoms of acute rhinitis and conjunctivitis developed after 6 months exposure.

In Italy, in what was described as a 'rare case of occupational asthma in an aquarium keeper following exposure to fish-feed', Resta *et al.* (1982) established that 'the larvae of a mosquito (Echinodorus Plasmosus)' was the responsible agent. Using skin-prick tests, specific IgE and bronchial challenge, they established the cause of the aquarist's occupational asthma as conjunctivitis and rhinitis. But what was the causative agent? There is no such mosquito as *Echinodorus plasmosus* (also spelled as *plamosus*), and besides, mosquitoes are not reared and used as fish-food. It seems likely to me that this name is a wild typographical error for *Chironomus plumosus* Linnaeus, a common component of commercial fish-food.

A further European record of allergy of exposure to midge larvae comes from Sweden, through Eriksson *et al.* (1984). Occupational exposure to chironomids undoubtedly occurs elsewhere than Europe, but the fact that most (perhaps all) medical reports derive from these countries probably relates to the high standards of occupational health and safety legislation.

One of the few documented observations of contact dermatitis from chironomids comes from Europe (Brasch *et al.*, 1992), in a German patient who handled bloodworms as an aquarist; this provoked eczema and, when scratch-tested, massive oedema. The patient was a leisure fisherman, who fished regularly in Swedish lakes, where he experienced similar allergic symptoms when exposed to the flying adult midges.

As far as can be established from the medical literature, occupational and hobbyist allergy to chironomids has been reported outside Europe only by Weil (1938, 1940) for one power station worker. However, the trade and use of chironomids as fish-food, and research on haemoglobins, involve people in many countries elsewhere in the world and the problems are likely to be more general than publications indicate.

14.5 THE EXTENT OF THE PROBLEM: EPIDEMIOLOGY

As the studies outlined above and more recent ones demonstrate, chironomids induce environmental allergies in exposed populations in Egypt, Italy, Sweden, UK, USA and particularly so in Japan (e.g. Cranston *et al.*, 1981, 1983; Kagen *et al.*, 1984; Giacomin and Tassi, 1988; Marcer *et al.*, 1990; Eriksson *et al.*, 1985; Sasa, 1985; Ito *et al.*, 1986, 1988; Kino *et al.*, 1987; Igarashi *et al.*, 1987; Murakami *et al.*, 1988; Ishii *et al.*, 1988). Furthermore, reports of allergic complaints grow in number amongst those who handle chironomids occupationally (e.g. Eriksson *et al.*, 1984; Baur *et al.*, 1982; Baur and Mazur, 1988). But just how many individuals suffer?

Epidemiological estimates were derived from a Sudanese study (Kay *et al.*, 1982) in which a community living beside the River Nile, and seasonally exposed to midge nuisance, were compared with a community living 8 km further from the midge source, in the upwind direction of the prevailing winds. In the exposed community the incidence of asthma amongst more than 5000 individuals surveyed was 4.9%, in contrast to 3.2% of more than 2500 controls. The disparity in rates of allergic rhinitis was greater, with rates of 6.7% for the exposed community compared with 1.5% in the control area. These statistically significant differences were ascribed to the seasonal exposure to allergenic midges. If the excess rates of asthma and rhinitis are all due to midge exposure, the incidence of chironomid-induced disease potentially number at least tens of thousands of cases, since such a substantial part of the population of Sudan and Egypt lives by the Nile, close to the allergen source (Figure 14.4).

Epidemiological estimates are difficult to acquire and are rarely available for other areas. However, the incidence amongst self-reported or referred patients to allergy clinics has been investigated. Kagen *et al.* (1984) found 45% of atopic individuals in Wisconsin heavily exposed to *C. plumosus* had immunological reactions to the midge. In Sweden, Eriksson *et al.* (1989) tested 2368 consecutive adult patients reporting to a clinic with asthma and or rhinitis: 26% of atopics and 4% of non-atopics (14% of patients) gave a positive skin-prick test to chironomid extract from fish-food. RAST on a subsample of 100 consecutive sera gave an 8% response in atopics, 4% overall. The authors concluded that chironomids might be allergens of clinical importance in Sweden.

In Japan, Ito *et al.* (1986) surveyed asthmatics in metropolitan Tokyo

(Figure 14.5). Although there had been no previous reports of chironomid hypersensitivity, they made the 'unexpected' finding that approximately 40% of >300 asthmatic patients were skin-prick positive to extracts of either larval or adult *T. akamusi*, with 32.4% of 105 individuals demonstrated by RAST to have specific IgE directed against the same species. In this study, midges were the second major allergen after house dust mites.

In Toyama, 119 asthmatic children (aged 1–15 years, mean 8.3) were tested with extracts of selected species of nuisance chironomids, and with house dust mite extracts (Igarashi *et al.*, 1987). Positive skin-test rates ranging between 7.6% and 23.4% were produced by different midge extracts, with nearly 95% positive for *Dermatophagoides farinae*, suggesting perhaps some cross-reactivity. RASTs performed on selected patients' sera demonstrated specific IgE directed against midges. This kind of evidence, together with demonstrations of the widespread nature of midge nuisance, encourages Japanese allergists to believe that chironomids are second only to house dust mite as environmental allergens.

14.6 THE ALLERGEN

Accurate diagnosis of allergy and prevention by means such as desensitization requires detailed knowledge of the allergen(s) responsible for the disease.

14.6.1 IDENTIFICATION OF ALLERGENS

In the chironomid studies two approaches have been taken to identify the allergens:

- For occupational allergy to the larvae, haemoglobins were identified rapidly as causative (e.g. Baur, 1982; Baur *et al.*, 1982) and no additional allergens were sought.
- For environmental allergy provoked by adult midges, a tedious search was undertaken to purify allergenic fractions from bulk extracts of adult flies (e.g. Gad El Rab *et al.*, 1980; Tee *et al.*, 1983).

By 1983, the evidence acquired by both approaches converged on the view that at least some of the allergens of adult *C. lewisi* in the Sudan were shared with some of those from larval *Chironomus riparius* Meigen (as *C. thummi thummi*, CTT) studied by Baur's group. This phenomenon, known as cross-reactivity, proved important in establishing the range of allergens present in Chironomidae.

Cross-reactive polymorphic haemoglobins were found within a single species of *Chironomus* (e.g. Baur *et al.*, 1983) as well as between other haemoglobin-possessing species (e.g. Cranston *et al.*, 1983; Tee *et al.*, 1985). The skin test results shown in Figure 14.2 provided some evidence that allergens were present in all stages of *C. lewisi*. Kagen *et al.* (1984) found both adults and larvae of *C. plumosus* contained the same major allergen and immunological studies showed that larval haemoglobin-based allergenicity could be retained through metamorphosis into the adult stage (e.g. Tee *et al.*, 1985, for *C. lewisi*; Prelicz *et al.*, 1986, for *Chironomus* species; Ito *et al.*, 1986, for *T. akamusi*).

In testing for the presence of haemoglobin-derived porphyrins in over 100 species in immature and adult life-history stages, Cranston (1988) found these chemicals were retained widely from larva to adult. Most species in those subfamilies in which larval red pigmentation occurs, namely the Tanypodinae and Chironominae, and exceptionally the Orthocladiinae (*Tokunagayusurika* Sasa), may retain haem to the adult (Cranston 1988, 1989a). Intepretation of these results should be treated with caution: the test identifies the presence of a porphyrin (haem) component, yet the allergenic determinant is the globin protein which may be dissociated from the haem. Using gel filtration and chromatography, Kawai and Konishi (1988) also found that some suspected haemoglobin did persist in small amounts into the adult. Furthermore, Kawai and Sakamoto (1992) found an immunological cross-reactivity between a larval haemoglobin component by mouse antibodies directed against a corresponding adult haemoglobin.

That chironomid haemoglobins are major allergens was demonstrated irrefutably by quantification of the immunological cross-reactivity between 33 different species using a wide range of immunological tests (Baur *et al.*, 1991). Earlier, Baur's group had been able to demonstrate the precise location of the antibody-binding site (epitope) on the different polymorphic haemoglobins (e.g. Baur *et al.*, 1986; Mazur *et al.*, 1987; Mazur *et al.*, 1988). These studies have been reviewed and expanded by Baur (1992), who reported the results of examination of 352 environmentally-exposed and 85 occupationally-exposed midge-allergy patients. With purified allergens, chironomid haemoglobin (*Chi t* I) was confirmed as the major allergen, with considerable cross-reactivity with other haemoglobins in the same and closely related species. Baur (*loc. cit.*) was able to identify T-cell and IgE-binding epitopes using peptides of *Chi t* I component III, and identified a genetic susceptibility to sensitization by *Chi t* I.

Taking a different approach, Kawai and Sakamoto (1992) sought (and found) certain common components amongst chironomid haemoglobins across the complete taxonomic spectrum of haemoglobin-possessing chironomids, using skin tests on mice to estimate cross-reactivity. Mouse IgE antibodies to a dimeric haemoglobin component from *Chironomus yoshimatui* Martin and Sublette reacted to dimeric components from tanytarsines, other chironomines and to *Tokunagayusurika*. An ever wider range of reactions was observed to a monomeric haemoglobin component of *Polypedilum nubifer* (Skuse), which reacted even to a tanypod component.

14.6.2 AIRBORNE ALLERGENS

A most interesting finding concerns the amount of insect-derived allergens in the air. The present dogma is that allergen particles must be less than 10 μm in diameter in order to reach the bronchi and lung parenchyma where they cause sensitization. Only Kino and colleagues (1987) appear to have addressed this question: how does the allergenic whole insect relate to airborne insect allergen?

Using a high-volume air sampler on the roof of their five-storey medical institute in Kyoto (Japan), airborne dust of diameter <10 μm was trapped on filter sheets. Examination of the eluates in different seasons allowed Kino and colleagues to establish airborne densities of allergens based on protein equivalents derived from caddisflies, silkworm wing and chironomid whole body. Chironomid equivalents were greater than the other two components, with a July (summer) peak of 5.2 ng m^{-3} and October value of 18 ng m^{-3}. The authors concluded that chironomid fragments of less than 10 μm were the most abundant allergenic particles of the three insects throughout the year, but were unable to say if these values of airborne insect allergens could cause asthmatic symptoms.

14.6.3 RETENTION OF HAEMOGLOBIN ALLERGENS FROM LARVA TO ADULT CHIRONOMID

The discovery that larval haemoglobins are retained into adult flies is of relatively recent origin. Schin *et al.* (1974) found that haemoglobins of *Chironomus pallidivittatus* (Malloch) were maximal in the fourth instar larva, declined through the pharate adult and were almost entirely absent in the mature adult. The mechanism of loss during metamorphosis was suggested to be through enzymatic degradation leading to haemoglobin breakdown products appearing in the meconium, the first excreta of the emerged adult.

Neonate adult midges frequently contain red pigments, presumed to be haemoglobin(s), and the darkening with age appears to be associated with haemoglobin degradation. However, Cranston (1988) found no evidence of excretion of haem in newly emerged individuals of two species and could not show that the meconium contained any haemoglobin or porphyrin derivate. Whether the allergenic determinant was excreted had not been tested but Kawai and Konishi (1988) did find allergenic components in the excreta of some species. Their method may not have distinguished between excretory products and saline-soluble allergenic components of deceased adult midges, and the matter remains an open question. Many allergists believe that it is the 'fishy' smelling excreta that is the means by which chironomid allergen is made available for inhalation.

14.6.4 CROSS-REACTIVITY WITH OTHER INSECTS

The causes of some cross-reactivity with more widespread insect proteins (Baldo and Panzani, 1988; Kino and Oshima, 1988; Eriksson *et al.*, 1989) have been little explored. There have been persistent reports of limited cross-reactivity between chironomid allergens (believed to include haemoglobins) and the allergens of other arthropods. Particularly persuasive evidence of cross-reactivity with house dust mite (*D. pteronyssinus*) through some shared antigens comes from the study of Nagano *et al.* (1992) involving human T-cells, which regulate response to allergens. However, there have been many publications suggesting allergen independent of chironomids, with no cross-reactivity against others, e.g. house dust mite, cockroach and locust. It is beyond the scope of this chapter to do more than suggest that the differing views may derive from the results of different immunological and biochemical techniques, which vary in sensitivity. Although it is possible that skin-prick tests may give indications of false positivity because of cross-reactivity, the problem is just as likely to result from faulty RAST tests that may incorrectly suggest low cross-reactivity (e.g. Yamashita *et al.*, 1989).

14.6.5 ALLERGENS OTHER THAN HAEMOGLOBINS

A question remains as to whether all allergenicity caused by chironomids is due to haemoglobins alone, as is indicated by the studies of Baur and colleagues (e.g. Baur *et al.*, 1991).

Kawai and Konishi (1988) immunized mice with adult extracts of three species of *Chironomus*, one *Glyptotendipes* species and one *Polypedilum*. A highly allergenic component occurring in a small volume was produced by all species. Its molecular weight of 15–18 kDa appears similar to the monomeric haemoglobins found in *C. lewisi* and *Chironomus* species studied by Baur's group. However, a higher molecular weight fraction of 110–374 kDa dominated in volume and was found to have twice the allergenicity of crude extract. A similar allergenic high molecular weight component was found by Tee and colleagues (1984) for *C. lewisi*. A corresponding component from *Chironomus tentans* Fabricius lacked allergenicity when tested by Prelicz *et al.* (1986).

Further evidence for an allergenic role for higher molecular weight fractions comes from the work of Matsuoka and colleagues (1988, 1990). The studies of this team indicate little cross-reactivity between *T. akamusi* and *Chironomus* species and a change in allergenicity between larval and adult stage. Using an enzyme-linked immunosorbent assay (the ELISA test, section 14.2.1(d)) for specific IgE, Matsuoka *et al.* (1990) inferred a role for the high molecular weight fraction that they argued could not be haemoglobin-associated. They speculated that the material was associated with egg production, perhaps vitellogenin, a substance with a molecular weight of greater than 500 kDa. The molecular weight, as the authors discuss, is greater than the usual maximum size of allergens of less than 100 kDa. This drawback, and the sometime failure of others to detect potent allergenicity associated with this high molecular weight fraction, leave its role in some doubt.

14.7 CURATIVE METHODS

Sadly, for the many Sudanese individuals allergic to chironomid midges, the answer remains as it has done since the 1930s: live at enough distance from the

River Nile to avoid the mass emergences of *C. lewisi*. This distance was deemed to be 5 km from Wadi Halfa, and studies of Kay *et. al.* (1982) imply that no immunological effects were detected at 8 km distance upwind from the river source.

In developed nations, avoidance must still be considered the best precaution, in the absence of the kind of control measures for the midge nuisance discussed in Chapter 13. As for immunotherapy, opinions differ – basically on a transatlantic basis. In North America, immunotherapy is widely prescribed and enacted. This involves challenging the sufferer of allergy with a series of increasing doses of allergen prior to natural exposure. This is particularly useful with seasonal allergens, such as grass pollens and chironomids. Kagen and colleagues investigated this technique for individuals exposed to *C. plumosus* and, although noting symptomatic improvement that correlated with specific immunoglobulin G (IgG), pointed out the lack of any double-blind trial of immunotherapy. Personal communication with American allergists suggests that midge immunotherapy has been attempted, at least in Florida, but the results have not been evaluated scientifically.

Generally the merits of immunotherapy are better recognized in North America and Europe than in Britain, where increased risks of anaphylaxis are stressed. The use of monoclonal antibodies to standardize extracts and development of cloned allergens alluded to by Kagen (1990) may be of benefit in immunotherapy, but the broad range of molecules involved in chironomid allergy may complicate development of simple immunotherapy.

14.8 SUMMARY AND CONCLUSIONS

The evidence is overwhelming for a major role for chironomid haemoglobins in allergen induction. Potentially there may be a role for higher molecular weight compounds which may be unassociated with haemoglobin, implying potential allergenicity even of midges that do not possess haemoglobin. Even if this latter suggestion is unconfirmed, the fact that so many nuisance Chironomidae (Chapter 13) have haemoglobins appears to justify some of the more sanguine predictions of medical significance. Thus the view of Prelicz *et al.* (1986) that 'all stages of Chironomidae must be seen as significant occupational and environmental allergens' has proved to be well founded, with the proviso that the potentially allergenic taxa possess larval haemoglobins. Recent studies by Baur and Liebers (1992) and Baur (1992) demonstrate that exposure to chironomid haemoglobins, whether through occupation, hobby or environmental exposure, give a high risk of sensitization with asthma, rhinitis and conjunctivitis.

However, great caution must be taken before inferring that all areas with nuisance midges must have concomitant allergy. Although predictions of the occurrence of allergy in several chironomid-exposed human populations have been fulfilled, there are exceptions, including the midge-exposed English community studied by McHugh *et al.* (1988) where there was no obvious increase in allergic disease. Other parts of the world with nuisance midge problems, such as the residential areas of Florida (Chapter 13) and the urban eutrophicated lakes of Perth, Western Australia, have little or no documented immunological problems associated with massive emergences of midges.

In view of this incomplete association between midge nuisance and immunological complaints, it is strongly advised that public (and press) speculation about any particular medical association should be restricted until epidemiological and clinical investigations demonstrate their existence.

15

Classification of water-bodies and pollution

C. Lindegaard

15.1 INTRODUCTION

The science of limnology was born some hundred years ago, and early in the 20th century freshwater biomonitoring used organisms as indicators of, primarily, organic pollution in streams (Kolkwitz and Marsson, 1909) and lakes (Thienemann, 1922). Chironomids were involved in the saprobic system and played an overwhelming role in the biological classification of lakes.

Although biomonitoring started 70–80 years ago and its necessity is evident, the state of the art remains primitive compared with other exact sciences. Why this is so is unclear but an important reason may be that our knowledge of taxonomy, zoogeography and ecology is very limited for aquatic invertebrates – and chironomids are no exception. However, biomonitoring has made considerable progress in recent years and many handbooks and reviews considering different aspects of human impact on freshwater emerge continually (e.g. Wiederholm, 1984a; Hellawell, 1986; Metcalfe, 1989; Rosenberg and Resh, 1992; Rosenberg, 1993).

The aim of this chapter is to describe the application of Chironomidae in classifying freshwater and in monitoring organic pollution and eutrophication and finally to review the rapidly growing area of new pollution types, where chironomids may be important monitoring organisms. In the latter case, particular reference is made to the use of chironomids in toxicity tests and in assessing deformities of chironomids when describing chemical pollution.

15.2 CLASSIFICATION AND EUTROPHICATION OF LAKES

Classification is an attempt to organize highly complex structures into a simpler characterization. Therefore, such procedures often raise more

The Chironomidae: Biology and ecology of non-biting midges
Edited by P.D. Armitage, P.S. Cranston and L.C.V. Pinder
Published in 1995 by Chapman & Hall. ISBN 0 412 45260 X

Table 15.1 Lake classification based on the profundal chironomid fauna (from Brundin, 1949, 1958; Sæther, 1975; Wiederhom, 1984a)

Lake type, trophic level	Indicator chironomids Europe	Indicator chironomids North America
Ultraoligotrophic	*Heterotrissocladius subpilosus* Brundin	*Heterotrissocladius oliveri* Sæther
Oligotrophic	*Tanytarsus lugens* Kieffer with *Heterotrissocladius grimshawi* (Edwards) and *H. scutellatus* (Goetghebuer)	*Tanytarsus* sp. with: *Monodiamesa tuberculata* Sæther and *Heterotrissocladius changi* Sæther
Mesotrophic	*Stictochironomus rosenschoeldi* (Zetterstedt) and *Sergentia coracina* (Zetterstedt)	*Chironomus atritibia* Malloch and *Sergentia coracina*
Moderately eutrophic	*Chironomus anthracinus* Zetterstedt	*Chironomus decorus* Johannsen
Strongly eutrophic	*Chironomus plumosus* L.	*Chironomus plumosus*
Dystrophic	*Chironomus tenuistylus* Brundin with: *Zalutschia zalutschicola* Lipina	*Chironomus* sp. with *Zalutschia zalutschicola*

questions than they solve and the long history of classification of waterbodies demonstrates this clearly. However, discussions of fundamental causal relations can be useful and the interaction between different aspects of limnology, as started by Thienemann (1922) and Naumann (1929), has been fruitful in producing the trophic classification scheme of lakes.

The history of trophic lake type systems is summarized by Brundin (1949) and Brinkhurst (1974). The predominance of chironomids in the profundal zone of lakes and their response to trophic conditions make them useful in classifying lakes, and a number of lake types are characterized by their chironomid assemblages (Table 15.1; Brundin, 1949, 1958; Sæther, 1975). Although this classification has wide validity in the northern hemisphere, Forsyth (1978) found difficulty in separating seven New Zealand lakes according to their chironomid fauna and Australian lakes likewise do not to conform (Timms, 1978).

The factors determining the chironomid assemblages in the profundal zone of regularly stratified lakes are an interaction primarily between the quality and quantity of food and the oxygen conditions. The available food originates exclusively from precipitating organic matter produced in the photic zone, except in relatively shallow oligotrophic lakes, which may also have a profundal production of phytobenthos. The chironomids characteristic of oligotrophic lakes are often free-living (Davies, 1975; Grimås and Wiederholm, 1979) and thus able to increase their range of foraging to seek scarce food at this end of the trophic spectrum (Wiederholm, 1984a). Food is more abundant in eutrophic lakes and the majority of chironomids live in fixed tubes from which they feed by browsing around and/or by filtration within the tube. The larval tubes allow respiration movements and facilitate compensation for the microstratification of oxygen occurring in these lakes in the sediment/water interface. Moreover these larvae can – due to the presence of haemoglobin – maintain a high oxygen uptake with decreasing ambient oxygen concentration, and they are able to survive up to several months without oxygen by switching to anaerobic metabolism combined with dormancy. Temperature, however, seems to be of minor importance as a controlling factor except in the ultraoligotrophic range, where some characteristic species are limited by warm water (Moore, 1978).

Deep lakes with a large hypolimnion maintain oligotrophic conditions (shortage of organic matter, high oxygen concentration) in the profundal zone even with a phytoplankton production, which in more shallow lakes causes eutrophic profundal conditions (high content of organic matter, oxygen depletion during stratification periods). Therefore, the correlation between trophic state (measured as phytoplankton standing crop or gross production) and profundal chironomid communities may be low.

The classical system was expanded by Sæther (1979c), who recognized 15 different profundal chironomid communities characteristic of six oligotrophic, three mesotrophic and six eutrophic lake types. These 15 chironomid communities are fairly well correlated with trophic conditions expressed as mean concentrations of total phosphorus/mean lake depth or as total chlorophyll a/mean lake depth in many North American and European lakes (Figure 15.1; Sæther, 1979c; Wiederholm, 1980). By using phosphorous and

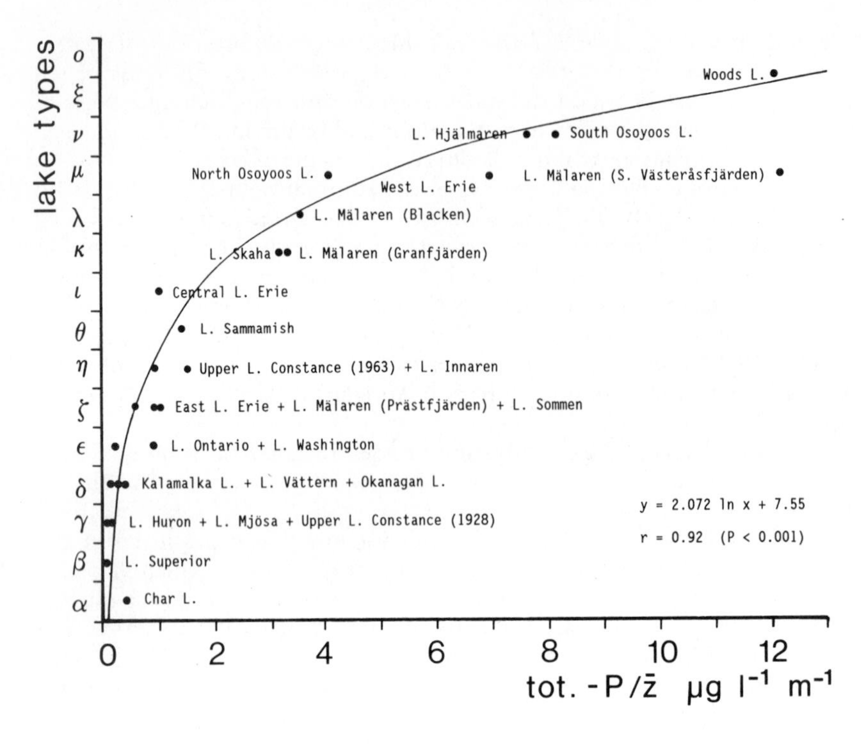
lake types
o
ξ
ν
μ
λ
κ
ι
θ
η
ζ
ε
δ
γ
β
α
Woods L.
L. Hjälmaren
South Osoyoos L.
North Osoyoos L.
West L. Erie
L. Mälaren (S. Västeråsfjärden)
L. Mälaren (Blacken)
L. Skaha
L. Mälaren (Granfjärden)
Central L. Erie
L. Sammamish
Upper L. Constance (1963) + L. Innaren
East L. Erie + L. Mälaren (Prästfjärden) + L. Sommen
L. Ontario + L. Washington
Kalamalka L. + L. Vättern + Okanagan L.
L. Huron + L. Mjösa + Upper L. Constance (1928)
L. Superior
Char L.
y = 2.072 ln x + 7.55
r = 0.92 (P < 0.001)
0
2
4
6
8
10
12
tot. -P/z̄ μg l⁻¹ m⁻¹

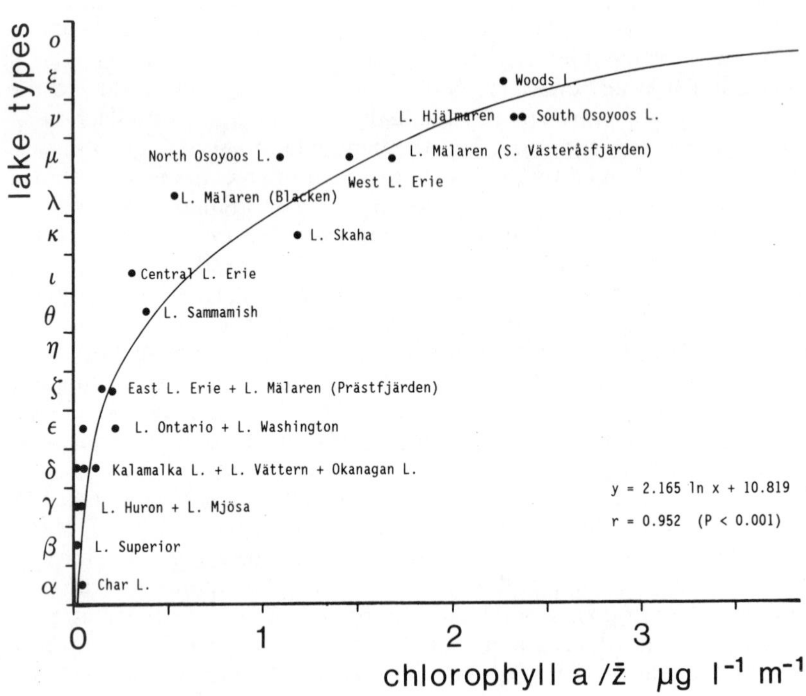
lake types
o
ξ
ν
μ
λ
κ
ι
θ
η
ζ
ε
δ
γ
β
α
Woods L.
L. Hjälmaren
South Osoyoos L.
North Osoyoos L.
L. Mälaren (S. Västeråsfjärden)
West L. Erie
L. Mälaren (Blacken)
L. Skaha
Central L. Erie
L. Sammamish
East L. Erie + L. Mälaren (Prästfjärden)
L. Ontario + L. Washington
Kalamalka L. + L. Vättern + Okanagan L.
L. Huron + L. Mjösa
L. Superior
Char L.
y = 2.165 ln x + 10.819
r = 0.952 (P < 0.001)
0
1
2
3
chlorophyll a/z̄ μg l⁻¹ m⁻¹

chlorophyll a divided by the lake mean depth, the problems of correlations as regards deep lakes are avoided (cf. Sæther, 1980b). Also, Aagaard (1986), Gerstmeier (1989a) and Lindegaard (1992) found that these chironomid assemblages reflected the trophic state of a number of lakes very well. However, Lindegaard and Mæhl (1993) pointed out that the correlations did not fit for a relatively shallow (mean depth 7 m) ultraoligotrophic lake in South Greenland. Also the high arctic Char Lake (mean depth 10 m) shows a higher phosphorous content than expected from the chironomid assemblage (Figure 15.1). This was true for a number of the oligotrophic lakes (maximum depth c. 20 m) studied by Aagaard (1986).

The creators of the lake type system regarded the various types as generalizations applying to certain sections of the continuum from oligotrophic to eutrophic conditions. Therefore, the system often failed to classify lakes having a mixture of indicator species. Wiederholm (1976, 1980) bypassed this problem by introducing the **benthic quality index** (BQI):

$$BQI = \sum_{i=0}^{5} \frac{k_i n_i}{N}$$

where k_i represents a value from 5 to 1 constant for each indicator species and 0 if these indicator species are absent, n_i represents the number of individuals of each indicator species, and N the total number of individuals of all indicator species present.

With only seven indicator taxa – *Heterotrissocladius subpilosus* (Kieffer) (k=5), *Micropsectra* spp. and *Paracladopelma* spp. (k=4), *Sergentia coracina* (Zetterstedt) and *Stictochironomus rosenschoeldi* (Zetterstedt) (k=3), *Chironomus anthracinus* (Zetterstedt) (k=2) and *Chironomus plumosus* (Linnaeus)(k=1) – Wiederholm (1980) applied this index and found a good correlation between total phosphorous/mean lake depth and the BQI of the great Swedish lakes. Aagaard (1986) found that the BQI agreed quite well with Sæther's (1979c) classification when applied on 50 Norwegian lakes, and Meriläinen (1987) used the BQI to describe a recovery of the profundal fauna in a Finnish lake due to a decrease in effluent loading over 15 years. Kansanen *et al.* (1990) found that the BQI expanded to 22 indicator taxa was useful to describe a pollution gradient in an oligotrophic Finnish lake heavily polluted by organic effluents from the wood-processing industry.

Wiederholm (1980) also suggested an oligochaete/oligochaete plus sedentary chironomid ratio (O/O+C) based on numbers of individuals. This ratio divided by the sampling depth showed a fairly good correlation with chlorophyll a content in the great Swedish lakes. This ratio seems to have been used in only a few studies (Bazzanti and Seminara, 1985; Seminara and Bazzanti, 1988) and its general application is still uncertain. However,

Figure 15.1 Correlation between different lakes based on profundal chironomid communities and total phosphorus/mean lake depth (upper panel) and chlorophyll a/mean lake depth (lower panel). (Redrawn from Sæther, 1979c, where also references to the different lakes are given.)

because a change from a chironomid-dominated community to an oligochaete-dominated community is often one of the first signs of eutrophication (Carter, 1978; Sæther, 1979c, 1980a), this ratio may be valuable as an early warning system.

The profundal chironomid communities have been valuable not only in describing the present trophic state but also in revealing the pollution history of this century through the analysis of subfossil chironomid remains in combination with older analyses of the profundal fauna as done by e.g. Warwick (1975), Wiederholm and Eriksson (1979), Kansanen and Aho (1981) and Kansanen (1985). For the use of fossil chironomid remains in interpreting whole lake histories, see Chapter 16.

Cluster analysis and ordination methods have been used to test the benthic lake type concept based on chironomid communities, but have not given rise to any change in the systems of Brundin and Sæther (Wiederholm, 1981; Kansanen *et al.*, 1984; Johnson, 1989; Johnson and Wiederholm, 1989). Discriminant analysis of possible factors controlling chironomid assemblages showed that the availability of food was most important and that oxygen concentration comes into effect only in eutrophic lakes (Kansanen *et al.*, 1984). Therefore, the system cannot describe the trophic level of, for example, the oligotrophic Banyoles Lake in which oxygen deficiency relates to morphology and subterranean inflow of water (Rieradevall and Prat, 1991).

The chironomids of open water areas in shallow non-stratified lakes do not clearly delimit trophic level, because both predation and the degree of resuspension of substrate influence zoobenthic conditions (e.g. Kajak, 1977; Jónasson and Lindegaard, 1979). However, in such lakes with high densities of zoobenthos, larvae of the *Chironomus plumosus* type often dominate (e.g. Iwakuma and Yasuno, 1981; Dévai, 1990). This may be due to their ability to feed both as surface collectors and as filterers (Walshe, 1951; McLachlan, 1977b). Nevertheless, Dévai and Moldován (1983) were able to correlate the chironomid fauna with a trophic gradient measured as organic content in the sediment of the shallow Hungarian Lake Balaton: the higher proportion of larvae of *C. plumosus* type [=*C. balatonicus* Dévai, Wülker et Scholl] the higher the carbon content in the sediment.

Although Sæther (1975, 1979c) lists chironomid species characteristic of littoral and sublittoral zones in both Nearctic and Palaearctic lakes, it seems that especially the littoral species, which characterize the meso- and eutrophic range, occur over a rather broad trophic gradient making them poor indicators, whereas littoral indicator species of oligotrophic lake types better delimit lake trophy (Hershey, 1985b). Also variations in, for example, geology and exposure to wind cause large differences in littoral fauna both within a single lake and between lakes, hampering a close relationship between littoral species or communities and lake trophy (Kansanen *et al.*, 1984; Harper and Cloutier, 1986; Dall *et al.*, 1990). However, as the majority of lakes are shallow, there is a clear need for a benthic lake typology based on littoral fauna.

15.3 CLASSIFICATION OF RUNNING WATERS

Many suggestions about classifying running waters have been proposed but the success from lake typology has never been followed up in a useful and practical running-water classification. A major problem is probably the obvious longitudinal zonation within the single river system, often resulting in a high number of zones in each river. A general world-wide classification system based on both physical and biological characteristics was presented by Illies and Botosaneanu (1963) by suggesting an upper (**rhithron**) and lower (**potamon**) region. These two reaches were mainly defined by temperature and substrate, but the fauna was also characterized. In order to make the classification more detailed a number of subdivisions was suggested. Realizing the gradual changes downstream in a river system, Vannote *et al.* (1980) launched the **river continuum** concept describing from headwaters to mouth the gradient of physical variables resulting in a continuum of biological responses. The biological continuum has been further discussed (e.g. Cushing *et al.*, 1983; Minshall and Petersen, 1985).

Chironomids were never involved in describing these systems but there have been some attempts to correlate chironomid communities with river zonation. Thienemann (1954) described a number of chironomid communities in primarily high altitude streams, but had at that time only a few data from lowland streams and very little from large rivers. His generalized view was that, in terms of species, Orthocladiinae (including Diamesinae and Prodiamesinae) decreased from about 70% in mountain streams to 37% in lowland cold streams and 29% in lowland rivers, while Chironomini simultaneously increased from 6 to over 16 to 43%, respectively. This general tendency has been confirmed by a number of subsequent studies. The overall conclusion has not changed, and briefly the zoocenoses of chironomids in different categories of northern hemisphere temperate running waters can be described in the following way (Table 15.2).

In a typical northern hemisphere glacial brook there are few species, a situation that is characterized by the cold-stenothermous Diamesinae making up 30–70% of the species. The remaining species are Orthocladiinae and a few Podonominae. At lower altitudes the water temperature in summer increases as does species richness and Orthocladiinae take over the dominance with 70–80%. Tanypodinae and Tanytarsini come in with up to c. 10% and a few Chironomini may occur. This high percentage of Orthocladiinae decreases gradually as the river reaches the lowland, where they constitute about 30–40%, while Chironomini increase to about the same level. Diamesinae decrease rapidly and only a few species persist into lowland rivers. When the mean water temperature in summer reaches c. 10 °C a few members of Prodiamesinae appear and remain stable members to the mouth except in very large rivers (e.g. the lower part of the River Rhine). The percentages of Tanypodinae and Tanytarsini are remarkably similar and vary from a few per cent to about 15–20%, but without any correlation to river zonation. Apparently, species richness declines in the large lowland rivers as shown by Wilson and Wilson (1983, 1985) in the lower Rhine. However, the

Table 15.2 Number of chironomid taxa and percentage distribution of chironomid subfamilies/tribes in a wide range of temperate running waters representing major sections of the river continuum. Rhithral groups are Pentaneurini, Diamesinae, Orthocladiinae, Chironominae connectentes (Wilson, 1980). Figures in brackets indicate ranges. References and more detailed review data are given in Lindegaard and Brodersen (in press)

River type	No. of studies	Podo-nominae	Tany-podinae	Buchono-myiinae	Diamesinae	Prodia-mesinae	Ortho-cladiinae	Chiro-nomini	Tanytarsini	No. of taxa	% 'rhithral groups'
Glacier brooks	4	0	0	0	56 (45–67)	0	44 (33–55)	0	0	7 (4–11)	100
Alpine streams	5	0	0	0	22 (14–35)	0	73 (59–81)	0	5 (0–12)	23 (17–31)	95 (88–100)
Subalpine streams	9	1 (0–5)	4 (0–8)	0	20 (10–32)	1 (0–4)	63 (51–73)	3 (0–12)	8 (4–11)	49 (37–58)	89 (85–96)
Mountain streams	9	<1 (0–2)	5 (2–10)	0	11 (4–18)	1 (0–2)	65 (50–73)	6 (2–12)	10 (4–16)	71 (26–144)	87 (74–93)
Lower mountain streams	20	<1 (0–1)	9 (2–18)	<1 (0–1)	6 (0–15)	1 (0–3)	59 (39–78)	10 (1–23)	14 (5–20)	95 (32–168)	80 (60–91)
Lowland cold streams	17	0	11 (2–25)	0	6 (1–11)	3 (0–6)	53 (44–62)	15 (0–23)	13 (3–24)	60 (28–134)	76 (66–86)
Lowland warm streams	8	0	15 (8–21)	0	4 (0–9)	1 (0–2)	46 (34–57)	19 (10–27)	15 (9–28)	78 (35–143)	67 (46–84)
Lowland warm rivers	23	0	9 (3–18)	<1 (0–1)	1 (0–5)	2 (0–8)	36 (24–47)	39 (26–54)	12 (5–20)	67 (18–148)	58 (46–75)

low species richness in the lower Rhine, as with other rivers in industrialized countries, might also be a consequence of both organic matter and a multitude of toxic substances in a river passing the most densely populated and highly industrial areas of western Europe (Klink, 1989).

The overall factors controlling this distribution pattern of chironomids are temperature and flow regime, which also to a large degree control availability of food and substrate. The importance of substrate and food is clearly correlated to the ratio of Orthocladiinae versus Chironomini. The latter group is suggested to be merely gathering collectors of detritus and algae (substrate feeders and filtrators), while Orthocladiinae often are scrapers preferring periphyton. Therefore the chironomid succession reflects very well the idea of the river continuum concept (Vannote *et al.*, 1980; Cummins *et al.*, 1984; Ward and Williams, 1986). However, as temperature and velocity influence the communities many exceptions from the observed pattern are possible. For example, small springfed lowland streams have low annual temperature amplitudes and often exhibit a greater slope than large rivers. Consequently the chironomid communities in such streams resemble those found in low mountain streams rather than communities in large lowland rivers.

The communities in tropical streams and rivers remain very poorly known, but apparently the higher average temperatures benefit Chironomini, which thus make up a higher percentage compared with otherwise similar temperate running waters. Chironomid studies from Australia and New Zealand demonstrate somewhat similar zonation as in the northern hemisphere, except that Aphroteniinae occur in low order streams (Cranston and Edward, 1992); in headwaters and cooler streams Podonominae are abundant; Diamesinae are species-poor although sometimes abundant; Prodiamesinae are absent and Pseudochironomini (*Riethia*) may be numerous in middle order streams. These patterns are evident in temperate south-eastern Australia (e.g. Metzeling *et al.*, 1984) and mediterranean south-west Australia even though the headwaters here are of low elevation (Bunn *et al.*, 1986; Storey and Edward, 1989). In tropical monsoonal northern Australia, where upland catchments are also of modest elevation, subfamily representation is restricted, with Diamesinae absent, Aphroteniinae rare only in upper reaches and Orthocladiinae frequent only in low order streams. In the middle sections and lowland flood-plains Chironominae dominate; Tanypodinae and Tanytarsini are distributed throughout (Cranston, 1992).

In New Zealand, the number of species appears to be low (e.g. Cowie, 1985; Boothroyd, 1987). Whether this reflects actuality or lack of sampling is discussed by Winterbourn *et al.* (1981) and Cowie (1983).

Until now we have dealt with chironomid communities at the level of subfamily or tribe only, but correlations between certain species assemblages and zonation have been suggested by e.g. Lehmann (1971), Kownacki and Zosidze (1980), Braukmann (1987), and Laville and Vinçon (1991). Further, Kownacka and Kownacki (1972) and Kownacki (1991) outline communities in a number of European glacier brooks in different mountain massifs. When detailed descriptions of spatial distribution and ecology at species level are used to develop a typology for running waters, they can be applied only within restricted areas with a uniform composition of species. However, Laville and Vinçon (1991) thought that their typology of Pyrenean running waters may be used in other European mountain massifs but that the knowledge of the exact species composition still is too small.

Wilson (1980) restricts himself to characterizing groups as 'rhithral' (Pentaneurini, Diamesinae, Orthocladiinae, Chironomini connectentes) or 'potamal' (Tanypodini, Macropelopiini, Prodiamesinae, Chironomini genuini, Tanytarsini). This approach shows a great similarity to the succession described on the basis of the Diamesinae/Orthocladiinae/Chironomini relationships (Table 15.2).

Attempts to use cluster analysis and ordination methods in classification of running-water communities have been made including all macroinvertebrates (e.g. Wright *et al.*, 1984; Braukmann, 1987) and for chironomids exclusively (Aagaard, 1993; Rossaro, 1991, 1993) but so far no practical classification system for a larger geographical area has been proposed.

15.4 ORGANIC ENRICHMENT OF RUNNING WATERS

The saprobic system developed by Kolkwitz and Marsson (1909) and continuously improved (e.g. Sládecek, 1973; Braukmann, 1987) has been widely used by monitoring services in Central Europe. The genus *Chironomus* (especially *C. riparius* (Meigen) = *C. thummi* Kieffer) is one of the characteristic taxa which by increasing density indicates a heavy loading of organic matter (e.g. Davies and Hawkes, 1981; Pinder and Farr, 1987; Ferrington and Crisp, 1989). However, chironomids indicating clean water have never been incorporated in the saprobic system due to difficulties in identifying larvae. Correctly used, the saprobic system is rather time-consuming, and to meet the authorities' demand of 'maximum information in minimum time' a number of rapid assessment systems (e.g. Hilsenhoff, 1982; Armitage *et al.*, 1983; De Pauw and Vanhooren, 1983; Metcalfe, 1989) have been developed. Chironomids enter these systems at family level or are, at most, identified to a reduced number of genera, which do not necessarily express much about pollution. In spite of this cursory identification used by water pollution control authorities, numerous papers describe detailed changes in chironomid communities caused by organic pollution (e.g. Murphy and Edwards, 1982; Armitage and Blackburn, 1985; Barclay and Harrel, 1985; Bazzanti and Bambacigno, 1987; Wilson, 1987; Ferrington and Crisp, 1989; Kownacki, 1989; Rae, 1989).

Recently, systems to assess water quality in running waters by using chironomids have been developed. These systems are based on samples of pupal exuviae obtained by drift nets or taken by net along the banks. Wilson and Bright (1973) and Wilson and McGill (1977) showed that samples obtained in this way reflect the chironomid communities in a 0.5 km upstream section of a river. These sampling methods also facilitate sampling large rivers, which have frequently been neglected in both faunistic and pollution surveys due to methodological problems. Further, because of the very long flying periods of lotic chironomids (e.g. Pinder, 1974; Lindegaard and Mortensen, 1988), a single collection during the period between late May and late September in northern hemisphere temperate rivers provides a representative picture of the species composition if at least 200 exuviae are analysed (McGill *et al.*, 1979; Ruse and Wilson, 1984).

To use the species composition in pollution assessment, Wilson and McGill (1982) allocated each genus of British Chironomidae to one of four pollution tolerance classes, based mainly on their tolerance to organic pollution. Later Ruse and Wilson (1984) simplified the four classes to two: tolerant or intolerant taxa. They suggested two indices: the percentage of intolerant taxa and the percentage of intolerant individuals.

The latter index in particular showed high correlations with routine quality monitoring methods (BMWP, Armitage *et al.*, 1983) in a single catchment area, although such agreements were not expected to be universal (Ruse and Wilson, 1984). Also Wilson (1989) found it difficult to describe the same reactions to organic pollution by chironomid communities in eight English lowland rivers, because it seemed that chironomid assemblages related more to the rivers themselves than to the influences of the sewage effluent. However, he concluded that the potential of chironomids as river quality indicators is excellent, and that with a better knowledge of the ecology of the individual chironomid species in different river types, it must be possible to predict the species assemblages which will characterize low-grade pollutional stress in rivers of different types. A general discussion of abiotic and biotic factors controlling communities in lotic habitats is given by Power *et al.* (1988). Further, Wright *et al.* (1985, 1988) and Moss *et al.* (1987) have predicted lotic macroinvertebrate communities in the British Isles using environmental data.

Using data on pupal exuviae from 60 stations in the River Garonne (France), Bazerque *et al.* (1989) established a 'Chironomid Index' based on only 26 species, each allocated to one of five pollution tolerance classes. This index provided similar information on pollution status when compared with some other biotic indices and the authors suggest it may be applicable to lowland streams and rivers of Central Europe.

15.5 ACIDIFICATION

Large areas of Europe and North America are poor in calcium and in such areas, with low buffered soils, acid precipitation lowers the pH of surface water to about 4.5. Whether by precipitation or by effluents of acid mine-wastewater, acidification has a detrimental effect on aquatic ecosystems. Organisms may be affected by:

- decreasing pH, disturbing for example the Ca and Na regulation;
- an increase in toxic elements (Al, Cd, Cu, Zn and probably Hg, Pb) leached from the surroundings;
- a possible change in nutrients and disruptions of food chains by, for

example, decreased microbial activity (e.g. Wiederholm, 1984a; Campbell and Stokes, 1985; Schindler *et al.*, 1985).

Recent studies on these aspects have improved our knowledge of zoobenthos responses to acidification (e.g. Økland and Økland, 1986; Raddum *et al.*, 1988). Chironomids have frequently been neglected in these applied studies, probably due to difficulties with identifications, and at family level they have often been considered to be tolerant of low pH. This is certainly true only in a limited number of species and in fact the traditional classification scheme of lakes comprises chironomid indicator species of natural acid lakes, namely the polyhumic lake type (Table 15.1). Therefore it has been possible to trace anthropogenic acidification through analyses of the subfossil chironomid remains in the upper sediment layer of a number of lakes (e.g. Henrikson and Oscarson, 1985; Johnson and McNeil, 1988; Brodin, 1990) and to describe or even classify the recent acidification of lakes by analysing their chironomid communities (e.g. Wiederholm and Eriksson, 1977; Raddum and Sæther, 1981; Walker *et al.*, 1985; Dermott *et al.*, 1986; Buskens, 1987; Johnson *et al.*, 1990). The trend is that, with increasing acidity, highly diverse chironomid assemblages are replaced by a few tolerant species, which at first become numerous but later show severe reductions with decreasing pH.

Members of the genus *Chironomus* especially seem to tolerate extremely low pH values. Thus, *C.* near *maturus* Johannsen occurred in an acid strip-mine lake with pH about 3 in Illinois (Zullo and Stahl, 1988); *C. riparius* lived in ponds with pH 2.8 at the Smoking Hills, North West Territory (Canada) (Havas and Hutchinson, 1982); *C. plumosus* was found in acid strip-mine lakes at pH down to 2.3 in central USA (Harp and Campbell, 1967); and *C. acerbiphilus* Tokunaga was recorded in a volcanic lake in Japan with pH at 1.4 (Yamamoto, 1986). Larvae of *C. riparius* from the Smoking Hills contained twice the amount of haemoglobin as that of the same species from neutral water and Jernelöv *et al.* (1981) suggested that the higher buffering capacity found in the haemolymph might be attributed to the higher content of haemoglobin. *C. riparius* larvae isolated from the fumigated pond at the Smoking Hills with pH 2.8 were more resistant to high levels of metals and had no or a minor loss of Na and Cl compared with other invertebrates from neutral water, when exposed to low pH (Havas and Hutchinson, 1982, 1983).

Effects on zoobenthos as a result of experimental acidification as well as of liming acid lakes have been followed over several years (e.g. Eriksson *et al.*, 1983; Raddum *et al.*, 1984; Schindler *et al.*, 1985; Bilyj and Davies, 1989). Littoral and sublittoral chironomid communities seemed to recover after lime treatment, in both quality and quantity, while profundal communities did not. Toxic effects of aluminium hydroxides precipitating with increasing pH and accumulating in the profundal sediment might be responsible for this (Raddum *et al.*, 1984, 1986). Further references to toxicity tests with pH and aluminium are given in Tables 15.3 and 15.4.

Chironomids in acidified streams have attracted much less attention, but apparently chironomid communities react in the same way as they do in lakes, namely by decreasing species richness and with a few tolerant species

Table 15.3 Examples of Chironomidae used in single species toxicity tests (further references in Rosenberg, 1993; Timmermans *et al.*, 1992)

Taxa	Toxicant(s)	Reference
Acute tests		
Chironomus decorus	B	Maier and Knight, 1991
Chironomus decorus	Cu	Kosalwat and Knight, 1987
Chironomus plumosus	DEHP (phthalate esters)	Streufert *et al.*, 1980
Chironomus riparius	eight herbicides	Buhl and Faerber, 1989
Chironomus riparius	lindane (insecticide)	Green *et al.*, 1986
Chironomus riparius	ammonia	Williams *et al.*, 1986a
Chironomus riparius	phenol	Green *et al.*, 1985
Chironomus riparius	Cd	Williams *et al.*, 1985, 1986b
Chironomus riparius	PCP (pentachlorophenol)	Fisher and Wadleigh, 1986
Chironomus riparius	carbaryl (insecticide)	Lohner and Fisher, 1990
Chironomus riparius	Ni	Powlesland and George, 1986
Chironomus tentans	10 heavy metals	Khangarot and Ray, 1989
Chironomus tentans	Cu	Gauss *et al.*, 1985
Chironomus tentans	methoxychlor (insecticide)	Sebastian and Lockhart, 1981
Chironomus tentans	acridine, quinoline (fuel products)	Cushman and McKamey, 1981
Chironomus tentans	4 synthetic fuels	Millemann *et al.*, 1984
Chironomus tentans, 3 pond species	aniline, phenol	Franco *et al.*, 1984
Chironomus yoshimatsui Martin et Sublette	Cd	Yamamura *et al.*, 1983
Chironomus sp.	trichlorfon (insecticide)	Neubert, 1986
Glyptotendipes pallens (Meigen)	Cd	Heinis *et al.*, 1990
Paratanytarsus dissimilis	10 different toxic chemicals	Thurston *et al.*, 1985
Chronic tests		
Chironomus anthracinus	pH, Al	Havas and Likens, 1985
Chironomus decorus	B	Maier and Knight, 1991
Chironomus decorus	Cu	Kosalwat and Knight, 1987
Chironomus riparius	pH and Al	Palawski *et al.*, 1989
Chironomus riparius	Cd	Pascoe *et al.*, 1989
Chironomus riparius	Ni	Powlesland and George, 1986
Chironomus riparius	Cd	Williams *et al.*, 1987
Chironomus riparius	Cd, Zn	Timmermans *et al.*, 1992
Chironomus tentans	Detroit River sediment	Giesy *et al.*, 1988
Chironomus tentans	sediment with heavy metals	Nebeker *et al.*, 1988
Polypedilum nubifer (Skuse)	Cu	Hatakeyama, 1988
Paratanytarsus parthenogeneticus Freeman (= *P. grimmii* (Schneider))	Cu	Hatakeyama and Yasuno, 1981
Paratanytarsus dissimilis	Cu, Cd, Zn, Pb	Anderson *et al.*, 1980

being abundant at moderate acidity. Oligotrophication in streams results in low densities which are further reduced with a decrease in pH (e.g. Simpson, 1983; Zischke *et al.*, 1983).

In many streams affected by mining effluents or drain water from soils with high content of metals, the effect of low pH is surpassed by the blanketing effect of precipitation of hydrous metal oxides, which greatly

Table 15.4 Examples of Chironomidae used in multiple species tests

Taxa	Toxicant(s)	Reference
Microcosms, 'ponds'		
Tanypodinae, Orthocladiinae Chironominae (*Paratanytarsus* sp.)	complex effluents (e.g. Pb, phenols)	Pontasch and Cairns, 1991
C. riparius, Daphnia magna	trifluralin, atrazine, fonofos (pesticides)	Huckins *et al.*, 1986
Chironomus salinarius Kieffer, *C. halophilus* Kieffer (= *aprilinus* Meigen) and *C. plumosus*	eight insecticides	Sinegre *et al.*, 1990
Microcosms, artificial streams		
Tanypodinae, Orthocladiinae, Chironomini, Tanytarsini	diflubenzuron (insecticide)	Hansen and Garton, 1982
Chironomidae, *Phaenopsectra* sp.	diflubenzuron (insecticide)	Rodrigues and Kaushik, 1986
Tanypodinae, Orthocladiinae Chironomini	Cu	Clements *et al.*, 1988
Mesocosms, ponds		
19 chironomid species	atrazine (herbicide)	Dewey, 1986
Tanypodinae, Chironominae	methyl parathion (insecticide)	Crossland, 1984
16 chironomid species	crude oil	Cushman and Goyert, 1984
Mesocosms, artificial streams		
23 chironomid species	pH	Zischke *et al.*, 1983
5 chironomid species	temephos, chlorphoxim (insecticides)	Yasuno *et al.*, 1985b
4 chironomid species	diflubenzuron, methoprene (insecticides)	Yasuno and Satake, 1990

decreases the abundance of periphyton, and consequently grazers such as many Diamesinae and Orthocladiinae disappear (e.g. McKnight and Feder, 1984; Rasmussen and Lindegaard, 1988).

15.6 TOXIC METALS AND CHEMICALS

Approximately 63 000 chemicals are in common use and may be released to nature but fewer than 500 have been tested for their toxicity to aquatic organisms (Giesy and Hoke, 1991). Further, individual chemicals are modified and integrated through physical, chemical and biological processes. Therefore aspects of contamination derive from a multitude of chemicals interacting in an unpredictable way and our knowledge about chemical pollution and its monitoring is fragmentary and kaleidoscopic, regardless of an exponential increase in the numbers of papers concerning effects of toxic

Table 15.5 Examples of Chironomidae used in toxicity manipulation of natural systems

Chironomid taxonomic level	Ecosystem type	Toxicant(s)	Reference
Short-term manipulation			
Family, *C. riparius*	stream	pH, Al	Ormerod *et al.*, 1987
Family, genera	stream	*Bacillus thuringiensis*	Back *et al.*, 1985
Family	stream	aminocarb (insecticide)	Eidt *et al.*, 1988
Species	streams	oil	Miller and Stout, 1986
Subfamilies, tribes	rivers	pirimiphos methyl (insecticide)	Dejoux and Troubat, 1982
Genera	river	TFM (lampricide)	Kolton *et al.*, 1986
Genera	ponds	deltamethrin (insecticide)	Morrill and Neal, 1990
Chironomus, Procladius	lake	toxaphene (piscicide)	Hilsenhoff, 1965
Long-term manipulation			
Family, genera	stream	methoxychlor ethane (insecticide)	Wallace *et al.*, 1989, 1991 Lugthart *et al.*, 1990
Subfamilies, species	lake	toxaphene (piscicide)	Webb, 1980

chemicals on freshwater ecosystems or freshwater organisms. Merely to review only those papers concerning chironomids is far beyond the scope of this chapter, and only a brief account is given here. Some references are given in Tables 15.3, 15.4 and 15.5; others can be found in Rosenberg (1993) and Timmermans *et al.* (1992).

Quite a number of studies describe the whole benthic fauna or parts of it in habitats polluted by one or (often) more toxic substances and compare the communities with those of undisturbed but otherwise similar habitats nearby. For example, Winner *et al.* (1980) recorded that chironomids numerically made up c. 80% of total fauna in a section of a small stream in Ohio heavily polluted by Cu, Cr, and Zn, while they constituted less than 10% in unpolluted sections. Precisely the same was found by Sheehan and Knight (1985) in a Californian stream polluted by Cu and Zn. However, in both studies the number of chironomid taxa were reduced to about one fourth in the polluted areas compared with reference stations. A large reduction in chironomid species was also recorded by Yasuno *et al.* (1985a) in three small Japanese streams polluted by heavy metals, but they noted that the most abundant species at the polluted stations were apparently lacking at the reference stations. Armitage and Blackburn (1985) found in the English Nent River system a gradual decrease in chironomid taxa with increasing zinc load, but a few taxa were tolerant to concentrations higher than 2 mg Zn l^{-1}. However, Wilson (1988) was unable to identify any characteristic 'zinc tolerant' species in three English streams, though he found a lower chironomid diversity at the zinc-loaded stations.

Yasuno *et al.* (1985a) found that some chironomids living at heavy-metal polluted stations accumulated metal ions (e.g. Cd 500–600 times compared to concentrations of Cd in water) but some species did not. However, a bioconcentration of toxicants is commonly found (e.g. Dixit and Witcomb, 1983;

Krantzberg and Stokes, 1989, 1990; Timmermans and Walker, 1989) and chironomids are able to spread the toxic effect to the terrestrial environment through the emergence of adults as shown for polychlorinated biphenyl (PCB) by Larsson (1984). Krantzberg and Stokes (1989) found that a *Chironomus (Camptochironomus) tentans* Fabricius population exposed to elevated concentrations of metals for several generations were able to control their uptake of Cu and Ni but not Pb and Cd in laboratory experiments. This was in contrast to experiments with *C. tentans* larvae from an undisturbed habitat, which showed bioaccumulation of all four metals. Timmermans and Walker (1989) found an elimination of Cd, Cu and Zn in *Stictochironomus sticticus* (Fabricius) during metamorphosis but not in *Chironomus anthracinus*.

15.6.1 TOXICITY TESTS

Studies of whole ecosystems are expensive and do not contain much information about toxic levels of pollutants to single organisms. Therefore, a true cascade of toxicity tests or bioassays have been performed. These may be classified in basically three groups:

- single species acute tests.
- single species chronic tests.
- multispecies tests.

Single species acute tests are short-term experiments and usually measure effects of toxicants as mortality during 24–96 hours. Single species chronic tests usually last longer – eventually the whole life span of the test organisms, which typically are easily cultured laboratory species. Single species tests are simplistic, but do not contain much information relevant to the environment. However, single species tests provide a simple method of screening for the toxicity of sediments, if standardized, as suggested by Giesy and Hoke (1991), who designed a battery of four different single species tests, one of which is a chronic growth test with *C. tentans* lasting for 10 days. Giesy *et al.* (1988) suggested that this test might be more sensitive in some cases than other commonly used tests, because assays with sediment or pore water from the Detroit River, which caused no lethality of *Daphnia magna* Straus, reduced the weight gain of *C. tentans* by up to 50%.

The general application of laboratory-cultured chironomids in single species tests is obvious from Table 15.3 and the most 'popular' species belong to the genus *Chironomus*. However, even between these related species, a considerable variation in sensitivity is observed. In addition, Timmermans *et al.* (1992) pointed out that there was a substantial difference in sensitivity to some heavy metals between larval instars. Generally first instar larvae are less resistant than fourth instar larvae, which are normally used in toxicity tests. Further, it seems that abiotic factors such as water hardness, pH, temperature and oxygen strongly influence toxicity (e.g. Wang, 1987; Lohner and Fisher, 1990). The widespread use of *Chironomus* probably also raises the toxic limit of pollutants obtained from these tests, because species of *Chironomus* all over the world seem to be more tolerant to toxicants and stressors of all kinds than nearly all other chironomid species.

Anderson (1980) mentioned a number of characteristics which make chironomids valuable for toxicity tests: easily identified life stages; short and well-known life history under laboratory conditions; and finally, that chironomids constitute a prominent part of benthic communities in all types of freshwater. Further, in the case of *Paratanytarsus dissimilis* (Johannsen) [probably *P. grimmii* Schneider], he described methods of culture and life cycle data under various laboratory conditions, and this species has later been used in toxicity tests (Tables 15.3–15.4). However, more sensitive species in the subfamilies Diamesinae and Orthocladiinae have not yet been included as a standard in single species tests.

Generally it is difficult to evaluate the sensitivity of chironomids in comparison with other freshwater organisms. Apparently an enormous range of thresholds to toxicity levels occurs within freshwater invertebrates. Thus, Brown and Pascoe (1988) demonstrated that among 22 species a range in cadmium LC_{50} values (96 h) from 0.03 mg l^{-1} for *Gammarus pulex* L. to 4600 mg l^{-1} for *Aphelocheirus aestivalis* (Fabricius) could be found in the literature. *Chironomus riparius* was located in the middle of this range with an LC_{50} of 300 mg l^{-1} (Timmermans *et al.*, 1992).

The shortage of useful ecological data obtained from single species tests may to some degree be remedied by using multispecies tests, which involve two or more species and may be conducted in **microcosms** ('small-scale systems in which colonization of the biotic assemblages either is completely controlled . . . or is restricted to a subset of the complete natural assemblages found in the system that is being modelled') or **mesocosms** ('large, outdoor experimental systems that resemble natural ecosystems more closely than microcosms') (Rosenberg, 1993). In particular, mesocosm tests, where semi-natural systems can be manipulated in a controlled way, may provide valuable ecological information. Unfortunately chironomids have been dealt with mostly on a family level in these experiments (Table 15.4) and the same is also the case in whole ecosystem manipulations (Table 15.5). However, if Chironomidae are identified to species level, valuable informations may be achieved as shown for example by Dewey (1986), who demonstrated that the herbicide atrazine in experimental ponds was toxic to the tanypodine *Labrundinia pilosella* (Loew) in concentrations more than one order of magnitude lower than indicated in single species experiments with *Chironomus tentans*.

15.6.2 CHIRONOMID DEFORMITIES AS AN INDEX OF POLLUTION

Long-term chronic tests for detecting sublethal effects may be supplemented by linking developmental abnormalities amongst the benthic fauna to chemical contamination. Chironomids are suitable for indexing defective morphology of larval head capsule structures. Deformities in antennae and mouthparts such as mentum/ligula, mandibles, premandibles and the epipharyngeal comb have been documented in the genera *Procladius, Chironomus* and *Cryptochironomus* (Warwick, 1985, 1988, 1989a, 1991; Warwick and Tisdale, 1988). Deformities range from mildly abnormal stages, such as the addition or deletion of a single tooth, to a grotesque, massive thickening and fusing of all head structures. The severity of deformities is ranked before calculating an index of morphological response from a population of a single species or of the whole chironomid community (Warwick, 1991). Deformities must be distinguished from wear or breakage of mouthparts, which occur due to feeding activities in coarse sediments (Janssens de Bisthoven and Ollevier, 1989). Deformities alter the symmetry of the mouthparts, while normal wear can be recognized by abrupt breaks and/or that the overall symmetry of the head capsule is maintained.

A number of studies show a positive correlation between incidence of deformity in chironomids and the degree of sediment contamination, but the correlations are only qualitative (e.g. Wiederholm, 1984b; Warwick, 1985; Klink, 1989; Pettigrove, 1989). It is notable that higher frequencies of deformities are reported for many contemporary communities compared with subfossil or fossil communities (Wiederholm, 1984b; Warwick, 1991). Studying a population of *Chironomus* gr. *thummi* in the polluted Dyle River (Belgium), Janssens de Bisthoven *et al.* (1992) found fourth instar larvae with deformed menta had significantly higher body tissue concentrations of Pb and Cu than larvae with normal menta, whereas no significant differences were found in concentrations of Cd and Zn. Only a few preliminary dose-response experiments to produce deformities have been published (Hamilton and Sæther, 1971; Cushman, 1984) but these have not revealed a correlation between toxicants and deformities. However, re-examination of Hamilton and Sæther's material by Warwick (1985) showed an unexpected decrease in antennal deformity as DDE concentration increased. Kosalwat and Knight (1987) demonstrated an initial positive correlation between copper and mouthpart abnormalities of *Chironomus decorus* Johannsen: 60% of larvae exposed to copper (up to 2.6 g Cu kg^{-1} dry weight of sediment) showed deformities in the epipharyngeal comb but at higher copper concentrations larvae failed to reach fourth instar. Other survey studies show that increased exposure to contaminants does not necessarily result in higher incidence of deformities. This inverse relationship may be a result of reduced survival of deformed larvae (Warwick, 1985; Dermott, 1991).

Though deformities in chironomids have been known for only 20 years and relatively few studies, primarily from the North American Great Lakes, have dealt with this phenomenon in a systematic way, it seems that deformities of chironomids have the potential to describe sublethal effects of contaminated sediment and particularly the history of the pollution sequence. However, before being a serious alternative or supplement to other chronic tests or to a simple description of community response, the methods of assessing the indices must be standardized and probably also simplified (Dermott, 1991; Rosenberg, 1993).

Table 15.6 Examples of responses of Chironomidae to different physical impacts

Chironomid taxonomic level	Ecosystem type	Stressor	Reference
Species	stream	flow regulation	Frank, 1991
Family	stream	winter warm water release	Raddum, 1985
Species	stream	thermal effluent	LeRoy Poff and Matthews, 1986
Species	experimental stream	elevated temperature	Nordlie and Arthur, 1981
Species	river	thermal effluent	Dusoge and Wisniewski, 1976
Species	river	thermal effluent	Rossaro, 1987
Species	stream	sediment addition	Rosenberg and Wiens, 1980
Species	stream	sediment release from reservoir	Gray and Ward, 1982
Species	lake	water level fluctuation	Bazzanti and Seminara, 1987
Species	lake	water level fluctuation	Hunt and Jones, 1972
Species	lake	water level fluctuation	Tikkanen, 1989
Genera	lake	water level fluctuation	Kaster and Jacobi, 1978
Subfamily, species	lake	water level fluctuation	Grimås, 1961
Species	reservoir	impoundment	Wiens and Rosenberg, 1984
Genera	reservoir	thermal effluent	Thorp and Chesser, 1983
Species	reservoir	thermal effluent	Stahl, 1986
Species	lake	thermal effluent	Wiederholm, 1971
Species	lake	thermal effluent	Webb, 1981
Species	lake	thermal effluent	Parkin and Stahl, 1981
Genera, species	lake	thermal effluent	Rasmussen, 1982

15.7 PHYSICAL IMPACTS

Physical impacts in this context include alterations such as damming for power plants or water supply reservoirs, cooling reservoirs for electricity plants, canalization for transport purposes or optimized water run-off, and disturbances such as road constructions or logging. These impacts are not directly toxic to freshwater organisms but nevertheless fundamentally influence ecosystem balance. The extensive literature (references in Armitage, 1984) on these subjects is mostly of a general nature but some studies, especially on water-level fluctuations and effects of elevated temperature, describe chironomid responses in detail (Table 15.6).

Elevated water temperatures may allow new species to colonize the habitat (Wiederholm, 1971), promote a dominance of Tanypodinae (Stahl, 1986), increase the number of generations (Stahl, 1986) and eventually result in adults emerging into terrestrial conditions to which they are not adapted (Raddum, 1985). Synergistic effects of heating water and organic or chemical pollution are also observed. For example, Dusoge and Wisniewski (1976) reported that a slight rise of only 2 °C in water temperature, together with sewage and industrial wastes, in a Polish lowland river resulted in a further decrease in number of chironomid species and in a considerable increase in the ratio of predatory species, accompanied by a violent fluctuation in abundance of most species.

Fluctuations in water level influence benthic fauna by changing water chemistry, eliminating macrophytes in the regulated zone, altering substrates by erosion and exposing substrates to air or ice-cover. Reservoirs therefore exhibit a reduced zoobenthic diversity compared with natural lakes. Kaster and Jacobi (1978) found that most fauna, including various chironomid species, were extinguished very soon in the exposed areas following water draw-downs in a Wisconsin reservoir. However, larvae of *Chironomus plumosus* not only survived for several weeks by burrowing in the sediment but also managed to increase in wet weight from 0.4 mg at 5 days before exposure to 37.4 mg at 14 days after being exposed to the atmosphere. Up to 80% of the fauna was found to survive if substrate was ice-covered during winter draw-downs (Grimås, 1961). Bazzanti and Seminara (1987) suggested that a mobile fauna (e.g. *Procladius* spp.) has an advantage in fluctuating waters compared with sedentary forms, which require stable conditions for survival. The survival of chironomids during draw-down situations and a later immigration from deeper water following the re-established water level are probably of little importance in recolonization ability of chironomids because a rapid repopulation of the littoral areas always occurs from egg depositions (Kaster and Jacobi, 1978). In contrast to natural lakes, the highest abundance and diversity of zoobenthos are found immediately below the draw-down limit. However, due to their rapid recolonization chironomids may show comparatively higher species richness above the draw-down limit (Fillion, 1967).

15.8 CONCLUDING REMARKS

Environmental factors result in effects at organism, population and community level. Examples of responses from each level have been given in this review. In spite of the extensive literature and comprehensive knowledge of responses at all three levels, the success of classifying freshwater and biomonitoring pollution using chironomids – with lake typology as the exception – has been limited. This may be due primarily to insufficient knowledge of factors creating and controlling other habitats. Relatively few, well known and clearly delimited factors operate in the profundal of stratified lakes, while littoral zone, shallow lake and running-water habitats are subjected to impacts from a large variety of factors interacting in an often unpredictable way. It is imperative that a genuine understanding of the ecology of natural habitats must be established before effects of disturbances from pollution can be predictable in more than general and qualitative terms. Therefore future approaches to ecology and pollution of freshwater ecosystems need to be more explanatory than descriptive and research should focus more on responses from organisms and populations than from communities.

16

Chironomids as indicators of past environmental change

I. Walker

16.1 INTRODUCTION

Compared with other insects, chironomids are well represented by their remains in aquatic sediments, allowing palaeoecologists to reconstruct past changes in their abundance. Together with diatoms, cladocera and ostracods, the chironomids have contributed immensely to our understanding of past changes in aquatic environments (Walker, 1993).

Although **palaeoecology** is often defined in terms of 'the relationship between *ancient* organisms and their environment' (e.g. Trowbridge, 1962), this definition is misleading; palaeoecological studies often focus on recent events including, for example, the anthropogenic eutrophication and acidification of lakes (e.g. Battarbee *et al.*, 1990; Charles *et al.*, 1990; Kansanen, 1985). These investigations have stimulated a rapidly developing branch of palaeoecological endeavour: **palaeolimnological biomonitoring** (Walker, 1993). Thus, palaeoecology is no longer a purely esoteric discipline; palaeoecological studies are now being used to address some of the most important environmental issues that we face today.

This chapter highlights what is known, unknown, or controversial in some areas of chironomid palaeoecology. It discusses case-histories that illustrate how chironomid assemblages respond to natural events, and it also presents case-histories that demonstrate how palaeoecological studies have contributed to our understanding of anthropogenically perturbed systems. Palaeolimnological studies are invaluable for impact assessment because historical data are rarely available, but are often necessary to prove deleterious impacts.

The intent is not to provide a thorough review of all chironomid palaeoecological research, nor to provide a comprehensive review of palaeolimnological methods. Comprehensive reviews of these topics are available (e.g. Berglund, 1986; Hofmann, 1988; Walker, 1987, 1993), and should be referred to where more detailed information is required by readers

The Chironomidae: Biology and ecology of non-biting midges
Edited by P.D. Armitage, P.S. Cranston and L.C.V. Pinder
Published in 1995 by Chapman & Hall. ISBN 0 412 45260 X

16.2 CHIRONOMID COMMUNITY RESPONSE TO NATURAL EVENTS – THE PREHISTORIC RECORD

To interpret recent changes in chironomid assemblage structure correctly, palaeoecologists must be fully aware of the natural variability which occurs in aquatic systems and must appreciate that natural events can produce dramatic changes in chironomid faunas. It is instructive, therefore, to examine prehistoric changes in assemblage structure, where the dominant impetus for change was natural rather than anthropogenic effects.

Our knowledge of the prehistoric fauna is still very limited, although the number of chironomid palaeoecological studies has grown rapidly in recent years. Almost all of the sites so far investigated are situated in north temperate regions, either in North America (e.g. Lawrenz, 1975; Pienitz *et al.*, 1992; Walker and Mathewes, 1987a, 1988, 1989a; Walker and Paterson, 1983; Walker *et al.*, 1991a; Wilson *et al.*, 1993; Worden, 1986) or western Europe (e.g. Andersen, 1938; Brodin, 1986; Bryce, 1962; Günther, 1983; Hofmann, 1971a, 1983a, 1983b, 1984, 1985, 1987, 1991; Hutchinson *et al.*, 1970; Keene, 1989; Ran, 1990; Schakau and Frank, 1984).

In recent years, apart from the observations of Tsukada (1967, 1972), no work has been published for Asia. Frey (1964) and Stahl (1969) summarize the findings of several early studies conducted in Kazakhstan and the Urals region (Konstantinov, 1951; Lastochkin, 1949).

Einarsson *et al.* (1988), Gardarsson *et al.* (1988) and Alm and Willassen (1993) have completed the first detailed analyses of midge fossils from the north polar region. Alpine lakes have yet to be studied.

In the southern hemisphere, chironomid palaeoecological studies have only been conducted within Australia (Paterson and Walker, 1974b), and New Zealand (Boubée, 1983; Deevey, 1955a; Schakau, 1986, 1991, 1993). Apart from one study conducted in north-west Sudan (Mees *et al.*, 1991), no detailed information is available on the palaeoecology of tropical Chironomidae. This dearth of information represents a tremendous research opportunity, especially for graduate students seeking original avenues for research.

Studies of past chironomid assemblages have been based almost entirely on subfossil head capsules. Recently, however, researchers have begun to recognize traces (i.e. evidence of larval trails and burrows) which may occasionally be preserved in lake sediments (Duck and McManus, 1984, 1987; Morrison, 1987).

16.2.1 LATE-GLACIAL CHANGES IN NORTH TEMPERATE LAKES

The most conspicuous and consistently observed phenomenon in studies of north temperate lakes is an abrupt change in chironomid assemblage composition shortly after deglaciation. Typically, there is an abrupt decline in the relative abundance of *Heterotrissocladius* and one or more associated taxa (for example, *Paracladopelma*, *Parakiefferiella nigra*, *Protanypus*: e.g. Günther, 1983; Hofmann, 1971a, 1983a; Lawrenz, 1975; Schakau and Frank, 1984; Walker and Mathewes, 1989a; Walker and Paterson, 1983; Walker *et al.*, 1993a). This abrupt change in assemblage composition appears to be centred on an interval of rapid climatic change c. 10 000 ^{14}C yr BP following the last glaciation. The change in assemblage composition is in some ways suggestive of an abrupt shift in trophic state – from oligotrophy to mesotrophy.

Deevey (1942, 1953, 1955b) felt that early post-glacial faunal shifts and the 'sigmoid growth phase' in lakes reflected a natural tendency for lakes to progress from oligotrophy to eutrophy, and made the preliminary assessment that lake assemblages were probably little affected by climatic change. In contrast, Livingstone (1957, p. 364) argued that 'the sigmoid increase is associated with extralacustrine changes and may be the result of the world-wide postglacial warming'. As Whiteside (1983) indicates, Deevey's view of the trophic ontogeny of lakes was uncritically accepted by later authors. Unfortunately Deevey's views are frequently presented as fact in introductory ecology courses and textbooks.

In recent years, more detailed stratigraphic analyses suggest that the changes in fauna and climate (and probably lake trophic state) were contemporaneous and might therefore be related (e.g. Walker and Mathewes, 1987a). Hofmann (1988, p. 507) notes:

> It is clear, for instance, that the rise of temperature at the end of the late-glacial period displaced the cold-stenothermic chironomids from the littoral zone and led to their extinction in shallow lakes (Hofmann, 1971a, 1978, 1983a; Walker and Paterson, 1983).

Despite this statement, there is no area in chironomid palaeoecology more controversial than how the late-glacial faunal discontinuity should be interpreted. Walker and co-workers (Walker and Mathewes, 1987a, 1987b, 1989a, 1989b, 1989c; Walker, 1991; Walker *et al.*, 1991a, 1991b, 1992; Wilson *et al.*, 1993) have argued that the changes are a predictable consequence of climatic change and are, in part at least, the direct result of temperature change. Several other authors make reference to the possible role of climate in shaping midge faunas (e.g. Brodin, 1985; Günther, 1983; Lawrenz, 1975; Sly, 1986).

Walker *et al.* (1991b) state that chironomids are excellent indicators of past climate and have even developed a quantitative model for palaeotemperature inference from fossil Chironomidae in unstratified lakes. Recent stratigraphic work in Atlantic Canada (Levesque *et al.*, 1993; Walker *et al.*, 1991a; Wilson *et al.*, 1993) indicates distinct oscillations in the composition of the late-glacial fauna (Figure 16.1). The changes do not conform to Deevey's unidirectional model of a lake shifting progressively from oligotrophy to eutrophy. The oscillations are believed to result from a major late-glacial climatic oscillation, correlative with the European Younger Dryas (Levesque *et al.*, 1993; Walker *et al.*, 1991a; Wilson *et al.*, 1993).

A second group (Hann *et al.*, 1992; Warner and Hann, 1987; Warwick, 1989b) vigorously opposes the notion that chironomids can serve as useful

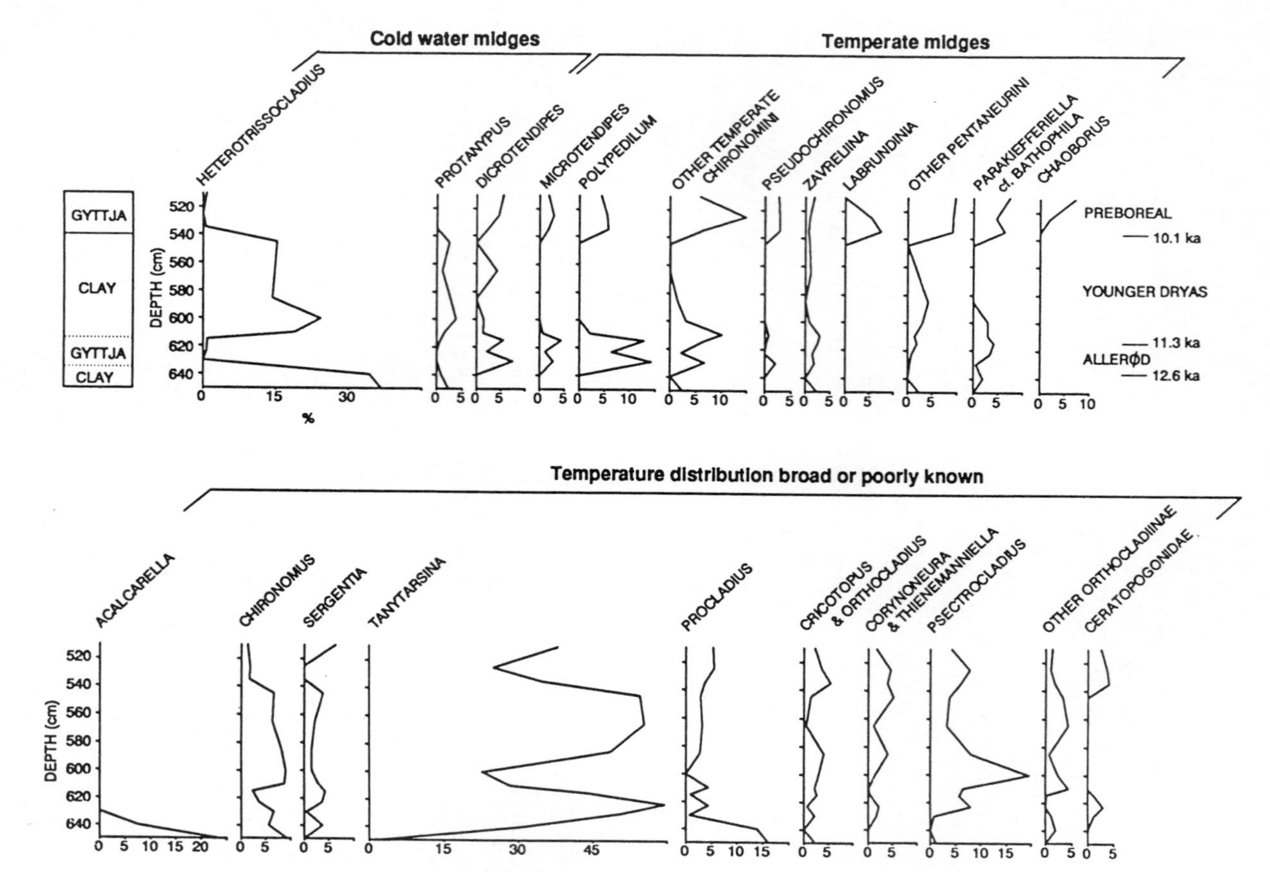

Figure 16.1 Late-glacial chironomid stratigraphy of Splan Pond, New Brunswick, Canada, showing distinct oscillations in the abundance of chironomid taxa during the Allerod/Younger Dryas climatic oscillation. (From Walker *et al*,, 1991a; reproduced with permission from *Science*; ©American Association for the Advancement of Science.)

palaeoclimatic indicators. This group favours indirect effects, especially abrupt changes in mineral sediment accumulation rates, as determinants of the late-glacial sequence.

For a full appreciation of both groups' arguments, the reader is referred to the many publications listed in the two preceding paragraphs. The controversy would benefit greatly from analyses examining how chironomid assemblages changed under different climatic scenarios (e.g. during periods of sustained climatic cooling). Hopefully the controversy will inspire research by a new generation of palaeoecologists.

16.2.2 OTHER RESEARCH PERTAINING TO CHIRONOMIDS AND CLIMATE

It is still far too early to make generalizations about how chironomid faunas have responded to climatic events outside the north temperate zone. For example, the record from Lake Taylor, New Zealand (Schakau, 1986) appears to be relatively complacent (i.e. the fauna remains more or less stable over time). Although there are changes in the relative abundances of taxa, the same species appear to be present throughout the core. It must be remembered, however, that this record spans only the last 3000 years, an interval in which climate (and therefore water temperature) has been relatively stable.

This situation is similar to that observed during the mid-Holocene in north temperate regions. For example, in Portey Pond in Atlantic Canada (Walker and Paterson, 1983), the fauna remained relatively stable throughout the Holocene (i.e. 0–10 000 y BP).

We have little information pertaining to those changes occurring in arctic regions, but clearly the climatic change that initiated deglaciation had an important impact on the fauna of Nedre Æråsvatn, Norway (Alm and Willassen, 1993). However, the discovery of 15 000-year-old *Corynocera ambigua* fossils in Endlevatn suggests that this species may have survived the last glaciation in northernmost Norway (Fjellberg, 1978).

The chironomids preserved in Plio-Pleistocene sediments from north-western Greenland are especially intriguing (Böcher, 1989); these fossils include taxa with distributions presently restricted in North America to low arctic/subarctic lakes (i.e. *C. ambigua*) or regions south of tree-line (i.e. *Endochironomus*). At present *C. ambigua* does not occur in lakes closer than southernmost Baffin Island, 1300 km south of the study site. Macrofossil evidence demonstrates that trees (e.g. *Larix*) were present in the area at this time, forming a part of the forest–tundra community.

In tropical lakes the maximum surface-water temperature attained in summer is similar to the summer surface-water temperature of temperate lakes; in both regions the Tanypodinae and Chironominae dominate the littoral fauna. Thus it may be more difficult to resolve changes in water temperature or climate at low-latitude, low-elevation sites. Much more research is necessary, however, to determine whether this difficulty is real or imagined.

One very promising avenue for study is the prospect of reconstructing the hydrological balance for closed-basin lakes, especially in arid and semi-arid regions (Smol *et al.*, 1991; Walker, 1991). Paterson and Walker (1974b), in their studies of an Australian lake, noted major fluctuations in the abundance of *Chironomus duplex, Procladius paludicola and Tanytarsus barbitarsis*. These oscillations appear to result from large variations in lake salinity, which were probably a consequence of long-term changes in evaporation/precipitation balance. Similarly Mees *et al.* (1991) and Pienitz *et al.* (1992) have inferred changes in climate from chironomid remains in saline lake sediments from two very different regions of the globe: north-western Sudan and subarctic Canada.

16.2.3 OTHER NATURAL PERTURBATIONS TO FAUNA

Although chironomid faunas often show distinct, sometimes abrupt, mid-Holocene changes in fauna, palaeoecologists are far from fully understanding their cause. For most of these changes, a climatic cause seems unlikely. Authors have invoked 'morphometric eutrophication' (*sensu* Deevey, 1955b), natural acidification, water level fluctuations, changes in mineral sediment accumulation rates or other phenomena as explanations (e.g. Boubée, 1983; Brodin, 1986; Lawrenz, 1975; Schakau, 1991; Walker and Paterson, 1983) but in few instances is the evidence strong enough to justify a firm conclusion. Palaeoecologists should also ponder whether chaotic processes (Gleick, 1987) are driving some of the unexplained variability.

To illustrate the problems in interpreting the palaeolimnological record, the patterns observed in two unpublished studies by Boubée (1983) and Lawrenz (1975) will be described.

(a) Lake Maratoto, New Zealand

In one particularly interesting study, Boubée examined the post-glacial record of Lake Maratoto, a naturally acidic lake in the Waikato basin, North Island, New Zealand (Boubée, 1983). The lake was formed about 17 000 years ago, at the end of the last ice age, and presumably had a circumneutral pH immediately subsequent to its formation. The development of the lake has been greatly affected by growth of nearby peatlands; the present pH of the lake varies between 4.1 and 5.5. The sediments include about 15 different tephra layers.

Boubée (1983) introduced surface sample data sets and multivariate statistics to chironomid palaeoecology. Analysing chironomid remains preserved in surficial sediments of 12 North Island lakes with principal components analysis, he determined that although most chironomid taxa were abundant in peaty waters, *Paratanytarsus grimmii* [as *agameta*] was characteristic of non-peaty sites. *P. grimmii*, *Polypedilum* and Orthocladiinae were characteristic of productive, turbid waters, but *Corynocera* and *Cladopelma curtivalva* distinguished shallow, clearwater lakes. Boubée (1983)

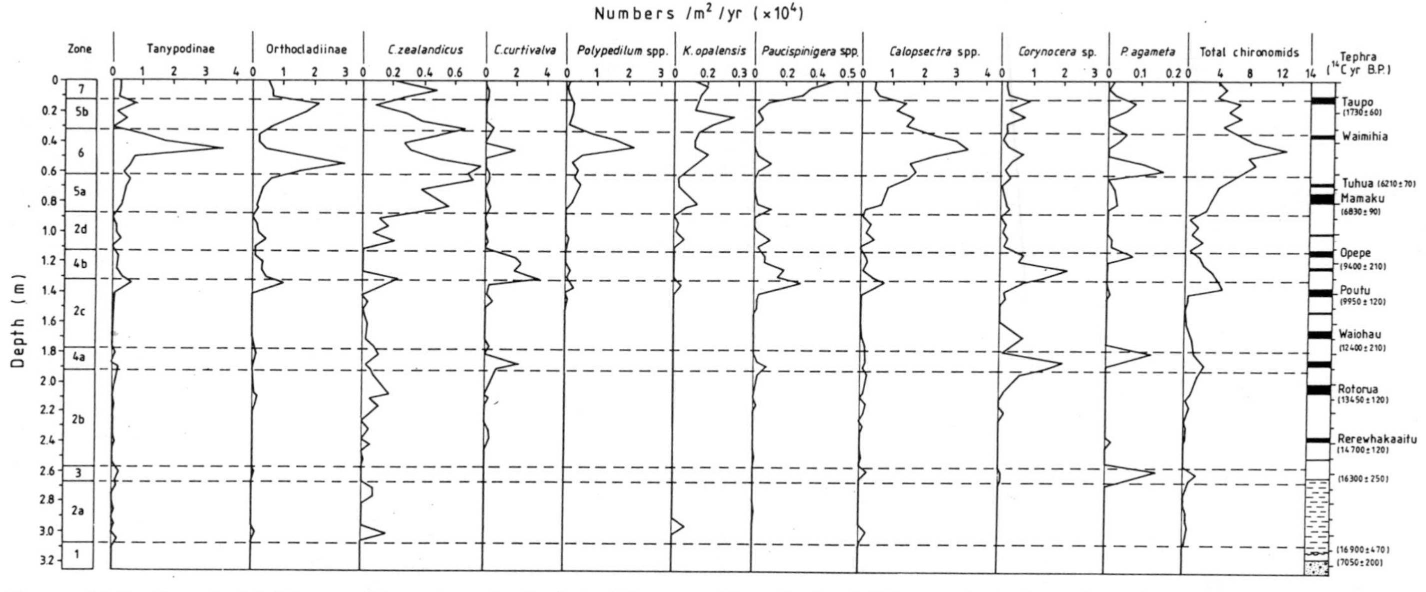

Figure 16.2 Postglacial chironomid stratigraphy for Lake Maratoto, New Zealand. Diagram includes radiocarbon dates, and positions of numerous tephra bands. (From Boubée, 1983; reproduced with the permission of J.A.T. Boubée.)

relied on these results to interpret the conditions represented in the Lake Maratoto core.

Boubée's (1983) results (Figure 16.2) indicate that Lake Maratoto was initially an unproductive, clearwater lake. With post-glacial warming and the shift to a more humid climate, Lake Maratoto developed into a more productive, dystrophic system. Superimposed on this gradual shift from oligotrophy to dystrophy are several major water level fluctuations. These may relate to changes in evaporation/precipitation balance, although the role of edaphic circumstances (i.e. damming of the lake by peat growth) is uncertain.

Although Boubée's (1983) core contains many tephra layers, he makes no reference to the possible effect of volcanic ash on the fauna. If these ash layers did have a significant effect, it is not easily recognizable in the chironomid stratigraphy (Figure 16.2). It is possible that the 5 cm (c. 500 y) sampling interval was too coarse to resolve any effect. In contrast, Tsukada (1967) noted that *Tanytarsus* 'genuinus' abundance decreased temporarily after two volcanic ashfalls in Lake Nojiri, Japan.

(b) Green Lake, Michigan

Another unpublished but highly significant investigation was performed by Lawrenz (1975), who examined the post-glacial record of Green Lake (16.2 ha), about 25 km east of the present Lake Michigan shoreline. In 1972, a core 4 m long was obtained from the centre of this basin and analysed for fossil pollen, loss on ignition, carbonate content and fossil Chironomidae.

One of the most striking features in the record was the close correspondence among the pollen and chironomid zones (Table 16.1). The earliest fauna, a *Heterotrissocladius*-dominated assemblage, prevailed during an interval (c. 13 000–12 000 BP) characterized by peak late-glacial concentrations of Cyperaceae (sedge) and *Fraxinus* (ash) pollen. *Phaenopsectra* (probably *Sergentia coracina*) prevailed throughout the remainder of the late-glacial (c. 12 000–10 000 BP), as relative abundances of *Picea* (spruce) and *Artemisia* (sage) peaked. *Phaenopsectra* abundance plummeted at the onset of the Holocene. Temperatures at this time were probably rising rapidly.

During the early Holocene (c. 10 000–8 500 BP), when *Pinus banksiana/resinosa* (jack and red pine) pollen was prevalent, *Microtendipes* and *Chironomus* larvae were very abundant. Between 8500 and 4000 BP, *Pinus strobus* (eastern white pine) pollen peaked, sediment organic content peaked, a minimum carbonate concentration was recorded and Tanytarsini came to dominate the fauna. During the late Holocene (c. 4000 BP to present), abrupt fluctuations in the abundance of Tanytarsini and Chironomini genera occurred, as the modern mixed coniferous/deciduous forest community was finally established.

The Green Lake record amply illustrates the difficulty of resolving which factors have been affecting the chironomid fauna over time. The close correspondence between chironomid and pollen zones is probably coupled to climatic change. However, because climate may have indirectly affected fauna through changes in lake productivity, sediment inputs from the

Table 16.1 Summary of the pollen and chironomid stratigraphies of Green Lake, Michigan (Lawrenz, 1975). Radiocarbon ages are inferred from regional pollen stratigraphy and three ^{14}C dates uncorrected for carbonate error. Interpretations reflect opinions of the chapter author

Age (^{14}C yr BP)	Pollen assemblage	Possible vegetation–climate interpretation	Chironomid asemblage
0			
	Betula–Fagus–Tsuga (Birch/Beech/Hemlock)	Warm, humid mixed forest	Tanytarsini–Chironomini
4000			
	Pinus strobus–Quercus (White Pine/Oak)	Maximum warmth, mesic forest	Tanytarsini–*Procladius*
8500			
	Pinus banksiana/resinosa–Quercus (Hard Pines/Oak)	Cool, dry forest (Southern Boreal Forest?)	Tanytarsini–*Microtendipes– Chironomus*
10 000			
	Picea–Artemisia (Spruce/Sage)	Cool, dry forest–tundra	*Phaenopsectra (Sergentia coracina?)* dominance
12 000			
	Cyperaceae-*Artemisia–Fraxinus* (Sedge/Sage/Ash)	Cold, dry steppe or tundra	*Heterotrissocladius* dominance
13 000			

catchment, lake stratification patterns and water levels, it is not possible to determine the lines of cause and effect from the palaeolimnological record. In addition, with glacial retreat, readvance and isostatic rebound, Lake Michigan water levels were in constant flux. Local climate, height of the water table and the volume of Green Lake may all have been influenced by the height and proximity of the Lake Michigan shoreline.

Ancient changes in chironomid faunas are often very difficult to interpret because we have no direct measurements of physico-chemical data. Studies of present biogeographical patterns via surface sample collections (e.g. Walker *et al.*, 1991b) are especially important, because they allow chironomid distributions to be directly related to physico-chemical data (P.H. Kansanen, *pers. comm.*)

16.3 RESPONSE OF CHIRONOMID ASSEMBLAGES TO ANTHROPOGENIC EVENTS

A common difficulty in lake management is differentiating between natural and perturbed conditions. Without good historical records, it is often difficult to know whether a particular lake was naturally eutrophic, naturally acidic, or whether the present state of the lake instead reflects the consequences of human activity. Although limnologists may be satisfied that major impacts have occurred, large industries wish to avoid costly clean-up operations, and are reluctant to accept responsibility. Industry may confront limnologists with a long list of alternative hypotheses, invoking natural processes to account for the present undesirable state of a particular lake.

As the lake acidification controversy has demonstrated, in the absence of sound historical data, there is little better evidence of human impact than that available in lake sediments (Battarbee *et al.*, 1990; Smol, 1990, 1992). At the same time the limnologist may wish to introduce mitigative measures to re-establish natural conditions. Palaeolimnology often provides the only means to determine what 'natural conditions' formerly existed. If a lake was naturally eutrophic, it makes little sense to try to 're-establish' an oligotrophic regime.

Many palaeoecological studies have focused on changes in chironomid fauna associated with anthropogenic perturbations of aquatic systems. Most of these studies address the problem of cultural eutrophication and all have been conducted at sites in temperate North America or Europe (e.g. Brodin, 1982; Carter, 1977; Kansanen, 1985; Räsänen *et al.*, 1992; Uutala, 1986; Walker *et al.*, 1993b; Wiederholm, 1979; Wiederholm and Eriksson, 1979). Palaeolimnological studies have tremendous potential as one component of biomonitoring programmes (Smol, 1992; Walker, 1993).

16.3.1 EUTROPHICATION

As would be expected, eutrophication of lakes is usually accompanied by a rapid increase in the relative abundance of *Chironomus* spp. (e.g. Warwick, 1980; Wiederholm and Eriksson, 1979), but with extreme eutrophication even *Chironomus anthracinus* and *C. plumosus* may be eliminated (e.g. Kansanen, 1985). Consequently the influx of chironomid remains may accelerate with primary productivity during the early stages of eutrophication, but the ultimate impact on the fauna may include an abrupt decline in *Chironomus* and other profundal taxa.

Palaeolimnological research, by providing a record of pre-impact conditions and anthropogenic perturbation, can provide a strong basis for water management decisions. This is illustrated by research conducted on Lake Mälaren, central Sweden (Wiederholm and Eriksson, 1979) and Prat and Daroca's (1983) studies of Spanish reservoir sediments.

(a) Lake Mälaren, Sweden

Wiederholm and Eriksson (1979) collected three undisturbed sediment cores from Ekoln Bay, an arm of Lake Mälaren near the city of Uppsala. The cores were all taken within a few hundred metres of each other, in 28 m of water. The cores, each approximately 70 cm long, are believed to represent changes spanning the past 150 years.

The chironomid remains recovered from the lowermost sediments

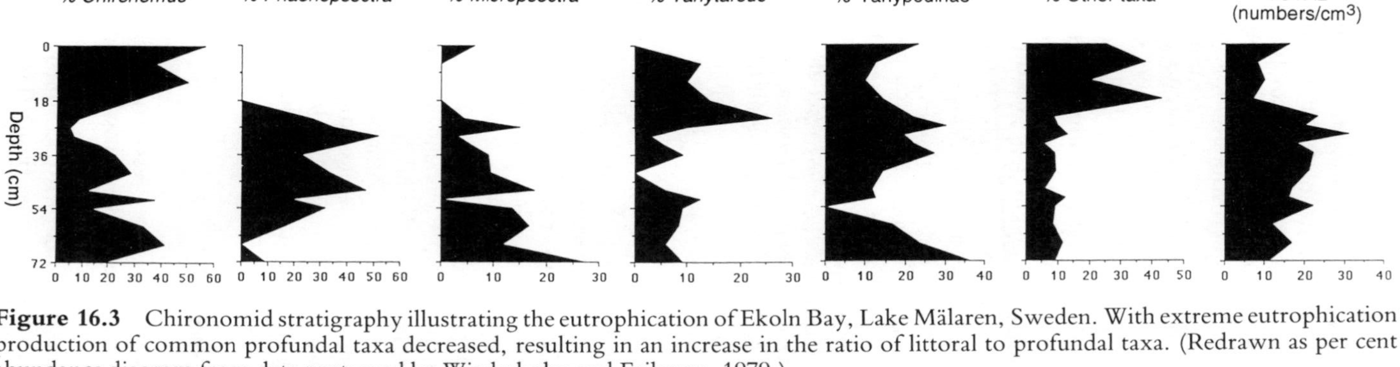

Figure 16.3 Chironomid stratigraphy illustrating the eutrophication of Ekoln Bay, Lake Mälaren, Sweden. With extreme eutrophication production of common profundal taxa decreased, resulting in an increase in the ratio of littoral to profundal taxa. (Redrawn as per cent abundance diagram from data portrayed by Wiederholm and Eriksson, 1979.)

(deposited c. AD 1830) portray an assemblage in which *Micropsectra* and Tanypodinae larvae dominate, although *Chironomus* larvae were also common (Figure 16.3). In subsequent sediments the relative abundance of *Micropsectra* gradually declines and *Phaenopsectra* (probably *Sergentia coracina*) larvae dominate in later sediments deposited prior to c. 1950.

Severe eutrophication in recent decades (c. 1950–1975) is reflected by:

- an abrupt drop in chironomid concentrations in the sediments;
- disappearance of *Micropsectra* and *Phaenopsectra*; and
- a rapid rise in the relative abundance of *Chironomus* and 'other taxa'.

Most of the 'other taxa' are genera usually collected from littoral habitats (e.g. *Corynoneura, Dicrotendipes, Microtendipes, Polypedilum, Psectrocladius*). Their remains probably originated in shallow water and were redeposited in the profundal sediments. Low oxygen levels, imposed by eutrophication, are no doubt responsible for these changes. As severe eutrophication gradually eliminated benthos from profundal habitats, littoral remains composed a larger proportion of those deposited.

This study demonstrates that recent events have had a tremendous impact on the fauna of Lake Mälaren. Furthermore, as Wiederholm and Eriksson (1979) indicate, their study also demonstrates the likely outcome of a lake restoration program. If attempts to reduce both point and non-point nutrient loading to natural levels are successful, re-establishment of the natural fauna should eventually occur. Indeed, the re-occurrence of *Micropsectra* in the uppermost sediments of the cores suggests that the lake mitigation program is having the desired effect (Figure 16.3).

The lake was naturally mesotrophic; thus it would be unrealistic to try to establish oligotrophic conditions in this lake. Wiederholm and Eriksson (1979) estimate, using the sedimentary record and a chironomid–phosphorus–depth relationship, that removal of all unnatural phosphorus inputs would allow the summer lake phosphorus concentrations to drop from 70 to 40 μg/l.

Similar eutrophication studies have been conducted by Brodin (1982), Carter (1977), Kansanen (1985), Warwick (1980) and Wiederholm (1979) for lakes in Sweden, Northern Ireland, Finland, Canada, and the United States, respectively. Modern radioisotopic dating methods (e.g. ^{210}Pb, ^{137}Cs) now provide highly reliable chronologies for such eutrophication studies (e.g. Kansanen, 1985; Kansanen and Jaakkola, 1985). These dates allow palaeoecologists confidently to correlate changes in the palaeolimnological record with historically documented events.

(b) Spanish reservoirs

Reservoirs also contain sediments preserving a detailed record of trophic changes. Unfortunately, these deposits have seldom been studied palaeolimnologically.

Prat and Daroca (1983) have recently used chironomid analyses to examine ontogenetic processes in Spanish reservoirs (Figure 16.4). Early sediments revealed diverse faunas, including *Cricotopus*, *Chironomus*, *Psectrocladius*, and

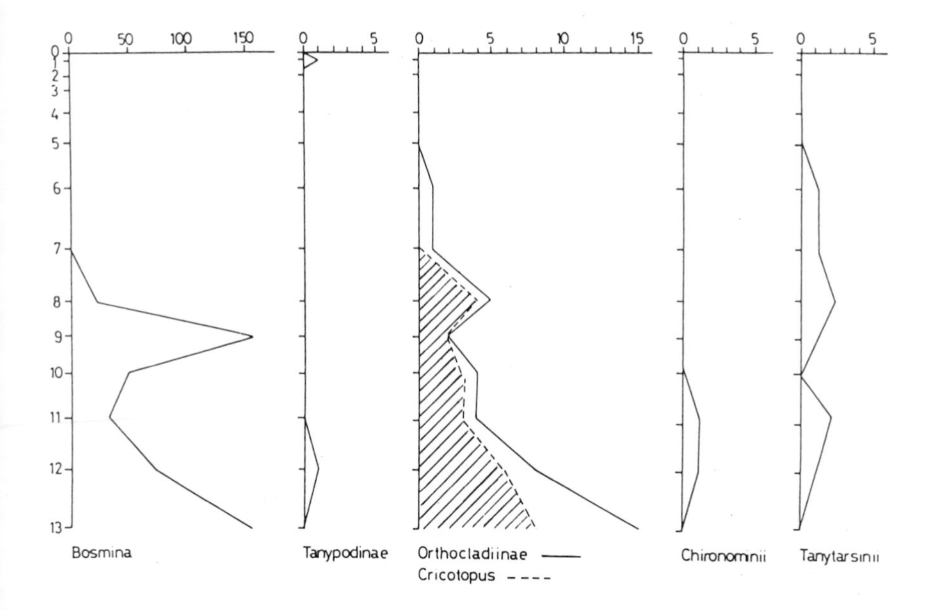

Figure 16.4 Invertebrate stratigraphy for Yasa Reservoir, Spain. Vertical axis represents depth (cm); horizontal axes portray number of individuals per g dry sediment. (From Prat and Daroca, 1983; reproduced with permission of Kluwer Academic Publishers.)

Tanytarsini. In one instance, basal samples included riverine chironomids (e.g. *Euorthocladius*). Chironomid remains were rare or completely absent in the uppermost sediments. These results suggest that the reservoirs were initially eutrophic but have become more oligotrophic during recent years.

Prat and Daroca (1983) provide the only detailed analyses of chironomid remains in reservoir sediments. Similar studies should be conducted in other areas to assess how representative these results are. Reservoirs provide a repository of sediments which has long been overlooked but may prove tremendously useful in future years.

16.3.2 ACIDIFICATION

Industrial emissions of SO_2 and NO_x are now well known for their role in generating acid precipitation and are responsible for the recent acidification of large numbers of lakes in northern Europe and eastern North America. During the 1980s, many palaeolimnological studies began to focus on the impact of acidification on lake benthos (Brodin, 1986; Buskens, 1989; Charles *et al.*, 1987; Dickman *et al.*, 1987; Henrikson *et al.*, 1982; Johnson and McNeil, 1988; Johnson *et al.*, 1990; Uutala, 1986).

Although little was known about the effects of low pH prior to 1980, it is now possible to make some generalizations. It is apparent that the kinds of

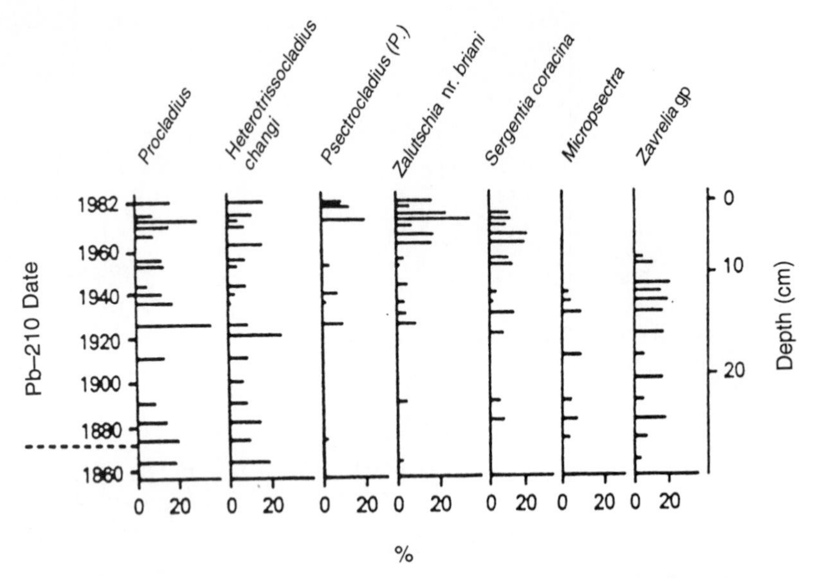

Figure 16.5 Chironomid stratigraphy prepared by A.J. Uutala for Big Moose Lake, New York, USA, portraying the impact of recent acidification. Significant changes in the relative abundance of *Zalutschia* nr *briani*, *Micropsectra* and *Zavrelia* are apparent. (From Charles *et al.*, 1987; reproduced with permission of Kluwer Academic Publishers.)

changes being observed in North America were not unlike those seen in European lakes.

The most detailed analyses were performed by Uutala (1986) in conjunction with the PIRLA (Paleoecological Investigations of Recent Lake Acidification) programme but, unfortunately, few of these results have appeared in print (Charles *et al.*, 1987; Uutala, 1990).

Uutala (1986) analysed recent sediments in five Adirondack lakes (northern New York State) for larval remains of *Chaoborus* and Chironomidae (Figure 16.5). Although one of these lakes has a circumneutral pH, the four remaining lakes are currently acidic.

Although the precise nature of the changes varied among the acidic lakes, declines in *Micropsectra* and abrupt increases in other Tanytarsina, *Psectrocladius* and *Zalutschia* nr *briani* were noted in sediments dated to recent decades. Diatom analyses provide a detailed record of changes in lakewater pH. The chironomid changes appear to coincide with the acidification of the lakes in the mid 1900s.

Significant changes were also apparent in Windfall Pond (the circumneutral lake) but these changes were distinctly different from those observed in the acidic systems. At Windfall Pond, a decrease in the *Tanytarsus glabrescens* group and an increased relative abundance of *Polypedilum sordens* occurred in recent sediments.

Uutala (1986) supplemented his research with careful statistical analyses.

For each core he provided a Bray-Curtis ordination, and a dendrogram constructed via cluster analysis. The Bray-Curtis ordinations provided an effective visual representation of the data, allowing intervals of abrupt faunal change to be identified.

Uutala's (1986) work clearly portrays the changes in fauna which occurred with acidification. It is unfortunate that our understanding of the pH preferences of Chironomidae is at present inadequate, for although quantitative models facilitate the reconstruction of past pH from fossil diatoms, the optima and tolerances of many chironomid taxa need to be determined before similar results can be provided from chironomid analysis.

Uutala's (1986) results are not unlike those obtained in other acidification studies. For example, Uutala (1986) indicates that the increases in *Psectrocladius*, *Zalutschia* and *Tanytarsus* and the declines in *Micropsectra* are consistent with the observations of Dermott *et al.* (1986), Walker *et al.* (1985) and other authors. These taxa may serve as useful biomonitors in future acidification research.

16.3.3 LOTIC ASSEMBLAGES

Lakes provide ideal traps for sediments and have long been favoured as study sites by palaeoecologists. In contrast, erosion prevails throughout much of the length of streams, and even in depositional areas the meandering habits of streams are likely to play havoc with the sedimentary record. Such considerations readily account for the dearth of attention that rivers and streams have received from palaeoecologists.

It is exciting, therefore, to ponder the results of a recent palaeoecological study, conducted by Klink (1989), which demonstrates that innovative strategies can overcome many of these obstacles. Indeed, much of the pessimism surrounding river and stream palaeoecology may prove unwarranted.

Rather than searching for a single study site with an uninterrupted sedimentary record, Klink (1989) pieced together the history of the Rhine from widely scattered deposits filling abandoned river channels in the Netherlands. The riverine sediments were easily distinguished from more recent lacustrine sediments by their coarse sandy texture; those sediments deposited at the time of channel abandonment provided a record of the stream fauna at that particular date. The date of abandonment was determined from historical records, including a series of dated maps.

Klink's (1989) analyses recorded the former presence of many stream insects no longer inhabiting the Rhine, as well as recent invaders. He noted a dramatic increase in the frequency of head capsule deformities, and the extinction of many Chironomidae and other insects. Both changes probably result from severe pollution during recent decades. In addition, thermal pollution and recent climatic changes may explain invasions of the Rhine by several southern taxa (e.g. *Rheocricotopus chalybeatus*, *Nanocladius rectinervis* and *Rheotanytarsus* spp.).

The construction of stone banks and jetties provided firm substrata –

habitats which were rare in the primeval Rhine. Klink (1989) suggests that these may explain the appearance of *Parachironomus longiforceps* and *Xenochironomus xenolabis*, which are indirectly dependent on the availability of firm substrata for colonization by bryozoans and sponges respectively.

Klink considered his report preliminary and notes that 'numerous questions arise from the analysed material and very few answers can be given' (Klink, 1989: p. 183). Nevertheless, Klink's analyses demonstrate that palaeoecologists should devote greater attention to stream deposits. The consequences of human activity are clearly depicted in Rhine sediments.

16.4 PROGNOSIS FOR FUTURE RESEARCH

Fifty years have passed since publication of Andersen's (1938) classic paper, describing the response of Chironomidae to Danish late-glacial climatic changes. In subsequent years, progress in chironomid palaeoecology has been slow; we have scarcely begun to understand how prehistoric events have influenced the fauna.

Progress has been hampered by three obstacles. Firstly, chironomid palaeoecologists have tended to work in isolation from other palaeoecologists. This circumstance has hampered the exchange of ideas, such that chironomid researchers have been unaware of progress being attained in related fields, and chironomid studies have seldom been integrated into major palaeoenvironmental research projects. Other palaeoecologists have not been made aware of the important contributions that chironomids can make to their studies.

Until chironomids are incorporated into major co-operative palaeoenvironmental projects, chironomid palaeoecologists will not have the advantage of comparing their results with independent evidence of change, including evidence of past climates, algal and zooplankton community structure, productivity or water level. Chironomid palaeoecology cannot be developed into a holistic framework if pursued in isolation from other research.

The second obstacle is that palaeoecologists require better biogeographical and physiological data for the modern fauna. As Hutton's principle of uniformitarianism ('the present is the key to the past') so aptly denotes, a thorough understanding of present systems is essential to interpreting past events. Despite years of ecological and physiological research, we still have only a very sketchy understanding of what environmental variables are instrumental in shaping chironomid distributions; optimal conditions for survival or reproduction have been determined for only a very few taxa. Even where good ecological data are available, we usually have little knowledge of the underlying physiological mechanisms which define lines of cause and effect.

As new ecological and physiological data become available, palaeoecologists will be able to refine their interpretations considerably. By recognizing the importance of extensive surface sample datasets (e.g. Boubée, 1983; Walker *et al.* 1991b) for determining species' environmental requirements, chironomid palaeoecologists are now emerging into the world of quantitative analysis and inference.

Finally, palaeoecologists require assistance from skilled taxonomists in order to refine further the level of identifications. Perhaps the most severe handicap faced by chironomid palaeoecologists is the difficulty of identifying larval remains. Although most specimens are readily identifiable to the generic level, in some instances groups of two or more genera must be considered as representing a single taxon. This difficulty is especially frustrating when working with the many genera composing the subtribe Tanytarsina. Tanytarsina often dominate the fossil fauna, yet the uncertainties in their identification are often so great that they yield no assistance when developing palaeoecological interpretations.

Publications which specifically address the problems of identifying subfossils (e.g. Hofmann, 1971b; Kokkinn, 1986; Walker *et al.*, 1993a) are especially welcome. As means become available for separating individual species of Chironomidae (e.g. Kokkinn, 1986; Walker *et al.*, 1993a), palaeoecologists will be able to provide reconstructions of prehistoric range expansions and contractions. Discussions of changing species distributions will become less speculative, being based to an increasing degree on the fossil record.

Despite these difficulties, chironomid palaeoecology is now being used to address key environmental issues. Palaeoecological studies have focused especially on the problems of eutrophication and acidification; studies are now under way which will tackle questions arising from global environmental change.

How will chironomid assemblages respond to future climatic change? Climatic change is not a new phenomenon. Palaeoecological studies have already revealed that chironomid assemblages responded rapidly to the changing environmental milieu during deglaciation. Late-glacial fluctuations in climate were abrupt, and similar in magnitude to those changes projected for coming decades. The late-glacial climatic amelioration was accompanied by a gradual increase in atmospheric CO_2. Thus, palaeoecological studies focusing on late-glacial changes will reveal how chironomids respond to climatic change and will provide the best available model for projecting future changes in chironomid distribution. Since abrupt changes in fauna are associated with tree-line environments, the impact of climatic change should be especially apparent in forest–tundra environments (Smol *et al.*, 1991; Walker, 1991). Changes in hydrological balance will have a dramatic impact on saline lake faunas.

Chironomid palaeoecological studies will also have an impact on new projections for future climatic trends. Although many models of global climate are now available, and are being used to predict future change, all models need to be tested adequately. Although models have been developed to portray accurately the earth's present climatic state, the ability of these models to simulate other climatic states can only be assessed by their ability to correctly model past climates. Modellers are therefore collating all available data on past climates and are anxious to see more data collected. Climatologists recognize that the key to projecting future climatic change is having a thorough understanding of past climatic events. All branches of palaeoenvironmental research will contribute to future projections of climatic change.

16.5 CONCLUSION

There is an exciting future for chironomid palaeoenvironmental research. Recent progress is attracting new investigators to the discipline and concerns about global change have attracted a surge of interest in all branches of palaeoenvironmental study. Chironomid palaeoecologists are now adopting the methods necessary to develop a quantitative framework for palaeoenvironmental reconstructions. Chironomid palaeoecology is a discipline that has scarcely escaped adolescence; so much work remains that future investigators will easily discover original avenues for research. Entire continents have yet to be investigated.

17

Chironomidae as food

P.D. Armitage

17.1 INTRODUCTION

Chironomidae in both aquatic and aerial stages are low in the food chain and contribute largely to the food of other groups of animals. This chapter reviews literature citing chironomids as food for a whole range of organisms and suggests how conditions can be altered by humans to increase production of chironomids as food for specific targeted organisms.

17.2 THE FOOD VALUE OF CHIRONOMIDS

The proximate composition and utilizable energy of chironomid larvae (species unknown) based on studies in the literature and their own analyses are presented by Yurkowski and Tabachek (1979). The rounded mean values (%) are: moisture content 86, protein 48, lipid 14, carbohydrate 23, chitin 4, ash 9; and the utilizable energy was 4.1 kcal g^{-1}. Other estimates of caloric content are given by Tubb and Doris (1965), Wissing and Hassler (1968, 1971), Cummins and Wuycheck (1971), Driver *et al.* (1974) and De la Noüe and Choubert (1985). Caloric content varies with season (Wissing and Hasler, 1971) and between species and life history stage (Driver *et al.*, 1974) but the range is from 4.6 to 6.1 kcal g^{-1}. The values reported in the literature are similar to those for other aquatic invertebrates.

Aquatic insects in general average 55% protein by dry weight (Moyle, 1961) and Sugden (1973) reports a value of 56% for chironomid larvae. Protein is the principal source of amino acids for insects, fish and other vertebrates. A comparison of several diets for fish found that chironomid larval biomass showed the highest total amino acid content (47.51 mg/ 100 mg dry weight: De la Noüe and Choubert, 1985). The relatively high protein content, the high digestibility (73.6%: De la Noüe and Choubert, 1985) and the apparent function in small quantities as a growth promoter in fish diets (Yashouv, 1956; Yashouv and Ben Shachar, 1967) make chironomid larvae a rich food for many organisms. This, in combination with high

The Chironomidae: Biology and ecology of non-biting midges
Edited by P.D. Armitage, P.S. Cranston and L.C.V. Pinder
Published in 1995 by Chapman & Hall. ISBN 0 412 45260 X

chironomid productivity, gives the group a key role in the trophic structure of communities. In addition a recent study on the nutrient composition of ground-up dried adult chironomids (mixed with *Chaoborus* and *Povilla*) also demonstrates the potential value of 'insect flour' as a food additive rich in digestible protein for human diets (Bergeron *et al.*, 1988).

17.3 CHIRONOMIDAE AS FOOD

The high production and ubiquitous occurrence of chironomid populations means that at some stage in their life history they will contribute to the food of a wide range of both small and large predators from *Hydra* (Grzybkowska, 1988) through to bats (Griffiths and Gates, 1985) and possibly humans.

17.3.1 INVERTEBRATES

Table 17.1 presents a series of invertebrate predators for which chironomids are cited as food. The list is not exhaustive but shows the wide range of invertebrates that depend to some extent on chironomids as a source of nutrients.

Sessile hydra may encounter the mobile free-swimming early instars of Chironomidae by chance, as may the predaceous oligochaete *Chaetogaster*. Reynoldson and Sefton (1976) found that *Planaria torva* fed on chironomids seasonally, in spring and early summer, whereas the species *Dugesia tigrina* took them more freely. A seasonal effect has also been noted in the feeding of the leech *Helobdella stagnalis* which preys upon chironomids mainly in April, with Oligochaeta becoming the major food item later in May and June (Davies *et al.*, 1979). The lake-dwelling amphipod *Pallasea quadrispinosa* feeds on chironomid larvae but they only become a major item of the diet in the deeper parts of the lake (Hill, 1988).

Table 17.1 A selection of references citing chironomids as food of invertebrate groups

Group	Reference
Hydrididae	Grzybkowska, 1988
Tricladida	Legner *et al.*, 1975b; Reynoldson and Sefton, 1976
Hirudinea	Davies, 1979; Keim, 1977; Young and Spelling, 1989
Crustacea	Hill, 1988; Reymond and Lagardere, 1990; Wilson, 1969
Odonata	Fischer, 1964; Folsom and Collins, 1984; Johnson, 1985; Thompson, 1978
Plecoptera	Johnson, 1983; Peckarsky *et al.*, 1990; Siegfried & Knight, 1976
Ephemeroptera	Soluk and Clifford, 1985
Corixidae	Ranta and Espo, 1989
Gerridae (marine)	Foster and Treherne, 1986
Megaloptera	Griffiths, 1973
Trichoptera	Crosby, 1975; Edington and Hildrew, 1981; Lavandier, 1984

Adult chironomids may also be taken by aquatic invertebrate predators at times of high density. Wilson (1969) notes that crayfish *Cambarus* spp. were the most common predators of dense aggregations of *Chironomus* adults along the shoreline of a lake. Midges were either pulled down from the surface or the crayfish would partially emerge from the water, seize a midge and retreat again. At 23.00 h crayfish densities of about 4 m^{-1} of shoreline were observed feeding on the chironomid aggregations.

It may be assumed from these observations that wherever or whenever chironomids are most available they are eaten and that there is little selection. However, some studies have shown high positive electivity coefficients for the aquatic life stages of chironomids. Such is the case in the stonefly *Acroneuria (Calineuria) californica* (Siegfried and Knight, 1976). *Hydropsyche* species in a New Zealand stream also appeared to select chironomids in preference to simuliids at certain times of the year (Crosby, 1975). Odonata also select Chironomidae and Thompson (1978) found that they were the most important food item (in terms of biomass) of the damselfly *Ischnura elegans*. Johnson (1985) notes that chironomid larvae were the major prey item (48.9%) of *Aeshna multicolor*.

In shifting sandy substrata the heptageniid mayfly *Pseudiron centralis* preys mainly on Chironomidae which are a major faunal component of this habitat (Soluk and Clifford, 1985). On rocky substrates the larval caddisfly *Rhyacophila tristis* feeds mainly on Chironomidae (Lavandier, 1984). On soft muddy bottoms in standing water the predominant food of large larvae of the megalopteran *Sialis lutaria* are chironomids and oligochaetes (Griffiths, 1973).

Thus it is clear that chironomids contribute to the food of many aquatic species in a wide range of aquatic habitats. Adult stages also contribute to the food of terrestrial invertebrates such as surface-feeding Hemiptera and Empididae. Predators and scavengers on adult *Chironomus* species include robber flies (Asilidae), ground beetles (Carabidae), ants (Formicidae) and spiders (Wilson, 1969).

17.3.2 FISH AND AMPHIBIA

(a) Fish

Chironomidae are major foods of both juvenile and adult fish and Table 17.2 lists a selection of fish species for which chironomids are cited as food. This section is concerned mainly with natural habitats – the manipulation of environments to increase food supply for fish is dealt with in section 17.4.

The seasonality and spatial variation of chironomids and their selection by fish is well illustrated in the work of Mackey (1979) in the River Thames, England. Chironomids were important in the diet of fish species, with peak numbers ingested in June. During the winter, fish fed on chironomids from the marginal *Acorus* zone and stony bed of the river. In summer, when large populations of chironomids were present in the *Nuphar* zone, fish utilized these. Seasonal and spatial variation in the use of the chironomid food resource were observed between fish species. Pope (*Gymnocephalus cernua*) fed all the year round on the bottom and only 6% of their diet came from the

Table 17.2 A selection of references citing chironomids as food of fish groups

Group	Reference
Acipenseridae (sturgeons etc.)	Baranova and Miroshnichenko, 1969
Lepisosteidae	Payne and Pearson, 1981
Clupeidae	Sprahamian, 1989
Salmonidae	Brown *et al.*, 1980; Davidowicz and Glivicz, 1983; Dudley, 1981; Elliott, 1973b; Glova, 1984; Huru, 1986; Johnsen and Ugedal, 1989; Kelly-Quinn and Bracken, 1990; Muir & Emmett, 1988; Vøllestad and Andersen, 1985
Thymallidae	Scott, 1985
Mormyridae	King, 1989; Olatunde and Moneke, 1985; Petr, 1968
Anguillidae	ter Heerdt and Nie, 1987; Lammens and Visser, 1989; Mann and Blackburn, 1991
Characidae	Bowmaker, 1969; Kenmuir, 1975; Petr, 1974
Cyprinidae	Debus, 1989; Jamet *et al.*, 1990; Kaeding and Zimmerman, 1983; Kokes, 1979; Lammens, 1989; Myllylä *et al.*, 1983; Rask, 1989
Cyprinidae (carp)	Adzimuradov, 1972; Barthelmes, 1970; Dimitrov, 1974; Martyniak, 1990; Matena, 1982
Cobitidae	Przybylski and Banbura, 1989; Perrin, 1980; Welton *et al.*, 1983
Siluroidea	Chilvers and Gee, 1974; Groenwald, 1964; Singh *et al.*, 1981
Cyprinodontidae	Durant *et al.*, 1979
Centrarchidae	Devries *et al.*, 1989; George and Hadley, 1979; Gomez, 1970; Shireman *et al.*, 1989
Percidae	Cordes and Page, 1980; Costa, 1979; Fox, 1989
Cichlidae	Spataru, 1976, 1978; Tudoranacea *et al.*, 1988
Mugilidae	Blaber and Whitfield, 1977
Cottidae	Hershey and McDonald, 1985; Snyder, 1984

Nuphar zone. Bream (*Abramis brama*) fed selectively on larvae, and pupae or adults comprised <1% of their diet. In contrast chironomid pupae and pharate adults were an important component of the diet of dace (*Leuciscus leuciscus*) and gudgeon (*Gobio gobio*) in the summer. Perch (*Perca fluviatilis*) were selective. In spring they ate many pupae, mainly *Procladius choreus*; however, in June 50% of the chironomid intake was *Parachironomus biannulatus* larvae which were nearly three times more frequent in perch stomachs than *Cricotopus* larvae, although the latter were 15 times more abundant in the river.

Most fish are opportunistic in their feeding and chironomids may be taken only when they are abundant and available (Costa, 1979). For example 'red chironomids' were the dominant food item of tench (*Tinca tinca*) in shallow experimental sites in a canal in which the vegetation had been depleted by the grazing activity of grass carp (*Ctenopharyngodon idella*). In the vegetated section, Crustacea (*Asellus* and *Eurycercus lamellatus*) and gastropods were the dominant food items (Petridis, 1990). In the River Tees, England, impoundment led to major changes in faunal composition (Armitage, 1976, 1977) with an increase in the number of chironomids. These changes were reflected in the food of trout (*Salmo trutta*), bullhead (*Cottus gobio*) and minnows (*Phoxinus phoxinus*), all of which increased the proportion of chironomids in their diets (Crisp *et al.*, 1978).

In the tropics similar responses to prey availability are observed. Mormyridae in the central African Lake Victoria relied heavily on the ephemeropteran nymph *Povilla adusta* which colonized submerged trees in these lakes (Petr, 1968). A similar reliance on *Povilla* was observed in Lake Kainjii, Nigeria (Blake, 1977). However, in the Volta Lake, Ghana, chironomids formed a major component of the diet of mormyrids (Petr, 1968), and in Zaria, Nigeria, Olatunde and Moneke (1985) found that chironomid larvae were the most important food of all four species of mormyrid studied.

Chironomidae may be an important food only during juvenile stages of a fish's life cycle. For example, Tudorançea *et al.* (1988) found that juvenile *Oreochromis (Tilapia) niloticus* of less than 25–30 mm in length fed largely on chironomid larvae. At the upper size-range they become predominantly herbivorous. This is an extreme example and for many fish species chironomids continue to form part of the diet as the fish increases in size but will be most important for the juvenile stages. Chironomids were the dominant food category for all sizes of larval grayling (Scott, 1985); and walleye (*Stizostedion vitreum vitreum*) fed mainly on chironomids and cyclopoid copepods in the first two weeks of life and chironomids thereafter (Fox, 1989).

Salmonids are economically important and their food has been the subject of much study (Table 17.2). In standing waters chironomids are a major item of the food of brown trout (*Salmo trutta*), rainbow trout (*Salmo gairdneri*) (Lien, 1978; Brown *et al.*, 1980) and char (Reimer, 1985). Free-living larval chironomids such as *Procladius* and some Orthocladiinae are more heavily predated than tube-dwelling species such as *Chironomus* and the Tanytarsini. Larvae which tend to leave their tubes when disturbed and enter the water column, such as *Endochironomus albipennis*, are more vulnerable to predation (Brown *et al.*, 1980). The relation between prey activity and consumption by salmonids is reflected in the large numbers of chironomid pupae and emerging adults taken in both running and standing waters (Williams, 1981; Vøllestad and Andersen, 1985; Reimer, 1985; Mundie *et al.*, 1991). Chironomidae emerge most frequently in crepuscular conditions and the ability of salmonids to detect prey under low light conditions (Brett, 1957) facilitates the exploitation of this resource.

Whitefish (Coregonidae) are another group of economic importance and Chironomidae form a high proportion of their food (Table 17.2). In Lough Neagh, Northern Ireland, *Coregonus pollan* feed on nauplii and copepodites initially but when the fry reach a weight of >1.4 g *Chironomus anthracinus* and *Chironomus* sp. larvae are among the dominant food items (Dabrowski *et al.*, 1984). In Hjarbaek Fjord, Denmark, the principal food items of *Coregonus lavaretus* of length group 30–50 cm were the larvae of chironomids, which made up more than 70% of the food volume during most of the growth season (Rasmussen, 1990). In the shallow Lake Suwa, Japan, in enclosures which included bottom sediment, *Coregonus lavaretus* fed selectively on the adults of a univoltine chironomid *Tokunagayusurika akamusi* during their emergence period. Thereafter the fish switched to bottom-feeding on the larvae of *T. akamusi* and *Chironomus plumosus*. In enclosures where access to the bottom sediment was prevented, the fish fed largely on zooplankters (Hanzato *et al.*, 1990).

The impact of fish feeding on benthic populations has been studied by a number of authors (Thorp and Bergey, 1981a,b; Allan, 1983; Healey, 1984; Gilliam *et al.*, 1989; Schlosser and Ebel, 1989 for references). When fish and turtle predators were excluded from areas of soft bottom in the littoral zone of a reservoir, Thorp and Bergey (1981b) found that vertebrate predation was not the fundamental parameter organizing the macroinvertebrate community and in general the abundance and average individual size remained constant between cage and control plots. Also the numbers of taxa in the dominant functional groups of Chironomidae were not significantly affected by predator treatment. Finding that reduction in trout density of 10–25% in a stream caused no increase in invertebrate density, Allan (1983) suggested that predators do not in fact limit the density of their benthic prey. Gilliam *et al.* (1989) noted reductions in the most common taxa in the benthos of small enclosures in a lowland stream but chironomids, which occurred in relatively low numbers, were not apparently affected. Schlosser and Ebel (1989) found that invertebrate densities were most affected by the hydrological regime and that cyprinid predation had weaker and more variable effects on the abundance of stream organisms. The role of predators in affecting stream communities is discussed further in Gilinsky (1984) and Thorp (1986).

In standing waters, Iwakuma and Yasuno (1983) estimated about 50% mortality of *T. akamusi* due to predation during the emergence period. Prior to emergence predation by benthic species (the goby *Tridentiger obscurus* and freshwater prawn *Macrobrachium nipponense*) accounted for 11–12%. Winkel and Davids (1987) found that Bream was a severe predator of chironomid larvae measuring >4 mm. However this species has two feeding modes (Lammens, 1985); in two lakes studied by Giles *et al.* (1990), bream fed either on zooplankton (Main Lake) or benthic invertebrates (Main Lake and St Peters). In 1986 bream fed mainly on *Daphnia hyalina* in Main Lake but in 1987 chironomid larvae were more important. Such dietary switching has been reported by Lammens (1982) and ten Winkel (1987). The removal of fish from these two lakes resulted in a substantial increase in macroinvertebrate biomass. Rasmussen (1990) calculated that 2–5% of the total chironomid production in Hjarbæk Fjord was consumed by introduced whitefish (*Coregonus lavaretus*). Hershey (1992) suggests that chironomids in a fertilized lake in Alaska were not food-limited and that their numbers were controlled by predation by the slimy sculpin.

(b) Amphibia

Chironomid larvae and pupae were found by Avery (1968) to be numerically the most important food of the newts *Triturus vulgaris helveticus* and *T. cristata*, though less so in the latter species. Strohmeier *et al.* (1989) examined the effects of red-spotted newts, *Notophthalmus viridescens*, on the densities of invertebrates in a permanent pond and found that the abundance of chironomid larvae was significantly reduced in an enclosure experiment. Frogs and toads were observed to feed nightly on adult *Chironomus* between 22.00 and 24.00 h along a lake shore during periods of high density (Wilson, 1969).

Table 17.3 A selection of references citing chironomids as food of birds

Family	Reference
Anatidae (waterfowl)	Austin *et al.*, 1990; Bagge *et al.*, 1973; Connelly and Chesemore, 1981; Danell and Sjöberg, 1977; Gardarsson, 1979; Giles, 1990; Owen and Thomas, 1979; Sjöberg and Danell, 1982; Woodin and Swanson, 1989
Charadriidae (waders)	Beltzer, 1988; Holmes, 1966; Moksnes, 1987
Laridae Sternidae (gulls and terns)	Cramp and Simmons (1983)
Apodidae (swifts)	Lack and Owen, 1955
Hirundinidae (swallows and martins)	Tokunaga, 1965; St Louis *et al.*, 1990
Motacillidae (wagtails and pipits)	Davies, 1977
Muscicapidae (flycatchers)	Briskie and Sealy, 1989

17.3.3 BIRDS

There have been numerous reports of birds feeding on chironomids (Table 17.3) with most research effort on the feeding of waterfowl and waders. However, many other groups of birds such as flycatchers, wagtails, swallows, martins, swifts, gulls and terns all take chironomids subject to their availability. In the pied and yellow wagtail, chironomids were an important food comprising over 80% of their diet in spring months (Davies, 1977). A review of the diets of Australian birds (Barker and Vestjens, 1989) showed that many birds, even some with only minor association with freshwater, including rails, quail and plovers, had chironomids in their digestive systems. (Table 17.3 lists Latin equivalents of English common names).

In northern latitudes Chura (1961) noted that the majority of invertebrates eaten by mallard ducklings were chironomids. The chironomid life-cycle stages of most importance to dabbling ducklings are the pupa, emerging sub-adult and adult insects on the surface. Adult chironomids were found in nearly every duckling examined by Bengtson (1975). High mortality of mallards in their first week of life was attributed by Street (1977) to the low availability of chironomids in an artificial lake. Giles (1990) found that the feeding efficiency of young tufted ducks on chironomid larvae increased with increased standing crop. Pochard (a diving duck) selected areas of relatively high larval chironomid densities (and biomass) to feed over, concentrating 77% of their feeding dives in one small bay of a single lake (Phillips, 1991). In the USA, ponds experimentally treated with a pesticide to control spruce bud-worm produced significantly fewer invertebrates. The ducklings of black duck, *Anas rubripes*, and mallard spent more time searching for food and less time resting and their rate of movement round the ponds was greater than that of ducklings from untreated ponds (Hunter *et al.*, 1984).

There is a strong contrast between northern temperate bird utilization of chironomids as food and those observed in Australia. In the temperate north, the patterns of northward migration of birds that rely on spring seasonal midge development is quite predictable. Breeding success or failure is determined directly by the adequacy of midge production (Danell and Sjöberg, 1977; Sjöberg and Danell, 1982; Street, 1977): in a poor year for midges, for example when ice cover is protracted, breeding failure is high. In contrast, conditions for the successful breeding of waterfowl in arid Australia are aseasonal and follow the unpredictable fall of sufficient rain. Nonetheless, success also depends on a chironomid food source, principally *Chironomus tepperi* Skuse, a species that tracks rainfall and is able to colonize temporary habitats rapidly and in high densities (Crome, 1986; Maher and Carpenter, 1984). Success of waterfowl breeding in these habitats depends very much on the extent of the previous area exposed by draw-down that is covered by new inundation to create *C. tepperi* habitat (Crome, 1986).

The competition between waterfowl and fish for invertebrate prey has been the subject of several studies (Eriksson, 1979; Hill *et al.*, 1987; Giles *et al.*, 1990) and five years of study of fish/duck interaction at the ARC Wildfowl Centre in England is reviewed by Giles *et al.* (1991). Midge larvae increased dramatically following removal of fish (perch, tench and bream) and brood sizes of tufted duck rose by 33%. The waterfowl in general have responded to fish removal both in summer and in winter by increased usage of the lakes as brood-rearing and wintering habitat.

Large swarms, however, may deter predators. Davenport (1978) observed gulls, swifts and house martins feeding on insects over water-storage reservoirs near London (England) but none ventured into the dense swarms above trees.

17.3.4 MAMMALS

With the exception of bats, few published accounts refer to mammals eating chironomids. Insectivorous bats will take chironomids but the proportions vary with age, sex and species of bat. Griffith and Gates (1985) found that, of two congeneric species feeding on the same night at the same location, only one fed on chironomids. At certain times Chironomidae were present in large numbers in stomachs and faecal samples and the authors suggest that the bats cue in on the humming sound emanating from large aggregations of swarming midges. Swarm feeding is considered to be energetically efficient and opportunistic.

Swarming dipterans are also used as food by humans living around some large African lakes. Most records are anecdotal and refer to chaoborids as being the main source (Beadle, 1974). The flies are attracted to lights and fall to the ground. They are collected, boiled and made into small cakes (Kungu cake) which are said to taste similar to caviar or salted locusts. Large numbers of chironomids emerge at the same time as chaoborids (MacDonald, 1956) and they also constitute a proportion of the midges collected for food. Bergeron *et al.* (1988) have analysed the flour made by grinding the dried insects and report high proportions of protein (67 g/100 g) with a high *in vitro* digestibility of 91%.

17.4 ENVIRONMENTAL MANIPULATION TO INCREASE CHIRONOMID DENSITIES

Chironomid larvae, as can be seen from the preceding sections, are important natural foods of many fish and birds and appear to function in small quantities as a growth promoter in fish diets (Yashouv, 1956; Yashouv and Ben Shachar, 1967). Chironomids are also used as fish food by aquarists in live and freeze-dried form and occasionally as bait by anglers. Although most larvae for these and other purposes are harvested from the natural environment, there have been several attempts to rear chironomids commercially on both large and small scales. The attempts to increase the number of chironomids available as food have been made by rearing in artificial habitats or by manipulating aspects of the natural environment (Catalan and White, 1982; Gersich *et al.*, 1989; Kiskaroly and Tafro, 1985; Maier *et al.*, 1990; Mundie *et al.*, 1983; Yashouv, 1970; Zieba, 1984). Biochemists, physiologists, geneticists and cytotaxonomists may require continuous cultures of chironomids for experimental purposes and these are reared under controlled conditions in the laboratory.

17.4.1 LABORATORY REARING

Most methods of culturing chironomids require regular attention and the availability of suitable food and may need frequent renewal of water and media (Beerman, 1952; Strenzke, 1959). The medium developed by Walter (1973), as modified for Australian conditions by Martin *et al.* (1980), has proved effective for laboratory rearing of *Chironomus*, *Kiefferulus*, *Polypedilum nubifer* (Skuse), most other lotic Chironomini and many Tanytarsini. Addition of thiamine HCl, use of commercial trout starter as food and continuous aeration with filtered air reduce the need for renewal of the culture medium.

If the substrate is a flat dish to which food is added, larvae distribute themselves in a monolayer with each larva requiring up to 5 cm^2 of bottom (*Chironomus tentans*, Beerman, 1952). Higher densities are obtained using a substrate of filamentous alga (Credland, 1973b). Martin *et al.* (1980) recommended shredded industrial Kleenex™ wipers to provide a three-dimensional medium for larval development of *Chironomus* species.

Biever (1971) considered aspects of the dietary requirements of laboratory cultures and found that the optimum food level of a proprietary pet food 'Dog Kisses' for the development of *Chironomus* '51' was 0.004 g per larva. In competition studies 0.004 g was also found to be optimal regardless of stocking rates. Feeding above or below this level increased both mortality and development time in intraspecific studies (*Chironomus* '51', *Goeldichironomus holoprasinus* (Goeldi), *Tanypus grodhausi* Sublette). Highest survival rates and fastest development time for all three species occurred at the lowest stocking rate of 50 larvae per unit and survival was lowest and development time most extended at stocking densities of 200 per unit. 'Dog Kisses' is not universally acceptable (J. Martin, *pers. comm.*) and Kawai and Konishi (1986) reared larvae at densities of 4–8 individuals cm^2 on a dosage of 7.5 mg of 'ground-up carp food' per individual. Diets were further investigated by Gersich *et al.* (1989) in their culturing of the parthenogenetic chironomid *Tanytarsus dissimilis* Johannsen (almost certainly *Paratanytarsus grimmii* (Schneider) (Langton *et al.*, 1988). They found that algae/alfalfa mix diets were the most effective, resulting in >80% survival consistently and approximately 120 eggs per adult. Photoperiod used in rearing varies from 12 h light/12 h dark, to 16 h light/8 h dark, with some laboratories using constant low light.

Although some chironomids can copulate on the ground (section 9.5.4), flight/swarming is a necessary precursor to mating in many species. For the continuous culture of sexual forms it is important to provide suitable conditions for adult mating. Details of lighting requirements are presented in Credland (1973) and Martin *et al.* (1980), who note the need for artificial twilight. Kawai and Konishi (1986), in addition to examining the effects of food dosage and larval density on the survival and development of larvae, also investigated the effects of illumination, temperature, humidity, space available for copulation and sex ratio on oviposition and fertilization of the adults. For *Chironomus samoensis* Edwards [as *flaviplumus*], they found that if 'dusk' illumination was introduced into the dark/light cycle, a considerable number of fertilized egg-masses were produced even in a relatively small space, when the sex ratio (m/f) was 2.

17.4.2 LARGE-SCALE REARING

Large-scale cultivation is necessary where chironomids are required as fish food. As early as 1935 *Chironomus tentans* was cultivated in concrete ponds in water enriched with sheep manure, soybean meal and sheep manure with superphosphate (Sadler, 1935). Soybean meal alone gave the best results. A breeding cage allowed adults to mate and oviposit and the eggs were collected to supplement natural spawning in the ponds. The most productive culture system for midge larvae recorded in the literature is that of Konstantinov (1952, 1954, 1958a). By supplementing the culture medium with egg-masses of *Chironomus dorsalis*, production rates of 400 g m^{-2} of floor space or 25 g m^{-2} of water surface per day were obtained. Without restocking with egg-masses, Yashouv (1956) harvested 250–375 g m^{-2} per week. Natural standing crops of benthic invertebrates cited in the literature are usually well under 15 g m^{-2} (Armitage, 1977). Using burlap sheets as a substratum for chironomids in fertilized ponds, McLarney *et al.* (1974) found that the average yield of the best ponds (11 g m^{-2} per week) did not approach that of Konstantinov or Yashouv (*op. cit.*). However, the authors suggest that the technique is convenient for practical fish culture because the burlap to which the chironomids adhere is simply removed from the culture pond and suspended in the fish pond and the fish allowed to graze.

The rearing of chironomids can provide a means of recycling farm manure and potentially can reduce water pollution. A chironomid farming technique which utilizes chicken manure is described by Shaw and Mark (1980). In Hong Kong, 13.5 ha are used for growing chironomids and 0.6% of the chicken manure produced annually is used as fertilizer. Chironomidae are used locally for fish-food or exported by air under refrigeration to western countries, mainly to North America for the aquarist market. Production was again lower than reported by Konstantinov and Yashouv (*op. cit.*) and on average 25 g m^{-2} were harvested per week. The authors consider that the development of chironomid farming using excess farm manure will reduce the organic pollution loading into streams draining farmland in Hong Kong. This link between chironomid production and pollution is explored further by Pandian (1987) who suggests that there is scope for using chironomids as 'biological filters' to remove bacteria, a view shared by Sokolova *et al.* (1993) who found that *Chironomus piger* at high larval densities was capable of assimilating 80–177 g wet weight m^{-2} d^{-1} of organic matter.

In France, a rearing system has been developed for *Chironomus riparius* on the outlet channel of the third basin of waste-lagoon purification works (Bouguenac and Giani, 1992). Between May and October, 51 kg wet weight of larvae were collected. Temperature, dissolved oxygen and sedimentation rate were the principal factors affecting production.

17.4.3 FISH REARING PONDS

Carp have been farmed in Europe at least since the Middle Ages in warm, shallow, sheltered ponds. Such conditions also suit chironomid larvae and the link between insect food and carp production has long been exploited (Thienemann, 1954, pp. 740–50; Barthelmes, 1970; Dimitrov, 1974). The techniques usually involve fertilization with farmyard manure (Schroeder, 1974) or organic liquors such as sugar factory wastes (Zieba, 1984). The enrichment of the nutrient content of the ponds lead to increases in bacterial and later algal production which is utilized by developing chironomid populations. Since intense manuring, especially in warm countries, may lead to severe depletion in oxygen resulting in fish kills, a balance has to be maintained between the benefits of increased production of natural food and the risks of oxygen depletion. Schroeder (1974) calculated the biological oxygen demand of manure at pond temperature to determine the maximum quantity of oxygen that organic decomposition would use and the maximum overnight reduction in dissolved oxygen in the pond water. From this it was possible to estimate the amount of manure that could be added to the fish ponds without causing anoxia. Schroeder notes increases in natural food in manured ponds ranging from 10 to 1000 and an increase in the growth of carp of 25–100%.

Chironomid production may also be increased by other means. In Bulgaria oats are sown in carp ponds in early March and the ponds are stocked and filled at the end of April (Dimitrov, 1980). The flooded vegetation acts as both substrate for chironomid larvae and food source, first as a substratum for the growth of epiphytic algae and then later as a source of bacteria as it decomposes. Very high production of chironomid larvae on flooded vegetation is frequently seen in the first years of new impoundments (Armitage, 1977).

17.4.4 STREAM ENRICHMENT TECHNIQUES

Knowledge of stream ecology can be used to raise fish species in high density (Mundie and Mounce, 1978) and running waters may also be fertilized to increase benthic production. Most studies in this area concern salmonids whose food in streams is almost entirely aquatic insects. The agents used to enhance production include sugar (Warren *et al.*, 1964) and cereal grain (Mundie *et al.*, 1983; Perrin *et al.*, 1987) and inorganic nitrogen (N) and phosphorus (P) have been added to streams to increase autotrophic production (Stockner and Shortreed, 1978; Perrin *et al.*, 1987). The input of inorganic nutrients can result in large diatom populations which are grazed by stream insects. Johnston *et al.* (1990) have shown three- to five-fold increases in the biomass of standing stock in treated areas.

Mundie *et al.* (1991) have examined in detail the responses of stream benthos to increases in algae in the absence of fish. The results show that insects respond to enrichment with a 2.2-fold increase in numbers of emerging insects over a 7-week period and a 1.75-fold increase in density of the benthos over the same period. Greatest responses were observed in Chironomidae and Ephemeroptera. Small species of chironomid (belonging to the genera *Corynoneura*, *Thienemanniella*, *Synorthocladius* and *Cladotanytarsus*) showed the greatest response to the addition of N and P. Changes in benthic density and biomass are considered to become most apparent after 3 or 4 months of nutrient additions to a stream as succeeding generations of insects respond to the sustained availability of food. Significantly, the smallest chironomids respond most to the nutrient input. Such small species are known to be heavily utilized by juvenile coho salmon and have high development rates (Mackey, 1977a; Tokeshi, 1986b). Their multivoltinism facilitates recruitment.

Stream enrichment appears to be a promising method of rearing fish, particularly in their young stages, but Mundie *et al.* (1990) did not find that the 'semi-natural' rearing of coho smolts was superior to standard hatchery procedures. However, semi-natural rearing has potential in certain areas where flows can be regulated in a simple manner, for example in the side channel of a braided river. This channel produces and transports invertebrates whose densities can be increased by means of the stream enrichment techniques described.

17.5 FUTURE POSSIBILITIES AND SUMMARY

Clearly chironomids are vital in trophic pathways of aquatic environments. They also lend themselves to both large- and small-scale rearing and have a potential role as biological filters of wastewaters. Future attempts at producing large quantities of chironomids may take account of the slow sand filter systems used to purify water for drinking purposes. These systems filter raw water though a bed of sand which develops a rich 'schmutzdecke' at the sand/water interface (Wotton *et al.*, 1993). This area is colonized by enormous numbers of chironomids which may reach standing crops of $250\,000\ m^{-2}$. Although the sampling of larvae would not be possible, as it disturbs the integrity of the filter surface, it may be possible to harvest the adults which emerge in vast numbers. The continual rain of nutrient particles onto the surface of the filter supports the large midge production. Alternatively geotextiles could be used to act as a false bottom in the filter beds. This could be removed together with its complement of chironomids and placed in fish ponds in much the same way as burlap was used by McLarney *et al.* (1974). The closer linking of necessary commercial undertakings such as the provision of clean water and waste disposal with chironomid farming is a logical future course and more trials are needed to assess the economic value of a wide range of techniques.

18

Conclusions

W.P. Coffman

18.1 INTRODUCTION

The preceding chapters of this book clearly document the far-reaching nature of chironomid study: from taxonomy, palaeontology and phylogenetics through aspects of autecology, genetics, physiology, development and immunology to the integrative subjects of community ecology, biogeography, water quality assessment and lake history. The review of such a broad field as chironomid studies in a limited space requires a very general treatment. Although the basic organization that follows is generally chronological, it is not a history nor an exhaustive review of the achievements of investigators. As it is difficult to dissect the development of various fields of chironomid study, from their beginnings to the present time, the discussion has been restricted to a few major themes, the interrelationships among them and needs for future work.

Before launching into the treatment of this subject, it seems worthwhile to ask why scientists of various casts have been interested in chironomids. In the beginning, i.e. at the time of Linnaeus and other pioneers, the interest stemmed from the early drive to describe all organisms including the relative great variety of midges. Only much later were other rationales called upon to justify other interests in this group. Four of these principal justifying arguments function as the main organizing themes of this chapter:

1. The **great species richness** of the family – at global, regional and local (within system) scales.
2. The **extensive geographical range** of the family.
3. The **great ecological diversity** exhibited by members of the family.
4. The **great abundance and biomass** of individuals within aquatic systems.

With the possible exception of studies in developmental and molecular genetics, biochemistry and physiology, all aspects of chironomid study can be thought of as having been stimulated by one or more of these organizing

The Chironomidae: Biology and ecology of non-biting midges
Edited by P.D. Armitage, P.S. Cranston and L.C.V. Pinder
Published in 1995 by Chapman & Hall. ISBN 0 412 45260 X

themes. It is a curious fact that at the same time these factors have played a major role in stimulating the study of chironomids in the minds of the few, the great richness and abundance, in particular, have played an equally large part in discouraging their study in the minds of the many. It is still not rare to encounter in the scientific literature exhaustive studies of lentic or lotic ecosystems in which the chironomids are either not included at all, or treated only at the subfamily or tribe level. Perhaps even worse are those studies in which the chironomids are treated at the generic and/or species level and include taxonomically incongruous and/or ecologically disharmonious lists of chironomid taxa – suggesting a severe disarticulation between the intent to be thorough in the treatment of the chironomids and the actual practice of sampling and identification.

It may not be an overstatement to suggest that one of the most important tasks facing contemporary and future workers in the field is the demonstration, through work in systematics, ecology and biogeography (to name a few), that a thorough treatment of the chironomids in studies of aquatic systems is not only necessary, but possible.

18.2 SPECIES RICHNESS OF THE CHIRONOMIDAE

Estimates of global species richness of the family range from about 8 000 to about 20 000 species. Considering the fact that only the western Palaearctic (with 1000 or more species) has been investigated with anything approaching thoroughness, an estimate toward the higher end of this range is certainly justifiable. The discrepancy in these global estimates reflects, quite simply, our relative ignorance of the species richness and levels of endemism to be found in the major biogeographical regions of the world. All indications suggest that a fauna of at least 2000–3000 species per biogeographical region (the majority endemic) is a realistic estimate. On a more solid basis is our knowledge of species richness in particular aquatic ecosystems – where more than 100 species are quite common, especially in lotic environments.

Over time an awareness has grown of high species richness on these different scales and the associated difficulties encountered in their pursuit. It cannot be emphasized too strongly that progress in the systematics of the Chironomidae (especially alpha taxonomy) is absolutely vital to advancements in many other disciplines involving chironomids. For this reason an important focus of this chapter is on the development, problems and future of chironomid systematics.

It is relatively easy to date the beginnings of formal chironomid systematics to the time of Linnaeus and his immediate followers (mid 18th to early 19th centuries). However, it is also clear that an awareness of chironomids predated these formal beginnings. The English word midge, for example, has been reliably traced back to an Old Teutonic root with surviving cognates in a number of current Germanic languages. It is even suggested that the original word was onomatopoeic – imitating the humming sound made by mating swarms of these insects. With less firm evidence some philologists trace the word midge to a proto Indo-European lexicon, from which Sanskrit, Greek and Latin, among others, have derived words relating to fly or to such sounds as those made by midges. A cursory examination of dictionaries from non-Indo-European languages reveals that many have a word for midge. However, it is not clear whether these words always refer to what we mean by midge or that the absence of the word in a dictionary means the absence of a word in the language. It would not be surprising to find that the languages of so-called primitive peoples have such a word. D.R. Oliver (*pers. comm.*) informs me that Canadian Eskimos do indeed have such a word.

Perhaps the safest thing to conclude is that adult chironomids (see Thienemann (1954) for a reference to the knowledge of larvae by Aristotle) have been recognized for a long time and that it is perfectly reasonable that humans living near suitable chironomid habitats (nearly everywhere) from the earliest times have known about them. It is not at all hard to imagine inhabitants of the great cities along the Tigris, Euphrates, Indus, Nile and Yangtze not only distinguishing these insects by their appearance and behaviour but also being able to refer to them by name.

Since the first 'scientific' descriptions of chironomid adults, several specific difficulties have arisen that have hindered the development of an awareness of the species-rich nature of the Chironomidae. Two of these – the development of stage specific systems of taxonomy and an inadequate concept of the morphological species – have been most responsible. These will be returned to below. Additionally, problems of nomenclature have, either directly or indirectly, affected our sense of the number of species in the family.

The replacement of the Meigen (1803) generic names by the Meigen (1800) names for the same genera at a relatively late date by Hendel (1908), with the consequence of changes in the names of some tribes and subfamilies, had a debilitating effect on potential students of chironomid taxonomy. This confusion resulted in a basically (although not entirely) transatlantic dichotomy in the use of these two sets of names for some taxa. Ecologists and potential systematists were put off by this internecine affair as well as by that engendered by the temporary shift of the genus name *Tanytarsus* to the tribe Chironomini. The confusion was only cleared away after years of consideration by the ICZN, when it ruled in favour of the principle of conservation of well-established names (1963) and placement (1961) in these cases. Additional problems in nomenclature, for example the use of the name Hydrobaeninae for Orthocladiinae by Townes (1945), further muddied the waters.

The significant degree of species-level synonymy has proved to be another factor in our understanding of species richness in the family. This has not only had the effect of temporarily inflating the number of species but also, together with a shifting sense of the morphological species definition in the family, has necessitated (although rarely achieved for most genera) constant revision.

As a result of these and other problems in nomenclature students of the Chironomidae have until very recently been faced with a library of taxonomic works which seem to be nomenclatorially irreconcilable and which may contain, even in the same work, species under more than one name. These facts, coupled with the great species richness of the Chironomidae, have clearly discouraged their study. Which of us has not heard the quip that 'you don't have to be crazy to work on midges, but it helps'?

With regard to the development of a dual system of taxonomy it should be noted that, in a strict sense, there were not really two systems but two levels of generic discrimination – one based on adults and one on immatures. This problem arose from two factors in the early decades of this century. First, chironomid taxonomy until that time was based almost exclusively on the adult stage and, second, ecological studies of aquatic systems were beginning to reveal a bewildering variety of larvae and pupae. Given this amazing diversity of immature stages, ecologists-cum-taxonomists were not altogether willing to force what they thought of as morphologically and ecologically distinct taxa into the relatively few genera recognized by the traditional taxonomists.

This duality continued to exist until recent decades. During this period there was, in a very piecemeal fashion, a constant effort to reconcile the two approaches in some groups. This has resulted in a gradual acceptance of a concept of the genus in the family that is quite a bit smaller than that held by the traditional taxonomists of old. It has also had the effect of stimulating the discrimination of species in which the principal morphological differences occur in the immature stages. However, the problem is not yet totally resolved. We still face large genera, albeit in a decreasing number, which are based on seemingly similar adults but which contain very heterogeneous immature stages. For example, presumed clades can be recognized from the immature stages in genera such as *Tanytarsus* and *Polypedilum*. Though progress has been made in this direction (namely, *Virgatanytarsus*), much yet remains to be done.

Although various views have been held, and to some degree still are, concerning the minimal conditions that warrant the description of new species, a general consensus has evolved: unless a taxon is clearly unique in the adult stage (male usually, but also the female) or originates from some ecological or geographical context that makes it unlikely that it has been described before, its description should include, in addition to the adult stage(s), at least one, and ideally both, of the immature stages. In a similar vein, although there is no stricture against describing species from the immature stage(s) alone, most taxonomists hesitate to do so because of the potential problem of synonymy with one of the many species already described from the adult stage(s) alone.

A less frequently discussed problem in the assessment of the number of species in the family concerns the shift that has occurred over time in the operational definition of the morphological species. In the beginning, when species were described without adequate quantitative data and slide preparation, descriptions were very general and rarely accompanied by adequate illustrations. They were often based on such phenotypically labile features as overall size and color pattern. This lack of precision in the description process, along with the fact that if museum specimens exist they may be fragmentary, renders much of the early taxonomic work essentially useless. Underlying this early work was the perception that there were relatively few species and that this level of distinction was adequate to separate them. Even a cursory examination of the chironomid taxonomic literature clearly shows a gradual progression in the completeness of species descriptions. Today's standards have been achieved by the recognition and remedying of the mistakes of the past. Careful slide preparation, accurate measurements (if possible involving a number of specimens representing different populations) and detailed and clear figures are now generally accepted as the ideal for adequate species level descriptions.

Yet several problems remain. Firstly, there is still a problem in the reconciliation of the life stages at the species level in that, for many genera, one life stage appears to be much better for the discrimination of species than do the others. With some possible exceptions (e.g. *Diamesa*), the life stage that seems to carry the most species-level taxonomic information is the pupa, with the features revealed most clearly in the pupal exuviae. However, this may be more an artifact of the clarity of morphological features (almost all being associated with the integument) in prepared pupal exuviae than the fact that the other life stages lack such species-level morphology.

Reasons have been suggested for the existence of greater species-level distinctiveness in the pupal stage. For example, the short pupal period may mean that natural selection has less time to operate on it through its phenotypic characters, allowing much greater morphological variability. This is coupled with the argument that selection may operate with less intensity on this stage since it is non-feeding and non-reproductive. However, there is no evidence that selection actually operates with less intensity on the phenotypic features of the pupa. If selection were really much less intense, it might be expected that intraspecific variation would be very great. But, with some exceptions, this does not appear to be the case. Langton (1989) has suggested that the phenomenon of species-specific morphology of the pupa may be a developmental one – perhaps related to the homeotic or other control genes and a reserve (bank) of unexpressed pupal genes. He suggests no firm mechanism for the sampling of this structural reserve, but thinks that it may be related to changes in larval form and/or habit. Coffman and Gottlieb (unpublished) have been developing a similar hypothesis that suggests species-specific pupal morphology is the secondary outcome of strong selection acting to maintain larval and adult morphologies even as the underlying structural and control genes for those features change as the result of various types of genetic events. The outcome would be conservation of structures in the larva and adult and, since many of the same genes may have effects on pupal structures, the production of species-level novelties in the pupa.

Another problem is the utility of the cytogenetic distinction of species. In considering this issue we must confront the shortcomings of the application of the Biological Species Definition. As in most other groups, chironomid taxonomy confronts the impracticality of actually demonstrating that species are reproductively isolated. It has been assumed that what appear to be stable morphological differences and inferred behavioural ones can be used as a measure of a species. Although not surprising, it is certainly disconcerting to learn from cytotaxonomists that what appear to be morphologically identical forms sometimes possess, for example, chromosomal inversions that may preclude reproductive compatibility. Since this phenomenon has been demonstrated for species within only a few genera (mostly in the Chironomini) it is unclear how pervasive it may be. Assuming for the moment that it is a family-wide phenomenon, what should be done about it? It might be argued, for example, that even if there are many such cytospecies, their recognition is a second-order problem facing taxonomists. For the present, it may not be advisable to concentrate on these species when there is clearly so much yet to be done on morphological species richness. Midge cytotaxonomy would gain much strength if it could be demonstrated convincingly that morphologically identical species have different ecological relationships. For the time being, the study of cytospecies should be on a wider taxonomic basis and should include a thorough search for accompanying morphological and/or ecological differences.

Numerous advances in recent years have made the study of chironomids much less daunting. All aspects of chironomid study, but particularly the identification process, have been positively affected by one or more of these advances:

- A clear improvement in the precision of species descriptions.
- A number of generic revisions and reviews.
- Well-written and illustrated up-to-date keys to all life stages for at least some parts of the world (e.g. Oliver, 1981; Coffman and Ferrington, 1984; Wiederholm, 1983, 1986, 1989), some at the species level (e.g. Pinder, 1978; Cranston, 1982b; Langton, 1984, 1989; Epler, 1992).
- A glossary of morphological terms (Sæther, 1980).
- A catalogue of generic and subgeneric names (Ashe, 1983).
- A synthesis of zoogeographic data (Ashe *et al.*, 1987).
- A bibliography of the Chironomidae (Fittkau *et al.*, 1976; Hoffrichter and Reiss, 1981).
- Regional catalogues of species (e.g. Freeman and Cranston, 1980; Oliver *et al.*, 1990; Ashe and Cranston, 1990).
- A periodically appearing newsletter.
- An ever-increasing sense of community among the students of the Chironomidae, engendered to a large extent by the success of the triannual Symposia.

There remain, however, several obvious needs in the area of systematics:

- More intensive study of extra-Holarctic regions.
- Association of adults with immature stages, including those from the Holarctic.
- World-wide revisions of genera, including redescriptions when necessary.

The recent trend of increasing use of pupal exuviae in faunistic studies has great promise for the advancement of our knowledge of species richness at all levels of distribution – global to local. Although first suggested by Thienemann (1910) and championed by his students and co-workers (e.g. Lenz, 1955), this approach did not gain widespread acceptance until relatively recently. The re-demonstration by Brundin (1966) of the ease with which collections of exuviae can be made and the amount of systematic information that can be extracted from them was critical to this resurgence. The nearly simultaneous publications of Coffman (1973) and Wilson and Bright (1973), in which the utility of studies of pupal exuviae for various aspects of lotic communities was demonstrated, played a major role as well. Since then, a number of workers (e.g. Laville, 1979, 1981) have made significant contributions to our understanding of assemblage composition, phenology and biogeography. In addition to the relative ease of species-level discrimination of pupal exuviae, their collection offers other distinct advantages:

- Large numbers of specimens can be collected in a short period.
- Sorting to the species level is often possible with a dissecting microscope.
- Sampling is non-destructive.

Finally, two questions concerning the species richness of the Chironomidae need to be treated:

- Why are there so many species of Chironomidae on all scales?
- Why is the family Chironomidae, and not some other group, so species rich?

We have barely begun to answer these questions, although a proximal answer can be given in ecological terms for high levels of local (within system) species richness (see below). The answers to these questions about the Chironomidae have significance in the fields of evolutionary and community ecology, systematics and biogeography. Possible answers involve high evolutionary plasticity, unusual susceptibility to geographical and/or ecological isolation and, low level of extinction. There is very fertile ground here for future study utilizing the Chironomidae which, compared with many other species-rich insect groups, offers the distinct advantage of a relatively advanced state in taxonomic and ecological knowledge of all life stages.

18.3 GEOGRAPHICAL RANGE OF THE CHIRONOMIDAE

Probably no other major family of aquatic macroinvertebrates occurs over as wide a geographical range as does the family Chironomidae. This is even more obvious when terrestrial, semi-terrestrial, brackish, hypersaline and marine forms are considered. In essence, chironomids are to be found wherever conditions permit and, given the extremely wide-ranging adaptations within the family, this means (with a few exceptions such as the deep ocean, the very highest latitudes and altitudes and hottest springs) essentially everywhere. Of course, not all taxa within the family occur everywhere, but the level of cosmopolitanism and/or endemism for the major biogeographical regions can only be estimated at the present time. One of the major tasks still before us is the exploration of most of the world and a description of regional chironomid faunas.

Although much has been accomplished the coverage of this work has often been rather narrow geographically and/or taxonomically, and work before the modern concept of species description was developed is often of questionable validity. Most striking among relatively recent evolutionary and biogeographical work is the description of the austral subfamilies Aphroteniinae and Chilenomyiinae and the localization of various clades of the Diamesinae and Podonominae in the southern hemisphere by Brundin (1966, 1983). The demonstration of Gondwanan distributions in most of these austral taxa via the use of Hennigian cladistics by Brundin has not only been a major advance in our knowledge of the evolutionary and biogeographical history of the Chironomidae, but has also served as a major influence in phylogeny and biogeography ranging far beyond this family. It seems clear from recent work (e.g. Edward, 1989; Cranston and Edward, 1992; P.S. Cranston and D.H.D. Edward, *unpubl.*) that other groups of Chironomidae may share in this Gondwanan distribution.

The clarification of the biogeographical interrelationships between the chironomid faunas of the Nearctic and Neotropical (e.g. Reiss, 1985b; Reiss and Sublette, 1985), the Nearctic and Palaearctic (e.g. Cranston and Oliver, 1987), the Palaearctic and Afrotropical (e.g. Fittkau, 1980) and Palaearctic, Oriental and Australasian regions await further study.

Superimposed on the pattern of endemic faunas in various parts of the world are a number of cosmopolitan genera and an unknown number of species which have multi-region distributions. Questions as to whether these represent phylogenetically old, phylogenetically new, exceptionally mobile or ecologically generalized taxa can only be clarified by additional systematic, phylogenetic and ecological analyses.

Although our knowledge of the specific ecological relationships of the Chironomidae is wanting in most extra-Holarctic settings (Australia perhaps being the principal exception), it is likely that widely distributed taxa have similar ecological relationships wherever they occur. It is also likely that the ecological organization (e.g. utilization of spatial and trophic resources) of local chironomid assemblages reflects the available opportunities and that all major ecological zones and functions are occupied by species. The ecological uniformity of chironomid assemblages in comparable ecological settings projects the family as a major candidate for comparative studies and for the development of universal water quality assessment schemes that utilize them.

The extensive geographical distribution of the Chironomidae may be due, in part, to the great age of the family – at least Jurassic (Brundin, 1966; Cranston and Edward, 1993; Cranston *et al.*, 1987). To the extent that the age of the family is coupled with endemism via vicariance events, many additional opportunities may exist for clarification of the evolutionary history of the family.

However, the question of monophyly of the family, which bears on the issue of phylogenetics and, therefore, biogeography, has not yet been satisfactorily addressed. Although various taxa within the family can be rather convincingly advanced as being monophyletic, unambiguous and convincing evidence of the monophyly of the family and the identity of the sister group has not yet been presented (e.g. Schlee, 1975). The resolution of this problem may well require the clarification of the phylogenetic relationships among all the families of the Nematocera. Attempts have certainly been made (e.g. Brundin, 1966; Sæther, 1977), but they have been clouded by uncertain polarities of characters and a very high frequency of homoplasy. In part, these problems led Brundin (1976) and Sæther (1979, 1983) to propose the concept of 'unique inside parallelisms' (underlying synapomorphies), which has been criticized (e.g. Farris, 1985; Cranston and Humphries, 1988).

It is conceivable, however, that there is a sound developmental argument involving control genes, similar to that raised by Langton (1989) and Coffman and Gottlieb (unpublished) concerning the origin of the structural richness of pupae, that speaks in favour of this idea. Consider that it has been demonstrated that a major part of the genome of *Drosophila* is never expressed. Since there are many developmental similarities between chironomids and drosophilids, it is likely that a similar phenomenon occurs in the Chironomidae. The question then concerns the nature of the unexpressed genetic material. It is possible that some of this material is in the form of intact, but unexpressed, structural genes (the gene bank of Langton) which are subject to expression, with relatively minor changes in the action of control genes. This would account for the seemingly sporadic occurrence of certain character states (homoplasy) within certain lineages. It follows that the more closely related lineages are, the greater the probability that they would possess the same intact unexpressed structural genes which would be subject to occasional activation – resulting in homoplasy. Therefore one could argue, unconventionally and in a purely speculative fashion, that the reoccurrence of a particular character state is a demonstration that the group within which such homoplasy occurs is a clade. Although ridden with difficulties, the use of underlying synapomorphies in groups such as the Chironomidae may be a necessary compromise until, if ever, appropriate non-homoplasic synapomorphies (perhaps at the molecular level) are identified.

18.4 ECOLOGICAL DIVERSITY OF THE CHIRONOMIDAE

Partially explaining, or at least correlated with, the great geographical distribution and global species richness of the Chironomidae is the great breadth of ecological conditions to which chironomids are adapted. Aquatic larvae (and pupae) are associated with all types of inorganic and organic substrates: from silts, sands, pebbles, cobble, boulders and bedrock to vascular plants, leaves, wood and a variety of animals (e.g. insects, sponges, bryozoans, molluscs and fish). They may occur on the surfaces of substrates, or where permissible, burrowed within them; deep in the hyporheic of streams, even in the water table; suspended from the surface film of Amazonian waters; skating on the surface of tropical lagoons; and towards the depths of ancient lakes like Baikal.

Regional chironomid faunas are likely to be composed of species which exhibit at least some of these specific adaptations and it is from the regional fauna that the specific assemblages of individual ecosystems must be drawn. The great species richness of the Chironomidae within particular aquatic ecosystems has almost become an axiom. Although often less pronounced in lentic than lotic systems, it is often the case that the number of chironomid species equals or exceeds that of all other benthic macroinvertebrate groups combined. Whether one is of the opinion that these assemblages represent species saturation resulting from competitive interactions, accidental co-occurrences or tightly-bound distribution units (e.g. Tokeshi, 1991b), to mention just a few of the controversial issues in community ecology, it is clear that one of the major tasks of chironomid ecology is to explain the phenomenon of high within-system species richness.

A possible key to the solution is the often-mentioned success of chironomid larvae in exploiting a wide range of trophic, temporal and micro-spatial aspects of the systems in which they occur – that is, via resource partitioning. Empirical data would suggest that this is certainly the case. It is not unusual to find, by trophic analysis, that lotic assemblages contain predaceous (including parasitic), herbivorous, detritivorous and fungivorous species. Collectively, these assemblages feed on a range of food materials (including a variety of invertebrates, algae, fine and coarse detritus, hyphae and spores) and fall into various functional groups (engulfers, piercers, scrapers, collectors, miners and filterers). Since the food resources and micro-habitats are patchy in time and/or space, so are the larvae.

But this is not a satisfactory explanation as to why this family and not some other group should be associated with this ecological diversity and degree of partitioning. The explanation must also involve the complex interrelationships of the effect of size, physiological tolerance, life-cycle strategies, labile mouthpart morphology, digestive processes, evolutionary age and reproductive behaviour, to mention just some of the factors that are likely to play a role.

The great ecological specificity of chironomid species has been exploited in a number of interesting ways to assess environmental change. Water quality assessment has often been linked to the use of chironomid indicator species. This is a potentially very useful approach but it is subject to misuse and misinterpretation. At the very least, it requires adequate sampling, identification and knowledge about the specific requirements and tolerances of the indicator species. A more promising approach is to use indicator assemblages which collectively respond to a broader spectrum of conditions. Lake classification along the trophic gradient has been effectively correlated with distinct profundal assemblages of chironomid species, and changing trophic status of lakes with time has been effectively revealed by the use of chironomid subfossils from cores.

Although we can infer the ecological significance of the Chironomidae from the number of species which occur in assemblages and the wide range of conditions to which they are adapted, it is clear that advancement in our understanding of chironomid ecology and the use of chironomids as indicators of environmental conditions will come only with a significant increase in the number of studies focused on individual species.

18.5 ABUNDANCE OF THE CHIRONOMIDAE

The first impression that one gets upon examining the benthos of lentic and lotic systems is not the great species richness, but the tremendous number of larvae – often as many as 50 000 m^{-2} or more. This observation was made by ecologists long before the taxonomy of the larval stages was perfected to the point where even generic identifications could be made with certainty, and was a major impetus for the development of interest in chironomid taxonomy. Even though the larvae might have been characterized as white, green and red species, and were often assigned exclusively to *Chironomus* and/or *Tanypus* (e.g. Forel, 1904; Forbes, 1925), these early workers were clearly aware of the important role that the midges play. This awareness stemmed in large part from the efforts of August Thienemann and his students. Thienemann, who at some point in his long career thought about and wrote about nearly everything of interest concerning chironomids, was not found wanting when it came to suggesting that any ecological study that passed over the chironomids was fragmentary at best (e.g. Thienemann, 1939). From the time of Lindeman (1941) to the present (e.g. Berg and Hellenthal, 1991; Caspers, 1983b; Ringe, 1974; Samietz, 1989) chironomids have been a major factor in the study of energy flow and production in aquatic ecosystems. A large number of workers in Russia and Eastern Europe, for example under the tutelage of Borutsky and Kajak, focused on the measurement of chironomid production in fish ponds and lakes. In North America, large scale studies of energy flow and production in aquatic ecosystems considered the chironomids in varying degrees of detail (e.g. Coffman *et al.*, 1971; Odum, 1957; Teal, 1957).

The great abundance of chironomid larvae, in addition to the obvious impact on system production, has been a major influence on trophic connections within the system. Chironomid larvae are responsible for the turnover of large quantities of many kinds of food materials, but most non-predaceous species depend on fine particles (detritus and/or algae) – what might be called the lowest common denominator of food materials. The universal distribution and quality of these foods, as well as coarse detritus (leaves, wood etc.) upon which some species feed, may play a role in the wide geographical distribution of the family and many of its substituent taxa. The large biomass of the Chironomidae, in turn, is available for utilization by a host of predator and parasitic species. Nearly every macroinvertebrate predator and most vertebrate predators, in most aquatic systems, utilize chironomid larvae (and, to a lesser extent, pupae and adults) at some point in their life cycle.

Abundance-related nuisance problems of adult midges have developed mainly in association with the construction of reservoirs and subsequent lake-shore exploitation for recreational, residential and commercial purposes. Larvae have occasionally been found to exist in nuisance numbers in water treatment facilities and even in drinking-water supplies. Immunological responses to proteins of adult chironomids are likely to occur wherever sensitive humans are repeatedly exposed.

18.6 OTHER AREAS OF RESEARCH ON THE CHIRONOMIDAE

Investigations on aspects of molecular and developmental genetics, biochemistry and physiology take advantage of particular features of the

Chironomidae that are more or less independent of their species richness, geographical extent, ecological diversity and abundance. Although midges were among the first organisms to be investigated with regard to the important discoveries of polytene chromosomes, chromosome puffs and the developmental consequences of egg polarity, they currently play a relatively minor part in the burgeoning fields of molecular and developmental genetics. This is largely because they are relatively difficult to rear in the laboratory, the typical generation period is relatively long, mutant variety is limited, chromosome maps are less complete and there is a lack of extensive knowledge of transmission genetic phenomena within the group. The wide range of morphological and ecological variation within the family, however, make the Chironomidae ideally suited for comparative studies in molecular and developmental genetics.

References

The numbers in square parentheses following each reference indicate the chapters in which the reference is cited.

Aagaard, K. (1978) The chironomids of Lake Målsjøen. A phenological, diversity, and production study. *Norwegian Journal of Entomology,* **25,** 21–37.[10].

Aagaard, K. (1982) Profundal chironomid populations during a fertilization experiment in Langvatn, Norway. *Holarctic Ecology,* **5,** 325–31.[10]

Aagaard, K. (1986) The chironomid fauna of North Norwegian lakes, with a discussion on methods of community classification. *Holarctic Ecology,* **9,** 1–12.[15]

Aagaard, K. (1993) Ordination or typology – The search for a stable classification of running water communities. *Netherlands Journal of Aquatic Ecology,* **26,** 441–5.[15]

Abul-Nasr, S., Isa, A.L., Kira, M.T. and El-Tantawy, A.M. (1970a) Biological studies on the bloodworms, *Chironomus* sp., injurious to rice seedlings in U.A.R. *Bulletin of the Entomological Society of Egypt,* **54,** 381–88.[7]

Abul-Nasr, S., Isa, A.L. and El-Tantawy, A.M. (1970b) Control measures of the bloodworms, *Chironomus* sp., in rice nurseries. *Bulletin of the Entomological Society of Egypt, Economic Series,* **4,** 127–33.[13]

Adams, A. (1983) Cryptobiosis in Chironomidae (Diptera). *Antenna,* **8,** 58–61.[6]

Adzimuradov, A. (1972) The food of juvenile carp *Cyprinus carpio* (L.) in early development stages in bodies of water of the Arakum (Terek River Delta). *Journal of Ichthyology,* **12,** 981–6.[17]

Alfred, J.R.B. (1974) On the food of *Chironomus costatus* (Chironomidae: Diptera) from a shallow freshwater pond in South India. *Freshwater Biology,* **4,** 337–42.[7]

Ali, A. (1980a) Diel adult eclosion periodicity of nuisance chironomid midges of central Florida, *Environmental Entomology,* **9,** 365–70.[13]

Ali, A. (1980b) Nuisance chironomids and their control: A review. *Bulletin of the Entomological Society of America,* **26,** 3–16.[9][13]

Ali, A. (1981a) *Bacillus thuringiensis* serovar. *israelensis* (ABG–6108) against chironomids and some nontarget aquatic invertebrates. *Journal of Invertebrate Pathology* **38,** 264–72.[13]

Ali, A. (1981b) Laboratory evaluation of organophosphate and new synthetic pyrethroid insecticides against pestiferous midges of central Florida. *Mosquito News,* **41,** 157–61.[13]

Ali, A. (1989) Chironomids, algae, and chemical nutrients in a lentic system in central Florida, USA. *Acta Biologica Debrecina, Oecologica Hungarica,* **3,** 15–27.[8, 14]

Ali, A. (1990) Seasonal changes of larval food and feeding of *Chironomus crassicaudatus* (Diptera: Chironomidae) in a subtropical lake. *Journal of the American Mosquito Control Association,* **6,** 84–8.[7][13]

REFERENCES

Ali, A. (1991a) Perspectives on management of pestiferous Chironomidae (Diptera), an emerging global problem. *Journal of the American Mosquito Control Association,* **7,** 260–81.[13]

Ali, A. (1991b) Activity of methoprene new formulations against midges (Diptera: Chironomidae) in experimental ponds. *Journal of the American Mosquito Control Association,* **7,** 616–20.[13]

Ali, A. and Chaudhuri, P.K. (1988) Evaluation of a new benzoylphenylurea insect growth regulator (UC–84572) against chironomid midges in experimental ponds. *Journal of the American Mosquito Control Association,* **4,** 542–4.[13]

Ali, A., Chaudhuri, P.K. and Guha, D.K. (1987) Description of *Stictochironomus affinis* (Johannsen)(Diptera: Chironomidae), with notes on its behavior. *Florida Entomologist,* **70,** 259–67.[7]

Ali, A. and Fowler, R.C. (1983) Prevalence and dispersal of pestiferous Chironomidae in a lakefront city of central Florida. *Mosquito News,* **43,** 55–9.[13]

Ali, A. and Lord, J. (1980) Experimental insect growth regulators against some nuisance chironomid midges of central Florida. *Journal of Economic Entomology,* **73,** 243–9.[13]

Ali, A. and Majori, G. (1984) A short-term investigation of chironomid midge (Diptera: Chironomidae) problem in saltwater lakes of Orbetello, Grosseto, Italy. *Mosquito News,* **44,** 17–21.[13]

Ali, A. and Mulla, M.S. (1976a) Insecticidal control of chironomid midges in the Santa Ana River water spreading system, Orange County, California. *Journal of Economic Entomology,* **69,** 509–13.[13]

Ali, A. and Mulla, M.S. (1976b) Chironomid larval density at various depths in a southern California water-percolation reservoir. *Environmental Entomology,* **5,** 1071–4.[13]

Ali, A. and Mulla, M.S. (1977a) Chemical control of nuisance midges in the Santa Ana River Basin, southern California. *Journal of Economic Entomology,* **70,** 191–5.[13]

Ali, A. and Mulla, M.S. (1977b) The IGR diflubenzuron and organophosphorus insecticides against nuisance midges in man-made residential-recreational lakes. *Journal of Economic Entomology,* **70,** 571–7.[13]

Ali, A. and Mulla, M.S. (1978a) Declining field efficacy of chlorpyrifos against chironomid midges and laboratory evaluation of substitute larvicides. *Journal of Economic Entomology,* **71,** 778–82.[13]

Ali, A. and Mulla, M.S. (1978b) Effects of chironomid larvicides and diflubenzuron on nontarget invertebrates in residential-recreational lakes. *Environmental Entomology,* **7,** 21–7.[13]

Ali, A. and Mulla, M.S. (1978c) Impact of the IGR diflubenzuron on invertebrates in a residential-recreational lake. *Archives of Environmental Contamination and Toxicology,* **7,** 483–91.[13]

Ali, A. and Mulla, M.S. (1980) Activity of organophosphate and synthetic pyrethroid insecticides against pestiferous midges in some southern California flood control channels. *Mosquito News,* **40,** 593–7.[13]

Ali, A. and Mulla, M.S. (1983) Evaluation of the planarian, *Dugesia dorotocephala,* as a predator of chironomid midges and mosquitoes in experimental ponds. *Mosquito News,* **43,** 46–9.[13]

Ali, A. and Nayar, J.K. (1986) Efficacy of *Bacillus sphaericus* Neide against larval mosquitoes (Diptera: Culicidae) and midges (Diptera: Chironomidae) in the laboratory. *Florida Entomologist,* **69,** 685–90.[13]

Ali, A. and Nayar, J.K. (1987) Laboratory toxicity of a new benzoylphenylurea insect growth regulator (UC–84572) against mosquitoes and chironomid midges. *Journal of the American Mosquito Control Association,* **3,** 309–11.[13]

REFERENCES

Ali, A. and Stanley, B. H. (1981) Effects of a new insect growth regulator, UC–62644, on target Chironomidae and some nontarget aquatic invertebrates. *Mosquito News,* **41,** 692–701.[13]

Ali, A., Baggs, R.D. and Stewart, J.P. (1981) Susceptibility of some Florida chironomids and mosquitoes to various formulations of *Bacillus thuringiensis* serovar. *israelensis, Journal of Economic Entomology,* **74,** 672–7.[13]

Ali, A., Mulla, M.S. and Pelsue, F.W. (1976) Removal of substrate for the control of chironomid midges in concrete lined flood control channels. *Environmental Entomology,* **5,** 755–8.[13]

Ali, A., Nayar, J.K. and Kok-Yokomi, M.L. (1987) Evaluation of new experimental insect growth regulators against mosquitoes (Diptera: Culicidae) and midges (Diptera: Chironomidae) in the laboratory. *Proceedings and Papers of the California Mosquito and Vector Control Association,* **55,** 81–6.[13]

Ali, A., Stafford, S.R. and Fowler, R.C. (1984) Attraction of adult Chironomidae (Diptera) to incandescent light under laboratory conditions. *Environmental Entomology,* **13,** 1004–9.[13]

Ali, A., Stanley, B.H. and Chaudhuri, P.K. (1986) Attraction of some adult midges (Diptera: Chironomidae) of Florida to artificial light in the field. *Florida Entomologist,* **69,** 644–50.[13]

Ali, A., Stanley, B.H. and Majori, G. (1985a) Daily abundance patterns of pestiferous Chironomidae (Diptera) in an urban lake front in central Florida. *Environmental Entomology,* **14,** 780–4.[13]

Ali, A., Stanley, B.H. and Stafford, S.R. (1983) Short-term daily emergence of adult Chironomidae from a natural lake and a man-made reservoir. *Environmental Entomology,* **12,** 765–7.[13]

Ali, A., Xue, R.D. and Lobinske, R. (1993) Efficacy of two formulations of the insect growth regulator, Pyriproxyfen (Nylar[R] or Sumilarv[R]) against nuisance midges (Chironomidae: Diptera) in man-made ponds. *Journal of the American Mosquito Control Association,* **9,** in press.[13]

Ali, A., Mulla, M. S., Federici, B. A. and Pelsue, F. W. (1977) Seasonal changes in chironomid fauna and rainfall reducing chironomids in urban flood control channels. *Environmental Entomology,* **6,** 619–22.[13]

Ali, A., Mulla, M.S., Pfuntner, A.R. and Luna, L. L. (1978) Pestiferous midges and their control in a shallow residential-recreational lake in southern California. *Mosquito News,* **38,** 528–35.[13]

Ali, A., Majori, G., Ceretti, G. *et al.* (1985b) A chironomid (Diptera: Chironomidae) midge population study and laboratory evaluation of larvicides against midges inhabiting the lagoon of Venice, Italy. *Journal of the American Mosquito Control Association,* **1,** 63–8.[13]

Ali, A., Barbato, L., Ceretti, G. *et al.* (1992) Efficacy of two temephos formulations against *Chironomus salinarius* (Diptera: Chironomidae) in the saltwater lagoon of Venice, Italy. *Journal of the American Mosquito Control Association,* **8,** 353–6.[13]

Ali, A., Ceretti, G., Barbato, L. *et al.* (1994) Attraction of *Chironomus salinarius* (Diptera: Chironomidae) to artificial light on an island in the saltwater lagoon of Venice, Italy. *Journal of the American Mosquito Control Association,* **10,** in press.[13]

Alimov, A.F., Boullion, V.V., Finogenova, N.P. *et al.* (1972) Biological productivity of Lakes Krivoe and Krugloe, in *Productivity Problems of Freshwaters* (eds Z. Kajak and A. Hillbricht-Ilkowska), PWN Polish Scientific Publishers, Warszawa-Krakow, Poland, pp.39–56.[11]

Allan, J.D. (1982) The effects of reduction in trout density on the invertebrate community of a mountain stream. *Ecology,* **63,** 1444–55.[12]

Allan, J.D. (1983) Food consumption by trout and stoneflies in a Rocky Mountain Stream with comparison to prey standing crop, in *Dynamics of Lotic Ecosystems,* (eds T.D. Fontaine, III and S.M. Bartell), Ann Arbor Science, Ann Arbor, pp.371–402.[17]

Allen, K.R. (1951) The Horokiwi Stream. *New Zealand Marine Department of Fisheries Bulletin*, **10,** 1–238.[11]

Alm, T. and Willassen, E. (1993) Late Weichselian Chironomidae (Diptera) stratigraphy of Lake Nedre Æråsvatn, Andøya, northern Norway. *Hydrobiologia*, **264**, 21–32.[16]

Alsterberg, G. (1922) Die respiratorischen Mechanismen der Tubificiden. *Kungliga Fysiografiske sällskapets i Lund Handlingar*, **33,** 1–175.[6]

Ambühl, H. (1959) Die Bedeutung der Strömung als Ökologischer Faktor. *Schweizer Zeitschrift für Hydrobiologie*, **21,** 133–264.[8]

Andersen, F.S. (1938) Spätglaciale Chironomiden. *Meddelelser Dansk geologisk Forening*, **9,** 320–6.[16]

Andersen, T. and Sæther, D.A. (in press) The first record of *Buchonomyia* Fittkav and the subfamily Buchomyiinae from the New World (Diptera: Chironomidae, in *Chironomids: from Genes to Ecosystems* (ed. P.S. Cranston), CSIRO, Melbourne.[3]

Anderson, J.F. and Hitchcock, S.W. (1968) Biology of *Chironomus atrella* in a tidal cove. *Annals of the Entomological Society of America*, **61,** 1597–603.[5]

Anderson, L.D., Bay, E.C. and Ingram, A.A. (1964) Studies on chironomid midge control in water spreading basins near Montebello, California. *California Vector Views*, **11,** 13–20.[13]

Anderson, N.H. (1982) A survey of aquatic insects associated with wood debris in New Zealand streams. *Mauri Ora*, **10,** 21–33.[7]

Anderson, N.H. (1989) Xylophagous Chironomidae from Oregon streams. *Aquatic Insects*, **11,** 33–45.[7]

Anderson, N.H. (1992) Influence of disturbance on insect communities in Pacific Northwest streams. *Hydrobiologia*, **248,** 79–92.[4]

Anderson, N.H. and Cummins, K.W. (1979) Influences of diet on the life histories of aquatic insects. *Journal of the Fisheries Research Board of Canada*, **36,** 335–42.[7]

Anderson, N.H. and Sedell, J.R. (1979) Detritus processing by macroinvertebrates in stream ecosystems. *Annual Review of Entomology*, **24** 351–77.[7]

Anderson, N.H., Sedell, J.R, Roberts, L.M. and Triska, F.J. (1978) The role of aquatic invertebrates in processing of wood debris in coniferous forest streams. *American Midland Naturalist*, **100,** 64–82.[7]

Anderson, N.H., Steedman, R.J. and Dudley, T. (1984) Patterns of exploitation by stream invertebrates of wood debris (xylophagy). *Internationale Vereinigung für Theoretische und Angewandte Limnologie Verhandlungen*, **22,** 1847–52.[6][7]

Anderson, R.L. (1980) Chironomidae toxicity tests – biological background and procedures, in *Aquatic Invertebrate Bioassays, ASTM STP 715*, (eds A. L. Buikema, Jr. and J. Cairns, Jr.), American Society for Testing and Materials, pp.70–80.[15]

Anderson, R.L., Walbridge, C.T. and Fiandt J.T. (1980) Survival and growth of *Tanytarsus dissimilis* (Chironomidae) exposed to copper, cadmium, zinc, and lead. *Archives of Environmental Contamination and Toxicology*, **9,** 329–35.[15]

Andronikowa, I.N., Drabkova, V.G., Kuzmenko, K.N. *et al.* (1972) Biological productivity of the main communities of the Red Lake, in *Productivity Problems of Freshwaters* (eds Z. Kajak and A. Hillbricht-Ilkowska), PWN Polish Scientific Publishers, Warszawa-Krakow, Poland, pp.57–71.[11]

Anonymous (1963) Annual Report and Statement of Accounts. *Auckland Metropolitan Drainage Board*, Auckland, New Zealand.[13]

Anonymous (1974) A Report on Midge in the Kanda River. *Report of the Department of Sewer Systems, Metropolitan Tokyo*, Tokyo.[13]

REFERENCES

Anonymous (1975) On the Important Factors for Habitation of the Midge Larvae in the Kanda River. *Report of the Department of Sewer Systems, Metropolitan Tokyo,* Tokyo.

Anonymous (1977) Economic Impact Statement, Blind Mosquito (Midge) Task Force. *Sanford Chamber of Commerce, Seminole County,* Florida.[13]

Anthony, D.W. (1975) Use of viruses for control of aquatic insect pests. *United States Environmental Protection Agency,* EPA-600/3–75–100.[13]

Armitage, P.D. (1968) Some notes on the food of the chironomid larvae of a shallow woodland lake in South Finland. *Annales Zoologici Fennici,* **5,** 6–13.[7][12]

Armitage, P. D. (1970) The Tanytarsini (Diptera: Chironomidae) of a shallow woodland lake in South Finland, with special reference to the effects of winter conditions on the larvae. *Annales Zoologici Fennici,* **7,** 313–22.[10]

Armitage, P.D. (1971) Some aspects of the ecology of *Pagastiella orophila* (Diptera: Chironomidae) in the Lake Kuusijärvi, in the south of Finland. *Canadian Entomologist,* **103,** 306–10.[7]

Armitage, P.D. (1976) A quantitative study of the invertebrate fauna of the River Tees below Cow Green Reservoir. *Freshwater Biology,* **6,** 229–40.[17]

Armitage, P.D. (1977) Development of the macro-invertebrate fauna of Cow Green reservoir (Upper Teesdale) in the first five years of its existence. *Freshwater Biology,* **7,** 441–54.[9][17]

Armitage, P.D. (1984) Environmental changes induced by stream regulation and their effect on lotic macroinvertebrate communities, in *Regulated Rivers,* (ed. A. Lillehammer and S. J. Saltveit), Universitetsforlaget, Oslo, pp.139–65.[15]

Armitage, P.D. and Blackburn, J.H. (1985) Chironomidae in a Pennine stream system receiving mine drainage and organic enrichment. *Hydrobiologia,* **121,** 165–72.[15]

Armitage, P.D., Moss, D., Wright, J.F. and Furse, M.T. (1983) The performance of a new biological water quality score system based on macroinvertebrates over a wide range of unpolluted running-water sites. *Water Research,* **17,** 333–47.[15]

Ashe, P. (1983) A catalogue of chironomid genera and subgenera of the world including synonyms (Diptera: Chironomidae). *Entomologica Scandinavica Supplement,* **17,** 1–68.[3][18]

Ashe, P. (1985) A larval diagnosis for the subfamily Buchonomyiinae with a description of the 1st instar of *Buchonomyia thienemanni* Fittk. (Diptera: Chironomidae). *Spixiana Supplement* **11,** 143–8.[3]

Ashe, P. (1990) Ecology, zoogeography and diversity of Chironomidae (Diptera) in Sulawesi with some observations relevant to other aquatic insects, Chapter 22, in *Insects and the Rain Forests of South East Asia (Wallacea)* (eds W.J. Knight and J.D. Holloway), The Royal Entomological Society, London, pp.261–8.[4]

Ashe, P. (1993) Corrections to the Chironomidae part of the Catalogue of Palaearctic Diptera. *Netherlands Journal of Aquatic Ecology,* **26,** 215–21.[3]

Ashe, P. and Cranston, P.S. (1990) Chironomidae, in *Catalogue of Palaearctic Diptera, Volume 2,* (eds A. Soos and L. Papp), Akademiai Kiado, Budapest and Elsevier Science Publishers, Amsterdam, pp.113–355.[3][18]

Ashe, P. and Murray, D.A. (1983) Observations on and descriptions of the egg-masses of *Buchonomyia thienemanni* Fitt. *Memoirs of the American Entomological Society,* **34,** 3–13.[5]

Ashe, P., Murray, D.A. and Reiss, F. (1987) The zoogeographical distribution of Chironomidae. *Annales de Limnologie,* **23,** 27–60.[3][4][18]

Austin, J.E., Serie, J.R. and Noyes, J.H. (1990) Diet of canvasbacks during breeding. *Prairie Naturalist,* **22(3),** 171–6.[17]

Avery, R.A. (1968) Food and feeding relations of three species of *Triturus* (Amphibia Urodela) during the aquatic phases. *Oikos,* **19,** 408–12.[17]

REFERENCES

Azzouzi, A., Laville, H. and Reiss, F. (1992) Nouvelles récoltes de Chironomidés du Maroc. *Annales de Limnologie,* **28,** 225–32.[4]

Back, C., Boisvert, J., Lacoursière, J.O. and Charpentier, G. (1985) High-dosage treatment of a Quebec stream with *Bacillus thuringiensis* serovar. *israelensis*: Efficacy against black fly larvae (Diptera: Simuliidae) and impact on non-target insects. *The Canadian Entomologist,* **117,** 1523–34.[15]

Bagge, P., Lemmetyinen, R. and Raitis, T. (1973) Spring food of some diving waterfowl in the southwestern Finnish archipelago. *Oikos,* **15,** 146–50.[17]

Baker, A.S. and McLachlan, A.J. (1979) Food preferences of Tanypodinae larvae (Diptera: Chironomidae). *Hydrobiologia,* **62,** 283–8.[7][12]

Baker, J.H. and Bradnam, L.A. (1976) The role of bacteria in the nutrition of aquatic detritivores. *Oecologia,* (Berlin) **24,** 95–104.[7]

Baldo, B.A. and Panzani, R.C. (1988) Detection of IgE antibodies to a wide range of insect species in subjects with suspected inhalant allergies to insects. *International Archives of Allergy and Applied Immunology,* **85,** 278–87.[14]

Baranova, V.P. and Miroshnichenko, M.P. (1969) Usloviya i perspektivy vyrashchivaniya molodi osetra na Volgogradskom Osetrovom Rybozavode. [Conditions and prospects for culturing sturgeon fry in the Volgograd sturgeon nursery]. *Gidrobiologicheskii Zhurnal,* **5,** 70–4.[17]

Barbato, L. C., Filia, L. A., Maraga, B. *et al.* (1990) Problematiche relative al controllo dei Chironomidi nella Laguna di Venezia, in *Chironomidi, Culicidi, Simulidi – Aspetti Sanitari ed Ecologici,* (eds F. D'Andrea and G. Marchese), Regione Veneto, ULSS 16, S.I.P., Venice.[13]

Barber, W.E. and Kevern, N.R. (1973) Ecological factors influencing macroinvertebrate standing crop distribution. *Hydrobiologia,* **43,** 53–75.[6]

Barclay, C.M. and Harrel, R.C. (1985) Effects of pollution effluents on two successive tributaries and Village Creek in southeastern Texas. *The Texas Journal of Science,* **37,** 175–88.[15]

Barthelmes, D. (1970) Untersuchungen zur Steigerung der Naturnahrungsproduktion und der Erträge im Karpfenteich durch Vervielfachung des Ersttespannungs – gzw. Neubespannungs effektes. *Zeitschrift für Fischerei und deren Hilfswissenshaften,* **18,** 87–95.[17]

Battarbee, R.W., Mason, J., Renberg, I. and Talling, J.F. (eds) (1990) Palaeolimnology and lake acidification. *Philosophical Transactions of the Royal Society of London (Series B),* **327,** 227–445.[16]

Batzer, D.P. and Resh, V.H. (1991) Trophic interactions among a beetle predator, a chironomid grazer, and periphyton in a seasonal wetland. *Oikos* **60,** 251–7.[7]

Baumgartner, D.L. (1986) Failure of mosquitoes to colonize teasel axils in Illinois. *Journal of the American Mosquito Control Association,* **2,** 371–3.[6]

Baur X. (1982) Chironomid hemoglobin. A major allergen for humans. *Chironomus,* **2,** 24–5.[14]

Baur X. (1992) Chironomid midge allergy. *Japanese Journal of Immunology,* **41,** 81–5.[14]

Baur, X. and Mazur, G. (1988) Structures of major chironomid allergens and epitopes: studies with allergen fragments, a synthetic peptide, monoclonal, and human antibodies, in *International Symposium on Mite and Midge Allergy,* (ed. T. Miyamoto), Ministry of Education, Science and Culture, Tokyo, pp.364–72.[14]

Baur, X., Dewair, M., Fruhmann, G. *et al.* (1982) Hypersensitivity to chironomids (non-biting midges): localization of the antigenic determinants within certain polypeptide sequences of hemoglobins (erythrocruorins) of *Chironomus thummi thummi* (Diptera). *Journal of Allergy and Clinical Immunology,* **68,** 66–76.[14]

Baur, X., Dewair, M., Haegele, K. *et al.* (1983) Common antigenic determinants of haemoglobin as basis of immunological cross-reactivity between chironomid species (Diptera, Chironomidae): studies with human and animal sera. *Clinical Experimental Immunology,* **54,** 599–607.[14]

Baur, X., Aschauer, H., Mazur, G. *et al.* (1986) Structure, antigenic determinants of some clinically important insect allergens: chironomid hemoglobins. *Science,* **233,** 351–4.[14]

Baur, X., Liebers, V., Mazur, G. *et al.* (1991) Immunological cross-reactivity of hemoglobins in the Diptera family Chironomidae. *Allergy,* **46,** 445–51.[14]

Bay, E.C. (1964) An analysis of the 'Sayule' (Diptera: Chironomidae) nuisance at San Carlos, Nicaragua, and recommendations for its alleviation. *World Health Organization,* WHO/EBL/20.[13]

Bay, E.C. (1974) Predator-prey relationships among aquatic insects. *Annual Review of Entomology,* **19,** 441–53.[13]

Bay, E.C. and Anderson, L.D. (1965) Chironomid control by carp and goldfish. *Mosquito News,* **25,** 310–6.[13]

Bay, E.C. and Anderson, L.D. (1966) Studies with the mosquito fish, *Gambusia affinis,* as a chironomid control. *Annals of the Entomological Society of America,* **59,** 150–3.[13]

Baz, L. G. (1959) Biology and morphology of the genus *Microtendipes* which live in the canals of the Utchinsk reservoir. *Trudy Vsesoyuznogo gidrobiologicheskogo Obshchestva,* **9,** 74–84.[5]

Bazerque, M. F., Laville, H. and Brouquet, Y. (1989) Biological quality assessment in two rivers of the Northern Plain of France (Picardie) with special reference to chironomid and diatom indices. *Acta Biologica Debrecina Supplementum Oecologica Hungarica,* **3,** 29–39.[15]

Bazzanti, M. and Bambacigno, F. (1987) Chironomids as water quality indicators in the river Mignone (Central Italy). *Hydrobiological Bulletin,* **21,** 213–22.[15]

Bazzanti, M. and Seminara, M. (1985) Seasonal changes of the profundal macrobenthic community in a polluted lake. *Schweizerische Zeitschrift für Hydrologie,* **47,** 57–63.[15]

Bazzanti, M. and Seminara, M. (1987) Environmental stress in a regulated eutrophic lake indicated by the profundal macrobenthic community. *Bolletino di Zoologia,* **54,** 261–6.[15]

Beadle, L.C. (1974) *The Inland Waters of Tropical Africa – an Introduction to Tropical Limnology.* Longman, London.[17]

Beattie, D.M. (1981) Investigations into the occurrence of midge plagues in the vicinity of Harderwijk. *Hydrobiologia,* **80,** 147–59.[9][13]

Beck, E. C. and Beck, Jr., W. M. (1969) Chironomidae of Florida. II. The nuisance species. *Florida Entomologist,* **52,** 1–11.[13]

Beermann, W. (1952) Chromomerenkonstanz und spezifische Modifikationen der Chromosomenstruktur in der Entwicklung und Organdifferenzierung von *Chironomus tentans. Chromosoma,* **5,** 139–98.[17]

Beermann, W. (1955) Cytologische Analyse eines *Camptochironomus*-Artbastards. I. Kreuzungsergebnisse und die Evolution des Karyotypus. *Chromosoma,* **7,** 198–259.[9]

Bellas, T. (1990) Occupational inhalant allergy to arthropods. *Clinical Reviews in Allergy,* **15,** 15–29.[14]

Belton, P. (1974) The analysis of direction finding in male mosquitoes, in *Experimental Analysis of Insect Behaviour,* (ed. L.B. Browne), Springer-Verlag, Berlin, pp.139–48.[9]

Beltzer, A.H. (1988) Feeding of *Calidris fuscicollis* Vieillot 1819 (Aves Scolopacidae) in Del Cristal Lagoon, Saladillo River Basin, Santa Fe, Argentina. *Iheringia serie Miscellanea,* **2,** 53–62.[17]

Benedict, P.R. and Fisher, G.T. (1972) Commensalistic relationships between

Plecopteracoluthus downesi (Diptera: Chironomidae) and *Chauliodes pectinicornis* (Megaloptera: Corydalidae). *Annales of the Entomological Society of America*, **65,** 109–11.[12]

Bengtson, S.A. (1975) Food of ducklings of surface feeding ducks at Lake Myvatn, Iceland. *Ornis Fennica,* **52,** 1–4.[12][17]

Benke, A.C. (1979) A modification of the Hynes method for estimating secondary production with particular significance for multivoltine populations. *Limnology and Oceanography,* **24,** 168–71.[11]

Benke, A.C., Van Arsdall, Jr., T., Gillespie, D.M. and Parrish, F.K. (1984) Invertebrate productivity in a subtropical blackwater river: the importance of habitat and life history. *Ecological Monographs,* **54,** 25–63.[7][11]

Benson, D.J., Fitzpatrick, L.C. and Pearson, W.D. (1980) Production and energy flow in the benthic community of a Texas pond. *Hydrobiologia,* **74,** 81–93.[11]

Berg, K. (1938) Studies on the bottom animals of Esrom Lake. *Kungelige Danske Videnskabernes Selskabs Skrifter,* 9, 1–255.[6]

Berg, M.B. and Hellenthal, R.A. (1991) Secondary production of Chironomidae (Diptera) in a north temperate stream. *Freshwater Biology,* **25,** 497–505.[7][10][11][18]

Berg, M.B. and Hellenthal, R.A. (1992) Life histories and growth of lotic chironomids (Diptera: Chironomidae). *Annals of the Entomological Society of America,* **85,** 578–89.[7][10]

Berg, M.B. and Hellenthal, R.A. (1993) Role of Chironomidae in energy flow of a lotic ecosystem. *Netherlands Journal of Aquatic Ecology,* **26,** 471–6.[7]

Bergeron, D., Bushway, R.J., Roberts, F.L. *et al.* (1988) The nutrient composition of an insect flour sample from Lake Victoria, Uganda. *Journal of Food Composition and Analysis,* **1,** 371–7.[17]

Berglund, B. E. (ed.) (1986) *Handbook of Holocene Palaeoecology and Palaeohydrology,* John Wiley and Sons, Chichester, England. pp.869.[16]

Berrie, A.D. (1976) Detritus, micro-organisms and animals in fresh water, in *The Role of Terrestrial and Aquatic Organisms in Decomposition Processes,* (eds J.M. Andersen and A. MacFadyen), Blackwell Scientific, Oxford, pp.323–38.[7]

Bickley, W. E. and Ludlam, K. W. (1968) Insecticidal control of *Glyptotendipes barbipes* in two Maryland sewage lagoons. *Journal of Economic Entomology,* **61,** 1454.[13]

Biever, K.D. (1971) Effect of diet and competition in laboratory rearing of chironomid midges. *Annals of the Entomological Society of America,* **64,** 1166–9.[7][17]

Bilyj, B. and Davies, I.J. (1989) Descriptions and ecological notes on seven new species of *Cladotanytarsus* (Chironomidae: Diptera) collected from an experimentally acidified lake. *Canadian Journal of Zoology,* **67,** 948–62.[15]

Bindloss, M.E. (1974) Primary productivity of phytoplankton in Loch Leven, Kinross. *Proceedings of the Royal Society of Edinburgh, Series B,* **74,** 157–181.[11]

Bjärnov, N. (1972) Carbohydrases in *Chironomus, Gammarus* and some Trichoptera larvae. *Oikos,* **23,** 261–3.[7]

Blaber, S.J.M. and Whitfield, A.K. (1977) The feeding ecology of juvenile mullet (Mugilidae) in south east African estuaries. *Biological Journal of the Linnean Society,* **9,** 277–84.[17]

Blake, B.F. (1977) Food and feeding of the mormyrid fishes of Lake Kainji, Nigeria, with special reference to seasonal variation and interspecific differences. *Journal of Fish Biology,* **11,** 315–28.[17]

Block, W., Burn, A.J. and Richard, K.J. (1984) An insect introduction to the maritime Antarctic. *Biological Journal of the Linnean Society,* **23,** 33–9.[9]

Böcher, J. (1989) Boreal insects in northernmost Greenland: palaeo-entomological evidence from the Kap Ko/benhavn Formation (Plio-Pleistocene), Peary Land. *Fauna norvegica, Series B,* **36,** 37–43.[16]

Boerger, H. (1981) Species composition, abundance and emergence phenology of midges (Diptera, Chironomide) in a brown-water stream of West-Central Alberta, Canada. *Hydrobiologia,* **80,** 7–30.[9][12]

Boerger, H., Clifford, H.F. and Davies, R.W. (1982) Density and microdistribution of chironomid larvae in an Alberta brown-water stream. *Canadian Journal of Zoology,* 60, 913–20.[6]

Boesel, M.W. (1937) Order Diptera: Family Chironomidae, in *Insects and arachnids from Canadian amber,* (ed. F.M. Carpenter), *University of Toronto Studies in Geology, Series,* **40,** 44–55.[4]

Boling, R.H., Jr., Goodman, E.D., VanSickle, J.A. *et al.* (1975) Toward a model of detritus processing in a woodland stream. *Ecology,* **56**, 141–51.[7]

Bonomi, G. (1962) La dinamica produttiva delle principali popolazioni macrobentoniche del lago di Varese. *Memorie dell Instituto Italiano di Idrobiologia,* **5,** 207–254.[11]

Booth, J.P. and Learner, M.A. (1978) The parasitization of chironomid midges (Diptera) by water–mite larvae (Hydracarina: Acari) in a eutrophic reservoir in South Wales. *Archiv für Hydrobiologie,* **84,** 1–28.[12]

Boothroyd, I.K.G. (1987) Taxonomic composition and life cycles of Chironomidae (Diptera) in a northern New Zealand stream. *Entomologica Scandinavica Supplement,* **29,** 15–22.[15]

Boothroyd, I.K.G. (1988) Temporal and diel emergence of Chironomidae (Diptera: Insecta) from a New Zealand Stream. *Verhandlungen der Internationalen Vereinigung für theoretische und angewandte Limnologie,* **23,** 1399–404.[9]

Borkent, A. (1984) The systematics and phylogeny of the *Stenochironomus* complex (*Xestochironomus, Harrisius,* and *Stenochironomus*) (Diptera: Chironomidae). *Memoirs of the Entomological Society of Canada,* **128,** 1–269.[7]

Borkent, A., Wirth, W.W. and Dyce, A.L. (1987) The newly discovered male of *Austroconops* (Ceratopogonidae: Diptera) with a discussion of the phylogeny of the basal lineages of the Ceratopogonidae. *Proceedings of the Entomological Society of Washington,* **89,** 587–606.[3]

Borob'ev, B.A. (1960) Dipterous larvae inhabiting water lying in the leaf axils of the teasel. *Entomological Review Washington,* **39,** 579–80.[6]

Borutski, E.V. (1939) Dinamika biomassy *Chironomus plumosus* profundali Belogo Ozera. *Trudy Limnologicheskoi Stantsii v Kosine,* **22,** 156–95.[11]

Borutski, E.V., Sokolova, N.Yu. and Yablonskaya, E.A. (1971) A review of Soviet studies into production estimates of chironomids. *Limnologica,* **8,** 183–91.[11]

Böttger, K. (1976) Types of parasitism by larvae of water mites (Acari: Hydrachnellae). *Freshwater Biology,* **6,** 497–500.[12]

Bottorff, R.L. and Knight, A.W. (1987) Ectosymbiosis between *Nanocladius downesi* (Diptera: Chironomidae) and *Acroneuria abnormis* (Plecoptera: Perlidae) in a Michigan stream, USA. *Entomologia Generalis,* **12,** 97–113.[12]

Botts, P.S. and Cowell, B.C. (1992) Feeding selectivity of two epiphytic chironomids in a subtropical lake. *Oecologia,* (Berl.) **89,** 331–7.[7]

Boubée, J.A.T. (1983) *Past and Present Benthic Fauna of Lake Maratoto with Special Reference to the Chironomidae.* Ph.D. thesis, University of Waikato, Hamilton, New Zealand. pp.251.[16]

Bouguenac, V. and Giani, N. (1992) Mise en place d'un élevation de *Chironomus riparius* Meigen (Diptera, Chironomidae) à l'aval d'une station d'épuration par lagunage. *Annales de Limnologie,* **28,** 233–43.[17]

Boulton, A.J. and Suter, P.J. (1986) Ecology of Temporary Streams – an Australian Perspective, in *Limnology in Australia* (eds P. De Deckker and W.D. Williams), CSIRO, Melbourne/Junk, Dordrecht, pp.313–27.[6]

Bowmaker, A.P. (1968) Preliminary investigations on some aspects of the biology of

Sinamwenda Estuary, Lake Kariba. Proceedings of Transactions of the Rhodesian Scientific Association, **53,** 3–8.[17]

Branch, H.E. (1923) Description of the early stages of *Tanytarsus fatigans* JOH. *Entomologists News,* **34,** 1–4.[5]

Bradshaw, W. E. and Creelman, R.A. (1984) Mutualism between the carnivorous purple pitcher plant and its inhabitants. *American Midland Naturalist,* **112,** 294–303.[6]

Brasch, J., Brüning, H. and Paulke, E. (1992) Allergic contact dermatitis from chironomids. *Contact Dermatitis,* **26,** 317–20.[6]

Braukmann, U. (1987) Zoozönologische und saprobiologische Beiträge zu einer allgemeinen regionalen Bachtypologie. *Archiv für Hydrobiologie Beiheft Ergebnisse der Limnologie,* **26,** 1–355.[15]

Brennan, A. (1976) *The Role of Particulate Material in the Ecology of Larval Chironomidae (Diptera) from an Upland River.* Ph.D. Dissertation, University of Newcastle upon Tyne.[7]

Brennan, A. (1981) *Chironomus. Biologist,* **28,** 133–8.[7]

Brennan, A. and McLachlan, A.J. (1979) Tubes and tube-building in a lotic chironomid (Diptera) community. *Hydrobiologia,* **67,** 173–8.[7]

Brennan, A. and McLachlan, A.J. (1980) Species of *Eukiefferiella* Thienemann (Dipt., Chironomidae) from a northern river with notes on larval dwellings. *Entomologist's Monthly Magazine,* **116,** 109–11.[7]

Brennan, A., McLachlan, A.J. and Wotton, R.S. (1978) Particulate material and midge larvae (Chironomidae: Diptera) in an upland river. *Hydrobiologia,* **59,** 67–73.[7]

Bretschko, G. (1974) The chironomid fauna of a high alpine lake (Vorderer Finstertaler See) Tyrol, Austria, 2237 m asl. *Entomologisk Tidskrift Supplement,* **95,** 22–33.[6]

Bretschko, G. (1982) *Pontomyia* Edwards (Diptera: Chironomidae), a Member of the Coral Reef Community at Carrie Bow Cay, Belize. *Smithsonian Contributions to the Marine Sciences,* **12,** 381–5.[6]

Brett, J.R. (1957) Salmon research and hydroelectric power development. *Bulletin of the Fisheries Research Board of Canada,* **114,** 1–26.[17]

Brinkhurst, R.O. (1974) *The Benthos of Lakes,* Macmillan Press, London.[15]

Brinkhurst, R.O. and Kennedy, C.R. (1965) Studies on the biology of the Tubificidae (Annelida, Oligochaeta) in a polluted stream. *Journal of Animal Ecology* **34,** 429–43.[7]

Briskie, J.V. and Sealy, S.G. (1989) Determination of clutch size in the least flycatcher. *Auk,* **106(2),** 269–78.[17]

Brock, E.M. (1960) Mutualism between the midge *Cricotopus* and the alga *Nostoc. Ecology,* **41,** 474–83.[7][12]

Brodin, Y.-W. (1982) Palaeoecological studies of the recent development of the Lake Växjösjön. IV. Interpretation of the eutrophication process through the analysis of subfossil chironomids. *Archiv für Hydrobiologie,* **93,** 313–26.[16]

Brodin, Y.-W. (1985) Lake history and climatic change in northern Europe interpreted from subfossil Chironomidae (Diptera). *Acta Universitatis Upsaliensis,* Abstracts of Uppsala Dissertations from the Faculty of Science, 787, Uppsala, pp.27.[16]

Brodin, Y.-W. (1986) The postglacial history of Lake Flarken, southern Sweden, interpreted from subfossil insect remains. *Internationale Revue der Gesamten Hydrobiologie,* **71,** 371–432.[16]

Brodin, Y.-W. (1990) Midge fauna development in acidified lakes in northern Europe. *Philosophical Transactions of the Royal Society London. Series B,* **327,** 295–8.[15]

Brook, A.J. (1954) The bottom living algal flora of slow sand filter beds of waterworks. *Hydrobiologia,* **6,** 333–51.[7]

Brooks, J.L. and Dodson, S.I. (1965) Predation, body size and composition of plankton. *Science,* **150,** 28–35.[12]

Brown, A.W.A., McKinley, D.J., Bedford, H.W. and Qutubuddin, M. (1961) Insecticidal operations against chironomid midges along the Blue Nile. *Bulletin of Entomological Research,* **51,** 789–801.[13]

Brown, A.E., Oldham, R.S. and Warlow, A. (1980) Chironomid larvae and pupae in the diet of brown trout (*Salmo trutta*) and rainbow trout (*Salmo gairdneri*) in Rutland Water, Leicestershire, in *Chironomidae: Ecology, Systematics, Cytology and Physiology* (ed. by D.A. Murray) Pergamon Press, Oxford. pp.307–14.[12][17]

Brown, A.F. and Pascoe, D. (1988) Studies on the acute toxicity of pollutants to freshwater macro-invertebrates. *Archiv für Hydrobiologie,* **114,** 311–9.[15]

Brues, C.T. (1924) Observations on animal life in the thermal waters of Yellowstone Park, with a consideration of the thermal environment. *Proceedings of the American Academy of Arts and Science.* **59,** 371–437.[6]

Brundin, L. (1949) Chironomiden und andere Bodentiere de südschwedischen Urgebirgseen. Ein Beitrag zur Kenntnis der bodenfaunistischen Charakterzüge schwedischer oligotropher Seen. *Report of the Institute of Freshwater Research, Drottningholm,* **30,** 1–914.[9][10][15]

Brundin, L. (1956) Die bodenfaunistischen Seetypen und ihre Anwendbarkeit auf die Südhalbkugel. Zugleich eine Theorie der produktionsbiologischen Bedeutung der glazialen Erosion. *Report of the Institute of Freshwater Research, Drottningholm,* **37,** 186–235.[12]

Brundin, L. (1958) The bottom faunistical lake type system and its application to the southern hemisphere. Moreover a theory of glacial erosion as a factor of productivity in lakes and oceans. *Verhandlungen der Internationalen Vereinigung für Theoretische und Angewandte Limnologie,* **13,** 288–97.[15]

Brundin, L. (1966) Transantarctic relationships and their significance, as evidenced by chironomid midges with a monograph of the subfamilies Podonominae and Aphroteniinae and the austral Heptagyiae. *Kunglica Svenska Vetenskapsakademiens Handlingar,* **11,** 1–472 + 30 plates.[2][3][4][6][7][18]

Brundin, L. (1976) A neocomian chironomid and Podonominae- Aphroteniinae (Diptera) in the light of phylogenetics and biogeography. *Zoologica Scripta* **5,** 139–60.[3][4][18]

Brundin, L. (1983a) *Chilenomyia paradoxa* gen. n. and Chilenomyiinae, a new subfamily amongst the Chironomidae. *Entomologica Scandinavica* **14,** 33–45.[2][3][18]

Brundin, L. (1983b) Two new aphrotenian larval types from Chile and Queensland, including *Anaphrotenia lacustris* n. gen., n.sp. (Diptera: Chironomidae). *Entomologica Scandinavica,* **14,** 415–33.[3]

Brundin, L. and Sæther, O.A. (1978) *Buchonomyia burmanica* sp. n. and Buchonomyiinae, a new subfamily among the Chironomidae (Diptera). *Zoologica Scripta* 7, 269–75.[3]

Bryce, D. (1962) Chironomidae (Diptera) from fresh water sediments, with special reference to Malham Tarn (Yorks.). *Transactions of the Society for British Entomology,* **10,** 41–54.[16]

Buchmann, W. (1932) Chironomidenschäden bei dem Belebt-Schlamm-Verfahren und ihre Verhütung und Behebung mit chemischen Mitteln. *Zeitschrift fuer Gesundheitstechnik und Städthygiene,* **1,** 31–8.[13]

Buckley, B.R. and Sublette, J.E. (1964) Chironomidae (Diptera) of Louisiana. II. The limnology of the upper part of Cane River Lake, Natchitoches Parish, Louisiana, with particular reference to the emergence of Chironomidae. *Tulane Studies in Zoology, New Orleans,* **11,** 151–66.[9]

Buhl, K. J. and Faerber, N. L. (1989) Acute toxicity of selected herbicides and surfactants to larvae of the midge *Chironomus riparius. Archives of Environmental Contamination and Toxicology,* **18,** 530–6.[15]

REFERENCES

Bund, W. van der (1993) Competition between *Stictochironomus histrio* and *Polypedilum agg. nubeculosum* in the littoral of Lake Maarsseveen I. in prep.[9]

Bunn, S.E., Edward, D.H.D. and Loneragan, N.R. (1986) Spatial and temporal variation in the macroinvertebrate fauna of streams of the northern jarrah forest, Western Australia: community structure. *Freshwater Biology,* **16,** 67–91.[6][15]

Burfield, I. and Williams, D.N. (1975) Control of parthenogenetic chironomid with pyrethrins. *Water Treatment and Examination,* **24,** 57–67.[13]

Burrett, C., Duhig, N., Berry, R. and Varne, R. (1991) Asian and south-western Pacific continental terranes derived from Gondwana, and their biogeographic significance, in *Austral Biogeography* (eds P.Y. Ladiges, C.J. Humphries and L.W. Martinelli), CSIRO, Melbourne, pp.13–24.[4]

Burtt, E.T., Perry, R.J.O. and McLachlan, A.J. (1986) Feeding and sexual dimorphism in adult midges (Diptera: Chironomidae). *Holarctic Ecology* **9,** 27–32.[2][3][9]

Bush, G.L. (1975) Modes of animal speciation. *Annual Reviews of Ecology and Systematics,* **6,** 339–54.[4]

Buskens, R.F.M. (1987) The chironomid assemblages in shallow lentic waters differing in acidity, buffering capacity and trophic level in the Netherlands. *Entomologica Scandinavica Supplement,* **29,** 217–24.[15]

Buskens, R.F.M. (1989) Monitoring of chironomid larvae and exuviae in the Beuven, a soft-water pool in the Netherlands, and comparisons with palaeolimnological data. *Acta Biologica Debrecina Oecologica Hungarica,* **3,** 41–50.[16]

Buskens, R.F.M. and Moller-Pillot, H.K.M. (1993) The Netherlands as an environment for chironomid fauna. *Netherlands Journal of Aquatic Ecology,* **26,** 223–8.[4]

Butler, M.G. (1980) Emergence phenologies of some arctic Alaskan Chironomidae, in *Chironomidae. Ecology, Systematics, Cytology and Physiology,* (ed. D.A. Murray), Pergamon Press, New York, pp.307–14.[9]

Butler, M.G. (1982a) Morphological and phenological delimitation of *Chironomus prior* new species and *Chironomus tardus* new species (Diptera: Chironomidae) sibling species from Arctic Alaska. *Aquatic Insects,* **4,** 219–35.[9]

Butler, M.G. (1982b) A 7-year life cycle for two *Chironomus* species in arctic Alaskan tundra ponds (Diptera: Chironomidae). *Canadian Journal of Zoology,* **60,** 58–70.[9][10]

Butler, M.G. (1982c) Production dynamics of some arctic *Chironomus* larvae. *Limnology and Oceanography,* **27,** 728–36.[10][11]

Butler, M.G. (1984) Life histories of aquatic insects, in *The Ecology of Aquatic Insects* (eds V.H. Resh and D.M. Rosenberg), Praeger, New York, pp.24–55.[9]

Campbell, P. G. C. and Stokes, P. M. (1985) Acidification and toxicity of metals to aquatic biota. *Canadian Journal of Fisheries and Aquatic Sciences,* **42,** 2034–49.[15]

Candolle, A.P. de (1820) Géographie botanique, in *Dictionaire des sciences naturelles* **XVII** Strasbourg, Paris [not seen, ex. Nelson, G., 1978].[4]

Candolle, A.P. de (1838) Statistique de la famille des composés, Paris [not seen, ex. Nelson, G., 1978].[4]

Cantrell, M.A. and McLachlan, A.J. (1982) Habitat duration and dipteran larvae in tropical rain pools. *Oikos,* **38,** 343–8.[6]

Carpenter, G.D.H. (1920) *A Naturalist on Lake Victoria.* Fisher Unwin Ltd, London.[14]

Carter, C.E. (1973) *A Study of the Chironomidae (Diptera) of Lough Neagh.* Ph.D thesis, New University of Ulster.[10]

Carter, C.E. (1977) The recent history of the chironomid fauna of Lough Neagh, from the analysis of remains in sediment cores. *Freshwater Biology,* **7,** 415–23.[16]

Carter, C.E. (1978) The fauna of the muddy sediments of Lough Neagh, with particular reference to eutrophication. *Freshwater Biology,* **8,** 547–59.[15]

Carter, C.E. (1980) The life cycle of *Chironomus anthracinus* in Lough Neagh. *Holarctic Ecology*, **3,** 214–7.[10]

Casas, J.J. and Vilchez-Quero, A. (1989) A faunistic study of the lotic chironomids (Diptera) of the Sierra Nevada (S.E. of Spain): changes in the structure and composition of the populations between spring and summer. *Acta Biologica Debrecen, Oecologica Hungarica*, **3,** 83–93.[4]

Caspary, V.G. and Downe, A.E.R. (1971) Swarming and mating of *Chironomus riparius* (Dipt., Chironomidae). *Canadian Entomologist*, **103,** 444–8.[9]

Caspers, N. (1951) Rhythmische Erscheinungen in der Fortpflanzung von *Clunio marinus* (Dipt., Chiron.) und das Problem der lunaren Periodizität bei Organismen. *Archiv für Hydrobiologie, Supplementband*, **18,** 415–594.[9]

Caspers, H. (1983a) Sukzessionsanalyse des Makrozoobenthos eines neu angelegten stehenden Gewassers. *Archiv für Hydrobiologie, Supplementband*, **65,** 300–70.[9]

Caspers, N. (1983b) Chironomiden-Emergenz zweier Lunzer Bäche, 1972. *Archiv für Hydrobiologie, Supplementband*, **65,** 484–549.[9][18]

Catalan, Z.B. and White, D.S. (1982) Creating and maintaining cultures of *Chironomus tentans* (Diptera: Chironomidae). *Entomology News*, **93,** 54–8.[17]

Cattaneo, A. (1983) Grazing on epiphytes. *Limnology and Oceanography*, **28,** 124–32.[7]

Cavanaugh, W.J. and Tilden, J.E. (1930) Algal food, feeding and casebuilding habits of the larvae of the midge fly, *Tanytarsus dissimilis*. *Ecology*, **11,** 281–7.[5]

Ceretti, G., Ferrarese, U., Francesion, A. and Barbaro, A. (1987) Chironomids (Diptera: Chironomidae) in the natural diet of gilthead seabream (*Sparus aurata* L.) farmed in the Venice lagoon. *Entomologica Scandinavica Supplement*, **29,** 289–92.[13]

Chapman, J. and Ecke, D. H. (1969) Study of a population of chironomid midges (*Tanytarsus*) parasitized by mermithid nematodes in Santa Clara County, California. *California Vector Views*, **16,** 83–8.[13]

Charles, D.F., Whitehead, D.R., Engstrom, D.R. *et al.* (1987) Paleolimnological evidence for recent acidification of Big Moose Lake, Adirondack Mountains, N. Y. (USA). *Biogeochemistry*, **3,** 267–96.[16]

Charles, D.F., Binford, M.W., Furlong *et al.* (1990) Paleoecological investigation of recent lake acidification in the Adirondack Mountains, NY. *Journal of Paleolimnology*, **3,** 195–241.[16]

Charles, W.N., East, K., Brown, D. *et al.* (1974) The production of larval Chironomidae in the mud at Loch Leven, Kinross. *Proceedings of the Royal Society of Edinburgh, Series B*, **74,** 241–58.[11]

Chaudhuri, P.K. and Guha, D.K. (1987) A conspectus of chironomid midges (Diptera: Chironomidae) of India and Bhutan. *Entomologica Scandinavica Supplement*, **29,** 23–33.[3][4]

Cheng, L. and Birch, M.C. (1978) Insect flotsam: an unstudied marine resource. *Ecological Entomology*, **3,** 87–97.[9]

Cheng, J.L. and Collins J.D. (1980) Observations on behaviour, emergence and reproduction of the marine midges *Pontomyia* (Diptera: Chironomidae). *Marine Biology (Berlin)*, **58,** 1–6.[9]

Chevrel, R. (1894) Sur un Diptère marin du genre *Clunio* Haliday. *Archives de Zoologie experimentale et génerale*, **3,** 583–98.[9]

Chiba, Y. (1967) Activity of mosquitoes, *Culex pipiens pallens* and *Aedes japonicus* under a step-wise decrease of light intensity. *Scientific Reports of Tohoku University, Series IV (Biology)*, **33,** 7–13.[9]

Chiba, Y., Kubota, M. and Nakamura, Y. (1982) Differential effects of temperature upon evening and morning peaks in the circadian activity of mosquitoes, *Culex pipiens pallens* and *C. pipiens molestus*. *Journal of Interdisciplinary Cycle Research* **13,** 55–60.[9]

Chilvers, R.M. and Gee, J.M. (1974) The food of *Bagrus docmac* (Forsk.) (Pisces: Siluriformes) and its relationship with *Haplochromis* (Pisces: Cichlidae), in Lake Victoria, East Africa. *Journal of Fish Biology*, **6.** 483–505.[17]

Chura, N.J. (1961) Food availability and preferences of Juvenile Mallards. *Transactions of the 26th North American Wildlife Conference*, **2,** 121–34.[17]

Claassen, P.W. (1922) The larva of a chironomid (*Trissocladius equitans* n. sp.) which is parasitic upon a mayfly nymph (*Rhithrogena* sp.). *University of Kansas Science Bulletin*, **14,** 395–405.[12]

Clement, S.L., Grigarick, A.A. and Way, M.O. (1977) Conditions associated with rice plant injury by chironomid midges in California. *Environmental Entomology*, **6,** 91–6.[7]

Clements, W.H., Cherry, D.S. and Cairns, J. Jr., (1988) Structural alterations in aquatic insect communities exposed to copper in laboratory streams. *Environmental Toxicology and Chemistry*, **7,** 715–22.[15]

Cloutier, L. and Harper, P.P. (1978) Phénologie de tanypodinae de ruisseaux des Laurentides (Diptera: Chironomidae). *Canadian Journal of Zoology*, **56,** 1129–39.[9][10]

Cobo, F. and Gonzáes, M.A. (1991) Étude de la dérive des exuvies nymphales de Chironomidés dans la rivière Sar (NO. Espagne) (Insecta, Diptera). *Spixiana* **14,** 193–203.[9]

Cocchi, F. (1966) Ricerche sui Ditteri Chironomidi dannosi al riso nella Bassa Bolognese. *Bollettino Osservatorio Malattie Piante Bologna*, **1,** 39–66.[13]

Codreanu, R. (1939) Recherches biologiques sur un chironomide *Symbiocladius rhithrogenae* (Zavr.) – ectoparasite 'cancérigène' des Éphémères torrenticoles. *Archives de Zoologie Expérimentale et Générale*, **81,** 1–283.[12]

Coffman, W.P. (1967) *Community Structure and Trophic Relations in a Small Woodland Stream, Linesville Creek, Crawford County, Pennsylvania*. Ph.D. Dissertation, University of Pittsburgh.[7]

Coffman, W.P. (1973) Energy flow in a woodland stream ecosystem. II. The taxonomic composition and phenology of the Chironomidae as determined by the collection of pupal exuviae. *Archiv für Hydrobiologie*, **71,** 281–322.[3][9][12][18]

Coffman, W.P. (1974) Seasonal differences in the diel emergence of a lotic Chironomidae community. *Entomologisk Tidskrift (Supplement)* **95,** 42–8.[9]

Coffman, W.P. (1979) Neglected characters in pupal morphology as tools in taxonomy and phylogeny of Chironomidae (Diptera). *Entomologica Scandinavica Supplement*. **10,** 37–46.[2][9]

Coffman, W.P. (1983) Thoracic chaetotaxy of chironomid pupae (Diptera: Chironomidae). *Memoirs of the American Entomological Society*, **34,** 61–70.[2]

Coffman, W.P. (1989) Factors that determine the species richness of lotic communities of Chironomidae. *Acta Biologica Debrecen, Oecologica Hungarica*, **3,** 95–100.[4]

Coffman, W.P. and Ferrington, L.C., Jr. (1984) Chironomidae, in *An Introduction to the Aquatic Insects of North America*, 2nd edn, (eds R.W. Merritt and K.W. Cummins), Kendall/Hunt, Dubuque, pp.551–652.[1][7][18]

Coffman, W.P., Yurasits, L.A. and de la Rosa, C. (1988) Chironomidae of South India. 1. Generic composition, biogeographical relationships and descriptions of two unusual pupal exuviae (Diptera: Chironomidae). *Spixiana Supplement*, **14,** 155–65.[4]

Coffman, W.P., de la Rosa, C., Cummins, K.W. and Wilzbach, M.A. (1993) Species richness in some Neotropical (Costa Rica) and Afrotropical (West Africa) lotic communities of Chironomidae (Diptera). *Netherlands Journal of Aquatic Ecology*, **26,** 229–37.[4]

Coffman, W.P., Cummins, K.W. and Wuycheck, J.C. (1971) Energy flow in a woodland stream ecosystem: I. Tissue support trophic structure of the autumnal community. *Archiv für Hydrobiologie,* **68,** 232–76.[7][18]

Coler, B.G. and Kondratieff, B.C. (1989) Emergence of Chironomidae (Diptera) from a delta-swamp receiving thermal effluent. *Hydrobiologia,* **174,** 67–78.[9]

Connelly, D.P. and Chesemore, D.L. (1981) Food habits of pintails, *Anas acuta,* wintering on seasonally flooded wetlands in the northern San Joaquin Valley, California. *California Fish and Game,* **66,** 233–7.[17]

Cook, S.F. Jr. (1962) Feeding studies of the *aeneus* catfish, *Corydoras aeneus* on aquatic midges. *Journal of Economic Entomology,* **55,** 155–7.[13]

Cook, S.F. Jr. (1964) The potential of two native California fish in the control of chironomid midges. *Mosquito News,* **24,** 332–3.[13]

Cook, S.F. Jr., Conners, J.D. and Moore, R.L. (1964) The impact of the fishery upon the midge populations of Clear Lake, Lake County, California. *Annals of the Entomological Society of America,* **57,** 701–7.[13]

Coope, G.R. (1979) Late Coenozoic Fossil Coleoptera: Evolution, biogeography and ecology. *Annual Review of Ecology and Systematics,* **10,** 247–267.[4]

Corbet, P.S. (1964) Temporal patterns of emergence in aquatic insects. *Canadian Entomologist,* **96,** 264–79.[10]

Corbet, P.S. (1967) Further observations on diel periodicities of weather factors near the ground at Hazen Camp, Ellesmere Island, N.W.T. Operation Hazen Number **31,** Defense Research Board of Canada.[10]

Cordes, L.E. and Page, L.M. (1980) Feeding chronology and diet composition of two darters (Percidae) in the Iroquois River system, Illinois. *American Midland Naturalist,* **104,** 202–6.[17]

Corliss, J.O. (1960) *Tetrahymena chironomi* sp. nov., a ciliate from midge larvae and the current status of facultative parasitism in the genus *Tetrahymena, Parasitology,* **50,** 111–53.[13]

Costa, H.H. (1979) The food and feeding chronology of yellow perch (*Perca flavescens*) in Lake Washington. *Internationale Revue der Gesamten Hydrobiologie,* **64,** 783–93.[17]

Cowie, B. (1980) *Community Dynamics of the Benthic Fauna in a West Coast Stream Ecosystem.* Ph.D. Thesis, University of Canterbury.[6]

Cowie, B. (1983) Macroinvertebrate taxa from a southern New Zealand montane stream continuum. *New Zealand Entomologist,* **7,** 439–47.[15]

Cowie, B. (1985) An analysis of changes in the invertebrate community along a southern New Zealand montane stream. *Hydrobiologia,* **120,** 35–46.[15]

Crafford, J.E. (1986) A case study of an alien invertebrate (*Limnophyes pusillus,* Diptera, Chironomidae) introduced on Marion Island: selective advantages. *South African Journal of Antarctic Research,* **16,** 115–7.[9]

Craig, D.A. (1969) The embryogenesis of the larval head of *Simulium venustum* Say (Diptera: Nematocera). *Canadian Journal of Zoology,* **47,** 495–503.[2]

Cramp, S. and Simmons, K.E.L. (eds) (1983) *Handbook of the birds of Europe, the Middle East and North Africa – The birds of the Western Palaearctic Volume III. Waders to Gulls.* Oxford University Press, Oxford.[17]

Cranston, P.S. (1982a) The metamorphosis of *Symposiocladius lignicola* (Kieffer) n. gen., n. comb., a wood-mining Chironomidae (Diptera). *Entomologica Scandinavica,* **13,** 419–29.[7]

Cranston, P.S. (1982b) A key to the larvae of the British Orthocladiinae (Chironomidae). *Scientific Publications of the Freshwater Biological Association,* **45,** 1–152.[6][18]

Cranston, P.S. (1984) The taxonomy and ecology of *Orthocladius (Eudactylocladius) fuscimanus* (Kieffer), a hygropetric chironomid (Diptera). *Journal of Natural History,* **18,** 873–95.[6]

REFERENCES

Cranston, P.S. (1985) *Eretmoptera murphyi* (Diptera: Chironomidae) an apparently parthenogenetic Antarctic midge. *British Antarctic Survey Bulletin,* **66,** 35–46.[6][9]

Cranston, P.S. (1987) A non-biting midge (Diptera: Chironomidae) of horticultural significance. *Bulletin of Entomological Research,* **77,** 661–8.[1][9][14]

Cranston, P.S. (1988) Allergens of non-biting midges (Diptera: Chironomidae): a systematic survey of chironomid haemoglobins. *Medical and Veterinary Entomology* **2,** 117–27.[14]

Cranston, P.S. (1989a) The biology of the Chironomidae, with particular reference to the phylogeny, ecology and occurrence of haemoglobin, in *International Symposium on Mite and Midge Allergy* (ed. T. Miyamoto), Ministry of Education, Science and Culture, Tokyo, pp.232–42.[14]

Cranston, P.S. (1989b) The adult males of Telmatogetoninae (Diptera: Chironomidae) of the Holarctic region – Keys and diagnoses. *Entomologica Scandinavica Supplement,* **34,** 17–21.[6]

Cranston, P.S. (1990) Biomonitoring and taxonomy. *Environmental Monitoring and Assessment,* **14,** 265–73.[1]

Cranston, P.S. (1991) *Immature Chironomidae of the Alligator Rivers Region* Open File Report 82. Supervising Scientist for the Alligator Rivers Region, pp.269.[3][6][15]

Cranston, P.S. and Armitage, P.D. (1988) The Canary Islands Chironomidae described by T. BECKER and by SANTOS-ABREU. *Deutsche Entomologische Zeitshrift N.F.,* **35,** 341–54.[4]

Cranston, P.S. and Edward, D.H.D. (1992) A systematic reappraisal of the Australian Aphroteniinae (Chironomidae) with dating from vicariance biogeography. *Systematic Entomology,* **17,** 41–54.[2][3][4][6][9][15][18]

Cranston, P.S. and Humphries, C.J. (1988) Cladistics and computers: a chironomid conundrum? *Cladistics* **4,** 72–92.[3][18]

Cranston, P.S. and Judd, D.D. (1987) *Metriocnemus* (Diptera; Chironomidae) – an ecological survey and description of a new species. *Journal of the New York Entomological Society,* **95,** 534–6.[6]

Cranston, P.S. and Kitching, R.L. (in press) The Chironomidae of Austro-oriental phytotelmata (plant-held waters): *Richea pandaniformis* Hook. f., in *Chironomids: from Genes to Ecosystems* (ed. P.S. Cranston), CSIRO, Melbourne.[6]

Cranston, P.S. and Judd, D.D. (1989) Diptera: Fam. Chironomidae of the Arabian Peninsula. *Fauna Saudi Arabia* **10,** 236–89.[2][3]

Cranston, P.S. and Martin, J. (1989) Family Chironomidae, in *Catalog of the Diptera of the Australasian and Oceanian Regions,* (ed. N.L. Evenhuis), Bishop Museum Press, Honolulu and E.J. Brill, Leiden, pp.252–74.[3]

Cranston, P.S. and Naumann, I. (1991) Biogeography, in *Insects of Australia,* (ed. I. Naumann), CSIRO and Melbourne University Press, pp.181–97.[4]

Cranston, P.S. and Oliver, D.R. (1987) Problems in Holarctic chironomid biogeography. *Entomologica Scandinavica Supplement,* **29,** 51–6.[4][18]

Cranston, P.S. and Oliver, D.R. (1988a) Additions and Corrections to the Nearctic Orthocladiinae (Diptera: Chironomidae). *Canadian Entomologist,* **120,** 425–62.[3][4]

Cranston, P.S. and Oliver, D.R. (1988b) Aquatic xylophagous Orthocladiinae – systematics and ecology (Diptera: Chironomidae). *Spixiana Supplement,* **14,** 143–54.[6][7]

Cranston, P.S., Gad el Rab, M.O. and Kay, A.B. (1981) Chironomid midges as a cause of allergy in the Sudan. *Transactions of the Royal Society of Tropical Medicine and Hygiene,* **75,** 1–4.[9][13][14]

Cranston, P.S., Gad El Rab, M.O., Tee, R.D. and Kay, A.B. (1983) Immediate-type skin reactivity to extracts of the 'green nimitti' midge (*Cladotanytarsus lewisi*) and other chironomids in asthmatic subjects in the Sudan and Egypt. *Annals of Tropical Medicine and Parasitology*, **77**, 527–33.[14]

Cranston, P.S., Oliver, D.R. and Saether, O.A. (1983) The larvae of Orthocladiinae (Diptera: Chironomidae) of the Holarctic Region – keys and diagnoses. *Entomologica Scandinavica Supplement*, **19**, 149–291.[7]

Cranston, P. S., Tee, R. M., Credland, P. F. and Kay, A. B. (1983) Chironomidae haemoglobins: their detections and role in allergy to midges in the Sudan and elsewhere. *Memoirs of the Entomological Society of America*, **34**, 71–87.[13]

Cranston, P.S., Edward, D.H.D. and Colless, D.H. (1987) *Archaeochlus* Brundin: a midge out of time (Diptera: Chironomidae). *Systematic Entomology*, **12**, 313–34.[3][4][18]

Cranston, P.S., Oliver, D.R. and Sæther, O.A. (1989) The adult males of Orthocladiinae (Diptera; Chironomidae) of the Holarctic region – Keys and diagnoses. *Entomologica Scandinavica Supplement*, **34**, 165–352.[6]

Cranston, P.S., Webb, C.J. and Martin, J. (1990) The saline nuisance chironomid '*Carteronica" longilobus* (Diptera: Chironomidae): a systematic reappraisal. *Systematic Entomology* **15:** 401–432.[3][6][15]

Credland, P.F. (1973a) A simple method of collecting eggs of some Chironomidae (Dipt.). *Entomologist's Monthly Magazine*, **109**, 126–7.[9]

Credland, P.F. (1973b) A new method for establishing a permanent laboratory culture of *Chironomus riparius Meigen (Diptera: Chironomidae)*. *Freshwater Biology*, **3**, 45–51.[17]

Crisp, D.T., Mann, R.H.K. and McCormack, J.C. (1978) The effects of impoundment and regulations upon the stomach contents of fish at Cow Green, Upper Teesdale. *Journal of Fish Biology*, **12**, 287–302.[17]

Croizat, L. (1958) *Panbiogeography*, privately published by the author, L. Croizat, Caracas, Vol.I. pp.1018., Vol.IIa + IIb, pp.1731.[4]

Croizat, L. (1964) *Space, Time, Form: The Biological Synthesis*, privately published by the author, L. Croizat, Caracas, pp.881.[4]

Croizat, L., Nelson, G.J. and Rosen, D.E. (1974) Centers of origin and related concepts. *Systematic Zoology*, **23**, 265–87.[4]

Crosby, T.K. (1975) Food of the New Zealand trichopterans *Hydrobiosis parumbripennis* McFarlane and *Hydropsyche colonica* McLachlan. *Freshwater Biology*, **5**, 105–14.[17]

Crossland, N. O. (1984) Fate and biological effects of methyl parathion in outdoor ponds and laboratory aquaria. II Effects. *Ecotoxicology and Environmental Safety*, **8**, 482–95.[15]

Crowder, L.B. and Cooper, W.E. (1982) Habitat structural complexity and the interaction between bluegills and their prey. *Ecology*, **63**, 1802–13.[12]

Cuker, B.E. (1983) Competition and coexistence among the grazing snail Lymnaea, Chironomidae, and microcrustacea in an arctic epilithic lacustrine community. *Ecology*, **64**, 10–5.[12]

Cummins, K.W. (1973) Trophic relations of aquatic insects. *Annual Review of Entomology*, **18**, 183–206.[7]

Cummins, K.W. (1974) Structure and function of stream ecosystems. *BioScience*, **24**, 631–41.[7][12]

Cummins, K.W. and Klug, M.J. (1979) Feeding ecology of stream invertebrates. *Annual Review of Ecology and Systematics*, **10**, 147–72.[7]

Cummins, K.W. and Wuycheck, J.C. (1971) Caloric equivalents for investigations in ecological energetics. *Mitteilungen der Internationalen Vereinigung für Theoretische and Angewandte Limnologie*, **18**, 1–158.[11][17]

Cummins, K.W., Minshall, G.W., Sedell, J.R. *et al.* (1984) Stream ecosystem theory. *Verhandlungen der Internationalen Vereinigung für Theoretische und Angewandte*

Limnologie, **22,** 1818–27.[15]
Cushing, C.E., McIntire, C.D., Cummins, K.W. *et al.* (1983) Relationships among chemical, physical, and biological indices along river continua based on multivariate analyses. *Archiv für Hydrobiologie,* **98,** 317–26.[15]
Cushman, R.M. (1984) Chironomid deformities as indicators of pollution from a synthetic, coal-derived oil. *Freshwater Biology,* **14,** 179–82.[15]
Cushman, R.M. and Goyert, J.C. (1984) Effects of a synthetic crude oil on pond benthic insects. *Environmental Pollution (Series A),* **33,** 163–86.[15]
Cushman, R.M. and McKamey, M.I. (1981) A *Chironomus tentans* bioassay for testing synthetic fuel products and effluents, with data on acridine and quinoline. *Bulletin of Environmental Contamination and Toxicology,* **26,** 601–5.[15]
Cyr, H. and Downing, J.A. (1988) The abundance of phytophilous invertebrates on different species of submerged macrophytes. *Freshwater Biology,* **20,** 365–74.[6]
Czeczuga, B. and Bobiatynska-Ksók, E. (1968) Ecological–biological aspects of the parasitic larvae *Cryptochironomus* ex. gr. *pararostratus* Harn. (Diptera: Chironomidae). *Internationale Revue der Gesamten Hydrobiologie,* **53,** 549–61.[12]
Dabrowski, K., Murawska, E., Terlecki, J. and Wielgosz, S. (1984) Studies on the feeding of *Coregonus pollan* (Thompson) alevins and fry in Lough Neagh. *Internationale Revue der Gesamten Hydrobiologie,* **69,** 529–40.[17]
Dall, P.C., Lindegaard, C. and Jónasson, P.M. (1990) In-lake variations in the composition of zoobenthos in the littoral of Lake Esrom, Denmark. *Verhandlungen der Internationalen Vereinigung für Theoretische und Angewandte Limnologie,* **24,** 613–20.[15]
Daneholt, B. and Edström, J.-E. (1967) The content of DNA in individual polytene chromosomes of *C. tentans*. *Cytogenetics,* **6,** 350–6.[3]
Danell, K. and Sjöberg, K. (1977) Seasonal emergence of chironomids in relation to egg-laying and hatching of ducks in a restored lake (northern Sweden). *Wildfowl,* **28,** 129–35.[17]
Danks, H.V. (1971a) Overwintering of some north temperate and arctic Chironomidae. II. Chironomid biology. *Canadian Entomologist,* **103,** 1875–910.[6][10]
Danks, H.V. (1971b) Life history and biology of *Einfeldia synchrona* (Diptera: Chironomidae). *Canadian Entomologist,* **103,** 1597–606.[7][10]
Danks, H.V. (1978) Some effects of photoperiod, temperature, and food on emergence in three species of Chironomidae (Diptera). *Canadian Entomologist,* **110,** 289–300.[9]
Danks, H.V. and Oliver, D.R. (1972a) Seasonal emergence of some high arctic Chironomidae (Diptera). *Canadian Entomologist,* 104, 661–86.[9]
Danks, H.V. and Oliver, D.R. (1972b) Diel periodicities of emergence of some high arctic Chironomidae (Diptera). *Canadian Entomologist,* **104,** 903–16.[9]
Danks, H.V. and Jones, J.W. (1978) Further observations on winter cocoons in Chironomidae (Diptera). *Canadian Entomologist,* **110,** 667–9.[6][10]
Darby, R.E. (1962) Midges associated with California rice fields, with special reference to their ecology (Diptera: Chironomidae). *Hilgardia,* **32,** 1–206.[7][9][13]
Davidowicz, P. and Gliwicz, Z.M. (1983) Food of brookchar *Salvelinus fontinalis* in extreme oligotrophic conditions of an alpine lake. *Environmental Biology of Fishes,* **8(1),** 55–60.[17]
Davies, B.R. (1973) *Field and Laboratory Studies on the Activity of Larval Chironomidae in Loch Leven, Kinross.* Ph.D. Thesis, Paisley College of Technology, Scotland.[5]
Davies, B.R. (1974) The planktonic activity of larval Chironomidae in Loch Leven, Kinross. *Proceedings of the Royal Society of Edinburgh, B,* **74,** 241–58.[5]
Davies, B.R. (1976a) Wind distribution of the egg masses of *Chironomus anthracinus*

(Zetterstedt) (Diptera: Chironomidae) in a shallow, wind-exposed lake (Loch Leven, Kinross). *Freshwater Biology,* **6,** 421–4.[5][9]

Davies, B. R. (1976b) The dispersal of chironomid larvae; A Review. *Journal of the Entomological Society of South Africa,* **39,** 39–62.[5]

Davies, I.J. (1975) Selective feeding in some arctic Chironomidae. *Internationale Vereinigung für Theoretische und Angewandte Limnologie Verhandlungen,* **19,** 3149–54.[7][15]

Davies, I.J. (1980) Relationships between dipteran emergence and phytoplankton production in the Experimental Lakes Area, Northwestern Ontario. *Canadian Journal of Fisheries and Aquatic Science,* **37,** 523–33.[9]

Davies, L.J. and Hawkes H. A. (1981) Some effects of organic pollution on the distribution and seasonal incidence of Chironomidae in riffles in the River Cole. *Freshwater Biology,* **11,** 549–59.[15]

Davies, N.B. (1977) Prey selection and social behaviour in wagtails (Aves: Motacillidae). *Journal of Animal Ecology,* **46,** 37–57.[17]

Davies, R.W., Wrona, F.J. and Linton, L. (1979) A serological study of prey selection by *Helobdella stagnalis* (Hirudinoidea). *Journal of Animal Ecology,* **48,** 181–94.[17]

Davis, C.C. (1966) A study of the hatching process in aquatic invertebrates XVI. Events of eclosion in *Calopsectra neoflavellus* Malloch. *Hydrobiologia,* **27,** 196–207.[5]

De Pauw, N. and Vanhooren, G. (1983) Method for biological quality assessment of watercourses in Belgium. *Hydrobiologia,* **100,** 153–68.[15]

Debus, L. (1989) Food composition of bream and roach from shallow brackish coastal waters of the Southern Baltic proper with comments on possible diet overlap. *Rapport et Proces-Verbaux des Reunions du Conseil Permanent International pour l'Exploration de la Mer, Copenhague,* **190,** 118–24.[17]

Deevey, E.S., Jr. (1942) Studies on Connecticut lake sediments. III. The biostratonomy of Linsley Pond. *American Journal of Science,* **240,** 233–64, 313–38.[16]

Deevey, E.S., Jr. (1953) Paleolimnology and climate, in *Climatic Change: Evidence, Causes, and Effects.* (ed. H. Shapley), Harvard University Press, pp.273–318.[16]

Deevey, E.S., Jr. (1955a) Paleolimnology of the Upper Swamp Deposit, Pyramid Valley. *Records of the Canterbury Museum,* **6,** 291–344.[16]

Deevey, E.S., Jr. (1955b) The obliteration of the hypolimnion. *Memorie dell'Istituto Italiano di Idrobiologia, Supplement,* **8,** 9–38.[16]

Dejoux, C. (1968) Le Lac Tchad et les Chironomides de sa partie Est. *Annales Zoologica Fennici,* **5,** 27–32.[6]

Dejoux, C. (1969) Les insectes aquatiques de lac Tchad Apercu systématique et bio-écologique. *Verhandlungen der Internationalen Vereinigung für theoretische und angewandte Limnologie,* **17,** 900–6.[9]

Dejoux, C. (1971) Recherches sur le cycle de développement de *Chironomus pulcher* (Diptera: Chironomidae). *The Canadian Entomologist,* **103,** 465–70.[5]

Dejoux, C. (1984a) Chironomidae in western Africa (Diptera: Nematocera). *Revue d'Hydrobiologie Tropicale,* **17,** 65–76.[4]

Dejoux, C. (1984b) Contribution to the knowledge of West African Chironomidae (Diptera-Nematocera). Chironomids from the Guinean Republic. *Aquatic Insects,* **6,** 157–67.[4]

Dejoux, C. and Troubat, J.-J. (1982) Toxicité pour la faune aquatique non cible de quelques larvacides antisimulidiens. II – L'Actellic^{OR} M20. *Revue Hydrobiologie Tropicale,* **15,** 151–6.[15]

De la Noüe, J. and Choubert, G. (1985) Apparent digestibility of invertebrate biomasses by rainbow trout. *Aquaculture,* **50,** 103–12.[17]

Delettre, Y.R. (1978) Biologie et écologie de *Limnophyes pusillus* Eaton, 1875 (Diptera, Chironomidae) aux Iles Kerguelen. I. Présentation générale et étude des populations larvaires. *Revue d'Écologie et de Biologie du Sol,* **15,** 475–86.[6][10]

Delettre, Y.R. (1984) *Recherches sur les Chironomides à Larves Édaphiques Biologie, Écologie, Mécanismes Adaptatifs*. Thése Doctorat d'Etat, Université de Rennes.[9]

Delettre, Y.R. (1986) La colonisation de biotopes multiples: une alternative a la resistance *in situ* aux conditions mesologoques defavorables. Cas de *Limnophyes minimus* (Mg.), Diptere Chironomidae a larves édaphiques des landes armoricaines. *Revue d'Écologie et de Biologie du Sol,* **23,** 29–38.[6]

Delettre, Y.R. (1988a) Chironomid wing length, dispersal ability, and habitat predictability. *Holarctic Ecology,* **11(3),** 166–70.[9]

Delettre, Y.R. (1988b) Flux d'évaporation corporelle et résistance à la dessication chez les larves de quelques Chironomidae terrestres (Diptera). *Revue Écologique et Biologique de Sol,* **25,** 129–38.[6][9][10]

Delettre, Y.R. (1993) Terrestrial Chironomidae: contribution of local emergence to global aerial flow in a heterogeneous environment. *Netherlands Journal of Aquatic Ecology,* **26,** 269–71.[9]

Delettre, Y.R. and Baillot, S. (1977) Sur la résistance de larves de Chironomidae Orthocladiinae à l'asséchement du sol. *Compte Rendu Hebdomadaire des Séances de l'Académie des Sciences. Paris, Series D,* **24,** 1717–9.[6]

Delettre, Y.R. and Cancela da Fonseca, J.P. (1979) Biologie et Écologie de *Limnophes pusillus* Eaton, 1875 (Diptera, Chironomidae), aux Iles Kerguelen. II: Étude des populations imaginales et discussion. *Revue d'Écologie et de Biologie du Sol,* **16,** 355–72.[9]

Dendy, J.S. (1973) Predation on chironomid eggs and larvae by *Nanocladius alternantherae* Dendy and Sublette (Diptera: Chironomidae, Orthocladiinae). *Entomological News,* **84,** 91–5.[7]

Dermott, R.M. (1991) Deformities in larval *Procladius* spp.and dominant Chironomini from the St. Clair River. *Hydrobiologia,* **219,** 171–85.[15]

Dermott, R.M. and Paterson, C.G. (1974) Determining dry weight and percentage dry matter of chironomid larvae. *Canadian Journal of Zoology,* **52,** 1243–50.[11]

Dermott, R.M., Kalff, J., Leggett, W.C. and Spence, J. (1977) Production of *Chironomus, Procladius,* and *Chaoborus* at different levels of phytoplankton biomass in Lake Memphremagog, Quebec–Vermont. *Journal of the Fisheries Research Board of Canada,* **34,** 2001–7.[11]

Dermott, R.M., Kelso, J.R.M. and Douglas, A. (1986) The benthic fauna of 41 acid sensitive headwater lakes in North Central Ontario. *Water, Air, and Soil Pollution,* **28,** 283–92.[16, 17]

Dévai, G. (1988) Emergence patterns of chironomids in Keszthely-basin of Lake Balaton (Hungary). *Spixiana Supplementum,* **14,** 201–11.[13]

Dévai, G. (1990) Ecological background and importance of the change of chironomid fauna (Diptera: Chironomidae) in shallow Lake Balaton. *Hydrobiologia,* **191,** 189–98.[15]

Dévai, G. and Moldován, J. (1983) An attempt to trace eutrophication in a shallow lake (Balaton, Hungary) using chironomids. *Hydrobiologia,* **103,** 169–75.[15]

Devonport, B.F. and Winterbourn, M.J. (1976) The feeding relationships of two invertebrate predators in a New Zealand river. *Freshwater Biology,* **6,** 167–76.[12]

Devries, D.R., Stein, R.A. and Chesson, P.L. (1989) Sunfish foraging among patches – the patch-departure decision. *Animal Behaviour,* **37(3),** 455–64.[17]

Dewey, S.L. (1986) Effects of the herbicide atrazine on aquatic insect community structure and emergence. *Ecology,* **67,** 148–61.[15]

Dickman, M.D., van Dam, H., van Geel, B. *et al.* (1987) Acidification of a Dutch moorland pool – a palaeolimnological study. *Archiv für Hydrobiologie,* **9,** 377–408.[16]

Dimitrov, M. (1974) Mineral fertilization of carp ponds in polycultural rearing. *Aquaculture,* **3,** 273–85.[17]

Dimitrov, M. (1980) Chironomid larvae growing over the oats sown for green fertilization of carp ponds. *Acta Universitatis Carolinae-Biologica,* **1978,** 43–8.[17]

Dinulesco, G. (1932) Sur la biologie d'un Chironomide nouveau *Cardiocladius leoni* Goetghebuer et Dinulesco ordinairement confundu avec la mouche de Golubatz *Simulium columbacensis. Diptera,* **6,** 1–9.[5][9]

Dittmar, H. (1955) Ein Sauerlandbach. *Archiv für Hydrobiologie,* 50, 305–552.[10]

Dixit, S.S. and Witcomb, D. (1983) Heavy metal burden in water, substrate and macroinvertebrate body tissue of a polluted River Irwell (England). *Environmental Pollution (Series B),* **6,** 161–72.[15]

Dodds, W.K. and Marra, J.L. (1989) Behaviors of the midge, *Cricotopus* (Diptera: Chironomidae) related to the mutualism with *Nostoc parmelioides* (Cyanobacteria). *Aquatic Insects,* **11,** 201–8.[7][12]

Dordel, H.-J. (1971) The process of copulation in the marine chironomid *Clunio marinus* (Diptera). *Canadian Entomologist,* **103,** 404–6.[9]

Dosdall, L.M. and Mason, P.G. (1981) A chironomid (*Nanocladius (Plecopteracoluthus) branchicolus*: Diptera) phoretic on a stonefly (*Acroneuria lycorias*: Plecoptera) in Saskatchewan. *Canadian Entomologist,* **113,** 141–7.[12]

Dosdall, L.M., Mason, P.G. and Lehmkuhl, D.M. (1986) First records of phoretic Chironomidae (Diptera) associated with nymphs of *Pteronarcys dorsata* (Say) (Plecoptera: Pteronarcyidae). *Canadian Entomologist,* **118,** 511–5.[12]

Downes, J.A. (1958) Assembly and mating in the biting Nematocera. *Proceedings of the 10th International Congress of Entomology, Montréal, 1956,* **2,** 425–34.[9]

Downes, J.A. (1969) The swarming and mating flight of Diptera. *Annual Review of Entomology,* **14,** 271–98.[9]

Downes, J.A. (1974) The feeding habits of adult Chironomidae. *Entomologisk Tidskrift (Supplement),* **95,** 84–90.[9]

Downes, J.A. and Colless, D. (1967) Mouthparts of the Biting and Blood-sucking Type in Tanyderidae and Chironomidae (Diptera). *Nature, London,* **21,** 1355–6.[9]

Downing, J.A. and Rigler, F.H. (1984) *A Manual on Methods for the Assessment of Secondary Productivity in Fresh Waters.* Blackwell Scientific Publications, Oxford.[11]

Drake, C.M. (1982) Seasonal dynamics of Chironomidae (Diptera) on the Bulrush *Schoenoplectus lacustris* in a chalk stream. *Freshwater Biology,* **12,** 225–240.[6][10][12]

Dratnal, E. (1979) *Eukiefferiella szczensnyi* sp. n. (Diptera, Chironomidae). *Bulletin Academie Polskie Science Cl. II Serie Science Biologique,* **27,** 183–93.[12]

Driver, E.A. (1977) Chironomid communities in small prairie ponds: characteristics and controls. *Freshwater Biology,* **7,** 121–33.[6]

Driver, E.A., Sugden, L.G. and Kovach, R.J. (1974) Calorific, chemical and physical values of potential duck foods. *Freshwater Biology,* **4,** 233–92.[17]

Duck, R.W. and McManus, J. (1984) Traces produced by chironomid larvae in sediments of an ice-contact proglacial lake. *Boreas,* **13,** 89–93.[16]

Duck, R. W. and McManus, J. (1987) Chironomid larvae trails in proglacial lake sediments: Comments. *Boreas,* **16,** 322.[16]

Dudley, T. and Anderson, N.H. (1982) A survey of invertebrates associated with wood debris in aquatic habitats. *Melanderia,* **39,** 1–21.[7]

Dudley, W.D. (1981) The 1st diets of postemergent brook trout (*Salvelinus fontinalis*) and Atlantic salmon (*Salmo salar*) alevins in a Quebec river, Canada.

Canadian Journal of Fisheries and Aquatic Sciences, **38,** 765–71.[17]
Durant, D.F., Shireman, J.V. and Gasaway, R.D. (1979) Reproduction, growth and food habits of seminole killifish, *Fundulus seminolis,* from two central Florida, USA, lakes. *American Midland Naturalist,* **102,** 127–33.[17]
Dusoge, K. (1980) The occurrence and role of the predatory larvae of *Procladius* Skuse (Chironomidae, Diptera) in the benthos of Lake Śniardwy. *Ekologia Polska,* **28,** 155–86.[7][12]
Dusoge, K. and Wisniewski, R. J. (1976) Effect of heated waters on biocenosis of the moderately polluted Narew River. Macrobenthos. *Polskie Archiwum Hydrobiologii,* **23,** 539–54.[15]
Edgar, W.D. and Meadows, P.S. (1969) Case construction, movement, spatial distribution and substrate selection in the larva of *Chironomus riparius* Meigen. *Journal of Experimental Biology,* **50,** 247–53.[7]
Edington, J.M. and Hildrew, A.G. (1981) A key to the caseless caddis larvae of the British Isles with notes on their ecology. *Scientific Publication of the Freshwater Biological Association,* **43,** 1–92.[17]
Edward, D.H.D. (1968) Chironomidae in temporary freshwaters. *Newsletter of the Australian Society for Limnology,* **6,** 3–5.[6]
Edward, D.H.D. (1986) Chironomidae (Diptera) of Australia, in *Limnology in Australia* (eds P. De Deckker and W.D. Williams), CSIRO, Melbourne/Junk, Dordrecht, pp.159–73.[6][9][10]
Edward, D.H.D. (1989) Gondwanaland elements in the Chironomidae (Diptera) of South-Western Australia. *Acta Biologica Debrecen, Oecologica Hungarica,* **2**, 181–7.[4][18]
Edward, D.H.D. and Colless, D.H. (1968) Some Australian parthenogenetic Chironomidae. *Journal of the Australian Entomological Society,* **7,** 158–62.[9]
Edwards, F.W. (1926a) On the British biting midges (Diptera, Ceratopogonidae). *Transactions of the Entomological Society of London* **74,** 389–426.[3]
Edwards, F.W. (1926b) The phylogeny of nematocerous Diptera: a critical review of some recent suggestions. *Verhandlung der II Internationaler Entomologen-Kongress, Zürich,* **II,** 111–30.[3]
Edwards, F.W. (1929) British non-biting midges (Diptera, Chironomidae). *Transactions of the Entomological Society of London,* **77,** 279–439.[3][9]
Edwards, F.W. (1931) Chironomidae. *Diptera of Patagonia and South Chile* **2,** 233–331.[3]
Edwards, J.S. and Baust, J. (1981) Sex ratio and adult behaviour of the Antarctic midge *Belgica antarctica* (Diptera, Chironomidae). *Ecological Entomology,* **6,** 239–43.[9]
Edwards, M. and Usher, M.B. (1985) The winged Antarctic midge *Parochlus steinenii* (Gerke) (Diptera: Chironomidae), in the South Shetland Islands. *Biological Journal of the Linnean Society,* **26,** 83–93.[1]
Edwards, R.W. (1958) The effect of larvae of *Chironomus riparius* Meigen on the redox potentials of settled activated sludge. *Annals of Applied Biology,* **46,** 457–64.[6]
Edwards, R.W., Egan, H., Learner, M.A. and Maris, P.J. (1964) The control of chironomid larvae in ponds, using TDE (DDD). *Journal of Applied Ecology,* **1,** 97–119.[13]
Eichenberger, E. and Schlatter, A. (1978) Effect of herbivorous insects on the production of benthic algal vegetation in outdoor channels. *Internationale Vereinigung für Theoretische und Angewandte Limnologie Verhandlungen,* **20,** 1806–10.[7]
Eidt, D.C., Bacon, G.B., DeGraeve, G.M. and Mallet, V.N. (1988) Fate and short-term persistence of the insecticide aminocarb in a New Brunswick

(Canada) headwater stream. *Archives of Environmental Contamination and Toxicology,* **17,** 817–29.[15]

Einarsson, A., Haflidason, H. and Oskarsson, H. (1988) Myvatn saga lífríkis og gjóskutímatal í Sydriflóa. *Náttúruverndarrad,* **17,** Reykjavik, Iceland. pp.96. [Icelandic with English summary].[16]

Elgmork, K. and Sæther, O.A. (1970) Distribution of invertebrates in a high mountain brook in the Colorado Rocky Mountains. *University of Colorado Studies Series in Biology,* **31,** 1–55.[6]

Elliott, J.M. (1973a) The diet activity pattern, drifting, and food of the leech (*Erpobdella octoculata*) in a Lake District stream. *Journal of Animal Ecology,* **42,** 449–59.[13]

Elliott, J.M. (1973b) The food of brown and rainbow trout (*Salmo trutta* and *S. gairdneri*) in relation to the abundance of drifting invertebrates in a mountain stream. *Oecologia,* **12,** 329–47.[17]

El Medzi, Z. and Guidicelli, J. (1986) Étude d'une écosystème limnique peu connu: les Khettaras de la région de Marrakech (Maroc), habitats et peuplements. *Sciences de l'eau,* **6,** 281–97.[4]

Elton, C.S. (1946) Competition and the structure of ecological communities. *Journal of Animal Ecology,* **15,** 54–68.[12]

Epler, J.H. (1986) A novel new neotropical *Nanocladius* (Diptera: Chironomidae), symphoretic on *Traverella* (Ephemeroptera: Leptophlebiidae). *Florida Entomologist,* **69,** 319–27.[12]

Epler, J.H. (1992) *Identification Manual for the Larval Chironomidae (Diptera) of Florida.* State of Florida Department of Environmental Regulation.

Eriksson, F., Hörnström, E., Mossberg, P. and Nyberg, P. (1983) Ecological effects of lime treatment of acidified lakes and rivers in Sweden. *Hydrobiologia,* **101,** 145–64.[15]

Eriksson, M.O.G. (1979) Competition between freshwater fish and goldeneye *Bucephala clangula* L. for common prey. *Oecologia,* **41,** 99–107.[17]

Eriksson, N.E., Vedal, S. and Belin, L. (1984) Röda mygglarver utlöst IgE-förmedlad allergi hos akvarieägare. *Läkartidningen,* **43,** 3951–3.[14]

Eriksson, N.E., Peterson, I., Vedal, S. *et al.* (1985) Allergy among farmers. Abstract 137 of Annual Meeting of the European Academy of Allergology and Clinical Immunology, Stockholm, **1985,** 199.[14]

Eriksson, N.E., Ryden, B, and Jonsonn, P. (1989) Hypersensitivity to larvae of chironomids (non-biting midges): Cross sensitization with crustaceans. *Allergy (Copenhagen),* **44,** 305–13.[14]

Faith, D.P. (1991) Cladistic permutation tests for monophyly and nonmonophyly *Systematic Zoology,* **40,** 366–75.[3]

Faith, D.P. and Cranston, P.S. (1991) Could a cladogram this short have arisen by chance alone?: on permutation tests for cladistic structure. *Cladistics* **7,** 1–28.[3]

Farris, J.S. (1981) Distance data in phylogenetic analysis, in *Advances in Cladistics: Proceedings of the First Meeting of the Willi Hennig Society,* (eds V.A. Funk and D.R. Brooks). New York Botanic Garden, New York, pp.3–23.[4]

Farris, J.S. (1985) On the boundaries of phylogenetic systematics. *Cladistics* **1,** 190–201.[18]

Federici, B.A., Hazard, E.I. and Anthony, D.W. (1973) A new cytoplasmic polyhedrosis virus from Chironomidae collected in Florida. *Journal of Invertebrate Pathology,* **22,** 136–8.[13]

Fellton, H.L. (1940) Control of aquatic midges with notes on the biology of certain species. *Journal of Economic Entomology,* 33, 252–64.[13]

Fellton, H.L. (1941) The use of chlorinated benzenes for the control of aquatic midges. *Journal of Economic Entomology,* **34,** 192–4.[13]

Ferrarese, U. (1993) Chironomids of Italian Rice Fields. *Netherlands Journal of Aquatic Ecology*, **26**, 341–46.[7]

Ferrington, L.C. Jr. (1984) Evidence for the Hyporheic Zone as a Microhabitat of *Krenosmittia* spp. Larvae (Diptera: Chironomidae). *Journal of Freshwater Ecology*, **2**, 353–8.[6]

Ferrington, L.C. Jr. (1993) Habitat and sediment preferences of *Axarus festivus* larvae. *Netherlands Journal of Aquatic Ecology*, **26**, 347–54.[7]

Ferrington, L.C. Jr. (in press) Utilization of anterior headcapsule structures in locomotion by larvae of *Constempellina* sp. (Diptera: Chironomidae), in *Chironomids: from Genes to Ecosystems* (ed. P.S. Cranston), CSIRO, Melbourne.[7]

Ferrington, L.C. Jr. and Sæther, O.A. (1987) Male, female, pupa and biology of *Oliveridia hugginsi* n.sp. (Chironomidae: Diptera) from Kansas. *Journal of the Kansas Entomological Society*, **60**, 451–61.[9]

Ferrington, L.C. Jr. and Crisp, N.H. (1989) Water chemistry characteristics of receiving streams and the occurrence of *Chironomus riparius* and other Chironomidae in Kansas. *Acta Biologica Debrecina Supplementum Oecologica Hungarica*, **3**, 115–26.[15]

Fillion, D.B. (1967) The abundance and distribution of benthic fauna of three mountain reservoirs on the Kananaskis River in Alberta. *Journal of Applied Ecology*, **4**, 1–11.[15]

Fischer, J. (1969) Zur Fortpflanzungsbiologie von *Chironomus nuditarsis* Str. *Revue Suisse Zoologie*, **76**, 23–55.[5]

Fischer, J. (1974) Experimentelle Beiträge zur Ökologie von *Chironomus* (Diptera). I. Dormanz bei *Chironomus nuditarsis* und *Ch. plumosus*. *Oecologia*, **16**, 73–95.[9]

Fischer, J. and Rosin, S. (1968) Einfluss von Licht und Temperatur auf die Schlüpf-Aktivität von *Chironomus nuditarsis*. *Revue Suisse de Zoologie*, **75**, 538–49.[8][9]

Fischer, J. and Rosin, S. (1969) Das larvale Wachstum von *Chironomus nuditarsis* Str. *Revue Suisse de Zoologie*, **76**, 727–34.[9]

Fischer, Z. (1964) Kilka uwag o odzywianiu sie larw wazek gatunkow *Erythromma najas* Hans. i *Coenagrion hastulatum* Charp. [Some observations concerning the food consumption of the dragonfly larvae of *Erythromma najas* Hans. and *Coenagrion hastulatum* Charp.)]. *Polskie Archiwum Hydrobiologii, Warszawa*, **12**, 253–64.[17]

Fisher, S.G. and Gray, L.J. (1983) Secondary production and organic matter processing by collector macroinvertebrates in a desert stream. *Ecology*, **64**, 1217–24.[11]

Fisher, S.W. and Wadleigh R.W. (1986) Effects of pH on the acute toxicity and uptake of [^{14}C]Pentachlorophenol in a midge, *Chironomus riparius*. *Ecotoxicology and Environmental Safety*, **11**, 1–8.[15]

Fittkau, E.J. (1955) *Buchonomyia thienemanni* n. gen. n. sp. Chironomidenstudien IV (Diptera; Chironomidae). *Beiträge zur Entomologie* 5, 403–14.[3]

Fittkau, E.J. (1962) Die Tanypodinae (Diptera: Chironomidae). (Die tribus Anatopynyiini, Macropelopiini und Pentaneurini). *Abhandlungen zur LarvenSystematik der Insekten*, **6**, 1–453.[3]

Fittkau, E.J. (1964) Remarks on limnology of central-Amazon rain-forest streams. *Verhandlungen der Internationalen Vereinigung für theoretische und angewandte Limnologie*, **15**, 1092–6.[6]

Fittkau, E.J. (1971a) Distribution and ecology of Amazonian chironomids (Diptera). *Canadian Entomologist*, **103**, 407–13.[6]

Fittkau, E.J. (1971b) Der Torsionsmechanismus beim Chironomiden-Hypopygium. *Limnologica* **8**, 27–34.[2][9]

Fittkau, E.J. (1974) *Ichthyocladius* n. gen., eine neotropische Gattung der Orthocladiinae (Chironomidae, Diptera) deren Larven epizoisch auf Welsen (Astroblepidae und Loricariidae) leben. *Entomologisk Tidskrift (Supplement)*, **95**, 91–106.[12]

Fittkau, E. J. (1980) Ein zoogeographischer Vergleich der Chironomiden der Westpalaearktis und der Aethiopis, in *Chironomidae. Ecology, Systematics, Cytology, and Physiology*, (ed. D.A. Murray), Pergamon Press, Oxford, pp.139–43.[4][18]

Fittkau, E.J. and Reiss, F. (1978) Chironomidae, in *Limnofauna Europaea* (ed. J. Illies), 2nd edition, Gustav Fischer Verlag, Stuttgart, pp.404–40.[4]

Fittkau, E.J. and Reiss, F. (1979) Die Zoogeographische Sonderstellung der neotropischen Chironomiden. *Spixiana,* **2,** 273–80.[3]

Fittkau, E.J., Reiss, F. and Hoffrichter, O. (1976) A bibliography of the Chironomidae. *Gunneria,* **26,** 1–177.[3][18]

Fjellberg, A. (1978) Fragments of a Middle Weichselian fauna on Andøya, north Norway. *Boreas,* **7,** 39.[16]

Flecker, A.S. (1984) The effects of predation and detritus on the structure of a stream insect community: a field test. *Oecologia (Berlin),* **64,** 300–5.[12]

Flentje, M.E. (1945a) Elimination of midge fly larvae with DDT. *Journal of the American Water Works Association,* **37,** 1053.[13]

Flentje, M.E. (1945b) Control and elimination of pest infestations in public water supplies. *Journal of the American Water Works Association,* **37,** 1194–203.[13]

Flössner, D. (1976) Biomasse und Produktion des Macrobentos der mittleren Saale. *Limnologica,* **10,** 123–53.[11]

Flynn, T. and Bolas, P. M. (1985) Simple method of chironomid control at water treatment works. *Journal of Institution of Water Engineers and Scientists,* **39,** 414–22.[13]

Folsom, T.C. and Collins, N.C. (1984) The diet and foraging behavior of the larval dragonfly *Anax junius* (Aeshnidae) with an assessment of the role of refuges and prey activity. *Oikos,* **42(1),** 105–13.[17]

Forbes, S.A. (1925) The lake as a microcosm. *Bulletin of the Illinois Natural History Survey,* **15**, 537–50.[18]

Ford, J.B. (1962) The vertical distribution of larval Chironomidae (Dipt.) in the mud of a stream. *Hydrobiologia,* **19,** 262–72.[6]

Forel, F.A. (1904) Le Léman. *Lausanne* **3**, pp.715.[18]

Forsyth, D.J. (1978) Benthic macroinvertebrates in seven New Zealand lakes. *New Zealand Journal of Marine and Freshwater Research,* **12,** 41–9.[15]

Forsyth, D.J. (1986) Distribution and production of *Chironomus* in eutrophic Lake Ngapouri. *New Zealand Marine and Freshwater Research,* **20,** 327–35.[11]

Forsyth, D.J. and James, M.R. (1988) The Lake Okaro ecosystem 2. Production of the chironomid *Polypedilum pavidus* and its role as food for two fish species. *New Zealand Journal of Marine and Freshwater Research,* **22,** 47–54.[11]

Forsyth, D.J. and McCallum, I.D. (1978) *Xenochironomus canterburyensis* (Diptera: Chironomidae) an insectan inquiline commensal of *Hyridella menziesi* (Mollusca: Lamellibranchia). *Journal of Zoology,* **186,** 331–4.[12]

Foster, W.A. and Treherne, J.E. (1986). The ecology and behaviour of a marine insect, *Halobates fijiensis* (Hemiptera: Gerridae). *Zoological Journal of the Linnean Society,* **86,** 391–412.[17]

Fox, M.G. (1989) Effect of prey density and prey size on growth and survival of juvenile walleye *Stizostedion vitreum vitreum. Canadian Journal of Fisheries and Aquatic Sciences,* **46(8),** 1323–8.[17]

Franco, P.J., Daniels, K.L., Cushman, R.M. and Kazlow, G.A. (1984) Acute toxicity of a synthetic oil, aniline and phenol to laboratory and natural populations of chironomid (Diptera) larvae. *Environmental Pollution (Series A),* **34,** 321–31.[15]

Frank, C. (1982) Ecology, production and anaerobic metabolism of *Chironomus plumosus* L. larvae in a shallow lake. I. Ecology and production. *Archiv für Hydrobiologie,* **94,** 460–91.[10]

Frank, C.A.P. (1991) Effects of flow regulation in a pre-alpine river on the chironomid community. *Verhandlungen der Internationalen Vereinigung für Theoretische und Angewandte Limnologie,* **24,** 1856–61.[15]

Frank, G.H. (1965) The hatching pattern of 5 species of chironomid from a small reservoir in the eastern Transvaal by a new type of trap. *Hydrobiologia* **25,** 52–68.[9]

Freeman, P. (1955) A study of the Chironomidae (Diptera) of Africa south of the Sahara, Part I. *Bulletin of the British Museum (Natural History), Entomology,* **4,** 1–67.[3]

Freeman, P. (1956) A study of the Chironomidae (Diptera) of Africa south of the Sahara, Part II. *Bulletin of the British Museum (Natural History), Entomology* **4,** 285–366.[3]

Freeman, P. (1957) A study of the Chironomidae (Diptera) of Africa south of the Sahara, Part III. *Bulletin of the British Museum (Natural History), Entomology* **5,** 321–426.[3]

Freeman, P. (1958) A study of the Chironomidae (Diptera) of Africa south of the Sahara, Part IV. *Bulletin of the British Museum (Natural History), Entomology* **6,** 261–363.[3]

Freeman, P. (1959) A study of the New Zealand Chironomidae (Diptera, Nematocera). *Bulletin of the British Museum of Natural History, Entomology,* **7,** 393–437.[3]

Freeman, P. (1961) The Chironomidae (Diptera) of Australia. *Australian Journal of Zoology,* **9,** 611–737.[3]

Freeman, P. and Cranston, P.S. (1980) Family Chironomidae, in *Catalogue of the Diptera of the Afrotropical Region',* (ed. R.W. Crosskey), British Museum (Natural History), London, pp.175–202.[3][18]

Freihofer, W.C. and Neil, E.H. (1967) Commensalism between midge larvae (Diptera: Chironomidae) and catfishes of the families Astroblepidae and Loricariidae. *Copeia,* **1,** 39–45.[12]

Frenzel, P. (1990) The influence of chironomid larvae on sediment oxygen microprofiles. *Archiv für Hydrobiologie,* **119,** 427–37.[6]

Frey, D. G. (1964) Remains of animals in Quaternary lake and bog sediments and their interpretation. *Ergebnisse der Limnologie,* **2,** 1–114.[16]

Frommer, S.I. and Sublette, J.E. (1971) The Chironomidae (Diptera) of the Philip L. Boyd Deep Canyon Desert Center, Riverside Co., California. *Canadian Entomologist,* **103,** 414–23.[9]

Fryer, G. (1959) Lunar rhythm of emergence, differential behaviour of the sexes, and other phenomena in the African midge *Chironomus brevibucca* (Kieff.). *Bulletin of Entomological Research,* **50,** 1–8.[9]

Fuchs, T. and Kleinhaus, D. (1982) Fischfutter-Allergie: Sensibilisierung gegen Chironomiden-Larven (Zuckmückenlarven). Kasuistik. *Allergologie* **5,** 81–2.[14]

Fukuhara, H. and Sakamoto, M. (1988) Ecological significance of bioturbation of zoobenthos community in nitrogen release from bottom sediments in a shallow eutrophic lake. *Archiv für Hydrobiologie,* **113,** 425–445.[11]

Fukuhara, H. and Yasuda, K. (1985) Phosphorus excretion by some zoobenthos in a eutrophic freshwater lake and its temperature dependency. *Japanese Journal of Limnology,* **46,** 287–96.[11]

Fukuhara, H. and Yasuda, K. (1989) Ammonium excretion by some freshwater zoobenthos from a eutrophic lake. *Hydrobiologia,* **173,** 1–8.[11]

Furnish, J.D., Belluck, D., Baker, D. and Pennington, B.A. (1981) Phoretic relationships between *Corydalus cornutus* (Megaloptera: Corydalidae) and Chironomidae in eastern Tennessee. *Annales of the Entomological Society of America* **74,** 29–30.[12]

Gad El Rab, M.O. and Kay, A.B. (1980) Widespread immunoglobulin E-mediated hypersensitivity in the Sudan to the 'green nimitti' midge *Cladotanytarsus lewisi* (Diptera: Chironomidae). *Journal of Allergy and Clinical Immunology,* **66,** 190–7.[14]

Gad El Rab, M.O., Thatcher, D.R. and Kay, A.B. (1980) Widespread IgE-mediated hypersensitivity in the Sudan to the 'green nimitti' midge *Cladotanytarsus lewisi* (Diptera: Chironomidae). II. Identification of a major allergen. *Clinical and Experimental Immunology,* **41,** 389–96.[14]

Gallepp, G.W. (1974) Behavioral ecology of *Brachycentrus occidentalis* Banks during the pupation period. *Ecology,* **55,** 1283–1294.[12]

Gallepp, G.W. (1979) Chironomid influence on phosphorus release in sediment-water microcosms. *Ecology,* **60,** 547–56.[6]

Gardarsson, A. (1979) Waterfowl populations of Lake Myvatn and recent changes in numbers and food habits. *Oikos,* **32,** 250–70.[17]

Gardarsson, A., Gíslason, G.M. and Einarsson, A. (1988) Long term changes in the Lake Myvatn ecosystem. *Aqua Fennica,* **18,** 125–35.[16]

Gardner, W.S., Nalepa, T.F., Quigley, M.A. and Malczyk, J.M. (1981) Release of phosphorus by certain benthic invertebrates. *Canadian Journal of Fisheries and Aquatic Science,* **38,** 978–81.[11]

Gardner, W.S., Nalepa, T.F., Slavens, D.R. and Laird G.A. (1983) Patterns and rates of nitrogen release by benthic Chironomidae and Oligochaeta. *Canadian Journal of Fisheries and Aquatic Science,* **40,** 259–66.[11]

Gardner, W.S. and Scavia, D. (1981) Kinetic examination of nitrogen release by zooplankters. *Limnology and Oceanography,* **26,** 801–10.[11]

Gauss, J. D., Woods, P. E., Winner, R. W. and Skillings, J. H. (1985) Acute toxicity of copper to three life stages of *Chironomus tentans* as affected by water hardness-alkalinity. *Environmental Pollution (Series A),* **37,** 149–57.[15]

Gendron, J.M. and Laville H. (1993) Diel emergence patterns of drifting chironomid (Diptera) pupal exuviae in the River Aude (Eastern Pyrenees, France). *Netherlands Journal of Aquatic Ecology,* **26,** 273–9.[9]

Gentry, A.W. and Sutcliffe, A.J. (1981) Pleistocene geography and mammal faunas, in *The Evolving Earth* (ed. R.L.M. Cocks), British Museum (Natural History), London, pp.273–51.[4]

George, E.L. and Hadley, W.F. (1979) Food and habitat partitioning between rock bass (*Ambloplites rupestris*) and smallmouth bass (*Micropterus dolomieui*) young of the year. *Transactions of the American Fisheries Society,* **108,** 253–61.[17]

Gersich, F.M., Millazzo, D.P. and Landenberger, B.D. (1989) A comparison of seven diets used to culture *Tanytarsus dissimilis,* in *Aquatic Toxicology and Hazard Assessment: 12th volume,* (eds U.M. Cowgill and L.R. Williams), ASTM Philadelphia, pp.392–401.[17]

Gerstmeier, R. (1989a) Lake typology and indicator organisms in application to the profundal chironomid fauna of Starnberger See (Diptera, Chironomidae). *Archiv für Hydrobiologie,* **116,** 227–34.[15]

Gerstmeier, R. (1989b) Phenology and bathymetric distribution of the profundal chironomid fauna in Starnberger See (F.R. Germany). *Hydrobiologia,* **184,** 29–42.[10]

Giacomin, C. and Tassi, G.C. (1988) Hypersensitivity to chironomid *Chironomus salinarius* (non-biting midge living in the Lagoon of Venice) in a child with serious skin and respiratory symptoms. *Bolletino dell'Instituto Sieroterapico Milanese,* **67,** 72–5.[14]

Gibson, N.H.E. (1945) On the mating of certain Chironomidae (Diptera). *Transactions of the Royal Entomological Society, London,* **95,** 263–94.[9]

Giesy, J.P. and Hoke, R.A. (1991) Bioassessment of the toxicity of freshwater sediment. *Verhandlungen der Internationalen Vereinigung für Theoretische und Angewandte Limnologie,* **24,** 2313–21.[15]

Giesy, J.P., Graney, R.L., Newsted, J.L. *et al.* (1988) Comparison of three sediment bioassay methods using Detroit River sediments. *Environmental Toxicology and Chemistry,* **7,** 483–98.[15]

Giles, N. (1990) Effects of increasing larval chironomid densities on the underwater feeding success of diving tufted ducklings *Aythya fuligula. Wildfowl,* **41,** 99–106.[17]

Giles, N., Street, M. and Wright, R.M. (1990) Diet composition and prey preference of tench, *Tinca tinca* (L.), common bream, *Abramis brama* (L.), perch, *Perca fluviatilis* L., and roach, *Rutilus rutilus* (L.), in two contrasting gravel pit lakes: potential trophic overlap with wildfowl. *Journal of Fish Biology,* **37,** 945–57.[17]

Giles, N., Street, M., Wright, R. *et al.* (1991) A review of the fish and duck research at Great Linford 1986–1990. *Game Conservancy Annual Review,* **22,** 129–33.[17]

Gilinsky, E. (1984) The role of fish predation and spatial heterogeneity in determining benthic community structure. *Ecology,* **65,** 455–68.[12][17]

Gilliam, J.F., Fraser, D.F. and Sabat, A.M. (1989) Strong effects of foraging minnows on a stream benthic invertebrate community. *Ecology,* **70,** 445–52.[12][17]

Gleick, J. (1987) *Chaos: Making a New Science,* Penguin, New York, pp.354.[16]

Glick, P.A. (1960) Collecting insects by airplane, with special reference to the dispersal of the potato leafhopper. *Technical Bulletin of the United States Department of Agriculture,* **1222,** 16.[9]

Glova, G.J. (1984) Management implications of the distribution and diet of sympatric populations of juvenile coho salmon *Oncorhynchus kisutch* and coastal cutthroat trout *Salmo clarki clarki* in small streams in British Columbia, Canada. *Progressive Fish Culturist,* **46(4),** 269–78.[17]

Glover, B. (1973) The Tanytarsini (Diptera: Chironomidae) of Australia. *Australian Journal of Zoology Supplement,* **23,** 403–78.[3]

Godbout, L. and Hynes, H.B.N. (1982) The three dimensional distribution of the fauna in a single riffle in a stream in Ontario. *Hydrobiologia,* **97,** 87–96.[6]

Goddeeris, B.R. (1987) The time factor in the niche space of *Tanytarsus*-species in two ponds in the Belgian Ardennes (Diptera: Chironomidae). *Entomologica Scandinavica Supplement,* **29,** 281–8.[9]

Goddeeris, B.R. (1990) Life cycle characteristics in *Tanytarsus sylvaticus* (van der Wulp, 1859) (Chironomidae, Diptera). *Annales de Limnologie,* **26,** 51–64.[10]

Goedkoop, W. and Johnson, R.K. (1993) Modelling the importance of sediment bacterial carbon for profundal macroinvertebrates along a lake nutrient gradient. *Netherlands Journal of Aquatic Ecology,* **26,** 477–83.[7]

Goetghebuer, M. (1932) Diptères (Nématocères). Chironomidae IV. Orthocladiinae, Corynoneurinae, Clunioninae, Diamesinae. *Faune de France,* **23,** 1–204.[9]

Goetz, G.P. (1980) Tracheal patterns in larval Chironomidae (Diptera, Nematocera). *Entomologica Scandinavica,* **11,** 291–6.[2]

Goff, A.M. (1972) Feeding of adult *Chironomus riparius* Meigen. *Mosquito News,* **32,** 243–4.[9]

Golini, V.I. and Sherry, J.P. (1979) *Chironomus plumosus* (Diptera: Chironomidae) from Lake Ontario parasitized by a mermithid nematode with subsequent colonization by a saprolegniaceous fungus. *Transctions of the American Microscopy Society,* **98,** 572–6.[12]

Gomez, R. (1970) Food habitats of young-of-the-year striped bass, *Roccus saxatilis* (Walbaum) in Canton reservoir. *Proceedings of the Oklahoma Academy of Science,* **50,** 79–83.[17]

Gordon, M.H., Swan, B.K. and Paterson, C.G. (1978) *Baeoctenus bicolor* (Diptera:

Chironomidae) parasitic in unionid bivalve mollusks and notes on other chironomid–bivalve associations. *Journal of Fisheries Research Board of Canada,* **35,** 154–7.[12]

Gotceitas, V. and Mackay, R.J. (1980) The phoretic association of *Nanocladius (Nanocladius) rectinervis* (Kieffer) (Diptera: Chironomidae) on *Nigronia serricornis* Say (Megaloptera: Corydalidae). *Canadian Journal of Zoology,* **58,** 2260–3.[12]

Gouin, F.J. (1959) Morphology of the larval head of some Chironomidae (Diptera: Nematocera). *Smithsonian Miscellaneous Collections,* **137,** 175–201.[2]

Goulden, C.E. (1971) Environmental control of the abundance and distribution of the chydorid Cladocera. *Limnology and Oceanography,* **16,** 320–31.[7]

Graham, A.A. and Burns, C.W. (1983) Production and ecology of benthic chironomid larvae (Diptera) in Lake Hayes, New Zealand, a warm-monomictic eutrophic lake. *Internationale Revue der Gesamten Hydrobiologie,* **68,** 351–77.[10][11]

Grandjean, F. (1964) Oribates mexicains (1'e série) *Dampfiella* Selln. et *Beckiella* n.g. *Acarologia,* **6,** 694–711.[3]

Grant, P.R. (1966) Ecological compatibility of bird species on islands. *American Naturalist,* **100,** 451–62.[12]

Gray, L.J. and Ward, J.V. (1982) Effects of sediment releases from a reservoir on stream macroinvertebrates. *Hydrobiologia,* **96,** 177–84.[15]

Green, D.W.J., Williams, K.A. and Pascoe, D. (1985) Studies on the acute toxicity of pollutants to freshwater macroinvertebrates. 2. Phenol. *Archiv für Hydrobiologie,* **103,** 75–82.[15]

Green, D.W.J., Williams, K.A. and Pascoe, D. (1986) Studies on the acute toxicity of pollutants to freshwater macroinvertebrates. 4. Lindane (- Hexachlorocyclohexane). *Archiv für Hydrobiologie,* **106,** 263–73.[15]

Gregory, S.V. (1983) Plant-herbivore interactions in stream systems, in *Stream Ecology: Application and Testing of General Ecological Theory* (eds J.R. Barnes and G.W. Minshall), Plenum Press, New York, pp.157–89.[7]

Griffith, L.A. and Gates, J.E. (1985) Food habits of cave-dwelling bats in the Central Appalachians. *Journal of Mammology,* **66,** 451–60.[17]

Griffiths, D. (1973) The food of animals in an acid moorland pond. *Journal of Animal Ecology,* **42,** 285–93.[17]

Grimås, U. (1961) The bottom fauna of natural and impounded lakes in northern Sweden (Ankarvattnet and Blåsjön). *Institute of Freshwater Research Drottningholm Report,* **42,** 183–237.[15]

Grimås, U. and Wiederholm, T. (1979) Biometry and biology of *Constempellina brevicosta* (Chironomidae) in a subarctic lake. *Holarctic Ecology,* **2,** 119–24.[15]

Grimm, O. von (1870) Die ungeschlechtliche Fortpflanzung einer *Chironomus* Art und deren Entwicklung aus dem unbefruchteten Ei. *Zapiski Imperatorskoi Akademii Nauk,* **7,** Série **17,** 1–20.[8]

Grisvold, C.E. (1991) Cladistic biogeography of Afromontane spiders, in *Austral Biogeography* (eds P.Y. Ladiges, C.J. Humphries and L.W. Martinelli), CSIRO, Melbourne, pp.73–89.[4]

Grodhaus, G. (1963) Chironomid midges as a nuisance. II. The nature of the nuisance and remarks on its control. *California Vector Views,* **10,** 27–37.[13]

Grodhaus, G. (1967) Identification of chironomid midges commonly associated with waste stabilizing lagoons in California. *California Vector Views,* **14,** 1–12.[13]

Grodhaus, G. (1971) Sporadic parthenogenesis in three species of *Chironomus* (Diptera). *Canadian Entomologist,* **103,** 338–40.[9]

Grodhaus, G. (1976) Two species of *Phaenopsectra* with drought-resistant larvae (Diptera: Chironomidae). *Journal of the Kansas Entomological Society,* **49,** 405–18.[6]

Grodhaus, G. (1980) Aestivating chironomid larvae associated with vernal pools in *Chironomidae. Ecology, Systematics, Cytology and Physiology,* (ed. D.A. Murray), Pergamon Press, New York, pp.315–22.[10]

Grodhaus, G. and Rotramel, G.L. (1980) Immature stages of *Polypedilum pedatum excelsius* (Diptera, Chironomidae) from seasonally flooded tree-holes. *Acta Universitatis Carolinae-Biologica,* **1978,** 69–76.[6]

Groenewald, A.A. (1964) Observations of the food habits of *Clarias gariepinus* Burchell, the South African freshwater barbel (Pisces: Clariidae) in Transvaal. *Hydrobiologia,* **23,** 287–91.[17]

Gruhl, K. (1924) Paarungsgewohnheiten der Dipteren. *Zeitschrift für wissenschaftliche Zoologie,* **122,** 205–80.[9]

Grzybkowska, M. (1988) Selective predation by *Hydra* sp. on the larvae of Chironomidae. *Przeglad Zoologiczny Wroclaw,* **32(4),** 605–10.[17]

Grzybkowska, M. (1989) Production estimates of the dominant taxa of Chironomidae (Diptera) in the modified River Widawka and the natural River Grabia, central Poland. *Hydrobiologia,* **179,** 245–59.[11]

Grzybkowska, M., Hejduk, J. and Zielinski, P. (1990) Seasonal dynamics and production of Chironomidae in a large lowland river upstream and downstream from a new reservoir in central Poland. *Archiv für Hydrobiologie,* **119,** 439–55.[11]

Guibé, J. (1943) Chironomes parasites de mollusques gastéropodes; *Chironomus varus limnaei* Guibé espèce jointive de *Chironomus varus varus* Gtgh. *Bulletin Biologique de la France et la Belgique,* **76,** 283–97.[12]

Günther, J. (1983) Development of Grossensee (Holstein, Germany): variations in trophic status from the analysis of subfossil microfauna. *Hydrobiologia,* **103,** 231–4.[16]

Hafiz, H. A. (1939) Observations on the bionomics of the midge *Chironomus (Limnochironomus) tenuiforceps* K. occurring on the filter beds of the Calcutta Corporation water works at Pulta, near Calcutta (Chironomidae: Diptera). *Records of Indian Museum,* **41,** 225–31.[13]

Hågvar, S. and Østbye, E. (1973) Notes on winter active chironomids. *Norsk Entomologisk Tidsskrift,* **73 (20),** 253–7.[9]

Halse, S.A. (1981) Faunal assemblages of some saline lakes near Marchagee, Western Australia. *Australian Journal of Marine and Freshwater Research,* **32,** 133–42.[6]

Hall, D.J., Cooper, W.E. and Werner, E.E. (1970) An experimental approach to the production dynamics and community structure of freshwater animal communities. *Limnology and Oceanography,* **15,** 839–928.[12]

Hall, R.J. (1990) Relative importance of seasonal, short-term pH disturbances during discharge variation on a stream ecosystem. *Canadian Journal of Fishery and Aquatic Sciences,* **47,** 2261–74.[12]

Hambrook, J.A. and Sheath, R.G. (1987) Grazing of freshwater Rhodophyta. *Journal of Phycology,* **23,** 656–62.[7]

Hamburger, K. and Dall, P.C. (1990) The respiration of common benthic invertebrate species from the shallow littoral zone of Lake Esrom, Denmark. *Hydrobiologia,* **199,** 117–30.[11]

Hamilton, A.L. (1965) *An Analysis of a Freshwater Benthic Community with Special Reference to the Chironomidae.* Ph.D. Thesis, University of British Columbia.[5]

Hamilton, A.L. (1969) On estimating annual production. *Limnology and Oceanography,* **14,** 771–82.[11]

Hamilton, A.L. and Sæther, O.A. (1971) The occurrence of characteristic deformities in the chironomid larvae of several Canadian lakes. *The Canadian Entomologist,* **103,** 363–8.[15]

Hanazato, T., Iwakuma, T. and Hayashi, H. (1990) Impact of white fish on an enclosure ecosystem in a shallow eutrophic lake: selective feeding of fish and predation effects on the zooplankton communities. *Hydrobiologia,* **200/201,** 129–40.[17]

Hann, B.J. (1991) Invertebrate grazer/periphyton interactions in a eutrophic marsh pond. *Freshwater Biology,* **26,** 87–96.[7]

Hann, B.J., Warner, B.G. and Warwick, W.F. (1992) Aquatic invertebrates and climate change: a comment on Walker et al. (1991). *Canadian Journal of Fisheries and Aquatic Sciences,* **49,** 1274–6.[16]

Hansen, S.R. and Garton, R.R. (1982) The effects of diflubenzuron on a complex laboratory stream community. *Archives of Environmental Contamination and Toxicology,* **11,** 1–10.[15]

Hare, L. and Carter J.C.H. (1987) Chironomidae (Diptera, Insecta) from the environs of a natural West African lake. *Entomologica Scandinavica Supplement,* **29,** 65–74.[4]

Harkrider, J.R. and Hall, I.M. (1978) The dynamics of an entomopoxvirus in a field population of larval midges of the *Chironomus decorus* complex. *Environmental Entomology,* **7,** 858–62.[13]

Harkrider, J.R. and Hall, I.M. (1979) The effect of an entomopoxvirus on larval populations of an undescribed midge species in the *Chironomus decorus* complex under laboratory conditions. *Environmental Entomology,* **8,** 631–5.[13]

Harnisch. O. (1954) Die physiologische Bedeutung der präanalen Tubuli der Larve von *Chironomus thummi. Zoologische Anzeiger,* **153,** 204–11.[2]

Harp, G.L. and Campbell, R.S. (1967) The distribution of *Tendipes plumosus* (Linné) in mineral acid water. *Limnology and Oceanography,* **12,** 260–3.[15]

Harper, P.P. and Cloutier, L. (1986) Spatial structure of the insect community of a small dimictic lake in the Laurentians (Québec). *Internationale Revue der Gesamten Hydrobiologie,* **71,** 655–85.[6][15]

Harrison, A.D. (1965) Geographical distribution of riverine invertebrates in Southern Africa. *Archiv für Hydrobiologie,* **61,** 380–6.[6]

Harrison, A.D. (1966) Recolonisation of a Rhodesian stream after drought. *Archiv für Hydrobiologie,* **62,** 405–21.[6]

Harrison, A.D. (1987) Chironomidae of five central Ethiopian Rift Valley lakes. *Entomologica Scandinavica Supplement* **29,** 39–43.[4]

Harrison, A.D. (1991) Chironomidae from Ethiopia. Part 1. Tanypodinae (Insecta, Diptera). *Spixiana,* **14,** 45–69.[4]

Harrison, A.D. (1992) Chironomidae from Ethiopia. Part 2. Orthocladiinae with two new species and a key to *Thienemanniella* Kieffer. *Spixiana,* **15,** 149–95.[4]

Hasegawa, H. and Sasa, M. (1987) Taxonomical notes on the chironomid midges of the tribe Chironomini collected from the Ryuku Islands, Japan, with descriptions of their immature stages. *Japanese Journal of Sanitary Zoology,* **38,** 275–95.[4]

Hashimoto, H. (1957) Peculiar mode of emergence in the marine chironomid *Clunio* (Diptera, Chironomidae). *Scientific Report, Tokyo Kyoiku Daig, Section B,* **8,** 217–26.[5][9]

Hashimoto, H. (1959) Notes on *Pontomyia natans* from Sado (Diptera, Chironomidae). *Scientific Report, Tokyo Kyoiku Daig., Section B,* **9,** 57–64.[6]

Hashimoto, H. (1962) Ecological significance of the sexual dimorphism in marine chironomids. *Scientific Reports of the Tokyo Kyoiku Daigaku, Section A,* **10,** 221–52.[9]

Hashimoto, H. (1965) Discovery of *Clunio takahashii* Tokunaga from Japan. *Japanese Journal of Zoology,* **14,** 13–29.[9]

Hashimoto, H. (1976) Non-biting midges of marine habitats (Diptera: Chironomidae), in *Marine Insects,* (ed. L. Cheng), North Holland Publishers, Amsterdam, pp.377–414.[6][9]

Hashimoto, H., Wongsiri, T., Wongsiri, N. *et al.* (1981) Chironomidae from rice fields of Thailand with descriptions of 7 new species. *Technical Bulletin, Entomology and Zoology Division, Department of Agriculture, Bangkok, Thailand,* **7,** 1–47.[4]

Haskell, P.T. (1966) Flight behaviour. *Symposium of the Royal Entomological Society, London,* **3,** 29–45.[9]

Hatakeyama, S. (1988) Chronic effects of Cu on reproduction of *Polypedilum nubifer* (Chironomidae) through water and food. *Ecotoxicology and Environmental Safety,* **16,** 1–10.[15]

Hatakeyama, S. and Yasuno, M. (1981) A method for assessing chronic effects of toxic substances on the midge, *Paratanytarsus parthenogeneticus* – effects of copper. *Archives of Environmental Contamination and Toxicology,* **10,** 705–13.[15]

Haufe, W.O. and Burgess, L. (1956) Development of *Aedes* (Diptera: Culicidae) at Fort Churchill, Manitoba, and prediction of dates of emergence. *Ecology,* **37,** 500–19.[9]

Hauer, F.R. and Benke, A.C. (1991) Rapid growth of snag-dwelling chironomids in a blackwater river: the influence of temperature and discharge. *Journal of the North American Benthological Society,* **10,** 154–64.[10]

Havas, M. and Hutchinson, T.C. (1982) Aquatic invertebrates from the Smoking Hills, N.W.T.: Effect of pH and metals on mortality. *Canadian Journal of Fisheries and Aquatic Sciences,* **39,** 890–903.[15]

Havas, M. and Hutchinson, T.C. (1983) Effect of low pH on the chemical composition of aquatic invertebrates from tundra ponds at the Smoking Hills, N. W. T., Canada. *Canadian Journal of Zoology,* **61,** 241–9.[15]

Havas, M. and Likens, G.E. (1985) Toxicity of aluminium and hydrogen ions to *Daphnia catawba, Holopedium gibberum, Chaoborus punctipennis,* and *Chironomus anthracinus* from Mirror lake, New Hampshire. *Canadian Journal of Zoology,* **63,** 1114–9.[15]

Hayashi, F. (1988) Prey selection by the dobsonfly larva, *Protohermes grandis* (Megaloptera: Corydalidae). *Freshwater Biology,* **20,** 19–29.[12]

Hayes, B.P. and Murray, D.A. (1987) Species composition and emergence of Chironomidae (Diptera) from three high arctic streams on Bathurst Island, Northwest Territories, Canada. *Entomologica Scandinavica Supplement,* **29,** 355–60.[9]

Hayne, D.W. and Ball, R.C. (1956) Benthic productivity as influenced by fish predation. *Limnology and Oceanography,* **1,** 162–75.[13]

Healey, M. (1984) Fish predation on aquatic insects, in *The Ecology of Aquatic Insects,* (eds V.H. Resh and D.M. Rosenberg), Praeger, New York, pp.255–88.[17]

Heinis, F. and Crommentuijn, T. (1989) The natural habitat of the deposit feeding chironomid larvae *Stictochironomus histrio* (Fabricius) and *Chironomus anthracinus* Zett. in relation to their responses to changing oxygen concentrations. *Acta Biologica Debrecen, Oecologica Hungarica,* **3**, 135–40.[6]

Heinis, F. and Crommentuijn, T. (1992) Behavioral responses to changing oxygen concentrations of deposit feeding chironomid larvae (Diptera) of littoral and profundal habitats. *Archiv für Hydrobiologie,* **124,** 173–85.[8]

Heinis, F. and Davids, C. (1993). Factors governing the spatial and temporal distribution of chironomid larvae in the Marsseveen lakes with special emphasis on the role of oxygen conditions. *Netherlands Journal of Aquatic Ecology,* **27,** 21–34.[6]

Heinis, F., Timmermans, K.R. and Swain, W.R. (1990) Short-term sublethal effects of cadmium on the filter feeding chironomid larva *Glyptotendipes pallens* (Meigen) (Diptera). *Aquatic Toxicology,* **16,** 73–86.[15]

Heinis, F., Van de Bund, W.J. and Davids, C. (1989) Avoidance of low oxygen and food concentrations by the larvae of *Tanytarsus* sp. *Acta Biologica Debrecen, Oecologica Hungarica,* **3**, 141–45.[6]

Hellawell, J.M. (1986) *Biological Indicators of Freshwater Pollution and Environmental Management,* Elsevier Applied Science Publishers, London and New York.[15]

Hely, P.C. (1958) Insect pests of the rice crop. *The Agricultural Gazette of N.S.W.,* **69,** 29–32.[13]

Hendel, F. (1908) Nouvelle classification des mouches à deux ailes (Diptera L.). D'aprés un plan tout nouveau par J.G. Meigen, Paris, an VIII (1800 v.s). Mit einem Kommentar herausgegeben von Friedrich Hendel (Wien). *Verhandlungen der Zoologische-Botanischen Gesellschaft in Wien* **58,** 43–69.[3][18]

Hennig, W. (1950) *Grundzüge einer Theorie der phylogenetischen Systematik.* Deutscher Zentralverlag, Berlin.[3]

Hennig, W. (1960) Die Dipteren-fauna von Neuseeland als systematisches und tiergeographisches problem. *Beiträge zur Entomologie,* **10,** 221–329.[4]

Hennig, W. (1966) *Phylogenetic Systematics.* Translated by D.D. Davis and R. Zangerl, University of Illinois Press, Urbana.[3]

Hennig, W. (1973) Ordnung Diptera (zweiflügler). *Handbuch der Zoologie,* **4(2)2/31,** 1–337.[3]

Henrikson, L. and Oscarson, H.G. (1985) History of the acidified Lake Gårdsjön: The development of chironomids. *Ecological Bulletin,* **37,** 58–63.[15]

Henrikson, L., Olofsson, J.B. and Oscarson, H.G. (1982) The impact of acidification on Chironomidae (Diptera) as indicated by subfossil stratification. *Hydrobiologia,* **86,** 223–9.[16]

Henson, E.B. (1966) Aquatic insects as inhalent allergens: A review of American literature. *Ohio Journal of Science,* **66,** 529–32.[14]

Herrmann, S.J., Sublette, J.E. and Sublette, M. (1987) Midwinter emergence of *Diamesa leona* Roback in the Upper Arkansas River, Colorado, with notes on other diamesines (Diptera: Chironomidae). *Entomologica Scandinavica Supplement,* **29,** 309–22.[10]

Hershey, A.E. (1985a) Effects of predatory sculpin on the chironomid communities of an arctic lake. *Ecology,* **66,** 1131–8.[12]

Hershey, A.E. (1985b) Littoral chironomid communities in an arctic Alaskan lake. *Holarctic Ecology,* **8,** 39–48.[15]

Hershey, A.E. (1986) Selective predation by Procladius in an arctic Alaskan lake. *Canadian Journal of Fisheries and Aquatic Sciences,* **43,** 252–328.[7]

Hershey, A.E. (1987) Tubes and foraging behavior in larval Chironomidae: implications for predator avoidance. *Oecologia* (Berl.), **73,** 236–41.[7][12]

Hershey, A.E. (1992) Effects of experimental fertilization on the benthic macroinvertebrate community of an arctic lake. *Journal of the North American Benthological Society,* **11,** 204–17.[17]

Hershey, A.E. and Dodson, S.I. (1985) Selective predation by a sculpin and a stonefly on two chironomids in laboratory feeding trials. *Hydrobiologia,* **124,** 269–73.[7][12]

Hershey, A.E. and McDonald, M.E. (1985) Diet and digestion rates of slimy sculpin, *Cottus cognatus,* in an Alaskan Arctic Lake. *Canadian Journal of Fisheries and Aquatic Sciences,* **42,** 483–7.[17]

Hildrew, A.G. and Townsend, C.R. (1976) The distribution of two predators and their prey in an iron rich stream. *Journal of Animal Ecology,* **45,** 41–57.[12]

Hildrew, A.G., Townsend, C.R. and Hasham, A. (1985) The predatory Chironomidae of an ironrich stream: feeding ecology and food web structure. *Ecological Entomology,* **10,** 403–13.[7][12]

Hill, C. (1988) Life cycle and spatial distribution of the amphipod *Pallasea quadrispinosa* in a lake in Northern Sweden. *Holarctic Ecology,* **11(4).** 298–304.[17]

Hill, D.A., Wright, R. and Street, M. (1987) Survival of mallard ducklings *Anas platyrhynchos* and competition with fish for invertebrates in a flooded gravel quarry in England. *Ibis,* **129,** 159–67.[17]

Hilsenhoff, W.L. (1959) The evaluation of insecticides for the control of *Tendipes plumosus* (Linnaeus). *Journal of Economic Entomology,* **52,** 331–2.[13]

Hilsenhoff, W.L. (1962) Granulated malathion as a possible control for *Tendipes plumosus* (Diptera: Tendipedidae). *Journal of Economic Entomology,* **55,** 71–8.[13]

Hilsenhoff, W.L. (1965) The effect of toxaphene on the benthos in a thermally-stratified lake. *Transactions of the American Fisheries Society,* **94,** 210–3.[15]

Hilsenhoff, W.L. (1966) The biology of *Chironomus plumosus* (Diptera: Chironomidae) in Lake Winnebago, Wisconsin. *Annals of the Entomological Society of America,* **59,** 465–73.[5][7][9][10]

Hilsenhoff, W.L. (1967) Ecology and population dynamics of *Chironomus plumosus* (Diptera: Chironomidae) in Lake Winnebago, Wisconsin. *Annals of the Entomological Society of America,* **60,** 1183–94.[10][12]

Hilsenhoff, W.L. (1968) Phoresy by *Plecopteracoluthus downesi* on larvae of *Nigronia serricornis*. *Annals of the Entomological Society of America,* **61,** 1622–3.[12]

Hilsenhoff, W.L. (1982) Using a biotic index to evaluate water quality in streams. *Department of Natural Resources, Wisconsin, Technical Bulletin,* **132,** 1–22.[15]

Hilsenhoff, W.L. and Lovett, O.L. (1966) Infection of *Chironomus plumosus* (Dipt. : Chir.) by a microsporidian (*Thelohania* sp.) in Lake Winnebago, Wisconsin. *Journal of Invertebrate Pathology,* **8,** 512–9.[12][13]

Hinton, H.E. (1951) A new chironomid from Africa, the larva of which can be dehydrated without injury. *Proceedings of the Zoological Society of London,* **121,** 371–80.[6]

Hinton, H.E. (1960a) A fly larva that tolerates dehydration and temperatures from −270°C to +102°C. *Nature,* **188,** 336–7.[6]

Hinton, H.E. (1960b) Cryptobiosis in the larva of *Polypedilum vanderplanki* Hint. (Chironomidae). *Journal of Insect Physiology,* **5,** 286–300.[6][10]

Hinton, H.E. (1968) Reversible suspension of metabolism and the origin of life. *Proceedings of the Royal Society, B.,* **171,** 43–57.[10]

Hinton, H.E. (1981) *The Biology of Insect Eggs.* Pergamon Press Ltd, Oxford.[5]

Hirvenoja, M. (1973) Revision der Gattung *Cricotopus* van der Wulp und ihrer Verwandten (Diptera: Chironomidae). *Annales Zoologici Fennici,* **10,** 1–363.[2]

Hitchcock, S.W. and Anderson, J.F. (1968) Field-plot tests with insecticides for control of *Chironomus atrella*. *Journal of Economic Entomology,* **61,** 16–9.[13]

Hodkinson, I.D. and Williams, K.A. (1980) Tube formation and distribution of *Chironomus plumosus* L. (Diptera: Chironomidae) in a eutrophic woodland pond, in *Chironomidae: Ecology, Systematics, Cytology and Physiology,* (ed. D.A. Murray), Pergamon Press, Oxford, pp.331–7.[7]

Hoffrichter, O. (1973) On the role of traffic as possible means of dispersal in chironomids. *Chironomus,* **1,** 102–4.[9]

Hoffrichter, O. and Reiss, F. (1981) Supplement 1 to 'A bibliography of the Chironomidae.' *Gunneria,* **37,** 1–68.[3][18]

Hofmann, W. (1971a) Die postglaziale Entwicklung der Chironomiden- und *Chaoborus*-Fauna (Dipt.) des Schöhsees. *Archiv für Hydrobiologie, Supplement,* **40,** 1–74.[English translation: Fisheries Research Board of Canada, Translation Series No. 2177] [16]

Hofmann, W. (1971b) Zur Taxonomie und Palökologie subfossiler Chironomiden (Dipt.) in Seesedimenten. *Ergebnisse der Limnologie,* **6,** 1–50.[16]

Hofmann, W. (1978) Analysis of animal microfossils from the Großer Segeberger See (F. R. G.). *Archiv für Hydrobiologie,* **82,** 316–46.[16]

Hofmann, W. (1983a) Stratigraphy of Cladocera and Chironomidae in a core from a shallow North German lake. *Hydrobiologia,* **103,** 235–9.[16]
Hofmann, W. (1983b) Stratigraphy of subfossil Chironomidae and Ceratopogonidae (Insecta: Diptera) in late-glacial littoral sediments from Lobsigensee (Swiss Plateau). Studies in the Late Quaternary of Lobsigensee 4. *Revue de Paléobiologie,* **2,** 205–9.[16]
Hofmann, W. (1984) Stratigraphie subfossiler Cladocera (Crustacea) und Chironomidae (Diptera) in zwei Sedimentprofilen des Meerfelder Maares. *Courier Forschunginstitut Seckenberg,* **65,** 67–80.[16]
Hofmann, W. (1985) Subfossile Cladocera (Crustacea) und Chironomidae (Diptera) aus Brackwassersedimenten des Silkteiches (Untere Trave). *Faunistisch -Ökologische Mitteilungen,* **5,** 431–42.[16]
Hofmann, W. (1987) Stratigraphy of Cladocera (Crustacea) and Chironomidae (Insecta: Diptera) in three sediment cores from the central Baltic Sea as related to paleo-salinity. *Internationale Revue der Gesamten Hydrobiologie,* **72,** 97–106.[16]
Hofmann, W. (1988) The significance of chironomid analysis (Insecta: Diptera) for paleolimnological research. *Palaeogeography, Palaeoclimatology, Palaeoecology,* **62,** 501–9.[16]
Hofmann, W. (1991) Weichselian chironomid and cladoceran assemblages from maar lakes. *Hydrobiologia,* **214,** 207–11.[16]
Holmes, R.T. (1966) Feeding ecology of the red-backed sandpiper (*Calidris alpina*) in Arctic Alaska. *Ecology,* **47,** 32–45.[17]
Holzapfel, E.P. and Perkins, B.D. (1969) Trapping of air-borne insects on ships in the Pacific, part 7. *Pacific Insects,* **11,** 455–76.[9]
Hooker, J.D. (1853) *The botany of the Antarctic voyage of H.M. discovery ships Erebus* and *Terror* in the years 1839–1843. II. Flora Novae-Zelandiae.[Introductory essay, pp.i-xxxxix.] Lovell Reeve, London, reprinted 1963, Weinheim: Cramer.[4]
Hopkins, C.L. (1976) Estimate of biological production in some stream invertebrates. *New Zealand Journal of Marine and Freshwater Research,* **10,** 629–40.[11]
Howarth, M.K. (1981) Palaeogeography of the Mesozoic, in *The Evolving Earth,* (ed. L.R.M. Cocks), British Museum (Natural History), London, pp.197–220.[4]
Huckins, J.N., Petty, J.D. and England, D.C. (1986) Distribution and impact of trifluralin, atrazine, and fonofos residues in microcosms simulating a northern prairie wetland. *Chemosphere,* **15,** 563–88.[15]
Hudson, P.L. (1987) Unusual larval habitats and life history of Chironomid (Diptera) genera. *Entomologica Scandinavica Supplement,* **29,** 369–73.[6]
Hudson, P.L., Lenat, D.R., Caldwell, B.A. and Smith, D. (1990) Chironomidae of the Southeastern United States: A checklist of species and notes on biology, distribution, and habitat. *Fish and Wildlife Research,* **7,** 1–46.[4]
Huger, A.M., Kreig, A., Emschermann, P. and Götz, P. (1970) Further studies on *Polypoxvirus chironomi,* an insect virus of the pox group isolated from the midge *Chironomus luridus. Journal of Invertebrate Pathology,* **15,** 253–61.[13]
Hughes, T.D. (1980) The imaginal ecdysis of the desert locust, *Schistocerca gregaria.* 1. A description of the behaviour. *Physiological Entomology,* **5,** 47–54.[8]
Humphries, C.F. (1938) The chironomid fauna of the Grosser Plöner See, the relative density of its members and their emergence period. *Archiv für Hydrobiologie,* **33,** 535–84.[9]
Humphries, F.C. (1937) Neue *Trichocladius*-Arten. *Stettiner Entomologische Zeitung,* **98,** 185–95.[8]
Humphreys, W.F. (1979) Production and respiration in animal populations. *Journal of Animal Ecology,* **48,** 427–53.[11]
Hunt, P.C. and Jones, J.W. (1972) The littoral fauna of Llyn Celyn, North Wales. *Journal of Fish Biology,* **4,** 321–31.[15]

Hunter, D.K. (1968) Response of populations of *Chironomus californicus* to a microspordian (*Gurleya* sp.). *Journal of Invertebrate Pathology,* **10,** 387–9.[12][13]

Hunter, M.L., Witham, J.W. and Dow, H. (1984) Effects of a carbaryl induced depression in invertebrate abundance on the growth and behaviour of American Black Duck and Mallard ducklings. *Canadian Journal of Zoology,* **62,** 452–6.[17]

Hurlbert, S. H., Fedler, J. and Fairbanks, D. (1972) Ecosystem alteration by mosquito fish (*Gambusia affinis*) predation. *Science,* **175,** 639.[12][13]

Huru, H. (1986) Diurnal variations in the diet of 0 to 3 years old Atlantic Salmon *Salmo salar* L. under semiarctic conditions in the Alta River, Northern Norway. *Fauna Norvegica, Series A,* **7(5),** 33–40.[17]

Huryn, A.D. (1990) Growth and voltinism of lotic midge larvae: patterns across an Appalachian mountain basin. *Limnology and Oceanography,* **35,** 339–51.[9][10]

Hutchinson, G.E. (1975) *A Treatise on Limnology, Volume 3. Limnological Botany,* Wiley-Interscience, New York. pp.660.

Hutchinson, G.E., Bonatti, E., Cowgill, U.M. *et al.* (1970) Ianula: an account of the history and development of the Lago di Monterosi, Latium, Italy. *Transactions of the American Philosophical Society,* **60 (4),** 1–178.[16]

Hynes, H.B.N. (1970) *The Ecology of Running Waters.* Liverpool University Press, Liverpool.[6]

Hynes, H.B.N. (1976) *Symbiocladius aurifodinae* sp. nov. (Diptera, Chironomidae), a parasite of nymphs of Australian Leptophlebiidae (Ephemeroptera). *Memoirs of the National Museum, Victoria,* **37,** 47–52.[12]

Hynes, H.B.N. (1980) A name change in the secondary production business. *Limnology and Oceanography,* **25,** 778.[11]

Hynes, H.B.N. and Williams, T.R. (1962) The effect of DDT on the fauna of a central African stream. *Annals of tropical Medicine and Parasitology* 56, 78–91.[6]

Hynes, H.B.N. and Coleman, M.J. (1968) A simple method of assessing the annual production of stream benthos. *Limnology and Oceanography,* **13,** 569–73.[11]

Hynes, J.D. (1975) Annual cycles of macroinvertebrates of a river in southern Ghana. *Freshwater Biology,* **5,** 71–83.[6]

Igarashi, T., Saeki, Y., Okada, T. *et al.* (1985) [Two cases of bronchial asthma induced by chironomid midges] *Chiryogaku,* **14,** 122–126 [in Japanese].[14]

Igarashi, T., Murakami, G., Adachi, Y. *et al.* (1987) Common occurrence in Toyama of bronchial asthma induced by Chironomid midges. *Japanese Journal of Experimental Medicine,* **57,** 1–9.[14]

Ikeshoji, T. (1981) Acoustic attraction of male mosquitoes in a cage. *Japanese Journal of Sanitary Zoology,* **32,** 7–15.[9]

Illies, J. (1961a) Phylogenie und Verbreitungsgeschichte der Ordnung Plecoptera. *Verhandlungen der Deutschen Zoologische Gesellschaft, Bonn/Rhein,* **1960,** 384–94.[4]

Illies, J. (1961b) Versuch einer allgemeinen biozonotischen Gliederung der Fliessgewasser. *Internationale Revue Hydrobiologie und Hydrographie,* 46, 205–13.[6]

Illies, J (1965) Phylogeny and zoogeography of the Plecoptera. *Annual reviews of Entomology,* **10,** 117–40.[4]

Illies, J. (1971) Emergenz 1969 im Breitenbach. Schlitzer produktionsbiologische Studien. 1. *Archiv für Hydrobiologie,* **69,** 14–59.[9]

Illies, J. (1978) *Limnofauna Europaea. A Checklist of the Animals Inhabiting European Inland Waters, with an Account of their Distribution and Ecology.* 2nd edn. Gustav Fischer Verlag, Stuttgart.[6]

Illies, J. and Botosaneanu, L. (1963) Problèmes et méthodes de la classification et de la zonation écologique des eaux courantes, considerées surtout du point de vue faunistique. *Mitteilungen der Internationalen Vereinigung für Theoretische and Angewandte Limnologie,* **12,** 1–57.[6][15]

Ineichen, H. Riesen-Willi, U. and Fischer, J. (1979) Experimental contributions to the ecology of *Chironomus* (Diptera). II. The influence of the photoperiod on the development of *Chironomus plumosus* in the 4th larval instar. *Oecologia,* **39,** 161–83.[9]

Inoue, Y. (1976) Changes of aquatic environments and chironomid midges. *Iden,* **30,** 25–31.[13]

Inoue, Y. and Mihara, M. (1973) On the unusual emergence of chironomid midges in the center of Tokyo city. *Japanese Journal of Sanitary Zoology,* **23,** 315.[13]

Inoue, Y. and Mihara, M. (1975) Studies on the Japanese chironomid midges as a nuisance. I. Larvicidal effects of some organophosphorus insecticides against the last instar larvae of *Chironomus yoshimatsui* Martin and Sublette. *Japanese Journal of Sanitary Zoology,* **26,** 135–8.[13]

International Commission on Zoological Nomenclature. (1961) Opinion 616. *Tanytarsus* van der Wulp, 1874 (Insecta, Diptera); designation of a type-species under the plenary powers. *Bulletin of Zoological Nomenclature,* **18**, 361–2.[18]

International Commission on Zoological Nomenclature. (1963) Opinion 678. The suppression under the plenary powers of the pamphlet published by Meigen, 1800. *Bulletin of Zoological Nomenclature,* **20,** 339–42.[18]

International Commission of Zoological Nomenclature (1985) *International Code of Zoological Nomenclature.* 3rd Edition, International Trust for Zoological Nomenclature, London, in association with British Museum (Natural History) and Berkeley, Los Angeles: University of California Press, pp.338.[3]

Iovino, A.J. and Miner, F.D. (1970) Seasonal abundance and emergence of Chironomidae of Beaver Reservoir, Arkansas (Insecta: Diptera). *Journal of the Kansas Entomological Society,* **43,** 197–216.[10]

Ishii, A., Matsuoka, H., Uchida, J.Y. *et al.* (1988) Chironomid midge and allergy around Lake Kojima, Okayama, in *International Symposium on Mite and Midge Allergy,* (ed. T. Miyamoto), Ministry of Education, Science and Culture, Tokyo, pp.284–317.[14]

Ito, K., Yamashita, N., Miyamoto, T. *et al.* (1985) [The role played by chironomid midge in bronchial asthma]. *Japanese Journal of Thoracic Disease,* **23,** 176 [in Japanese].[14]

Ito, K., Miyamoto, T., Shibuya, T. *et al.* (1986) Skin test and radioallergosorbent test with extracts of larval and adult midges of *Tokunagayusurika akamusi* Tokunaga (Diptera: Chironomidae) in asthmatic patients of the metropolitan area of Tokyo. *Annals of Allergy,* **57,** 199–204.[14]

Ito, K., Yamashita, N., Morita, Y. *et al.* (1988) Allergenicity and allergenic independency of chironomid midges, in *International Symposium on Mite and Midge Allergy,* (ed. T. Miyamoto), Ministry of Education, Science and Culture, Tokyo, pp.318–37.[14]

Iversen, T.M. (1988) Secondary production and trophic relationships in the invertebrate community of a Danish spring. *Limnology and Oceanography,* **33,** 582–92.

Iwakuma, T. (1986a) Ecology and production of *Tokunagayusurika akamusi* (Tokunaga) and *Chironomus plumosus* (L.) (Diptera: Chironomidae) in a shallow eutrophic lake. Ph. D thesis, Kyushu University, Japan.[10][11][12]

Iwakuma, T. (1986b) Factors controlling the secondary productivity of benthic macroinvertebrates in freshwaters: a review. *Japanese Journal of Ecology,* **36,** 169–187. (In Japanese).[11]

Iwakuma, T. (1987) Density, biomass, and production of Chironomidae (Diptera) in Lake Kasumigaura during 1982–1986. *Japanese Journal of Limnology,* **48,** 559–75.[11]

Iwakuma, T. and Yasuno, M. (1981) Chironomid populations in highly eutrophic Lake Kasumigaura. *Verhandlungen der Internationalen Vereinigung für Theoretische und Angewandte Limnologie,* **21,** 664–74.[11][15]

Iwakuma, T. and Yasuno, M. (1983) Fate of the univoltine chironomid *Tokunagayusurika akamusi* (Diptera: Chironomidae), at emergence in Lake Kasumigaura, Japan. *Archiv für Hydrobiologie,* **99,** 37–59.[17]

Iwakuma, T., Sugaya, Y. and Yasuno, M. (1989) Dependence of the autumn emergence of *Tokunagayusurika akamusi* (Diptera: Chironomidae) on water temperature. *The Japanese Journal of Limnology,* **50,** 281–8.[9]

Izvekova, E.I. (1971) On the feeding habits of chironomid larvae. *Limnologica,* **8,** 201–2.[7]

Izvekova, E.I. and Lvova-Katchanova, A.A. (1972) Sedimentation of suspended matter by *Dreissena polymorpha* Pallas and its subsequent utilization by Chironomidae larvae. *Polskie Archiwum Hydrobiologii,* **19,** 203–10.[7]

Jackson, J.K. (1988) Diel emergence, swarming and longevity of selected adult aquatic insects from a Sonoran Desert stream. *American Midland Naturalist,* **119,** 344–52.[9]

Jackson, J.K. and Fisher, S.G. (1986) Secondary production, emergence, and export of aquatic insects of a Sonoran desert stream. *Ecology,* **67,** 629–38.[11]

Jacobsen, R.E. (1993) Description of the larvae of four nearctic species of *Epoicocladius* (Diptera: Chironomidae) with a redescription of *Epoicocladius ephemerae* (Kieffer). *Netherlands Journal of Aquatic Ecology,* **26,** 145–55.[12]

Jamet, J.L., Gres, P., Lair, N. and Lasserre, G. (1990) Diel feeding cycle of roach (*Rutilus rutilus,* L.) in eutrophic Lake Aydat (Massif Central, France). *Archiv für Hydrobiologie,* **118(3),** 371–82.[17]

Jankovič, M. (1971) Anzahl der Generationen der Art *Chironomus plumosus* in den Karpfenteichen Serbiens. *Limnologica (Berlin),* **8,** 203–20.[10]

Jankovič, M. (1974) Feeding and food assimilation in larvae of *Prodiamesa olivacea. Entomologica Tidskrift Supplement,* **95,** 116–9.[7]

Janssens de Bisthoven, L. and Ollevier, F. (1989) Some experimental aspects of sediment stress on *Chironomus* gr. *thummi* larvae (Diptera: Chironomidae). *Acta Biologica Debrecina Supplementum Oecologica Hungarica,* **3,** 147–55.[15]

Janssens de Bisthoven, L.G., Timmermans, K.R. and Ollevier, F. (1992) The concentration of cadmium, lead, copper and zinc in *Chironomus* gr. *thummi* larvae (Diptera, Chironomidae) with deformed *versus* normal menta. *Hydrobiologia,* **239,** 141–9.[15]

Jernelöv, A., Nagell, B. and Svenson, A. (1981) Adaption to an acid environment in *Chironomus riparius* (Diptera, Chironomidae) from Smoking Hills, NWT, Canada. *Holarctic Ecology,* **4,** 116–9.[15]

Johannsson, O.E. (1980) Energy dynamics of the eutrophic chironomid *Chironomus plumosus* f. *semireductus* from the Bay of Quinte, Lake Ontario. *Canadian Journal of Fisheries and Aquatic Sciences,* **37,** 1254–65.[7][10][11]

Johannsson, O.E. and Beaver, J.L. (1983) Role of algae in the diet of *Chironomus plumosus* f. *semireductus* from the Bay of Quinte, Lake Ontario. *Hydrobiologia,* **107,** 237–47.[7][11]

Johnsen, B.O. and Ugedal, O. (1989) Feeding by hatchery-reared brown trout *Salmo trutta* L. released in lakes. *Aquaculture and Fisheries Management,* **20(1),** 97–104.[17]

Johnson, A.A. (1955) Life history studies on *Hydromermis contorta* (Kohn), a nematode parasite of *Chironomus plumosus* (L.). Ph.D. thesis, University of Illinois, Urbana. USA.[13]

Johnson, G.D. and Mulla, M.S. (1980) Investigations on nuisance midges in water percolation basins, Montclair, California. *Proceedings and Papers of the California Mosquito and Vector Control Association,* **48,** 118–20.[13]

Johnson, G.D. and Mulla, M.S. (1981) Chemical control of aquatic nuisance midges in residential-recreational lakes. *Mosquito News*, **41,** 495–501.[13]

Johnson, G.D. and Mulla, M.S. (1982) Suppression of nuisance aquatic midges with a urea insect growth regulator. *Journal of Economic Entomology, 75,* 297–300.[13]

Johnson, G.D. and Mulla, M.S. (1983) An aquatic macrophyte affecting nuisance chironomid midges in a warm-water lake. *Environmental Entomology,* **12,** 266–9.[13]

Johnson, J.H. (1983) Diel food habits of two species of setipalpian stone-flies (Plecoptera) in tributaries of the Clearwater River, Idaho. *Freshwater Biology,* **13(2),** 105–12.[17]

Johnson, J.H. (1985) Diel feeding ecology of the nymphs of *Aeshna multicolor* and *Lestes unguiculatus* (Odonata). *Freshwater Biology,* **15(6),** 749–56.[17]

Johnson, M.G. and Brinkhurst, R.O. (1971) Production of benthic macroinvertebrates of Bay of Quinte and Lake Ontario. *Journal of the Fisheries Research Board of Canada,* **28,** 1699–714.[11]

Johnson, M.G. and McNeil, O.C. (1988) Fossil midge associations in relation to trophic and acidic state of the Turkey Lakes. *Canadian Journal of Fisheries and Aquatic Sciences,* **45** (Supplement 1), 136–44.[15][16]

Johnson, M.G., Kelso, J.R.M., McNeil, O.C. and Morton, W.B. (1990) Fossil midge associations and the historical status of fish in acidified lakes. *Journal of Paleolimnology,* **3,** 113–27.[16]

Johnson, R.K. (1985) Feeding efficiencies of *Chironomus plumosus* (L.) and *C. anthracinus* Zett. (Diptera: Chironomidae) in mesotrophic Lake Erken. *Freshwater Biology,* **15,** 605–12.[7][11]

Johnson, R.K. (1987) Seasonal variation in diet of *Chironomus plumosus* (L.) and *C. anthracinus* Zett. (Diptera: Chironomidae) in mesotrophic Lake Erken. *Freshwater Biology,* **17,** 525–32.[7]

Johnson, R.K. (1989) Classification of profundal chironomid communities in oligotrophic/humic lakes of Sweden using environmental data. *Acta Biologica Debrecina Supplementum Oecologica Hungarica,* **3,** 167–75.[15]

Johnson, R.K., Bostrom, B. and van de Bund, W. (1989) Interactions between *Chironomus plumosus* (L.) and the microbial community in surficial sediments of a shallow, eutrophic lake. *Limnology and Oceanography,* **34,** 992–1003.[7][12]

Johnson, R.K. and Goedkoop, W. (1993) *Monoporeia* interference with larval Chironomidae: time-series and mesocosm studies. *Netherlands Journal of Aquatic Ecology,* **26,** 491–7.[9]

Johnson, R.K. and Pejler, B. (1987) Life histories and coexistence of the two profundal *Chironomus* species in Lake Erken, Sweden. *Entomologica Scandinavica Supplement,* **29,** 233–8.[7][9][10][11][12]

Johnson, R.K. and Wiederholm, T. (1989) Classification and ordination of profundal macroinvertebrate communities in nutrient poor, oligo- mesohumic lakes in relation to environmental data. *Freshwater Biology,* **21,** 375–86.[15]

Johnson, R.K., Wiederholm, T. and Eriksson, L. (1990) The influence of season on the classification and ordination of profundal communities of nutrient poor, oligo-mesohumic Swedish lakes using environmental data. *Verhandlungen der Internationalen Vereinigung für Theoretische und Angewandte Limnologie,* **24,** 646–52.[15]

Johnston, N.T., Perrin, C.J., Slaney, P.A. and Ward, B.R. (1990) Increased juvenile salmonid growth by whole-river fertilization. *Canadian Journal of Fisheries and Aquatic Sciences,* **47,** 862–72.[17]

Jónasson, P.M. (1961) Population dynamics in *Chironomus anthracinus* Zett. in the profundal zone of Lake Esrom. *Verhandlungen der Internationalen Vereinigung für Theoretische und Angewandte Limnologie,* **14,** 196–203.[10]

Jónasson, P.M. (1965) Factors determining population size of *Chironomus anthracinus* in Lake Esrom. *Mitteilungen der Internationalen Vereinigung für Theoretische and Angewandte Limnologie,* **13,** 139–62.[10]

Jónasson, P.M. (1972) Ecology and production of the profundal benthos in relation to phytoplankton in Lake Esrom. *Oikos Supplement,* **14,** 1–148.[7][10][11]

Jónasson, P.M. (1978) Zoobenthos of lakes. *Verhandlungen der Internationalen Vereinigung für theoretische und angewandte Limnologie* **20,** 13–37.[6]

Jónasson, P.M. and Kristiansen, J. (1967) Primary and secondary production in Lake Esrom. Growth of *Chironomus anthracinus* in relation to seasonal cycles of phytoplankton and dissolved oxygen. *Internationale Revue der Gesamten Hydrobiologie,* **52,** 163–217.[6][7][10]

Jónasson, P.M. and Lindegaard, C. (1979) Zoobenthos and its contribution to the metabolism of shallow lakes. *Archiv für Hydrobiologie Beiheft Ergebnisse der Limnologie,* **13,** 162–80.[15]

Jones, E.L. (1968) Bloodworms, pests of rice. *The Agricultural Gazette of N.S.W.,* **79,** 477–8.[13]

Jones, R.E. (1974) The effects of size–selective predation and environmental variation on the distribution and abundance of a chironomid, *Paraborniella tonnoiri* Freeman. *Australian Journal of Zoology,* **22,** 71–89.[6][12]

Jones, R.E. (1975) Dehydration in an Australian rockpool chironomid larva, *Paraborniella tonnoiri. Proceedings of the Royal Entomological Society of London, Series A General Entomology,* **49,** 111–9.[10]

Jónsson, E. (1985) Population dynamics and production of Chironomidae (Diptera) at 2 m depth in Lake Esrom, Denmark. *Archiv für Hydrobiologie Supplement,* **70,** 239–78.[10][11]

Jutting, T.v.B. (1938) A freshwater pulmonate (*Physa fontinalis*), inhabited by larva of a non-biting midge (*Tendipes* (*Parachironomus*) *varus* Gtgh.). *Archiv für Hydrobiologie,* **32,** 693–96.[12]

Kaeding, L. and Zimmerman, M.A. (1983) Life history and ecology of the humpback chub *Gila cypha* in the Little Colorado and Colorado rivers of the Grand Canyon, USA. *Transactions of the American Fisheries Society,* **112(5),** 577–94.[17]

Kagen, S.L. (1990) Inhalent Allergy to Arthropods, Insects, Arachnids, and Crustaceans. *Clinical Reviews in Allergy,* **8,** 99–125.[14]

Kagen, S.L., Yunginger, J.W. and Johnson, R. (1984) Lake fly allergy: incidence of chironomid sensitivity in an atopic population. *Journal of Allergy and Clinical Immunology,* **73,** 187 (abstract).[14]

Kajak, Z. (1963) The effect of experimentally induced variations in the abundance of *Tendipes plumosus* L. larvae on intraspecific and interspecific relations. *Ekologia Polska (Series A),* **11,** 355–67.[10]

Kajak, Z. (1977) Factors influencing benthos biomass in shallow lake environments. *Ekologia Polska,* **25,** 421–9.[7][15]

Kajak, Z. (1980) Role of invertebrate predators (mainly *Procladius* sp.) in benthos, in *Chironomidae: Ecology, Systematics, Cytology and Physiology,* (ed. D.A. Murray), Pergamon Press, Oxford, pp.339–48.[7][12]

Kajak, Z. and Dusoge, K. (1970) Production efficiency of *Procladius choreus* Mg (Chironomidae, Diptera) and its dependence on the trophic conditions. *Polskie Archiwum Hydrobiologii,* **17,** 217–24.[7][11]

Kajak, Z. and Dusoge, K. (1976) Benthos of Lake Sniardwy as compared to benthos of Mikolajskie Lake and Lake Taltowisko. *Ekologia Polska,* **24,** 77–101.[11]

Kajak, Z. and Rybak, J.I. (1966) Production and some trophic dependences in benthos against primary production and zooplankton production of several Masurian lakes. *Verhandlungen der Internationalen Vereinigung für Theoretische und Angewandte Limnologie,* **16,** 441–451.

Kajak, Z. and Pieczynski, E. (1966) The influence of invertebrate predators on the abundance of benthic organisms (chiefly Chironomidae). *Ekologia Polska,* **12,** 175–9.[7]

Kajak, Z. and Warda, J. (1968) Feeding of benthic nonpredatory Chironomidae in lakes. *Annales Zoologici Fennici,* **5,** 57–64.[7]

Kajak, Z., Dusoge, K. and Stanczykowska, A. (1968) Influence of mutual relations of organisms, especially Chironomidae, in natural benthic communities, on their abundance. *Annales Zoologici Fennici,* **5,** 49–56.[7][12]

Kalugina, N.S. (1959) Changes in morphology and biology of chironomid larvae in relation to growth (Diptera Chironomidae). *Fisheries Research Board of Canada Translation Series* No. 1160 (1968), pp 48.[5]

Kalugina, N.S. (1976) Komary-zvontsy podsemeistva Diamesinae (Diptera: Chironomidae) iz verkhnego mela Taimyra. *Palaeontologicheskii Zhurnal,* **1,** 87–93.[4]

Kalugina, N.S. (1980) Cretaceous Aphroteniinae from North Siberia (Diptera: Chironomidae). *Electrotenia brundini* gen. nov., sp. nov. *Acta Universitatis Carolinensis-Biologica* **1978,** 89–93.[4]

Kamei, M., Shimada, H., Okubo, S. and Ishii, T. (1982) Effects of Altosid^R 10F on the chironomid midge, *Chironomus yoshimatsui* Martin et Sublette (Diptera: Chironomidae) of Tokushima City. *Japanese Journal of Sanitary Zoology,* **33,** 355–61.[13]

Kangasniemi, B.J. and Oliver, D.R. (1983) Chironomidae (Diptera) associated with *Myriophyllum spicatum* in Okanagan Valley Lakes, British Columbia. *Canadian Entomologist,* **115,** 1545–46.[7]

Kansanen, P.H. (1985) Assessment of pollution history from recent sediments in Lake Vanajavesi, southern Finland. II. Changes in the Chironomidae, Chaoboridae and Ceratopogonidae (Diptera) fauna. *Annales Zoologici Fennici,* **22,** 57–90.[15]

Kansanen, P.H. and Aho, J. (1981) Changes in the macrozoobenthos associations of polluted Lake Vanajavesi, southern Finland, over a period of 50 years. *Annales Zoologici Fennici,* **18,** 73–101.[15][16]

Kansanen, P.H., Aho, J. and Paasivirta, L. (1984) Testing the benthic lake type concept based on chironomid associations in some Finnish lakes using multivariate statistical methods. *Annales Zoologici Fennici,* **21,** 55–76.[15]

Kansanen, P.H., Paasivirta, L. and Väyrynen, T. (1990) Ordination analysis and bioindices based on zoobenthos communities used to assess pollution of a lake in southern Finland. *Hydrobiologia,* **202,** 153–70.[15]

Kansanen, P.H. and Jaakkola, T. (1985) Assessment of pollution history from recent sediments in Lake Vanajavesi, southern Finland. I. Selection of representative profiles, their dating and chemostratigraphy. *Annales Zoologici Fennici,* **22,** 13–55.[16]

Kaster, J.L. and Jacobi, G.Z. (1978) Benthic macroinvertebrates of a fluctuating reservoir. *Freshwater Biology,* **8,** 283–90.[15]

Kaufman, M.G. and King, R.H. (1987) Colonization of wood substrates by the aquatic xylophage *Xylotopus par* (Diptera: Chironomidae) and a description of its life history. *Canadian Journal of Zoology,* **65,** 2280–86.[7]

Kaufman, M.G., Pankratz, H.S. and Klug, M.J. (1986) Bacteria associated with the ectoperitrophic space in the midgut of the larva of the midge *Xylotopus par* (Diptera: Chironomidae). *Applied Environmental Microbiology,* **51,** 657–60.[7]

Kawai, K. and Konishi K. (1986) Fundamental studies on chironomid allergy I. Culture methods of some Japanese chironomids (Chironomidae, Diptera). *Japanese Journal of Sanitary Zoology,* **37,** 47–57.[17]

Kawai, K. and Konishi, K. (1988) Fundamental studies on chironomid allergy. III. Allergen analyses of some adult Japanese chironomid midges (Chironomidae: Diptera). *Japanese Journal of Allergology,* **37,** 944–51.[13][14]

Kawai, K. and Sakamoto, K. (1988) Cross-reactivities of murine IgE-inducing larval hemoglobins among various chironomid species. *Japanese Journal of Sanitary Zoology,* **43,** 95–103.[14]

Kawecka, B. and Kownacki, A. (1974) Food conditions of Chironomidae in the River Raba. *Entomologica Tidskrift Supplement,* **95** 120–8.[7]

Kawecka, B., Kownacki, A. and Kownacka, M. (1978) Food relations between algae and bottom fauna communities in glacial streams. *Internationale Vereinigung für Theoretische und Angewandte Limnologie Verhandlungen,* **20,** 1527–30.[7]

Kay, A.B., Gad el Rab, M.O., Stewart, J. and Erwa, H.H. (1978) Widespread IgE-mediated hypersensitivity in Northern Sudan to the chironomid *Cladotanytarsus lewisi* ('green nimitti'). *Clinical and Experimental Immunology,* **34,** 106–10.[14]

Kay, A.B., MacLean, U., Wilkinson, A.H. and Gad El Rab, M.O. (1982) The prevalence of asthma and rhinitis in a Sudanese community seasonally exposed to a potent airborne allergen (the 'green nimitti' midge *Cladotanytarsus lewisi*). *Journal of Allergy and Clinical Immunology,* **71,** 345–52.[14]

Keene, J.A. (1989) *Holocene environmental changes interpreted from arthropod remains in British lake sediments with especial reference to Malham Tarn, Yorkshire.* Ph.D. thesis, University of Birmingham, Birmingham, England.[16]

Keim, A. (1977) Electrophoretic analyses of the crop contents of *Helobdella stagnalis* (L.) (Hirudinea). *Zeitschrift für Naturforschung, Wiesbaden,* **32c,** 739–42.[17]

Kelly-Quinn, M. and Bracken, J.J. (1990) A seasonal analysis of the diet and feeding dynamics of brown trout *Salmo trutta* L. in a small nursery stream. *Aquaculture and Fisheries Management,* **21(1),** 107–24.[17]

Kenmuir, D.H.S. (1975) The diet of fingerling tigerfish, *Hydrocynus vittatus* Cast., in Lake Kariba, Rhodesia. *Arnoldia (Rhodesia),* **7,** 1–8.[17]

Kennedy, J.S. (1940) The visual responses of flying mosquitoes. *Proceedings of the Zoological Society of London, Series A,* **109,** 221–42.[9]

Kesler, D.H. (1981) Grazing rate determination of *Corynoneura scutellata* Winnertz (Chironomidae: Diptera). *Hydrobiologia,* **80,** 63–6.[7][11]

Khangarot, B.S. and Ray, P.K. (1989) Sensitivity of midge larvae of *Chironomus tentans* Fabricius (Diptera, Chironomidae) to heavy metals. *Bulletin of Environmental Contamination and Toxicology,* **42,** 325–30.[15]

Kieffer, J.J. (1923) Nouvelles contribution à l'étude des Chironomides de la Nouvelle-Zemble. *Reports of the Scientific Results of the Norwegian Expedition to Nova Zemlya,* **9,** 3–11.[3]

Kimball, J.H. (1968) Carp for chironomid midge control. *Mosquito News* **28**: 147–8.[13]

Kimerle, R.A. and Anderson, N.H. (1971) Production and bioenergetic role of the midge *Glyptotendipes barbipes* (Staeger) in a waste stabilization lagoon. *Limnology and Oceanography,* **16,** 646–59.[11]

King, R.P. (1989) Distribution abundance size and feeding habits of *Brienomyrus brachyistius* Gill 1862 (Teleostei: Mormyridae) in a Nigerian rainforest stream. *Cybium,* **13(1),** 25–36.[17]

Kino, T., Chihara, J., Fukuda, K. *et al.* (1987) Allergy to insects in Japan. III. High frequency of IgE antibody responses to insects (moth, butterfly, caddis fly, and chironomid) in patients with bronchial asthma and immunochemical quantitation of the insect-related airborne particles smaller than 10 μm in diameter. *Journal of Allergy and Clinical Immunology,* **79,** 857–66.[14]

Kino, T. and Oshima, S. (1988) Determination of chironomid midge airborne allergen, in comparison with moth, butterfly and caddis fly allergens, in *International Symposium on Mite and Midge Allergy,* (ed. T. Miyamoto), Ministry of Education, Science and Culture, Tokyo, pp.338–63.[14]

Kiskaroly, M. and Tafro, A. (1985) Artificial production of large forms of zooplankton as additional living food in the nutrition of some kinds of young fish and fish fry. *Veterinarski Glasnik Beograd Zagreb,* **39,** 515–22.[17]

Kitching, R.L. (1972) Population studies of the immature stages of the tree-hole midge *Metriocnemus martinii* Thienemann Diptera: Chironomidae). *Journal of Animal Ecology,* **41,** 53–62.[6]

Klink, A. (1989) The Lower Rhine: Palaeoecological analysis, in *Historical Change of Large Alluvial Rivers: Western Europe,* (eds G. E. Petts with H. Möller and A. L. Roux), John Wiley & Sons, pp.183–201.[15][16]

Kluge, A. (1989) A concern for evidence and a phylogenetic hypothesis of relationships amongst *Epicrates* (Boidae, Serpentes). *Systematic Zoology,* **38,** 7–25.[3]

Kobayashi, T. (1993) *Eurycnemus* sp. (Diptera: Chironomidae) larvae ectoparasitic on pupae of *Goera japonica* (Trichoptera), newly recorded in Japan. *Japanese Journal of Sanitary Zoology,* **44,** 401–404.[12]

Kobayashi, T. (in press) *Eurycnemus* sp. (Diptera: Chironomidae) larvae ectoparasitic on pupae of *Goera japonica* (Trichoptera: Limnephilidae), in *Chironomids: from Genes to Ecosystems* (ed. P.S. Cranston), CSIRO, Melbourne.[12]

Kohshima, S. (1984). A novel cold-tolerant insect found in a Himalayan glacier. *Nature,* **310,** 225–7.[1][6]

Kohshima, S. (1985) Migration of the Himalayan wingless glacier midge (*Diamesa* sp.): slope direction assessment by sun-compassed straight walk. *Journal of Ethology,* **3,** 93–104.[9]

Kokes, J. (1979) Food eaten by the fry of chub, *Leuciscus cephalus* in the Rokytna River, Czechoslovakia. *Folia Zoologica,* **28,** 361–70.[17]

Kokkinn, M.J. (1986) Identification of two Australian salt-lake chironomid species from subfossil larval head capsules. *Palaeogeography, Palaeoclimatology, Palaeoecology,* **54,** 317–28.[16]

Kokkinn, M.J. (1990) Is the Rate of Embryonic Development a Predictor of Overall Development Rate in *Tanytarsus barbitarsis* Freeman (Diptera : Chironomidae). *Australian Journal of Marine and Freshwater Research,* **41,** 575–9.[5]

Kokkinn, M.J. and Williams, W.D. (1988) Adaptations to life in a hypersaline water-body: Adaptations at the egg and early embryonic stage of *Tanytarsus barbitarsis* Freeman. *Aquatic Insects,* **4,** 205–14.[5]

Kokkinn, M.J. and Williams, W.D. (1989) An experimental study of phototactic responses of *Tanytarsus barbitarsis* Freeman (Diptera: Chironomidae). *Australian Journal of Marine and Freshwater Research,* **40 (6),** 693–702.[9][13]

Kolkwitz, R. and Marsson, M. (1909) Ökologie der tierischen Saprobien. Beiträge zur Lehre von der biologischen Gewässerbeurteilung. *Internationale Revue der Gesamten Hydrobiologie,* **2,** 126–52.[15]

Kolton, R.J., MacMahon, P.D., Jeffrey, K.A. and Beamish, F.W.H. (1986) Effects of the lampricide, 3-trifluoromethyl–4-nitrophenol (TFM) on the macroinvertebrates of a hardwater stream. *Hydrobiologia,* **139,** 251–67.[15]

Kon, M. (1984) Swarming and mating of *Chironomus yoshimatsui* (Diptera: Chironomidae): seasonal change in the timing of swarming and mating. *Journal of Ethology,* **2,** 37–45.[9]

Kon, M. (1985) Activity patterns of *Chironomus yoshimatsui* (Diptera: Chironomidae) 1. Effects of temperature conditions on the adult activity patterns. *Journal of Ethology,* **3,** 131–4.[9]

Kon, M. (1986) Swarming and insemination of *Chironomus yoshimatsui* (Diptera: Chironomidae). *Memoirs of the Faculty of Science, Kyoto University (Series of Biology),* **11,** 43–6.[9]

Kon, M. (1987) The mating system of chironomid midges (Diptera: Chironomidae): a review. *Memoirs of the Faculty of Science, Kyoto University,* **12,** 129–34.[9]

Kon, M. (1989) Swarming and mating behaviour of *Chironomus flaviplumus* (Diptera: Chironomidae), compared with a sympatric congeneric species *C. yoshimatsui*. *Journal of Ethology*, **7**, 125–31.[9]

Kon, M., Otsuka, K. and Hidaka, T. (1986) Mating system of *Tokunagayusurika akamusi* (Diptera: Chironomidae) I Copulation in the air by swarming and on the ground by searching. *Journal of Ethology*, **4**, 49–58.[9]

Kondo, S., Hamashima, S. and Hashimoto, H. (1989) Life history and seasonal occurrence of *Pentapedilum tigrinum* Hashimoto associated with *Nymphoides indica* O. Kuntze in an irrigation reservoir. *Acta Biologica Debrecina Oecologica Hungarica*, **2**, 237–45.[7]

Konstantinov, A.S. (1951) Istoriya fauny khironomid nekotorykh ozer sapovednika 'Borovoye' (Severniy Kazakhstan). *Trudy Laboratorii Sapropelevykh Otlozheniy*, **5**, 91–107.[16]

Konstantinov, A.S. (1952) Poluproizvodstvennoe razvedenie lichinok khironomid [Semicommercial propagation of Chironomid larvae]. *Rybnoje Choziajstvo*, **1**, 31–33 [in Russian].[17]

Konstantinov, A.S. (1954) Opyt poluproizvodostvennogo razvedeniya motylya [An experience in semicommercial midge propagation]. *Rybnoje Choziajstvo*, **11**, 41–3 [in Russian].[17]

Konstantinov, A.S. (1958a) Biologiya khironomid i ikh razvedenie [*Biology of the Chironomidae and their Cultivation*]. (Vniorch), Saratov, USSR, pp.362.[in Russian].[17]

Konstantinov, A.S. (1958b) Influence of temperature on the rate of development and growth of chironomids. *Doklady Akadamii Nauk SSSR*, **120**, 1362–65. (In Russian).[10]

Konstantinov, A.S. (1961) O biologii komarov semeistva Chironomidae. [On the biology of the midge family Chironomidae]. *Nauchnye doklady vysshei shkoly Biologicheskie nauki*, **4**, 20–3.[9]

Konstantinov, A.S. (1971a) Ecological factors affecting respiration in chironomid larvae. *Limnologica (Berlin)*, 8, 127–34.[6][11]

Konstantinov, A.S. (1971b) Feeding habits of the chironomid larvae and certain ways to increase the food content of the water basins. *Fisheries Research Board of Canada Translation Series* No. 1853.[7]

Koreneva, T.A. (1959) Egg deposition by females of Pelopiinae in the Uchinsk Reservoir. *Trudy Vsesoyuznogo gidrobiologicheskogo Obshchestva*, **9**, 108–20. (In Russian).[5]

Kornijów, R. (1992) Seasonal migration by larvae of an epiphytic chironomid. *Freshwater Biology*, **27**, 85–9.[7][10]

Kosalwat, P. and Knight, A. W. (1987) Chronic toxicity of copper to a partial life cycle of the midge, *Chironomus decorus*. *Archives of Environmental Contamination and Toxicology*, **16**, 283–90.[15]

Koskenniemi, E. and Paasivirta, L. (1987) The chironomid (Diptera) fauna in a Finnish reservoir during its first four years. *Entomologica Scandinavica Supplement*, **29**, 239–46.[9]

Koskinen, R. (1969) Observations on the swarming of *Chironomus salinarius* Kieff. (Diptera: Chironomidae). *Annales Zoologici Fennici.*, **6**, 145–9.[9]

Kowalyk, H. E. (1985) The larval cephalic setae in the Tanypodinae (Diptera: Chironomidae) and their importance in generic determinations. *Canadian Entomologist* **117**, 67–106.[2]

Kownacka, M. and Kownacki, A. (1972) Vertical distribution of zoocenoses in the streams of the Tatra, Caucasus and Balkans Mts. *Verhandlungen der Internationalen Vereinigung für Theoretische und Angewandte Limnologie*, **18**, 742–50.[15]

Kownacka, M. and Kownacki, A. (1975) Gletscherbach-Zuckmücken der Ötztaler Alpen in Tirol (Diptera: Chironomidae: Diamesinae). *Entomologica Germanica,* **2,** 35–43.[6]

Kownacki, A. (1989) Taxocenes of Chironomidae as an indicator for assessing the pollution of rivers and streams. *Acta Biologica Debrecina Supplementum Oecologica Hungarica,* **3,** 219–30.[15]

Kownacki, A. (1991) Zonal distribution and classification of the invertebrate communities in high mountain streams in South Tirol (Italy). *Verhandlungen der Internationalen Vereinigung für Theoretische und Angewandte Limnologie,* **24,** 2010–4.[15]

Kownacki, A. and Zosidze, R. S. (1980) Taxocens of Chironomidae (Diptera) in some rivers and streams of the Adzhar ASSR (Little Caucasus Mts). *Acta Hydrobiologica,* **22,** 67–87.[6][15]

Krantzberg, G. and Stokes, P. M. (1989) Metal regulation, tolerance, and body burdens in the larvae of the genus *Chironomus. Canadian Journal of Fisheries and Aquatic Sciences,* **46,** 389–98.[15]

Krantzberg, G. and Stokes, P. M. (1990) Metal concentrations and tissues distribution in larvae of *Chironomus* with reference to X-ray microprobe analysis. *Archives of Environmental Contamination and Toxicology,* **19,** 84–93.[15]

Krecker, F.H. (1939) A comparative study of the animal population of certain submerged aquatic plants. *Ecology,* **20,** 553–62.[6]

Krueger, C.C. and Waters, T.F. (1983) Annual production of macroinvertebrates in three streams of different water quality. *Ecology,* **64,** 840–50.[11]

Kugler, J. and Wool, D. (1968) Chironomidae (Diptera) from the Hula Nature Preserve, Israel. *Annales Zoologici Fennici,* **5,** 76–83.[9]

Kugler, J. and Chen, H. (1968) Distribution of chironomid larvae in Lake Tiberias and their occurrence in the food of fish in the lake. *Israel Journal of Zoology,* **17,** 95–115.[13]

Kukichi. M. and Sasa, M. (1990) Studies on the chironomid midges (Diptera: Chironomidae) of the Lake Toba area, Sumatra, Indonesia. *Japanese Journal of Sanitary Zoology,* **41,** 291–329.[4]

Kullberg, A. (1988) The case, mouthparts, silk and silk formation of *Rheotanytarsus muscicola* Kieffer (Chironomidae: Tanytarsini). *Aquatic Insects,* **10,** 249–55.[7]

Kureck, A. (1966) Schlüpfrhythmus von *Diamesa arctica* (Diptera: Chironomidae) auf Spitzbergen. *Oikos,* **17,** 276–7.[9]

Kureck, A. (1979) Two circadian eclosion times in *Chironomus thummi* (Diptera), alternately selected with different temperatures. *Oecologia (Berlin),* **40,** 311–23.[9]

Kureck, A. (1980) Circadian eclosion rhythm in *Chironomus thummi*; ecological adjustment to different temperature levels and the role of temperature cycles, in Chironomidae, in *Chironomidae, Ecology, Systematics, Cytology and Physiology* (ed. D.A. Murray), Pergamon Press, New York, 73–80.[9]

Lack, D. and Owen, D.F. (1955) The food of the swift. *Journal of Animal Ecology,* **24,** 120–36.[17]

Ladle, M. (1982) Organic detritus and its role as a food-source in chalk streams. *Freshwater Biological Association of the United Kingdom Fiftieth Annual Report*: 30–7.[7][9]

Ladle, M. and Bass, J.A.B. (1981) The ecology of a small chalk stream and its responses to drying during drought conditions. *Archiv für Hydrobiologie,* **90,** 448–66.[6]

Ladle, M., Cooling, D.A., Welton, S. and Bass, J.A.B. (1985) Studies on Chironomidae in experimental recirculating stream systems. II. The growth, development and production of a spring generation of *Orthocladius (Euorthocladius) calvus* Pinder. *Freshwater Biology,* **15,** 243–55.[10]

Ladle, M., Welton, J.S. and Bass, J.A.B. (1984) Larval growth and production of three species of Chironomidae from an experimental recirculating stream. *Archiv für Hydrobiologie,* **102,** 201–14.[10][11]

Laird, M. (1961) New American locality records for the species of *Coelomomyces* (Blastocladiales, Coelomomycetaceae). *Journal of Insect Pathology,* **3,** 249–53.[12]

Lake, P.S., Barmuta, L.A., Boulton, A.J. *et al.* (1986) Australian streams and Northern Hemisphere stream ecology: comparisons and problems. *Proceedings of the Ecological Society of Australia,* **14,** 61–82.[6]

Lake, P.S., Doeg, T. and Morton, D.W. (1985) The macroinvertebrate community of stones in an Australian upland stream. *Verhandlungen der Internationalen Vereinigung für theoretische und angewandte Limnologie,* **22,** 2141–7.[6]

Lamberti, G.A. and Moore J.W. (1984) Aquatic insects as primary consumers, in *The Ecology of Aquatic Insects,* (eds V.H. Resh and D.M. Rosenberg), Praeger, New York, pp.164–95.[7]

Lamberti, G.A. and Resh, V.H. (1983) Geothermal effects on stream benthos: Separate influences of thermal and chemical components on periphyton and macroinvertebrates. *Canadian Journal of Fisheries and Aquatic Science,* **40,** 1995–2009.[6]

Lammens, E.H.R.R. (1982) Growth, condition and gonad development of Bream *Abramis brama* L. in relation to its feeding conditions in Tjeukemeer. *Hydrobiologia,* **95,** 311–20.[17]

Lammens, E.H.R.R. (1985) A test model for planktivorous filter feeding by bream *Abramis brama* L. *Environmental Biology of Fishes,* **13,** 289–96.[17]

Lammens, E.H.R.R. (1989) Causes and consequences of the success of bream in Dutch eutrophic lakes. *Hydrobiological Bulletin,* **23(1),** 11–8.[17]

Lammens, E.H.R.R. and Visser, J.T. (1989) Variability of mouth width in European eel *Anguilla anguilla* in relation to varying feeding conditions in three Dutch lakes. *Environmental Biology of Fishes,* **26(1),** 63–76.[17]

Lancaster, J. (1990) Predation and drift of lotic macroinvertebrates during colonization. *Oecologia (Berlin),* **85,** 48–56.[12]

Langton, P.H. (1974) On the biology of the parthenogenetic *Paratanytarsus* breeding in the Essex Water Company distribution system. *Water Treatment and Examination,* **23,** 230–2.[8]

Langton, P.H. (1980) *Taxonomy and Biology of some British Chironomidae.* Ph.D. thesis, C.N.A.A.[8]

Langton, P.H. (1984) *A key to pupal exuviae of British Chironomidae.* Privately published by P.H. Langton, 1 Brooks Road, March, Cambridgeshire.[3][8][10][12][18]

Langton, P.H. (1989) Functional and phylogenetic interpretation of chironomid pupal structure. *Acta Biologica Debrecen, Oecologica Hungarica,* **2,** 247–52.[8][18]

Langton, P.H. (1991) A key to pupal exuviae of West Palaearctic Chironomidae. Privately published by P.H. Langton, 3 St Felix Road, Ramsey Forty Foot, Cambridgeshire.[3]

Langton, P.H. (1994) Adhesion marks on the abdomen of pupal Chironomidae (Diptera). *British Journal of Entomology and Natural History,* **7,** 89–91.[8]

Langton, P.H., Cranston, P.S. and Armitage, P.D. (1988) The parthenogenetic midge of water supply systems *Paratanytarsus grimmii* Schneider (Diptera: Chironomidae). *Bulletin of Entomological Research,* **78(2),** 317–28.[5][8][9][13][17]

Larsson, P. (1984) Transport of PCBs from aquatic to terrestrial environments by emerging chironomids. *Environmental Pollution,* Series A, **34,** 283–9.[13][15]

Lastochkin, D. A. (1949) Ocherki po paleolimnologii Urala. *Trudy Laboratorii Sapropelevykh Otlozheniy,* **3,** 101–35.[16]

Lavandier, P. (1984) Dynamique des populations larvaires et régime alimentaire de *Rhyacophila tristis* Pictet (Trichoptera: Rhyacophilidae) dans un ruisseau de haute montagne. *Annales de Limnologie,* **20,** 209–14.[17]

Laville, H. (1972) Recherches écologiques sur les Chironomides (Diptera) des Lacs de Montagne. Doctoral thesis, L'Université Paul Sabatier de Toulouse.[10][12]

Laville, H. (1975) Production d'un chironomide semivoltin (*Chironomus commutatus* Str.) dans le lac de Port–Bielh. (Pyrénées centrales) *Annales de Limnologie,* **11,** 67–77.[11]

Laville, H. (1979) Étude de la dérive des exuvies nymphales de Chironomides au niveau du confluent Lot-Tryuère. *Annales de Limnologie,* **15 (2)**, 155–80.[18]

Laville, H. (1981) Récoltes dexuvies nymphales de Chironomides (Diptera) dans le Haut-Lot, de la source (1 295 m) áu confluent de la Truyère (223 m). *Annales de Limnologie,* **17 (3)**, 255–89.[18]

Laville, H. and Reiss, F. (1993) The Chironomid fauna of the Mediterranean region reviewed. *Netherlands Journal of Aquatic Ecology,* **26,** 239–45.[4]

Laville, H. and Serra-Tosio, B. (1987) Chironomidés (Diptera) du Massif Central et des basses région avoisinantes. *Annales de Limnologie,* **3,** 185–204.[4]

Laville, H. and Tourenq, J.N. (1975) Contribution à la connaissance de trois chironomides de Camargue et des marismas du Guadalquivir. *Annales de Limnologie,* **3,** 185–204.[6]

Laville, H. and Vinçon, G. (1986) Inventaire 1986 des Chironomidés (Diptera) connus des Pyrénées. *Annales de Limnologie,* **22,** 239–51.[4]

Laville, H. and Vinçon, G. (1991) A typological study of Pyrenean streams: Comparative analysis of the Chironomidae (Diptera) communities in the Ossau and Aure Valleys. *Verhandlungen der Internationalen Vereinigung für Theoretische und Angewandte Limnologie,* **24,** 1775–84.[15]

Lawrenz, R. W. (1975) *The developmental paleoecology of Green Lake, Antrim County, Michigan,* M.S. thesis, Central Michigan University, Mount Pleasant, Michigan. pp.78.[16]

Layton, R.J. and Voshell Jr., J.R. (1991) Colonization of new experimental ponds by benthic macroinvertebrates. *Environmental Entomology,* **20 (1),** 110–7.[9]

Learner, M.A. and Edwards, R.W. (1966) The distribution of the midge *Chironomus riparius* in a polluted river system and its environs. *Air and Water Pollution,* **10,** 757–68.[9][10]

Learner, M.A. and Potter, D.W.B. (1974) The seasonal periodicity of emergence of insects from two ponds in Hertfordshire, England, with special reference to the Chironomidae (Diptera: Nematocera). *Hydrobiologia,* **44,** 495–510.[10]

Learner, M., Wiles, R. and Pickering, J. (1990) Diel emergence patterns of chironomids. *Internationale Revue der Gesamten Hydrobiologie,* **75,** 569–81.[9]

Leavitt, P.R., Carpenter, S.R. and Kitchell, J.F. (1989) Whole-lake experiments: the annual record of fossil pigments and zooplankton. *Limnology and Oceanography,* **34,** 700–17.[7]

Legner, E.F. and Medved, R.A. (1973) Predation of mosquitoes and chironomid midges in ponds by *Tilapia zillii* (Gervais) and *T. mossambica* (Peters) (Teleosteii: Cichlidae). *Proceeding and Papers of the California Mosquito Control Association,* **41,** 119–21.[13]

Legner, E.F., Medved, R.A. and Hauser, W.J. (1975a) Predation of the desert pupfish, *Cyprinodon macularius,* on *Culex* mosquitoes and benthic chironomid midges. *Entomophaga,* **20,** 23–30.[13]

Legner, E.F., Yu, H.S., Medved, R.A. and Badgley, P.M.E. (1975b) Mosquito and chironomid midge control by planaria. *California Agriculture,* **29,** 3–6.[17]

Lehmann, J. (1971) Die Chironomiden der Fulda (Systematische, ökologische und faunistische Untersuchungen). *Archiv für Hydrobiologie Supplement,* **37,** 466–555.[10][15]

Lehmann, J. (1979) Chironomidae (Diptera) aus Fliessgewässern Zentralafrikas. Teil I: Kivu-Gebiet, Ostzaire. *Spixiana Supplement* **3,** 1–144.[4]

Lehmann, J. (1981) Chironomidae (Diptera) aus Fliessgewässern Zentralafrikas. II. Die Region um Kisangami, Zentralzaire. *Spixiana Supplement,* **5,** 1–85.[4]

Lellak, J. (1968) Positive Phototaxis der Chironomiden-larvulae als regulierender Faktor ihrer Verteilung in stehenden Gewässer. *Annales Zoologici Fennici,* **5,** 84–7.[5]

Lenat, D.R. (1987) The macroinvertebrate fauna of the Little River, North Carolina: taxa list and seasonal trends. *Archiv für Hydrobiologie,* **110,** 19–43.[12]

Lenz, F. (1951) Neue Beobachtungen zur Biologie der Jugendstadien der Tendipedidengattung *Parachironomus* Lenz. *Zoologische Anzeiger,* **147,** 95–111.[8]

Lenz, F. (1955) Der Wert der Exuviensammlung für die Beurteilung der Tendipedidenbesiedlung eines Sees. *Archiv für Hydrobiologie Supplement,* **22**, 415–21.[18]

Leonard, S.D. (1972) The natural history of *Paraclunio alaskensis* and *Paraclunio trilobatus* (Diptera: Chironomidae), two intertidal flies. M.S. thesis, Humbolt State College, Humbolt, USA.[8]

LeRoy Poff, N. and Matthews, R.A. (1986) Benthic macroinvertebrate community structural and functional group response to thermal enhancement in the Savannah River and a coastal plain tributary. *Archiv für Hydrobiologie,* **106,** 119–37.[15]

LeSage, L. and Harrison, A.D. (1980a) The biology of Cricotopus (Chironomidae: Orthocladiinae) in an algal enriched stream: 1. Normal biology. *Archiv für Hydrobiologie Supplementband,* **57,** 375–418.[9][10][12]

LeSage, L. and Harrison, A.D. (1980b) The biology of Cricotopus (Chironomidae: Orthocladiinae) in an algal enriched stream: 2. Effects of parasitism. *Archiv für Hydrobiologie, Supplementband,* **58,** 1–25.[9]

Leuchs, H. and Neumann, D. (1990) Tube texture, spinning and feeding behaviour of *Chironomus* larvae. *Zoologische Jahrbucher Abteilung für Systematik, Okologie und Geographie der Tiere,* **117,** 31–40.[7]

Levesque, A.J., Mayle, F.E., Walker, I.R. and Cwynar, L.C. (1993) A previously unrecognized late-glacial cold event in eastern North America. *Nature,* **361,** 623–6.[16]

Lewis, D.J. (1954) Nimitti and some other small annoying flies in the Sudan. *Sudan Notes and Records,* **35,** 76–89.[14]

Lewis, D.J. (1956) Chironomidae as a pest in the Northern Sudan. *Acta Tropica,* **13,** 142–58.[14]

Lewis, D.J. (1957) Observations on Chironomidae at Khartoum. *Bulletin of Entomological Research,* **48,** 155–84.[13]

Lien, L. (1978) The energy budget of the brown trout (*Salmo trutta*) population of Øvre Heimdalsvatn, Norway. *Holarctic Ecology,* **1,** 279–300.[17]

Lindeberg, B. (1958) A parthenogenetic race of *Monotanytarsus boreoalpinus* Th. (Dipt., Chironomidae) from Finland. *Suomen hyönteistietellinen Aikakauskirja,* **24,** 35–8.[5][8][9]

Lindeberg, B. (1964) The swarm of males as a unit for taxonomic recognition in the chironomids (Diptera). *Annales Zoologici Fennici,* **1,** 72–6.[9]

Lindeberg, B. (1971) Parthenogenetic strains and unbalanced sex ratios in Tanytarsini (Diptera, Chironomidae). *Annales Zoologici Fennici,* **8,** 310–7.[8][9]

Lindeberg, B. (1974) Parthenogenetic and normal populations of *Abiskomyia virgo* Edw. (Diptera, Chironomidae). *Entomologisk Tidskrift (Supplement),* **95,** 157–61.[9]

Lindegaard, C. (1989) A review of secondary production of zoobenthos in freshwater ecosystems with special reference to Chironomidae (Diptera). *Acta Biologica Debrecen, Oecologica Hungarica,* **3** 231–240.[11]

Lindegaard, C. (1992) Zoobenthos ecology of Thingvallavatn: vertical distribution, abundance, population dynamics and production. *Oikos,* **64,** 257–304.[6][10][11][15]

Lindegaard, C. and Brodersen, K.P. (in press) Distribution of Chironomidae (Diptera) in the river continuum, in *Chironomids: from Genes to Ecosystems* (ed. P.S. Cranston), CSIRO, Melbourne.[15]

Lindegaard, C. and Jónasson, P.M. (1979) Abundance, population dynamics and production of zoobenthos in Lake Myvatn, Iceland. *Oikos,* **32,** 202–7.[6][9][10][11]

Lindegaard, C. and Jónsson, E. (1987) Abundance, population dynamics and high production of Chironomidae (Diptera) in Hjarbaek Fjord, Denmark, during a period of eutrophication. *Entomologica Scandinavica Supplement,* **29,** 293–302.[10][11][13]

Lindegaard, C. and Mortensen, E. (1988) Abundance, life history and production of Chironomidae (Diptera) in a Danish lowland stream. *Archiv für Hydrobiologie Supplement,* **81,** 563–87.[10][11][15]

Lindegaard, C. and Mæhl, P. (1993) Abundance, population dynamics and production of Chironomidae (Diptera) in an ultraoligotrophic lake in South Greenland. *Netherlands Journal of Aquatic Ecology,* **26,** 297–308.[10][11][15]

Lindegaard-Petersen, C. (1971) An ecological investigation of the Chironomidae (Diptera)) from a Danish lowland stream (Linding Å). *Archiv für Hydrobiologie,* 69, 465–507.[6]

Lindeman, R. L. (1941) Seasonal food-cycle dynamics in a senescent lake. *American Midland Naturalist,* **26,** 636–73.[18]

Linevich, A.A. (1963) K biologii komarov semeistva Tendipedidae 'Biologiya bespozvonochnykh Baikala'. *Trudy Limnologicheskogo Instituta,* **1,** 3–48.[1]

Livingstone, D.A. (1957) On the sigmoid growth phase in the history of Linsley Pond. *American Journal of Science,* **255,** 364–75.[16]

Lloyd, D.G. and Wells M.S. (1992) Reproductive biology of a primitive angiosperm, *Pseudowintera colorata* (Winteraceae) and the evolution of pollination systems in the Anthophyta. *Plant Systematics and Evolution,* **181,** 77–95.[9]

Loden, M.S. (1974) Predation by chironomid (Diptera) larvae on oligochaetes. *Limnology and Oceanography,* **19,** 156–9.[7]

Lodge, D.M. (1991) Herbivory on freshwater macrophytes. *Aquatic Botany,* **41,** 195–224.[7]

Lohner, T.W. and Fisher, S.W. (1990) Effects of pH and temperature on the acute toxicity and uptake of carbaryl in the midge, *Chironomus riparius. Aquatic Toxicology,* **16,** 335–54.[15]

Luferov, V.P. (1958) Investigations on the biology of predatory larvae of Tendipedidae (Diptera) in the Rybinsk reservoir. Ph.D. Dissertation. Moscow.[In Russian].[7]

Luferov, J. (1971) The role of light in the populating of water bodies by epibiotic chironomid larvae. *Limnologica,* **8,** 139–40.[5]

Lugthart, G.J., Wallace, J.B. and Huryn, A.D. (1990) Secondary production of chironomid communities in insecticide-treated and untreated headwater streams. *Freshwater Biology,* **24,** 417–27.[15]

MacArthur, R.H. (1957) On the relative abundance of bird species. *Proceedings of the National Academy of Sciences, USA* **43,** 293–5.[12]

MacArthur, R.H. and Wilson, E.O. (1967) *The Theory of Island Biogeography,* Princeton University Press, Princeton.[4]

Macchiusi, F. and Baker, R.L. (1991) Prey behaviour and size selective predation by fish. *Freshwater Biology,* **25,** 533–8.[7][12]

MacDonald, W.W. (1956) Observations on the biology of chaoborids and chironomids in Lake Victoria and on the feeding habits of the 'elephant-snout fish' (*Mormyrus kannume* Forsk.) *Journal of Animal Ecology,* **25,** 36–53.[5][9][17]

Mackey, A.P. (1976a) Quantitative studies on the Chironomidae (Diptera) of the Rivers Thames and Kennet. 1. The *Acorus* zone. *Archiv für Hydrobiologie,* **78,** 240–76.[6]

Mackey, A.P. (1976b) Quantitative studies on the Chironomidae (Diptera) of the Rivers Thames and Kennet. II. The Thames flint zone. *Archiv für Hydrobiologie,* **78,** 310–8.[6]

Mackey, A.P. (1977a) Growth and development of larval Chironomidae. *Oikos,* **28,** 270–75.[6][10][11][17]

Mackey, A.P. (1977b) Quantitative studies on the Chironomidae (Diptera) of the Rivers Thames and Kennet. III. The *Nuphar* zone. *Archiv für Hydrobiologie,* **79,** 62–102.[7][10][11]

Mackey, A.P. (1979) Trophic dependencies of some larval Chironomidae (Diptera) and fish species in the River Thames. *Hydrobiologia,* **62,** 241–7.[7][12][17]

Macquart, J. (1838) Diptéres exotique nouveaux ou peu connus. *Mémoires du Société (Royale) des Sciences, de l'Agriculture et des Artes à Lille,* **1838 (2),** 9–225.[3]

Macrae, I.V., Winchester, N.N. and Ring, R.A. (1989) An evaluation of *Cricotopus myriophylli* as a potential biocontrol for Eurasian Watermilfoil (*Myriophyllum spicatum*). *Acta Biologica Debrecina Oecologica Hungarica* **3,** 241–8.[7]

Macrae, I.V., Winchester, N.N. and Ring, R.A. (1990) Feeding activity and host preference of the milfoil midge, *Cricotopus myriophylli* Oliver (Diptera: Chironomidae). *Journal of Aquatic Plant Management,* **28,** 89–92.[7]

Madder, M.C.A., Rosenberg, D.M. and Wiens, A.P. (1977) Larval cocoons in *Eukiefferiella claripennis* (Diptera; Chironomidae). *Canadian Entomologist,* **109,** 891–2.[6]

Magy, H.I. (1968) Vector and nuisance problem emanating from man-made recreational lakes. *Proceedings and Papers of the California Mosquito Control Association,* **36,** 36–7.[13]

Maier, K.J. and Knight, A.W. (1991) The toxicity of waterborne boron to *Daphnia magna* and *Chironomus decorus* and the effects of water hardness and sulfate on boron toxicity. *Archives of Environmental Contamination and Toxicology,* **20,** 282–7.[15]

Maitland, P.S. and Hudspith, P.M.G. (1974) The zoobenthos of Loch Leven, Kinross, and estimates of its production in the sandy littoral area during 1970 and 1971. *Proceedings of the Royal Society of Edinburgh, Series B,* **74,** 219–39.[11]

Maitland, P.S., Charles, W.N., Morgan, N.C. *et al.* (1972) Preliminary research on the production of Chironomidae in Loch Leven, Scotland, in *Productivity Problems of Freshwaters* (eds J. Kajak and Hillbricht-Ilkowska, A.), PWN Polish Scientific Publishers, Warszawa-Krakow, Poland, pp.795–812.[10]

Majori, G., Ali, A., Donelli, G. *et al.* (1986) The occurrence of a virus of the pox group in a field population of *Chironomus salinarius* Kieffer in Italy. *Florida Entomologist,* **69,** 418–21.[13]

Makarchenko, E.A. (1989) A review of the Diamesinae (Diptera, Chironomidae) from the USSR, with notes on systematics of *Pseudodiamesa* G. and *Pagastia* Ol. *Acta Biologica Debrecen, Oecologica Hungarica,* **2,** 265–74.[4]

Malloch, J.R. (1917) A preliminary classification of Diptera, exclusive of Pupipara, based upon larval and pupal characters, with keys to the imagines in certain families *Bulletin of the Illinois State Laboratory of Natural History (1918),* **12,** 161–409.[3]

Mancini, E.R. (1979) A phoretic relationship between a chironomid larva and an operculate stream snail. *Entomological News,* **90,** 33–6.[12]

Mann, K.H. (1957) The breeding, growth, and age structure of a population of the leech, *Helobdella stagnalis* (L.). *Journal of Animal Ecology*, **26,** 171–7.[13]

Mann, R.H.K. and Blackburn, J.H. (1991) The biology of the eel *Anguilla anguilla* (L.) in an English chalk stream and interactions with juvenile trout *Salmo trutta* L. and salmon *Salmo salar* L. *Hydrobiologia*, **218,** 65–76.[17]

Marcer, G., Saia, B., Zanetti, C. *et al.* (1990) Aspetti sanitari dell'infestazione da Chironomidi, in *Chironomidi, Culicidi, Simulidi – Aspetti Sanitari et Ecologici* (eds F. D'Andrea and G. Marchese), Regione Veneto, ULSS 16, S.I.P., Venice. pp.89–99.[14]

Marchant, R. (1982) Seasonal variation in the macroinvertebrate fauna of billabongs along Magela Creek, Northern Territory. *Australian Journal of Marine and Freshwater Research*, **33,** 329–42.[6]

Marchant, R., Metzeling, L., Graesser, A. and Suter, P. (1984) The organisation of macroinvertebrate communities in the major tributaries of the La Trobe River, Victoria, Australia. *Freshwater Biology*, **15,** 315–31.[6]

Margolina, G.L. (1971) Contribution to the problem on feeding *Tendipes plumosus* in the Rybinsk water reservoir. *Fisheries Research Board of Canada Translation Series* No. 1798.[7]

Marker, A.F.H. (1976) The benthic algae of some streams in southern England. I. Biomass of the epilithon in some small streams. *Journal of Ecology*, **64,** 343–58.[7]

Marker, A.F.H., Clarke, R.T. and Rother, J.A. (1986) Changes in epilithic population of diatoms, grazed by chironomid larvae, in an artificial recirculating stream. *Proceedings of the 9th Diatom Symposium, 1986,* 143–9.[7]

Martin, J. (1963) The cytology and larval morphology of the Victorian representatives of the subgenus *Kiefferulus* of the genus *Chironomus*. *Australian Journal of Zoology*, **11,** 301–22.[7]

Martin, J. (1979) Chromosomes as tools in taxonomy and phylogeny of Chironomidae (Diptera). *Entomologica Scandinavica, Supplement*, **10,** 67–74.[3]

Martin, J. and Lee, B.T.O. (1989) Indirect evidence for multiple insemination in *Chironomus oppositus* Walker (Diptera: Chironomidae). *Journal of the Australian Entomological Society*, **28,** 77–80.[9]

Martin, J. and Porter, D.L. (1977) Laboratory biology of the rice midge, *Chironomus tepperi* Skuse (Diptera: Nematocera): mating, behaviour, productivity and attempts at hydridization. *Journal of the Australian Entomological Society*, **16,** 411–6.[9]

Martin. J., Kuvangkadilok, C., Peart, D.H. and Lee, B.T.O. (1980). Multiple sex determining regions in a group of related *Chironomus* species (Diptera: Chironomidae). *Heredity*, **44,** 367–82.[17]

Martyniak, A. (1990) Feeding of carp *Cyprinus carpio* in the artificially aerated Lake Mutek, Czechoslovakia. *Folia Zoologica*, **39,** 279–84.[17]

Mason, C.F. (1977) Populations and production of benthic animals in two contrasting shallow lakes in Norfolk. *Journal of Animal Ecology*, **46,** 147–72.[11]

Mason, C.F. and Bryant, R.J. (1975) Periphyton production and grazing by chironomids in Alderfen Broad, Norfolk. *Freshwater Biology*, **5,** 271–7.[7]

Matěna, J. (1982) Benthos development in planktonic and nursery ponds. *Buletin Vyzkumny Ustav Rybarsky a Hydrobiologicky Vodnany*, **18,** 10–5.[17]

Matěna, J. (1990) Succession of *Chironomus* Meigen species (Diptera, Chironomidae) in newly filled ponds. *Internationale Revue der Gesamten Hydrobiologie*, **75,** 45–57.[9]

Matěna, J. and Soldán, T. (1982) Taxonomy and ecology of the subgenus Symbiocladius s. str. (Diptera, Chironomidae). *Facultatis Scientiarum Naturalium Universitatis Purkynianae Brunensis*, **23,** 83–6.[12]

Matěna, J. and Soldán, T. (1986) New findings of larvae of the genus *Epoicocladius* (Diptera, Chironomidae). *Dipterologica Bohemoslovaca,* **4,** 39–41.[12]
Mathews, K.P. (1989) Inhalant Insect-derived Allergens, in *Airborne Allergens,* Immunology and Allergy Clinics of North America. 9. pp.321–38.[14]
Matsuoka, H., Ishii, A. and Noono, S. (1988) Detection of IgE antibodies to larvae and adults of chironomids by enzyme-linked immunosorbent assay. *Allergy,* **43,** 425–9.[14]
Matsuoka, H., Ishii, A., Kimura, J.Y. and Noono, S. (1990) Developmental change of chironomid allergen during metamorphosis. *Allergy,* **45,** 115–20.[14]
Mattingly, R.L., Cummins, K.W. and King, R.H. (1981) The influence of substrate organic content on the growth of a stream chironomid. *Hydrobiologia,* **77,** 161–5.[7][10]
May, R.M. (1975) Patterns of species abundance and diversity, in: *Ecology and Evolution of Communities* (ed. M.L. Cody and J.M. Diamond), Belknap/Harvard University Press, Cambridge, Mass., pp.81–120.[12]
Maynard-Smith, J. (1971) What use is sex. *Journal of Theoretical Biology,* **30,** 319–35.[9]
Mazur, G., Becker, W.-M. and Baur, X. (1987) Epitope mapping of major insect allergens (chironomid hemoglobins) with monoclonal antibodies. *Journal of Allergy and Clinical Immunology,* **80,** 876–83.[14]
Mazur, G., Baur, X., Modrow, S. and Becker, W.-M. (1988) A common epitope on major allergens from non-biting midges (Chironomidae). *Molecular Immunology,* **25,** 1005–10.[14]
McAlpine, J.F. (1981) Morphology and terminology – adults, in *Manual of Nearctic Diptera,* (ed. J.F. McAlpine), Agriculture Canada Monograph 27, pp.9–63.[2]
McAuliffe, J.R. (1984) Competition for space, disturbance, and the structure of a benthic stream community. *Ecology,* **65,** 894–908.[12]
McGill, J.D., Wilson, R.S. and Brake, A.M. (1979) The use of chironomid pupal exuviae in the surveillance of sewage pollution within a drainage system. *Water Research,* **13,** 887–94.[15]
McGowan, L.M. (1975) The occurrence and behaviour of adult *Chaoborus* and *Procladius* (Diptera: Nematocera) from Lake George, Uganda. *Zoological Journal of the Linnean Society,* **57,** 321–34.[9]
McHugh, S.M., Credland, P.F., Tee, R.D. and Cranston, P.S. (1988) Evidence of allergic hypersensitivity to chironomid midges in an English village community. *Clinical Allergy,* **18,** 275–85.[14]
McKnight, D. M. and Feder, G. L. (1984) The ecological effect of acid conditions and precipitation of hydrous metal oxides in a Rocky Mountain stream. *Hydrobiologia,* **119,** 129–38.[15]
McLachlan, A.J. (1969) Substrate preferences and invasion behaviour exhibited by larvae of *Nilodorum brevibucca* Freeman (Chironomidae) under experimental conditions. *Hydrobiologia,* **33**: 237–49.[6]
McLachlan, A.J. (1970) Submerged trees as a substrate for benthic fauna in the recently created Lake Kariba (Central Africa). *Journal of Applied Ecology,* **7,** 253–66.[5]
McLachlan, A.J. (1976a) Variation of 'gill' size in larvae of the African midge *Chironomus transvaalensis* Kieffer. *Journal of the Limnological Society of Southern Africa* **2,** 55–6.[2]
McLachlan, A.J. (1976b) Factors restricting the range of *Glyptotendipes paripes* Edwards (Diptera: Chironomidae) in a bog lake. *Journal of Animal Ecology,* **45,** 105–13.[6]
McLachlan, A.J. (1977a) Density and distribution in laboratory populations of midge larvae (Chironomidae: Diptera). *Hydrobiologia,* **55,** 195–9.[7]

McLachlan, A.J. (1977b) Some effects of tube shape on the feeding of *Chironomus plumosus* L. (Diptera: Chironomidae). *Journal of Animal Ecology,* **46,** 139–46.[7][15]

McLachlan, A.J. (1981a) Food sources and foraging tactics in tropical rain pools. *Zoological Journal of the Linnean Society,* **71,** 265–77.[7]

McLachlan, A.J. (1981b) Interaction between insect larvae and tadpoles in tropical rain pools. *Ecological Entomology,* **6,** 175–82.[6][7]

McLachlan, A.J. (1983a) Life-history strategies of rain-pool dwellers. *Journal of Animal Ecology,* **52,** 545–62.[9]

McLachlan, A.J. (1983b) Habitat distribution and body size in rain pool dwellers. *Zoological Journal of the Linnean Society,* **79(4),** 399–408.[9][10]

McLachlan, A.J. (1985a) The relationship between habitat predictability and wing length in midges (Chironomidae). *Oikos,* **44(3),** 391–7.[9]

McLachlan, A.J. (1985b) What determines the species present in a rain-pool? *Oikos,* **45,** 1–7.[6]

McLachlan, A.J. (1986a) Sexual dimorphism in midges strategies for flight in the rain-pool dweller *Chironomus imicola* (Diptera: Chironomidae). *Journal of Animal Ecology,* **55,** 261–8.[9]

McLachlan, A.J. (1986b) Survival of the smallest: advantages and costs of small size in flying animals. *Ecological Entomology,* **11** 237–40.[9]

McLachlan, A.J. (1988) Refugia and habitat partitioning among midges (Diptera: Chironomidae) in rain–pools. *Ecological Entomology,* **13,** 185–93.[12]

McLachlan, A.J. and Allen, D.F. (1987) Male mating success in Diptera: advantages of small size. *Oikos,* **48,** 11–14.[9]

McLachlan, A.J. and Cantrell, M.A. (1976) Sediment development and its influence on the distribution and tube structure of *Chironomus plumosus* L. (Chironomidae, Diptera) in a new impoundment. *Freshwater Biology,* **6,** 437–43.[6][7]

McLachlan, A.J. and Cantrell, M.A. (1980) Survival strategies in tropical rain pools. *Oecologia (Berlin),* **47,** 344–51.[10]

McLachlan, A.J., Brennan, A. and Wotton, R.S. (1978) Particle size and chironomid (Diptera) food in an upland river. *Oikos,* **31,** 247–52.[7]

McLachlan, A.J., Pearce, L.J. and Smith, J.A. (1979) Feeding interactions and cycling of peat in a bog lake. *Journal of Animal Ecology,* **48,** 851–61.[7]

McLachlan, A.J. and Neems, R. (in press) Swarm based mating systems, in *Insect Reproduction* (eds S.R. Leather and J. Hardy), CRC Press.

McLarney, W.O., Henderson, S. and Sherman, M.M. (1974) A new method for culturing *Chironomus tentans* Fabricius larvae using burlap substrate in fertilized pools. *Aquaculture,* **4,** 267–76.[17]

McLusky, D.S. and McFarlane, A. (1974) The energy requirements of certain larval chironomid populations in Loch Leven, Kinross. *Proceedings of the Royal Society of Edinburgh, Series B,* **74,** 259–64.[11]

McNeill, S. and Lawton, J.H. (1970) Annual production and respiration in animal populations. *Nature,* **225,** 472–4.[11]

Mees, F., Verschuren, D., Nijs, R. and Dumont, H. (1991) Holocene evolution of the crater lake at Malha, Northwest Sudan. *Journal of Paleolimnology,* **5,** 227–53.[16]

Meier, M. (1987) Lebenszyklus und Parasit–Wirt–Beziehung von *Parachironomus varus* (Diptera: Chironomidae) und *Radix ovata* (Pulmonata: Lymnaeidae) in einem Weiher in Süddeutschland. *Archiv für Hydrobiologie,* **109,** 367–76.[12]

Meigen, J.W. (1800) *Nouvelle classification des mouches à deux ailes (Diptera L.) d'aprés un plan tout nouveau.* Perronneau, Paris, pp.40.[3][18]

Meigen, J.W. (1803) Versuch einer neuen Gattungseinteilung der europäischen zweiflügeligen Insekten. *Magazin für Inseketenkunde (Illiger)* **2,** 259–81.[3][18]

Menzie, C.A. (1981) Production ecology of *Cricotopus sylvestris* (Fabricius) (Diptera: Chironomidae) in a shallow esturine cove. *Limnology and Oceanography*, **26,** 467–81.[10][11]

Meriläinen, J.J. (1987) The profundal zoobenthos used as an indicator of the biological condition of Lake Päijänne. *Biological Research Report University of Jyväskylä,* **10,** 87–94.[15]

Merritt, R.W. and Cummins, K.W. (eds) (1978) *An Introduction to the Aquatic Insects of North America* 1st Edition. Kendall/Hunt, Dubuque, Iowa.[12]

Merritt, R.W. and Cummins, K.W. (eds) (1984) *An Introduction to the Aquatic Insects of North America,* 2nd Edition. Kendall/Hunt, Dubuque, Iowa.[7]

Merritt, R.W., Dadd, R.H. and Walker, E.D. (1992) Feeding behavior, natural food, and nutritional relationships of larval mosquitoes. *Annual Review of Entomology,* **37,** 349–76.[7]

Merritt, R.W., Olds, E.J. and Walker, E.D. (1990) Natural food and feeding behavior of *Coquillettidia perturbans* larvae. *Journal of the American Mosquito Control Association,* **6,** 35–42.[7]

Metcalfe, J.L. (1989) Biological water quality assessment of running waters based on macroinvertebrate communities: history and present status in Europe. *Environmental Pollution,* **60,** 101–39.[15]

Metzeling, L., Graesser, A., Suter, P. and Marchant, R. (1984) The distribution of aquatic macroinvertebrates in the upper catchment of the La Trobe River, Victoria. *Occasional Papers from the Museum of Victoria,* **1,** 1–62.[6][15]

Mezger, E. G. (1967) Insecticidal and naturalistic control of chironomid larvae in Lake Dalwigk, Vallejo, California. *Proceeding and Papers of the California Mosquito Control Association,* **35,** 125–8.[13]

Miall, L.C. (1895) *The Natural History of Aquatic Insects.* MacMillan and Co., London. pp.395.[5]

Miall, L.C. and Hammond, A.R. (1900) *The Structure and Life-history of the Harlequin Fly (Chironomus).* Clarendon Press, Oxford.[2][5][8][9]

Michaelova, P.V. (1989) The polytene chromosomes and their significance to the systematics of the family Chironomidae. *Acta Zoologica Fennica,* **186,** 1–107.[3]

Miehlbradt, J. and Neumann, D. (1976) Reproduktive Isolation durch optische Schwarmmarken bei den sympatrischen *Chironomus thummi* und *Ch. piger. Behaviour,* **58,** 272–97.[9]

Millemann, R.E., Birge, W.J., Black, J.A. *et al.* (1984) Comparative acute toxicity to aquatic organisms of components of coal-derived synthetic fuels. *Transactions of the American Fisheries Society,* **113,** 74–85.[15]

Miller, R.B. (1941) A contribution to the ecology of the Chironomidae of Costello Lake, Algonquin Park, Ontario. *University of Toronto Studies,* **49,** 1–63.[9]

Miller, M.C. and Stout, J.R. (1986) Effects of a controlled under-ice oil spill on invertebrates of an arctic and a subarctic stream. *Environmental Pollution (Series A),* **42,** 99–132.[15]

Minshall, G.W. (1968) Community dynamics of the benthic fauna in a woodland springbrook. *Hydrobiologia,* **32,** 305–39.[6]

Minshall, G. W. and Petersen, R. C. (1985) Towards a theory of macroinvertebrate community structure in stream ecosystems. *Archiv für Hydrobiologie,* **104,** 49–76.[15]

Mizukami, Y., Watanabe, H., Igarashi, T. *et al.* (1986) [A case of bronchial asthma induced by direct inhalation of a chironomid midge, *Tanytarsus oyamai* Sasa, 1979]. *Japanese Journal of Thoracic Disease,* **24,** 287–91.[14]

Moksnes, A. (1987) Food biology of ringed plover *Charadrius hiaticula* and Temminck's stint *Calidris temminckii* in the regulation zone of a hydroelectric power reservoir. *Fauna Norvegica Series C,* **10,** 103–13.[17]

Monakov, A.V. (1972) Review of studies on feeding of aquatic invertebrates conducted at the Institute of Biology of Inland Waters, Academy of Science, USSR. *Journal of the Fisheries Research Board of Canada,* **29,** 363–83.[7]

Moore, J.W. (1978) Some factors influencing the diversity and species composition of benthic invertebrate communities in twenty arctic and subarctic lakes. *Internationale Revue der Gesamten Hydrobiologie,* **63,** 757–71.[15]

Moore, J.W. (1979a) Factors influencing algal consumption and feeding rate in *Heterotrissocladius changi* Saether and *Polypedilum nebeculosum* (Meigen)(Chironomidae: Diptera). *Oecologia* (Berlin), **40,** 219–27.[7]

Moore, J.W. (1979b) Some factors influencing the distribution, seasonal abundance and feeding of subarctic Chironomidae (Diptera). *Archiv für Hydrobiologie,* **85,** 302–25.[7][10]

Moore, J.W. (1980) Factors influencing the composition, structure and density of a population of benthic invertebrates. *Archiv für Hydrobiologie,* **88,** 202–18.[6]

Morduchai-Boltovskoy, F.D. and Shilova, A.I. (1955) On the planktonic life of *Glyptotendipes* larvae (Diptera, Tendipedidae) *Doklady Akademii nauk SSSR,* **105,** 163–5 (in Russian).[5]

Morgan, N.C. and Waddell, A.B. (1961) Diurnal variation in the emergence of some aquatic insects. *Transactions of the Royal Entomological Society, London,* **113,** 123–37.[9]

Morley, R.L. and Ring, R.A. (1972) The inter-tidal Chironomidae (Diptera) of British Columbia. II. Life history and population dynamics. *Canadian Entomologist,* **104,** 1099–121.[9]

Morrill, P.K. and Neal, B.R. (1990) Impact of deltamethrin insecticide on Chironomidae (Diptera) of prairie ponds. *Canadian Journal of Zoology,* **68,** 289–96.[15]

Morrison, A. (1987) Chironomid larvae trails in proglacial lake sediments. *Boreas,* **16,** 318–21.[16]

Mortensen, E. and Simonsen, J.L. (1983) Production estimates of the benthic invertebrate community in a small Danish stream. *Hydrobiologia,* **102,** 155–62.[11]

Moss, D., Furse, M. T., Wright, J. F. and Armitage, P. D. (1987) The prediction of the macro-invertebrate fauna of unpolluted running-water sites in Great Britain using environmental data. *Freshwater Biology,* **17,** 41–52.[15]

Motomura, I. (1932) On the statistical treatment of communities. *Zoological Magazine, Tokyo* **44,** 379–83 (In Japanese).

Moubayed, Z. (1987) Complément à l'inventaire des Chironomides (Diptera) du Liban. *Bulletin de la Société d'Histoire Naturelle de Toulouse,* **123,** 51–2.[4]

Moubayed, Z. (1988) Chironomidae (Diptera) de Thaïlande récoltés par l'expédition Thaï 87. *Expéditions de l'APS en Asie du Sud-Est, travaux scientifique,* **1,** (1988), 41–2.[4]

Moubayed, Z. and Laville, H. (1983) Les Chironomides (Diptera) du Liban I. Premier inventaire faunistique. *Annales de Limnologie,* **19,** 219–28.[4]

Moyle, J.B. (1961) Aquatic invertebrates as related to larger plants and waterfowl. *Minnesota Department of Conservation Investigative Report,* **233,** 1–24.[17]

Mozley, S.C. (1970) Morphology and ecology of the larva of *Trissocladius grandis* (Kieffer) (Diptera, Chironomidae), a common species in lakes and rivers of northern Europe. *Archiv für Hydrobiologie,* **67,** 433–51.[10]

Mozley, S.C. (1971) Maxillary and premental patterns in Chironominae and Orthocladiinae (Diptera: Chironomidae). *Canadian Entomologist,* **103,** 298–305.[2]

Mozley, S.C. (1979) Neglected characters in larval morphology as tools in taxonomy and phylogeny of Chironomidae (Diptera). *Entomologica Scandinavica Supplement,* **10,** 27–36.[2][3][5]

Müh, C. (1985) Kopfmorphologie der Larven der Tanypodinae (Chironomidae: Diptera) am Beispiel von *Macropelopia nebulosa* (Meigen). *Zoologische Jahrbücher, Anatomie,* **113,** 331–62.[2]

Muir, W.D. and Emmett, L.R. (1988) Food habits of migrating salmonid smolts passing Bonneville dam in the Columbia River, Washington, 1984. *Regulated Rivers: Research and Management,* **2(1),** 1–10.[17]

Mulla, M.S. (1974) Chironomids in the residential-recreational lakes. An emerging nuisance problems – measures for control. *Entomologisk Tidskrift (Supplement),* **95,** 172–6.[9]

Mulla, M. S. and Darwazeh, H. A. (1975) Evaluation of insect growth regulators against chironomids in experimental ponds. *Proceedings and Papers of the California Mosquito Control Association,* **43,** 164–8.[13]

Mulla, M.S. and Khasawinah, A.M. (1969) Laboratory and field evaluations of larvicides against chironomid midges. *Journal of Economic Entomology,* **62,** 37–41.[13]

Mulla, M.S., Barnard, D.R. and Norland, R.L. (1975) Chironomid midges and their control in Spring Valley Lake, California. *Mosquito News,* **35,** 389–95.[13]

Mulla, M.S., Kramer, W.L. and Barnard, D.R. (1976) Insect growth regulators for the control of chironomid midges in residential-recreational lakes. *Journal of Economic Entomology,* **69,** 285–91.[13]

Mulla, M.S., Navvab-Gojrati, H.A. and Darwazeh, H.A. (1978a) Biological activity and longevity of synthetic pyrethroids against mosquitoes and some nontarget insects. *Mosquito News,* **38,** 90–6.[13]

Mulla, M.S., Navvab-Gojrati, H.A. and Darwazeh, H.A. (1978b) Toxicity of mosquito larvicidal pyrethroids to four species of freshwater fishes. *Environmental Entomology,* **7,** 428–30.[13]

Mulla, M.S., Norland, R.L., Ikeshoji, T. and Kramer, W.L. (1974) Insect growth regulators for the control of aquatic midges. *Journal of Economic Entomology,* **67,** 165–70.[13]

Mulla, M.S., Norland, R.L., Fanara, D.M. *et al.* (1971) Control of chironomid midges in recreational lakes. *Journal of Economic Entomology,* **64,** 300–7.[13]

Mulla, M.S., Norland, R.L., Westlake, W.E. *et al.* (1973) Aquatic midge larvicides, their efficacy and residues in water, soil and fish in a warm water lake. *Environmental Entomology,* **2,** 58–65.[13]

Mundie, J.H. (1957) The ecology of Chironomidae in storage reservoirs. *Transactions of the Royal Entomological Society of London,* **109,** 149–232.[9]

Mundie, J.H. (1959) The diurnal activity of the larger invertebrates at the surface of Lac la Ronge, Saskatchewan. *Canadian Journal of Zoology,* **37,** 945–56.[9]

Mundie, J.H. and Mounce, D.E. (1978) Application of stream ecology to raising salmon smolts in high density. *Verhandlung der Internationale Vereiningung für Theoretische und Angewandte Limnologie,* **20,** 2013–8.[17]

Mundie, J.H., McKinnell, S.M. and Traber, R.E. (1983) Responses of stream zoobenthos to enrichment of gravel substrates with cereal grain and soy bean. *Canadian Journal of Fisheries and Aquatic Sciences,* **40,** 1702–12.[17]

Mundie, J.H., Mounce, D.E. and Simpson, K.S. (1990) Semi-natural rearing of coho salmon, *Oncorhynchus kisutch* (Walbaum), smolts, with an assessment of survival to the catch and escapement. *Aquaculture and Fisheries Management,* **21,** 327–45.[17]

Mundie, J.H., Simpson, K.S. and Perrin, C.J. (1991) Responses of stream periphyton and benthic insects to increases in dissolved organic phosphorus in a mesocosm. *Canadian Journal of Fisheries and Aquatic Sciences,* **48,** 2061–72.[17]

Munsterhjelm, G. (1920) Om Chironomidernas Äggläggning och Äggrupper. *Acta Societatis pro Fauna et Flora Fennica,* **47,** 1–174.[5]

Murakami, G., Igarashi, T., Adachi, Y.S. *et al.* (1988) Common occurrence in Toyama of bronchial asthma induced by chironomid midges, in *International Symposium on Mite and Midge Allergy,* (ed. T. Miyamoto), Ministry of Education, Science and Culture, Tokyo, pp.261–83.[14]

Murphy, P.M. and Edwards, R.W. (1982) The spatial distribution of the freshwater macroinvertebrate fauna of the River Ely, South Wales, in relation to pollutional discharges. *Environmental Pollution (Series A),* **29,** 111–24.[15]

Murray, D.A. and Ashe, P. (1981) A description of the adult female of *Buchonomyia thienemanni* FITTKAU, with notes on its ecology and on the phylogenetic position of the subfamily Buchonomyiinae (Diptera: Chironomidae). *Spixiana,* **4,** 55–68.[3][6]

Murray, D.A. and Ashe, P. (1985) A description of the adult female of *Buchonomyia thienemanni* Fittkau and a reassessment of the phylogenetic position of the subfamily Buchonomyiinae. *Spixiana Supplement* **11,** 149–60.[3]

Murray, D.W. (1976) *Tilapia* fish for midge control in a sewage treatment pond. *Proceedings and Papers of the California Mosquito Control Association,* **44,** 122.[13]

Muthukrishnan, J. and Palavesam, A. (1992) Secondary production and energy flow through *Kiefferulus barbitarsis* (Diptera: Chironomidae) in tropical ponds. *Archiv für Hydrobiologie,* **125,** 207–26.[7]

Myers, A.A. and Giller, P.S. (eds) (1988) *Analytical Biogeography. An integrated approach to the study of animal and plant distributions,* Chapman & Hall, London, New York.[4]

Myllylä, M., Torssonen, M., Pulliainen, E. and Kuusela, K. (1983) Biological studies on the minnow *Phoxinus phoxinus* in northern Finland. *Aquilo, Serie Zoologica,* **22(0),** 149–56.[17]

Naeem, S. (1988a) Predator-prey interactions and community structure: chironomids, mosquitoes and copepods in *Heliconia imbricata* (Musaceae). *Oecologia (Berlin),* **77,** 202–9.[7][12]

Naeem, S. (1988b). Resource heterogeneity fosters coexistence of a mite and a midge in pitcher plants. *Ecological Monographs,* **58,** 215–27.[12]

Nagano, T., Ohta, N., Okano, M. *et al.* (1992) Analysis of antigenic determinants shared by two different allergens recognized by human T cells: House dust mite (*Dermatophagoiudes pteronyssinus*) and chironomid midge (*Chironomus yoshimatsui*). *Allergy,* **47,** 554–9.[14]

Nagell, B. and Landahl, C.C. (1978) Resistance to anoxia of *Chironomus plumosus* and *Chironomus anthracinus* (Diptera) larvae. *Holarctic Ecology,* **1,** 333–6.[6]

Nagell, B. and Orrhage L. (1981) On the structure and function of the ventral tubuli of some Chironomus larvae. *Hydrobiologica* **78,** 11–6.[2]

Naumann, E. (1929) Einige neue Gesichtspunkte zur Systematik der Gewässertypen. Mit besonderer Berücksichtigung der Seetypen. *Archiv für Hydrobiologie,* **20,** 191–8.[15]

Nebeker, A.V., Onjukka, S.T. and Cairns, M.A. (1988) Chronic effects of contaminated sediment on *Daphnia magna* and *Chironomus tentans. Bulletin of Environmental Contamination and Toxicology,* **41,** 574–81.[15]

Neems, R., McLachlan, A.J. and Chambers, R. (1990) Body size and lifetime mating success of male midges (Diptera: Chironomidae). *Animal Behaviour,* **40,** 648–52.[9]

Neems, R., Lazarus, J. and McLachlan, A.J. (1992) Swarming behavior in male chironomid midges: a cost-benefit analysis. *Behavioural Ecology,* **3,** 285–290.[9]

Nelson, G. (1978) From Candolle to Croizat: Comments on the History of Biogeography. *Journal of the History of Biology,* **11,** 269–305.[4]

Nelson, G. and Platnick, N.I. (1978) The perils of plesiomorphy: widespread taxa, dispersal, and phenetic biogeography. *Systematic Zoology,* **27,** 474–7.[4]

Nelson, G. and Platnick, N.I. (1984) *Biogeography*, Carolina Biological Supply Co., Burlington, North Carolina, 16pp.[4]

Neubert, J. (1986) Zur akuten Toxizität von Trichlorfon gegenüber ausgewählten Wasserorganismen. *Acta Hydrochimie Hydrobiologie*, **14**, 643–51.[15]

Neumann, D. (1961) Osmotische Resistenz und Osmoregulation aquatischer Chironomidenlarven. *Biologisches Zentralblatt*, **6**, 693–715.[6]

Neumann, D. (1966) Die lunare und tägliche Schlüpfperiodik der Mücke *Clunio*. Steuerung und Abstimmung auf die Schlüpfperiodik. *Zeitschrift für vergleichende Physiologie*, **53**, 1–61.[9]

Neumann, D. (1976) Adaptations of chironomids to intertidal environments. *Annual Review of Entomology*, **21**, 387–414.[6][9]

Neumann, D. and Honegger, H.W. (1969) Adaptions of the intertidal midge *Clunio* to arctic conditions. *Oecologia*, **3**, 1–13.[9]

Neumann, D. and Kruger, M. (1985) Combined effects of photoperiod and temperature on the diapause of an intertidal chironomid *Clunio marinus*. *Oecologia*, **67**, 154–6.[9][10]

Newman, L.J. (1977) Chromosomal evolution of the Hawaiian *Telmatogeton* (Chironomidae: Diptera). *Chromosoma*, **64**, 349–69.[3]

Newman, R.M. (1991) Herbivory and detritivory on freshwater macrophytes by invertebrates: a review. *Journal of the North American Benthological Society*, **10**, 89–114.[7]

Neves, R.J. (1979) Secondary production of epilithic fauna in a woodland stream. *American Midland Naturalist*, **102**, 209–24.[11]

Nickle, W.R. (1972) A contribution to our knowledge of the Mermithidae (Nematoda). *Journal of Nematology*, **4**, 113–46.[12]

Nielsen, E.T. and Greve, H. (1950) Studies on the swarming habits of mosquitoes and other Nematocera. *Bulletin of Entomological Research*, **41**, 227–58.[9]

Nielsen, E.T. and Haeger, J.S. (1960) Swarming and mating in mosquitoes. *Miscellaneous Publications of the Entomological Society of America*, **1**, 71–95.[9]

Nielsen, H.T, and Nielsen, E.T. (1962) Swarming of mosquitoes. Laboratory experiments under controlled conditions. *Entomologia Experimentalis et Applicata (Amsterdam)*, **5**, 14–32.[9]

Nilsson, C. (1984) Filtrera – ett sätt att äta hos vatteninsekter. *Fauna och Flora*, **79**, 227–38.[7]

Noda, H. and Miyazaki, M. (1986) Injury to rice leaves by chironomid larvae (Diptera: Chironomidae). *Japanese Journal of Applied Entomology and Zoology*, **30**, 66–8.[7]

Noda, H., Miyazaki, M. and Hashimoto, H. (1986) Injury to rice leaves by chironomid larvae (Diptera: Chironomidae). *Japanese Journal of Applied Entomology and Zoology*, **30**, 66–8.[7]

Noll, W. (1952) Es regnete Zuckmückeneier. *Nachrichten naturwissenschaftlichen Museums der Stadt Aschaffenburg*, **34**, 71–4.[9]

Nolte, U. (1988) Small water colonization in pulse stable varzea and constant terra firma biotopes on the Neotropics. *Archiv für Hydrobiologie*, **113(4)**, 541–50.[9]

Nolte, U. (1990) Chironomid biomass determination from larval shape. *Freshwater Biology*, **24**, 443–51.[11]

Nolte, U. (1993) Egg masses of Chironomidae (Diptera). A review, including new observations and a preliminary key. *Entomologica Scandinavica Supplement*, **43**, 1–75.[2][5][9]

Nolte, U. (in press) From egg to imago in less than seven days: *Apedilum elachistum* (Chironomidae), in *Chironomids: from Genes to Ecosystems* (ed. P.S. Cranston), CSIRO, Melbourne.[5]

Nolte, U. and Hoffman, T. (1992) Fast life in cold water: *Diamesa incallida* (Chironomidae). *Ecography* **15,** 25–30.[2][5]

Nolte, U. and Hoffmann, T. (1993) Life cycle of *Pseudodiamesa branickii* (Chironomidae) in a small upland stream. *Netherlands Journal of Aquatic Ecology,* **26,** 309–314.[10]

Nordlie, K.J. and Arthur, J.W. (1981) Effect of elevated water temperature on insect emergence in outdoor experimental channels. *Environmental Pollution (Series A),* **25,** 53–65.[9][15]

Økland, J. and Økland K.A. (1986) The effects of acid deposition on benthic animals in lakes and streams. *Experientia,* **42,** 471–86.[15]

Odum, H. (1957) Trophic structure and productivity of Silver Springs, Florida. *Ecological Monographs,* **27**, 55–112.[18]

Ogawa, K. (1992) Field trapping of male midges *Rheotanytarsus kyotoensis* (Diptera: Chironomidae) by sounds. *Japanese Journal of Sanitary Zoology,* **43,** 77–80.[9]

Ohkura, T. and Tabaru, Y. (1975) Control of chironomids breeding in eel culture ponds. *Suisan Zoshoku,* **23**: 1–7.[13]

Ohno, M. and Okamoto, H. (1980) Test on the susceptibility of the last instar larvae of *Chironomus yoshimatsui* Martin and Sublette (Diptera: Chironomidae) collected at the Kanda River to two organophosphorus insecticides. *Tokyo Metropolitan Public Health Research Laboratory Annual Report,* **31,** 261–4.[13]

Ohno, M. and Shimizu, K. (1982) On changes in numbers of larvae and imagines of *Chironomus yoshimatsui* Martin and Sublette (Diptera: Chironomidae), and drift of the larvae after applications of temephos in the Kanda River. *Tokyo Metropolitan Public Health Research Laboratory Annual Report,* **33,** 314–21.[13]

Oka, H. and Hashimoto, H. (1959) Lunare Periodizität in der Fortpflanzung einer spezifischen Art von *Clunio*. *Biologische Zentralblatt,* **78,** 545–59.[9]

Olafsson, J.S. (1992) A comparative study on mouthpart morphology of certain larvae of Chironomini (Diptera: Chironomidae) with reference to the larval feeding habits. *Journal of Zoology, London,* **228,** 183–204.[2][3]

Olander, R. and Palmén, E. (1968) Taxonomy, ecology and behaviour of the northern Baltic *Clunio marinus* Halid. (Dipt., Chironomidae). *Annales Zoologici Fennici,* **5,** 97–110.[9]

Olatunde, A.A. and Moneke, C.C. (1985) The food habits of four mormyrid species in Zaria, Nigeria. *Archiv für Hydrobiologie,* **102,** 503–17.[17]

Oliver, D.R. (1959) Some Diamesini (Chironomidae) from the Nearctic and Palaearctic. *Entomologisk Tidskrift,* **80,** 48–64.

Oliver, D.R. (1968) Adaptations of arctic Chironomidae. *Annales Zoologici Fennici,* **5,** 111–8.[5][9][10]

Oliver, D.R. (1971) Life history of the Chironomidae. *Annual Review of Entomology,* **16,** 211–30.[6][7][9][10]

Oliver, D.R. (1979) Contribution of life history information to taxonomy of aquatic insects. *Journal of the Fisheries Research Board of Canada,* **36,** 318–21.[1]

Oliver, D.R. (1981) Chapter 29: Chironomidae, in *Manual of Nearctic Diptera,* Vol. 1 (eds J. F. McAlpine *et al.*), Research Branch, Agriculture Canada, **27,** 1–674.[1][18]

Oliver, D.R. (1984) Description of a new species of *Cricotopus* Van der Wulp (Diptera: Chironomidae) associated with *Myriophyllum spicatum. Canadian Entomologist,* **116,** 1287–92.[7]

Oliver, D.R. and Corbet, P.S. (1966) Aquatic habitats in a high arctic locality: the Hazen camp study area, Ellesmere Island, N.W.T. Operation Hazen No. 26, Defense Research Board of Canada.[10]

Oliver, D.R. and Sinclair, B.J. (1989) Madicolous Chironomidae (Diptera), with a review of *Metriocnemus hygropetricus* Kieffer. *Acta Biologica Debrecina Oecologica Hungarica,* **2,** 285–93.[6]

Oliver, D.R., Dillon, M.E. and Cranston, P.S. (1990) *A Catalog of Nearctic Chironomidae.* Research Branch, Agriculture Canada, Ottawa, pp.89.[3][18]

Olsson, T.I. (1981) Overwintering of benthic macroinvertebrates in ice and frozen sediment in a North Swedish river. *Holarctic Ecology,* **4,** 161–6.[10]

Ormerod, S.J., Boole, P., McCahon, C.P. *et al.* (1987) Short-term experimental acidification of a Welsh stream: comparing the biological effects of hydrogen ions and aluminium. *Freshwater Biology,* **17,** 341–56.[15]

Ostrovskii, I.S. (1982) Growth and production of *Chironomus plumosus* (Chironomidae, Diptera) larvae in Lake Sevan. *Soviet Journal of Ecology,* **14,** 54–60.[11]

Ostrofsky, M.L. and Zettler, E.R. (1986) Chemical defences in aquatic plants. *Journal of Ecology,* **74,** 279–87.[7]

Otsuka, K. Kon, M. and Hidaka, T. (1986) The mating system of *Tokunagayusurika akamusi* (Diptera: Chironomidae): II. Experimental analysis of male mating behaviour at the resting place. *Journal of Ethology,* **4,** 147–52.[9]

Otto, C. (1983) Adaptations to benthic freshwater herbivory, in *Periphyton of Freshwater Ecosystems,* (ed. R.G. Wetzel), Dr. W. Junk, The Hague, pp.199–205.[7]

Outridge, P.M. (1988) Seasonal and spatial variations in benthic macroinvertebrate communities of Magela Creek, Northern Territory. *Australian Journal of Marine and Freshwater Research,* **39,** 211–23.

Owen, H.G. (1981) Constant dimensions or an expanding Earth ?, in *The Evolving Earth* (ed. L.R.M. Cocks), British Museum (Natural History), London, pp.179–192.[4]

Owen, M. and Thomas, G.J. (1979) The feeding ecology and conservation of wigeon wintering at the Ouse Washes, England, UK. *Journal of Applied Ecology,* **16,** 795–810.[17]

Paasivirta, L. (1972) Taxonomy, ecology and swarming behaviour of *Tanytarsus gracilentus* Holm. (Diptera, Chironomidae) in Valasaaret, Gulf of Bothnia, Finland. *Annales Zoologici Fennici,* **9,** 255–64.[9]

Pagast, F. (1947) Systematik und Verbreitung der um die Gattung Diamesa gruppierten Chironomiden. *Archiv für Hydrobiologie,* **61,** 435–596.[3]

Palavesam, A. and Muthukrishnan, J. (1992) Influence of food quality and temperature on fecundity of *Kiefferulus barbitarsis* (Kieffer) (Diptera: Chironomidae). *Aquatic Insects,* **14,** 145–52.[7]

Palawski, D.U., Hunn, J.B., Chester, D.N. and Wiedmeyer, R.H. (1989) Interactive effects of acidity and aluminium exposure on the life cycle of the midge *Chironomus riparius* (Diptera). *Journal of Freshwater Ecology,* **5,** 155–62.[15]

Palmén, E. (1955) Diel periodicity of pupal emergence in natural populations of some chironomids (Diptera). *Suomalaisen eläin-ja kasvitieteellisen seuran vanamon kasvitieteellisia julkaisuja,* **17,** 1–30.[9]

Palmén, E. (1962) Studies on the ecology and phenology of the chiromonids (Dipt.) of the Northern Baltic. 1. *Allochironomus crassiforceps* K. *Suomen hyönteistietellinen Aikakauskirja,* **28,** 137–68.[9]

Palmén, E. and Lindeberg, B. (1959) The marine midge *Clunio marinus* Hal. (Dipt., Chironomidae) found in brackish water in the Northern Baltic. *Internationale Revue der Gesamten Hydrobiologie,* **44,** 383–94.[6][9]

Palomäki, R. (1987) The Chironomidae of some lakes and rivers in Nicaragua. *Entomologica Scandinavica, Supplement* **29,** 45–9.[13]

Pandian, T.J. (1987) Sustainable clean water and aquaculture. *Ergebnisse der Limnologie,* **28,** 333–42.[17]

Paris, O.H. and Jenner, C.E. (1959) Photoperiodic control of diapause in the pitcher-plant midge, *Metriocnemus knabi* in *Photoperiodism and Related Phenomena in Plants and Animals* (ed. R.B. Withrow). AAAS, Washington, pp.601–24.[10]

Parker, C.R. and Voshell, J.R. (1979) *Cardiocladius* (Diptera: Chironomidae) larvae ectoparasitic on pupae of Hydropsychidae. *Environmental Entomology,* **8,** 357–61.[7][12]

Parkin, R.B. and Stahl, J.B. (1981) Chironomidae (Diptera) of Baldwin Lake, Illinois, a cooling reservoir. *Hydrobiologia,* **76,** 119–28.[15]

Parlato, S.J. (1929) A case of coryza and asthma due to sand flies (caddis flies). *Journal of Allergy,* **1,** 35–42.[14]

Parlato, S.J. (1930) The sand fly (caddis fly) as exciting cause of allergic coryza and asthma. II. Its relative frequency. *Journal of Allergy,* **1,** 307–12.[14]

Parlato, S.J. (1938) The mayfly as exciting cause of seasonal allergic coryza and asthma. *Journal of Allergy,* **10,** 56.[14]

Parma, S. and Krebs, B.P.M. (1977) The distribution of chironomid larvae in relation to chloride concentration in a brackish water region of the Netherlands. *Hydrobiologia,* **52,** 117–26.[6]

Pascoe, D., Williams, K.A. and Green, D.W.J. (1989) Chronic toxicity of cadmium to *Chironomus riparius* Meigen – effects upon larval development and adult emergence. *Hydrobiologia,* **175,** 109–15.[15]

Paterson, C.G. and Cameron, C.J. (1982) Seasonal dynamics and ecological strategies of the pitcher plant chironomid, *Metriocnemus knabi* Coq. (Diptera: Chironomidae), in southeast New Brunswick. *Canadian Journal of Zoology,* **60,** 3075–83.[7][9][10]

Paterson, C.G. and Walker, K.F. (1974a) Seasonal dynamics and productivity of *Tanytarsus barbitarsis* Freeman (Diptera: Chironomidae) in the benthos of a shallow, saline lake. *Australian Journal of Marine and Freshwater Research,* **25,** 151–65.[11]

Paterson, C.G. and Walker, K.F. (1974b) Recent history of *Tanytarsus barbitarsis* Freeman (Diptera: Chironomidae) in the sediments of a shallow, saline lake. *Australian Journal of Marine and Freshwater Research,* **25,** 315–25.[10][16]

Patterson, R.S. (1964) Recent investigations on the use of BHC and EPN to control chironomid midges in central Florida. *Mosquito News* **24,** 294–9.[13]

Patterson, R.S. (1965) Control of chironomid midges in Florida. *Florida Anti-Mosquito Association Annual Report,* **36,** 35–9.[13]

Patterson, R.S. and Wilson, F.L. (1966) Fogging and granule applications are teamed to control chironomid midges on Florida lake fronts. *Pest Control,* **34,** 26–31.[13]

Patterson, R.S., von Windeguth, D.L., Glancy, B.M. and Wilson, F.L. (1966) Control of the midge *Glyptotendipes paripes* with low-volume aerial sprays of malathion. *Journal of Economic Entomology,* **59,** 864–6.[13]

Payne, S.L. and Pearson, W.D. (1981) Feeding preferences of postlarval longnose gar (*Lepidosteus osseus*) of the Ohio River. *Transactions of the Kyoto Academy of Science,* **42,** 119–131.[17]

Peckarsky, B.L. (1984) Predator–prey interactions among aquatic insects, in *The Ecology of Aquatic Insects* (eds V.H. Resh and D.M. Rosenberg), Praeger, New York., pp.196–254.[12]

Peckarsky, B.L. (1985) Do predaceous stoneflies and siltation affect the structure of stream insect communities colonizing enclosures? *Canadian Journal of Zoology,* **63,** 1519–30.[12]

Peckarsky, B.L., Hom, S.C. and Statzner, B. (1990) Stonefly predation along a hydraulic gradient a field test of the harsh-benign hypothesis. *Freshwater Biology,* **24(1),** 181–92.[17]

Peckham, V. (1971) Notes on the chironomid midge *Belgica antarctica* Jacobs at Anvers Island in the maritime Antarctic. *Pacific Insects Monograph,* **25,** 145–66.[9]

Pelsue, F.W. and McFarland, G.C. (1971) Laboratory and field studies on a new chironomid species in the Southeast Mosquito Abatement District. *Proceedings and Papers of the California Mosquito Control Association,* **39,** 74–9.[13]

Pelsue, F.W., McFarland, G.C. and Beesley, C. (1974) Field evaluation of two insect growth regulators against chironomid midges in water spreading basins. *Proceedings and Papers of the California Mosquito Control Association,* 42, 157–63.[13]

Pennak, R.W. (1978) *Fresh-Water Invertebrates of the United States,* 2nd ed. Wiley, New York.[6]

Pereira, C.R.D., Anderson, N.H. and Dudley, T. (1982) Gut content analysis of aquatic insects from wood substrates. *Melanderia,* **39,** 23–33.[7]

Perrin, C.J. (1980) Structure and functioning of French Upper Rhone ecosystems: 14. Alimentary preferences of the common loach (*Noemacheilus barbatulus*) by a modified points method. *Hydrobiologia,* **71,** 217–24.[17]

Perrin, C.J., Bothwell, M.L. and Slaney, P.A. (1987) Experimental enrichment of a coastal stream in British Columbia: effects of organic and inorganic additions on autotrophic periphyton production. *Canadian Journal of Fisheries and Aquatic Sciences,* **44,** 1247–56.[17]

Petr, T. (1968) Distribution, abundance and food of commercial fish in the Black Volta and the Volta man-made lake in Ghana during its first period of filling (1964–1966). I. Mormyridae. *Hydrobiologia,* **32,** 417–48.[17]

Petr, T. (1974) Distribution, abundance and food of commercial fish in the Black Volta and the Volta man-made lake in Ghana during the filling period (1964–1968). II. Characidae. *Hydrobiologia,* **45,** 303–37.[17]

Petridis, D. (1990) The influence of grass carp on habitat structure and its subsequent effect on the diet of tench. *Journal of Fish Biology,* **36(4),** 533–44.[17]

Pettigrove, V. (1989) Larval mouthpart deformities in *Procladius paludicola* Skuse (Diptera: Chironomidae) from the Murray and Darling Rivers, Australia. *Hydrobiologia,* **179,** 111–7.[15]

Pflüger, W. (1973) Die Sanduhrsteuerung der gezeitensynchronen Schlüpfrhythmik der Mücke *Clunio marinus* im arktischen Mittsommer. *Oecologia,* **11,** 113–50.[9]

Pflüger, W. and Neumann, D. (1971) Die Steuerung einer gezeitenparallelen Schlüpfrhythmik nach dem Sanduhrprinzip. *Oecologia,* **7,** 262–6.[9]

Phillipp, P. (1938) XX. Studien über den jahres- und tagezeitlichen Insektenflug über Teichen. *Zeitschrift für Fischerei und deren Hilfswissenschaften,* **35,** 731–75.[9]

Phillips, V.E. (1991) Pochard *Aythya ferina* use of chironomid-rich flowing feeding habitat in winter. *Bird Study,* **38,** 118–22.[17]

Pienitz, R., Walker, I.R., Zeeb, B.A, Smol, J.P. and Leavitt, P.R. (1992) Biomonitoring past salinity changes in an athalassic subarctic lake. *International Journal of Salt Lake Research,* **1,** 91–123.[16]

Pinder, A.M., Trayler, K.M. and Davis, J.A. (1991) Chironomid control in Perth wetlands. Final report and recommendations. *School of Biological and Environmental Sciences, Murdoch University,* Australia.[13]

Pinder, A.M., Trayler, K.M., Mercer, J.W. *et al.* (1993) Diel periodicities of adult emergence of some chironomids (Diptera: Chironomidae) and a mayfly (Ephemeroptera: Caenidae) at a Western Australian wetland. *Journal of the Australian Entomological Society,* **32,** 129–135.[9]

Pinder, L.C.V. (1974) The Chironomidae of a small chalk-stream in southern England. *Entomologisk Tidskrift (Supplementum),* **95,** 195–202.[9][10][12][15]

Pinder, L.C.V. (1977) The Chironomidae and their ecology in chalk streams. *Report of the Freshwater Biological Association* **45,** 62–9.[7]

Pinder, L.C.V. (1978) A key to adult males of the British Chironomidae (Diptera). *Scientific Publications of the Freshwater Biological Association,* **37**, 1–169.[18]

Pinder, L.C.V. (1980) Spatial distribution of Chironomidae in an English chalk stream, in *Chironomidae: Ecology, Systematics, Cytology and Physiology,* (ed. D.A. Murray), Pergamon Press, Oxford, pp.153–61.[6][7]

Pinder, L.C.V. (1983) Observations on the life-cycles of some Chironomidae in southern England. *Memoirs of the American Entomological Society,* **34,** 249–65.[10]

Pinder, L.C.V. (1986) Biology of freshwater Chironomidae. *Annual Review of Entomology,* **31,** 1–23.[6][7]

Pinder, L.C.V. (1992) Biology of epiphytic Chironomidae (Diptera: Nematocera) in chalk streams. *Hydrobiologia,* **248,** 39–51.[5][6][7]

Pinder, L.C.V. and Farr, I.S. (1987) Biological surveillance of water quality – 3. The influence of organic matter enrichment on the macroinvertebrate fauna of small chalk streams. *Archiv für Hydrobiologie,* **109,** 619–37.[15]

Plante, C. and Downing, J.A. (1990) Empirical evidence for differences among methods for calculating secondary production. *Journal of the North American Benthological Society,* **9,** 9–16.[11]

Poinar, G.O. (1993) Insects in amber. *Annual Reviews of Entomology,* **46,** 145–59.[4]

Polls, I., Greenberg, B. and Lue-Hing, C. (1975) Control of nuisance midges in a channel receiving treated municipal sewage. *Mosquito News,* **35,** 533–7.[13]

Pontasch, K.W. and Cairns, J. Jr., (1991) Multispecies toxicity tests using indigenous organisms: predicting the effects of complex effluents in streams. *Archives of Environmental Contamination and Toxicology,* **20,** 103–12.[15]

Potter, D.W.B. and Learner, M.A. (1974) A study of the benthic macroinvertebrates of a shallow eutrophic reservoir in South Wales with emphasis on the Chironomidae (Diptera); their life-histories and production. *Archiv für Hydrobiologie,* **74,** 186–226.[9][10][11]

Power, M.E. (1991) Shifts in the effects of tuftweaving midges on filamentous algae. *American Midland Naturalist,* **125,** 275–85.[7]

Power, M.E., Stout, R.J., Cushing, C.E. *et al.* (1988) Biotic and abiotic controls in river and stream communities. *Journal of the North American Benthological Society,* **7,** 456–79.[15]

Powlesland, C. and George, J. (1986) Acute and chronic toxicity of nickel to larvae of *Chironomus riparius* (Meigen). *Environmental Pollution (Series A),* **42,** 47–64.[15]

Prat, N. and Daroca, M.V. (1983) Eutrophication processes in Spanish reservoirs as revealed by biological records in profundal sediments. *Hydrobiologia,* **103,** 153–8.[16]

Prelicz, H., Baur, X., Dewair, M. *et al.* (1986) Persistence of hemoglobin allergenicity and antigenicity during metamorphosis of Chironomidae (Insecta: Diptera). *International Archives of Allergy and Applied Immunology,* **79,** 72–6.[14]

Pringle, C.M. (1985) Effects of chironomid (Insecta: Diptera) tube-building activities on stream diatom communities. *Journal of Phycology,* **21,** 185–94.[7]

Pritchard, G. (1964) The prey of dragonfly larvae (Odonata: Anisoptera) in ponds in northern Alberta. *Canadian Journal of Zoology,* **42,** 785–800.[13]

Pritchard, G. (1983) Biology of Tipulidae. *Annual Reviews of Entomology,* **28,** 1–22.[10]

Provost, M.W. and Branch, N. (1959) Food of chironomid larvae in Polk County lakes. *Florida Entomologist,* **42,** 49–62.[7][13]

Przybylski, M. and Banbura, J. (1989) Feeding relations between the gudgeon *Gobio gobio* L. and stone loach *Noemacheilus barbatulus* L. *Acta Hydrobiologica, Kraków,* **31(1–2),** 109–20.[17]

Raddum, G.G. (1985) Effects of winter warm reservoir release on benthic stream invertebrates. *Hydrobiologia,* **122,** 105–11.[15]

REFERENCES

Raddum, G.G. and Sæther, O.A. (1981) Chironomid communities in Norwegian lakes with different degrees of acidification. *Verhandlungen der Internationalen Vereinigung für Theoretische und Angewandte Limnologie*, **21,** 399–405.[15]

Raddum, G.G., Hagenlund, G. and Halvorson, G.A. (1984) Effects of lime treatment on the benthos of Lake Sondre Boksjo. *Institute of Freshwater Research Drottningholm Report*, **61,** 167–76.[15]

Raddum, G.G., Brettum, P., Matzow, D. *et al.* (1986) Liming the acid Lake Hovvatn, Norway: A whole-ecosystem study. *Water, Air, and Soil Pollution*, **31,** 721–63.[15]

Raddum, G.G., Fjellheim, A. and Hesthagen, T. (1988) Monitoring of acidification by the use of aquatic organisms. *Verhandlungen der Internationalen Vereinigung für Theoretische und Angewandte Limnologie*, **23,** 2291–7.[15]

Rae, J.G. (1985) A multivariate study of resource partitioning in soft bottom lotic Chironomidae. *Hydrobiologia*, **126,** 275–85.[12]

Rae, J.G. (1989) Chironomid midges as indicators of organic pollution in the Sciotio River Basin, Ohio. *The Ohio Journal of Science*, **89,** 5–9.[15]

Rainey, R.C. (1976) Insect Flight. *Symposium of the Royal Entomological Society*, **7,** Blackwell, Oxford.[9]

Ran, E.T.H. (1990) Dynamics of vegetation and environment during the middle pleniglacial in the Dinkel Valley (The Netherlands). *Mededelingen van's Rijks Geologische Dienst*, **44–3,** 141–205.[16]

Ranta, E. and Espo, J. (1989) Predation by the rock-pool insects *Arctocorisa carinata*, *Callicorixa producta* (Heteroptera: Corixidae) and *Potamonectes griseostriatus* (Coleoptera: Dytiscidae). *Annales Zoologici Fennici*, **26(1),** 53–60.[17]

Räsänen, M, Salonen, V.-P., Salo, J. *et al.* (1992) Recent history of sedimentation and biotic communities in Lake Pyhäjärvi, SW Finland. *Journal of Paleolimnology*, **7,** 107–26.[16]

Rask, M. (1989) A note on the diet of roach *Rutilus rutilus* L. and other cyprinids at Tvärminne, northern Baltic Sea. *Aqua Fennica*, **19 (1),** 19–28.[17]

Rasmussen. J.B. (1982) The effect of thermal effluent, before and after macrophyte harvesting, on standing crop and species composition of benthic macroinvertebrate communities in Lake Wabamun, Alberta. *Canadian Journal of Zoology*, **60,** 3196–205.[15]

Rasmussen, J.B. (1984a) Comparison of gut contents and assimilation efficiency of fourth instar larvae of two coexisting chironomids, *Chironomus riparius* Meigen and *Glyptotendipes paripes* (Edwards). *Canadian Journal of Zoology*, **62,** 1022–6.[7][11]

Rasmussen, J.B. (1984b) The life-history, distribution, and production of *Chironomus riparius* and *Glyptotendipes paripes* in a prairie pond. *Hydrobiologia*, **119,** 65–72.[7][10][11]

Rasmussen, J.B. (1985) Effects of density and microdetritus enrichment on the growth of chironomid larvae in a small pond. *Canadian Journal of Fisheries and Aquatic Sciences*, **42,** 1418–22.[7][10]

Rasmussen, K. (1990) Some positive and negative effects of stocking whitefish on the ecosystem redevelopment of Hjarbaek Fjord, Denmark. *Hydrobiologia*, **200/201,** 593–602.[13][17]

Rasmussen, K. and Lindegaard, C. (1988) Effects of iron compounds on macroinvertebrate communities in a Danish lowland river system. *Water Research*, **22,** 1101–8.[15]

Rathburn, C.B. (1988) Insecticides labeled for the control of adult and larval mosquitoes by ground and aerial application methods. *Journal of the Florida Anti-Mosquito Association*, **59,** 27–36.[13]

Rawson, D.S. and Moore, J.E. (1944) The saline lakes of Saskatchewan. *Canadian Journal of Research, D,* **22,** 141–201.[6]

Reice, S.R. (1985) Experimental disturbance and the maintenance of species diversity in a stream community. *Oecologia,* **67,** 90–7.[12]

Reice, S.R. and Edwards, R.L. (1986) The effect of vertebrate predation on lotic macroinvertebrate communities in Quebec, Canada. *Canadian Journal of Zoology,* **64,** 1930–6.[12]

Reimer, G. (1985) Contributions on the feeding habits of arctic char *Salvelinus alpinus* in Austria. *Archiv für Hydrobiologie,* **105,** 229–38.[17]

Reisen, W.K., Milby, M.M., Meyer, R.P. and Reeves, W.C. (1983) Population ecology of *Culex tarsalis* (Diptera: Culicidae) in a foothill environment in Kern County, California: temporal changes in male relative abundance and swarming behavior. *Annals of the Entomological Society of America,* **76,** 809–15.[9]

Reiss, F. (1966) Zum Kopulationmechanismus bei Chironomiden (Diptera). Chironomidenstudien IV. *Zoologischer Anzieger,* **176,** 440–9.[9]

Reiss, F. (1968) Ökologische und systematische Untersuchungen an Chironomiden (Diptera) des Bodensees. Ein Beitrag zur lakustrischen Chironomidenfauna des nördlichen Alpenvorlandes. *Archiv für Hydrobiologie,* **64,** 176–323.[9][10]

Reiss, F. (1971) Zum Kopulations-Mechanismus bei Chironomiden (Diptera) II. *Limnologica,* **8,** 35–42.[9]

Reiss, F. (1977) Qualitative and quantitative investigations on the macrobenthic fauna of Central Amazon lakes. *Amazoniana,* **6,** 203–35.[6]

Reiss, F. (1985a) A contribution to the zoogeography of the Turkish Chironomidae (Diptera). *Israel Journal of Entomology,* **19,** 161–70.[4]

Reiss, F. (1985b) Die panamerikanisch verbreitete Tanytarsini-Gattung *Skutzia* gen. nov. (Diptera, Chironomidae). *Spixiana Supplement,* **11,** 173–8.[18]

Reiss, F. (1986) Ein Beitrag zur Chironomidenfauna Syriens (Diptera: Chironomidae). *Entomofauna,* **7,** 153–66.[4]

Reiss, F. (1989) Erster Beitrag zur Chironomidenfauna Portugals. *Nachrichtenblatt der Bayerischen Entomologen,* **38,** 46–50.[4]

Reiss, F. and Kohmann, F. (1982) Die Chironomidenfauna (Diptera, Insecta) des unteren Inn. *Mitteilungen der Zoologischen Gesellschaft Braunau,* **4,** 77–88.[6]

Reiss, F. and Sublette, J.E. (1985) *Beardius* new genus with notes on additional Pan-American taxa (Diptera, Chironomidae). *Spixiana Supplement,* **11**, 179–93.[18]

Remmert, H. (1955) Untersuchungen über das tageszeitlich gebundene Schlüpfen von *Pseudosmittia arenaria. Zeitschrift für vergleichende Physiologie,* **37,** 338–54.[9]

Remmert, H. (1960) Uber die Eiablage von *Trichocladius vitripennis* (Meigen). *Kieler Meeresforschungen,* **16,** 236–7.[9]

Remmert, H. (1962) *Der Schlüpfrhythmus der Insekten.* Franz Steiner Verlag, Wiesbaden.[9]

Remmert, H. (1965) Über den Tagesrhythmus Arktischer Tiere. *Zeitschrift für Morphologie und Ökologie der Tiere,* **55,** 142–60.[9]

Rempel, J.G. (1936) The life-history and morphology of *Chironomus hyperboreus. Journal of the Biological Board of Canada,* **2,** 209–21.[5]

Rempel, R.S. and Harrison, A.D. (1987) Structural and functional composition of the community of Chironomidae (Diptera) in a Canadian Shield stream. *Canadian Journal of Zoology,* **65(10),** 2545–54.[9]

Resta, O., Forschino-Barbaro, M.P., Carnimeo, N. *et al.* (1982) Occupational asthma from fish-feed (Echinodorus plamosus larva). *La medicina del lavoro,* **3,** 234–6.[14]

Reymond, H. and Lagardere, J.P. (1990) Feeding rhythms and food of *Penaeus japonicus* Bate (Crustacea: Penaeidae) in salt marsh ponds: role of halophilic entomofauna. *Aquaculture,* **84(2),** 125–44.[17]

Reynoldson, T.B. and Sefton, A.D. (1976) The food of *Planaria torva* (Müller) (Turbellaria – Tricladida), a laboratory and field study. *Freshwater Biology,* **6,** 229–40.[17]

Richardson, J.S. (1991) Seasonal food limitation of detritivores in a montane stream: an experimental test. *Ecology,* **72,** 873–87.[11]

Rieradevall, M. and Prat, N. (1991) Benthic fauna of Banyoles Lake (NE Spain). *Verhandlungen der Internationalen Vereinigung für Theoretische und Angewandte Limnologie,* **24,** 1020–3.[15]

Ring, R. (1989) Intertidal Chironomidae of B.C., Canada. *Acta Biologica Debrecina, Oecologica Hungarica,* **3, 1989** 275–88.[2]

Ringe, F. (1974) Chironomiden-Emergenz, 1970 in Breitenbach und Rohrwiesenbach. Schlitzer Produktionsbiologische Studien (10). *Archiv für Hydrobiologie, Supplementband,* **45,** 212–304.[9][18]

Ripley, M.P. (1980) The relation of dry weight and temperature to respiration in some benthic chironomid species in Lough Neagh, in *Chironomidae: Ecology, Systematics, Cytology and Physiology* (ed. D.A. Murray), Pergamon, New York, pp.81–8.[11]

Risbec, J. (1951) Les Diptéres nuisibles au riz de Camargue au début de son développment. *Revue de Pathologie Vegetale et d'Entomologie Agricole de France,* **30,** 211–7.[13]

Roback, S.S. (1965) A new subgenus and species of *Symbiocladius* from South America (Diptera, Tendipedidae). *Entomological News,* **76,** 113–22.[12]

Roback, S.S. (1969a) Notes on the food of Tanypodinae larvae. *Entomological News,* **80,** 13–8.[7][12]

Roback, S.S. (1969b) The immature stages of the genus *Tanypus* Meigen (Diptera: Chironomidae). *Transactions of the American Entomological Society,* **94,** 407–28.[7]

Roback, S.S. (1976) Note on the feeding habits of *Clinotanypus pinguis* (Loew) (Diptera: Chironomidae: Tanypodinae). *Entomological News,* **87,** 243–4.[7]

Roback, S.S. (1977a) The immature chironomids of the eastern United States II. Tanypodinae – Tanypodini. *Proceedings of the Academy of Natural Sciences of Philadelphia,* **128,** 55–87.[7]

Roback, S.S. (1977b) First record of a chironomid larva living phoretically on an aquatic hemipteran (Naucoridae). *Entomological News,* **88,** 192.[12]

Roback, S.S. (1979) Numerical taxonomic methodology and its application to chironomid classification. *Entomologica Scandinavica Supplement,* **10,** 75–9.[3]

Roback, S.S. (1989) The larval development of *Djalmabatista pulcher* (Joh.) (Diptera: Chironomidae: Tanypodinae). *Proceedings of the Academy of Natural Sciences of Philadelphia,* **141,** 73–84.[3]

Roback, S.S. and Moss, W.W. (1978) Numerical taxonomic studies on the congruence of classifications for the genera and subgenera of Macropelopiini and Anatopyniini (Diptera: Chironomidae: Tanypodinae). *Proceedings of the Academy of Natural Sciences of Philadelphia,* **129,** 125–50.[3]

Roback, S.S. and Coffman, W.P. (1983) The results of the Catherwood Bolivian-Peruvian Altiplano Expedition. Part II. Aquatic Diptera including montane Diamesinae and Orthocladiinae (Chironomidae) from Venezuela. *Proceedings of the Academy of Natural Sciences, Philadelphia,* **135,** 9–79.[3]

Roback, S.S., Bereza, D.J. and Vidrine, M.F. (1979) Description of an *Ablabesmyia* (Diptera: Chironomidae: Tanypodinae) symbiont of unionid fresh-water mussels (Mollusca: Bivalvia: Unionacea), with notes on its biology and zoogeography. *Transactions of the American Entomological Society,* **105,** 577–620.[12]

Robles, C. (1984) Coincidence of agonistic larval behaviour, uniform dispersion, and unusual pupal morphology in a genus of marine midges (Diptera: Chironomidae). *Journal of Natural History,* **18,** 897–904.[8]

Rodcharoen. J., Mulla, M.S. and Chaney, J.D. (1991) Microbial larvicides for the control of nuisance aquatic midges breeding in mesocosms and man-made lakes in California. *Journal of the American Mosquito Control Association,* **7,** 56–62.[13]

Rodina, A.G. (1971) The role of bacteria in the feeding of the tendipedid larvae. *Fisheries Research Board of Canada Translation Series* No. 1848.[7]

Rodova, R.A. (1978) Opredelitelj samok komarov-zvonzov Tribyj Chironomini (Diptera: Chironomidae). *Institut Akad Nauk, SSSR,* 1–145.[2]

Rodrigues, C.S. and Kaushik, N.K. (1986) Laboratory evaluation of the insect growth regulator diflubenzuron against black fly (Diptera: Simuliidae) larvae and its effects on nontarget stream invertebrates. *The Canadian Entomologist,* **118,** 549–58.[15]

Römer, F. and Rosin, S. (1969) Untersuchungen über die Bedeutung der Flugtöne beim Schwärmen von *Chironomus plumosus* L. *Revue Suisse de Zoologie,* **76,** 734–40.[9]

Römer, F. and Rosin, S. (1971) Einfluss von Licht und Temperatur auf die Schwärmzeit von *Chironomus plumosus* L. im Jahresverlauf. *Revue Suisse de Zoologie,* **78,** 851–67.[9]

Rosenberg, D.M. (1993) Freshwater biomonitoring and Chironomidae. *Netherlands Journal of Aquatic Ecology,* **26,** 101–122.[15]

Rosenberg, D.M. and Resh, V.H. (eds) (1992) *Freshwater Biomonitoring and Benthic Macroinvertebrates,* Chapman & Hall, New York.[15]

Rosenberg, D.M. and Wiens, A.P. (1980) Responses of Chironomidae (Diptera) to short-term experimental sediment additions in the Harris River, Northwest Territories, Canada. *Acta Universitatis Carolinae-Biologica,* **1978,** 181–92.[15]

Ross, P.E. and Kalff, J. (1975) Phytoplankton production in Lake Memphremagog, Quebec (Canada) – Vermont (USA). *Verhandlungen der Internationalen Vereinigung für Theoretische und Angewandte Limnologie,* **19,** 760–9.[11]

Rossaro, B. (1979) Confronto tra classifizione filetica e fenetica nel genere *Cricotopus* van der Wulp. *Bolletino della Societé Entomologica Italiana,* **111,** 76–82.[3]

Rossaro, B. (1981) Utilizazione della tassonomia numerica per una revisione delle Diamesinae (Diptera: Chironomidae). *Bolletino della Zoologiae Supplementa,* **48,** 97.[3]

Rossaro, B. (1987) Chironomid emergence in the Po River (Italy) near a nuclear power plant. *Entomologica Scandinavica Supplement,* **29,** 331–8.[15]

Rossaro, B. (1988) A contribution to the knowledge of chironomids in Italy. *Spixiana Supplement,* **14,** 191–200.[4]

Rossaro, B. (1989) A numerical taxonomy study of Orthocladiinae (Diptera: Chironomidae). *Acta Biologica Debrecen, Oecologica Hungarica,* **2,** 315–324.[3]

Rossaro, B. (1991) Chironomids of stony bottom streams: a detrended correspondence analysis. *Archiv für Hydrobiologie,* **122,** 79–93.[15]

Rossaro, B. (1993) Ordination methods and chironomid species in stony bottom streams. *Netherlands Journal of Aquatic Ecology,* **26,** 447–56.[12][15]

Ruse, L.P. and Wilson, R.S. (1984) The monitoring of river water quality within the Great Ouse Basin using the chironomid exuvial analysis technique. *Water Pollution Control,* **83,** 116–35.[15]

Ruzickova, J. (1987) Cohort analysis and production estimate of Chironomus larvae (Diptera, Chironomidae) in a carp pond in southwest Bohemia. *Věstnik Ceské Společnosti Zoologické,* **51,** 140–51.[11]

Rzóska, J. (1964) Mass outbreaks of insects in the Sudanese Nile Basin. *Verhandlungen der Internationalen Vereinigung für theoretische und angewandte Limnologie,* **15,** 194–200.[9]

Sadler, W.O. (1935) The biology of the midge *Chironomus tentans* Fabricius, and methods for its propagation. *Memoirs of Cornell University Agricultural Experimental Station,* **173,** 1–25.[5][17]

Samietz, R. (1989) Der Einfluss des Sammelrhythmus auf die Ausbeute der Chironomiden-Emergenz (Diptera). *Acta Biologica Debrecen, Oecologica Hungarica,* **2,** 325–33.[18]

Sand-Jensen, K. and Madsen, T.V. (1989) Invertebrates graze submerged rooted macrophytes in lowland streams. *Oikos,* **55,** 420–3.[7]

Sandberg, G. (1969) A quantitative study of chironomid distribution and emergence in Lake Erken. *Archiv für Hydrobiologie, Supplementband,* **35,** 119–201.[9][12]

Sasa, M. (1979) A morphological study of adults and immature stages of 20 Japanese species of the family Chironomidae (Diptera). *National Institute of Environmental Studies Research Report,* **7,** 1–148.[13]

Sasa, M. (1985) [Chironomid midges as the cause of bronchial asthma] *Kankyo Eisei,* **32,** 8–14 [In Japanese].[13][14]

Sasa, M. (1989) Chironomidae of Japan: checklist of species recorded, key to males and taxonomic notes. *Research Report from the National Institute for Environmental Studies, Japan,* **125,** 1–177.[3][4]

Sasa, M. (1990) Studies on the chironomid midges (Diptera: Chironomidae) of the Nansei Islands, southern Japan. *Japanese Journal of Experimental Medicine,* **60,** 111–65.[4]

Sasa, M. and Hasegawa, H. (1983) Chironomid midges of the tribe Chironomini collected from sewage ditches, eutrophicated ponds, and some clean streams in the Ryuku Islands, southern Japan. *Japanese Journal of Sanitary Zoology,* **34,** 305–41.[4]

Sæther, O.A. (1962) Larval overwintering cocoons in *Endochironomus tendens* Fabricius *Hydrobiologia,* **20,** 377–81.[6][10]

Sæther, O.A. (1968) Chironomids of the Finse Area, Norway, with special reference to their distribution in a glacier brook. *Archiv für Hydrobiologie,* **64,** 426–83.[6]

Sæther, O.A. (1971) Notes on general morphology and terminology of the Chironomidae. *Canadian Entomologist,* **103,** 1237–60.[2]

Sæther, O.A. (1974) Morphology and terminology of female genitalia in Chironomidae (Diptera). *Entomologisk Tidskrift Supplement,* **95,** 216–24.[2]

Sæther, O.A. (1975) Nearctic chironomids as indicators of lake typology. *Verhandlungen der Internationalen Vereinigung für Theoretische und Angewandte Limnologie,* **19,** 3127–33.[12][15]

Sæther, O.A. (1976) Revision of *Hydrobaenus, Trissocladius, Zalutschia, Paratrissocladius,* and some related genera (Diptera: Chironomidae). *Bulletin of the Fisheries Research Board of Canada,* **195,** 1–287.[3]

Sæther, O.A. (1977a) Female genitalia in Chironomidae and other Nematocera: morphology, phylogenies, keys. *Bulletin of the Fisheries Research Board of Canada,* **197,** 1–204.[2][3][18]

Sæther, O.A. (1977b) Taxonomic studies on Chironomidae: *Nanocladius, Pseudochironomus* and the *Harnischia* complex. *Bulletin of the Fisheries Research Board of Canada,* **196,** 1–143.[3]

Sæther, O.A. (1979a) Underlying synapomorphies and anagenetic analysis. *Zoologica Scripta,* **8,** 305–12.[3][18]

Sæther, O.A. (1979b) Hierarchy of the Chironomidae with special emphasis on the female genitalia (Diptera). *Entomologica Scandinavica Supplement,* **10,** 17–26.[3]

Sæther, O. A. (1979c) Chironomid communities as water quality indicators. *Holarctic Ecology,* **2,** 65–74.[15]

Sæther, O.A. (1980a) Glossary of chironomid morphology terminology (Diptera, Chironomidae). *Entomologica Scandinavica Supplement,* **14,** 1–51.[2][18]

Sæther, O.A. (1980b) The influence of eutrophication on deep lake benthic invertebrate communities. *Progressive Water Technology,* **12,** 161–80.[15]

Sæther, O.A. (1981) Orthocladiinae (Chironomidae: Diptera) from British West Indes with descriptions of *Antillocladius* n. gen., *Lipurometriocnemus* n.gen., *Compterosmittia* n.gen. and *Diplosmittia* n.gen. *Entomologica Scandinavica Supplement,* **16,** 1–46.[4]

Sæther, O.A. (1983) The canalized evolutionary potential: inconsistencies in phylogenetic reasoning. *Systematic Zoology,* **32,** 343–59.[3][18]

Sæther, O.A. (1986) The myth of objectivity – post Hennigian deviations. *Cladistics,* **2,** 1–13.[3]

Sæther, O.A. (1989) Phylogenetic trends and their evaluation in chironomids with special reference to orthoclads. *Acta Biologica Debrecen, Oecologica Hungarica,* **2,** 53–75.[2][3]

Sæther, O.A. (1990a) Midges and the electronic Ouija board. The phylogeny of the *Hydrobaenus* group (Chironomidae: Diptera) revised. *Zeitschrift für Systematische Evolutions-forschung* **28,** 107–36.[3]

Sæther, O.A. (1990b) A review of the genus *Limnophyes* Eaton from the Holarctic and Afrotropical regions. *Entomologica Scandinavica Supplement* **35,** 1–135.[3]

Sæther, O.A. and Willassen, E. (1987) Four new species of *Diamesa* Meigen, 1835 (Diptera: Chironomidae) from the glaciers of Nepal. *Entomologica Scandinavica Supplement,* **29,** 189–203.[1][6]

Sato, H. and Yasuno, M. (1979) Tests on Chironomidae larval susceptibility to various insecticides. *Japanese Journal of Sanitary Zoology,* **30,** 361–6.[13]

Saunders, L.G. (1928) Some marine insects of the Pacific coast of Canada. *Annals of the Entomological Society of America,* **21,** 521–45.[9]

Säwedal, L. (1982) Distribution of leg sensilla chaetica in male Chironomidae (Diptera) and its phylogenetic significance. *Entomologica Scandinavica,* **13,** 1–22.[2]

Säwedal, L. and Hall, R. (1979) Flight tone as a taxonomic character in Chironomidae (Diptera). *Entomologica Scandinavica Supplement,* **10,** 139–43.[9]

Schakau, B. (1986) Preliminary study of the development of the subfossil chironomid fauna (Diptera) of Lake Taylor, South Island, New Zealand, during the younger Holocene. *Hydrobiologia,* **143,** 287–91.[16]

Schakau, B. (1991) Stratigraphy of the fossil Chironomidae (Diptera) from Lake Grasmere, South Island, New Zealand, during the last 6000 years. *Hydrobiologia,* **214,** 213–21.[16]

Schakau, B. (1993) *Palaeolimnological studies on sediments from Lake Grasmere, South Island, New Zealand, with special reference to the Chironomidae, Diptera.* Ph. D. thesis, University of Canterbury, Christchurch, N.Z.[16]

Schakau, B. and Frank, C. (1984) Die Entwicklung der Chironomiden-Fauna (Diptera) des Tegeler Sees im Spät- und Postglazial. *Verhandlungen der Gesellschaft für Ökologie,* **12,** 375–82.[16]

Schin, K.S., Poluhowich, J.J., Gamo, T. and Laufer, H. (1974) Degradation of hemoglobin in *Chironomus* during metamorphosis. *Journal of Insect Physiology,* **20,** 561–71.[14]

Schindler, D.W., Mills, K.H., Malley, D.F. *et al.* (1985) Long-term ecosystem stress: The effects of years of experimental acidification on a small lake. *Science,* **228,** 1395–1401.[15]

Schlee, D. (1968) Vergleichende Merkmalsanalyse zur Morphologie und Phylogenie der *Corynoneura*-Gruppe (Diptera: Chironomidae) Zugleich eine Allgemeine Morphologie der Chironomiden-Imago. *Stuttgarter Beiträge zur Naturkunde,* **180,** 1–150.[2]

Schlee, D. (1970) Insektenführender Bernstein aus der Unterkreide des Libanon.

Neues Jahrbuch für Geologie und Paläontologie, Stuttgart. Monatshefte, **1970,** 40–50.[4]

Schlee, D. (1975) Das Problem der Podonominae-Monophylie; Fossiliendiagnose und Chironomidae-Phylogenetik (Diptera). *Entomologica Germanica,* **1,** 316–51.[3][4][18]

Schlee, D. (1977) Florale und extraflorale Nektarien sowie Insektenkot als Nahrungsquelle für Chironomiden-Imagines (und andere Diptera). *Stuttgarter Beitrage zur Naturkunde aus dem Staatlichen Museum für Naturkunde in Stuttgart, Series A. (Biologie),* **300,** 1–16.[9]

Schlosser, I.J. and Ebel, K.K. (1989) Effects of flow regime and cyprinid predation in a headwater stream. *Ecological Monographs,* **59(1),** 41–57.[17]

Schmid, P.E. (1992a) Population dynamics and resource utilization by larval Chironomidae (Diptera) in a backwater area of the River Danube. *Freshwater Biology,* **28,** 111–27.[10][12]

Schmid, P.E. (1992b) Community structure of larval Chironomidae (Diptera) in a backwater area of the River Danube. *Freshwater Biology,* **27,** 151–67.[12]

Schmid, P.E. (1993a) Habitat preferences as patch selection of larval and emerging chironomids (Diptera) in a gravel brook. *Netherlands Journal of Aquatic Ecology,* **26,** 419–29.[10]

Schmid, P.E. (1993b) Random patch dynamics of larval Chironomidae (Diptera) in the bed sediments of a gravel stream. *Freshwater Biology,* **30,** 239–55.[12]

Schmidt, E.R. (1989) Molecular Biology of *Chironomus*: state-of-the-art. *Acta Biologica Debrecen, Oecologica Hungarica,* **2,** 151–63.[1]

Schnell, Ø. (1988) Twenty eight Chironomidae (Diptera) new to Norway. *Fauna norwegica Series B,* **35,** 1–4.[4]

Schnell, Ø. (1991) New records of Chironomidae (Diptera) from Norway (II), with two new species synonyms. *Fauna Norvewica, series B,* **38,** 5–10.[4]

Schofield, K., Townsend, C.R. and Hildrew, A.G. (1988) Predation and the prey community of a headwater stream. *Freshwater Biology,* **20,** 85–95.[12]

Schroeder, G.L. (1974) Use of fluid cowshed manure in fish ponds. *Bamidgeh,* **26,** 84–96.[17]

Sclater, W.L. and Sclater, P.L. (1899) *The Geography of Mammals,* Kegan, Paul, Trench, Trubner, London.[4]

Scott, A. (1985) Distribution, growth, and feeding of post emergent grayling *Thymallus thymallus* in an English River. *Transactions of the American Fisheries Society,* **114,** 525–31.[17]

Scott, W. and Opdyke, D.F. (1941) The emergence of insects from Winona Lake. *Investigations of Indiana Lakes and Streams,* **2,** 4–14.[9]

Sebastian, R.J. and Lockhart, W.L. (1981) The influence of formulation on toxicity and availability of a pesticide (methoxychlor) to black fly larvae (Diptera: Simuliidae), some non-target aquatic insects and fish. *The Canadian Entomologist,* **113,** 281–93.[15]

Seminara M. and Bazanti, M. (1988) Trophic level assessment of profundal sediments of the artificial lake Campotosto (Central Italy), using midge larval community (Diptera: Chironomidae). *Hydrobiological Bulletin,* **22,** 183–93.[15]

Sendstad, E., Solem, J.O. and Aagaard, K. (1977) Studies of terrestrial chironomids (Diptera) from Spitsbergen. *Norwegian Journal of Entomology,* **24,** 91–8.[6]

Sephton, T.W. (1987) Some observations on the food of larvae of *Procladius bellus* (Diptera: Chironomidae). *Aquatic Insects,* **9,** 195–202.[7]

Sephton, T.W. and Paterson, C.G. (1986) Production of the chironomid *Procladius bellus* in an annual drawdown reservoir. *Freshwater Biology,* **16,** 721–33.[11]

Sephton, T.W., Paterson, C.G. and Fernando, C.H. (1980) Spatial interrelationships of bivalves and nonbivalve benthos in a small reservoir in New Brunswick, Canada. *Canadian Journal of Zoology,* **58,** 852–9.[12]

Serra-Tosio, B. (1971) *Contribution a l'étude Taxonomique, Phylogénétique, Biogéographique, et écologique des Diamesini (Diptera, Chironomidae).* Doctoral thesis, University of Grenoble, France.[3]

Serra-Tosio, B. (1989a) Révision des espèces ouest-paléactiques et néarctiques de *Boreoheptagyia* Brundin avec des clés pour les larves, les nymphs et les imagos (Diptera, Chironomidae). *Spixiana,* **11,** 133–73.[3]

Serra-Tosio, B. (1989b) Chironomidés (Diptera) des Alpes françaises et des basses régions avoisinantes. *Annales de Limnologie,* **25,** 159–175.[4]

Serra-Tosio, B. (1989c) Ecologie et Biogeographie des *Boreoheptagyia* (Diptera, Chironomidae, Diamesinae). *Acta Biologica Debrecen, Oecologica Hungarica,* **3,** 289–94.[4]

Serra-Tosio, B. and Laville, H. (1991). List annotée des Diptéres Chironomidés de France continentale at de Corse. *Annales de Limnologie,* **27,** 37–74.[4]

Shaw, P.C. and Mark, K.-K. (1980) Chironomid farming – a means of recycling farm manure and potentially reducing water pollution in Hong Kong. *Aquaculture,* **21,** 155–63.[17]

Sheehan, P.J. and Knight, A.W. (1985) A multilevel approach to the assessment of ecotoxicological effects in a heavy metal polluted stream. *Verhandlungen der Internationalen Vereinigung für Theoretische und Angewandte Limnologie,* **22,** 2364–70.[15]

Sheldon, A.L. (1969) Size relationships of *Acroneuria californica* (Perlidae: Plecoptera) and its prey. *Hydrobiologia,* **34,** 85–94.[12]

Shilova, A.I. (1960) The metamorphosis of *Cryptochironomus burganadzeae* Tshern. *Transactions of the Biological Institute of Inland Waters of the Academy of Science, USSR,* **5,** 71–80.[In Russian].[7]

Shilova, A.I. (1965) The metamorphosis of *Parachironomus vitiosus* Goetgh. and some data on its biology. *Transactions of the Biological Institute of Inland Waters of the Academy of Science, USSR,* **8,** 102–9.[In Russian].[7]

Shilova, A.I. (1966) The metamorphosis of *Odontomesa fulva* Kieff. *Transactions of the Biological Institute of Inland Waters of the Academy of Science, USSR,* **12,** 239–50.[In Russian].[7]

Shilova, A.I. (1968) The information on biology of genus *Parachironomus* Lenz. *Transactions of the Biological Institute of Inland Waters of the Academy of Science, USSR,* **17,** 104–23.[In Russian].[7]

Shilova, A.I. and Zelentsov, N.I. (1972) Vliyanie fotoperiodizma na diapauzu u khironomid.[The influence of photoperiodism on diapause in Chironomidae]. *Informatsionnyi Byulleten Biologiya. Vnutrennikh Vod,* **13,** 37–42.[9]

Shimizu, K. (1978) On the nuisance midge in Kanda River. *Seikatsu to Kankyo* **23**: 25–38.[13]

Shiozawa, D.K. and Barnes, J.R. (1977) The microdistribution and population trends of larval *Tanypus stellatus* Coquillett and *Chironomus frommeri* Atchley and Martin (Diptera: Chironomidae) in Utah Lake, Utah. *Ecology,* **58,** 610–8.[10]

Shireman, J.V., Opuszynski, K. and Okoniewska, G. (1989) Food and growth of hybrid bass fry *Morone saxatilis* x *Morone chrysops* under intensive culture conditions. *Polskie Archiwum Hydrobiologii,* **35(1),** 109–18.[17]

Siegfried, C.A. (1984) The benthos of a eutrophic mountain reservoir: influence of reservoir level on community composition, abundance, and production. *California Fish and Game,* **70,** 39–52.[10][11]

Siegfried, C.A. and Knight, A.W. (1976) Prey selection by a setipalpian stonefly nymph, *Acroneuria (Calineuria) californica* Banks (Plecoptera: Perlidae). *Ecology,* **57,** 603–8.[12][17]

Silina, Y.P. (1959) Systematics and morphology of the chironomid genus *Limno-*

chironomus Kieffer. *Trudy Vsesoyuznogo gidrobiologicheskogo Obshchestva,* **9,** 121–8. (In Russian).[5]

Simpson, K.W. (1983) Communities of Chironomidae (Diptera) from an acid-stressed headwater stream in the Adirondack Mountains, New York. *Memoirs of the American Entomological Society,* **34,** 315–27.[15]

Sinegre, G., Babinot, M., Vigo, G. and Tourenq, J. N. (1990) Sensibilité de trois espèces de *Chironomus* (Diptera) à huit insecticides utilisés en démoustication. *Annales de Limnologie,* **26,** 65–71.[13][15]

Singh, M.P. and Harrison, A.D. (1982) Diel periodicities of emergence of midges (Diptera: Chironomidae) from a wooded stream on the Niagara Escarpment, Ontario. *Aquatic Insects,* **4,** 29–38.[9]

Singh, M.P. and Harrison, A.D. (1984) The chironomid community (Diptera: Chironomidae) in a southern Ontario stream and the annual emergence patterns of common species. *Archiv für Hydrobiologie,* **99,** 221–53.[10]

Singh, S.P., Malhotra, J.C., Seth, R.N. *et al.* (1981) Observations on rearing of hatchlings of catfish *Mystus seenghala. Aquaculture,* **26,** 161–6.[17]

Sjöberg, K. and Danell, K. (1982) Feeding activity of ducks in relation to diel emergence of chironomids. *Canadian Journal of Zoology,* **60(6),** 1383–7.[17]

Sládecek, V. (1973) System of water quality from the biological point of view. *Archiv für Hydrobiologie Beiheft Ergebnisse der Limnologie,* **7,** 1–218.[15]

Sly, P. G. (1986) Review of postglacial environmental changes and cultural impacts in the Bay of Quinte. *Canadian Special Publications in Fisheries and Aquatic Sciences,* **86,** 7–26.[16]

Smit, H., Klaren, P. and Snoek, W. (1991) *Lipiniella arenicola* Shilova (Diptera: Chironomidae) on a sandy flat in the Rhine-Meuse estuary: distribution, population structure, biomass and production of larvae in relation to periodical drainage. *Internationale Vereinigung für Theoretische und Angewandte Limnologie Verhandlungen,* **24,** 2918–23.[7]

Smit, H., Heinis, F., Bijkerk, R. and Kerkum, F. (1993) *Lipiniella arenicola* (Chironomidae) compared with *Chironomus muratensis* and *Ch. nudiventris*: distribution patterns related to depth and sediment characteristics, diet and behavioural response to reduced oxygen concentrations. *Netherlands Journal of Aquatic Ecology,* **26,** 431–40.[6]

Smith, E.L. (1969) Evolutionary morphology of external insect genitalia. 1. Origin and relationships to other appendages. *Annals of the Entomological Society of America,* **62,** 1051–260.[2]

Smith, I.M. and Oliver, D.R. (1976) The parasitic associations of larval water mites with imaginal aquatic insects, especially Chironomidae. *Canadian Entomologist,* **108,** 1427–42.[12]

Smith, L.C. and Smock, L.A. (1992) Ecology of invertebrate predators in a Coastal Plain stream. *Freshwater Biology,* **28,** 319–29.[7]

Smith, V.G.F. and Young, J.O. (1973) The life histories of some Chironomidae (Diptera) in two ponds on Merseyside, England. *Archiv für Hydrobiologie,* **72,** 333–55.[10]

Smock, L.A. (1980) Relationships between body size and biomass of aquatic insects. *Freshwater Biology,* **10,** 375–83.[11]

Smol, J. P. (1990) Paleolimnology – recent advances and future challenges, in *Scientific Perspectives in Theoretical and Applied Limnology* (eds R. de Bernardi, G. Giussani and L. Barbanti). *Memorie dell'Istituto Italiano di Idrobiologia,* **47,** 253–76.[16]

Smol, J.P. (1992) Paleolimnology: an important tool for effective ecosystem management. *Journal of Aquatic Ecosystem Health,* **1,** 49–58.[16]

Smol, J.P., Walker, I.R. and Leavitt, P.R. (1991) Paleolimnology and hindcasting climatic trends. *Internationale Vereinigung für Theoretische und Angewandte Limnologie Verhandlungen,* **24,** 1240–6.[16]

Sneath, P.H.A. and Sokal, R.R. (1973) *Numerical Taxonomy. The Principles and Practice of Numerical Classification.* W.H. Freeman and Company, San Francisco, pp.573.[3]

Snyder, R.J. (1984) Seasonal variation in the diet of the threespine stickleback, *Gasterosteus aculeatus* in Contra Costa County, California. *California Fish and Game,* **70(3),** 167–72.[17]

Sokolova, N.Yu. (1971) Life cycles of chironomids in the Uchinskoye Reservoir. *Limnologica,* **8,** 151–5.[11]

Sokolova, N.Yu., Palij, A.V. and Izvekova, E.I. (1993) Biology of *Chironomus piger* Str. and its role in the self-purification of the river. *Netherlands Journal of Aquatic Ecology,* **26,** 509–12.[17]

Soldán, T. (1978) Die Wirtsspezifizität und Verbreitung von *Symbiocladius rhithrogenae* (Diptera, Chironomidae) in der Tschechoslowakei. *Acta Entomologica Bohemoslovaca,* **75,** 194–200.[12]

Soldán, T. (1979) The effect of *Symbiocladius rhithrogenae* (Diptera, Chironomidae) on the development of reproductive organs of *Ecdyonurus lateralis* (Ephemeroptera, Heptageniidae). *Folia Parasitologica (Praha),* **26,** 45–50.[12]

Soldán, T. (1988) Distributional patterns, host specificity and density of an epoictic midge, *Epoicocladius flavens* (Diptera, Chironomidae) in Czechoslovakia. *Věstnik Ceské Společnosti Zoologické,* **52,** 278–89.[12]

Soltz, D.B., Hilsenhoff, W.L. and Stitch, H.F. (1968) A virus disease of *Chironomus plumosus. Journal of Invertebrate Pathology,* **12,** 118–28.[13]

Soluk, D.A. (1985) Macroinvertebrate abundance and production of psammophilous Chironomidae in shifting sand areas of a lowland river. *Canadian Journal of Fisheries and Aquatic Sciences,* **42,** 1296–302.[7][10][11]

Soluk, D.A. and Clifford, H.F. (1985) Microhabitat shifts and substrate selection by the psammophilous predator *Pseudiron centralis* McDunnough (Ephemeroptera: Heptageniidae). *Canadian Journal of Zoology,* **63,** 1539–43.[17]

Soponis, A.R. (1977) A revision of the Nearctic species of *Orthocladius (Orthocladius)* van der Wulp (Diptera: Chironomidae). *Memoirs of the Entomological Society of Canada,* **102,** 1–187.[2]

Soponis, A.R. and Russell, C.L. (1984) Larval Drift of Chironomidae (Diptera) in a North Florida Stream. *Aquatic Insects,* **6,** 191–9.[5]

Southwood, T.R.E. (1966) *Ecological Methods.* Chapman & Hall, London.[10]

Spataru, P. (1976) Natural feed of *Tilapia aurea* Steindachner in polyculture, with supplementary feed and intensive manuring. *Bamidgeh,* **28,** 57–63.[17]

Spataru, P. (1978) Food and feeding habits of *Tilapia zillii* (Gervais) (Cichlidae) in Lake Kinneret (Israel). *Aquaculture,* **14,** 327–38.[17]

Sprahamian, M.W. (1989) The diet of juvenile and adult twaite shad *Alosa fallax fallax* Lacepede from the rivers Severn and Wye, Britain UK. *Hydrobiologia,* **179(2),** 173–82.[17]

Sprules, W.M. (1947) An ecological investigation of stream insects in Algonquin Park, Ontario. *University of Toronto Studies. Biological Series,* **56,** 1–81.[9]

Stahl, J.B. (1969) The uses of chironomids and other midges in interpreting lake histories. *Mitteilungen Internationale Vereinigung für Theoretische und Angewandte Limnologie,* **17,** 111–25.[16]

Stahl, J.B. (1986) A six-year study of abundance and voltinism of Chironomidae (Diptera) in an Illinois cooling reservoir. *Hydrobiologia,* **134,** 67–79.[15]

Stark, J.D. (1989) Chironomidae (non-biting midges) in *Guide to the Aquatic Insects of New Zealand* (eds M.J. Winterbourne and K.L.D. Gregson), *Bulletin of the Entomological Society of New Zealand,* **9,** pp.72–81.[3]

Steffan, A.W. (1965) *Plecopteracoluthus downesi* gen. et sp. nov. (Diptera, Chironomidae), a species whose larvae live phoretically on larvae of Plecoptera. *Canadian Entomologist,* **97,** 1323–44.[12]

Steffan, A.W. (1968) Zur Evolution und Bedeutung epizoischen Lebensweise bei Chironomidae–Larven (Diptera). *Annales Zoologici Fennici,* **5,** 144–50.[12]

Steffan, A.W. (1971) Chironomid (Diptera) biocoenoses in Scandinavian glacier brooks. *Canadian Entomologist,* **103,** 477–86.[6]

Steinhaus, E.A. and Brinley, F.J. (1957) Some relationships between bacteria and certain sewage-inhabiting insects. *Mosquito News,* **17,** 299–302.[13]

Stites, D.L. and Benke, A.C. (1989) Rapid growth rates of chironomids in three habitats of a subtropical blackwater river and their implications for P:B ratios. *Limnology and Oceanography,* **34,** 1278–89.[10]

St Louis, V.L., Breebaart, L. and Barlow, J.C. (1990) Foraging behavior of tree swallows over acidified nonacidic lakes. *Canadian Journal of Zoology,* **68(11),** 2385–92.[17]

Stockner, J.G. and Shortreed, K.R.S. (1978) Enhancement of autotrophic production by nutrient addition in coastal rainforest stream on Vancouver Island. *Journal of the Fisheries Research Board of Canada,* **35,** 28–34.[17]

Storey, A.W. (1986) *Population Dynamics, Production and Ecology of Three Species of Epiphytic chironomid.* Ph.D. Dissertation. University of Reading.[5][7]

Storey, A.W. (1987) Influence of temperature and food quality on the life history of an epiphytic chironomid. *Entomologica Scandinavica Supplement,* **29,** 339–7.[10]

Storey, A.W. and Edward, D.H.D. (1989) Longitudinal variation in community structure of Chironomidae (Diptera) in two south-western Australian river systems. *Acta Biologica Debrecina Supplementum Oecologica Hungarica,* **3,** 315–28.[15]

Storey, A.W. and Pinder, L.C.V. (1985) Mesh size and efficiency of sampling of larval Chironomidae. *Hydrobiologia,* **124,** 193–7.[5]

Stout, R.J. and Taft, W.H. (1985) Growth patterns of a chironomid shredder on fresh and senescent tag alder leaves in two Michigan streams. *Journal of Freshwater Ecology,* **3,** 147–53.[7][10]

Street, M. (1977) The food of Mallard ducklings in a wet gravel quarry, and its relation to duckling survival. *Wildfowl,* **28,** 113–25.[17]

Street, M. and Titmus, G. (1979) The colonization of experimental ponds by Chironomidae (Diptera). *Aquatic Insects,* **1,** 233–44.[9]

Strenzke, K. (1950a) Die Pflanzengewässer von *Scirpus sylvaticus* und ihre Tierwelt. *Archiv für Hydrobiologie,* **44,** 123–70.[6]

Strenzke, K. (1950b) Systematik, Morphologie und Ökologie der terrestrischen Chironomiden. *Archiv für Hydrobiologie, Supplement,* **18,** 207–414.[6]

Strenzke, K. (1959) Revision der Gattung *Chironomus* Meig. 1. Die Imagines von 15 norddeutschen Arten und Unterarten. *Archiv für Hydrobiologie,* **56,** 1–42.[17]

Strenzke, K. (1960a) Metamorphose und Verwandtschaftsbeziehungen der Gattung *Clunio* Hal.(Dipt.). *Suomalaisen Eläin-ja Kasvitieteellisen Seuran Vanamon Tiedonannot,* **22,** 1–30.[2][3]

Strenzke, K. (1960b) Die systematische und ökologische Differenzierung der Gattung *Chironomus. Annales Zoologici Fennici,* **26,** 111–38.[9]

Strenzke, K. (1960c) Terrestrische Chironomiden. XIX–XXIII (Diptera: Chironomidae). *Deutsche Entomologische Zeiting,* **7,** 414–41.

Strenzke, K. and Neumann, D. (1960) Die Variabilität der abdominalen Körperanhänge aquatischer Chironomidenlarven in Abhängigkeit von der Ionenzusammensetzung des Mediums. *Biologisches Zentralblatt,* **79,** 199–225.[2][6]

Streufert, J.M., Jones, J.R. and Sanders, H.O. (1980) Toxicity and biological effects of phthalate esters on midges (*Chironomus plumosus*). *Transactions, Missouri Academy of Science,* **14,** 33–40.[15]

Strohmeier, K.L., Crowley, P.H. and Johnson, D.M. (1989) Effects of red-spotted newts *Notophthalmus viridescens* on the densities of invertebrates in a permanent fish-free pond: a one-month enclosure experiment. *Journal of Freshwater Ecology,* **5(1),** 53–66.[17]

Strong, D.R., Simberloff, D., Abele, L.G. and Thistle, A.B. (1984) *Ecological Communities: Conceptual Issues and the Evidence.* Princeton University Press, Princeton.

Stuart, T.A. (1942) Chironomid larvae of Millport shore pools. *Transactions of the Royal Society of Edinburgh,* **60,** 475–502.[5]

Sublette, J.E. (1979) Scanning electron microscopy as a tool in taxonomy and phylogeny of the Chironomidae. *Entomologica Scandinavica Supplement,* **10,** 47–65.[2]

Sublette, J.E. and Sublette, M.S. (1973) Family Chironomidae, in *Catalogue of the Diptera of the Oriental Region, Part 1,* (eds M. Delfinado and E.D. Hardy), University Press of Hawaii, Honolulu, pp.389–422.[3]

Sublette, J. E. and Sublette, M. S. (1974) A review of the genus *Chironomus* (Diptera: Chironomidae). V. The *maturus* – complex. *Studies of Natural Sciences,* Portales, New Mexico, **1,** 1–41.[13]

Sublette, J. E. and Sublette, M. S. (1988) An overview of the potential for Chironomidae (Diptera) as a world-wide source for potent allergens, in *International Symposium on Mite and Midge Allergy,* (ed. T. Miyamoto), Ministries of Education, Science and Culture, Tokyo.[13]

Sugden, L.G. (1973) Feeding ecology of Pintail, Gadwall, American Wigeon and Lesser Scaup ducklings. *Canadian Wildlife Services Report,* **24,** pp.45.[17]

Sugg, P., Edwards, J.S. and Baust, J. (1983) Phenology and life history of *Belgica antarctica,* an Antarctic midge (Diptera: Chironomidae) *Ecological Entomology,* **8,** 105–13.[10]

Sullivan, R.T. (1980) Insect swarming and mating. *Florida Entomologist,* **64,** 44–65.[9]

Suomalainen, E., Saura, A. and Lokki, J. (1976) Evolution of parthenogenetic insects. *Evolutionary Biology,* **9,** 209–57.[9]

Sutcliffe, D.W. (1960) Osmotic regulation in the larvae of some euryhaline Diptera. *Nature,* **187,** 331–2.[6]

Svensson, B.S. (1976) The association between *Epoicocladius ephemerae* Kieffer (Diptera: Chironomidae) and *Ephemera danica* Müller (Ephemeroptera). *Archiv für Hydrobiologie,* **77,** 22–36.[12]

Svensson, B.S. (1979) Pupation, emergence and fecundity of phoretic *Epoicocladius ephemerae* (Chironomidae). *Holarctic Ecology,* **2,** 41–50.[9][12]

Svensson, B.S. (1986) *Eukiefferiella ancyla* sp. n. (Diptera: Chironomidae) a commensalistic midge on *Ancylus fluviatilis* Müller (Gastropoda: Ancylidae). *Entomologica Scandinavica,* **17,** 291–98.[12]

Syrjämäki, J. (1960) Humidity reactions and water balance of *Allochironomus crassiforceps* Kieff. (Dipt., Chironomidae). *Suomen hyönteistiellinen Aikakauskirja,* **26,** 138–56.[9]

Syrjämäki, J. (1963) The change in the humidity reaction of *Cricotopus sylvestris* Fabr. (Dipt., Chironomidae) during the swarming period. *Annales Entomologici Fennici,* **29,** 147–51.[9]

Syrjämäki, J. (1964) Swarming and mating behaviour of *Allochironomus crassiforceps* Kieff. (Dipt., Chironomidae). *Annales Zoologici Fennici,* **2,** 145–52.[9]

Syrjämäki, J. (1965) Laboratory studies on the swarming behaviour of *Chironomus strenzkei* Fittkau in litt. (Dipt., Chironomidae). I. Mechanism of swarming and mating. *Annales Zoologici Fennici,* **2,** 145–52.[9]

Syrjämäki, J. (1966) Dusk swarming of *Chironomus pseudothummi* Strenzke (Dipt., Chironomidae). *Annales Zoologici Fennici,* **3,** 20–8.[9]

Syrjämäki, J. (1968a) Diel patterns of swarming and other activities of two arctic dipterans (Chironomidae and Trichoceridae) on Spitsbergen. *Oikos,* **19,** 250–8.[9]

Syrjämäki, J. (1968b) A peculiar swarming mechanism of an Arctic chironomid (Diptera) at Spitzbergen. *Annales Zoologici Fennici,* **5,** 151–2.[9]

Tabaru, Y. (1975) Outbreak of chironomid midge in a polluted river and chemical control of the larvae. *Japanese Journal of Sanitary Zoology,* **26**: 247–51.[13]

Tabaru, Y. (1985a) Studies on chemical control of a nuisance chironomid midge (Diptera: Chironomidae). 1. Larvicidal activity of organophosphorus insecticides against *Chironomus yoshimatsui. Japanese Journal of Sanitary Zoology,* **36,** 289–94.[13]

Tabaru, Y. (1985b) Studies on chemical control of nuisance chironomid midge (Diptera: Chironomidae). 2. Chemical control of the midge larvae in a sewage treatment plant by dripping technique and effects of the chemicals on microorganisms in activated sludge. *Japanese Journal of Sanitary Zoology,* **36,** 295–302.[13]

Tabaru, Y. (1985c) Studies on chemical control of a nuisance chironomid midge (Diptera: Chironomidae). 3. Susceptibility of *Chironomus yoshimatsui* collected from various rivers to the three kinds of organophosphorus larvicides and larval control trials with chlorpyrifos methyl in the field. *Japanese Journal of Sanitary Zoology,* **36,** 303–8.[13]

Tabaru, Y. (1985d) Studies on chemical control of a nuisance chironomid midge (Diptera: Chironomidae). 4. Efficacy of two insect growth regulators to *Chironomus yoshimatsui* in laboratory and field. *Japanese Journal of Sanitary Zoology,* **36,** 309–13.[13]

Tabaru, Y. (1986) Some problems on the chemical control of the midges. *Seikats to Kankyo,* **31,** 55–61.[13]

Tabaru, Y., Matsunaga, H. and Sato, A. (1978) Chemical control of the chironomid larvae (Diptera: Chironomidae) in the polluted river in reference to the controlled area and the mortality of the intoxicated larvae carried by the stream. *Japanese Journal of Sanitary Zoology,* **29,** 87–91.[13]

Tabaru, Y., Moriya, K. and Ali, A. (1987) Nuisance midges (Diptera: Chironomidae) and their control in Japan. *Journal of American Mosquito Control Association,* **3,** 45–9.[13][14]

Tait-Bowman, C.M. (1978) Chironomid larvae in the Shropshire meres: the relationship between distribution and tracheal patterns. *Acta Universitatis Carolinae-Biologica,* **1978,** 245–51.[2]

Tait-Bowman, C. (1980) Emergence of chironomids from Rostherne Mere, England, in *Chironomidae: Ecology, Systematics, Cytology and Physiology* (ed. D.A. Murray), Pergamon, New York, pp.291–5.[10]

Tang, Y. and Chen, T. (1959) Control of chironomid larvae in milkfish ponds. *Chinese-American Joint Commission on Rural Reconstruction, Fish series,* **4,** 1–36.[13]

Tarwid, M. (1969) Analysis of the contents of the alimentary tract of predatory Pelopiinae larvae (Chironomidae). *Ekologia Polska Series A,* **17,** 125–31.[7]

Taylor, L.R. (1974) Insect migration, flight periodicity and the boundary layer. *Journal of Animal Ecology,* **43,** 225–38.[9]

Teal, J. M. (1957) Community metabolism in a temperate cold spring. *Ecological Monographs,* **27**, 283–302.[18]

Tee, R.D., Gad El Rab, M.O., Cranston, P.S. and Kay, A.B. (1983) Allergens of the 'green nimitti' midge. *Proceedings of the XI International Congress of Allergy and Clinical Immunology,* **1983,** 541–4.[14]

Tee, R.D., Cromwell, O., Longbottom, J.L. *et al.* (1984) Partial characterisation of the allergens associated with hypersensitivity to the green nimitti midge

(*Cladotanytarsus lewisi,* Diptera, Chironomidae). *Clinical Allergy,* **14,** 117–27.[14]

Tee, R.D., Cranston, P.S., Dewair, M. *et al.* (1985) Evidence for haemoglobins as common allergenic determinants in IgE-mediated hypersensitivity to chironomids (non-biting midges). *Clinical Allergy,* **15,** 335–43.[14]

ten Winkel, E.H. (1987) *Chironomid Larvae and Their Food Web Relations in the Littoral Zone of Lake Maarsseveen.* PhD Thesis, University of Amsterdam, Amsterdam, The Netherlands. pp.145.[5][17]

ten Winkel, E.H. and Davids, C. (1987) Population dynamic aspects of chironomid larvae of the littoral zone of Lake Maarsseveen I. *Hydrobiological Bulletin,* **21,** 81–94.[12][17]

ten Winkel, E.H., Davids, C. and De Nobel, J.G. (1989) Food and feeding strategies of water mites of the genus *Hygrobates* and the impact of their predation on the larval population of the chironomid *Cladotanytarsus mancus* (Walker) in Lake Maarsseveen. *Netherlands Journal of Zoology,* **39,** 246–63.[12]

Terek, J. and Losos, B. (1979) Zivotné cykly a produkcia dominantnych foriem cel'ade Chironomidae vo vel'kom vihorlatskom jazere. *Biológia (Bratislava),* **34,** 851–60.[10][11]

ter Heerdt, G. and de Nie, H.W. (1987) A note on the feeding behaviour of the eel *Anguilla anguilla* (L.) in aquaria. *Archiv für Hydrobiologie,* **109(3),** 471–75.[17]

Terry, F.W. (1913) On a new genus of Hawaiian chironomids. *Proceedings of the Hawaiian Entomological Society,* **2,** 291–5.[5][9]

Theron, J.G. (1972) Chironomidae (Diptera) causing damage to motor cars (by ruining fresh paint). *Journal of the Entomological Society of South Africa,* **35,** 361.[9]

Thienemann, A. (1910) Das Sammeln von Puppenhäuten der Chironomiden. *Archiv für Hydrobiologie,* **6,** 213–4.[18]

Thienemann, A. (1922) Die beiden *Chironomus*-arten der Tiefenfauna der norddeutschen Seen. Ein hydrobiologisches Problem. *Archiv für Hydrobiologie,* **13,** 609–46.[15]

Thienemann, A. (1934) Chironomiden-Metamorphosen. *Stettiner Entomologischer Zeitung,* **95,** 1–23.[3]

Thienemann, A. (1937) Podonominae, eine neue Unterfamilie der Chironomiden (Chironomiden aus Lappland I). Mit einem Beitrag: Edwards, F.W.: On the European Podonominae (adult stage). *Internationale Revue der Gesamten Hydrobiologie, Hydrographie,* **35,** 65–112.[3]

Thienemann, A. (1939) Die Chironomiden forschung in ihrer Bedeutung für Limnologie und Biologie. *Biologische Jaarbäch,* **6,** 107–54.[18]

Thienemann, A. (1950) Verbreitungsgeschichte der Süsswassertierwelt Europas. Versuch einer historischen Tiergeographie der europäischen Binnengewässer. *Die Binnengewässer,* **18,** 1–809.[10]

Thienemann, A. (1952) Bestimmungstabelle für die Larven der mit *Diamesa* nächste verwandten Chironomiden. *Beiträge zur Entomologie,* **2,** 244–56.[3]

Thienemann, A. (1954). Chironomus. Leben, Verbreitung und wirtshaftliche Bedeutung der Chironomiden. *Binnengewässer* **20,** pp.834.[2][5][6][7][9][10][15][17][18]

Thienemann, A. and Strenzke, K. (1940a) Terrestrische Chironomiden III-IV: Zwei parthenogenetische Formen. *Zoologischer Anzeiger,* **132,** 24–40.[5]

Thienemann, A. and Strenzke, K. (1940b) Terrestrische Chironomiden I: *Pseudosmittia holsata,* eine neue Art mit fakultativer Parthenogenese. *Zoologischer Anzeiger,* **132,** 238–44.[5]

Thienemann, A. and Zavřel, J. (1916) Die Metamorphose der Tanypinen. *Archiv für Hydrobiologie und Planktonkunde,* **2,** 566–654.[3]

Thompson, D.J. (1978) Prey size selection by larvae of the damselfly, *Ischnura elegans* (Odonata). *Journal of Animal Ecology*, **47.** 769–85.[17]

Thornhill, R. and Alcock, J. (1983) *The Evolution of Insect Mating Systems.* Harvard University Press, London.[9]

Thornton, I.W.B. and New, T.R. (1988) Krakatau invertebrates: the 1980s fauna in the context of a century of recolonisation. *Philosophical Transactions of the Royal Society*, **322,** 493–522.[4]

Thorp, J.H. (1986) Two distinct roles for predators in freshwater assemblages. *Oikos*, **47,** 75–82.[17]

Thorp, J.H. and Bergey, E.A. (1981a) Field experiments on responses of a freshwater, benthic macroinvertebrate community to vertebrate predators. *Ecology*, **62,** 365–75.[17]

Thorp, J.H. and Bergey, E.A. (1981b) Field experiments on interactions between vertebrate predators and larval midges (Diptera: Chironomidae) in the littoral zone of a reservoir. *Oecologia (Berlin)*, **50,** 285–90.[12][17]

Thorp, J.H. and Chesser, R. K. (1983) Seasonal responses of lentic midge assemblages to environmental gradients. *Holarctic Ecology*, **6,** 123–32.[12][15]

Thurston, R.V., Gilfoil, T.A and Meyn, E.L. *et al.* (1985) Comparative toxicity of ten organic chemicals to ten common aquatic species. *Water Research*, **19,** 1145–55.[15]

Tikkanen, P. (1989) The chironomid larvae of a flow-through lake in northern Finland: the situation after 22 years of intense regulation. *Acta Biologica Debrecina Supplementum Oecologica Hungarica*, **3,** 351–9.[15]

Timmermans, K.R. and Walker, P.A. (1989) The fate of trace metals during metamorphosis of chironomids (Diptera, Chironomidae). *Environmental Pollution*, **62,** 73–85.[15]

Timmermans, K.R., Peeters, W. and Tonkes, M. (1992) Cadmium, zinc, lead and copper in *Chironomus riparius* (Meigen) larvae (Diptera; Chironomidae): uptake and effects. *Hydrobiologia*, **241,** 119–34.[15]

Timms, B.V. (1978) The benthos of seven lakes in Tasmania. *Archiv für Hydrobiologie*, **81,** 422–44.[6][15]

Titmus, G. (1979) The emergence of midges (Diptera: Chironomidae) from a wet gravel-pit. *Freshwater Biology*, **9,** 165–79.[9]

Titmus, G. and Badcock, R.M. (1980) Production and emergence of chironomids in a wet gravel pit, in *Chironomidae: Ecology, Systematics, Cytology and Physiology* (ed. D.A. Murray), Pergamon, Oxford. pp.299–305.[11]

Titmus, G. and Badcock, R.M. (1981) Distribution and feeding of larval Chironomidae in a gravelpit lake. *Freshwater Biology*, **11,** 263–71.[7]

Tjönneland, A. (1962) The nocturnal flight activity and the lunar rhythm of emergence in the African midge *Conochironomus acutistilus* (Freeman). *Contributions from the Faculty of Science, University College of Addis Ababa, Series Zoology*, **4,** 1–21.[9]

Tokeshi, M. (1985) Life cycle and production of the burrowing mayfly *Ephemera danica*: a new method for estimating growth. *Journal of Animal Ecology*, **54,** 919–30.[12]

Tokeshi, M. (1986a) Resource utilization, overlap and temporal community dynamics: a null model analysis of an epiphytic chironomid community. *Journal of Animal Ecology*, **55,** 491–506.[7][11]

Tokeshi, M. (1986b) Population dynamics, life histories and species richness in an epiphytic chironomid community. *Freshwater Biology*, **16,** 431–42.[7][10][17]

Tokeshi, M. (1986c) Population ecology of the commensal chironomid *Epoicocladius flavens* on its mayfly host *Ephemera danica*. *Freshwater Biology*, **16,** 235–44.[10][12]

Tokeshi, M. (1988) Two commensals on a host: habitat partitioning by a ciliated protozoan and a chironomid on the burrowing mayfly, *Ephemera danica. Freshwater Biology,* **20,** 31–40.[12]

Tokeshi, M. (1990a). Niche apportionment or random assortment: species abundance patterns revisited. *Journal of Animal Ecology,* **59,** 1129–1146.[12]

Tokeshi, M. (1990b). Density – body size allometry does not exist in a chironomid community on *Myriophyllum. Freshwater Biology,* **24,** 613–8.[12]

Tokeshi, M. (1991a) On the feeding habits of *Thienemannimyia festiva* (Diptera: Chironomidae). *Aquatic Insects,* **13,** 9–16.[7][12]

Tokeshi, M. (1991b). Faunal assembly in chironomids (Diptera): generic association and spread. *Biological Journal of the Linnean Society,* **44,** 353–367.[12][18]

Tokeshi, M. (1992) Dynamics of distribution in animal communities: theory and analysis. *Researches on Population Ecology,* **34,** 249–73.[12]

Tokeshi, M. (1993a). The structure of diversity in an epiphytic chironomid community. *Netherlands Journal of Aquatic Ecology,* **26,** 461–70.[12]

Tokeshi, M. (1993b). On the evolution of commensalism in the Chironomidae. *Freshwater Biology,* **29,** 481–489.

Tokeshi, M. (1993c). Species abundance patterns and community structure. *Advances in Ecological Research,* **24,** 111–86.[12]

Tokeshi, M. (1994) Community ecology and patchy freshwater habitats, in *Aquatic Ecology: Scale, Pattern and Process* (eds P.S. Giller, A.G. Hildrew and D.G. Raffaeli), Blackwell Scientific Publications. (in press).[12]

Tokeshi, M. and Pinder, L.C.V. (1985) Microhabitats of stream invertebrates on two submersed macrophytes with contrasting leaf morphology. *Holarctic Ecology,* **8,** 313–9.[6][12]

Tokeshi, M. and Pinder, L.C.V. (1986) Dispersion of epiphytic chironomid larvae and the probability of random colonization. *Internationale Revue der Gesamten Hydrobiologie, Hydrographie,* **71,** 613–20.[12]

Tokeshi, M. and Townsend, C.R. (1987) Random patch formation and weak competition: coexistence in an epiphytic chironomid community. *Journal of Animal Ecology,* **56,** 833–45.[12]

Tokunaga, M. (1932) Morphological and biological studies on a new marine chironomid fly, *Pontomyia pacifica* from Japan. Part I: Morphology and taxonomy. *Memoirs of the College of Agriculture, Kyoto University,* **19,** 1–56.[5][6][9]

Tokunaga, M. (1935) Chironomidae from Japan (Diptera). IV. The early stages of a marine midge, *Telmatogeton japonicus* Tokunaga. *Philippine Journal of Science,* **57,** 491–511.[5][9]

Tokunaga, M. (1965) Chironomids as winter bait of the over-wintering swallows. *Akitu,* **12,** 39–41.[17]

Tokunaga, M. and Kuroda, M. (1936) *Stenochironomus* midge from Japan (Diptera), with notes on controlling methods of a leaf mining midge. *Transactions of Kansai Entomological Society,* **7,** 1–6.[13]

Tompkins, A.M. and Gee, J.H. (1983) Foraging behavior of brook stickleback, *Culaea inconstans* (Kirtland): optimization of time, space, and diet. *Canadian Journal of Zoology,* **61,** 2482–90.[12]

Tonnoir, A. (1923) Le cycle évolutif de *Dactylocladius commensalis* sp. nov. *Annales de Biologie Lacustre,* **21,** 279–91.[12]

Toscano, R.J. and McLachlan, A.J. (1980) Chironomids and particles: micro-organisms and chironomid distribution in a peaty upland river, in *Chironomidae: Ecology, Systematics, Cytology and Physiology,* (ed. D.A. Murray), Pergamon Press, Oxford, pp.171–7.[7]

Townes, H.K. (1938) VI. Studies on the food organisms of fish. A biological survey of the Allegheny and Chemung watersheds. *Annual Report of the New York State Conservation Department, Supplement,* **27,** 162–75.[10]

Townes, H. K. (1945) The nearctic species of Tendipedini (Diptera, Tendipedidae (= Chironomidae)). *American Midland Naturalist,* **34**, 1–206.[18]

Tracy, B.H. and Hazelwood, D.H. (1983) The phoretic association of *Urnatella gracilis* (Entoprocta: Urnatellidae) and *Nanocladius downesi* (Diptera: Chironomidae) on *Corydalus cornutus* (Megaloptera: Corydalidae). *Freshwater Invertebrate Biology,* **2,** 186–91.[12]

Treverrow, N. (1985) Susceptibility of *Chironomus tepperi* (Diptera: Chironomidae) to *Bacillus thuringiensis* serovar *israelensis*. *Journal of the Australian Entomological Society,* **24,** 303–4.[7][13]

Trowbridge, A.C. (ed.) (1962) *Dictionary of Geological Terms,* Dolphin Books. pp.545.[16]

Tsai, S.C. and Legner, E.F. (1977) Exponential growth in culture of the planarian mosquito predator, *Dugesia dorotocephala* (Woodworth). *Mosquito News,* **37,** 474–8.[13]

Tsukada, M. (1967) Successions of Cladocera and benthic animals in Lake Nojiri. *Japanese Journal of Limnology,* **28,** 107–23.[16]

Tsukada, M. (1972) The history of Lake Nojiri, Japan. *Transactions of the Connecticut Academy of Arts and Science,* **44,** 339–65.[16]

Tsumuraya, T., Moriya, K., Matsumoto, T. and Mori, T. (1982) The ecology and control of *Chironomus yoshimatsui* in the sewage water treatment plant. *Yosui to Haisui,* **24,** 1356–62.[13]

Tubb, R.A. and Doris, T.C. (1965) Herbivorous insect populations in oil refinery effluent-holding pond series. *Limnology and Oceanography,* **10,** 121–34.[17]

Tudorançea, C., Fernando, C.H. and Paggi, J.C. (1988) Food and feeding ecology of *Oreochromis niloticus* Linnaeus 1758 juveniles in Lake Awassa Ethiopia. *Archiv für Hydrobiologie Supplementband,* **79(2/3),** 267–89.[17]

Tuiskunen, J. and Lindeberg, B. (1986) Chironomidae (Diptera) from Fennoscandia north of 68°N, with a description of ten new species and two new genera. *Annales Zoologici Fennici,* **23,** 361–93.[4]

Tuxen, S.L. (1944) The hot springs, their animal communities and their zoogeographical significance. *Zoology of Iceland,* **1,** 1–206.[6]

Ueno, M. (1930) Mayfly nymph and chironomid larva. *Transactions of the Kansai Entomological Society,* **1,** 46–48. (In Japanese).[12]

Usher, M.B. and Edwards, M. (1984) A dipteran from south of the Antarctic Circle: *Belgica antarctica* (Chironomidae), with a description of its larva. *Biological Journal of the Linnean Society,* **23,** 19–31.[1]

Uutala, A.J. (1981) Composition and secondary production of the chironomid (Diptera) communities in two lakes in the Adirondack mountain region, New York, in *Effects of Acidic Precipitation on Benthos, Proceedings of the Symposium on the effects of Acidic Precipitation on Benthos, 1980* (ed. R. Singer), North American Benthological Society, Hamilton, N.Y., pp.139–54.[11]

Uutala, A.J. (1986) *Paleolimnological Assessment of the Effects of Lake Acidification on Chironomidae (Diptera) Assemblages in the Adirondack Region of New York.* Ph.D. thesis, State University of New York College of Environmental Science and Forestry, Syracuse, N.Y., pp.156.[16]

Uutala, A. J. (1990) *Chaoborus* (Diptera: Chaoboridae) mandibles – paleolimnological indicators of the historical status of fish populations in acid-sensitive lakes. *Journal of Paleolimnology,* **4,** 139–51.[16]

Vaillant, F. (1956) Recherches sur la faune madicole (hygropetrique s.l.) de France, de Corse et d'Afrique de Nord. *Mémoires du Muséum d'Histoire Naturelle, Paris,* **11,** 1–258.[6]

Van Der Velde, G. and Hiddink, R. (1987) Chironomidae mining in *Nuphar lutea* (L.) Sm. (Nymphaeaceae). *Entomologica Scandinavica Supplement,* **29,** 255–64.[6][7]

Vannote, R.L., Minshall, G.W., Cummins, K.W. *et al.* (1980) The river continuum concept. *Canadian Journal of Fisheries and Aquatic Sciences,* **37,** 130–7.[6][7][15]

Vareschi, E. and Jacobs, J. (1984) The ecology of Lake Nakuru (Kenya) V. Production and consumption of consumer organisms. *Oecologia,* **61,** 83–98.[11]

Vepsäläinen, K. (1986) Chironomid wing length: a measure of habitat duration and predictability? *Oikos,* **46,** 269–71.[9]

Vilchez-Quero, A. and Casas, J.J. (1987) Quironomidos (Diptera) de los rios de Sierra Nevada (Grenada, Espana). *Actas IV Congresso Espanol de Limnologia,* 223–232.[4]

Vilchez-Quero, A. and Lavandier, P. (1986) Composition et rythme journalier de la dérive des exuvies nymphales de Chironomidés dans le Guadalquivir (Sierra de Cazorla – Espagne). *Annales de Limnologie,* **22,** 253–60.[9]

Vinikour, W.S. (1982) Phoresis between the snail *Oxytrema* (=*Elima*) *carnifera* and aquatic insects, especially *Rheotanytarsus* (Diptera: Chironomidae). *Entomological News,* **93,** 143–51.[12]

Vodopich, D.S. and Cowell, B.C. (1984) Interaction of factors governing the distribution of a predatory aquatic insect. *Ecology,* **65,** 39–52.[7][12]]

Vøllestad, L.A. and Andersen, R. (1985) Resource partitioning of various age groups of brown trout *Salmo trutta* in the littoral zone of Lake Selura, Norway. *Archiv für Hydrobiologie,* **105,**

von Zilah, G.V. (1932) Chironomiden-Studien. *Arbeiten des ungarischen biologischen Forschungsinstitutes*, **5,** 77–84.[5]

Walde S.J. (1986) Effect of an abiotic disturbance on a lotic predator–prey interaction. *Oecologia (Berlin),* **69,** 243–7.[12]

Walde, S.J. and Davies, R.W. (1984) Invertebrate predation and lotic prey communities: evaluation of *in situ* enclosure/exclosure experiments. *Ecology,* 1206–13.[7][12]

Walentowicz, A.T. and McLachlan, A.J. (1980) Chironomids and particles: a field experiment with peat in an upland stream, in *Chironomidae: Ecology, Systematics, Cytology and Physiology* (ed. D.A. Murray), Pergamon Press, Oxford, pp.179–85.[7]

Walker, I.R. (1987) Chironomidae (Diptera) in paleoecology. *Quaternary Science Reviews,* **6,** 29–40.[16]

Walker, I.R. (1991) Modern assemblages of arctic and alpine Chironomidae as analogues for late-glacial communities. *Hydrobiologia,* **214,** 223–7.[16]

Walker, I.R. (1993) Paleolimnological biomonitoring using freshwater benthic macroinvertebrates. in *Freshwater Biomonitoring and Benthic Macroinvertebrates,* (eds D.M. Rosenberg and V.H. Resh), Routledge, Chapman & Hall, Inc., New York, pp.306–343.[16]

Walker, I.R. and Mathewes, R.W. (1987a) Chironomidae (Diptera) and postglacial climate at Marion Lake, British Columbia, Canada. *Quaternary Research,* **27,** 89–102.[16]

Walker, I.R. and Mathewes, R.W. (1987b) Chironomids, lake trophic status, and climate. *Quaternary Research,* **28,** 431–7.[16]

Walker, I.R. and Mathewes, R.W. (1988) Late-Quaternary fossil Chironomidae (Diptera) from Hippa Lake, Queen Charlotte Islands, British Columbia, with special reference to *Corynocera* Zett. *The Canadian Entomologist,* **120,** 739–751.[16]

Walker, I.R. and Mathewes, R.W. (1989a) Early postglacial chironomid succession in southwestern British Columbia, Canada, and its paleoenvironmental significance. *Journal of Paleolimnology,* **2,** 1–14.[16]

Walker, I.R. and Mathewes, R.W. (1989b) Much ado about dead Diptera. *Journal of Paleolimnology,* **2,** 19–22.[16]

Walker, I.R. and Mathewes, R.W. (1989c) Chironomidae (Diptera) remains in surficial lake sediments from the Canadian Cordillera: analysis of the fauna across an altitudinal gradient. *Journal of Paleolimnology,* **2,** 61–80.[16]

Walker, I.R. and Paterson, C.G. (1983) Post-glacial chironomid succession in two small, humic lakes in the New Brunswick – Nova Scotia (Canada) border area. *Freshwater Invertebrate Biology,* **2,** 61–73.[16]

Walker, I.R., Fernando, C.H. and Paterson, C.G. (1985) Associations of Chironomidae (Diptera) of shallow, acid, humic lakes and bog pools in Atlantic Canada, and a comparison with an earlier paleoecological investigation. *Hydrobiologia,* **120,** 11–22.[15][16]

Walker, I.R., Mott, R.J. and Smol, J.P. (1991a) Allerød Younger Dryas lake temperatures from midge fossils in Atlantic Canada. *Science,* **253,** 1010–2.[16]

Walker, I.R., Smol, J.P., Engstrom, D.R. and Birks, H.J.B. (1991b) An assessment of Chironomidae as quantitative indicators of past climatic change. *Canadian Journal of Fisheries and Aquatic Sciences,* **48,** 975–87.[16]

Walker, I.R., Smol, J.P., Engstrom, D.R. and Birks, H.J.B. (1992) Aquatic invertebrates, climate, scale, and statistical hypothesis testing: a response to Hann, Warner, and Warwick. *Canadian Journal of Fisheries and Aquatic Sciences,* **49,** 1276–80.[16]

Walker, I.R., Oliver, D.R. and Dillon, M.E. (1993a) The larva and habitat of *Parakiefferiella nigra* Brundin (Diptera: Chironomidae). *Netherlands Journal of Aquatic Ecology* **26,** 527–531.[16]

Walker, I.R., Reavie, E.D., Palmer, S. and Nordin, R.N. (1993b) A paleoenvironmental assessment of human impact on Wood Lake, Okanagan Valley, British Columbia, Canada. *Quaternary International,* **20,** 51–70.[16]

Wallace, A.R. (1876) *The Geographical Distribution of Animals,* Macmillans, London.[4]

Wallace, J.B. and Merritt, R.W. (1980) Filter-feeding ecology of aquatic insects. *Annual Review of Entomology,* **25,** 103–32.[7]

Wallace, J.B., Lugthart, G.J., Cuffney, T.F. and Schurr, G.A. (1989) The impact of repeated insecticidal treatments on drift and benthos of a headwater stream. *Hydrobiologia,* **179,** 135–47.[15]

Wallace, J.B., Huryn, A.D. and Lugthart, G.J. (1991) Colonization of a headwater stream during three years of seasonal insecticidal applications. *Hydrobiologia,* **211,** 65–76.[15]

Walshe, B.M. (1951) The feeding habits of certain chironomid larvae (subfamily Tendipedinae). *Proceedings of the Zoological Society of London,* **121,** 63–79.[15]

Walter, L. (1973) Syntheseprozesse an den Riesenchromosomen von *Glyptotendipes. Chromosoma,* **41,** 327–60.[17]

Wang, W. (1987) Factors affecting metal toxicity to (and accumulation by) aquatic organisms – overview. *Environment International,* **13,** 437–57.[15]

Wang, X. and Zheng, L. (1993) Checklist of Chironomidae records from China. *Netherlands Journal of Aquatic Ecology,* **26,** 247–55.[4]

Ward, A.F. and Williams, D.D. (1986) Longitudinal zonation and food of larval chironomids (Insecta: Diptera) along the course of a river in temperate Canada. *Holarctic Ecology,* **9,** 48–57.[7][15]

Ward, A.K., Dahm, C.N. and Cummins, K.W. (1985) *Nostoc* (Cyanophyta) productivity in Oregon stream ecosystems: invertebrate influences and differences between morphological types. *Journal of Phycology,* **21,** 223–7.[7][12]

Ward, G.M. and Cummins, K.W. (1978) Life history and growth pattern of *Paratendipes albimanus* in a Michigan headwater stream. *Annals of the Entomological Society of America,* **71,** 272–284.[10]

Ward, G.M. and Cummins, K.W. (1979) Effects of food quality on growth of a stream detritivore, *Paratendipes albimanus* (Meigen)(Diptera: Chironomidae). *Ecology,* **60,** 57–64.[10]

Ward, J.V. (1992) *Aquatic Insect Ecology 1. Biology and Habitat,* John Wiley and Sons, New York.[6]

Ward, J.V. and Dufford, R.G. (1979) Longitudinal and seasonal distribution of macroinvertebrates and epilithic algae in a Colorado springbrook-pond system. *Archiv für Hydrobiologie,* **86,** 284–321.[6]

Warner, B.G. and Hann, B.J. (1987) Aquatic invertebrates as paleoclimatic indicators? *Quaternary Research,* **28,** 427–30.[16]

Warren, C.E., Wales, J.H., Davis, G.E. and Doudoroff, P. (1964) Trout production in an experimental stream enriched with sucrose. *Journal of Wildlife Management,* **28,** 617–60.[17]

Wartinbee, D.C. (1979) Diel emergence patterns of lotic Chironomidae. *Freshwater Biology,* **9,** 147–56.[9]

Warwick, W.F. (1975) The impact of man on the Bay of Quinte, Lake Ontario, as shown by the subfossil chironomid succession (Chironomidae, Diptera). *Verhandlungen der Internationalen Vereinigung für Theoretische und Angewandte Limnologie,* **19,** 3134–41.[15]

Warwick, W.F. (1980) Palaeolimnology of the Bay of Quinte, Lake Ontario: 2800 years of cultural influence. *Canadian Bulletin of Fisheries and Aquatic Sciences,* **206,** 1–117.[16]

Warwick, W.F. (1985) Morphological abnormalities in Chironomidae (Diptera) larvae as measures of toxic stress in freshwater ecosystems: indexing antennal deformities in *Chironomus* Meigen. *Canadian Journal of Fisheries and Aquatic Sciences,* **42,** 1881–914.[15]

Warwick, W.F. (1988) Morphological deformities in Chironomidae (Diptera) larvae as biological indicators of toxic stress, in *Toxic Contaminants and Ecosystem Health: a Great Lakes Focus* (ed. M. S. Evans). Advances in Environmental Science and Technology, Vol. 21, Wiley & Sons, New York, pp.281–320.[15]

Warwick, W.F. (1989a) Morphological deformities in larvae of *Procladius* Skuse (Diptera: Chironomidae) and their biomonitoring potential. *Canadian Journal of Fisheries and Aquatic Sciences,* **46,** 1255–70.[15]

Warwick, W.F. (1989b) Chironomids, lake development and climate: a commentary. *Journal of Paleolimnology,* **2,** 15–7.[16]

Warwick, W.F. (1991) Indexing deformities in ligulae and antennae of *Procladius* larvae (Diptera: Chironomidae): application to contaminant- stressed environments. *Canadian Journal of Fisheries and Aquatic Sciences,* **48,** 1151–66.[15]

Warwick, W.F. and Tisdale, N.A. (1988) Morphological deformities in *Chironomus, Cryptochironomus,* and *Procladius* larvae (Diptera: Chironomidae) from two differentially stressed sites in Tobin Lake, Saskatchewan. *Canadian Journal of Fisheries and Aquatic Sciences,* **45,** 1123–44.[15]

Waters, T.F. (1969) The turnover ratio in population ecology of freshwater invertebrates. *American Naturalist,* **103,** 173–85.[11]

Waters, T.F. (1977) Secondary production in inland waters. *Advances in Ecological Research,* **10,** 91–164.[11]

Waters, T.F. and Crawford, G.W. (1973) Annual production of a stream mayfly population: a comparison of methods. *Limnology and Oceanography,* **18,** 286–96.[11]

Waters, T.F. and Hockenstrom, J.C. (1980) Annual production and drift of the stream amphipod *Gammarus pseudolimnaeus* in Valley Creek, Minnesota. *Limnology and Oceanography,* **25,** 700–10.[11]

Watrous, L.E. and Wheeler, Q.D. (1981) The out-group comparison method of character analysis. *Systematic Zoology*, **30,** 1–11.[3]

Watson, C.N. and Heyn, M.W. (1993) A preliminary survey of the Chironomidae (Diptera) of Costa Rica, with emphasis on the lotic fauna. *Netherlands Journal of Aquatic Ecology*, **26,** 257–62.[3]

Way, M.O. and Wallace, R.G. (1989) First record of midge damage to rice in Texas. *Southwestern Entomologist*, **14,** 27–33.[7][13]

Webb, C.J. and Scholl, A. (1987) Comparative morphology of the larval ventromental plates of European species of *Einfeldia* Kieffer and *Chironomus* Meigen (subgenera *Lobochironomus* and *Camptochironomus* (Diptera: Chironomidae). *Entomologica Scandinavica Supplement*, **29,** 75–86.[2]

Webb, C.J., Cranston, P.S. and Martin, J. (1989) Congruence between larval ventromental plate ultrastructure and immature morphology in *Yama* Sublette & Martin and some Oceanian species of *Chironomus* Meigen (Diptera: Chironomidae). *Biological Journal of the Linnean Society* **97:** 81–100.[3]

Webb, D.W. (1980) The effects of toxaphene piscicide on benthic macroinvertebrates. *Journal of the Kansas Entomological Society*, **53,** 731–44.[15]

Webb, D.W. (1981) The benthic macroinvertebrates from the cooling lake of a coal-fired electric generating station. *Illinois Natural History Survey Bulletin*, **32,** 358–77.[15]

Weil, C.K. (1938) Hay-fever due to the scales of the *Tanytarsus*. *International Correspondence Society of Allergists*, **1938,** 63.[14]

Weil, C.K. (1940) Summer hay fever of unknown origin in the Southwest. *Journal of Allergy*, **11,** 361–75.[14]

Weiser, J. (1948) Zwei neuartige Erkrankungen bei Insekten. *Experientia (Basel)*, **4,** 317–8.[13]

Weiser, J. (1976) The intermediary host for the fungus *Coelomomyces chironomi*, *Journal of Invertebrate Pathology*, **28,** 273–4.[13]

Weiser, J. and McCauley, V.J.E. (1971) Two *Coelomomyces* infections of Chironomidae (Diptera) larvae in Marion Lake, British Columbia. *Canadian Journal of Zoology*, **49,** 65–8.[12][13]

Weiser, J. and McCauley, V.J.E. (1974) *Bertramia marionensis* n. sp., a fungal parasite of Chironomidae (Diptera) larvae. *Zeitschrift für Parasitenkunde*, **43,** 299–304.[13]

Weiser, J. and Vávra, J. (1964) Zur Verbreitung der *Coelomomyces*–Pilzen in europäischen Insekten. *Tropenmedezin und Parasitologie*, **15,** 38–42.[12]

Welch, H.E. (1973) Emergence of Chironomidae (Diptera) from Char Lake, Resolute, Northwest Territories. *Canadian Journal of Zoology*, **51,** 1113–23.[9]

Welch, H.E. (1976) Ecology of Chironomidae (Diptera) in a polar lake. *Journal of the Fisheries Research Board of Canada*, **33,** 227–47.[9][10][11]

Welch, H.E., Jorgenson, J.K. and Curtis, M.F. (1988) Emergence of Chironomidae (Diptera) in fertilized and natural lakes at Saqvaqjuac, N.W.T. *Canadian Journal of Fisheries and Aquatic Science*, **45,** 731–7.[9][11]

Welton, J.S., Mills, C.A. and Rendle, E.L. (1983) Food and habitat partitioning in two small benthic fishes, *Noemacheilus barbatulus* (L.) and *Cottus gobio* L. *Archiv für Hydrobiologie*, **97,** 434–54.[17]

Welton, J.S., Ladle, M., Bass, J.A.B. and Clarke, R.T. (1991) Grazing of epilithic chironomid larvae at two different water velocities in recirculating streams. *Archiv für Hydrobiologie*, **121,** 405–18.[7]

Wensler, R.J.D. and Rempel, J.G. (1962) The morphology of the male and female reproductive systems of the midge, *Chironomus plumosus* L. *Canadian Journal of Zoology*, **40,** 199–229.[2][5]

Werner, E.E., Gilliam, J.F., Hall, D.J. and Mittlebach, G.G. (1983) An experimental test of the effects of predation risk on habitat use in fish. *Ecology*, **64,** 1540–8.[12]

Wesenberg-Lund, C. (1913) Fortpflanzungsverhältnisse: Paarung und Eiablage der Süsswasserinsekten. *Fortschritte der naturwissenschaftlichen Forschung,* **8,** 161–286.[9]

Wetzel, R.G. (1983) *Limnology,* 2nd edn, Saunders, Philadelphia.[7]

White, T.R. and Fox, R.C. (1979) Chironomid (Diptera) larvae and hydroptilid (Trichoptera) pupae in a phoretic relationship on a macromiid (Odonata) nymph. *Notulae Odonatologica,* **1,** 76–7.[12]

White, T.R., Weaver, J.S. III and Fox, R.C. (1980) Phoretic relationships between Chironomidae (Diptera) and benthic macroinvertebrates. *Entomological News,* **91,** 69–74.[12]

Whiteside, M. C. (1983) The mythical concept of eutrophication. *Hydrobiologia,* **103,** 107–11.[16]

Wiederholm, T. (1971) Bottom fauna and cooling water discharges in a basin of Lake Mälaren. *Institute of Freshwater Research Drottningholm Report,* **51,** 197–214.[15]

Wiederholm, T. (1976) Chironomids as indicators of water quality in Swedish lakes. *Naturvårdsverkets Limnologiska Undersökningar Information,* **10,** 1–17.[15]

Wiederholm, T. (1979) Chironomid remains in recent sediments of Lake Washington. *Northwest Science,* **53,** 251–6.[16]

Wiederholm, T. (1980) Use of benthos in lake monitoring. *Journal of the Water Pollution Control Federation,* **52,** 537–47.[15]

Wiederholm, T. (1981) Associations of lake-living Chironomidae. A cluster analysis of Brundin's and recent data from Swedish lakes. *Schweizerische Zeitschrift für Hydrologie,* **43,** 140–50.[15]

Wiederholm, T. (ed.)(1983) Chironomidae of the Holarctic region. Keys and Diagnoses. Part 1. Larvae. *Entomologica Scandinavica Supplement,* **19,** 1–457.[1][2][3][18]

Wiederholm, T. (1984a) Responses of aquatic insects to environmental pollution, in *The Ecology of Aquatic Insects,* (eds V. H. Resh and D. M. Rosenberg), Praeger, New York, pp.508–57.[15]

Wiederholm, T. (1984b) Incidence of deformed chironomid larvae (Diptera: Chironomidae) in Swedish lakes. *Hydrobiologia,* **109,** 243–9.[15]

Wiederholm, T. (ed.)(1986) Chironomidae of the Holarctic region. Keys and Diagnoses. Part II. Pupae. *Entomologica Scandinavica Supplement,* **28,** 1–482.[1][2][3][18]

Wiederholm, T. (ed.)(1989) Chironomidae of the Holarctic region. Keys and Diagnoses. Part III. Adults. *Entomologica Scandinavica Supplement,* **34,** 1–532.[1][2][4][18]

Wiederholm, T. and Eriksson, L. (1977) Benthos of an acid lake. *Oikos,* **29,** 261–7.[15]

Wiederholm, T. and Eriksson, L. (1979) Subfossil chironomids as evidence of eutrophication in Ekoln Bay, Central Sweden. *Hydrobiologia,* **62,** 195–208.[15][16]

Wiens, A.P. and Rosenberg, D.M. (1984) Effect of impoundment and river diversion on profundal macrobenthos of Southern Indian Lake, Manitoba. *Canadian Journal of Fisheries and Aquatic Sciences,* **41,** 638–48.[15]

Wiens, A.P., Rosenberg, D.M. and Evans, K.W. (1975) *Symbiocladius equitans* (Diptera: Chironomidae), an ectoparasite of Ephemeroptera in the Martin River, Northwest Territories, Canada. *Entomologica Germanica,* **2,** 113–20.[12]

Wiggins, G.B. and Mackay, R.J. (1978) Some relationships between systematics and trophic ecology in nearctic aquatic insects, with special reference to Trichoptera. *Ecology,* **59,** 1211–20.[7]

Wiklund, C. and Fagerström, T. (1977) Why do males emerge before females? A hypothesis to explain the incidence of protandry in butterflies. *Oecologia,* **31,** 153–8.[9]

Wilda, T.J. (1984) The production of five genera of Chironomidae (Diptera) in Lake Norman, a North Carolina reservoir. *Hydrobiologia,* **108,** 145–52.[11]

Wiles, P. R. (1982) A note on the watermite *Hydrodroma despiciens* feeding on chironomid egg masses. *Freshwater Biology,* **12,** 83–7.[5]

Wiley, M.J. (1978) The biology of some Michigan trout stream chironomids (Diptera: Chironomidae) *Michigan Academician,* **1978,** 193–209.[11]

Wiley, M. and Kohler, S.L. (1984) Behavioral adaptations of aquatic insects, in *The Ecology of Aquatic Insects,* (eds V.H. Resh and D.M. Rosenberg), Praeger, New York, pp.101–33.[9]

Wiley, M.J. and Warren, G.L. (1992) Territory abandonment, theft, and recycling by a lotic grazer: a foraging strategy for hard times. *Oikos,* **63,** 495–505.[7]

Willassen, E. and Cranston, P.S. (1986) Afrotropical montane midges (Diptera, Chironomidae, *Diamesa*). *Zoological Journal of the Linnean Society,* **87,** 91–123.[4][6]

Williams, C.B. (1964) *Patterns in the Balance of Nature.* Academic Press, London.[12]

Williams, C.J. (1982) The drift of some chironomid egg masses. *Freshwater Biology,* **12,** 573–8.[5]

Williams, D.A. (1974) An infestation by a parthenogenetic chironomid. *Water Treatment and Examination,* **23,** 215–29.[8][13]

Williams, D.D. (1981) The first diets of postemergent Brook Trout (*Salvelinus fontinalis*) and Atlantic Salmon (*Salmo salar*) alevins in a Quebec River. *Canadian Journal of Fisheries and Aquatic Sciences,* **38,** 765–71.[17]

Williams, D.D. (1982) Emergence pathways of adult insects in the upper reaches of a stream. *Internationale Revue der Gesamten Hydrobiologie,* **67,** 223–4.[9]

Williams, D.D. and Feltmate, B.W. (1992) *Aquatic Insects,* C.A.B. International, Wallingford, UK.[6]

Williams, G.C. (1979) The question of adaptive sex-ratio in outcrossed vertebrates. *Proceedings of the Royal Society of London, Series B,* **205,** 567–80.[9]

Williams, K.A. (1981) *Population dynamics of epiphytic chironomid larvae in a chalk stream.* Ph.D. Dissertation. University of Reading.[5][6][7]

Williams, K.A., Green, D.W.J. and Pascoe, D. (1985) Studies on the acute toxicity of pollutants to freshwater macroinvertebrates. 1. Cadmium. *Archiv für Hydrobiologie,* **102,** 461–71.[15]

Williams, K.A., Green, D.W.J. and Pascoe, D. (1986a) Studies on the acute toxicity of pollutants to freshwater macroinvertebrates. 3. Ammonia. *Archiv für Hydrobiologie,* **106,** 61–70.[15]

Williams, K.A., Green, D.W.J., Pascoe, D. and Gower, D.E. (1986b) The acute toxicity of cadmium to different larval stages of *Chironomus riparius* (Diptera: Chironomidae) and its ecological significance for pollution regulation. *Oecologia (Berlin),* **70,** 362–6.[15]

Williams, K.A., Green, D.W.J., Pascoe, D. and Gower, D.E. (1987) Effect of cadmium on oviposition and egg viability in *Chironomus riparius* (Diptera: Chironomidae). *Bulletin of Environmental Contamination and Toxicology,* **38,** 86–90.[9][15]

Wilson, H. (1913) Preliminary report of a case of sensitization to the May fly (Ephemera). *Journal of the American Medical Association,* **61,** 1648.[14]

Wilson, L.F. (1969) Shoreline aggregation behaviour of adults of a midge *Chironomus* sp. (Diptera: Chironomidae) at Solberg Lake, Wisconsin. *Michigan Entomologist,* **2,** 14–9.[9][17]

Wilson, R.S. (1977) Chironomid pupal exuviae in the River Chew. *Freshwater Biology,* **7,** 9–17.[12]

Wilson, R.S. (1980) Classifying rivers using chironomid pupal exuviae, in *Chironomidae – Ecology, Systematics, Cytology and Physiology* (ed. D.A. Murray), Pergamon Press, Oxford, pp.209–16.[15]

Wilson, R.S. (1987) Chironomid communities in the River Trent in relation to water chemistry. *Entomologica Scandinavica Supplement,* **29,** 387–93.[15]

Wilson, R.S. (1988) A survey of the zinc-polluted River Nent (Cumbria) and the East and West Allen (Northumberland), England, using chironomid pupal exuviae. *Spixiana Supplement,* **14,** 167–74.[15]

Wilson, R.S. (1989) The modification of chironomid pupal exuvial assemblages by sewage effluent in rivers within the Bristol Avon catchment, England. *Acta Biologica Debrecina Supplementum Oecologica Hungarica,* **3,** 367–76.[15]

Wilson, R.S. and Bright, P.L. (1973) The use of chironomid pupal exuviae for characterizing streams. *Freshwater Biology,* **3,** 283–302.[15][18]

Wilson, R.S. and McGill, J.D. (1977) A new method of monitoring water quality in a stream receiving sewage effluent, using chironomid pupal exuviae. *Water Research,* **11,** 959–62.[15]

Wilson, R.S. and McGill, J.D. (1982) *A Practical Key to the Genera of Pupal Exuviae of the British Chironomidae (Diptera, Insecta).* University of Bristol Printing Office, Bristol.[15]

Wilson, R.S. and Wilson, S.E. (1983) A reconnaissance of the river Rhine using Chironomidae pupal exuviae (Insecta: Diptera). *Memoirs of the American Entomological Society,* **34,** 361–85.[15]

Wilson, R.S. and Wilson, S.E. (1985) A survey of the distribution of Chironomidae (Diptera, Insecta) of the river Rhine by sampling pupal exuviae. *Hydrobiological Bulletin,* **18,** 119–32.[15]

Wilson, S.E., Walker, I.R., Mott, R.J. and Smol, J.P. (1993) Climatic and limnological changes associated with the Younger Dryas in Atlantic Canada. *Climate Dynamics,* **8,** 177–87.[16]

Winberg, G.G., Patalas, K., Hillbricht-Ilkowska, A. *et al.* (1971) Methods for calculating productivity, in *A Manual on Methods for the Assessment of Secondary Productivity in Fresh Waters* (eds W.T. Edmonson and G.G. Winberg), Blackwell, pp.296–317.[11]

Winberg, G.G., Babitsky, V.A., Gavrilov, S.I. *et al.* (1972) Biological productivity of different types of lakes, in *Productivity Problems of Freshwaters* (eds Z. Kajak and A. Hillbricht–Ilkowska), PWN Polish Scientific Publishers, Warszawa–Krakow, Poland, pp.383–404.[11]

Winberg, G.G., Alimov, A.F., Boullion, V.V. *et al.* (1973) Biological productivity of two subarctic lakes. *Freshwater Biology,* **3,** 177–97.[11]

Winfield, D.K. (1991) The ecology of the overwintering diving ducks of Lough Neagh, with particular reference to their interactions with the fish populations. M.Phil. Thesis, University of Ulster.[12]

Winner, R.W., Boesel, M.W. and Farrel, M.P. (1980) Insect community structure as an index of heavy-metal pollution in lotic ecosystems. *Canadian Journal of Fisheries and Aquatic Sciences,* **37,** 647–55.[15]

Winterbourn, M.J. (1969) The distribution of algae and insects in hot spring thermal gradients at Waimangu, New Zealand. *New Zealand Journal of Marine and Freshwater Research,* **3,** 459–65.[6]

Winterbourn, M.J., Rounick, J.S. and Cowie, B. (1981) Are New Zealand stream ecosystems really different? *New Zealand Journal of Marine and Freshwater Research,* **15,** 321–8.[15]

Wirth, W.W. (1947) A review of the Genus *Telmatogeton* Schiner, with descriptions of three new Hawaiian species (Diptera: Tendipedidae). *Proceedings of the Hawaiian Entomological Society,* **13,** 145–91.[8][9]

Wirth, W. W. (1949) A revision of the Clunioninae midges with descriptions of a new genus and four new species (Diptera: Tendipedidae). *University of California Publications in Entomology,* **8,** 151–82.[3]

Wirth, W.W. (1975) The species of *Cricotopus* midges living in the blue–green alga *Nostoc* in California (Diptera: Tendipedidae). *Pan-Pacific Entomology,* **33,** 121–6.[12]

Wirtz, R.A. (1984) Allergic and toxic reactions to non-stinging arthropods. *Annual Review of Entomology,* **24,** 47–69.[14]

Wishart, G. and Riordan, D.F. (1959) Flight responses to various sounds by adult males of *Aedes aegypti* (L.). *Canadian Entomologist,* **91,** 181–91.[9]

Wissing, T.E. and Hassler, A.D. (1968) Calorific values of some invertebrates in Lake Mendota, Wisconsin. *Journal of the Fisheries Research Board of Canada,* **25,** 2515–8.[17]

Wissing, T.E. and Hassler, A.D. (1971) Intraseasonal change in caloric content of some freshwater invertebrates. *Ecology,* **52,** 371–3.[17]

Wood, D.M. and Borkent, A. (1989) Phylogeny and classification of the Nematocera, in *Manual of Nearctic Diptera, Volume 3,* (ed. J.F. McAlpine), Research Branch, Agriculture Canada, Ottawa. Agriculture Canada Monograph **32,** pp.1333–70.[3]

Woodin, M.C. and Swanson, G.A. (1989) Foods and dietary strategies of prairie-nesting ruddy ducks and redheads. *Condor,* **91(2),** 280–7.[17]

Wool, D. and Kugler, J. (1969) Circadian rhythm in chironomid species (Diptera) from the Hula Nature Preserve, Israel. *Annales Zoologici Fennici,* **6,** 94–7.[9]

Worden, D. M. (1986) *Chironomid paleoecology and the paleolimnology of Otsego Lake, New York,* M.S. thesis, University of Massachussetts, Amherst, Massachussetts, pp.116.[16].

Wotton, R.S., Armitage, P.D., Aston, K. *et al.* (1993) Colonization and emergence of midges (Chironomidae: Diptera) in slow sand filter beds *Netherlands Journal of Aquatic Ecology,* **26,** 331–9.[17]

Wright, J.F., Moss, D., Armitage, P.D. and Furse, M.T. (1984) A preliminary classification of running-water sites in Great Britain based on macro-invertebrate species and the prediction of community type using environmental data. *Freshwater Biology,* **14,** 221–56.[15]

Wright, J.F., Armitage, P.D., Furse, M.T. and Moss, D. (1985) The classification and prediction of macroinvertebrate communities in British rivers. *Annual Report Freshwater Biological Association,* **53,** 80–93.[15]

Wright, J.F., Armitage, P.D., Furse, M.T. and Moss, D. (1988) A new approach to the biological surveillance of river quality using macroinvertebrates. *Verhandlungen der Internationalen Vereinigung für Theoretische und Angewandte Limnologie,* **23,** 1548–52.[15]

Wrubleski, D.A. and Ross, L.C.M. (1989) Diel periodicties of adult emergence of Chironomidae and Trichoptera from the Delta Marsh, Manitoba, Canada. *Journal of Freshwater Ecology,* **5,** 163–70.[9]

Wülker, W. (1958) Parasitäre Intersexualität bei Chironomiden des Schluchsees. *Gewässer und Abwässer,* **20,** 62–7.[12][13]

Wülker, W. (1960) Neue Ergebnisse der parasitären Intersexualität der Chironomiden. *Naturwissenschaften,* **47,** 21–2.[12]

Wülker, W. (1961) Untersuchungen über die Intersexualität der Chironomiden (Dipt.) nach *Paramermis* Infektion. *Archiv für Hydrobiologie Supplementband,* **25,** 127–81.[13]

Wülker, W. (1964) Parasite-induced changes of internal and external sex characters in insects. *Experimental Parasitology,* **15,** 561–97.[12]

Wülker, W. (1970) Mechanism of parasitic gonad depression in insects. *Journal of Parasitology,* **56,** 484–5.

Yamagishi, H. and Fukuhara, H. (1971) Ecological studies on chironomids in Lake Suwa. I. Population dynamics of two large chironomids. *Oecologia,* **7,**

309–27.[10][12]

Yamagishi, H. and Fukuhara, H. (1972) Vertical migration of *Spaniotoma akamushi* larvae (Diptera: Chironomidae) through the bottom deposits of Lake Suwa. *Japanese Journal of Ecology*, **22**, 226–7.[10]

Yamamoto, M. (1986) Study of the Japanese *Chironomus* inhabiting high acidic water (Diptera, Chironomidae) I. *Kontyû, Tokyo*, **54**, 324–32.[15]

Yamamura, M., Suzuki, K.T., Hatakeyama, S. and Kubota, K. (1983) Tolerance to cadmium and cadmium-binding proteins induced in the midge larva, *Chironomus yoshimatsui* (Diptera, Chironomidae). *Comparative Biochemistry and Physiology*, **75C**, 21–4.[15]

Yamashita, N., Ito, K., Miyamoto, T. *et al.* (1989) Allergenicity of Chironomidae in asthmatic patients. *Annals of Allergy*, **63**, 423–6.[14]

Yashouv, A. (1956) Problems in carp nutrition. *Bamidgeh*, **8**, 79–87.[17]

Yashouv, A. (1970) Propagation of chironomid larvae as food for fish fry. *Bamidgeh*, **22**, 101–5.[17]

Yashouv, A. and Ben Shachar, R. (1967) Breeding and growth of Mugilidae II. Feeding experiments under laboratory conditions with *Mugil cephalus* L. and *M. capito* (Cuvier). *Bamidgeh*, **19**, 50–66.[17]

Yasuno, M. and Satake, K. (1990) Effects of diflubenzuron and methoprene on the emergence of insects and their density in an outdoor experimental stream. *Chemosphere*, **21**, 1321–35.[15]

Yasuno, M., Hasegawa, J., Iwakuma, T. *et al.* (1982) Effects of temephos on chironomid and plankton populations in eel culture ponds. *Japanese Journal of Sanitary Zoology*, **33**, 207–12.[13]

Yasuno, M., Hatakeyama, S. and Sugaya, Y. (1985a) Characteristic distribution of chironomids in the rivers polluted with heavy metals. *Verhandlungen der Internationalen Vereinigung für Theoretische und Angewandte Limnologie*, **22**, 2371–7.[15]

Yasuno, M., Sugaya, Y. and Iwakuma, T. (1985b) Effects of insecticides on the benthic community in a model stream. *Environmental Pollution (Series A)*, **38**, 31–43.[15]

Yokogi, K. and Ueno, R. (1971) Wasabi (Japanese horseradish, *Wasabia japonica*). *Nohson-Gyoson Bunka Kyoukai*, Tokyo.[13]

Young, J.O. and Spelling, S.M. (1989) Food utilization and niche overlap in three species of lake-dwelling leeches (Hirudinea). *Journal of Zoology (Oxford)*, **219(2)**, 231–44.[17]

Young, M.R. (1973) Seasonal variation in the occurrence of the parasitic larvae of *Cryptochironomus* sp. near *pararostratus* Kieffer (Diptera) in various freshwater molluscs. *Entomologist's Monthly Magazine*, **109**, 143–6.[12]

Yurkowski, M. and Tabachek, J.L. (1979) Proximate and amino acid composition of some natural fish foods, in *Proceedings of the World Symposium on Finfish Nutrition and Fish Feed Technology vol. 1* (eds J. Halver and K. Tiews), Heinemann, Berlin, pp.435–48.[17]

Zavřel, J. (1918) Über die Atmung und Respirations-organe der Chironomidenlarven. *Archiv für Hydrobiologie*, **12**, 202–7.[2]

Zavřel, J. (1926) '*Tanytarsus connectens*'. *Spisy vydavane Prirodovedeckov fakultou Masarykovy University*, **65**, 1–47.[5]

Zavřel, J. (1941a) Chironomidarum larvae et nymphae iii (*Pseudokiefferiella*). *Entomologické Listy*, **4**, 1–6.[3]

Zavřel, J. (1941b) Vergleichend-morphologische Untersuchungen an den Podonomidenlarven (Diptera, Chironomidae) I. Labrum und Praemandibeln. *Zoologische Anzeiger*, **134**, 105–15.[2]

Zelinka, M., Helan, J., Opravilova, V. *et al.* (1977) Production conditions of the polluted trout brook. *Folia Přírodovědecké Fakulty University J.E. Purkyně v Brně,* **18(7),** 5–105.[11]

Zieba, J. (1984) The food of some Chironomini in a carp pond enriched with organic sewage. *Polskie Archiwum Hydrobiologii,* **31,** 257–76.[17]

Zischke, J.A., Arthur, J.W., Nordlie, K.J. *et al.* (1983) Acidification effects on macroinvertebrates and fathead minnows (*Pimephales promelas*) in outdoor experimental channels. *Water Research,* **17,** 47–63.[15]

Zullo, S.J. and Stahl, J.B. (1988) Abundance, distribution and life cycles of midges (Chironomidae: Diptera) in an acid strip-mine lake in southern Illinois. *American Midland Naturalist,* **119,** 353–65.[15]

Zur, O. (1980) The importance of chironomid larvae as natural feed and as biological indicator of soil condition in ponds containing common carp (*Cyprinus carpio*) and *Tilapia (Sarotherodon aureus). Bamidgeh,* **32,** 66–77.[13]

Index

Lakes and rivers are indexed according to their proper names, e.g. Thames, River for River Thames. (Bot.) = Botanical name.

INDEX